Photophysical and Photochemical Tools in Polymer Science

Conformation, Dynamics, Morphology

NATO ASI Series

Advanced Science Institutes Series

A series presenting the results of activities sponsored by the NATO Science Committee, which aims at the dissemination of advanced scientific and technological knowledge, with a view to strengthening links between scientific communities.

The series is published by an international board of publishers in conjunction with the NATO Scientific Affairs Division

A Life Sciences B Physics	Plenum Publishing Corporation London and New York
C Mathematical and Physical Sciences	D. Reidel Publishing Company Dordrecht, Boston, Lancaster and Tokyo
D Behavioural and Social Sciences E Engineering and Materials Sciences	Martinus Nijhoff Publishers The Hague, Boston and Lancaster
F Computer and Systems Sciences G Ecological Sciences	Springer-Verlag Berlin, Heidelberg, New York and Tokyo

Photophysical and Photochemical Tools in Polymer Science

Conformation, Dynamics, Morphology

edited by

Mitchell A. Winnik

Department of Chemistry, University of Toronto,
Ontario, Canada

D. Reidel Publishing Company

Dordrecht / Boston / Lancaster / Tokyo

Published in cooperation with NATO Scientific Affairs Division

Proceedings of the NATO Advanced Study Institute on
Photophysical and Photochemical Tools in Polymer Science:
Conformation, Dynamics, Morphology
San Miniato, Italy
13-24 October, 1985

Library of Congress Cataloging in Publication Data

NATO Advanced Study Institute on Photophysical and Photochemical Tools in Polymer
    Science: Conformation, Dynamics, Morphology (1985: San Miniato, Italy)
    Photophysical and photochemical tools in polymer science.
    (NATO ASI series. Series C, Mathematical and physical sciences; vol. 182)
    "Proceedings of the NATO Advanced Study Institute on Photophysical and Photo-
chemical Tools in Polymer Science: Conformation, Dynamics, Morphology, San Miniato, Italy,
October 1985"—Verso t.p.
    "Published in cooperation with NATO Scientific Affairs Division."
    Includes index.
    1. Polymers and polymerization—Analysis—Congresses. 2. Luminescence spec-
troscopy—Congresses. I. Winnik, Mitchell A. II. North Atlantic Treaty Organization.
Scientific Affairs Division. III. Title. IV. Series: NATO ASI series. Series C, Mathematical and physical sciences; vol. 182.
QD381.8.N37   1985   547.7'046   86–15469

ISBN-13: 978-94-010-8601-1   e-ISBN-13: 978-94-009-4726-9
DOI: 10.1007/978-94-009-4726-9

Published by D. Reidel Publishing Company
P.O. Box 17, 3300 AA Dordrecht, Holland

Sold and distributed in the U.S.A. and Canada
by Kluwer Academic Publishers,
101 Philip Drive, Assinippi Park, Norwell, MA 02061, U.S.A.

In all other countries, sold and distributed
by Kluwer Academic Publishers Group,
P.O. Box 322, 3300 AH Dordrecht, Holland

D. Reidel Publishing Company is a member of the Kluwer Academic Publishers Group

CONTENTS

PREFACE

In 1980 the New York Academy of Sciences sponsored a three-day conference on luminescence in biological and synthetic macromolecules. After that meeting, Professor Frans DeSchryver and I began to discuss the possibility of organizing a different kind of meeting, with time for both informal and in-depth discussions, to examine certain aspects of the application of fluorescence and phosphorescence spectroscopy to polymers. Our ideas developed through discussions with many others, particularly Professor Lucien Monnerie. By 1983, when we submitted our proposal to NATO for an Advanced Study Institute, the area had grown enormously.

It is interesting in retrospect to look back on the points which emerged from these discussions as the basis around which the scientific program would be organized and the speakers chosen. We decided early on to focus on applications of these methods to provide information about polymer molecules and polymer systems: The topics would all relate to the conformation and dynamics of macromolecules, or to the morphology of polymer-containing systems. Another important decision was to expand the scope of the ASI to include certain photochemical techniques, particularly laser flash photolysis. These applications were at the time quite new, but full of promise as important sources of information about polymers.

We made a conscious decision not to distinguish between synthetic and biological macromolecules. Scientists working with these different types of polymers often have different focal points, which tend to inhibit their interaction, even though they often use the same techniques to extract similar kinds of information about their respective systems.

An Advanced Study Institute should be organized as a school, with the teaching function taking precedence over the announcement of new research findings. The lecturers for this ASI were asked to prepare their presentations from a pedagogic point of view and to write their chapter contributions from a similar perspective. The individual chapters were photocopied and distributed to all participants at the start of the meeting, thus serving also as detailed lecture notes. These chapters were modestly revised after the meeting for inclusion in this book. In this way we hoped to overcome some of the problems inherent in achieving a balanced presentation when preparing a text in which different authors contribute different chapters.

The meeting itself was very exciting, from many points of view. Good food, good weather, and good science make a rather heady combination. As editor of this volume, I have tried to convey some of the flavor as well as the substance of the meeting. Ideally, a book of

this sort should serve a dual function, presenting the scientific
material in such a way that it would be accessible both to polymer
scientists interested in new techniques, and to photochemists and photophysicists concerned with how their methods might be applied to polymer
systems.   In practice, such objectives are too broad to be achieved in a
single book.   This book is in fact aimed at the polymer scientist.   The
ultimate success of the meeting and of the book will be judged in part
by the extent to which the techniques described here are adopted by
scientists whose primary interests are with polymers, and not with the
techniques themselves.

Mitchell A. Winnik

ACKNOWLEDGEMENTS

There are many people and organizations who contributed to the success of the meeting in San Miniato. First and foremost I would like to thank NATO for its sponsorship, and Dr. Craig Sinclair and Mr. John A.S. Walker of the NATO Scientific Committee for their help and understanding. All non-scientific aspects of the meeting itself, from the facilities and the food, to the perfect weather, were arranged by Professor F. Ciardelli of the University of Pisa, with the help of his colleagues Professors C. Carlini and O. Pieroni. Dr. Irena Bronstein of Allied Instrument Laboratories (USA) made important contributions to the early organization of the meeting, and then, unfortunately, was unable to attend.

The Advanced Study Institute was sponsored by NATO under its double-jump program. This program is intended to encourage interactions among the academic community, government laboratories, and industry. To help us in this regard, we were fortunate to receive direct support from the University of Pisa, The U.S. Army European Research Office, and the following industries: Allied Instrument Laboratories (USA), C.D.F. Chimie (France), CIL Ltd. (Canada), ICI SA (Belgium), ICI Chemical Industries (UK), Rhone-Poulenc (France).

Much of the editing of the book was done while I was on sabbatical leave in the laboratory of Professor S. Tazuke at the Tokyo Institute of Technology in Yokohama, Japan. I would like to express my appreciation for his hospitality and helpfulness. Many administrative contributions were made by Ms. Yolanda Sivasankaran and by Ms. Susan Arbuckle. The meeting would not have been organized nor the book produced without their able assistance.

FLUORESCENCE PHENOMENA USEFUL FOR THE STUDY OF POLYMERS.

H. Morawetz
Department of Chemistry
Polytechnic Institute of New York
333 Jay Street
Brooklyn,NY U.S.A.

ABSTRACT. The development of fluorescence techniques for the study of
macromolecules, with special emphasis to those of biological origin,
is reviewed. Fluorescent labels may "report" the polarity or rigidity
of the medium. Fluorescence quenching depends on the spatial distribution of the emitting and quenching species. The polarization of the
emitted light may be used to characterize rotational diffusion and conformational mobility. Excimer emission from polymer solutions reflects
energy transfer and need not involve neighboring monomer residues along
the contour of the polymer chain. Nonradiative energy transfer may be
used to characterize a distance within a single macromolecule or distances between constituents of an association complex of macromolecules.
The fluorescence photobleaching recovery technique is a powerful method
for measuring minute diffusion coefficients.

1. Introduction.

Fluorescence is associated with a surprising variety of phenomena which
can be utilized to study properties of systems containing macromolecules.
Historically, many of these methods were explored by biochemists long
before they were used for the study of synthetic polymers and I shall
concentrate in this account on some characteristic work in this area
since it is less well known to the polymer community.
The emission from a fluorophore, the quantum yield and the spectrum of
the emitted radiation, frequently depend on the properties of the medium, its polarity and rigidity. Since emission from a rigorously defined excited state follows exponential decay, a deviation from this
simple pattern may be an indication that the excited chromophores are
not equivalent. Collisional quenching depends on the spatial distribution of the fluorescing and quenching species. The polarization of the
emitted light depends on rotational diffusion of the fluorophore; this
may be characterized by a single relaxation time or may exhibit a more
complex pattern of behavior. Energy transfer between "donor" and
"acceptor" chromophores over relatively long distances may be used as a
"spectroscopic ruler" to estimate distances either within a molecule or
between molecules. A sandwich complex, a so-called excimer, may form

1

M. A. Winnik (ed.), Photophysical and Photochemical Tools in Polymer Science, 1–13.
© *1986 by D. Reidel Publishing Company.*

between an excited and a ground state aromatic molecule with a characteristic shift in the emission spectrum. Since the intramolecular formation of such excimers from two aromatic residues requires a conformational transition, this phenomenon may be taken advantage of for the
study of conformational mobility. Finally, in the "fluorescence photobleaching recovery" (FPR) technique, the fluorescing species is destroyed in a small area by a sharply focussed intense light beam, and the
recovery of fluorescence is used as a measure of the diffusion of emitting chromophores from the surrounding areas. This is a powerful method
for the measurement of extremely small diffusion coefficients.

2. Polarization of fluorescence.

It had been known for many years that the polarization of fluorescence
increases with the viscosity of the medium containing the emitttng species and a quantitative theory connecting the phenomenon with the rotational diffusion of the fluorophore was formulated by Perrin in 1929
(1). The utilization of this effect for the study of macromolecules was
first suggested by Weber (2) who proposed the determination of the
rotary diffusion coefficient of globular proteins from the polarization
of fluorescence of a dye rigidly attached to the protein, using

$$P = (I_{\shortparallel} - I_{\perp})/(I_{\shortparallel} + I_{\perp})$$

$$[P^{-1} - (1/3)]/[P_o^{-1} - (1/3)] = [1 + (3\tau_e/\varphi)] \tag{1}$$

where $I_{\shortparallel}$ and $I_{\perp}$ are intensities of the emitted light (at right angles
to the incident beam) polarized parallel and perpendicular to the plane
of polarization of the exciting radiation, τ is the lifetime of the excited state and φ is the rotational relaxation time of the chromophore.
For spherical particles of volume V, $\varphi = 3\eta V/kT$ where η is the viscosity of the medium, while for ellipsoidal particles φ is the harmonic
mean of relaxation times for rotation around the equatorial and the
symmetry axis. Weber demonstrated that for serum albumin φ varies with
pH as would be expected from the denaturation of the protein in acid
and basic media and that the increase in the polarization of the fluorescence of the protein conjugate can be used to follow the denaturation of ovalbumin. If the protein carries more than one fluorescent
label,every energy transfer between them is equivalent to a rotation
from the orientation of the donor to that of the acceptor. This effect
was studied (3) on serum albumin tagged with varying numbers of fluorophores. With insulin conjugates, the change in the polarization of
fluorescence when organic solvents were added to an aqueous solution of
the protein showed that an insulin dimer dissociated to isolated insulin molecules (4).
    The power of this method was greatly increased when instrument-
ation became available which made it possible to follow the polarization
of the emitted light as a function of time after excitation by a light
flash. In this case it is convenient to express data by the emission
anisotropy, $r = (I_{\shortparallel} - I_{\perp})/(I_{\shortparallel} + 2I_{\perp})$, where r should exhibit simple

exponential decay if all emitting moieties are equivalent and if rotational diffusion is characterized by a single relaxation time. In the case of an immunoglobulin associated with a fluorescent hapten, the biphasic decay of the emission anisotropy was interpreted as due to a flexible "hinge" in the immunoglobulin, whose motion is much faster than the tumbling of the molecule (5).

When a flexible chain is attached to a fluorescent dye, the rotational diffusion of the dye is slowed down, but as the chain is extended, ρ approaches an asymptotic limit, since the whole of the chain does not participate in the motion of the dye (6 ,7). The extent to which the rotational diffusion of the dye is impeded is then a measure of the rigidity of the chain. If the middle of the chain is tagged with a fluorescent label, the rotational relaxation will reflect, in principle, both the local conformational transition and the rotation of the molecule as a whole. With ρ_{ct} and ρ_{rot} characterizing these two processes, the observed relaxation time will be (8)

$$\rho_{obs} = 1/(\rho_{ct}^{-1} + \rho_{rot}^{-1}) \qquad (2)$$

If the polymer chain behaves as an impermeable coil with the volume of the hydrodynamically equivalent sphere given by $V_e = (2/5)[\eta]M/\bar{N}$ where $\bar{N}$ is Avogadro's number, then $\rho_{rot} = 3\eta V/kT = (6/5)\eta[\eta]M/RT$. It is instructive to evaluate this quantity for a typical polymer chain of moderate molecular weight. Cyclohexane, a θ-solvent for polystyrene at 34°C, has $\eta = 0.006P$ and polystyrene with M = 10,000 has $[\eta]$ 18mL/g. Thus, ρ_{rot} = 300 ns and since this relaxation time for the rotation of the whole molecule is much longer than the relaxation time for a conformational transition, we may safely approximate $\rho_{obs} \approx \rho_{ct}$.

The interpretation of the polarization of fluorescence of labeled polymer chains requires a knowledge of the orientation of the transition moment of the label relative to the chain backbone (9). In copolymers of 9-p-vinylphenyl-10-phenylanthracene (10) these orientations are perpendicular to each other, while with the incorporation of an anthracene residue into the chain backbone (11) the transition moment of the label is parallel to the chain backbone:

Anufrieva and her colleagues have exploited the emission anisotropy of labeled polymers in a variety of ways (8, 12). For instance, they have shown that the conformational mobility is much smaller in polymethacrylates and polystyrene than in polyacrylates, that the rotational relaxation time of the label decreases with increasing solvation of the polymer and that the cooperative transition of poly(methacrylic acid) from a contracted to an expanded form at a critical charge density produces an abrupt decrease in $\wp$. A similar result has been obtained for the helix-coil transition of a labeled polypeptide (13). An increase in the emission anisotropy of labeled poly(methacrylic acid) may also be used to monitor complex formation with polyoxyethylene (12) and the decrease in the polarization of the emitted light after the addition of of unlabeled poly(methacrylic acid) to a solution of this complex allows the study of the kinetics of "complex interchange" (14)

$$PMA^*\text{-}POE + PMA = PMA\text{-}POE + PMA^*$$

In another type of application a low molecular fluorescent probe is added to a system containing macromolecules. As would be expected, the rotation of a small species is insensitive to the molecular weight of high polymers, but depends on tne "microscopic viscosity" which is a function of free volume. For instance, Nishijima has shown that the microscopic viscosity of liquid paraffin hydrocarbons levels off for molecular weights above 1000 and that the microscopic viscosity of polystyrene containing 10 volume% benzene is only 200 times as high as that of benzene (15). Nishijima also showed that the emission anisotropy is a useful index of molecular orientation. Since both the excitation and the emission are anisotropic, the method yields the fourth moment of the distribution function of orientations, while other optical properties (dichroism, birefringence) depend on the second moment (15).

3. Effect of the microenvironment on the emission characteristics.

The emission characteristics of fluorophores (the quantum yield and the wavelength of the emission maximum) are sensitive to their immediate environment and this leads to pronounced changes of fluorescence during protein denaturation (16). The high sensitivity of fluorescence to environmental effects was demonstrated in a particularly effective way in a study in which the decay of tryptophan fluorescence was followed in a number of proteins containing a single tryptophan residue (17). Deviations from simple exponential decay were interpreted as demonstrating that these proteins have a variable conformations, with the rate of their interconversion slow compared with the lifetime of the excited tryptophan.
      A number of chromophores which fluoresce strongly in organic media emit very weakly when surrounded by water molecules (18). When adsorbed or covalently bound to proteins, they "report" on the local polarity. With their use, the sensitivity with which even minor changes in protein conformation can be detected is greatly magnified. For instance, with such a dye adsorbed on chymotrypsinogen, the fluorescence intensity increases dramatically when the zymogen is converted to the active enzyme (19). The same principle has been used in employing the dansyl label for the study of the transition of maleic acid-vinyl ether

copolymers from the contracted state, in which the label is shielded
from the aqueous environment, to the expanded state (20).

In the case of some dyes, fluorescence is quenched by internal
rotation of the excited molecule and the emission intensity increases
then with the rigidity of the medium which keeps the chromophore in the
planar configuration (21). This effect has been used to demonstrate that
well below the glass transition point the rigidity of polystyrene decreases with increasing temperature (22). Complexation of such a dye
with a polymer apparently also restricts substantially the internal
rotation, so that the dye which does not fluoresce when dissolved in a
solvent of low viscosity becomes fluorescent on complexation (23). More
recently, a dye of this type has been used to monitor the increase in
microviscosity during a polymerization (24). It has also been claimed
(25) that the fluorescence of such a dye incorporated into polystyrene
increased sharply when the molecular weight of the polymer was raised
from 10,000 to 100,000. Since the glass transition temperature (which
reflects the free volume) is no longer sensitive to the chain length in
this range, this result was surprising and was rationalized as due to
"an abrupt change of morphology". It would be desirable to confirm such
a phenomenon.

4. Fluorescence quenching.

In systems containing low molecular weight species, the relation between
the fluorescence intensities, I and I_o, in the presence and absence of
a quencher, is commonly expressed by the Stern-Volmer equation

$$I_o/I = 1 + K_{SV}(Q) \qquad\qquad (3)$$

where (Q) is the quencher concentration. In solutions containing polymers, complications may arise from the non-uniform distribution of the
fluorescing and quenching species. If the polymer carries quenching
groups distributed at random along the chain molecule and the quencher
is a small molecule (26) quenching is retricted to the polymer domains
and the dependence of I_o/I on (Q) may be interpreted in terms of the
critical concentration at which the polymer chains are sufficiently entangled so that the quencher concentration becomes uniform. If the polymer carries the fluorescent residues and the quenchers are small molecules, the quenchers behave like cosolvents which may be concentrated
or diluted in the polymer domains and this will determine their quenching efficiency (27). If a polymer carries both fluorescing and quenching groups, then any quenching in dilute solution takes place within
the domain of a single polymer coil and reflects the probability that
the polymer-bound emitting and quenching groups can encounter each other
within the lifetime of the excited state. Kirsh et al. (28) carried out
an experiment of this type and concluded that the efficiency of quenching was only smaller by a factor of four compared with what it would
have been if the quenching species had been free to diffuse within the
polymer domain. They also carried out experiments in which different

chain molecules carried the fluorescing and the quenching substituents. In that case no quenching was observed below a critical concentration at which the molecular coils were presumably forced to overlap. This concentration was close to the reciprocal of the intrinsic viscosity, as would have been expected.

The presence of a polymer may also profoundly affect the frequency of encounters between two small molecules, one fluorescing, the other quenching. This will be particularly the case if the polymer bears a large ionic charge and if the two interacting species carry charges of the opposite sign (29). In that case the emitter and quencher are concentrated in the polymer domain and the quenching efficiency is sharply increased at low quencher concentrations. However, at larger quencher concentrations an unusual effect is observed: The quencher now displaces the fluorescing species from the environment of the polyion into a portion of space where the quencher concentration is low. As a result, the fluorescence intensity increases sharply. Under such conditions the fluorescence decay is biphasic, reflecting the behavior of emitters in the polymer domain, where the quencher concentration is high, and the rets of the system, where quenching is much less effective (30).

5. Excimer fluorescence.

In 1954 Förster and Kasper discovered that when the concentration of pyrene solutions was increased a new structureless emission band appeared, red-shifted to the normal pyrene emission, which decreased in intensity (31). Since the absorption spectrum was found to be independent of concentration, the new emission was ascribed to a complex of an excited and a ground-state pyrene molecule. Similar phenomena were soon noted with other aromatic molecules. In the crystalline state, an emission characteristic of the excited dimer ("excimer") was observed only when the molecules lay face-to-face at a distance of about 0.35 nm (32) and it was concluded that this is the geometry required for excimer formation.

Of particular interest was the finding that 1,3-diphenylpropane exhibits typical excimer emission independent of its concentration, so that excimer formation had to be intramolecular (33). No other α,ω-diphenylalkane behaved in this way, so that it was first believed that intramolecular excimer formation is only possible if the two chromophores are linked by a three atom bridge. Later it was found that such a restriction is not valid with all chromophores and that α,ω-dipyrenylalkanes, in particular, form intramolecular excimers even for derivatives of long paraffin hydrocarbons (34). Since the conformation of diarylpropanes in which the two aromatic residues lie face-to-face would correspond to a high potential energy in the ground state, a conformational transition had to take place after excitation of one chromophore to form the excimer. Denoting by k_{DM} and k_{MD} rate constants for the conformational transitions by which the excimer is formed and dissociated, by k_{EM} and k_{ED} rate constants for emission from the "monomer" and dimer and by k_{IM} and k_{ID} the corresponding rate constants for radiationless deactivation, we may represent the physical situation by the following scheme:

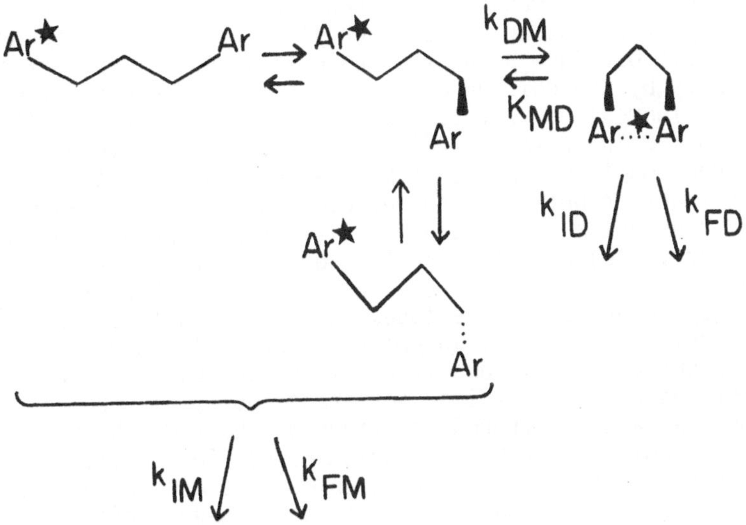

Since excimer formation is exothermic (35), k_{MD} is negligible at low temperature and the ratio of emission intensity of excimer and monomer, I_D/I_M, depends on $f(k_{DM}/k_{FM})[k_{FD}/(k_{ID}+ k_{FD})]$, where f is the fraction of the monomer in the tg conformation. Thus, at low temperature excimer emission increases with increasing temperature reflecting an increase in k_{DM}. An increasing viscosity of the medium will reduce excimer fluorescence since it hinders conformational transitions. However, the excimer yield correlates with the viscosity of the medium only for chemically similar solvents (36); for diaryl compounds in methanol - ethylene glycol mixtures the sensitivity of excimer formation to viscosity increases with the bulk of the chromophore (37). At higher temperatures, where $k_{MD} \gg k_{ID}+k_{FD}$, the I_D/I_M ratio reflects the monomer - excimer equilibrium.

Excimer fluorescence will be discussed in detail by other contributors to this volume and I shall limit myself to a few comments. Polystyrene was the first polymer investigated (38) and its excimer fluorescence was not surprising, since neighboring phenyl groups have the same steric relationship to one another as in 1,3-diphenylpropane. It was noted that in the isotactic polymer I_D/I_M is larger than in conventional polystyrene and the stereoisomeric 2,4-diphenylpentanes exhibit an analogous effect (39), I_D/I_M being seven times larger in the meso isomer. It was also soon recognized (40) that the excitation energy migrates from chromophore to chromophore before it is trapped by excimer formation; such energy transfer is particularly important in glassy polymers where no conformational transitions can occur.

For excimer formation in polymer solutions two fundamental questions should be considered: (1) When can it be assumed that excimers are formed (or formed exclusively) from contiguous monomer residues along the polymer chain ? (2) Does energy transfer take place exclusively along the polymer chain, or can it take place between chromophores attached at some distance along the chain backbone but brought into close proximity by the coiling of the chain ? The decrease of excimer emission with increasing solvation observed with polystyrene (41) and poly(1-naphthyl methacrylate) (42) could be due either to excimer formation from non-contiguous pairs of monomer residues, or to a more efficient energy transfer between such residues in the less expanded chain. With poly(2-naphthyl methacrylate), poly(1-vinyl naphthalene) and poly-(1-naphthyl acrylate) no solvent effect on excimer formation was observed (43) and this was interpreted as suggesting that excimer formation from adjoining monomer residues is so easy in these cases that it becomes the dominant process. In polystyrene, I_D/I_M increases with increasing concentration (41) and if excimers can form from phenyl groups of different chains, then excimer formation should also be possible from phenyl groups attached different sections of the same chain. In the very stiff chain of polyacenaphthylene, which cannot form excimers from nearest neighbor residues, excimer formation was assumed to involve next-to-nearest neighbor interaction (44) and this has been confirmed by the strong excimer emission from alternating acenaphthylene copolymers (45):

(The I_D/I_M is highly sensitive to the chain substituents X and Y) Alternating copolymers of 1-vinyl naphthalene

have not been studied but it is hard to believe that the less rigidly bound chromophore has a smaller tendency to form excimers by next-to-nearest group interaction than in the alternating acenaphthylene copolymers.

In polymers in which excimer formation from neighboring chain residues is difficult or impossible and where the driving force towards

excimer formation is relatively small, excimer emission may be observed
only at higher solution concentrations. This was the behavior of a styrene-butadiene copolymer with a low styrene content (46) and of poly-
(benzyl methacrylate) (47). A particularly interesting result was obtained with the polyester

$$[-OOC-CH=CH-\langle O \rangle-CH=CH-COO-CH_2-\langle \rangle-OCH_2CH_2-]_n$$

In spite of the large spacing of the fluorescent residues, an intense
excimer emission was observed (48). The I_D/I_M ratio increased with concentration beyond a sharply defined point, corresponding presumably to
the onset of chain interpenetration. However, the lifetime of monomer
emission was concentration independent and it was concluded that a cooperative effect of evenly spaced chromophores favored chromophore dimer
formation even in the ground state. The high intensity of excimer emission observed even in very dilute solutions seemed to imply a folded
chain conformation.
         Energy transfer in polymers and copolymers of aromatic monomers
should depend on the characteristic energy transfer distance R_o between
a pair of aromatic residues. For a random orientation of such residues
the R_o value of low molecular weight analogs, toluene (0.72 nm), 1-me-
thyl naphthalene (0.83 nm) and N-methyl carbazole (2.1 nm) may be cited
as a measure of energy transfer to be found in polymers carrying such
chromophores (49). It is then not surprising that fluorescence from an
alternating N-vinyl carbazole-fumaronitrile copolymer is fully depolarized, proving efficient energy transfer between non-neighboring monomer
residues (50)For copolymers of 1-vinyl naphthalene, 2-vinyl naphthalene
and styrene it has been suggested that I_D/I_M is proportinal to $f_{aa}\bar{\ell}_a$,
where f_{aa} is the fraction of aromatic pair in the copolymer and $\bar{\ell}_a$
is the mean length of uninterrupted sequences of aromatic comonomers
(51). This expression is based on the assumption that energy transfer
occurs only between neighboring aromatic residues along the chain. It
would be desirable to check this assumption by depolarization of fluorescence of various alternating copolymers. At any rate, the observation
that for any given $f_{aa}\bar{\ell}_a$, the I_D/I_M ratio is six times higher for
1-vinyl naphthalene-methyl acrylate than for 1-vinyl naphthalene-methyl
methacrylate copolymers (52) cannot be understood unless energy transfer
can occur between non-neighboring aromatic residues. Since the efficiency of energy transfer is highly sensitive to the mutual orientation
of donor and acceptor, such transfer between aromatic monomers should
depend on the nature of the intervening comonomer.

6. Nonradiative energy transfer as a measure of the donor-acceptor
distance.

         When the emission spectrum of one molecule overlaps the absorption
spectrum of another, dipole-dipole interaction leads to energy transfer
from "donor" to "acceptor". The theory of this effect was developed by
Förster (53) who showed that the efficiency of energy transfer is

given by

$$Eff = R_o^6/(R_o^6 + r^6)$$

$$R_o^6 = (8.8 \times 10^{-25}) \, Jn^{-4}k^2\phi_o \qquad (4)$$

where r is the distance betwwen donor and acceptor, J is the overlap
integral , n the refractive index, k^2 a factor depending on the mutual
orientation of donor and acceptor transition moments and ϕ_o is the fluorescence quantum yield of the donor in the absence of acceptors. With
large J, the characteristic distance R_o may be as large as 5 nm, and
with R_o known, the efficiency of energy transfer may be used as a measure of donor-acceptor separation (provided their mutual orientation is
well defined and known, or can be assumed to be random, corresponding
to $k^2 = 2/3$).
    In an early application of this technique, it was shown that the
energy transfer from the tryptophan residue of trypsin to a dansyl
group attached to the enzyme decreased when the enzyme associated with
its specific inhibitor (54). This was taken to imply a conformational
transition during complex formation. More recently, Cantor and his
associates have exploited this technique in a spectacular manner for a
study of biologically important structures formed by the association
of a large number of macromolecules. The ribosome "s-30 subunit" contains 21 distinct protein molecules and a molecule of RNA. The proteins
can be isolated, labeled with fluorophores and reassembled into a functional ribosome particle. The distance between various pairs of proteins
can then be "measured" by energy transfer and the information can be
used to suggest some features of the ribosome architecture (55). A similar study was carried out on nucleosomes, the chromosome fragments
whose core is a compact aggregate of eight protein molecules, which is
surrounded by a DNA fragment wound around the core. Here one of the proteins and the end of the DNA were labeled, and the dependence of the
structure on the solvent medium was investigated (56).

7. Fluorescence recovery after photobleaching.

    In 1974 Peters et al. (57) reported an experiment in which they
labeled proteins in the erythrocyte "ghost" with fluorescein, destroyed
the label in a portion of the "ghost" by an intense light flash and observed the distribution of the fluorescent species for the enxt 20 minutes. They could not detect any change and concluded that the diffusion coefficient, D, of the tagged protein in the erythrocyte membrane
must lie below 3×10^{-12} cm^2 s^{-1}. Since the mean square displacement in a
given direction of a molecule engaged in Brownian motion is 2Dt, the
smallest D which can be characterized in an experimental time t is inversely proportional to the square of the spatial resolution; thus,if a
small area can be bleached, the fluorescence recovery by diffusion of
the emitting species from the surrounding area can be used to measure
minute diffusion coefficients.
    Originally, spherical areas with a diameter of the order of 1μm

were bleached, allowing diffusion coefficients as small as $10^{-12} - 10^{-11}$ cm^2s^{-1} to be characterized. This range was found to be typical of proteins specifically attached to a cell membrane surface (58). Later it was shown that the recovery after bleaching of a periodic pattern offers experimental and computational advantages (59). By using a pattern periodic in two dimensions, diffusion anisotropy could be studied (60).

The measurements of very small diffusion coefficients is of special interest to polymer science. Smith (61) used the fluorescence recovery method to study the diffusion of a fluorescent dye in slightly swollen poly(methyl methacrylate), measuring D values as small as 10^{-13} cm^2s^{-1}. A spectacular study of this type was reported from Sillescu's laboratory (62) for polystyrene melts. The data were so precise that the dependence of the diffusion of short chains ($M_w = 4000$) on the molecular weight of the surrounding matrix or the dependence of the diffusion of long chains on the number of attached labels could be demonstrated. The data yielded D proportional to $M^{-2.2}$, close to the theoretical predictions for the chain length dependence of the diffusion coefficient (63, 64).

References.

(1) J. Perrin, Ann.Phys.(Paris), [10], 12,169 (1929)
(2) G. Weber, Biochem.J.,51,145, 155 (1951)
(3) S.R. Anderson and G. Weber, Biochemistry,8,371 (1969)
(4) R.F. Steiner and A. McAllister, J.Colloid Sci.,12,80 (1957)
(5) P. Yguerabide, H.F. Epstein and L. Stryer,J.Mol.Biol.,51,573 (1970)
(6) M. Frey, P. Wahl and H. Benoit, J.Chim.Phys.,61,1005 (1964)
(7) Y. Nishijima, A. Teramoto, M. Yamamoto and S. Hiratsuka, J.Polym.Sci.,A-2,5,23 (1967)
(8) E.V. Anufrieva and Yu. Ya. Gotlib, Adv.Polym.Sci.,40,1 (1981)
(9) Yu.Ya. Gotlib and P. Wahl, J.Chim.Phys.,60,849 (1963)
(10) D. Biddle and T. Nordstrom, Arkiv Kemi,32,359 (1970)
(11) E.V. Anufrieva, M.V. Volkenstein, Yu.Ya. Gotlib, M.G. Krakovyak, S.S. Skorokhodov and T.V. Sheveleva, Dokl.Akad.Nauk SSSR,194,1108 (1970)

(12) E.V. Anufrieva, Yu.Ya. Gotlib, M.G. Krakovyak and S.S. Skorokhodov, Vysokomol.Soedin.,A 14,1430 (1972)
(13) T.J. Gill,Jr. and G.S. Omenn, J.Am.Chem.Soc.,87,4188 (1965)
(14) E.V. Anufrieva, V.D. Pautov, I.M. Papisov and V.A. Kabanov, Dokl.Akad.Nauk SSSR,232,1096 (1977)
(15) Y. Nishijima, J.Polym.Sci.,C 31,353 (1970)
(16) G. Weber,Biochem.J.,75,345 (1960)
(17) A. Grinwald and I.Z. Steinberg, Biochim.Biophys. Acta,427,663(1976)
(18) R.A. Kenner and A.A. Aboderin, Biochemistry,10,4433 (1971)
(19) W.O. Mcclure and G.M. Edelman, Biochemistry,6,567 (1967)
(20) U.P. Strauss and G. Vesnaver, J.Phys.Chem.,79,2426 (1975)
(21) G. Oster and Y. Nishijima, J.Am.Chem.Soc.,78,589 (1957)
(22) G. Oster and Y. Nishijima, Fortschr.Hochpolym.Forsch.,3,313 (1964)
(23) G. Oster, Compt.Rend.,232,1708 (1951)
(24) R.O. Loufty, Macromolecules,14,270 (1981)

(25) R.O. Loufty, Macromolecules,16,678 (1983)
(26) G. Duportail, D. Frolich and G. Weill, Eur.Polym.J.,7,977 (1971)
(27) L. Moldavan and G. Weill, Eur.Polym.J.,7,1023 (1971)
(28) Yu.E. Kirsh, N.R. Pavlova and V.A. Kabanov, Eur.Polym.J.,11,495
     (1973)
(29) I.A. Taha and H. Morawetz, J.Am.Chem.Soc.,93,829 (1971);
     J.Polym.Sci.,A-2,9,1668 (1971)
(30) C.D. Jonah, M.S. Mathieson and D. Meisel, J.Phys.Chem.,83,257(1979)
(31) T. Förster and K. Kasper, Z.Phys.Chem.(Frankfurt),1,275 (1954)
(32) B. Stevens, Spectrochim.Acta,18,439 (1962)
(33) F. Hirayama,J.Chem.Phys.,42,3163 (1965)
(34) K. Zachariasse and W. Kuhnle, Z.Phys.Chem.(Frankfurt),101,267(1976)
(35) J.B. Birks, "Photophysics of Aromatic Molecules",Wiley, New York,
     1970, pp.354-359
(36) G.E. Johnson, J.Chem.Phys.,63,4047 (1963)
(37) M. Goldenberg, J.Emert and H. Morawetz, J.Am.Chem.Soc.,100,7171
     (1978)
(38) M.T. Vala, J. Haebig and S.A. Rice, J.Chem.Phys.,43,886 (1965)
(39) L. Bokobza, B. Jasse and L. Monnerie, Eur.Polym.J.,13,921 (1977)
(40) W. Klöpffer, J.Chem.Phys.,50,2337 (1969)
(41) F. Nishihara and M. Kaneko, Makromol.Chem.,124,84 (1969)
(42) A.C. Sommersall and J.E. Guillet, Macromolecules,6,218 (1973)
(43) J.S. Aspler and J.E. Guillet, Macromolecules,12,1082 (1979)
(44) C. David, M. Lempereur and G. Geuskens, Eur.Polym.J.,8,417 (1972)
(45) Y.-C. Wang and H. Morawetz, Makromol.Chem.,Suppl.1,283 (1975)
(46) R. Qian, in "Macromolecules,Main Lectures Presented at the 27th
     International Symposium on Macromolecules", H.Benoit and P. Rempp,
     Eds.,Pergamon,Oxford,1982, p.139
(47) J. Jachowicz and H. Morawetz, Macromolecules,15,828 (1982)
(48) M. Grayley, A. Reiser, A.J. Roberts and D. Phillips, Macromolecules
     14,1752 (1981)
(49) I.B. Berlman,"Energy Transfer Parameters of Aromatic Molecules",
     Academic Press,New York, 1973
(50) M. Yokoyama, T. Tamamura, M. Atsumi. M. Yoshimura, Y. Shirota
     and H. Mikawa, Macromolecules,8,101 (1975)
(51) R.F. Reid and I. Soutar, J.Polym.Sci.,Polym.Phys.Ed.,16,231 (1978)
(52) R.A. Anderson, R.F. Reid and I. Soutar, Eur.Polym.J.,15,925 (1979)
(53) T. Förster, Naturwiss.,35,166 (1946); Ann.Physik,2,55 (1948);
     Disc.Faraday Soc.,22,7 (1959)
(54) H. Edelhoch and R.F. Steiner, J.Biol.Chem.,240,2877 (1965)
(55) K.-H. Huang, R.H. Fairclough and C.R. Cantor, J.Mol.Biol.,97,443
     (1975)
(56) H.E. Shaghpour, A.E. Dieterich, C.R. Cantor and D.M. Crothers,
     Biochemistry,19,443 (1975)
(57) R. Peters, J. Peters, K.H. Tews and W. Bahr, Biochim.Biophys.Acta,
     363,282 (1974)
(58) J. Schlessinger, D.E. Koppel, D. Axelrod, K. Jacobson, W.W. Webb
     and E.L. Elson, Proc.Natl.Acad.Sci.,U.S.,73,2409 (1976)
(59) L.R. Smith, J.W. Parce, B.A. Smith and H.M. McConnell,
     Proc.Natl.Acad.Sci.,U.S.,76,4177 (1979)

(60) B.A. Smith, W.R. Clark and H.M.McConnell,
     Proc.Natl.Acad.Sci.,U.S.,76,5649 (1979)
(61) B.A. Smith, Macromolecules ,15,469 (1982)
(62) M. Antonietti, J. Contandin, R. Grütter and H. Sillescu,
     Macromolecules,17,798 (1984)
(63) P.G. de Gennes, J.Chem.Phys.,55,572 (1971)
(64) M. Doi and S.F. Edwards, J.Chem.Soc.,Faraday Trans.2,74,1789 (1978)

FLUORESCENCE AND PHOSPHORESCENCE SPECTROSCOPY IN POLYMER SYSTEMS: A GENERAL INTRODUCTION

Shigeo Tazuke
Tokyo Inst. of Technology, Research Lab. of Resources Utilization
5259 Nagatsuta, Midori-ku, Yokohama 227, Japan

Mitchell A. Winnik
Department of Chemistry and Erindale College
Univiversitiy of Toronto
Toronto, Ontario M5S 1A1, Canada

ABSTRACT. This chapter provides a general introduction to the techniques of fluorescence and phosphorescence spectroscopy. Emphasis is focussed on aspects particularly relevant to the study of polymer samples, both in solution and in the solid state. Techniques are discussed, as are common artifacts. We attempt to establish guidelines to show how various problems of current interest in polymer science might be studied by these luminescence techniques.

1. INTRODUCTION

The applications of fluorescence and phosphorescence spectroscopy as tools to study synthetic polymers are enjoying rapid growth. This is a recent phenomena. It follows from, and benefits from, the extensive applications of these techniques in the biological sciences (1). The factor, we believe, which has in the past limited the application of these tools to synthetic polymer systems, has been a misapprehension that they involve difficult techniques, requiring optically perfect samples, expensive equipment, and specialized training. A clear sign that an analytical technique is widely appreciated is that it is used by groups who do not specialize in studying the technique itself. This situation exists in various biological fields but is not yet the case in the study of synthetic polymers. It is our hope that this text and this chapter will serve to make these tools better appreciated and more accessible.

Polymer scientists use photophysical measurements for two quite distinct purposes (2). On one hand there are those whose primary interest is in the intrinsic photophysical properties of polymers which contain fluorescent and phosphorescent groups. Their samples normally

15

M. A. Winnik (ed.), Photophysical and Photochemical Tools in Polymer Science, 15–42.

contain many fluorophores, often one per repeat unit. Their major
interest is in interfluorophore excited state interactions. Early
experiments focussed on readily available polymers such as polystyrene,
poly(vinylnaphthalene), poly(vinylcarbazole) and various aromatic methacrylate homopolymers (2). Steady-state fluorescence measurements were
used to study excimer formation, energy migration/transfer, and fluorescence quenching behavior of each individual polymer. Simple models were
presented. These in turn made specific predictions about the fluorescence decay behavior of the systems.

As more and more research groups gained access to the time-
correlated single photon counting instrument for measuring fluorescence
decay profiles, it became clear that these models presented a serious
oversimplification of the phenomena. Even now with picosecond sources
and even more complex models, data interpretation remains controversial.
This subject will be examined in more detail in chapters by Soutar,
Phillips, and Frank. It may be that a new phase in this area of research
will be opened when one has access to a series of polymer samples of
controlled end groups, known stereochemistry, and low molecular weight
polydispersity.

The second application of luminescence spectroscopy in polymer
science has been as a tool to study polymer systems themselves. Here a
fluorescent or phosphorescent dye is introduced into a polymer environment as a molecular sensor of the environment. One chooses the dye with
a knowledge of its spectroscopy in the hopes that changes in its emission
spectrum, or, in a pulsed experiment, its emission decay profile, will
convey detailed molecular level information about the polymer system
itself. These are the experiments which mimic applications of
luminescent sensor techniques in biology, where these dyes provide
information about hydrophobicity in proteins, local polarity at
water-membrane interfaces, distances in antibody-antigen interactions,
and a wide variety of other issues concerning system morphology and
dynamics.

It is common to use the term "probe" to describe a small dye molecule added passively to a system for study. By contrast, a "label" is
attached covalently to some component of the system. Fluorescent probe
experiments are easy to carry out, but the onus rests with the experi-
mentor to prove where in the system the probe is located. Labelling
experiments are more demanding because of the need to synthesize the
labelled component. Data interpretation is often easier because the
dilemma of dye location is less severe. We like the phrase "luminescent
sensor" to describe the general case where a dye is used either as a
label or as a probe.

In all sensor experiments, one has to be concerned about whether the
dye perturbs its environment enough to affect the measurement. This is
not a concern when the chromophore is intrinsic to the system: e.g., a
tryptophan in a protein or an adventitiously fluorescent curing agent in
an epoxy resin. Nevertheless, one tries wherever possible to use
complementary techniques to establish confidence in the meaningfullness
of the sensor experiments.

The ultimate impact of fluorescence and phosphorescence spectroscopy
on polymer science is going to depend upon the ability of these sensor

techniques to provide answers to questions of conformation, dynamics, and morphology in polymer systems. These are the application areas with the fastest current growth in activity and where the future seems most promising. It is therefore timely to discuss the strengths and weaknesses of luminescence spectroscopy, and to explore the scope and limitations when these techniques are applied to the study of synthetic polymer systems.

2. BRIEF DESCRIPTION OF FLUORESCENCE AND PHOSPHORESCENCE PROCESSES (3)

The absorption of light by a molecule produces an electronically excited state. Since both the ground and excited states possess multiple vibrational levels, the transition from the ground state to the excited state (photoabsorption) or vice versa (photoemission) takes place to various vibrational levels either in the excited state or in the ground state. The absorption and emission spectra therefore exhibit vibrational structures reflecting vibrational levels of the excited state and the ground state, respectively. Absorption or emission to a state with a dissociative energy surface will be broad and structureless in shape. These transitions are non-adiabatic processes and occur within a time scale of $<10^{-15}$s so that the nuclei in the molecule cannot change their arrangement during the transition (Franck-Condon principle). Subsequently, the excess vibrational energy is dissipated by collision with surrounding molecules, during which the molecule relaxes to its equilibrium geometry. In fluid condensed phases, rearrangement of the surrounding solvent molecules takes place on the same time scale (10^{-13} to 10^{-12}s). In the case where a molecule is excited to higher electronic levels, transitions from the higher energy state (S_n) to the lowest excited state (S_1) (internal conversion) occur with dissipation of the excess vibrational energy. This process occurs within a much shorter time than the fluorescence lifetime, particularly in condensed phases, so that fluorescence is usually observed only from the lowest singlet excited state. In polymer systems, emission from higher excited singlet states has never been observed.

2.1. Singlets and triplets

The photophysical pathways which dissipate the energy of the S_1 state are classified into radiative and non-radiative processes. For our purposes, the most important and informative unimolecular decay process is fluorescence ($S_1 \rightarrow S_0 + h\nu$). Other processes include intersystem crossing to the triplet state ($S_1 \rightarrow T_1 + heat$), radiationless decay and intramolecular chemical reaction pathways. The lowest triplet excited states decay by a radiative path (phosphorescence, $T_1 \rightarrow S_0 + h\nu$) or radiationlessly (intersystem crossing, $T_1 \rightarrow S_0 + heat$) to the ground state. These photophysical processes are illustrated in the Jablonski state diagram in Figure 1.

Since the transition probability of phosphorescence is much smaller than that of fluorescence, phosphorescence has a much longer lifetime.

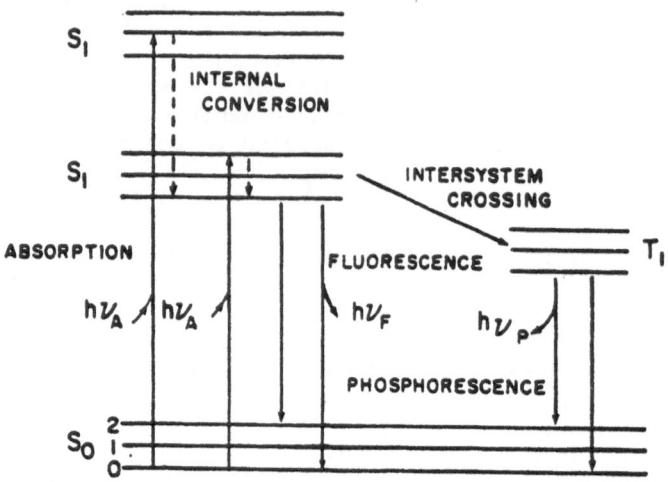

Figure 1. A Jablonski diagram showing the lowest
singlet and triplet states of a typical molecule
and identifying the various transitions.

This facilitates the relative importance of radiationless decay by
internal conversion or by quenching through collision with traces of
impurities. Consequently, phosphorescence is rarely observable in fluid
media. An important exception is in the case of ketones which have
lowest energy $^3(n\pi^*)$ triplet excited states (4). Here photon emission
occurs at rates of 10^2 to 10^3 sec^{-1}, fast enough to compete with solvent
or impurity quenching if care is taken to deoxygenate the samples and
purify the solvents. For molecules such as acetone, acetophenone, benzophenone, biacetyl and benzil, phosphorescence is readily observed in
fluid solution at ordinary temperatures with (1/e) lifetimes of
50--500 μs. Heavy atoms promote phosphorescence rates. Dibromoacetonaph-
thone (5), with a lowest $^3(\pi\pi^*)$ triplet state is a useful phosphorescence
probe of micellar systems. There is a whole literature on heavy--atom
induced room-temperature phosphorescence applications in analytical
chemistry (6).
    Nevertheless phosphorescence of simple aromatics can normally be
observed only in rigid matrices. Here impurity diffusion is suppressed
so that emission from triplet states with lifetimes of 1 to 10 seconds
competes effectively with diffusive quenching processes. Many of the
early spectroscopic studies of the phosphorescence of these aromatics
were carried out in films of glassy polymers such as poly(methyl metha--

crylate) [PMMA] (7). More recently there has been a recognition that impurity diffusion effects in the matrix, which are parasitic to a purely spectroscopic study of the dye, are rich in information about relaxation processes in the polymer matrix which affect diffusion processes (8).

2.2. Interactions leading to different energy levels of excited states

Besides radiative and non-radiative unimolecular decay pathways to the ground state, bimolecular processes may participate, resulting either in total deactivation of the excited state or formation of another excited state species. The latter case is of particular interest since the new state, and its generation and dissipation processes, provide information about the molecular environment of the chromophore. Various phenomena fall into this category, including excimer/exciplex formation and energy transfer. These bimolecular interactions will be discussed in detail in latter sections of this chapter. An intramolecular process, twisted intramolecular charge transfer (TICT) emission is another phenomenon useful to probe molecular environments and molecular motion in polymer systems.

2.3. Orientation of transition moments

For a molecule to absorb light, not only must the energy of the light correspond to the energy of the transition, but also a component of the electric vector of light must be parallel to the transition moment of the molecule. One of the consequences of this requirement is that exciting a sample of randomly oriented molecules with plane-polarized light produces an optical selection of which molecules are excited. If molecular motion is slow compared to emission, the emitted light will also be polarized. Furthermore, the rate of molecular motion can be estimated by determining the degree of polarization. When the molecular motion is fast enough compared with luminescence lifetime, the polarization (P) or the anisotropy (r) of the emission fall to zero.

The terms, the emission polarization $\underline{P}$ and the emission anisotropy $\underline{r}$, defined in equations 1 and 2, are the experimentally and theoretically significant measures of the extent to which the emitted light is polarized after excitation of the sample with plane polarized light. In these expressions $I_{\parallel}$ and $I_{\perp}$ denote the measured intensities when the observing polarizer is parallel or perpendicular to the direction of the polarized excitation. Polarization measurements are treated in much greater detail in the chapter by Monnerie.

$$P = \frac{I_{\parallel} - I_{\perp}}{I_{\parallel} + I_{\perp}} \qquad (1)$$

$$r = \frac{I_{\parallel} - I_{\perp}}{I_{\parallel} + 2I_{\perp}} \qquad (2)$$

Rotation of emission dipoles can also occur by energy migration among randomly oriented molecules. Molecular motion and energy migration can be separated by measurements in a glassy matrix or in very viscous solution, where molecular motion can be suppressed. Polarization measurements are also informative about the orientational distribution of fluorophores in solid polymers (10). This particular application is also discussed in detail in the chapter by Monnerie (00).

In concluding this section, we would like to point out that there are nine different experiments that one can in principle carry out on any given sample. These are listed in Table 1. Every dye can emit fluorescence from its lowest singlet state and phosphorescence from its lowest triplet state. Both emissions may not, however, always be observable. For each of these emissions one can run an excitation spectrum by setting the emission monochromator to the appropriate wave-length and scanning the excitation monochromator. Using pulsed sources one can measure the decay profile of fluorescence and of phosphores-cence. By exciting the sample with plane polarized light, one can measure the steady state polarization, or with a pulsed source, the polarization decay of the sample.

Table 1

TYPES OF LUMINESCENCE MEASUREMENTS

Steady State	Transient
fluorescence spectra	fluorescence decay time resolved fluorescence spectra
phosphorescence spectra	phosphorescence decay time resolved fluorescence spectra
excitation spectra	
polarization	polarization decay

3. MERITS OF THE METHOD

3.1. High sensitivity

Luminescence can be measured with great sensitivity. Micromolar quantities of dye can be studied with ease; nanomolar concentrations are within the scope of the method. One can also work with tiny amounts (micrograms) of sample if necessary. This is why fluorescence methods have been such powerful tools in biological systems where the available amount of sample is often extremely limited (1,9). High sensitivity is essential to reduce the amount of externally added probe or label since these are, of course, contaminants in the system. Luminescence methods allow one to do meaningful experiments with dyes added at the 1-100 ppm level in the sample.

3.2. Versatility of information

Luminescence methods can provide a broad spectrum of information about a particular system. The versatility of the method derives from three separate factors. First there are the nine different luminescence detection experiments described above. Second, different dyes are capable of sensing different aspects of any system into which they are introduced. Third and perhaps most important, bimolecular quenching processes provide a battery of high resolution tools for measuring distances and studying dynamics in macromolecular systems. These processes will be described in more detail in a subsequent section of this chapter.

3.3. Commercially available instrumentation

Good fluorescence spectrometers are commercially available at reasonable prices. The most important consideration in choosing an instrument is the quality of the monochromators. Polymer samples are frequently highly scattering (powders, turbid solutions). Scattered light is the most serious source of interference in these measurements; one should pay careful attention to the stray light rejection characteristics of the monochromators in choosing an instrument. With double monochromators one gains an important enhancement in stray light rejection at the expense of lower light throughput and hence lower sensitivity. Another considera-tion is a sample compartment which will easily accommodate front-face (reflectance) mesurements. A sample compartment with good temperature control is very important. Solid polymer samples often exhibit discon-tinuous changes in luminescence behavior at transition temperatures other than T_g. Under ideal circumstances, one could carry out measurements at any temperature from 77°K to 500°K.

Most spectrometers can be purchased already interfaced to a small computer. One uses the software to carry out rather standard manipula-tions of the spectra. These are incorporated into most data analysis packages. Common spectra manipulations include substracting background, normalizing spectra, converting wavelength to frequency (including the λ^2 correction), integrating areas, and correcting emission and excitation spectra for the lamp output and the spectral response of the monochromators and photomultiplier tube. It is also a convenience to be able to calculate polarization values from direct data input to the computer.

Phosphorescence is a delayed emission often much weaker than fluorescence from the same chromophore. In a standard spectrum, phosphorescence often appears as small ripples in the tail of the fluorescence spectrum. To remove the fluorescence from the spectrum, two approaches are taken. Traditionally one used a mechanical shutter or a rotating cam with slits that block the exciting light while passing the emitted light to allow only delayed emission to reach the photomultiplier tube. Mechanical choppers have millisecond resolution and permit decay

time measurements by varying the chopping rate. More recently, intense
pulsed lamp attachments have become available for commercial spectrometers. These, coupled with gated detection circuitry permit microsecond
resolution, increased sensitivity, and better precision in measuring
phosphorescence decays.

New types of detectors, such as OMA (optical multichannel analyser),
SMA (spectrum multichannel analyser) and MCPD (multichannel photodiode
array), are able to record the total emission spectrum by a single shot
after flash excitation, so that acquisition of time-resolved emission
spectra becomes much easier. One uses appropriate gating techniques,
synchronized to the excitation flash, to control the time scale in the
data acquisition. At the moment semiconductor based detectors are still
less sensitive than photomultiplier (PMT) detectors.

Nanosecond fluorescence decays can be measured either by the
phase-shift method (9) or with the time-correlated single photon counting
technique (11). Few people working with synthetic polymer systems
currently use the phase-shift method. Most fluorescence decays in
synthetic polymers are non-exponential. One needs a technique that
allows one to see the entire form of the decay curve. In the
time-correlated single photon counting technique one obtains a signal
which is a convolution of the instrument response (the excitation profile
of the flash lamp and the temporal response of the detector) with the
true decay curve. Computer methods are used to separate these, most
effectively by assuming a form for the true decay, and iteratively
reconvoluting it with the instrument response function, separately
determined. These methods are discussed in more detail later in this
volume and elsewhere (11). Here we wish to make the point that data
analysis is model dependent. When one uses this technique, one has to
write new software for every new model one develops. It is important
when purchasing a commercial instrument to make certain that software
purchased is fully documented so that changes can be made as needed.

4. PROBES AND LABELS

In some experiments, such as the fluorescence-recovery-after-pattern-
photobleaching method and the forced Rayleigh scattering method used to
measure diffusion rates (12), dyes serve only as tracers. The experiment
does not depend much on the spectroscopic characteristics of the dye
used. In most other experiments using luminescence detection in
polymers, one designs the experiment to take advantage of a particular
spectroscopic property of the chosen dye. The broad choice of dyes and
the variety of spectroscopic interactions is the source of the power of
luminescence techniques as a tool. Nevertheless, this variety can
seem overwhelming to someone just entering the field. In the following
paragraphs we try to summarize the most important of these spectroscopic
properties.

4.1. Single molecule sensors

Most dyes have different charge distributions in their ground and excited states. Hence absorption and emission spectra are often sensitive to solvent polarity. Dyes which have particularly large differences are very sensitive probes of local polarity. There are many such dyes used in the biomembrane field to probe the local environment of the membrane-water interface. Recent papers by Morawetz (13a), Osada (13b), and by Turro (13c) have shown that such dyes are also useful for examining aqueous solutions of polymers.

Molecules with twisted intramolecular charge transfer states (14a) are also very sensitive to the effective polarity of the local environment. These dyes undergo rotation in the excited state to produce the charge separated species. This motion is retarded by local friction imposed by the medium. There are strong indications that this motion is sensitive to local free volume and its distribution in solid polymer samples (14b).

The shape of the fine structure in the absorption and fluorescence spectra of pyrene is very sensitive to local polarity (15a). The so-called "Py-scale" of solvent polarities (15b) is a manifestation of the Hamm effect (15c), whereby locally anisotropic electric fields relax the forbiddenness of the $(0,0)$ band in $S_1 \rightarrow S_0$ transitions of symmetrical aromatic chromophores. Thomas has used pyrene fluorescence to probe local polarity in aqueous micelles and in aqueous solutions of polyelectrolytes (15d).

Molecules which undergo large scale motion after excitation are sensitive to friction effects of the medium traditionally associated with bulk solvent viscosity. For example, the extent and rate of fluorescence depolarization of diphenylhexatriene, dissolved in membrane bilayers, is often used to infer local or microscopic fluidity (9). Similarly, intramolecular excimer formation rates in molecules of the form $ArCH_2XCH_2Ar$ (Ar is an aromatic group; $X = O$, CH_2, $NHCOCH_3$) reflect frictional effects in their immediate microenvironment (16). It is well documented that in homologous series of similar liquids, these rates decrease in proportion to the viscosity increase. This is an active area of research with many questions still to be answered. There are strong indications that at very high viscosities, rates of conformational change are limited by the dynamics of free volume distribution associated with the glass transition at the frequency of the conformational change in the sensor (17).

One application of single molecular sensors of particular interest to polymer scientists is the ability of polarization measurements to detect local orientation of polymer molecules under elongational stress (10).

4.2. Quenching processes

Important bimolecular processes in the excited state include complex formation and energy transfer processes. When these processes produce non-fluorescent species, fluorescence or phosphorescence quenching alone is observed. Quenching is the general word used to describe any

bimolecular process which decreases emission intensity or increases the
emission decay rate. There are many such mechanisms. The most common
ones are presented in Table 2. The important feature of Table 2 is that
each particular quenching mechanism involves interactions between groups
over different interaction distances. For some processes, particularly
bond formation, adjacency and orbital overlap is necessary. For other
processes such as electron transfer and Förster energy transfer,
communication can occur over substantial distances.

Table 2

Bimolecular Excited State Quenching Processes

	interaction mechanism	effective distance[a, b]
1.	Energy transfer by	
	dipole coupling	10 Å to 100 Å
	electron exchange	4 Å to 15 Å
	reabsorption	as far as emission reaches
2.	Electron transfer	4 Å to 25 Å
3.	Exciplex formation	4 Å to 15 Å
4.	Excimer formation	ca. 4 Å
5.	Non-emissive self-quenching	4 Å to 15 Å
6.	Heavy atom effect	ca. 4 Å
7.	Chemical bond formation	ca. 2 to 4 Å

[a]The mimimum interaction distance is arbitrarily taken to be 4
Å except where new chemical bonds are formed.
[b]Each pair of chromophores, for each interaction mechanism, has
its own characteristic distance R_0 at which the interaction
rate equals the decay rate of the unquenched excited state.
These values are estimates of the range of R_0 for randomly
oriented non-diffusing pairs of species.

Quenching rates, for non-diffusing species, vary with chromophore
separation and orientation. This situation, where an interaction is
governed by a distance-dependent rate constant $k(r)$, is potentially rich

in information. For simplicity, one normally describes each pair of
chromophores for each interaction mechanism by its own characteristic Ro
value. Ro is defined as the distance between pairs of chromophores
(assuming random orientation) where the rate of quenching is equal to the
unimolecular decay rate of the excited donor. For some processes such as
energy transfer by the Förster mechanism (18), good theory, expressed in
terms of readily accessible spectroscopic information, allows Ro values
to be calculated. For electron transfer processes and energy transfer by
the exchange (Dexter) mechanism, there is good theory, but Ro has to be
obtained from experiments on model systems. Sometimes, when the details
of the quenching mechanism are only poorly understood, one can only guess
at the range of Ro values.

For the polymer scientist, consideration of these Ro distances is
one of the most important aspects of experimental design. Emission
quenching is the tool of choice for studying problems of interfaces and
interphases. One labels one component with a dye A and the other
component with Q. Studying the rate and efficiency whereby Q quenches A*
provides a powerful measure of molecular intimacy in the overlap region.
There are other applications pertinent to studying polymer conformation
and dynamics.

From a qualitative point of view, a large Ro makes experiments
easier in disordered systems. One needs fewer labels. With the coarser
resolution, the signal due to quenching is stronger. For quantitative
data interpretation, one has to refer to the appropriate theory from
which Ro was evaluated.

In the following paragraphs, we comment briefly on each of the
various bimolecular quenching mechanisms cited in Table 2 and provide
references to the literature for those interested in a more detailed
discussion of the theory and phenomena:

4.2.1. Energy transfer

Energy transfer describes the process

$$A^* + Q \rightarrow A + Q^* \qquad\qquad [7a]$$

whereby the excitation energy localized on A is transferred to Q to
produce electronically excited Q. Both singlet and triplet energy can be
transferred, and for the transfer to be most effective, the excited state
of Q should be lower than that of A. A special case of energy transfer
common to polymer systems containing a chromophore in the repeat unit is
the processs of energy migration (19)

$$A^* + A \rightarrow A + A^* \qquad\qquad [7b]$$

Energy transfer to an identical dye conserves the concentration of
[A*]. Unless energy migration leads to more rapid deactivation of A*,

this process is very difficult to study. Polarization measurements
provide one method to study this reaction since energy migration leads to
fluorescence depolarization (20).

The Förster or "long range" energy transfer mechanism (17) operates
via a coupling of the transition dipoles of the donor A* and the acceptor
Q. Since the oscillator strength of the transition enters into the
expression for the rate constant k_{ET} for energy transfer, this mechanism
is most common to singlet energy transfer. In the expression for the
efficiency ϕ_{ET} of energy transfer the donor lifetime has a compensating
effect. If there is good overlap between the emission spectrum of A* and
the absorption spectrum of Q, and the extinction coefficient of Q_o is
large in the overlap region, R_o can attain values as large as 60 Å.

This energy transfer mechanism is the basis of the "spectroscopic
ruler" concept (1c): k_{ET} and ϕ_{ET} vary sensitively with the separation
(r) between the centers of the transition moments of A and Q:

$$k_{ET}, \quad \phi_{ET} \quad \sim \quad (\frac{R_o}{r})^6 \qquad\qquad\qquad [8]$$

Consequently, measurements of rates and efficiencies of energy transfer
in non-diffusing systems can determine the distances between groups. If
A and Q are attached to specific sites in polymers, important high
resolution distance information can be obtained.

We have not discussed the problems of molecular orientation in this
process. The orientation factor involved in the proportionality constant
of eq. 8 is taken to be $K^2 = 2/3$ for randomly oriented dipoles. This
assumption is probably acceptable for amorphous polymers, but doubtful
for highly oriented systems. Long-range energy transfer/migration in
organized molecular aggregates is becoming more and more important.
Oriented molecular aggregates are also within the scope of polymer
science. For example, there have been recent reports of polymerized
Langmuir-Blodgett films, which are highly oriented (22).

Energy transfer by the exchange mechanism requires orbital overlap
between A* and Q. The original theory is due to Dexter (23a). Most
examples of triplet energy transfer occur in this way. The application
of this theory to experiments in rigid media has been developed by
Inokuti and Hirayama (23b). Typical R_o values are 10 Å to 15 Å. One
problem with the data interpretation is that the theoretical treatment of
experiments in rigid media makes assumptions about an isotropic distribution of chromophores in the system. There are many possible applications
of this process to glassy polymer systems, virtually none of which have
been exploited.

There are two other possible mechanisms of energy transfer. When
excimer formation and dissociation to the excited monomer are fast and
efficient, the partner can carry away the excitation energy. The second
process, called "trivial" energy transfer, is based upon reabsorption of
donor emission by an acceptor. The condition of emission-absorption
overlap is identical with that for the Förster mechanism. Although this
mechanism is uninteresting and trivial from a mechanistic point of view,

actually correcting for the contribution of this mechanism is not
trivial. Furthermore, this process is effective over any distance that
the emitted photon can traverse. The trivial process is often a dominant
mechanism of energy transfer in dilute solution (21).

4.2.2. Electron transfer

Electronically excited states are more easily oxidized as well as more
easily reduced than ground states. In consequence, photochemically
driven electron transfer processes are quite common (24). The general
expression for such a reaction between a donor molecule D and an acceptor
A, where either may be excited, is

$$A* \ + \ D \ \text{->} \ A^- \ + \ D^+ \ \text{<--} \ A \ + \ D* \tag{9}$$

The topic has received considerable attention of late, stimulated in part
by the interest in solar energy conversion. Recent papers have examined
the theory of electron transfer rate constant $k_{et}(r)$, which in non-
diffusing systems decreases exponentially with distance, and the
influence of energetics and diffusion on electron transfer rates (24b).
Electron transfer can occur over distances of tens of Angstroms.
Singlet and triplet states are characterized by different R_0 values
(24c).
There are many subtle features associated with these processes and
one has to be careful not to oversimplify the relation between electron
transfer quenching of excited states and subsequent ion-radical
formation. Coulombic interaction between A^- and D^+ plays a vital role in
back electron transfer to produce D and A. A recent paper discusses some
of these problems both in polyelectrolyte and small molecule systems
(25). The so called "inverted Marcus region" has been another point of
dispute. Marcus (26) predicted not only that the rate of electron
transfer would increase as, ΔG goes from positive to negative values (the
normal region), but also that for $\Delta G << 0$, the electron transfer rates
would again decrease (the inverted region). There are some cases (24c)
where this effect is observed, and others, where it is not. Mataga (27a)
has recently proposed a model to explain both phenomena in terms of
factors modifying the shape of the potential energy surfaces. Masuhara's
chapter examines these phenomena in more detail (27b).

4.2.3. Exciplex formation

Exciplexes are electronically excited binary complexes which have only
weakly associating or dissociative ground state interactions. Weller
(28) treats exciplexes as contact radical ion pairs ($[A^-/D^+]$ <-->
$[A/D]*$). In sufficiently polar solvents, the ion pairs dissociate to
free ions. In non-polar solvents one can frequently observe a new
emission, broad and structureless, and red-shifted from that of A* or
D*. The mechanism of exciplex formation is obviously closely tied to
that of excited state electron transfer reactions. There is an important

difference in the behavior of exciplex forming groups in polymers and in small molecules. In the latter the ground state donor-acceptor forces are never sufficient to cause significant association when both species are present in low concentration. If these groups are attached to polymers with a distribution that permits multiple D/A interactions, the cooperative nature of these interactions can lead to polymer-polymer association even at very low polymer concentration (29). This inter-polymer association depends sensitively on the regularity of the D and A arrangement along the polymer chain. Polymers with a regular alternating arrangement of D and A, or with fixed spacing of paired DA groups exhibit efficient polymer association. Other factors, such as the size of the polymer and solvent effects, also affect polymer association (30).

The energy of the exciplex state is dependent upon solvent polarity. Exciplexes can, therefore, act as sensors of the micropolarity around a polymer chain. Intramolecular exciplex forming groups, such as 1-(1'-pyrenyl)-3-(4'-N,N-dimethylaminophenyl)propane attached to a polymer chain are useful probes of local mobility. Very poor solvents for polystyrene containing this group seem to retard the extent of exciplex formation, suggesting that local motion of the side groups is reduced (31).

4.2.4. Excimer formation

Excimers are bound excited dimers from species which are non-associating in the ground state. The characteristic reaction can be written

$$A^* \; + \; A \; \rightarrow \; (AA)^* \qquad\qquad\qquad \text{[10a]}$$

Unlike exciplexes, the primary forces holding (AA)* together are exciton resonance (3), although charge resonance can become important. Excimers seem to have much more stringent orientational requirements than exciplexes. In some systems, the best face-to-face sandwich geometry is prevented by constraints imposed by an intervening chain connecting the two chromophores. These excimers have different radiative and non-radiative properties than excimers formed from the same chromophores in systems free from these constraints (32).

$$A^* \; + \; A \; \rightarrow \; 2A \qquad\qquad\qquad\qquad \text{[10b]}$$

An extreme case is sometimes seen in polymer systems. Here A* and A can interact to lead to the net deactivation of A*, eq. 10b, but where no excimer emission is observed (33). This process is referred to as self-quenching, and the detailed mechanism of this process is unknown.

4.2.5. Heavy atom effect

Spin-orbit coupling, which promotes rapid singlet-triplet interconversion, increases with atomic number to the fourth power. Heavy atoms promote this intersystem crossing, both internally as a substituent on a dye molecule, and externally. If Q contains a heavy atom (Hg, I, etc.) quenching of A* might be due to stimulated singlet-triplet interconversion. In many of these systems, electron transfer can also occur; and this, frequently, is the dominant interaction mechanism.

4.2.6. Chemical bond formation

Most bimolecular photochemical reactions require close proximity of reactants. Some processes, especially those involving concerted formation of two or more bonds, also have strict orientational requirements, and these processes are often closely related to excimer or exciplex formation.

5. PROBLEMS AND DIFFICULTIES

Every experimental technique suffers from artifacts. Luminescence spectroscopy is no exception. These quickly become known to the practitioner, and each artifact has its own particular folklore. New people entering the luminescence field are fortunate that good texts exist on the spectroscopic theory and on the practice of fluorescence and phosphorescence spectroscopy and photochemistry. Several of the older texts have become classics. These include books by Calvert and Pitts (34) (photochemical techniques), Parker (35) (luminescence measurement techniques), Birks (3) (spectroscopy of aromatic molecules) and McGlynn (36) (phosphorescence). These are supplemented by really excellent new volumes on fluorescence decay techniques (11) and its applications to biological systems. The Lakowicz text (9) on fluorescence is particularly useful.

5.1. Artifacts

An artifact is a distortion of a true result caused by a feature of the experimental technique. Encountering artifacts can be very discouraging, especially for the skilled scientist using a new technique for the first time. Our objective here is to point out the most common artifacts associated with fluorescence and phosphorescence spectroscopy. Recognition of the symptoms is frequently more than half the battle. Many of these problems and the appropriate measures one takes to avoid them are discussed in detail by Lakowicz (9).

Two kinds of spectral distortions arise from the use of right angle measurement geometry, where the incident light and collection lens are focussed typically at the centre of a 1 cm square cell. Both effects appear only when the optical density of the sample exceeds a value of 0.1. Self-absorption occurs when photons emitted by the dye have a

significant probability of being reabsorbed by the sample. This occurs
at the blue edge of the fluorescence spectrum where the absorption and
emission bands overlap. Self-absorption distorts not only the
fluorescence spectrum but leads as well to an artificial increase in the
fluorescence lifetime.

The most troublesome artifact when first encountered is the inner
filter effect. In samples of low optical density, luminescence intensity
increases with sample concentration. At higher concentrations the
optical density becomes sufficiently large that only a fraction of the
incident light reaches the midpoint of the cell where the collection
optics normally are focussed. One finds that fluorescence intensity
decreases as concentration increases. More disconcerting are distortions
caused to excitation spectra, which should in principle resemble absorption spectra. Since the extinction coefficient ε of the sample can vary
by orders of magnitude with a change in wavelength, the inner filter
effect can be prominent at wavelengths of large ε and absent elsehwere.
For proper determination of an excitation spectrum, the optical density
at the maximum absorption should be much less than 0.1. Also, for solid
polymer samples the problem of optical inhomogeneity is often
encountered.

In some applications, one requires corrected spectra. Corrections
in excitation spectra have to be made to account for the wavelength
dependence of the lamp source and the efficiency of the excitation monochromator throughput, as well as for temporal fluctuations in lamp
intensity (9). These corrections are commonly made within the fluorescence spectrometer by use of a quantum counter. A quantum counter is a
substance such as rhodamine B which has a fluorescence quantum efficiency
independent of the excitation wavelength. A small fraction of the
incident light is directed onto the quantum counter, whose emission
intensity serves as a reference for the emission intensity of the sample
itself. The ratio of the two intensities is independent of distortions
arising from the excitation source.

Emission spectra are also distorted by the wavelength sensitivities
of the photomultiplier tube and the emission monochromator (9). For many
routine applications, correcting for these effects is unimportant. For
example, in quenching experiments, the shape of the spectrum remains
identical but the area changes. The quenching efficiency is proportional
to the ratio of the peak areas, irrespective of any correction factors.
When, however, one has to compare intensities in different regions
of the spectrum, or determine the location of peak maxima, particularly
in energy units (ν, cm^{-1}), or measure absolute quantum yields, these
corrections have to be made.

While it seems to be widely appreciated that one should correct for
the wavelength-dependent response of the emission monochromator and
photomultiplier tube, it is less obvious that one must introduce an
additional correction for converting spectra from units of wavelength to
units of wavenumber. The bandpass of grating monochromators is constant
in $\Delta\lambda$ which corresponds to a different $\Delta\nu$ at different regions of the
spectrum. For spectra obtained in the usual form of intensity per
wavelength interval, conversion to the wavenumber scale requires that
each intensity be multiplied by λ^2

$$I(\nu) \;=\; \lambda^2 I(\lambda) \qquad\qquad [11]$$

This correction and (in most cases) the correction for spectral response both enhances the intensity contribution at low energies and shifts the peak maximum to the red.

Scattering is an ubiquitous problem. Rayleigh scattering, at the excitation wavelength λ_0, is particularly prominent with turbid or powder samples. The parasitic contribution to emission spectra and to fluorescence decay measurements is due to stray light. It is best removed in steady state measurements with good monochromators or in decay measurements with a proper choice of filters. Raman scattering is much less intense but becomes important when working with weakly emitting samples. The most prominent Raman bands are due to O-H and C-H stretching vibrations. The former is particularly prominent when water is used as the solvent; the latter is present when hydrocarbons are used. In alcohol solvents both may appear. These bands show up as peaks in the spectrum, shifted by 3600 cm^{-1} and 3000 cm^{-1} to the red of λ_0 respectively. By varying the excitation wavelength, the Raman peak also shifts. One can, therefore, minimize its consequences. It is worth remembering that monochromators pass harmonics of the chosen wavelength. A sample excited at 280 nm (e.g., naphthalene) will show an apparent emission at 560 nm which is really due to Rayleigh scattering of light passed by the excitation monochromator. Passing the excitation light through a band pass filter eliminates the problem.

Polarization measurements are sensitive to four types of artifacts. Misalignment of the polarizers will lead to apparent depolarization. Light scattering can lead to anomalously large depolarization. Sample reabsorption at high optical densities leads to depolarization. Anomalous results will also be obtained if the sample is embedded in a birefringent matrix. This problem can arise in experiments on dyes in polymer films.

Pulsed transient decay measurements are in many ways less sensitive to light scattering than steady state measurements. Scattering occurs during the time scale of the excitation and has the same temporal profile. Many research groups carry out measurements on powder samples. An obvious difficulty here is in distinguishing an important contribution to the fluorescence decay on a time scale shorter (e.g., picoseconds) than the width of the excitation pulse. More problematic in all such measurements is lamp stability (i.e., drift of flash timing and change in the lamp profile during measurement), which affects one's ability to deconvolute or reconvolute the data with the lamp profile. This is a necessary step for interpreting the measured I(t) decay curve.

There are a variety of well-known problems associated with the time-correlated single photon counting technique (11), the technique most commonly used for measuring fluorescence decays in synthetic polymer systems. Some are seemingly trivial (some filters fluoresce) until one stumbles across them. Others require rather delicate judgment. Among the latter, two problems are particularly noteworthy: First, it

frequently happens that one has to measure the decay of one of two
signals which have overlapping emissions. This occurs in excimer and
exciplex studies as well as in many energy transfer experiments. One has
to exercise considerable care in executing the measurement, as well as
some restraint in interpreting the data. Second, judgment is particularly delicate in choosing an appropriate time scale for a given
measurement. In many polymer systems, fluorescence decays are non-
exponential. Too short a time scale and the fractional decay of the
fluorescence is too short for one to assess the proper decay form. Too
long a time scale and the signal decays into the noise level with too few
points along the time axis for proper data analysis. A good rule is that
one should choose a time scale which gives two to three decades of decay
$I(t)$ [e.g., from 20,000 counts in the peak channel to 200 counts in the
tail, corresponding to 6 to 20 lifetimes for a single exponential decay].

Polymers are not homogeneous in a microscopic scale and a number of
perturbed states for a dye molecule are expected. As a matter of fact,
non-exponential decay of luminescence in polymer systems is a common
phenomenon. For some reaction processes (e.g., excimer and exciplex
formation), one tries to fit the decay curve to sums of two or three
exponential terms, since this kind of functional form is predicted by
kinetic models. Here one has to worry about the uniqueness of the fit
and the reliability of the parameters. Other processes can not be
analyzed in this way. Examples include transient effects in diffusion-
controlled processes, energy transfer in rigid matrices, and processes
which occur in a distribution of different environments, each with its
own characteristic rate. This third example is quite common when solvent
relaxation about polar excited states occurs on the same time scale as
emission from those states. Careful measurement of time-resolved
fluorescence spectra is an approach to this problem. These problems and
many others are treated in detail in recent books (9,11), including
various aspects of data analysis.

Many of us and our students have had problems in luminescence
measurements which we would like to forget, because in retrospect their
origin seems so obvious. We therefore recount without comment some facts
worth remembering: Some filters fluoresce, as do occasional samples of
quartz and pyrex. Stopcock grease is intensely fluorescent. Many
solvents (e.g., ketones, aromatics) have an intrinsic fluorescence if
they absorb at the excitation wavelength. Dyes must absorb light to
emit: Pyrex does not transmit light below 300 nm. Optical glass lenses
cut off light transmission at longer wavelengths. Many solvents absorb
strongly in the region 200 nm to 300 nm and each has a characteristic
cut-off wavelength where the optical density exceeds 2. Cylindrical
tubes make inexpensive and useful cells if the experiment is not
sensitive to their lensing effects. Many solvents can act to quench
fluorescence or phosphorescence. Samples may be photodecomposed by the
exciting light. Strict temperature control is often necessary. A cell
containing a volatile solvent, without a tight cap, is to be avoided.
The cap is a potential source of impurities.

5.2. Sample Impurities

Sample impurities can be a serious source of problem in luminescence experiments if they emit light at the wavelengths one intends to study, if they absorb light in competition with the probe, or if they quench the probe fluorescence. The first problem is the most serious since a trace substance with a high quantum yield of emission can generate a signifi- cant signal. The simplest method to detect the contribution of impurity fluorescence is to measure excitation spectra carefully as a function of monitoring wavelength. If a single emitting species is present, the excitation spectra should be identical irrespective of the monitoring wavelength. Oxygen, although not itself luminescent, is ubiquitous and quenches both fluorescence and phosphorescence at the diffusion- controlled rate.

It is sometimes a healthy attitude for one to consider whether impurity fluorescence in a sample can itself be used as an intrinsic sensor of molecular events within the polymer, and whether parasitic quenching might not provide a useful means of studying molecular relaxation processes which control impurity (oxygen) diffusion. There are numerous reports in the literature of how such studies of emission quenching can provide interesting information relevant to mechanical relaxation processes in polymers (8).

Commercial polymers frequently contain numerous additives such as fillers, antioxidants and plasticizers which may fluoresce or act as fluorescence quenchers. Little is known about the effectiveness of fluorescence and phosphorescence techniques in studying systems of such complexity. The situation is far from hopeless. One of us had the misfortune to find a fluorescence from a contaminant in a study of phenanthrene-labelled colloidal polymer particles (37). The impurity fluorescence seemed to be due to the choice of benzoyl peroxide as an initiator in the preparation. The problem could be overcome in three different ways: (i) use of a computer to substract away the impurity fluorescence, Figure 2, and to accommodate its contribution ($\tau^{-1} = 1$ ns) to the sample fluorescence decay; (ii) use of a different fluorescent label (e.g., pyrene) which could be excited where the impurity did not absorb; and (iii) removal of the offending impurity by modifying the synthesis of the sample.

5.3. Qualitative data analysis and interpretation

The most difficult kind of experimental problem is one in which quantita- tive data analysis is model dependent. Kinetics experiments are notori- ous in this regard, since one first assumes a mechanism, uses the mechanism to derive a rate law in terms of the rate constants, and then fits the raw data to the rate law. Any problems with the assumed mechanism carries over into the conclusions drawn from the data. This problem is particularly delicate for fluorescence decay measurements for several reasons:

First, with the time-correlated single photon counting technique one reconvolutes an assumed decay profile (obviously model dependent) with an experimental lamp decay profile. The latter varies from day to day and

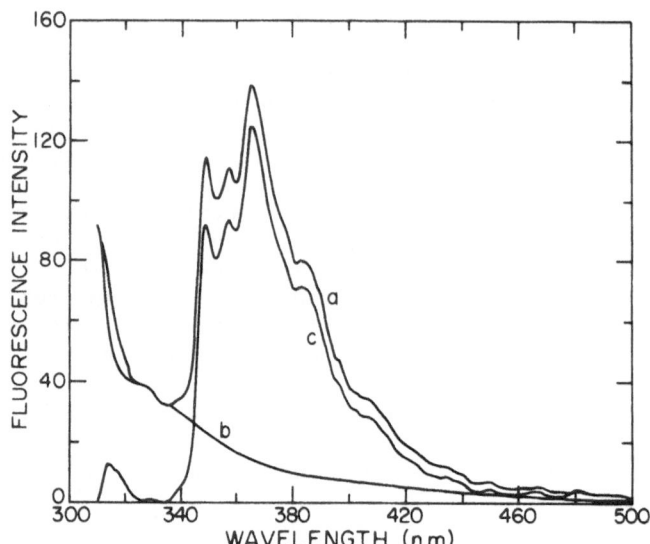

Figure 2. Fluorescence spectrum of a colloidal particle
composed of poly(vinyl acetate) [PVAc] and poly(2-ethylhexyl
methacrylate) labelled with phenanthrene in the PVAc phase,
curve a. Curve b is the fluorescence spectrum of an identical
unlabelled particle. Curve c is obtained by subtracting b
from a. The excitation wavelength was 290 nm.

from instrument to instrument. Raw data are not easily reanalyzed except
in the laboratory where they were obtained.

Second, one often uses values of quenching rate constants to
describe dynamic processes in polymers. Aside from the rate
law/mechanism problem mentioned above, diffusion controlled processes
offer another complication due to the transient contributions to the
quenching rate (38). These in turn affect the form of the decay profile
one uses to fit fluorescence decay data.

Third, the interpretation of polarization measurements normally
requires assumptions about the nature of the motion leading to emission
depolarization. Isotropic motion is easy to interpret. In some
instances one can get a very detailed description of anisotropic
rotational diffusion by combining nmr, light scattering and fluorescence
depolarization measurements.

Fourth, as with any technique, it is tempting to forget the simplifying assumptions behind the theory one uses to interpret observations.
Dye fluorescence may be very sensitive to local polarity or hydrophobicity, but what does that mean about the membrane–water interface or the
surface of an aqueous polymer layer? The absence of hydrophobicity
demonstrated by a fluorescence probe does not necessarily mean the
absence of hydrophobic domains in a system under study. Such domains may
be either too small to bind the hydrophobic probe or sterically
hindered. When a hydrophobic probe fluoresces strongly with the proper

emission, it is safe to conclude the presence of hydrophobic regions. But when it does not, one cannot say much.

There are other related problems. How does one assure a random distribution and random orientation of probes and labels, if that it is what is is desired? How does one determine the location of a probe in a complex system or insure that it does not perturb its own microenvironment sufficiently to distort the measurement? What about aggregation affects? These can change the spectroscopy of the dye.

These are difficult questions with no easy answers. Fortunately, some answers are available through carrying out different luminescence experiments, particularly time-resolved measurements, on any one labelled system. In addition one can change sensors to see if the information is sensor-dependent. Luminescence represents just one set of tools for studying polymer systems. Other tools (nmr, scattering techniques, etc.) provide other information all of which one should assemble into a self-consistent picture of the system.

6. PRESENT STATUS OF LUMINESCENCE METHODS IN POLYMER SCIENCE

Other chapters in this text will present a detailed evaluation of many of the various ways photophysical and photochemical processes can be used as a tool to examine polymer systems. In order to introduce this material we provide here a brief overview of the capabilities of these methods from a problem-oriented point of view. We have compiled in Table 3 a list of many important phenomena in polymer systems that one would like to examine. Next to each phenomenon we indicate photophysical and photochemical methods that have been, or might be, used to study these phenomena.

There are of course many other techniques with which to study polymer systems. These include, for example, stress-strain curve analysis for the study of mechanical properties, thermal methods such as DSC and DTA, which focus on bulk properties of the system. In addition there are an ever growing variety of techniques which provide information at the molecular level. Surface phenomena can be studied by SIMS, ESCA and other electron spectroscopies as well as ATR-infrared techniques. Other properties can be studied by ESR, NMR, X-ray, light, and neutron scattering.

In this volume we focus on analytical techniques based upon fluorescence and phosphorescence spectroscopy as well as flash photolysis methods. We emphasize the power and versatility of these methods for studying polymers in solution, in the solid state and in colloidal dispersion. We have not forgotten that the practitioner in polymer science is involved with solving problems. Normally a combination of methods is necessary to obtain a desired result. We hope that this chapter and this volume helps to make luminescence methods more accessible to the polymer scientist.

Table 3

HOW TO EXAMINE VARIOUS PHENOMENA OCCURRING IN POLYMERIC SYSTEMS

Phenomena	Techniques	Underlying Principles
rotational mobility changes	fluorescence (ns) and phosphorescence (μs, ms) depolarization of isolated chromophores	rotation of excited chromophore emission polarization
local chain dynamics	excimer/exciplex formation between groups closely spaced on the polymer backbone, TICT phenomena of dye bonded to polymer	local motion is rate-limiting for complex formation and charge separation
	emission depolarization of bound chromophores	backbone motion or side group motion causes reorientation of the transition dipole
large amplitude chain dynamics	fluorescence or phorphorescence quenching between pairs of groups far removed along the chain contour	chain motion is rate-limiting in cyclization needed for quenching
	phosphorescence depolarization of labelled proteins	in globular proteins, center-of-mass rotation rather than local motion leads to depolarization
translational diffusion coefficient	pattern photobleaching recovery and forced Rayleigh scattering	photochemical reaction causes a pattern of dye-rich and dye-poor areas, and diffusion leads to pattern fading
	energy transfer between donor-labelled and acceptor-labelled polymers	diffusion leads to changes in the extent of energy transfer

collapse transition	phosphorescence of labelled chains	sufficiently compact chain conformations inhibit impurity diffusion into the polymer interior
interpolymer association	excimer/exciplex formation between pairs of groups far removed along the chain contour	decrease in chain dimensions decreases distance between groups
	rapid and extensive intermolecular energy transfer or excimer/exciplex formation from groups on different chains under high dilution conditions	polymer association in the ground state leads to close proximity between groups on different chains
solvent relaxation	changes in the time-resolved fluorescence spectra and wavelength dependence of fluorescence lifetime measurements	temporal (sub-nanosecond to nanosecond) changes in fluorescence spectra reflect molecular motion and solvation
polymer conformation	polarization anistropy r(t) of polymer-bound chromophores dissolved in viscous media	separation of chromophores affects r(t) via energy migration (no polymer motion assumed)
	intramolecular excimer/exciplex formation between pairs of chromophores along the chain	cyclization kinetics and equilibria depend on the spatial separation of the chromophores
chain orientation under stress	measured polarization varies with sample orientation	orientation of polymer chains induces orientation of dyes attached to or dissolved in the matrix
glass transition at 1 Hz	temperature dependence of fluorescence of dyes which twist in the excited states	free volume at time of excitation permits or retards twisting, which leads to a non-fluorescent states

glass transition at high frequency	phosphorescence lifetime in the presence of air, or	diffusion of quenchers, including oxygen, over large distances is greatly enhanced above T_g
	fluorescence quenching by mobile probes	
other mechanical relaxation processes	intramolecular excimer/exciplex formation in small probe molecules	motion necessary for cyclization requires molecular relaxation of the matrix
	fluorescence depolarization of bound chromophores	polymer motion requires adequate free volume
	fluorescence and phosphorescence intensity and their decay profiles as a function of temperature	probe motion, as well as impurity diffusion, are sensitive to these relaxation processes
volume changes (swelling/contraction) in the glassy state	fluorescence and phosphorescence decay profiles of chains labelled with a significant mole fraction (ca. 2%) of chromophore, or	the mean separation of groups and their distribution controls self-quenching and energy migration rates (no diffusive motion)
	depolarization of fluorescence from polymer bound groups at sufficient concentration that energy migration can occur	
polymer-polymer compatibility and phase separation	excimer/exciplex formation or energy transfer between pairs of chromophores on a single chain	smaller mean dimensions of the labelled polymer increases interaction efficiency between chromophores
	excimer/exciplex formation, energy transfer, or emission quenching between pairs of chromophores on two different chains	mean separation of chromophores is decreased by polymer interpenetration
energy migration along polymer chain	polarization measurement in a frozen matrix	energy migration is exclusively responsible for external depolarization in the frozen state

8. REFERENCES

1. (a) E.L. Wehry, editor, "Modern Fluorescence Spectroscopy," Plenum Press, New York (1976).

    (b) R.F. Chen and H. Edelhoch, editors, "Biochemical Fluorescence," Marcel Dekker, New York (1985).

    (c) J. Schlessinger and E.L. Elson, Methods of Experimental Physics, 30, 197 (1982).

2. (a) D. Phillips and A.J. Roberts, editors, "Photophysics of Synthetic Polymers," Science Review, Northwood, UK, 1982.

    (b) Polymer Photochemistry, 5, (1984), special issue, "Photochemistry and Photophysics in Polymers."

3. J.B. Birks, "Photophysics of Aromatic Molecules," Wiley–Interscience, New York, 1971.

4. (a) U. Maharaj and M.A. Winnik, J. Amer. Chem. Soc., 103, 2328 (1981).

    (b) R.O. Loutfy, R.W. Yip, and S. Dogra, Tetrahedron Lett., 2843 (1977).

5. (a) J.D. Bolt and N.J. Turro, Photochem. Photobiol., 35, 305 (1982).

    (b) N.J. Turro and M. Aikawa, J. Amer. Chem. Soc., 102, 4866 (1980).

6. T. Vo-Dinh, "Room Temperature Phosphorescence for Chemical Analysis," J. Wiley and Sons, New York, 1984.

7. P.F. Jones and S. Siegel, J. Chem. Phys., 50, 1134 (1969).

8. (a) A.C. Sommersall, E. Dan, and J.E. Guillet, Macromolecules, 7, 233 (1974).

    (b) H. Rutherford and I. Soutar, J. Polym. Sci. Polym. Phys. Ed., 18, 1021 (1980).

    (c) F.E. El-Sayed, J.R. MacCallum, P.J. Pomery, and T.M. Shepperd, J. Chem. Soc. Faraday Trans. II, 75, 79 (1979).

    (d) W.H. Melhuish, Trans. Faraday Soc., 62), 3384 (1966).

    (e) W.E. Graves, R.H. Hofeldt, and S.P. McGlynn, J. Chem. Phys., 56, 1309 (1972).

(f) O. Pekcan, M.A. Winnik, and M.D. Croucher, Can. J. Chem.,
     63, 129 (1985).

9. R. Lakowicz, "Principles of Fluorescence Spectroscopy," Plenum
    Press, New York, 1983.

10. (a) Y. Nishijima, J. Polym. Sci., Polym. Sym., 31, 353 (1970).

    (b) S.W. Beavan, J.S. Hargreaves, and D. Phillips, Advances
         in Photochemistry, 11, 207 (1979).

11. (a) J.N. Demas, "Excited State Lifetime Measurements," Academic
         Press, New York, 1983.

    (b) D.V. O'Connor and D. Phillips, "Time Correlated Single Photon
         Counting," Academic Press, New York, 1984.

    (c) R.B. Cundall and R.E. Dale, editors, "Time Resolved
         Fluorescence Spectroscopy in Biochemistry and Biology,"
         Plenum Press, New York, 1982.

12. (a) H. Hervet, L. Leger, and F. Rondelez, Phys. Rev. Letters,
         42, 1681 (1979).

    (b) B.A. Smith, E.T. Samulski, L.-P. Yu, and M.A. Winnik,
         Macromolecules, in press, 1985; Phys. Rev. Letters, 52,
         45 (1984).

13. (a) H.L. Chen and H. Morawetz, Eur. Polym. J., 19, 923 (1983).

    (b) Y. Osada, M. Koike, and E. Katsumura, Chem. Letters, 809
         (1981).

    (c) N.J. Turro and T. Okubo, J. Amer. Chem. Soc., 104, 2985
         (1982).

14. (a) Z.R. Grabowski, K. Rotkiewcz, W. Rubaszewska, and
         E. Kirkor-Kominska, Acta Phys. Pol., A 54, 767 (1978).

    (b) R.O. Loutfy, Macromolecules, 16, 687 (1983).

15. (a) A. Nakajima, Bull. Chem. Soc. Jpn., 44, 3272 (1977).

    (b) D.C. Dong and M.A. Winnik, Can. J. Chem., 62, 2560 (1984).

    (c) G. Durocher and C. Sandorfy, J. Mol. Spectrosc.. 20, 410
         (1966).

16. (a) M. Goldenberg, J. Emert, and H. Morawetz, J. Amer. Chem.
         Soc., 100, 7171 (1978).

(b) K.A. Zachariasse, G. Duveneck, and R. Busse, J. Amer. Chem., 106, 3328 (1983).

17. R.O. Loutfy, J. Polym. Sci. Polym. Phys. Ed., 20, 825 (1982).

18. (a) Th. Förster, Ann. Physik, 2, 55 (1948).

(b) I.B. Berlman, "Energy Transfer Parameters of Aromatic Compounds," Academic Press (1973).

19. J.E. Guillet, "Polymer Photochemistry and Photophysics," Cambridge University Press, Cambridge, UK, 1985.

20. S. Tazuke, H. Tomono, N. Kitamura, K. Sato and N. Hayashi, Chem. Lett., 85 (1979).

21. Ref. 3, p. 522.

22. C.M. Paleos, Chem. Soc. Rev., 14, 45 (1985).

23. (a) D.L. Dexter, J. Chem. Phys., 21, 836 (1953).

(b) M. Inokuti and F. Hirayama, J. Chem. Phys., 43, 1978 (1965).

24. (a) D. Rehm and A. Weller, Israel J. Chem., 8, 259 (1970).

(b) J.R. Miller, J.V. Beitz, and R.K. Huddleston, J. Amer. Chem.Soc., 106, 5057 (1984).

(c) J.R. Miller, K.W. Hartman, and S. Abrash, J. Amer. Chem. Soc., 104, 4296 (1982).

25. (a) S. Tazuke, R. Takasaki, Y. Iwaya and Y. Suzuki, A.C.S. Symp. Ser., 266, 187, (1984).

(b) S. Tazuke, N. Kitamura and Y. Kawanishi, J. Photochem., in press (1985).

(c) N. Kitamura, Y. Kawanishi and S. Tazuke, Chem. Lett., 1185, (1983).

26. R.A. Marcus, DISC Faraday Soc., 29, 21 (1960).

27. (a) T. Kakitani and N. Mataga, J. Phys. Chem., 89, 8 (1985).

(b) H. Masuhara, Chapter 4 in this volume.

28. (a) M. Gordon and W. Ware, editors, "The Exciplex," Academic Press, New York, 1975.

(b) A. Weller, Z. Phys. Chem. (Wiesbaden), 130, 129 (1982);
     ibid, 133, 93 (1982).

29. S. Tazuke and H.L. Yuan, Macromolecules, 17, 1878 (1984).

30. S. Tazuke, Makromol. Chem., Supplement (1985) in press.

31. S. Tazuke, R. Iwasaki and T. Ikeda, Japan-US Polymer Symp. 1985.

32. K. Zachariasse and W. Kühnle, Z. Phys. Chem. (Wiesbaden), 101,
     267 (1976).

33. L.S. Egan, M.A. Winnik, and M.D. Croucher, Polym. Eng. and Sci., in
     press (1985).

34. J. Calvert and J. Pitts, "Photochemistry," J. Wiley, New York,
     1965.

35. C.A. Parker, "Photoluminescence of Solutions," Elsevier, New York
     (1968).

36. S.P. McGlynn, P. Azumi, M. Kinoshita, "Molecular Spectroscopy
     of the Triplet State," Prentice Hall, Englewood Cliffs, New Jersey
     (1969).

37. L.S. Egan, M.A. Winnik, and M.D. Croucher, to be published; cf.
     reference 33.

38. (a) R.M. Noyes, Prog. React. Kinet., 1, 129 (1961).

     (b) M.H. Hui and W.R. Ware, J. Amer. Chem. Soc., 98, 4712 (1976).

AN INTRODUCTION TO TRANSIENT ABSORPTION SPECTROSCOPY AND NONLINEAR PHOTOCHEMICAL BEHAVIOR OF POLYMER SYSTEMS

Hiroshi Masuhara
Kyoto Institute of Technology
Department of Polymer Science and Engineering
Matsugasaki, Kyoto 606
Japan

ABSTRACT.   Transient absorption spectral measurements in the ns and ps time regions are described and their importance as photophysical and photochemical tools are discussed.   Various artifacts such as spatial stray light effects, depletion effects on the ground state molecules, transient inner filter effects, nonlinear refractive index changes, thermal lensing effects, Resonance Raman scattering due to solvent, and two-photon excitation of solvent have to be examined in order to obtain a reliable and accurate absorption spectrum. Some of these factors are nonlinear with respect to excitation intensity and result in a characteristic behavior of polymers.   Intrapolymer interactions between the excited singlet states, formation of a transient polyelectrolyte, and intrapolymer multiphoton charge separation are demonstrated and discussed by correlating the results with the polymer structure.   Simultaneous two-photon absorption of the ps excitation pulse makes it possible to measure the absorption spectrum of the excited state of polymer solids. A combination of a polymer with pendant aromatic groups and a laser pulse with high intensity leads to new dynamic behavior and brings about a new methodology in polymer science.

1.   TRANSIENT ABSORPTION SPECTROSCOPY

The flash photolysis method was first developed by Norrish and Porter (1) and has made a great contribution to the fields of chemical reaction and relaxation phenomena.   Its temporal resolution has been improved from μs to tens of fs by introducing pulsed lasers as excitation and monitoring light sources.   The dynamic behavior of unstable radicals and molecular triplet states was the main subject of concern for studies in the μs time range.   Ns laser flash photolysis made it possible to observe dynamics of excited singlet (S_1)

43

M. A. Winnik (ed.), Photophysical and Photochemical Tools in Polymer Science, 43–63.
© *1986 by D. Reidel Publishing Company.*

states, n-π^* triplet states and short-lived intermediates.
Two of the most active areas of investigation in the ns
time range have been studies of electron and charge transfer phenomena in chemistry (2), and S_1-S_1 annihilation
processes in solution, molecular crystals and semiconductors
(3). Laser photolysis methods in the ps and fs time domains
are now available and are recognized as indispensable
techniques in chemistry, physics, biology, and engineering
(4).

In these measurements, dynamic processes have been analyzed
primarily by probing a transient absorption at one wavelength.
In general, however, absorption spectra of excited states
and chemical intermediates overlap each other. Furthermore,
conformational change and orientational relaxation of the
surrounding solvent molecules result in a time-dependence
of the spectral band shape. For example, intramolecular
exciplex systems give an absorption spectrum, the band shape
of which is a function of solvent properties and delay times
(5). Examples of phenomena which have been studied by analyzing absorption spectral changes in the ps time region are
isomerization of trans-stilbene from the S_1 state (6) and
vibrational relaxation of the triplet benzophenone (7) as
well as the excited singlet anthracene (8). It is

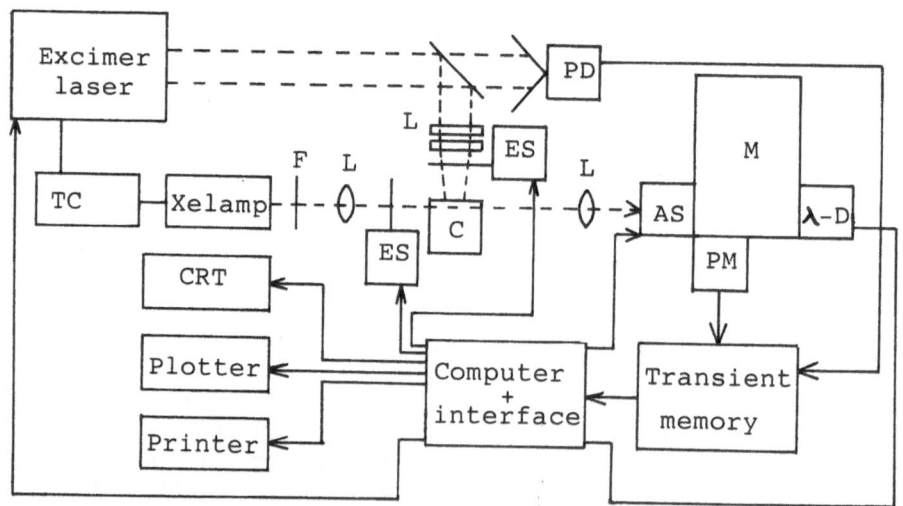

Figure 1. Schematic diagram of the microcomputer-
controlled ns excimer laser photolysis system. PD,
photodiode; PM, photomultiplier; ES, electromagnetic
shutter; AS, automatic slit; λ-D, wavelength driver;
L, lens; C, sample cell; F, filter; TC, timing circuit;
M, monochromator.

indispensable in these kinds of experiments to measure the
absorption spectra over a wide range of wavelengths for
elucidating the photophysical and photochemical processes.

From this viewpoint we have constructed several laser
photolysis systems which are controlled and processed by
a microcomputer. In Figs. 1 and 2, ns excimer and ps Nd^{3+}:
YAG laser photolysis systems are shown (9-11). The monitoring lamp is a 150 W Xe lamp (Wacom) which is operated in
the DC and pulsed modes. This is synchronized to an excimer
laser (Lumonix 432T) through a home-made timing circuit. A
time-profile of the transmitted monitoring light is digitized
by a transient memory (Kawasaki Electronica M-50E), transferred to a microcomputer (NEC PC9801F2) and processed. In
the ps photolysis the monitoring beam is a ps continuum
produced by focussing the fundamental pulse (1064 nm) into a
quartz cell of 10 cm length with a lens of 15 cm focal length.
The material for generating this continuum is D_2O (Merk
UVASOL, 99.95%) which makes it possible to measure the full

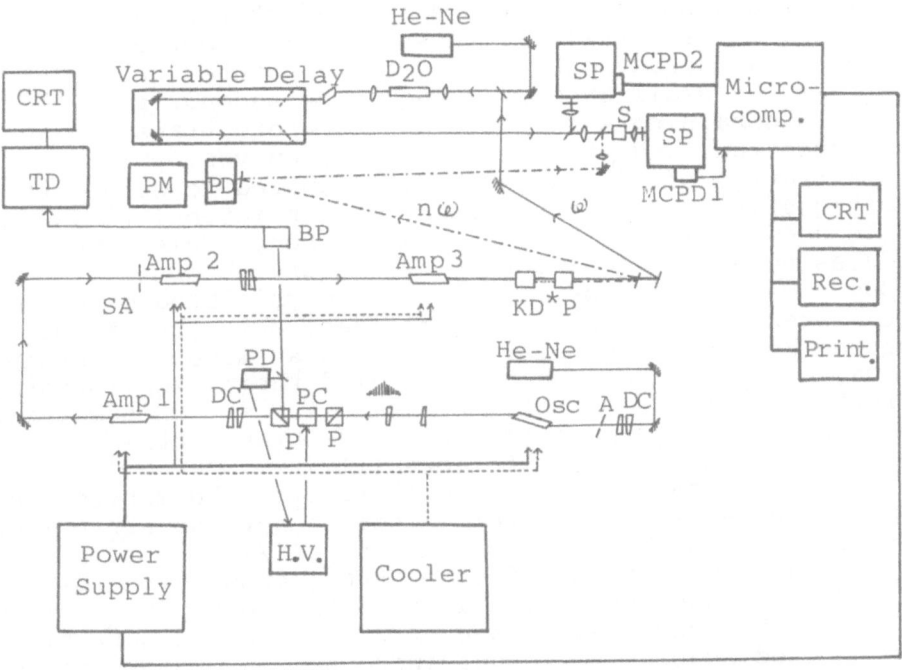

Figure 2. Schematic diagram of the microcomputer-
controlled ps Nd^{3+}:YAG laser photolysis system. DC,
dye cell; A, aperture; P, polarizer; PC, Pockels cell;
PD, photodiode; SA, soft aperture; BP, biplanar
photodiode; SP, spectrograph; S, sample; Rec., recorder; PM, power meter; TD, transient digitizer.

range of the visible region. The excitation pulse is either
the second (532 nm), the third (355 nm), or the fourth (266
nm) harmonic of the Nd^{3+}:YAG laser. The spectrum of the ps
continuum is detected by a multichannel photodiode array
(Matsushita MEL 1024 KV). Two diode arrays are attached to
each spectrograph (f=25 cm, Jovin Yvon grating 300 lines/nm),
covering a wavelength range of 380 nm. The output of each
diode array is sent to a microcomputer (SORD M223 Mark II),
whose processing principle is the double-beam method. This
one-shot measurement of spectra is also possible in ns
photolysis by using a gated photodiode array. These optical
and electronic arrangements are considered to be standard
techniques at the present state of investigation.

Transient absorption spectroscopy, of course, can be applied
to nonluminescent species, which makes it possible to
investigate a wider range of systems compared to fluorescence techniques. This spectroscopy probes several electronic transitions, while the fluorescence measurement follows only one. This indicates that more detailed information
on electronic structure of the excited state, and the factors
affecting it, are available in absorption spectroscopy.
However, two disadvantages have to be pointed out. The
first is its small dynamic range. The second is the difficulty in obtaining a reliable and accurate transient absorption spectrum, and its rise and decay curves. Since a
concentration of 10^{-4} to 10^{-5} M is required to detect the
excited states or intermediates, a high intensity laser has
to be used as an excitation light source. This sometimes
induces nonlinear behavior which affects absorption spectroscopy. We have performed laser photolysis studies on
exciplexes, charge transfer complexes, polymers, aromatic
liquids and molecular aggregates such as micelles, organic
films, and vesicles, and have studied their high intensity
effects. On the basis of these results, we can summarize
the important phenomena affecting the transient absorption
spectral measurements as follows (11).
    (i)   A spatial stray light effect. The monitoring
beam which passes the unexcited volume acts as stray light
for transient absorption spectroscopy (12).
    (ii) A depletion effect on the ground state molecules.
If the excitation condition is such that most of the molecules in the ground state are excited, the concentration of
the excited species no longer increases in proportion to
the increasing excitation intensity. Lachish et al. reported
a qualitative criterion for this depletion effect as $\sigma h \geq 1$,
where σ and h are the absorption cross-section in cm^2, and
the excitation intensity in photons cm^{-2}, respectively (13,
14).
    (iii) A transient inner filter effect. The excitation
pulse may be absorbed by the excited singlet state or by

other photoinduced chemical intermediates, suppressing the effective number of excitation photons (14,15).

(iv) A nonlinear refractive index change. Since the excitation photon density per unit area per unit time is very high, a change of the refractive index of the solvent may be induced (16). At its worst, an excitation pulse produces a non-uniform distribution of the refractive index which leads to scattering of the ps continuum, resulting in an apparent transient absorption.

(v) A thermal lensing effect. A similar refractive index change is also induced by local heating around the absorbing molecule (17).

(vi) Resonance Raman scattering due to solvents. In the case of aliphatic solvents, the Stokes-Raman scattering induced by a pulsed laser disturbs the absorption spectral measurements (18).

(vii) Two-photon excitation of solvents. It has been demonstrated that the ps 355 nm photolysis of aromatic liquids and solvents containing double bonds gives $S_n \leftarrow S_1$ absorption spectra with appreciable intensity (18,19), which may modify the absorption spectral shape.

All these phenomena are considered to induce an apparent spectral change and to affect the absorption rise and decay curves. Therefore, it is indispensable to examine experimentally or to evaluate numerically these effects in order to obtain reliable and accurate data. Among the above described phenomena, effects (vi) and (vii) can be excluded by selecting appropriate solvents and increasing the solute concentration. The effect of spatial stray light (i) can also be decreased to a great extent by adjusting the optical alignment carefully. The contribution of effects (ii) and (iii) in transient absorbance can be estimated numerically if molar extinction coefficients of the ground and the excited states at the excitation wavelength are known. Along this line we have examined photophysical and photochemical processes of molecules in condensed phases.

In dilute solution of polymers with pendant aromatic groups, the local concentration of the chromophore is high while its mean concentration is low. A combination of this chromophore distribution and an intense laser excitation results in interesting nonlinear photochemical behavior under appropriate experimental conditions. We have succeeded in demonstrating phenomena characteristic of polymers excited by a laser, which we summarize in the following sections.

2. INTRAPOLYMER INTERACTIONS BETWEEN EXCITED SINGLET STATES

In experiments carried out in our laboratory, an intrapolymer S_1-S_1 annihilation process was found to occur in the carbazole polymers shown in Fig. 3. This process is quite a familiar phenomenon for concentrated solutions of molecules, for semiconductors and for molecular crystals (20). We demonstrated it for the first time for polymer solutions (14). Several excited states are produced in one polymer chain by laser excitation. These migrate along the chain, and interact with each other, leading to an efficient deactivation. A depletion of the ground state chromophores, which was summarized in (ii) in the above section, is the origin of this phenomenon. In actual fact, our experimental conditions of carbazole excitation by a Ruby laser (347 nm) and by a N_2 gas laser (337 nm) satisfy the criterion proposed by Lachish et al. A depletion effect of N-ethylcarbazole was confirmed by examining the relation between the excitation intensity and its transmitted one for sample solutions with different absorbances (21).

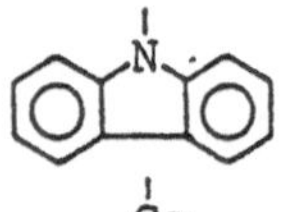

Here we summarize the experimental results confirming the S_1-S_1 annihilation as follows (14,22):

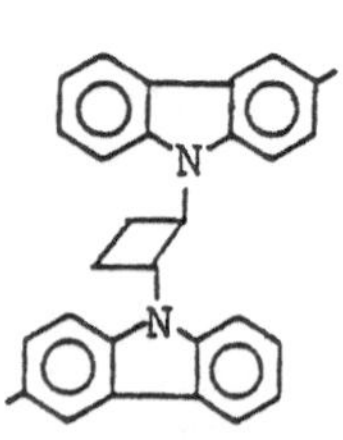

Figure 3. Molecular structures and abbreviations of the polymers studied and their reference compounds. The notation n is the mean degree of polymerization.

(i) Excitation intensity dependence of fluorescence spectra (14,23). In the case of PVCz, two kinds of excimers are observed, and their relative intensity shows an interesting dependence upon excitation intensity. The so-called second excimer increases relatively with increasing excitation intensity, which is more pronounced in the case of PVCz with higher n-values. This behavior can be ascribed to the fact that the sandwich excimer may be produced from the monomer excited state and all these fluorescent states interact with each other.

(ii) Excitation intensity dependence of fluorescence intensity (24). Here we compare the excitation intensity

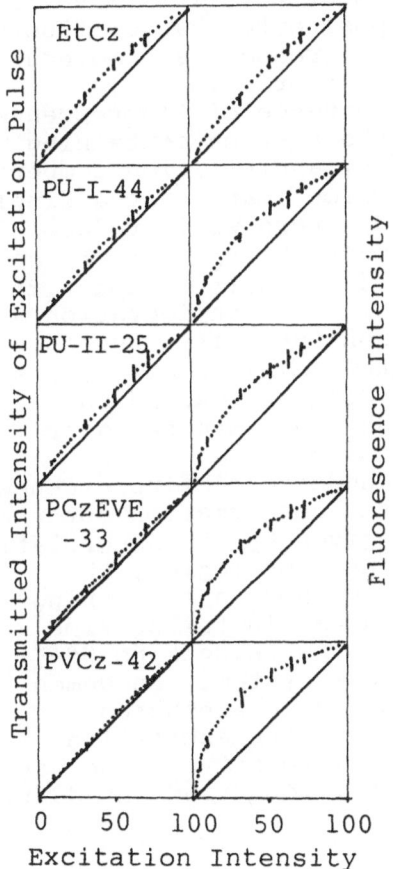

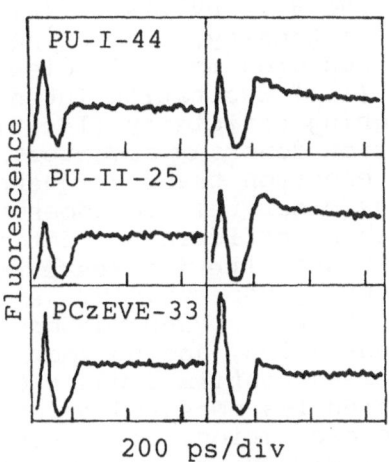

200 ps/div

Figure 5. Fluorescence decay curves observed by the low (left) and high (right) excitation. The first sharp pulse is a reference pulse used for calibration and data processing.

Figure 4. Variations of the fluorescence intensity (420 nm) and the transmitted intensity of the excitation pulse (337 nm) with the excitation intensity in DMF. The excitation intensity 100 corresponds to $\sim 1 \times 10^{16}$ photons/cm^2. The lines connecting the origin and the value measured at intensity 100 are given to have a slope of 45°, as a measure for linearity.

dependences of the time-integrated fluorescence intensity
and of the transmitted intensity of N_2 laser pulse for some
polymers. As shown in Fig. 4, the fluorescence intensity
increases with the excitation intensity, but a deviation
from a linear relation is observed for all the systems. In
the case of polymers the observed curves approach saturation rather rapidly at high excitation intensity, and the
degree is in the order of EtCz < Pu-I-44 ∿ PU-II-25 <
PCzEVE-33 < PVCz-42. This tendency in the polymer corresponds to a decreasing order of the mean distance between
neighbouring chromophores. On the other hand, the deviation
of the transmitted intensity of the excitation pulse from a
linear relation is in the order of EtCz > PU-I-44 ∿ PU-II-25
> PCzEVE-33 > PVCz-42. This is opposite to that of fluores-
cence intensity, suggesting that the ground state molecules
are recovered more efficiently in this order.

(iii) Excitation intensity dependence of fluorescence
quenching efficiency (14,25). In the systems containing
electron donors or acceptors, the S_1-S_1 annihilation competes
with electron transfer quenching, which means that a simple
Stern-Volmer relation does not hold. Here we introduce
quenching efficiency Q defined as $Q=(F_0-F)/F_0$ where F_0 and
F represent the fluorescence intensities without and with
quencher, respectively. A simple kinetic consideration
predicts that Q depends on excitation intensity, quencher
concentration, and polymer structure. The inequality, $Q(s)>$
$Q(\ell)$, was confirmed, where $Q(s)$ and $Q(\ell)$ are quenching
efficiencies measured by steady light and intense laser
pulse excitation, respectively.

(iv) Excitation intensity dependence of fluorescence
decay (14,24). Direct experimental confirmation for the
S_1-S_1 annihilation process has been given by the accerelation
of the fluorescence decay. In the ns time region, the
sandwich excimer emission of PVCz-400 shows a shortening
from 35 ns to 14 ns. A more rapid decay in the ps time
region was observed for the first time by using a ps Nd^{3+}:
YAG pulse (355 nm) and a streak camera system. As demonstrated in Fig. 5, the present systems give a constant
intensity in the 100 ps time range when the excitation
intensity is low. As the excitation intensity is increased
gradually, the rapid-decay component is observed in addition
to the plateau value.

In order to explain these observations, we considered that
the ground state molecule or the triplet state is formed by
the S_1-S_1 annihilation process as given in the following
scheme:

$$S_1 + S_1 \longrightarrow S_n + S_0 \longrightarrow S_1 + S_0 \qquad (1)$$

$$\nearrow T_1 + T_1 \qquad (2)$$

The precise confirmation of process (1) via the Förster
dipole-dipole interaction was given for pyrene in a rigid
solvent (26), while the process (2) was invoked to explain
the ps fluorescence decay curve of the in vivo photosyn-
thetic unit (27). The possibility of process (2) can be
examined in our system by measuring the excitation intensity
dependence of the $T_n \leftarrow T_1$ absorbance. In the case of EtCz,
the latter dependence is the same as that of the time-
integrated fluorescence intensity, which is consistent with
the usual intersystem crossing process $S_1 \rightarrow T$. If the process (1) occurs exclusively in the polymer systems, the
excitation intensity dependence of the $T_n \leftarrow T_1$ absorbance
would be identical to that of the fluorescence. However,
the deviation of the $T_n \leftarrow T_1$ absorbance from a linear relation is not as large as that of the time-integrated
fluorescence intensity. This indicates that another formation process (2) also occurs. This is the first experimental indication of triplet state formation obtained
by absorption spectroscopy, although no quantitative analysis of the data has been carried out. We consider that
both processes (1) and (2) operate within one polymer chain.

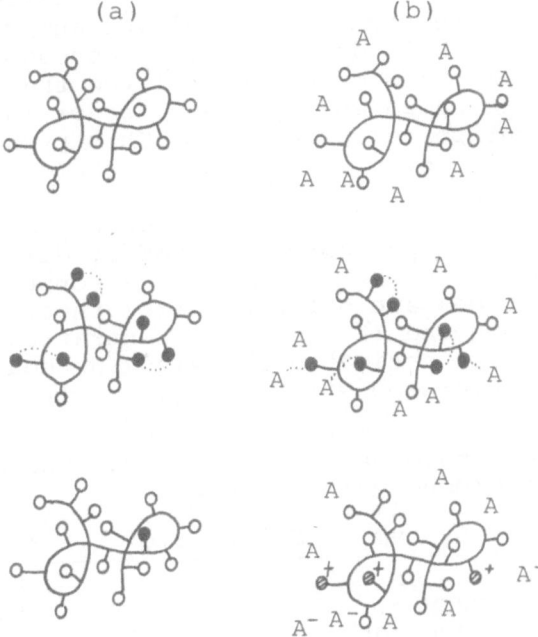

Figure 6. Schematic diagram showing intrapolymer
S_1-S_1 annihilation (a) and transient polyelectrolyte
formation (b). Carbazole (O), the excited carbazole
(●), carbazole cation (∅⁺), electron acceptor (A),
and acceptor anion (A⁻).

The present S_1-S_1 annihilation process is schematically
shown in Fig. 6.

3. FORMATION OF A TRANSIENT POLYELECTROLYTE

It is well established that electron transfer quenching of
the excited singlet state in polar solvents results in the
formation of a donor cation and an acceptor anion. In the
laser photolysis of polymer-quencher systems, this electron
transfer competes with the S_1-S_1 annihilation. By increasing the quencher concentration, the quenching is able
to overcome the latter process, and some of the excited
states in one polymer chain are converted to the corresponding ion radicals. The net effect of this process is
that polymers having several ionic chromophores are formed
by one excitation pulse of the laser. Energy migration and
interaction between the excited states are determined by
dipole-dipole interactions, while electron migration along
the polymer is due to an exchange interaction. Since the
latter interaction is a short-range one and electron transfer requires   solvent reorientation and segment motion,
the charge may migrate slowly or may be localized on one
chromophore. Recombination of the polymer and quencher ion
radicals determines their decay time. In consequence, ionic
states in the polymer are much longer lived than the excited
singlet states. We named this process "laser-induced formation of a transient polyelectrolyte". This process is shown
schematically in Fig. 6.

We have confirmed for the first time this characteristic
behavior for carbazole polymers quenched by dimethyl
terephthalate (DMTP) in DMF. Since the absorption spectra
of the polymer cations are different from those of the
monomeric model compound (28), the concentrations of ionic
chromophores had to be estimated by using the counter DMTP
anion (25,29). Conversion efficiencies of the carbazole

TABLE 1. Conversion efficiency of carbazole
          chromophore into ionized one

Excitation	System	Delay time	Efficiency
A Ruby laser	EtCz	0 μs	0.8
(347 nm)	PVCz-17	0	0.25
	PVCz-400	0	0.18
	PVCz-400	1	0.11
a N_2 laser	EtCz	1	0.14
(337 nm)	PVCz-400	1	0.06
	PU-I-44	0	0.13
	PMA-65	0	0.05

chromophore into an ionized one were calculated and listed
in Table 1. Immediately after excitation, PVCz-17 has four
carbazole cations, whereas 18% of chromophores are converted
into cationic species in the case of PVCz-400. The efficiency of PU-I-44 is higher than that of PMA-65, although
both have similar n-values. In the case of polymers with a
large distance between chromophores, solvent reorientation
and diffusion of ion radicals may occur easily, while energy
migration leading to the S_1-S_1 annihilation is inefficient.
PU-I-44 satisfies these conditions, but the chromophores of
PMA-65 are very crowded. On the basis of these considerations, the conditions favoring the laser-induced formation
of transient polyelectrolytes are summarized as follows.
       (i)   The absorption cross section of the chromophore
bound to the polymer is large at the excitation wavelength.
       (ii)   Nonradiative processes other than the formation
of free ions occur with small rate constants.
       (iii) The mean distance between chromophores is long.
Condition (i) is determined by the electronic structure of
the chromophore, and (ii) is characteristic of the donor-
acceptor pair. The present pair of carbazole-DMTP satisfies these conditions to some extent, and PU-I and PU-II
are the polymers having a suitable structure. Finally, it
has to be mentioned that the residence time of the polyelectrolyte will be lengthened by preventing its recombination with the counter ion radical.

4.   INTRAPOLYMER MULTIPHOTON CHARGE SEPARATION PROCESS

Successive two-photon absorption is quite a general phenomenon in laser spectroscopy and one of inner filter effects
in transient absorption measurements . This leads to the
formation of the higher excited states whose spin states are
the singlet in the case of ps excitation and the triplet
in the case of tens of ns excitation. In nonpolar solvents,
the higher excited singlet states of aromatic hydrocarbons
are deactivated rapidly to the lowest excited singlet state,
while electron ejection is sometimes induced in polar solvents. One believes that this multiphoton absorption should
operate in polymers having aromatic groups. There is, however, no spectroscopic evidence for this process. Recently,
we have succeeded in measuring a new multiphoton behavior
characteristic of polymers with a special structure during
our systematic laser photolysis studies on polymers and
related systems (30,31). This process is detected not by ns
but by ps excitation, because the transient species have a
short lifetime. Here we introduce the experimental results
and try to establish a relationship between the multiphoton
dynamics and the polymer structure.

Poly(1-pyrenylalanine):P(L-PyA)-n, P(DL-PyA)-n
  -(NHCH(CH$_2$-Py)CO)$_n$-(NHCH(C$_2$H$_4$CO$_2$CH$_2$φ)CO)$_{150}$-NH(CH$_2$)$_5$CH$_3$

N-Acetyl-L-1-pyrenylalaninemethylester:L-PyA
  CH$_3$CONHCH(CH$_2$-PY)COOCH$_3$

Monomer ester model
  CH$_3$COOCH$_2$CH(CH$_2$-Py)CH$_2$OCOCH$_3$

Dimer ester model
  CH$_3$COOCH$_2$CH(CH$_2$Py)CH$_2$OCO(CH$_2$)$_m$COOCH$_2$CH(CH$_2$-Py)CH$_2$OCOCH$_3$

Polyester
  -(OCH$_2$CH(CH$_2$-Py)CH$_2$OCO(CH$_2$)$_m$CO)$_n$-

Poly(1-vinylpyrene)
  -(CH$_2$CHPy)$_n$-

Dipeptide model
  CH$_3$CONHCH(CH$_2$-Py)CONHCH(CH$_2$-Py)COOCH$_3$

     -Py; 1-Pyrenyl,

Figure 7. Molecular structures and abbreviations of
polymers used and their reference compounds. The
notation n is the mean degree of polymerization.

Geometrical structures of the polymers studied are shown in
Fig.7. The polymer P(L-PyA)-n has a helical structure (32).
The same 1-pyrenyl group is connected to polypeptide, poly-
ester, and polyvinyl chains, so that effects of higher-order
structure, mean chromophore distance, and/or local concentration upon the photochemical processes can be elucidated.
The ps 355 nm photolysis was performed in DMF with the Nd^{3+}:
YAG laser photolysis system described above. As shown in
Fig.8, the absorption spectra of P(L-PyA)-20 at 33 ps and
P(DL-PyA)-20 at 100 ps consist of three bands at 420, 460
and 500 nm. These are quite different from the $S_n \leftarrow S_1$
absorption spectra of the monomer model, L-PyA, which was
measured under the same experimental conditions. In order
to assign these three peaks, transient absorption spectra of
L-PyA quenched with O$_2$, an electron acceptor and a donor were
measured. Upon quenching, intersystem crossing and electron
transfer are enhanced, giving the triplet, cationic and
anionic states of L-PyA. They are compared with each other
in Fig.8, and here we learn that the triplet and ionic states
are formed in this polypeptide immediately after excitation.

It appears that intrapolymer charge separation is induced quantitatively by ps laser excitation. The maximum absorbance of these bands was attained at 20 ps, and their rise was faster than that of the $S_n \leftarrow S_1$ band of L-PyA. This strange behavior suggests that all three states are formed by multiphoton processes. This idea is supported by examing an excitation intensity effect upon the transient absorption spectra. As shown in Fig.9, the spectrum of P(DL-PyA)-20 at low excitation intensity loses its structure and can be interpreted in terms of a monomer-excimer equilibrium. In the case of P(L-PyA)-20, the triplet absorbance at 420 nm is higher than the absorbance of ionic bands at high excitation intensity. Upon lowering the excitation intensity to a tenth of that of normal laser photolysis, the ionic absorptions were still observed whereas the triplet band was suppressed. This detailed examination shows that the

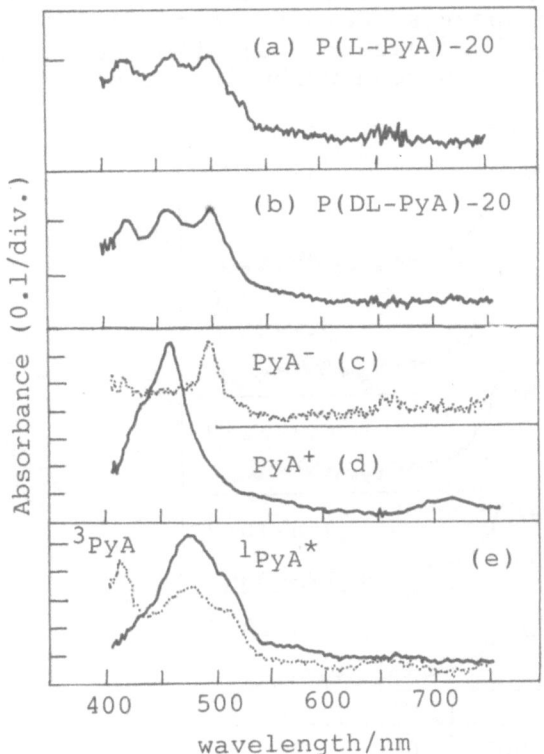

Figure 8. Transient spectra of the polypeptide and reference spectra of the model compound. (a) P(L-PyA)-20 at 33 ps, (b) P(DL-PyA)-20 at 100 ps, (c) L-PyA anion, (d) L-PyA cation, and (e) the excited singlet and triplet PyA.

dependence of the triplet band upon the excitation intensity
is larger than that of the ionic bands. We believe that the
intrapolymer charge separation resulting in the pyrenyl
cation and anion is induced by a two-photon excitation, and
that the triplet may be formed via a three-photon process.
In other words, the triplet state may be formed from a
higher excited state of the charge-separated state.

There are several possible interpretations for the present
intrapolymer charge separation. In one explanation, intrapolymer S_1-S_1 annihilation is followed by charge separation
to cationic and anionic chromophores. As discussed in the
second section, this process is related to a depletion
effect of the ground state chromophores whose qualitative
criterion is given as $\sigma h \geq 1$. However, this inequality
can not be satisfied by the present condition of ps photolysis. Furthermore, the mean degree of **polymerization is**
rather low and the possibility that one polypeptide has
a few excited chromophores seems to be small. The second
interpretation is that an electron ejection occurs from the
higher excited state of pyrene which is formed by a

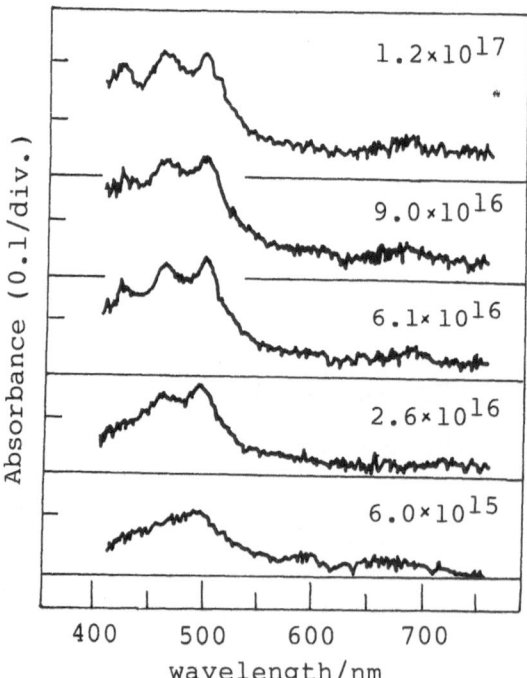

Figure 9. Excitation intensity dependence of the
transient spectrum of P(DL-PyA)-20 in DMF at 100
ps. Excitation intensity values in photons/cm^2
are given in the figure.

successive two-photon absorption. The energy level of such
a state can be estimated to be 6.8 eV, and lower than the
ionization potential of pyrene in DMF. The ejected electron
could be captured by the neighbouring pyrenyl chromophore,
leading to intrapolymer charge separation. However, ps
excitation of the monomer model, L-PyA, did not result in
the two-photon ionization, which may be due to the fact that
the solvent DMF is an electron donor. The third probable
interpretation is to consider the role of Rydberg states
in processes occurring from higher excited states.
The Rydberg state of pyrene in solution can be estimated to
be around 5 eV above the ground state according to the rule
that the lowest Rydberg state has an energy level lower
than the gas-phase ionization potential by 2.5 eV (33).
The electron density of the molecule in this state is more
expanded compared to other electronic configurations, so
interchromophoric interaction is considered to be more
easily induced. Even if the Rydberg state is not reached
directly by two-photon excitation, a mixing of this state
with the higher excited states results in a similar kind
of interaction. In the case that the neighbouring chromophore, in its ground state, is within the long interaction
radius of the Rydberg state, electron transfer from the
latter to the former may be brought about. This idea is
similar to the discussions on the photoionization of some
amines (34) and can explain the present characteristic behavior of polypeptides.

Although the history of photophysics of π-electronic molecules is very long, a charge separation between identical
molecules is a rare phenomenon. An exceptional case was
reported for some dyes and for octaethylporphyrins (35). No
report has been given for aromatic hydrocarbons. This
suggests that the geometrical structure characteristic of
polypeptides has an important role in the present primary
process. The higher-order helix structure seems not to be
a determining factor, since P(DL-PyA)-20 gives similar
ionic bands to P(L-PyA). This DL-Polymer designation does
not imply alternative copolymerization of D- and L-alanines,
but suggests possible block copolymer substructure. Namely, some of the local structures of P(DL-PyA)-n are still
identical to that of P(L-PyA)-n. The specific relative
structure of pyrenyl chromophores is probably the most
important criterion for the present process. This structural factor can be elucidated further by measuring transient absorption spectra of the related pyrenyl systems under
the same experimental condition. Polyesters, where the
interchromophoric distance is larger, showed the similar
dynamics. At normal laser intensity, ionic bands were
observed, whereas a monomer-excimer equilibrium could
explain the spectra obtained upon decreasing the excitation

intensity. On the other hand, no intramolecular charge
separation was observed in N-acetyl-bis(1-pyrenylalanine)-
methylester (36), a diester model, a diethyl ether model (37),
and poly(1-vinylpyrene) (38). Neither the high local
concentration nor the parallel sandwich structure of pyrenyl
chromophores seems sufficient to obtain the kind of results
described above in the present process. We conclude that a
loose structure of relevant chromophores is an important
factor for inducing electron transfer between identical
moieties and that their strict mutual orientation is not
required.

5. TWO-PHOTON PHOTOLYSIS OF POLYMERS

The wide applicability and usefulness of transient absorption spectroscopy in solution photochemistry are well
recognized, but reports on solids are rare. In the latter
case the chromophore concentration is very high, leading to
the fact that the excited states are produced only in the
surface part of the solid. The S_1-S_1 annihilation is
induced efficiently, resulting in rapid local heating.
This thermal energy cannot be dissipated as in solution,
causing cracking or optical damage of the solid. This
phenomenon is considered to be the main factor for making
it difficult to perform laser photolysis studies on solids.
If excited states were formed, not densely in the surface,
but homogeneously throughout the bulk, the S_1-S_1 annihilation would be suppressed and no cracking would be induced.
This condition is satisfied simply when the molar extinction
coefficient of the molecule at the excitation wavelength
is sufficiently small. Actually, the $T_n \leftarrow T_1$ absorption spectrum of a single crystal of benzophenone was measured by N_2
gas laser photolysis, where the excitation photon (337 nm)
is absorbed into the weak n-π^* transition. We came to the
conclusion that a simultaneous two-photon excitation is an
excitation method to produce the excited states homogeneously in bulk. We (18) and Hamanoue et al. (19) have
reported independently that simultaneous two-photon excitation of neat aromatics is so efficient that the absorption
spectrum of the excited singlet state can be measured under
the normal ps photolysis condition. This nonlinear effect,
summarized as (vii) in the first section, provides a new
spectroscopic method for studying the solid state. In the
following discussion we introduce briefly the experimental
results and discuss its possibilities.

Neat liquids of several benzene derivatives in a quartz
cell with a 1 cm path length were excited by the 347 and 355
nm ps pulses (18,39). Note that there is no one-photon
absorption at these wavelengths. Measured absorption

spectra and their time-dependence were interpreted in terms
of excimer formation and decay dynamics. After examining
various experimental conditions, we concluded that these
excimers are formed via a simultaneous two-photon absorption.
Just the same excimer absorption, of course, could be
observed by using a ps 266 nm pulse as a one-photon excitation. Here the decay is very rapid due to the S_1-S_1 annihilation in the front part of the cell.

On the basis of these results, we consider that transient
absorption spectral measurements of solids are possible with
the two-photon excitation method. Here our recent experimental demonstration of two typical polymer systems is
presented in order to confirm our idea. One example is the
ps 355 nm photolysis of a polystyrene plate with a thickness
of 1 mm (40). As in the case of benzenes, solid polystyrene gave an excimer absorption band whose peak position
was the same as those of liquid toluene and cumene. Another
system we examined is a very viscous, glass-like solution
of poly(N-vinylcarbazole) in THF. This sample was used
instead of a block of a solid polymer because the latter
was very diffucult to prepare. The ps 532 nm photolysis of
this sample gave the absorption spectrum shown in Fig.10

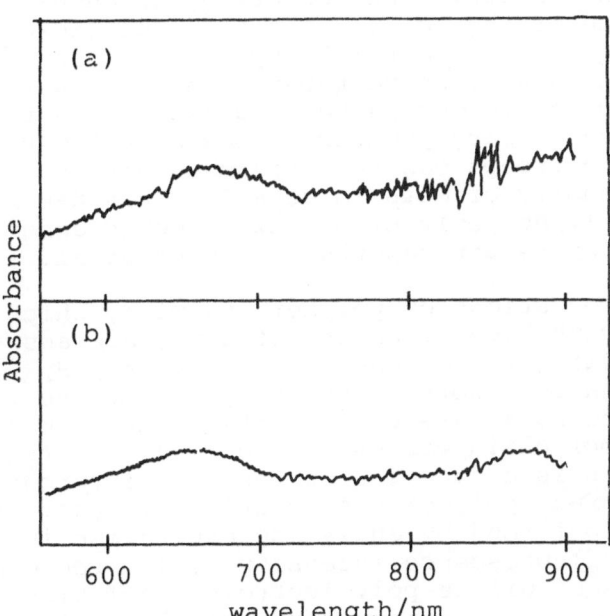

Figure 10. Transient absorption spectra of poly(N-
vinylcarbazole) at 100 ps. (a) the 532 nm two-photon
photolysis of the glass-like solution. (b) the 355
nm one-photon photolysis of the dilute solution.

(41). Since the carbazolyl chromophore has no one-photon
absorption at this wavelength, the present result is due
to simultaneous two-photon excitation. A comparison with
the spectrum of a dilute solution of the same polymer,
measured by the normal 355 nm photolysis, is also shown in
Fig.10. Although some differences are observed for the
band around 670 nm, the electronic structure of the polymer
glass seems to be very similar to that of the isolated
polymer in dilute solution. This measurement suggests that
an absorption spectral study on the photoconductive mechanism of solid polymers can be performed by this method.

6. SUMMARY

We have described the characteristic behavior of polymers
excited by an intense laser pulse and interpreted the
results in terms of nonlinear photochemical phenomena.
These are commonly observed in transient absorption spectroscopy, and make it difficult to analyze one-photon
photoprocesses of polymers. An exceptional case involves
studies on polymers having carbonyl groups reported by
Schnabel et al. (42). No distinct contribution of intrapolymer S_1-S_1 annihilation was observed, which was ascribed
to a small value of the molar extinction coefficient at the
laser wavelength and to fast intersystem crossing. Simultaneous as well as successive two-photon processes might be
suppressed compared to one-photon processes. On the other
hand, studies on primary photoprocesses of polymers with
pendant aromatic groups can be performed only by examining
excitation intensity effects in detail. This has been demonstrated clearly not only by our experiments described
here, but also by recent reports by Webber et al. (43).

Some important questions on polymer dynamics, which have
nothing to do with high intensity effects, are answered
by studying nonlinear photochemical behavior. By
analyzing the rapid fluorescence decay curves due to the
S_1-S_1 annihilation, a rate of excitation energy migration
along the polymer chain was estimated (24). As this annihilation process is a function of the mean interchromophoric
distance, different polymer conformations in poor and rich
solvents were confirmed by analyzing excitation intensity
dependences of fluorescence intensity (24). Concerning the
dynamic aspects of dilute polyelectrolyte solutions, we
believe that the study of transient polyelectrolytes will
yield fruitful information. Furthermore, laser photolysis
studies on polymers have made interesting contributions to
photochemistry and spectroscopy. Electron transfer between
identical chromophores is a rare process, and a new role of
the higher excited states has been demonstrated.

Simultaneous two-photon photolysis makes it possible to measure transient absorption spectra of solids and should now open a new aspect of photochemistry to experimental investigation.

Before closing this chapter, we emphasize that nonlinear photochemical behavior is quite general, and occurs during any interaction between an intense laser pulse and a condensed organic material. Not only polymer systems but also any molecularly associated ones such as crystals, amorphous solids, films, powders, and so on, will show similar phenomena upon excitation with an intense laser pulse.

ACKNOWLEDGEMENTS

The author wishes to express his sincere thanks to Prof. N. Mataga (Osaka University) for his discussion and support. Thanks are due to Dr. N. Ikeda, Dr. H. Miyasaka, Mr. S. Ohwada, and Mr. J. A. Tanaka.

REFERENCES
(1)   "Fast Reactions and Primary Processes in Chemical Kinetics" ed. by S. Claesson, Almqvist and Wiksell, Uppsala, 1967.
(2)   N. Mataga and M. Ottolenghi, in "Molecular Association, Vol. 2" ed. by R. Forster, Academic Press, London, 1979, p.1; N. Mataga, in "Molecular Interactions, Vol.2", ed. by H. Ratajczak and W. J. Orville-Thomas, John Wiley & Sons, London, 1981, p.509; H. Masuhara and N. Mataga, Accounts Chem. Res. 14, 312 (1981).
(3)   For example, C. E. Swenberg and N. F. Giacintov, in "Organic Molecular Photophysics, Vol.1", ed. by J. B. Birks, John Wiley & Sons, London, 1973, p.489; von der Linde, in "Ultrashort Light Pulses", ed. by S. L. Shapiro, Springer-Verlag, Berlin, 1977, p.203.
(4)   "Picosecond Phenomena III", ed. by K. B. Eisenthal, R. M. Hochstrasser, W. Kaiser, and A. Laubereau, Springer-Verlag, Berlin, 1982; "Ultrafast Phenomena IV", ed. by D. H. Auston and K. B. Eisenthal, Springer-Verlag, Berlin, 1984.
(5)   T. Okada, M. Migita, N. Mataga, Y. Sakata, and S. Misumi, J. Am. Chem. Soc., 103, 4715 (1981) and papers cited theirin.
(6)   B. I. Greene, R. M. Hochstrasser, and R. B. Weisman, Chem. Phys., 48, 289 (1980) and papers cited theirin.
(7)   B. I. Greene, R. M. Hochstrasser, and R. B. Weisman, J. Chem. Phys., 70, 1247 (1979).
(8)   R. W. Anderson, in "Picosecond Phenomena II", ed. by R. M. Hochstrasser, W. Kaiser, and C. V. Shank, Springer-Verlag, Berlin, 1980, p.163.

(9)   K. Imagi, N. Ikeda, and H. Masuhara, to be published.

(10) H. Masuhara, N. Ikeda, H. Miyasaka, and N. Mataga, J.
      Spectrosc. Soc. Jpn., 31, 19 (1982).

(11) H. Miyasaka, H. Masuhara, and N. Mataga, Laser Chem.,
      1, 357 (1983).

(12) C.R. Goldschmidt, in "Lasers in Physical Chemistry
      and Biophysics", ed. by J. Joussot-Dubien, Elsevier.
      Amsterdam, 1975, p.499.

(13) U. Lachish, A. Schafferman, and G. Stein, J. Chem.
      Phys., 64, 4205 (1976).

(14) H. Masuhara, S. Ohwada, N. Mataga, A. Itaya, K.
      Okamoto, and S. Kusabayashi, J. Phys. Chem., 84, 2363
      (1980).

(15) For example, M. M. Fischer, B. Veyret, and K. Weiss,
      Chem. Phys. Lett., 28, 60 (1974).

(16) D. H. Auston, in "Ultrashort Light Pulses", ed. by S.
      L. Shapiro, Springer-Verlag, Berlin, 1977, Chap.4.

(17) D. S. Kliger, Accounts Chem. Res., 13, 129 (1980);
      K. Fuke, private communication (1982).

(18) H. Masuhara, H. Miyasaka, N. Ikeda, and N. Mataga,
      Chem. Phys. Lett., 82, 59 (1981).

(19) K. Hamanoue, T. Hidaka, T. Nakayama, and H. Teranishi,
      Chem. Phys. Lett., 82, 55 (1981).

(20) N. A. Tolstoi and A. P. Abramov, Sov. Phys. -Solid
      State (Eng. Transl.) 9, 255 (1967); N. Nakashima and
      N. Mataga, J. Phys. Chem., 79, 1788 (1975) and papers
      cited theirin.

(21) H. Masuhara, K. Inoue, N. Tamai, and N. Mataga,
      Nippon Kagaku Kaishi, 14 (1984).

(22) H. Masuhara and N. Mataga, J. Luminescence, 24/25,
      511 (1981).

(23) H. Masuhara, S. Ohwada, Y. Seki, N. Mataga, K. Sato
      and S. Tazuke, Photochem. Photobiol., 32, 9 (1980).

(24) H. Masuhara, N. Tamai, K. Inoue, and N. Mataga, Chem.
      Phys. Lett., 91, 109 (1982).

(25) H. Masuhara, H. Shioyama, N. Mataga, T. Inoue, N.
      Kitamura, T. Tanabe, and S. Tazuke, Macromolecules,
      14, 1738 (1981).

(26) N. Nakashima, Y. Kume, and N. Mataga, J. Phys. Chem.,
      79, 1788 (1975).

(27) G. S. Beddard and G. Porter, Biochim. Biophys. Acta,
      462, 63 (1977).

(28) H. Masuhara, in this book, Chap. 4 ; H. Masuhara, K.
      Yamamoto, N. Tamai, K. Inoue, and N. Mataga, J. Phys.
      Chem. 88, 3971 (1984)

(29) H. Masuhara, S. Ohwada, K. Yamamoto, N. Mataga, A.
      Itaya, K. Okamoto, and S. Kusabayashi,Chem. Phys.
      Lett., 70, 276 (1980).

(30) J.A. Tanaka, H. Masuhara, N. Mataga, S.Egusa, M.
      Sisido, and Y. Imanishi, Abs. for Jpn. Symposium on

Molecular Structure, 710 (1984); J. A. Tanaka, Mc Thesis, Osaka University (1985); H. Masuhara, J. A. Tanaka, N. Mataga, M. Sisido, S. Egusa, Y. Imanishi, submitted to J. Phys. Chem.

(31) J. A. Tanaka, H. Masuhara, N. Mataga, Y. Higuchi, and S. Tazuke, in preparation.

(32) S. Egusa, M. Sisido, and Y. Imanishi, Chem. Lett., 1307 (1983); Macromolecules, 18, 882, 890 (1985).

(33) M. B. Robin, "Higher Excited States of Polyatomic Molecules,Vols. 1 and 2", Academic Press, New York (1974, 1975).

(34) Y. Nakato, J. Am. Chem. Soc., 98, 7203 (1976).

(35) M. Koizumi, S. Kato, N. Mataga, T. Matsuura, and Y. Usui, "Photosensitized Reactions", Kagaku Dojin, 1977, Chap.6; Y. Kurabayashi, K. Kikuchi, and H. Kokubun, Abstract for the 24th Jpn. Symposium on Photochem., Tsukuba, 1983, p.43.

(36) J.A. Tanaka, H. Masuhara, N. Mataga, R. Goedeweck, and F. C. De Schryver, to be submitted.

(37) H. Masuhara, J. A. Tanaka, N. Mataga, F. C. De Schryver, and P. Collart, Polym. J., 15, 915 (1983).

(38) J. A. Tanaka, H. Masuhara, and N. Mataga, Polym. J., submitted.

(39) H. Miyasaka, H. Masuhara, and N. Mataga, J. Phys. Chem., 89, 1631 (1985).

(40) H. Miyasaka, private communication (1984).

(41) H. Masuhara, unpublished results (1982).

(42) W. Schnabel, Makromol. Chem., 180, 1487 (1979); in "Developments in Polymer Photochemistry, Vol. 3", ed by N. S. Allen, Applied Science (1982), Chap. 2.

(43) J. S. Hargreaves and S. E. Webber, Macromolecules, 17, 235 (1984).

ELECTRON TRANSFER DYNAMICS IN THE EXCITED POLYMER AND RELATED SYSTEMS IN SOLUTION

Hiroshi Masuhara
Kyoto Institute of Technology
Department of Polymer Science and Engineering
Matsugasaki, Kyoto 606
Japan

ABSTRACT. An overview is presented of electron transfer quenching of aromatic hydrocarbons in solution. In nonpolar solvents, emissive exciplexes are formed, whereas in polar solvents free ions can be detected. The so-called inverted region, predicted by using a formula given by Marcus, whereby quenching with large negative value of free enthalpy change occurs slower than that with small negative one, is examined from other theoretical points of view. In polymers, electron transfer processes lead to exciplexes or ion radicals with complex structures, due to chromophore-chromophore interactions. The exciplexes formed in polymer systems are exterplexes composed of two donor and one acceptor moieties. Electronic structure and dynamics of polymer ion radicals are considered by comparing their transient absorption spectral data with those of model bichromophoric compounds.

1. INTRODUCTION

Electron transfer, energy transfer, proton transfer, dissociation, isomerization etc. are important primary photoprocesses and their detailed studies have provided a dynamic viewpoint of molecules, molecular aggregates, and polymers. Electron transfer, in particular, is a very interesting elementary step because it polarizes the surrounding environment, induces solvent reorientation, results in complex formation, and triggers chemical reactions. In nonpolar solvents fluorescence quenching of aromatic hydrocarbon by an electron donor or acceptor results in a new emission in the long wavelength region. As an example, results on N-isopropylcarbazole-1,3-dicyanobenzene in di-n-butyl ether are shown in Fig. 1 (1). As the quencher concentration is increased, the monomer fluorescence decreases and a weak emission with a peak at 410 nm appears. The broad and structureless band shape indicates that the formed

M. A. Winnik (ed.), Photophysical and Photochemical Tools in Polymer Science, 65–84.
© *1986 by D. Reidel Publishing Company.*

species giving this emission is dissociative in the ground
state. This type of emission was first called heteroexcimer
by Mataga and is now well known as an exciplex (2). This
new fluorescence is very sensitive to solvent polarity. As
the dielectric constant increases a little, the fluorescence
maximum shifts to the red and the fluorescence yield as well
as the lifetime decrease. From the analysis of this behavior,
it has been concluded that the electronic structure of
exciplexes of aromatic hydrocarbon-electron donor (or
acceptor) pairs is quite polar and is deemed to be a contact
ionpair. In polar solvents ionic species such as free ions
and ionpairs of the donor cation and acceptor anion are
formed instead of exciplexes. These processes are summarized as follows:

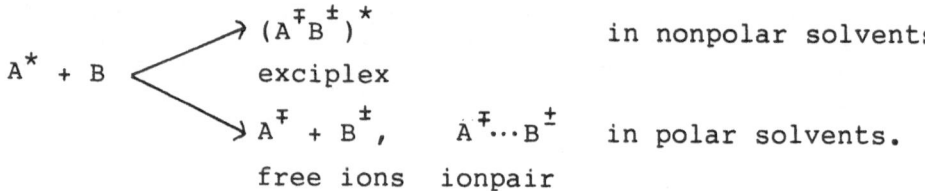

$$A^* + B \begin{cases} \longrightarrow (A^{\mp}B^{\pm})^* & \text{in nonpolar solvents} \\ & \text{exciplex} \\ \longrightarrow A^{\mp} + B^{\pm}, \quad A^{\mp}\cdots B^{\pm} & \text{in polar solvents.} \\ & \text{free ions} \quad \text{ionpair} \end{cases}$$

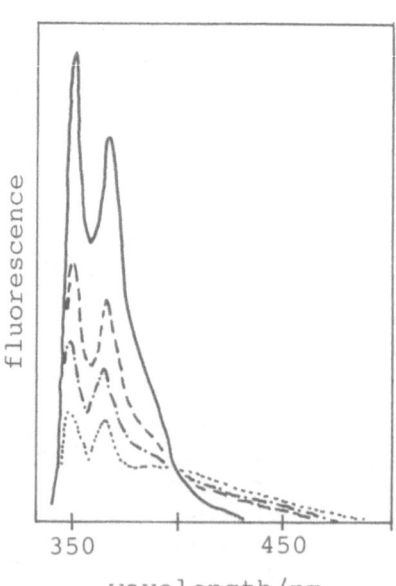

Figure 1. Fluorescence spectra of 10^{-4} M N-iso-
propylcarbazole-1,3-dicyanobenzene in di-n-butyl
ether at room temperature. Concentrations of 1,3-
dicyanobenzene are 0(——), 4.7×10^{-3}(---), 9.7×
10^{-3}(—·—·), and 1.8×10^{-2}(·······) M.

Here, A^* and B are fluorescer and quencher, respectively.

Energetic and kinetic aspects of these electron transfer processes have been considered in detail and correlated with each other, by which microscopic elucidation of these processes has been made possible. Environmental conditions in polymers such as micropolarity and microviscosity, molecular motions, and interchromophoric interactions are different from those in typical small molecule electron donor-acceptor systems. As a consequence, the structure and dynamics of polymer exciplexes and ion radicals are more complicated compared to those of aromatic hydrocarbons. In the following sections these topics are described and discussed.

2. ENERGETIC AND KINETIC ASPECTS OF PHOTOINDUCED ELECTRON TRANSFER IN POLAR SOLVENTS

A formation of donor cation and acceptor anion upon fluorescence quenching indicates that the free enthalpy change of electron transfer (ΔG_{23}) is negative. This value in polar solvents can be estimated according to the following equation

$$\Delta G_{23} = E(D/D^+) - E(A^-/A) - \frac{e^2}{a\varepsilon} - \Delta E_{00} \qquad (1)$$

Here, $E(D/D^+)$ and $E(A^-/A)$ are oxidation potential of the donor and reduction potential of the acceptor, respectively. The third term represents the coulombic attractive force in $A^{\mp} \cdots B^{\pm}$. a and ε are their intermolecular distance and the solvent dielectric constant, respectively. ΔE_{00} is the electronic excitation energy of the aromatic hydrocarbon.

The ΔG_{23}-value can be correlated with the rate constant of fluorescence quenching by using the following scheme (3,4).

$$A^* + B \underset{k_{21}}{\overset{k_{12}}{\rightleftharpoons}} A^* \cdots B \underset{k_{32}}{\overset{k_{23}}{\rightleftharpoons}} A^{\mp} \cdots B^{\pm} \xrightarrow{k_{30}} {}^3A + B, \; A + B, \; A^{\mp} + B^{\pm}, \qquad (2)$$

Here, k_{12} and k_{21} correspond to diffusion in solution, k_{23} and k_{32} represent the rate constants of electron transfer in the respective encounter complex, and k_{30} is the sum of those of rapid deactivation processes. The fundamental theory for elucidating these electron transfer processes was proposed by Marcus (3). It considers that the steps described by k_{23}, k_{32}, and k_{30} are outer-sphere electron transfer. The theory pictures very weak intermolecular

interaction in a nonequilibrium state with a fluctuating
solvation structures, resulting in a non-adiabatic electron
transfer which followed by relaxation to the newly solvated
ionpair. The experimental rate constant of fluorescence
quenching is related to these rate parameters in eq.(2)
through the following equation.

$$k_q = \frac{k_{12}}{1+(k_{21}/k_{23})(1+(k_{32}/k_{30})} \tag{3}$$

Marcus theory predicts the bell-shaped relationship
between log k_q and ΔG_{23} shown in Fig. 2. In other
words, for $\Delta G_{23} > 0$, k_q is predicted to increase
with decreasing ΔG_{23}; but, as ΔG_{23} becomes increasingly negative, k_q is predicted to pass through
a maximum and again decrease.

The first experimental test of this relationship between log
k_q and ΔG_{23} was given by Rehm and Weller (4). As shown in
Fig. 2, k_q decreases as ΔG_{23} increases in the endothermic
region, in a manner almost identical with the theoretical
prediction. In other regions, however, k_q is constant, and
a large difference from the theory is observed in the region
with $\Delta G_{23} < -0.7$ eV

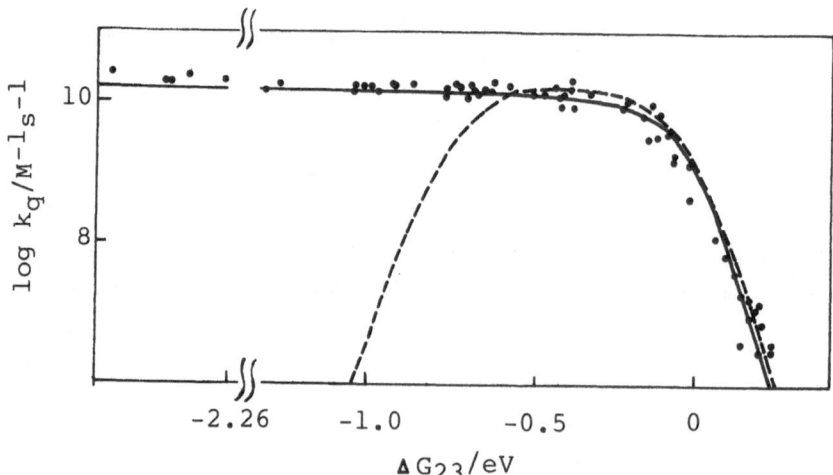

Figure 2. A relation between the fluorescence
quenching rate constant k_q and the free enthalpy
change ΔG_{23}. Experimental point (•), theoretical
curve calculated by Rehm and Weller (——) and one
calculated with a formula derived by Marcus (----).
Cited from Ref.4.

A decrease of the theoretical value of k_q in this exothermic region is due to a decrease of the k_{23}-value which is ascribed to small values of Franck-Condon factors in this process. A similar theoretical consideration was given by Levich and Dogonadze, using the polaron model (5). A quantum mechanical treatment of electron transfer was developed by Kestner et al. (6). These results indicated a similar bell-shaped curve for the relation between log k_q and ΔG_{23}. None of the treatments can interpret the experimental results.

Some attempts have been made to account for the experimental results and to try to elucidate the nature of electron transfer in solution. The first, by Rehm and Weller (4), was an empirical equation which assumed an appropriate monotonic dependence of the activation free energy on ΔG_{23}. It provides no new theoretical insight, and the equation has no distinct physical meaning. The second explanation is that there are several pathways of electron transfer in the inverted region which result in an excited state of $A^{\mp} \cdots B^{\pm}$. A relation between the k_q and the ΔG_{23} values may be reproduced by an overlap of some bell-shape curves ascribed to competing pathways (4,7). The third explanation considers a contribution of exciplex formation (8,9), which corresponds to the inner-sphere electron transfer in terms of Marcus theory. In the region where ΔG_{23} is largely negative and the ionpair state is distinctly lower in energy than the initial neutral states, both fluorescer and quencher may collide many times before electron transfer is induced, since the electron transfer has a smaller rate constant compared to the diffusion-limited one. The forth explanation invokes a detailed consideration on the Franck-Condon factors for electron transfer. Because of the change of electronic structure before and after electron transfer, intramolecular vibrational modes should be different from each other. From this viewpoint, Efrima and Bixon performed a theoretical analysis by including contribution of the higher excited modes of vibration (10). All these ideas seem to explain the anomaly in the inverted region, however, a determing experimental confirmation has not been given and it is still unclear what is the most probable explanation.

Although it is well known that electron transfer phenomena are very sensitive to   solvent polarity (2), effects of solvent upon electron transfer processes have never been considered explicitly. Kakitani and Mataga formulated a new expression for the rate constant of these processes, taking into consideration   the solvent role (11). They considered that solvent orientation around the ionpair produced by electron transfer is quite different from that around the   neutral fluorescer-quencher pair. Solvent molecules are strongly bound to the ions, which results in the steeper

potential surface for the solvent vibrational mode for the
ions compared to the neutral state. This is schematically
drawn in Fig. 3 where several final states with different
energies are given. Since the initial state is very flat
compared to the final one, the energy level of the crossing
point where electron transfer occurs is almost independent
of the energy of its final states. The flatness of the
$(A^*\cdots B)$ surface suggests a small activation energy and a
similar Franck-Condon factor for the pathways to any final
states. Therefore, the rate constant of the electron
transfer in the region of large negative value of ΔG_{23} is
expected to be almost constant. In order to represent the
shape of the energy surface, they used the force constants
and defined the following parameter.

$$\beta = k_i/(k_f - k_i) \tag{4}$$

Here, k_i and k_f represent the force constants of the initial

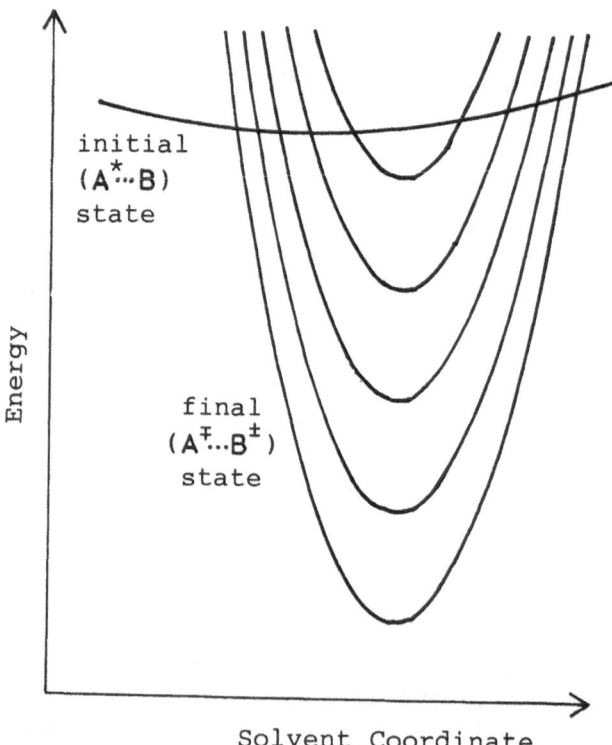

Figure 3. Schematic diagram of potential enegy curves
of electron transfer. Several final states with different minimum energies are given. Cited from Ref.11.

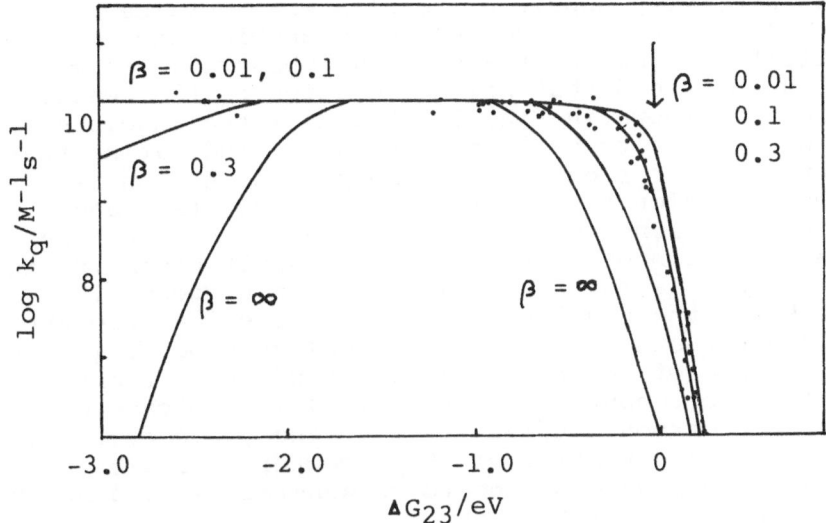

Figure 4.   Calculated relation between the fluorescence quenching rate constant k_q and the free enthalpy
change ΔG_{23} as a function of β given in eq.(4).   Experimental values ($\cdot$).   Cited from Ref.11.

and the final states, respectively. The k_q value was
numerically evaluated by changing β, and a good agreement
between experimental and calculated data was obtained in
the case of small β. As k_f approaches to k_i, β becomes
large, which gives a bell-shape curve for log k_q and ΔG_{23}.
These are shown schematically in Fig. 4. Therefore, the
large difference of the solvation structure before and
after electron transfer was concluded to be an origin for
the so-called anomaly in the inverted region.

3.   FORMATION YIELD OF ION RADICALS UPON PHOTOINDUCED
     ELECTRON TRANSFER

One would think that electron transfer quenching in polar
solvents would lead to the efficient formation of ion radicals. This event was first demonstrated in 1961 by flash
photolysis studies of the perylene-N,N-dimethylaniline
system (12). The transient absorption spectrum obtained
was the superposition of that of the perylene anion and the
amine cation. Since then several experiments were reported
(13) and the time resolution of measurement has been improved
to ns by using lasers (14). When these experiments were
carried out in a quantitative way so that ion yield could

be determined, very suprising results were obtained.
Under conditions where fluorescence quenching should lead
to quantitative electron transfer, we observed a remarkable
scatter in ion yield that had no relation to the reaction
energetics, as shown in Fig. 5 (15). All the absolute
yields were proved to be less than unity, indicating that
in electron-transfer quenching some processes such as
internal conversion to the original neutral state and
intersystem crossing to the triplet state are competing
with the formation of free ions. The result given in Fig.
5 is summarized as follows. (i) There is no general relation between the yield and ΔG_{23}. The chemical properties
such as substituents on the quencher molecule determine the
formation yield of ion radicals, independent of the redox
potential of component molecules. (ii) In the case of
aromatic hydrocarbon-nitrile systems, the yield decreases
as the reduction potential of the acceptor increases.
These conclusions were proved to be general by studies on
many similar systems (16).

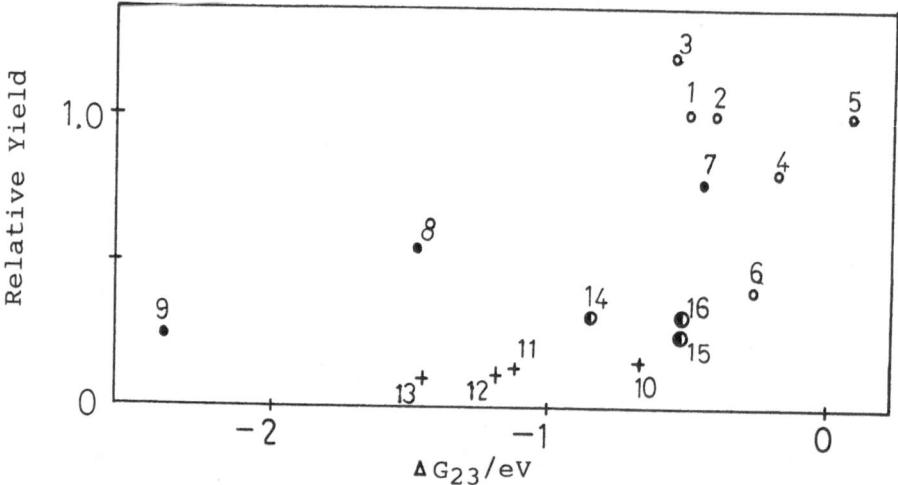

Figure 5. A relation between the relative yield of
free ion radicals and the free enthalpy change ΔG_{23}.
The fluorescer is pyrene and the quenchers are as
follows. (1) N,N-dimethylaniline, (2) N,N-diethyl-
aniline, (3) N,N-dimethyl-m-toluidine, (4) N-methyl-
aniline, (5) aniline, (6) triethylamine, (7) 1,4-
dicyanobenzene, (8) 1,2,4,5-tetracyanobenzene, (9)
tetracyanoethylene, (10) phthalic anhydride, (11)
tetrachlorophthalic anhydride, (12) maleic anhydride,
(13) pyromellitic dianhydride, (14) diethyl iso-
phthalate, (15) diethyl phthalate, (16) diethyl
terephthalate. Cited from Refs.9 and 15.

A possible interpretation for these ns laser photolysis
studies was that fluorescence quenching leads to the formation of an ionpair of the donor cation and the acceptor
anion with unit quantum yield (9,15). From this state
dissociation occurs,competing with other nonradiative
processes.

$$A^{\ast}\cdots B \longrightarrow A^{\overline{+}}\cdots B^{\pm} \begin{cases} \xrightarrow{k_d} A^{\overline{+}} + B^{\pm} \\ \xrightarrow{k_n} A + B, \ ^3A + B \end{cases} \quad (5)$$

ionpair

Provided that the rate constants of the dissociation and of
other pathways are k_d and k_n, respectively, the yield was
given as $\phi = k_d/(k_d+k_n)$. As the first approximation the k_d-
value was assumed to be common to all the systems with
different quenchers, since their molecular volume is almost
the same and no specific interaction between quencher and
solvent molecules operates. Therefore, the yield was considered to be mainly determined by the k_n-value.

Theoretically this k_n-value is again given by an electronic
matrix element and a Franck-Condon factor. Systems with the
same free enthalpy change should have the similar Franck-
Condon factors, since the latter is a function of the former.
However, the yields are scattered, which should be reduced to the difference of the electronic matrix elements.
Small yields were observed for the systems with a component
molecule having carbonyl group or heteroatom. In these
molecules, an n-$\pi^{\ast}$ triplet level is placed between or near
the initial electron transfer and the final triplet states,
which may accelerate the intersystem crossing. The k_n-
values increase and the formation yield of ion radicals is
reduced. This interpretation is consistent with the result
that the yield is determined by the chemical properties of
the component molecules.

A direct demonstration of the mechanism discussed above was
given for CT complexes stable in the ground state (17).
In this kind of simple system, the complexities of encounter
and dissociation processes, which complicate the data analysis, are avoided. The ps excitation of the CT absorption
band of 1,2,4,5-tetracyanobenzene complexes gave an absorption spectrum identical to that of the 1,2,4,5-tetracyano-
benzene anion, independent of the delay time. The species
was formed immediately after excitation and its decay up to a
few ns consisted of the fast and plateau components as shown
in Fig.6. The latter corresponds to free ion radicals, while
the former obeys first order decay kinetics. It is our

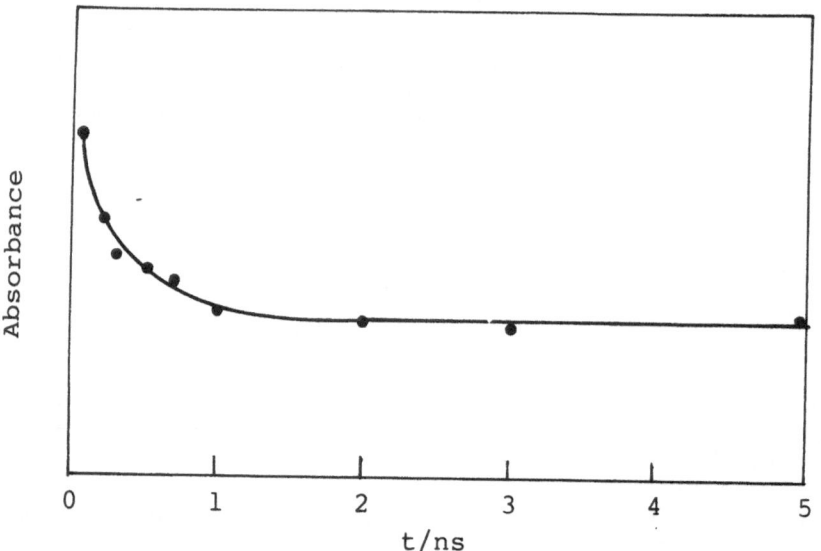

Figure 6. A decay curve of 1,2,4,5-tetracyanobenzene anion band which was induced by exciting the CT absorption of 1,2,4,5-tetracyanobenzene-toluene complex in acetonitrile. Cited from Ref.17.

interpretation that this fast component, with a time-constant of a few hundreds ps, is due to a contact ionpair of the donor cation and the acceptor anion. We determined the decay time of this ionpair and the formation yield of the free ions for these 1,2,4,5-tetracyanobenzene complexes. The values of k_d and k_n were calculated to be in the order of 10^9 sec^{-1} for substituted benzene donors. On the other hand, the k_n-values larger than 3×10^{11}s^{-1} were obtained in the case of donor molecules containing nitrogen or oxygen atom.

4. BICHROMOPHORIC MODELS FOR EXCIPLEX FORMATION IN POLYMERS

Structure and dynamics of exciplexes in polymers are in many ways even more complex than the typical aromatic hydrocarbon systems. First, solvation structure and solvent properties around the polymer such as microviscosity and micropolarity are different from those of simple solutions. Second, in the case of polymers with pendant aromatic groups, the mean distance between chromophores is very

short so that interchromophoric interaction is favored.
Third, configurational and conformational structures of
polymer chains may result in a plural geometrical structures
of exciplexes, and thus in complicated behavior. We believe
that detailed studies on bichromophoric models are fruitful
for understanding polymer exciplexes. In this section, two
topics are discussed. The first is on intramolecular
exciplex forming compounds such as $A-(CH_2)_n-D$ where A and D
are acceptor and donor moieties connected to each other by a
polymethylene chain. The second describes the structure and
dynamics of exterplexes. These are a kind of three-body
exciplex composed of two donor and one acceptor moieties.

Ns laser photolysis measurements on $(1-pyrenyl)-(CH_2)_n-$
(p-N,N-dimethylaminophenyl) (abbreviated as $P-(CH_2)_n-D$)
were performed in some solvents (18). Their absorption
spectra are reproduced by the superposition of bands of
the donor cation and the acceptor anion. Only the bandwidth is dependent upon solvent polarity and the number of
CH_2 groups, which is reduced to the relative geometrical
structure of the donor and the acceptor. Exciplex formation
dynamics of these compounds has been established by ps
transient absorption spectroscopy (19). The absorption
spectrum of $P-CH_2-D$ in 2-propanol is almost independent of
the delay time. On the other hand, $P-(CH_2)_3-D$ gives a
similar spectrum at about 1 ns after excitation, and its
spectral shape broadens with a time-constant of 1.2 ns.
These kinds of behavior were affected also by solvent polarity, viscosity, and the number of methylene chain units.
We interpreted these results by assigning the sharp absorption bands to loose exciplex geometries and the broad bands
to the sandwich structure. In our view, electron transfer
occurs within the extended structure, resulting in a loose
exciplex, which is followed by a conformational change to the
sandwich structure. This may be due to a coulombic interaction between two ionic moieties. In acetonitrile, the
sharp ionic bands were observed up to 5 ns, indicating no
formation of the sandwich exciplex. In nonpolar solvents,
exciplex formation is considered to be possible only in
the sandwich structure because $P-(CH_2)-D$ did not give
exciplex emission (20). On the basis of these results,
we concluded that the electronic and geometrical structures
are sensitive to the surrounding environments and several
structures are probable. This conclusion is just the basis
for understanding polymer exciplexes.

Poly(N-vinylcarbazole)(abbreviated as PVCz) quenched by
dimethyl terephthalate (DMTP) was first analyzed in terms of
exciplexes formed between the excited polymers and monomer
quenchers (21). The emission maximum of this pair is red-
shifted by 60 nm compared to the reference monomer model

systems, and the decay curves are significantly different
(22).   Examining various possibilities, we concluded that
the results characteristic of polymer exciplexes should be
ascribed to the exterplex formation. Geometric aspects of
these exterplexes, including their formation dynamics are
best understood through studies on bichromophoric model
compounds.   Thus we examined exciplex emissions of N-ethyl-
carbazole(N-EtCz),rac-2,4-di(N-carbazolyl)pentane (r-DCzPe),
and meso-2,4-di(N-carbazolyl)pentane (m-DCzPe) quenched by
DMTP were examined (1). Their spectra had peaks at 410, 460
and 510 nm, respectively. In the case of the 1,2-trans-di(N-
carbazolyl)cyclobutane (DCzB)-DMTP system, the exciplex
emission observed was similar to that of the EtCz system.
All the spectra showed red-shifts as the solvent polarity
was increased. These results indicate that an overlap between two carbazolyl groups plays an important role in the
exciplex, and the exciplex formed is quite polar. The red-
shifts are consistent with exterplexes of the form DDA, D and
A being carbazole and DMTP, respectively. This conclusion
seems to be quite general for exterplex structure as far
as we know (23). Direct confirmation of this structure
was given for the first time in the studies on 1,2,4,5-
tetracyanobenzene-toluene charge transfer complex (24).
The $S_n \leftarrow S_1$ absorption spectrum was composed of the bands re-
sembling the 1,2,4,5-tetracyanobenzene anion and toluene dimer
cation. Similar results were observed also in the systems

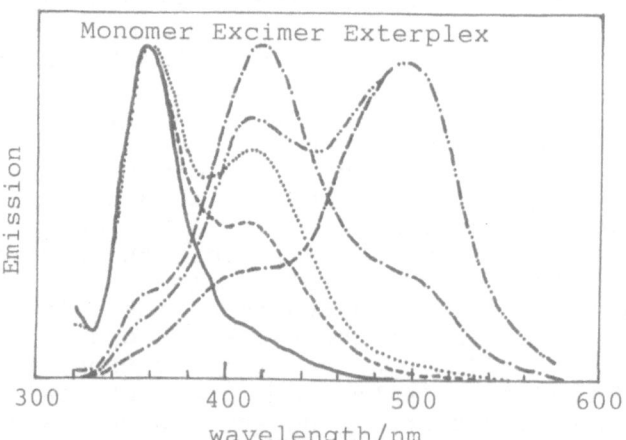

Figure 7. Normalized time-resolved spectra of meso-
2,4-di(N-carbazolyl)pentane quenched by 6×10^{-3}M 1,3-
dicyanobenzene in di-n-butyl ether at room temperature.   Gated time-width: 0-4.9 ns (———), 15.1-20.0
ns (----), 31.1-41.7 ns (·········), 40.4-60.2 ns (—·—·),
75.8-89.3 ns (—··—), 110-140 ns (——·). Cited from
Ref. 1.

of this acceptor in liquid benzene derivatives and 1,4-
dicyanobenzene in α-methylnaphthalene. In the present
carbazole systems, more detailed information on exterplex
structure could be obtained. The difference of exterplex
emission between r-DCzPe and m-DCzPe may be reduced to the
different geometries of the two carbazolyl groups. The
excimer as well as the dimer cation of r-DCzPe and m-DCzPe
take the partial overlap and the sandwich structures, respectively (25), and these structures are considered to be
responsible for their DDA structures.

The exterplex formation processes of the m- and r-DCzPe systems were investigated by measuring time-resolved fluorescence spectra. In both compounds excimers were formed first
and then quenched by 1,3-dicyanobenzene, leading to exterplexes. This sequence might be reduced to a rather low concentration of the quencher. Namely, the rate of the normal
exciplex formation is distinctly smaller than that of intramolecular excimer formation (25). The result on the m-DCzPe
system is shown in Fig. 7 (1). The relative contribution of
the excimer emission decreased gradually and did not show an
equilibrium with the exterplex emissions, indicating the
absence of the return from the exterplex to the excimer. If
we increase the quencher concentration and examine the compound having a smaller rate-constant of excimer formation,
the pathway of monomer → exciplex → exterplex may operate.
Actually this has been confirmed directly for 1,3-di(N-
carbazolyl)propane quenched by 1,2-dicyanobenzene, using ps
transient absorption spectroscopy (26). It seems
that there is no appreciable energy barrier in the steps
from exciplex to exterplex and from excimer to exterplex.
In the PVCz systems energy migration should be involved,
and accelerate the pathway of monomer → excimer → exterplex.

5. ELECTRONIC STRUCTURE AND DYNAMICS OF POLYMER ION
   RADICALS

In order to understand the nature of electron transfer
dynamics in polymers, it is necessary to understand the electronic structure of polymer ion radicals and their interaction with the counter ion. These are closely related to
configurational and conformational structures of polymers.
Similar concerns occur in fluorescence studies of polymers
where one finds that the electronic properties of the excited singlet state are sensitive to the structure of the
main chain, the molecular weight of the polymer, and the
position of the chromophore in the polymer structure. These
factors determine the statistical distributions of interchromophoric distance and relative orientation resulting in
characteristic fluorescence behavior. A similar relationship

between electronic and geometrical structures was expected
for ion radicals, but few detailed studies have been reported. One of the pioneering experiments described absorption spectra of polystyrene and poly(2-vinylnaphthalene)
cation radicals. Irie et al. reported that their spectra
are different from those of   corresponding model compounds and pointed out the importance of configurational
and conformational structures (27). Here we present our
laser photolysis studies on PVCz and related systems. This
carbazole system has certain advantages, because the relationship between absorption spectra and geometrical as well
as electronic structures is simple compared to pyrenyl and
naphthyl systems.

Cation radicals of carbazolyl chromophore can be formed by
quenching its excited state by the electron acceptor, DMTP,
in DMF. The measured absorption spectra were composed of
the carbazole cation and the counter acceptor anion. The
former is observed in the wavelength region above 600 nm,
while the latter is seen below 600 nm. The molecular structures of PVCz, Poly(N-carbazolylethyl vinyl ether) and polyurethanes studied and the absorption spectra of their cation
radicals are shown in Fig. 8. The spectra of the polyvinyl-
ethers and polyurethanes are very broad and show a peak at
780 nm (25, 28-30). PVCz has a maximum at 770 nm and no
shoulder is detected. It is worth noting that this spectral shape is common to all PVCz polymer cations with different n-values from 4 to 1100 and is observed even for PVCz-4.

Studies on bichromophoric model compounds are useful for
establishing how the shape of the cation radical spectrum
depends upon geometrical structure. Absorption spectra and
geometrical structures of some compounds in the cationic
state are shown in Fig. 9. The conformations of DCzP, r-
DCzPe, and m-DCzPe in the excited and ionic states were
determined by analyzing NMR data and time-resolved absorption
spectra (25, 29). On the basis of these results, the measured spectra of their cationic states were classified into
three groups due to open, sandwich and partial overlap structures. In other words, the positive charge is stabilized
not only in the sandwich form, but also in rather loose structures, in which the interactions are of the long-range
Coulombic type.

The spectra of the polymer cations except PVCz are similar to
that of the open dimer cation but not to that of the sandwich and
partial overlap ones. This result indicates that the positive
charge in these polymers is trapped in a loose dimer site
and no polymer conformation giving other dimer structures
formed between nonadjacent chromophores is possible.

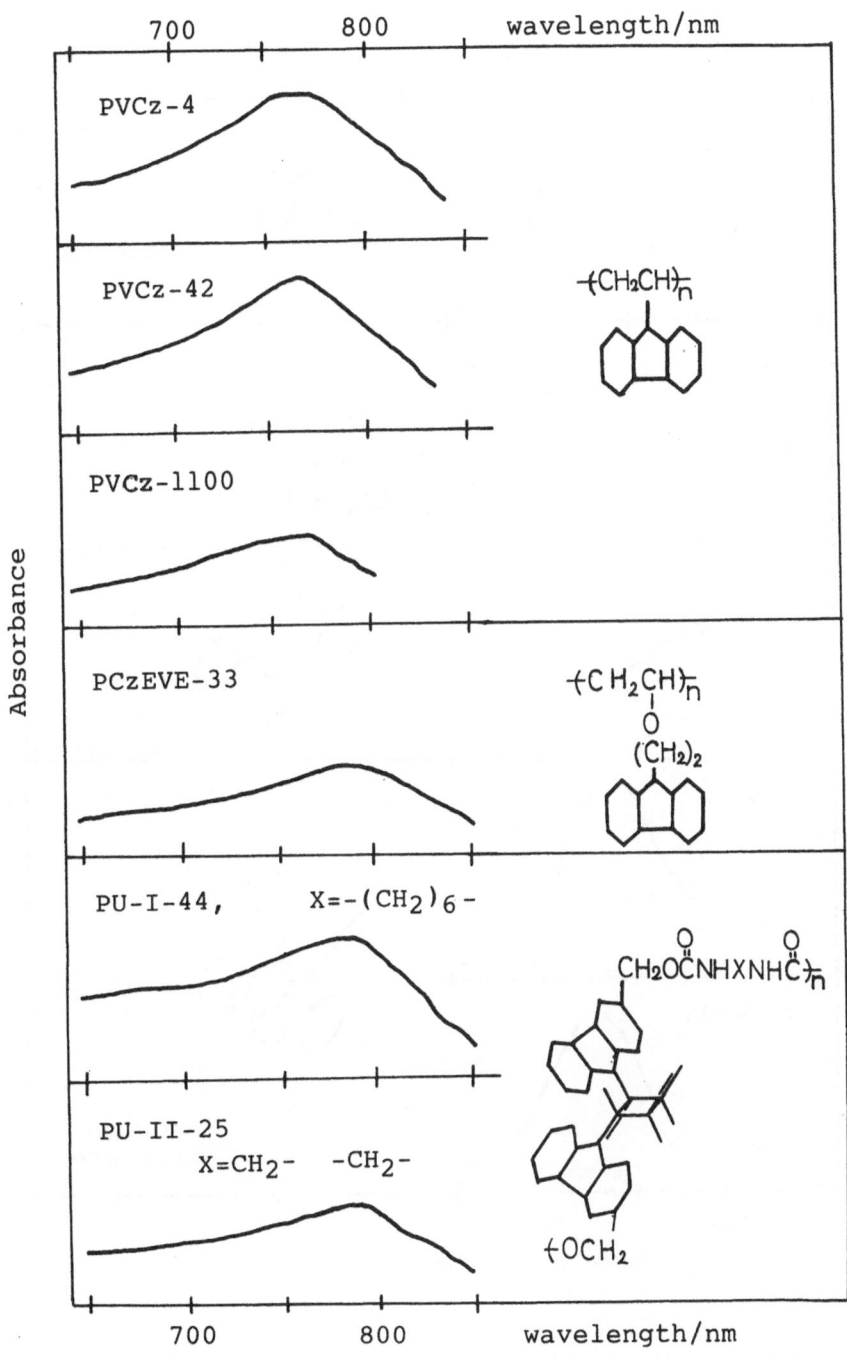

Figure 8. Absorption spectra of polymer cation radicals.

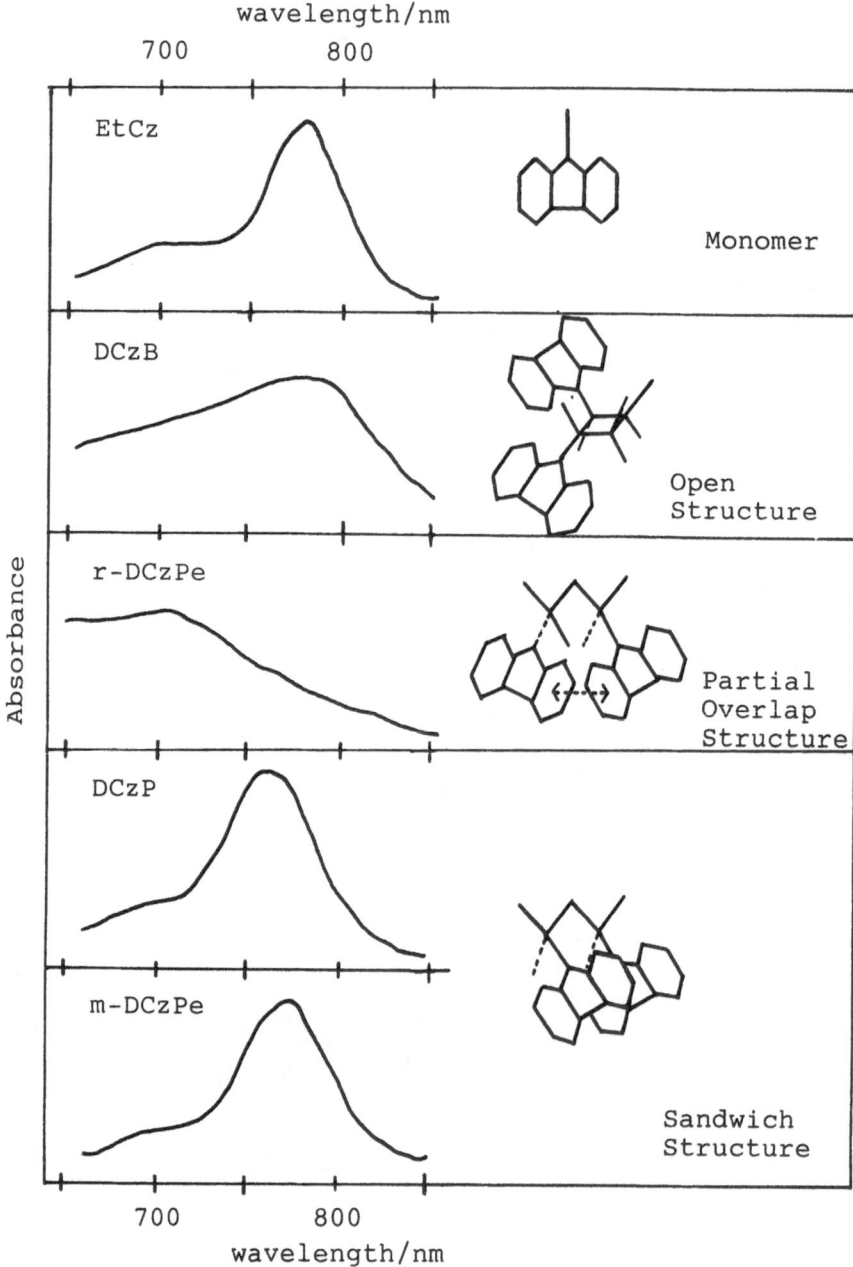

Figure 9. Absorption spectra of dimer model compounds in the cationic state.

The spectra of PVCz-n cations are different from those of
all monomer and dimer model compounds and characteristic
of the polymer itself. One possible explanation is that the
positive charge is delocalized over a number of chromophores
or trapped in a new kind of site produced only in the polymer
with high n-value. However, the present absorption spectra
are common to all PVCz-n systems including PVCz-4. Another
and more probable explanation is based on the assumption
that the positive charge is localized in dimer sites. There
should be some distribution of configurational and conformational structures even for PVCz-4. The measured spectrum
is a superposition of absorption bands of the different
dimer cation radicals as classified above. If one tries to
simulate the shape of these absorption spectra, one finds
that at least three kinds of dimer cations contribute to
the PVCz cation spectra.

The carbazole polymer anions behave quite differently. Electron transfer quenching of the present polymers by 1,4-
diazabicyclo[2,2,2]octane (Dabco) results in the formation
of the carbazole anion and the Dabco cation. The measured
spectrum is common to all model compounds and polymers, and
no interchromophoric interaction operates between anionic
and neutral chromophores, even in the polymer system. The
electron is trapped as a monomer anion radical.

This kind of information on the electronic and geometrical
structures of polymers is considered to be indispensable for
elucidating electron transfer and exciplex dynamics. The
kinetic aspects of formation of polymer ion radicals are
very complicated compared to those in monomer donor-acceptor
systems. There are distributions of configurational and
conformational structures, molecular weight, and solvation
structures around chromophores. Furthermore, diffusion of
polymers and energy migration along the main chain should be
involved in electron transfer processes leading to ion
radicals. We believe that some of these problems can be
overcome by studying a series of polymers containing the
same aromatic pendant groups in which local and higher
order-structures are variables.

The most interesting aspect of decay dynamics of polymer ion
radicals is that slow geminate recombination is observed even
in polar solvents. During our systematic studies of carbazole
polymers quenched by DMTP in N,N-dimethylformamide, we found
that the decay of the ion radicals produced had both of the
fast and slow components. Although only normal second-
order decay kinetics could be observed in the cases of
small molecule donor-acceptor systems and of polymers with
low degree of polymerization, an additional first-order
decay was detected for PVCz with n-values of 400 and

1100.   After examining various possibilities, we ascribe
this new component to the geminate ionpair of the polymer
and quencher ion radicals.   Some of the quencher ion radicals fail to diffuse out from the microdomain around polymers into the bulk and recombine with the original counter
ions (31). The relative   yield and the lifetime of this
ionpair increase as the n-value does, independent upon the
kind of quencher.   In the case of PVCz-3700, the whole
decay was analyzed as the first order decay and its time
constant was obtained to be 42 μs (32).   A similar behavior
was also observed in a polymer having crowded chromophores
(33).   We concluded that the geminate recombination is a
quite general behavior of polymer systems.

ACKNOWLEDGEMENTS

The author wishes to express his sincere thanks to Prof. N.
Mataga (Osaka University) for many helpful discussions and
support.

REFERENCES

(1)   H. Masuhara, J. Vandendriessche, K. Demeyer, N.
      Boens, and F. C. De Schryver, Macromolecules, 15,
      1471 (1982).

(2)   N. Mataga and M. Ottolenghi, in "Molecular Associ-
      ation, Vol. 2", ed. by R. Foster, Academic Press,
      London, 1979, p.1; N. Mataga, in "Molecular Inter-
      actions, Vol. 2", ed. by H. Ratajczak and W. J.
      Orville-Thomas, John-Wiley & Sons, NY, 1981, p.509.

(3)   R. A. Marcus, J. Chem. Phys., 24, 966 (1956); Dis-
      cussion Farad. Soc., 29, 21 (1960); Ann. Rev. Phys.
      Chem., 15, 155 (1964).

(4)   D. Rehm and A. Weller, Ber. Bunsenges. Phys. Chem.,
      73, 834 (1969); Israel J. Chem., 8, 259 (1970).

(5)   V. G. Levich and R. R. Dogonadze, Collection Czech.
      Chem. Commun., 26, 193 (1961).

(6)   N. R. Kestner, J. Logan, and J. Jortner, J. Phys.
      Chem., 78, 2148 (1974).

(7)   N. Mataga, Bull. Chem. Soc. Jpn., 43, 3623 (1970).

(8)   A. Weller, in "Light-Induced Charge Separation in
      Biology and Chemistry", ed. by H. Gerischer and J.
      Katz, Verlag Chem., Weinheim, 1979, p.132.

(9)   H. Masuhara and N. Mataga, Accounts Chem. Res., 14,
      312 (1981).

(10) S. Efrima and M. Bixon, Chem. Phys. Lett., 25, 34
      (1974); Chem. Phys., 13, 447 (1976).

(11) T. Kakitani and N. Mataga, J. Phys. Chem., 89, 8
      (1985); Chem. Phys., 93, 381 (1985).

(12) H. Leonhardt and A. Weller, Z. Phys. Chem. N. F.,
      29, 277 (1961); Ber. Bunsenges. Phys. Chem., 67,
      791 (1963).

(13) H. Knibbe, D. Rehm, and A. Weller, Ber. Bunsenges.
      Phys. Chem., 72, 257 (1968); M. Koizumi and H.
      Yamashita, Z. Phys. Chem. N. F., 57, 103 (1968);
      K. Kawai, N. Yamamoto, and H. Tsubomura, Bull. Chem.
      Soc. Jpn., 42, 369 (1969); T. Okada, H. Oohari, and
      N. Mataga, ibid., 43, 2750 (1970).

(14) Y. Taniguchi and N. Mataga, Chem. Phys. Lett., 13,
      596 (1972); Y. Taniguchi, Y. Nishina, and N. Mataga,
      Bull. Chem. Soc. Jpn., 45, 764 (1972).

(15) T. Hino, H. Akazawa, H. Masuhara, and N. Mataga, J.
      Phys. Chem., 80, 33 (1976); H. Masuhara, T. Saito,
      Y. Maeda, and N. Mataga, J. Mol. Struct., 47, 243
      (1978).   •

(16) H. Masuhara, M. Shimada, N. Tsujino, and N. Mataga,
      Bull. Chem. Soc. Jpn., 44, 3310 (1971); J. Hinatu,
      F. Yoshida, H. Masuhara, and N. Mataga, Chem. Phys.
      Lett., 59, 80 (1978).

(17) T. Uemiya, Mc Thesis, Osaka University (1984); T.
      Uemiya, H. Miyasaka, H. Masuhara, and N. Mataga, to

be published; N. Mataga, Radiat. Phys. Chem. 21, 83 (1983); N. Mataga, Pure Appl. Chem., 56, 1255 (1984).

(18) J. Hinatu, H. Masuhara, N. Mataga, Y. Sakata, and S. Misumi, Bull. Chem. Soc. Jpn. 51, 1032 (1978).

(19) T. Okada, M. Migita, N. Mataga, Y. Sakata, and S. Misumi, J. Am. Chem. Soc., 103, 4715 (1981).

(20) T. Okada, T. Saito, N. Mataga, Y. Sakata, and S. Misumi, Bull. Chem. Soc. Jpn., 50, 331 (1976).

(21) A. Itaya, K. Okamoto, and S. Kusabayashi, Bull. Chem. Soc. Jpn., 49, 2082 (1976); A. Itaya, Dc Thesis, Osaka University (1980).

(22) C. E. Hoyle and J. E. Guillet, Macromolecules, 11, 221 (1978); 12, 956 (1979).

(23) H. Beens and A. Weller, Chem. Phys. Lett., 2, 140 (1968); T. Mimura and M. Itoh, J. Am. Chem. Soc., 98, 1095 (1976); Bull. Chem. Soc. Jpn., 50, 1739 (1977).

(24) N. Tsujino, H. Masuhara, and N. Mataga, Chem. Phys. Lett., 21, 301 (1973).

(25) H. Masuhara, N. Tamai, N. Mataga, F. C. De Schryver, and J. Vandendriessche, J. Am. Chem. Soc., 105, 7256 (1983).

(26) H. Masuhara, T. Mito, and N. Mataga, unpublished data (1982).

(27) S. Irie, H. Horii, and M. Irie, Macromolecules, 13, 1355 (1980).

(28) H. Masuhara, K. Yamamoto, N. Tamai, K. Inoue, and N. Mataga, J. Phys. Chem., 88, 3971 (1984).

(29) H. Masuhara, J. Mol. Struct., 126, 145 (1985).

(30) H. Masuhara, Makromol. Chem. Supplement, in press (1985).

(31) H. Masuhara, S. Ohwada, N. Mataga, A. Itaya, K. Okamoto, and S. Kusabayashi, Kobunshi Ronbunshu, 37, 275 (1980); H. Masuhara and N. Mataga, J. Luminescence, 24/25, 511 (1981).

(32) U. Lachish, R. W. Anderson, and D. J. Williams, Macromolecules, 13, 1143 (1980).

(33) H. Masuhara, H. Shioyama, N. Mataga, T. Inoue, N. Kitamura, T. Tanabe, and S. Tazuke, Macromolecules, 14, 1738 (1981).

STUDIES OF POLYMERS CARRYING MEDIUM-SENSITIVE FLUOROPHORES.

H. Morawetz
Department of Chemistry
Polytechnic Institute of New York
333 Jay Street
Brooklyn,NY 11201

ABSTRACT. The fluorescence intensity of the dansyl group is strongly reduced when the group is in contact with water molecules. When the label
is attached to to polymeric acids which undergo on ionization a transition from a contracted to an expanded form, the transition can be monitored by the decrease in fluorescence. When solutions of dansyl-labeled poly(methacrylic acid) are subjected to pH jumps, part of the conformational transition is slow enough to be followed in a stopped-flow
apparatus. Association of dansyl-labeled poly(acrylic acid) in aqueous
solution with polymers which are hydrogen bond acceptors results in a
change in the emission intensity and the location of the emission peak.
Kinetics have been studied of the complex formation, complex dissociation and complex interchange with polyoxyethylene (POE). The dependence of the emission characteristics of the complex on the POE chain
length exhibit some unexpected features.

1. Introduction.

The use of 1-dimethylamino-5-naphthyl sulphonate ("dansyl") as a label
was first introduced in 1952 for the determination of the rotational
diffusion of protein molecules from the anisotropy of the dansyl fluorescence (1). Somewhat later, (2) it was found that the emission intensity of fluorophores of this type is strongly increased when they are
adsorbed from water solution on proteins. This increase in quantum yield
of emission (accompanied by a blue-shift of the emission spectrum) could
be correlated with a decreasing polarity of the medium (3). The effect
can be explained as follows: If the dipolar interaction of the excited
chromophore with the solvent is smaller than this interaction in the
ground state, the emitted quantum will be increased; at the same time,
the probability of nonradiative deactivation will be reduced, increasing
the quantum yield of fluorescence (4). If the fluorescent label can be
attached to a well-defined site on a protein, its emission characteristics will "report" about the polarity of this site (5) and the use of

M. A. Winnik (ed.), Photophysical and Photochemical Tools in Polymer Science, 85–95.
© *1986 by D. Reidel Publishing Company.*

such "reporters" has been widespread in physical biochemistry.

In the case of synthetic polymers, the utility of medium-sensitive fluorescent labels has been demonstrated in studies of two types:(a)Certain ionizable polymers exhibit a sharp transition from a contracted to an expanded state at a critical charge density along the chain molecule. This expansion increases the contact between dansyl groups attached to the polyion with the aqueous medium and this is reflected in a decreased fluorescence intensity. (b) When an aqueous solution of a dansyl-labeled polymer forms an association complex with another polymer, the label is shielded from contact with water molecules and such associations can, therefore, be monitored by fluorimetry.

2. Studies of conformational transitions of polymeric acids.

When a polymer carrying a large number of carboxyl groups is ionized,the mutual repulsion of the ionic charges leads to an expansion of the molecular coil, as shown most dramatically by a sharp increase in solution viscosity. Also, because of the increasing free energy required for the removal of a hydrogen ion from the highly charged polyion, the effective acidity of the carboxyl groups decreases with an increasing charge density of the polyion. The situation becomes more complex if some force - such as the hydrophobic cohesion of nonpolar polyion substituents - opposes chain expansion. This effect was first described for alternating maleic acid copolymers (6). In such copolymers, pairs of carboxyls are separated from each other by the comonomer residues and up to half-neutralization only one carboxyl in each pair is ionized. At higher degrees of ionization the acidity of the carboxyl groups in a styrene copolymer is much smaller than in a copolymer with methyl vinyl ether, since the hydrophobic forces between the phenyl residues resist chain expansion and thus ensure a greater concentration of the anionic charge.

A difference similar to that between the two maleic acid copolymers was later demonstrated (7) in a comparison of the titration behavior of poly(acrylic acid)(PAA) and poly(methacrylic acid)(PMA). In the case of PAA, the pK of the carboxyls increases smoothly with an increasing degree of ionization α while for PMA a plot of pK against α exhibits a plateau region. This has been interpreted (8) as resulting from a transition between a contracted and an expanded state of the polyion which behaves in a manner analogous to a phase transition. Since addition of methanol to the aqueous PMA solution eliminates the anomaly in the titration curve, it was proposed (9) that the force opposing chain expansion results from hydrophobic cohesion of the α-methyl groups. However, a calorimetric study of the PMA titration (10) revealed an endothermic peak in the transition range, contrary to the expected enthalpy for the dissociation of a hydrophobic bond. Also, a spectroscopic study of the titration of PAA and PMA in an aqueous methanol solution in which no hydrophobic bonding would be expected still revealed differences, suggesting that other factors must contribute to the characteristic PMA transition (11).

The use of a dansyl label for the study of the transition from the contracted to the expanded form of polymeric acids was first described by Strauss and his students (12,13) for alternating copolymers of maleic acid with vinyl alkyl ethers. They found that in the copolymer with vinyl methyl ether the emission intensity of the label was low no matter what the degree of ionization. By contrast,with the vinyl butyl ether copolymer the dansyl label emitted strongly at low degrees of ionization indicating that the label was effectively shielded by the highly contracted chain molecule from contact with water molecules. The fluorescence intensity decreased precipitously at higher degrees of ionization and the range within which this change occurred corresponded to the transition to the expanded form as deduced from titration data. As expected, when the length of the alkyl side chain or the ionic strength were increased, a higher charge density along the polyion chain was required before the contracted form was destabilized as indicated by the drop in fluorescence intensity of the label.

In our laboratory we used the changes in the fluorescence of dansyl labels attached to PMA to follow in a stopped-flow apparatus the kinetics of the chain expansion or contraction after a pH jump (14). The excitation wavelength of 330 nm used in this work was close to the absorption maximum of the basic form of the dansyl group while the absorption by the protonated form was negligible. As can be seen in Figure 1, which compares the emission characteristics of the dansyl group when

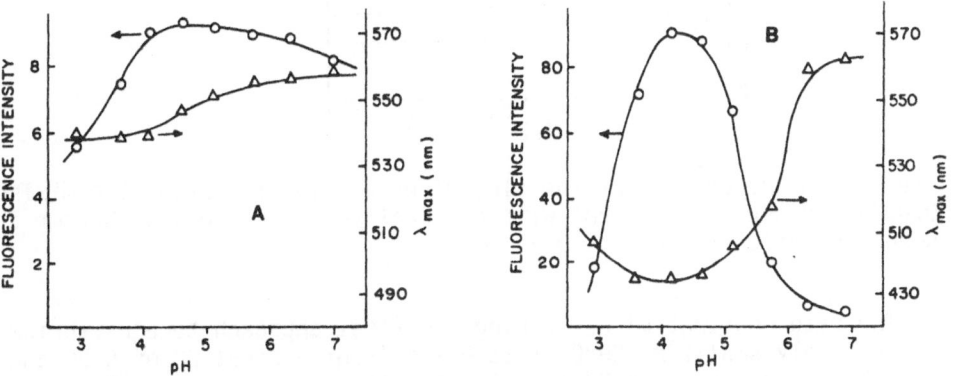

Figure 1. The pH dependence of the emission characteristics of (A) dansyl-labeled PAA and (B) dansyl-labeled PMA.

attached to PAA and to PMA, an increase of the pH to about 4.5 leads to an increased emission intensity as expected from the deprotonation of the dansyl group (pK~4). At higher pH, the expansion of the chain leads to an increasing exposure of the label to water molecules resulting in the quenching of fluorescence, but this effect is much less with PAA than with PMA which exists in a highly contracted state at low pH values The position of the emission maximum is highly revealing since it is insensitive to the state of ionization of the label and reflects only the polarity of its microenvironment. The results show that this polar-

ity changes much less during the ionization of PAA - which is expanded
even at low charge density - than with PMA which is highly contracted at
low pH. In fact, the blue-shift of the emission observed when the pH of
the labeled PMA is raised from 3 to 4 suggests that the initial ionization of PMA leads to a contraction, possibly due to hydrogen bonding
between ionized and unionized carboxyls.

A typical trace of the time dependence of the emission intensity
after the labeled PMA solution is mixed with a buffer inducing a polymer
expansion is shown in Figure 2. Here I* stands for the emission intensity which was observed when the PMA was mixed with a salt solution instead of the buffer, I_o is the emission 1 ms after mixing and I_∞ the
emission after 90 s, sufficient to attain equilibrium. It may be seen
that about half of the fluorescence decrease takes place too fast to be
followed in the stopped-flow apparatus. The remainder can be represented
by biphasic kinetics:

$$(I_t - I_\infty)/(I_o - I_\infty) = q \exp(-kt) + (1-q) \exp(-k't) \qquad (\kappa > k')$$

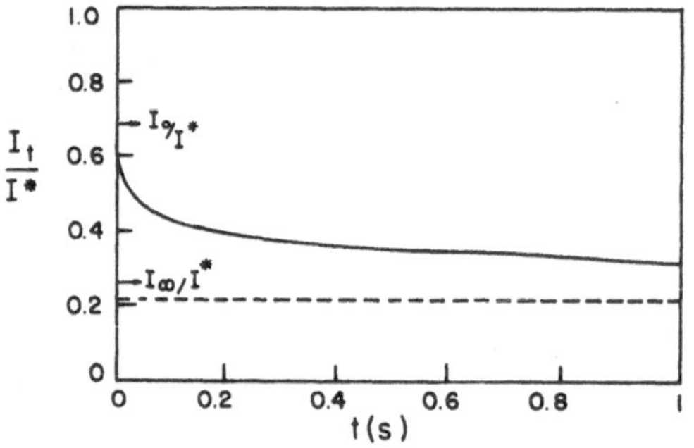

Figure 2. Time dependence of the fluorescence intensity after the mixing
of 0.0023 N dansyl-labeled PMA with an equal volume of pH 5.52 buffer at
20°C and an ionic strength of 0.4.

The rate constant k' describing the final approach to equilibrium
is surprisingly small. At 20°C it is 0.7 s^{-1} for a final pH of 5.52 and
32 s^{-1} for a final pH of 6.83.The slower process at the lower pH may reflect chain entanglements, hydrogen bonds and hydrophobic bonding, all
constituting impediments to conformational rearrangements in the relatively contracted chain at pH 5.52. Similar results were obtained for
the rate at which the fully ionized PMA collapses when it is mixed with
buffers in the pH range 4 - 5.

There have been, to my knowledge, three other attempts to follow
the kinetics of a conformational transition of a polymeric acid after a
pH jump. The first was made by Ohno et al. (15) who took advantage of a
change in the extinction coefficient of phenyl groups in the contracted
and expanded form of an alternating styrene-maleic acid copolymer to monitor chain expansion. They reported a mean relaxation time passing
through a maximum of 0.2 s at the midpoint of the conformational transition and suggested that their data can be interpreted on the basis of

the theory derived for a cooperative helix-coil transition of polypeptides (16). This approach is unconvincing since the polypeptides undergo an order-disorder change, while the conformation of the maleic acid copolymer is disordered in both the contracted and the expanded form. (Strauss and Barbieri (17, 18) modeled the cooperative transitions in maleic acid-vinyl alkyl ether copolymers by a chain containing micelles of constant size separated by randomly coiled sequences; the transition corresponds then to a gradual reduction in the number of micellized sequences). Another stopped-flow study was reported by Okubo (19) who interperted the time dependence of conductivity after mixing a PAA solution with dilute alkali as reflecting chain expansion. He found rate constants decreasing with an increasing degree of ionization from $100s^{-1}$ to $1s^{-1}$ at 25°C. However, this method does not allow the investigator to determine the fraction of the change which is too fast to be followed by the experimental method, since there is no way of measuring the dependence of conductivity on chain expansion at a constant charge of the polyion. Even so, this study is of great interest in that it shows that even PAA, which exhibits no cooperative transition from a contracted to an expanded state, exhibits a relatively slow approach to conformational equilibrium. More recently, Irie and Schnabel (20) have generated hydroxide ions photochemically in a methanol solution of PMA and interpreted the change in conductivity and light scattering after a light flash as a result of chain expansion. Their experiment extended over a time of about 0.2 ms, so that they would not have observed much slower changes such as studied in our work, in that of Ohno et al. and in that of Okubo

3. Studies of polymer complex formation. Equilibrium characteristics.

Polycarboxylic acids and water-soluble polymers which can function as hydrogen bond acceptors form association complexes of sharply reduced water solubility. This phenomenon was first described for the complexation of poly(acrylic acid) with polyoxyethylene (21) and has been studied by changes in solution viscosity (21,22) by a rise in pH (20,23) a drop of conductivity (19, 22), an enthalpic change (24,25) and by ultracentrifuge sedimentation (26). The association equilibrium has been shown to be highly sensitive to the length of the interacting chain molecules, suggesting a cooperative process.
    We have found (27) that the fluorescence change resulting from the mixing of solutions of dansyl-labeled PAA with solutions of polyoxyethylene (POE) or poly(N-vinyl pyrrolidone) (PVP) provide evidence concerning this type of macromolecular association which constitutes an important addition to the information deduced from the older methods. On addition of increasing amounts of PVP with a molecular weight ranging from 10,000 to 412,000 to a solution of labeled PAA, the emission intensity approached a five-fold enhancement with the initial slope of a plot of the fluorescence intensity against the PVP/PAA ratio becoming steeper with an increasing molecular weight of PVP. This suggests that the composition and structure of the complex are independent of the chain

length, although the driving force toward complex formation increases when the PVP molecule is extended. When solutions of PAA nad PVP were mixed at different pH, the data represented in Figure 3 were obtained. They show that as the PAA becomes partially ionized, its affinity for the PVP is reduced. Also, the limiting enhancement of the emission intensity at large PVP/PAA declines suggesting a swelling of the complex with an increasing contact between the dansyl label and water molecules.

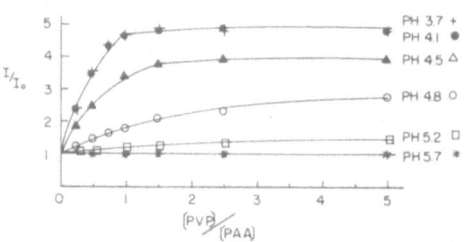

Figure 3. Increase of the fluorescence intensity of 2.5 x 10^{-4} N dansyl-labeled PAA on addition of PVP with a molecular weight of 57,000 at various pH values.

The association of PAA with POE proved to be a much more complex phenomenon. Our first experiments, represented in Figure 4, indicated the expected dependence of the complex stability on the length of the

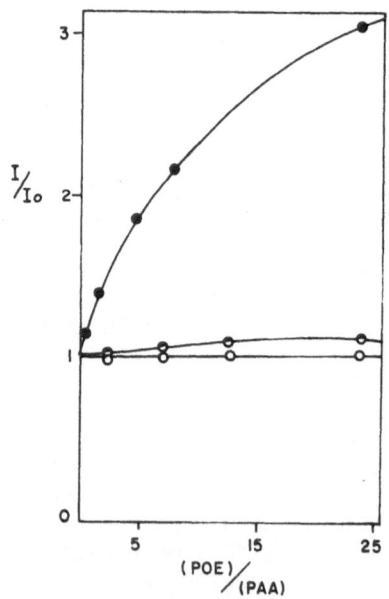

Figure 4. Enhancement of the fluorescence of 7.1 x 10^{-3} N dansyl labeled PAA on addition of POE of different molecular weights at pH 3.7. POE-3400 (O); POE-8000 (◑); POE-24000 (●).

POE chain: With a POE molecular weight of 3,400, there was no evidence of complex formation, only a marginal increase in fluorescence of the dansyl labeled PAA was observed on addition of POE with a molecular weight of 8,000, while addition of POE with a molecular weight of 24,000 led to a threefold increase in fluorescence intensity. This seemed consistent with the pronounced cooperativity of the process: The association depends on the small difference between the strength of the carboxyl-ether hydrogen bond and the strength of the hydrogen bonds of carboxyl and ether groups with the aqueous medium, so that many inter-polymer hydrogen bonds are required to produce a stable complex. However, it was surprising that the emission intensity of the dansyl labeled PAA increased up to a 25-fold excess of POE-24,000 with no indication that a limit had been approached. Even more troubling was the evidence represented in Figure 5 that the enhancement of the emission intensity was the same, for any given POE/PAA, at two PAA concentrations differing by a factor of 12. This result is clearly inconsistent with the specification of a simple association constant to describe the complexation equilibrium as proposed by Osada (25).

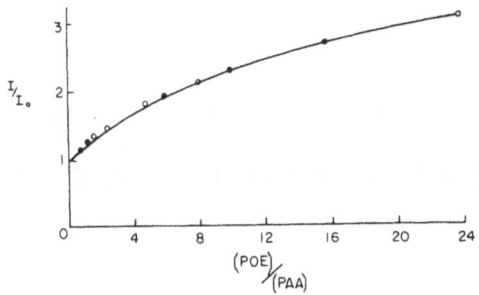

Figure 5. Increase in the fluorescence intensity of dansyl labeled PAA as a function of the POE/PAA ratio. Molecular weight of POE 24,000. Normality of PAA: 0.0071 (O); 0.00057 (●).

More unexpected results were obtained in a study of the fluorescence of dansyl labeled PAA in the presence of POE of much higher molecular weight (28). The data plotted in Figure 6 show that the addition of POE with molecular weights ranging from 100,000 to 4,000,000 produced much smaller enhancements of emission intensities than the POE with a molecular weight of 24,000. In addition, it was found (Figure 7) that whereas complexation with POE-24,000 produced the blue-shift of the emission spectrum which would be expected when the dansyl label is removed from contact with water molecules, the presence of the very high molecular weight POE shifted the emission maximum in the opposite direction. I know of no previous observation where a change in the microenvironment of the dansyl group led to an increase in the quantum yield of fluorescence and a red-shift of the emission spectrum and the interpretation of this result is uncertain at this time. In any case, the data show that the nature of the POE-PAA association is distinctly different from the simpler pattern characterizing the PAA association with

PVP,where the emission intensity of the dansyl label is independent of
the chainlength of the PVP and the emission spectrum exhibits invari-
ably the expected red-shift.

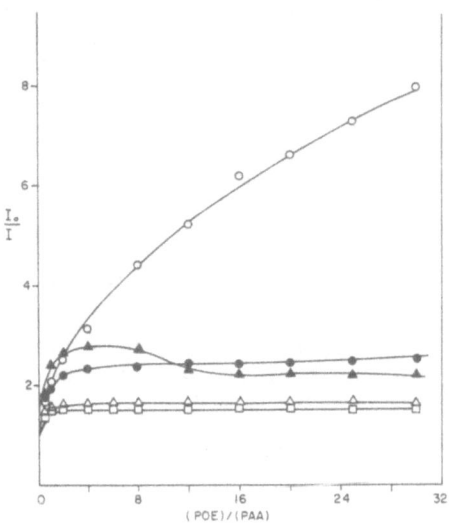

Figure 6. Enhancement of fluorescence intensity on addition of POE to a
solution of dansyl labeled 0.0063 N PAA. Molecular weight of POE:
14,000 (●);24,000 (O);100,000 (▲);300,000 (△);4,000,000 (□).

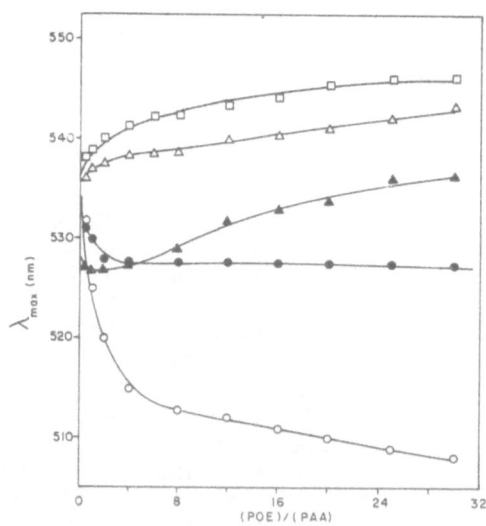

Figure 7. Emission maxima of solutions containing dansyl labeled PAA
and POE of various molecular weights. Significance of symbols as in
Figure 6.

4. Polymer complexation kinetics.

The use of dansyl labeled PAA allowed us to study in a stopped flow apparatus with fluorescence detection the kinetics of the association of PAA with POE (29). Similarly, when a solution containing the PAA-POE complex was mixed with a buffer at a higher pH, the dissociation of the complex could be monitored by the decrease in the fluorescence of the label. The complex formation could be described by biphasic kinetics while the complex decomposition was a first-order process. The most interesting part of the results concerned the activation parameters: Although the activation enthalpy of both complex formation and complex decomposition (3.7 kcal/mole and 2.8 kcal/mole, respectively) were very low, characterizing both processes as diffusion controlled, the rates were low enough to be followed by the stopped flow technique because of the extraordinarily large negative entropies of activation (-35.9 eu and -43 eu, respectively). We have interpreted the results as follows: The initial step in complex formation involves the formation of the first hydrogen bond between the interacting polymers - this would be expected to be diffusion controlled. However, to produce a stable complex requires the formation of a large number of hydrogen bonds and this can omly be achieved by an extensive conformational transition of the molecular chains, a process leading to a large entropy loss. A similar argument can be made to explain the small enthalpy of activation and the large negative entropy of activation of complex decomposition.

Previously, Okubo (19) has studied the kinetics of the association of PAA and PVP in a stopped flow apparatus using the change in conductivity to follow the process. Since the hydrogen bonding of carboxyl groups with the basic oxygens of the pyrrolidone residues inhibits carboxyl dissociation, the decrease in the hydrogen ion concentration (and the resulting drop in conductivity) was interpreted as a measure of the number of interacting groups in contact with each other. This treatment neglects the possible effects on the conductivity due to a change in the hydrodynamic properties of the polymers accompanying complex formation. In his published account, Okubo specified only one rate constant, but in a recent private communication he stated that the stopped flow data yielded biphasic kinetics and that a process slower by a factor of 10,000 can be followed by light scattering. Since Okubo believed that the formation of the polymer complex can be described by a simple equilibrium constant, he estimated rates of complex decomposition from the rate constant of complex formation and and the equilibrium constant. A direct measurement of the rate of complex decomposition after a pH jump, such as used in our work, would probably not be feasible with the conductance technique.

The use of a medium-sensitive fluorescent label enabled us also to study the rate at which the partners in a polymer complex can be interchanged. Thus, when an excess of unlabeled PAA is added to a solution containing a complex of dansyl labeled PAA with POE or PVP, the displacement of the labeled by the unlabeled PAA results in an increased exposure of the label to contact with water molecules, leading to a reduced fluorescence inyensity. As would be expected, such a process is much slower than complex formation and it can be easily followed without the use of a stopped flow apparatus (27). The complex interchange is faster with POE than with PVP, reflecting the stronger hydrogen bond with the pyrrolidone residue.

Exchange of the components of the PAA-POE complex has been investigated previously by Anufrieva et al. (30) who used the depolarization of fluorescence to monitor the displacement of the labeled PAA. It is instructive to compare the significance of results obtained in studies using the emission anisotropy of fluorescent labels and medium sensitive labels to monitor polymer association. In the first case the results reflect the rotational diffusion constant of the labeled species which is reduced when a polymer complex is formed but would not be expected to be sensitive to the nature of the association. With the second technique we sample the microenvironment of the label, so that a loose and a tight association can be distinguished. This enabled us to demonstrate the swelling of the PAA-PVP complex when the pH was raised and to show that that the nature of the PAA-POE complex depends on the molecular weight of the POE.

References.

(1) G. Weber, Biochem. J.,51,155 (1952)
(2) G. Weber and D.J.R. Laurence, Biochem. J.,56,XXXI,(1954)
(3) L. Stryer, J.Mol.Biol.,13,482 (1965)
(4) E. Lippert, Z.Elektrochem.,61,962 (1967)
(5) M. Burr and D.E. Koshland,Jr.,Proc.Natl.Acad.Sci.U.S.,52,1017 (1964)
(6) J.D. Ferry,D.C. Udy, F.C. Wu, G.E. Heckler and D.B. Fordyce, J.Colloid Sci.,6,425 (1951)
(7) R. Arnold, J.Colloid Sci.,12,549 (1957)
(8) J.C. Leyte and M. Mandel, J.Polymer Sci.,A-2,1879 (1964)
(9) E.V. Anufrieva, T.M. Birstein, T.N. Nekrasova, O.B. Ptitsyn and T.V. Sheveleva, J.Polym.Sci.,Polym.Symposia,16,3519 (1968)
(10) V. Crescenzi, F. Quadrifoglio and F. Delben, J.Polym.Sci.,A-2, 10,357 (1972)
(11) M. Mandel, J.C. Leyte and G. Stadhouder, J.Phys.Chem.,71,603 (1967)
(12) U.P. Strauss and G. Vesnaver, J.Phys.Chem.,79,2426 (1975)
(13) U.P. Strauss and M.S. Schlesinger, J.Phys.Chem.,82,1627 (1978)
(14) B. Bednar, H. Morawetz and J.A. Shafer, Macromolecules,in press.
(15) N. Ohno, K. Nitta and S. Sugai, J.Polym.Sci.,Polym.Phys.Ed., 16,513 (1978)
(16) G. Schwartz, J.Mol.Biol.,11,64 (1965)
(17) U.P. Strauss and B.W.Barbieri, Macromolecules,15,1347 (1982)
(18) B.W. Barbieri and U.P. Strauss, Macromolecules,18,411 (1985)
(19) T. Okubo, Biophys.Chem.,11,425 (1980)
(20) M. Irie and W. Schnabel, Makromol.Chem.,Rapid Commun.,5,413 (1984)
(21) F. Bailey, R.D. Lundberg and R.W. Collard, J.Polym.Sci.,A2,845 (1964)

(22) A.D. Antipina, V.Yu. Baranovskii, I.M. Papisov, and V.A. Kabanov, Vysokomol.Soedin.,A 14,1941 (1972)

(23) L.A. Bimedina, V.V. Roganov and E.A. Bakturov, J.Polym.Sci., Polym.Symposia ,44,65 (1974)

(24) I.M. Papisov, V.Yu. Baranovskii, I. Sergeeva, A.D. Antipina and V.A. Kabanov, Vysokomol.Soedin., A 16,1133 (1974)

(25) Y. Osada, J.Polym.Sci.,Polym.Chem.Ed.,17,3485 (1979)

(26) I.M. Papisov, V.Yu. Baranovskii and V.A. Kabanov, Vysokomol.Soedin. A 17,2104 (1975)

(27) H.-L. Chen and H. Morawetz, Eur.Polym.J.,19,923 (1983)

(28) B. Bednar, Z. Li, Y. Huang and H. Morawetz, Macromolecules,in press.

(29) B. Bednar, H. Morawetz and J.A. Shafer, Macromolecules,17,1634 (1984)

(30) E.V. Anufrieva, V.O. Pautov, I.M. Papisov and V.A. Kabanov, Dokl.Akad.Nauk SSSR,232,1096 (1977)

SINGLET ENERGY MIGRATION, TRAPPING AND EXCIMER FORMATION IN POLYMERS

Ian Soutar
Chemistry Department
Heriot-Watt University
Riccarton
Currie
Edinburgh   EH14 4AS
UK

and

David Phillips
The Royal Institution
21, Albemarle Street
LONDON   W1X 4BS
UK

ABSTRACT.   Intramolecular excimer formation in synthetic polymers is
reviewed.   The antenna effect in macromolecules is examined and evidence
for singlet energy migration and trapping is discussed.

1.   INTRODUCTION

The problem of intramolecular singlet energy migration in synthetic
polymers has evoked much interest and debate.   The interest of research
workers in the phenomenon derives from the implications in terms of the
efficiency of population of low concentration energy traps through an
act of mimicry of the processes whereby chlorophyll molecules in the
chloroplast of green plants populate the active centres in the process
of photosynthesis.   In a synthetic polymer consisting of repeat units
containing chromophoric groups, a very high local chromophore concentration is confined within the dimensions of the polymer coil which results
in efficient energy transfer to low energy traps.   The "energy
harvesting" behaviour of polymers has been christened aptly by Guillet
as "the antenna effect".   It is the purpose of this article to review
some aspects of energy trapping, particularly through excimer formation,
in antenna polymers and to discuss the current mechanisms which have
been proposed to model the kinetics displayed by such systems.
      The degree of debate which has arisen concerning the very exist-
ence of singlet energy migration has resulted in the main from conflict-
ing data in certain experiments and the absence of a definitive

M. A. Winnik (ed.), Photophysical and Photochemical Tools in Polymer Science, 97–127.

demonstration of the migrative act. The problem in the latter context
is that unlike the existence of triplet energy migration which is
observable through the occurrence of triplet-triplet annihilation, the
singlet energy migration which has been proposed by most workers is
believed to be much less extensive (in terms of mean diffusion length)
and only observable somewhat vicariously through its effect on the
kinetics of trap population. However, the observed photophysical behaviour of most aromatic polymers is consistent at least with the occurrence of energy migration.

2. EXCIMER FORMATION

In the absence of impurities or other energy traps introduced into a
macromolecule during synthesis, excimer forming sites constitute the
natural energy trap in aromatic polymers.
    First observed in concentrated solutions of pyrene [1], excimers
are now known to be formed by the majority of aromatic molecules.
Kinetic scheme 1, suggested by Forster and the kinetic derivations
developed by Birks [2] et al have been validated in a host of systems
involving intermolecular interactions between low molecular weight
species.

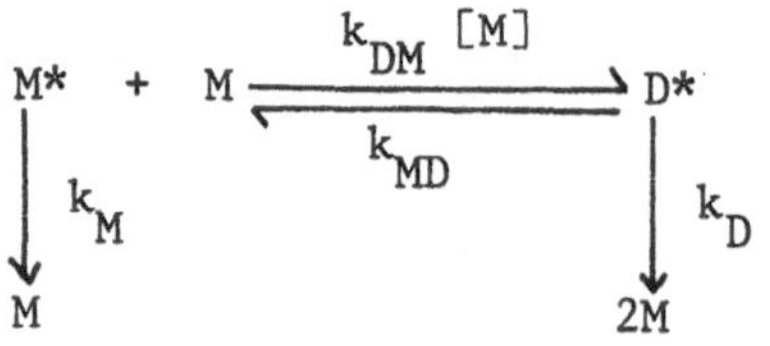

$$M^* \; + \; M \underset{k_{MD}}{\overset{k_{DM}\,[M]}{\rightleftharpoons}} D^*$$

$$\downarrow k_M \qquad\qquad\qquad\qquad \downarrow k_D$$

$$M \qquad\qquad\qquad\qquad\qquad 2M$$

    Scheme 1.

Studies of intramolecular excimer formation in bichromophoric systems,
as described by Professor De Schryver in this publication, have been
invaluable in provision of information relevant to understanding of the
photophysics of macromolecules and in the early years led to the
generalisation of Hirayama [3] known as the "n = 3 rule". Whilst the
n=3 rule is not strictly obeyed it has focussed the thoughts of workers
seeking elucidation of macromolecular photophysics especially for the
case of vinyl aromatic polymers.

3. STEADY-STATE FLUORESCENCE STUDIES OF INTRAMOLECULAR EXCIMER
       FORMATION IN POLYMERS

In order to gain information about the parameters governing excimer
formation in macromolecules, several approaches have been adopted.
Early investigations were concerned primarily with considerations of
the thermodynamic characteristics of the interaction through a study of
the temperature and solvent dependence of the excimer to monomer emission intensity ratio. In addition, some kinetic information was

derived from transient excitation experiments which assumed a validity
of the Forster [1] mechanism and direct applicability of Birks' [2]
kinetic derivations to the intramolecular phenomenon in macromolecules.
The extension of these methods to encompass studies of copolymeric
species introduced a new dimension to the investigations, namely the
ability to examine the intramolecular concentration dependence of the
photophysical behaviour. More recently, developments in sophisticated
instrumental techniques and the advent of stable, subnanosecond excitation sources has resulted in doubts regarding the validity of direct
application of the Forster/Birks [1,2] kinetic scheme to macromolecular
photophysics.

3.1. Temperature Dependence of Excimer Formation

In general, studies of the temperature dependence of the excimer-
monomer population in polychromophoric macromolecular species have been
treated assuming that a Birks [2] kinetic analysis might be capable of
modification and applied to polymer photophysics. The pertinence of
the consequent data must be viewed within this context and is rendered
questionable by the validity of the assumptions implicit in the treatments of Stevens and Ban [4] and Birks et al. [5] for the case of
polymers. The basis of the approach is outlined below.
     Treatment of kinetic scheme 1, assuming photostationary excited
state conditions, results in the following expression for the ratio of
fluorescence quantum yields Φ_D and Φ_M of excimer and monomer respectively;

$$\frac{\Phi_D}{\Phi_M} = \frac{k_{FD} \, k_{DM} [M]}{k_{FM} (k_{MD} + k_D)} \tag{1}$$

where [M] is the concentration of chromophoric species; k_{FX} and k_{IX}
are the rate coefficients governing fluorescence and non-
radiative decay of a species X; $k_D = k_{FD} + k_{ID}$. For adoption in
polymer photophysics, it is assumed that intramolecular excimer formation occurs via a two-state model in which a single type of excimer is
kinetically linked to a single population of excited states and that
[M] represents some local concentration of potential excimer forming
species within the polymer coil. (In section 4.1 we shall discuss how
time-resolved emission data have changed the views of photophysicists
regarding the validity of a two-state model and the kinetic models which
have evolved to better explain the observed behaviour). In early
studies, the nature of the concentration term [M], as dictated by the
mechanism of intramolecular excimer formation, was little questioned.
It was recognised that intramolecular excimer configurations might be
expected to be attained as a result of interactions between adjacent
chromophores upon the polymer backbone which, in head to tail placements, satisfy the "n=3 rule" of Hirayama [3] (which states that excimer
formation is expected between aromatic substituents attached to an
alkane chain only when the chromophores are separated by three carbon
atoms. The limitations of the rule are briefly noted in section 2).
In addition, excimer formation might be achieved through "long-range
interactions" involving juxtaposition of fluorophores, in distant

Table I   Activation energies and binding energies of excimer formation

Polymer	Solvent	E_{DM} (kJ mol^{-1})	$-\Delta H$ (kJ mol^{-1})	Reference
Polystyrene	THF/E	5.9	34.7	6
	DCE	12.8	–	7
	DCM	15.9	–	8
isotactic Polystyrene	DCM	8.4	–	8
Poly(α-methyl styrene)				
90% syndiotactic	DCM	10.5	–	8
70% syndiotactic	DCM	7.5	–	8
Poly(1-vinyl-naphthalene)	THF/E	8.8	28.9	6
	MTHF	11.3	15.1	9
Poly(2-vinyl-naphthalene)	THF/E	8.4	27.2	6
	PB	15.1	30.9	10
	MTHF	5.4	–	11
Poly(acenaphthylene)	MTHF	3.4	8.8	9
Poly(1-naphthyl methacrylate)	DCM	–	11.7	12
	C	–	15.5	12
	EA	–	19.2	12
Poly(2-naphthyl methyl methacrylate)	THF	–	11.3	13
	EA	–	14.2	13
Poly(2-naphthyl ethyl methacrylate)	THF	–	10.5	13
	EA	–	13.8	13
Poly(2-naphthyl propyl methacrylate)	THF	–	11.7	13
	EA	–	16.3	13
Poly(4-vinyl biphenyl)	PB	8.0	–	14
Poly(vinyl PPO)	MTHF	5.9	17.0	15

THF=tetrahydrofuran;   E=ether;   DCE=1,2 dichloroethane;

DCM=dichloromethane;   MTHF=2-methyltetrahydrofuran;

PB=n-propylbenzene;   C=chloroform;   EA=ethyl acetate.

placements along the polymer backbone, through the agency of chain
coiling. Attempts to elucidate the relative importance of these mechanisms are reviewed in section 3.2.

The conventional approach to the analysis of the temperature dependence of intramolecular excimer formation in fluid solutions assumes
that the quantum yield ratio Φ_D/Φ_M in equation (1) may be approximated
by the intensity ratio I_D/I_M. The variation of I_D/I_M with temperature is treated in terms of establishment of "high" and "low temperature"
kinetic regions which can yield estimates of the binding energy $(-\Delta H)$
and activation energy (E_{DM}) of excimer formation respectively.

Low temperature regime

At low temperatures, if the rate of excimer dissociation is sufficiently low relative to alternative means of excimer state depopula-
tion $(k_{MD} \ll k_D)$ and provided it may be assumed that k_{FM}, k_{FD} and k_D are
invariant with temperature, equation (1) becomes

$$I_D/I_M = kk_{DM}[M] \qquad (1a)$$

If it is further assumed that the concentration term [M] (which in the
case of a polymer represents a "local chromophore concentration") remains constant over the temperature range it is apparent that a plot of
$\ln(I_D/I_M)$ as a function of T^{-1} (a Stevens-Ban plot) should be linear
and yield a value of the "activation energy of excimer formation", E_{DM},
from its slope.

High temperature regime

If the temperature dependence of the rate coefficient for excimer
dissociation, k_{MD}, is sufficiently greater than that representative of
the thermal dependence of k_D, it is to be expected that at sufficiently
elevated temperatures k_D may be regarded as negligible relative to k_{MD}
and I_D/I_M may be expressed as

$$I_D/I_M = \frac{k_{FD}}{k_{FM}} \cdot \frac{k_{DM}}{k_{MD}} [M] = k \frac{k_{DM}}{k_{MD}} \qquad (1b)$$

where k is a constant provided k_{FD}, k_{FM} and [M] are regarded as temperature invariant. Given the validity of these assumptions, a state of
equilibrium is obtained in the high temperature region and the dependence of $\ln(I_D/I_M)$ upon T^{-1} should yield an estimate of the binding
energy of the excimer.

Table I presents typical data obtained in such a manner.

Since most of the experiments cited in Table I were performed when
the vast majority of the work on macromolecular luminescence was confined to steady state excitation the aspirations of the investigators
seems reasonable. However, the validity of the data in representation
of the functions E_{DM} and $-\Delta H$ is open to considerable doubt not only in
view of the assumptions discussed above in derivation of the low and
high temperature relationships from the Birks [2] scheme but also as a
result of doubts recently expressed [16,20] regarding the applicability
of the basic kinetic scheme itself to macromolecules.

Evidence for the failure of the assumptions inherent in the treatment discussed above to withstand empirical evaluation may be found both from steady state excitation and transient data. For example Nishijima [21] and David et al [22] have presented data (for poly 2- and 1-vinylnaphthalenes, respectively) which indicate that I_M does not increase in the manner expected in the high temperature region if excimer dissociation were dominant. More emphatically, Holden and Guillet [23] have tabulated individual rate parameters derived from transient decay data for poly(1-vinylnaphthalene) [23,24] and poly(1-naphthylmethacrylate) [12] in dichloromethane [12,24] or THF [23] at 298K. (Under these conditions, steady state dependence of I_D/I_M as a function of T^{-1} would be classified as being above the region characterized by "low temperature behaviour"). In no case was $k_{MD} \gg k_D$ as required by Birks et al [5] for high temperature behaviour. (It is worth noting however that the latter data have been derived from kinetic data obtained at a single chromophore concentration (that dictated by the equivalent local concentration within the polymer coil dimensions) from inter-relationships between the rate parameters which assume validity of the basic Birks [2] scheme).

More recently, using a revised kinetic scheme [17,26] (cf. section 4). Phillips et al. [25] have examined the temperature dependence of intramolecular excimer formation in poly(1-vinylnaphthalene) in THF under transient excitation conditions. It was shown that at no temperature within the region defined as "low temperature" in a Stevens-Ban presentation is the tenet $k_{MD} \ll k_D$ valid. Furthermore, since the approach adopted allows, within the limits of the kinetic scheme, derivation of the individual rate constants dissociated from the local chromophore concentration dependence (as described in section 4.2) estimates of "true" activation energies, E_{DM} may be made. For poly (1-vinylnaphthalene) in THF a value of 4.8 kJ mol^{-1} was obtained.

3.2. Influence of Microcomposition upon Excimer Formation

3.2.1. Intramolecular concentration dependence
The study of photophysical phenomena in homopolymers alone precludes variation of one of the most powerful parameters available to a photochemist, namely concentration. The local concentration experienced by the constituent chromophores of a polymer may be varied to some extent by altering the thermodynamic compatibility of the solvent and more dramatically by the use of copolymers containing the luminophore of interest and a spectroscopically inactive comonomer.

Following the early observations of excimer emissions from a variety of macromolecules it was necessary to try to establish the mechanisms whereby excimer sites were created. Amongst other considerations of interest were

a) the extent to which Hirayama's [3] n = 3 rule was obeyed and what fraction of excimer sites was generated by longer-range interactions

b) the extent to which energy migration served to populate potential excimer sites

and c) the influence of chain flexibility upon the overall photophysics.

It soon became apparent that Hirayama's n = 3 rule is not strictly held in polymer systems even for chromophores such as naphthalene which obey the generalization in low molar mass systems. As examples, excimer formation is observed in polyacenaphthylene, [6,16,27-29] in which adjacent repeat units can not adopt the correct geometry for excimer formation; and in poly(naphthyl methacrylates [30-32] in which the minimum spacing between naphthalenes involves seven carbon atoms. Indeed, chromophores such as pyrene which have a great affinity for excimer formation have been adopted to study and model cyclization dynamics. [33-36] However, the question arises as to the extent to which long range interactions are important in polymers in which excimer formation between nearest neighbours is feasible for chromophores of moderate excimer forming tendency.

The enhancement of excimer emission relative to monomer as the thermodynamic compatibility of solvent and polymer is reduced has been cited as evidence of the influence of long range interactions. [6,12,30] However, since the coil dimensions will affect the extent of energy migration and, to a lesser extent, the concentration of conformations appropriate to excimer formation, caution is advised. Solvent effects upon excimer to monomer emission intensity ratio have certainly been observed in polymers [37] in which nearest neighbour interactions are thought to dominate the photophysics. For vinylaromatic polymers such as polystyrene and poly(vinyl naphthalene) with relatively weakly interacting chromophores, evidence for the dominance of nearest-neighbour effects has been furnished by

a)   the absence of excimer fluorescence from alternating copolymer systems [38,39]

b)   the absence of excimer emission from the spectra of head to head polystyrene [40,41]

c)   studies of random copolymers as described below.

The influence of the intramolecular distribution of chromophores upon the spectral characteristics of polymer luminescence is evident in the works of Nishijima [42] and David et al [43]. The latter workers were the first to attempt to characterize the extent of excimer formation in polymers in terms of the microcomposition of the macromolecule. It was argued that in styrene-methyl methacrylate copolymers excimer formation required energy migration from the site of absorption to that suitable for excimer formation. It was proposed that excimer formation was dominated by interactions between adjacent chromophores upon the polymer chain and that the excimer site concentration was consequently proportional to the fraction of pairs of styrene residues in the copolymer f_{aa} .

Reid and Soutar [44] claimed that the linear dependence of I_D/I_M upon f_{aa} was not maintained over an extended composition range and suggested [44,45] that in polymers in which energy migration occurs to a limited extent and serves to populate potential excimer sites, description of the excimer-forming capabilities requires adoption of a function characteristic of the extent to which energy migration may populate the potential excimer sites in addition to that quantifying the concentration of such sites. Accordingly in copolymers incorporating vinylnaphthalenes and styrenes with methyl methacrylate

[44] it was shown that the intensity ratio I_D/I_M was adequately described by a function of the form:

$$I_D/I_M = \frac{k_{FD}\, k_{DM}[M]}{k_{FM}(k_D + k_{MD})} \tag{1}$$

in which f_{aa} is the fraction of bonds between aromatic species and $\overline{l}_a$ is the mean sequence length of chromophore.

Development of the model assumed that
a) the Birks [2] scheme or, rather, the resultant steady state expression of intermolecular concentration influence upon diffusion limited excimer formation namely

$$I_D/I_M = k\, \overline{l}_a f_{aa} \tag{2}$$

is capable of being modified to describe the intramolecular phenomenon in polymers.
b) excimer formation is resultant upon interactions between nearest neighbour chromophores
c) the function, $\overline{l}_a$, descriptive of energy migration as sensed by fluorescence anisotropy measurements on dilute glassy solutions is capable of description of the extent to which potential excimer sites may be sampled in fluid solution by the migrating excitation.

The functional dependence of I_D/I_M upon copolymer microcomposition described by equation (1) remained valid in description of excimer formation in the 1-vinyl naphthalene-co-methyl methacrylate system as both the temperature and thermodynamic nature of the solvent was varied [37].

The general approach involving adoption of a term to describe the energy migration characteristics and a function descriptive of excimer site concentration has furnished functions descriptive of I_D/I_M in situations where the copolymers studied were of enhanced flexibility[46] (due to variation of comonomer) or in which the chromophores were altered such as to change either the compositional dependence of the energy migration or that of the excimer site [37]. For example, in copolymers of acenaphthylene with methylacrylate [47] or methyl methacrylate [29] the energy migration term was f_a, the mole fraction of acenaphthylene in the copolymer, and the excimer site concentration term a summation over the different pentad microcompositional situations for the acenaphthylene residues which reflected the differential excimer forming capabilities of sites of the type AMA and AAA (where A = acenaphthylene and M = comonomer). The adoption of such a function relies on earlier suggestions by David et al [9] that next-to-nearest neighbours dominate excimer formation in poly(acenaphthylene) and gains support from the work of Wang and Morawetz [27] on acenaphthylene copolymers of highly alternating character.

It is relevant to stress that the implementation of two functions, one descriptive of energy migration and one characteristic of excimer site concentration, is anticipated to be necessary only in cases where

energy migration is a determinant in excimer site generation. Two extremes in which this will not be the case are those of:
a) negligible energy migration, and,
b) migration sufficiently extensive to ensure complete statistical
sampling of the distribution of excimer sites.
     One example of a copolymer system in which a single term descriptive
of the excimer site concentration is adequate in description of excimer
formation is that of styrene-acrylonitrile. In this system I_D/I_M is
controlled by the fraction of styrene pairings in the polymer [48-50]
     Further studies that report on the relationship between the
luminescence properties of copolymers and the chromophore microcomposition include those of Majumdar et al [51] and Nakahira et al [52,53].
     It should be stressed that it is important to generate functions
descriptive of excimer formation from a given type of chromophore if it
is intended that the influence of factors such as chain mobility and
comonomer geometry are to be assessed by studying the emission characteristics of a series of copolymers of varying structure. It is not
sufficient to merely compare the spectral data obtained from copolymers
of equal aromatic content [46]. Furthermore, the intramolecular
concentration terms derived from studies under steady state conditions
have assumed importance in derivation of rate parameters within the
context of one model which has been applied to transient emission data
as will be described in section 4.

3.3. Influence of conformation and configuration

The importance of conformational considerations in bichromophoric
systems has been recognized by several workers [10,20,54-59]. The influence of ground-state conformational control, which has been invoked
in explanation of intramolecular exciplex photophysics in α-aryl-ω-N-
alkylakanes [60,61] and excimer formation in 1,1'-di-2-naphthyldiethyl-
ether [20], has been suggested to be of importance too in macromolecular
photophysics [20,59].
     Configurational aspects are apparent in publications concerning the
fluorescence characteristics of polymers of differing tacticity and of
model compounds. It has been shown that the intensity ratio of excimer
to monomer fluorescence is greater in fluid solutions of isotactic
polystyrene [62,63] and poly(p-methyl-styrene) [64,65] relative to that
of the atactic polymers. This phenomenon has been attributed in the
case of poly(p-methylstyrene) to the existence of a lesser energy
barrier to excimer formation in meso dyads compared to the racemic dyad
[65]. Similar conclusions of direct relevance to excimer formation in
polystyrene were made by Bokobza et al [55] in studies of intramolecular
excimer formation in model compounds.
     A dependence of I_D/I_M upon tacticity in fluid solutions of poly(1-
and 2-naphthyl methacrylate)s has also been reported [31]. The implications of these observations are less apparent, since it has been suggested [12,30] that excimer formation in such poly (methacrylate)s is
dominated by long-range interactions.

3.4. Summary

The picture which evolved from stationary state experiments complement-
ed by a few early time-resolved measurements invoked a modified Birks
kinetic scheme. This notion was accepted virtually without question in
studies prior to 1980. Further, the majority of authors cited the
occurrence of singlet energy migration and subsequent sampling of
potential excimer sites in the absence of concrete evidence for migration except for that afforded by emission depolarization [66,67]
     Development of a more detailed understanding of energy migration
and trapping via excimer formation evolved from access within the
polymer community to sophisticated time-resolved emission spectroscopy has altered our concepts of polymer photophysics is the subject
of the following section of this chapter.

4. TIME-RESOLVED FLUORESCENCE STUDIES OF INTRAMOLECULAR EXCIMER FORMATION IN POLYMERS

Initial investigations analysed the time dependence of the fluorescences
of a variety macromolecular systems through a force-fitting to dual
exponential functions of the type depicted in equations (3) and (4)
resultant upon Birks [2] treatment of intermolecular excimer formation

$$i_m(t) = A_1 \exp(-\lambda_1 t) + A_2 \exp(-\lambda_2 t) \tag{3}$$

$$i_d(t) = A_3[\exp-\lambda_1 t - \exp(-\lambda_2 t)] \tag{4}$$

The means of data acquisition in single photon counting allied to
modern data processing techniques allows an extremely critical inter-
rogation of the adequacy of a mathematical function to describe the
decay behaviour over several decades of intensity. It was generally
realized that functions of the type (3) and (4) were not judged statistically to be competent descriptors of the temporal characteristics of
fluorescences analysed in either the monomer, $i_m(t)$, or excimer, $i_d(t)$,
spectral regions. However, such complications were frequently dismissed
in the belief that the complexities of polymeric species were not
conducive to the production of "ideal" behaviour. However, increasing
disquiet has resulted in the expression of serious doubts concerning
the validity of the predictions of the Birks [2] analysis as applied to
intramolecular excimer formation in polymers [16-20]. The foundation
of these doubts was laid upon observations of the following type:
1. The recognition that the inability to describe emission decays in
certain polymers in terms of dual exponential functions was
photophysically significant [16,17,20].
2. The observation that in dual exponential fitting to the decay data
obtained in the spectral region of excimer emission, pre-exponential
factors of equal magnitude but opposite sign were not obtained if all
four fitting parameters were allowed to vary freely [20,68].
3. The observation that the decay times evaluated in the regions of
monomer and excimer emission, respectively, do not coincide [18,68].

Accordingly models have been proposed as alternatives to the
Birks [2] mechanism in attempts to explain the complex fluorescence
decays obtained from aromatic polymers. Generally it has been assumed
that the decays may be described by multiexponential functions which
result from the existence of kinetically distinct species within the
polymer microcomposition. Such an approach, adopting the principle of
Occam's razor, is common practice amongst kineticists and allows, within the framework of the mechanism postulated, derivation of rate coefficients descriptive of the influence of concentration terms upon the
physical behaviour of the system. This point will bear some elaboration
in our following chapter. Recently, two publications have appeared which
question the applicability of this approach to macromolecular photophysics [69,70]

4.1 Kinetic Models

The diversity of polymer types which have been subjected to photophysical
scrutiny precludes expectation of the ability of a single general
kinetic scheme to be applied in description of polymer photophysics.
The models which have been proposed are relevant to the particular
systems concerned. However, it is possible to discuss proposals generated to explain the photophysics of polymers containing chromophores which
show similar strengths of interaction and excimer formation over similar
diffusive distances. The first group of models pertain to systems in
which the chromophores are naphthyl or phenyl groups, constituting or
contained in substituents capable of a reasonable amount of rotation
independent of the macromolecular backbone.

(a) <u>Polymers Containing Naphthyl and Phenyl Chromophores</u> Holden et al[18]

have proposed kinetic scheme 2 to describe the photophysics of dilute
solutions of polymers containing naphthalenes whose chromophores are
separated by more than three carbon atoms. Analysing the decay curves
of some poly(naphthyl alkyl methacrylate)s and their model compounds

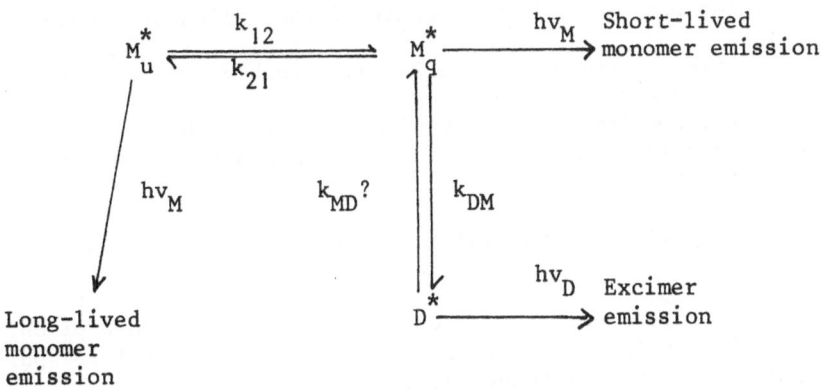

Scheme 2

both in fluid solution and rigid matrix in terms of dual exponential functions the authors noted that the lifetime, $\tau_1 (= \frac{1}{\lambda_1})$, of the longer lived component measured in the region of monomer emission was very different to that characteristic of the excimer fluorescence. Within experimental error, the monomer τ_1 values were equal to those of the model compounds. Consequently it was proposed that this component arises not as a result of excimer dissociation but through the existence of an unquenched monomer species (M_u in scheme 2). It was supposed that such monomers exist in configuration too far removed from that of the excimer to be quenched within the excited state lifetime. The similarity of lifetime obtained for this species in solid matrices to that in fluid solution lends credence to the proposal.

A somewhat similar kinetic scheme to that of Holden et al [18] was proposed independently by Phillips, Roberts and Soutar as a consequence of studies of polymers and copolymers of the vinylnaphthalenes [17,25, 68,71], naphthyl methacrylate [68], acenaphthylene [16] and styrene [72,73] in fluid solutions. Variation of the intramolecular chromophore concentration in series of copolymers and observation of the consequences in the decay kinetics has resulted in the proposal of scheme (3) for application to macromolecules containing the chromophores listed above.

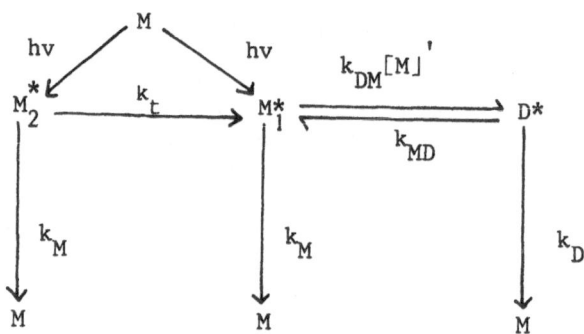

Scheme 3

This scheme was adopted as a result of the following general observations:

1. The Birks scheme is generally inapplicable to description of intramolecular excimer formation in polymers.
2. For the copolymer series studied, decay in the monomer region could only be described adequately if three exponential terms are adopted, as in Equation (5):

$$i_m(t) = A_1 \exp(-\lambda_1 t) + A_2 \exp(-\lambda_2 t) + A_3 \exp(-\lambda_3 t). \tag{5}$$

3. Time-resolved emission spectra reveal the existence of only two spectrally distinct species as illustrated in Fig. (1) for polystyrene in degassed dichloromethane [72].
4. The decay parameters, λ_i, assigned as resultant upon the existence of three kinetically distinct species, vary with the aromatic content of the polymers.

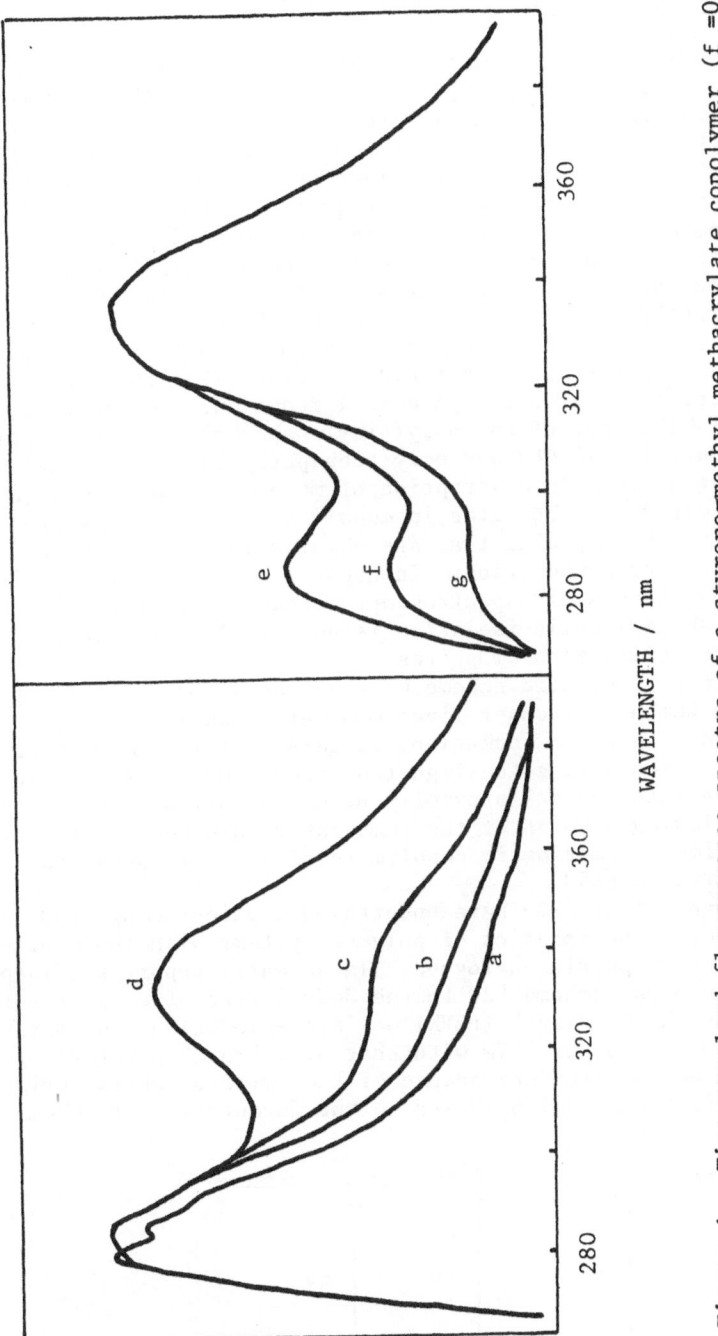

Figure 1. Time-resolved fluorescence spectra of a styrene-methyl methacrylate copolymer (f_a =0.75; dichloromethane solution; 298K) recorded at delays of: (a) 0 (b) 3.8 (c) 7.7 (d) 11.5 (e) 15.4 (f) 19.3 and (g) 28.8ns. Excitation at 257.3nm; gate width 3.2nm (After Soutar et al. [72])

5.   Time-resolved emission spectra suggest that dissociation of excimers to create excited state monomer does occur in the copolymers studied including polymers containing styrene [72] in contrast to previous reports [24,74].

In scheme (3), it is envisaged that D^*, the excimer, is populated from M_1^*, by an 'exciton sampling' mechanism; M_2^* is a chromophore that enjoys a certain degree of kinetic isolation from the normal distribution of quenched excited monomer. The M_2^* sites are not unquenched as supposed in the kinetic scheme of Holden et al [18], but suffer an intramolecular concentration dependent quenching that is supposed to originate from energy transfer into the reservoir of M_2^* sites.

The relative importance of M_2^* in the photophysical behaviour of macromolecules appears to depend upon the type of chromophore and macro-molecular structure involved. For example, there is no detectable in-fluence from isolated monomeric groups in the decays of the monomeric emission band in polystyrene in fluid solution [72]; the decay is ad-equately described by a dual exponential function. In contrast, the photophysical behaviour of the poly(vinyl-naphthalene)s [68], poly(1-naphthyl methacrylate) [68] and poly(acenaphthylene) [68] requires a third exponential term in description of monomer decay. Consequently it may be inferred that M_2^* sites in copolymers containing styrene are associated with chromophores that are physically isolated within the macromolecular microcomposition. In naphthalene polymers, M_2^* sites are evident in the absence of 'spectroscopic spacers' and can be associated with species whose kinetic isolation is not solely consequent upon separation from similar chromophores.

An alternative kinetic scheme was considered by Phillips et al [17,26,71] in which the longer lived monomeric species M_2^* was considered to be quenched by excimer formation, largely as a result of long-range segmental diffusion acting to align two chromophores distant along the chain backbone into spatial proximity by chain coiling. This model was discarded following studies of the temperature dependence of intra-molecular excimer formation in copolymers of 1-vinylnaphthalene and methyl methacrylate [25].

De Schryver et al [20] have undertaken a painstaking study of the comparison of the photophysics of polymer systems with those of low molar mass bichromophoric analogues. In an early report the inapplic-ability of the Birks scheme [2] to the description of excimer formation in poly(2-vinyl naphthalene) (P2VN) was expressed [19]. Comparison of the emission behaviour of P2VN with that of 1,3-di(2-naphthyl)propane combined with an observed dependence of the emission characteristics of P2VN upon excitation wavelength led to the formulation of scheme 4 [19].

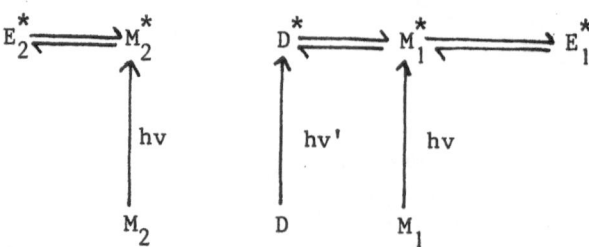

Scheme 4

The scheme invokes the existence of different local environments that can not interconvert on the lifetime of the excited states (M^* and M_1^*). It is further supposed that all emission in the monomer region is due to one type of localized excited state (M^*) whereas excimer formation occurs very rapidly from (M_2^*). Deviations in the excitation spectra and excimer decay kinetics are explained by the existence of a subset (D) of M_1 responsible for the 'deviation of 304 nm' constituting an excited-state rapid equilibrium between unperturbed (M^*) and perturbed (D^*) alkylnaphthalene species. The temperature dependence of the decay of monomer emission was rationalized in terms of two excimeric species E_1^* and E_2^*. At low temperatures, decay from M^* is biexponential, the longer-living component resulting from dissociation of E_1^* whereas the greater complexity of the decay kinetics at higher temperatures is a consequence of the influence of the excimer E_2^*.

Amplification of the concepts involved in scheme 4 resulted from studies of the photophysics of the meso and racemic isomers of 1,1'-di-2-naphthyl diethyl ether (D2NEE) [20]. These isomers serve as models for the isotactic and syndiotactic sequences in P2VN respectively. It was shown that:
1. The ratio I_D/I_M of excimer to monomer fluorescence in both P2VN and meso D2NEE is dependent upon excitation wavelength, whereas no such dependence is evident in the racemic isomer of D2NEE. It is therefore supposed that the observed excitation wavelength dependence, P2VN emission characteristics is due to isotactic sequences in the polymer.
2. The decay kinetics of P2VN are temperature dependent and complex (compared, for example, to the predictions of the Birks [2] scheme). Comparison with the kinetic behaviour of meso D2NEE led to the proposal that isotactic sequences in the polymer were responsible for two exponential decay terms as contribution to the total excimer emission in P2VN.
3. The decay processes in P2VN are considered to be super-position of excimer forming processes from different conformational sites. Further, the efficiency of excimer formation at different conformational sites differs substantially. The dual decay process in P2VN, if indicative of the existence of two excimer sites, is supposed to result from ground state conformational control of excimer formation.

Subsequent to these publications several alternative mechanisms have appeared in the literature to explain the complexity of the fluorescence decay behaviour in macromolecular systems which exhibit excimer formation. Some of these proposals are briefly discussed below.

MacCallum [66,67], addressing the particular case of styrene containing polymers, has suggested that kinetic discrimination within the microcomposition might be consequent upon the dependence of excited monomer quenching through formation of excimer as a result of the location of styrenes in triads of the type −MS*M−, −MS*S− and −SS*S−. The existence of three kinetically distinct species according to the triad distribution would be consistent with the necessity for triple exponential fitting to the data obtained in the monomer emission region of styrene copolymers [72]. However it has been claimed [73,75] that such a model is inconsistent with observations regarding excimer dissociation in time resolved spectroscopy [72], decay data for styrene block copolymers [73] and transient intramolecular quenching studies [75]

Itagaki et al [76] have proposed that the decay kinetics observed for fluid solutions of poly(1-methoxy-4-vinylnaphthalene) (PMVN) may be explained in terms of two excimer species in addition to excited monomer. The proposal is based on variations in steady-state 1,3-di(4-methoxy-1-naphthyl)propane (BMNP) [77,78], presumes that the lower energy excimer comprises the normal structure with fully overlapped aromatic rings, whereas the second excimer is thought to comprise a partially overlapped structure. Steady state spectroscopy indicated a reasonable degree of evidence that three excited state species do indeed exist in the sterically hindered naphthalene systems. However the spectral and numerical deconvolution procedures adopted make the separation of intensity components used in the kinetic analysis, necessary for excited state assignment, rather optimistic.

Gupta et al [79] have proposed a similar scheme to describe the photophysics of poly(1-vinylnaphthalene) involving two excimers differing in the degree of overlap between the aromatic rings. However, the decays were described in terms of dual and single exponential functions according to the wavelength of analysis which is undoubtedly oversimplified. The subsequent assignment of the lifetimes concerned to individual photophysical entities required an assumption of negligble spectral overlap which was not justified.

(b) <u>Poly(N-vinylcarbazole)</u> Poly(N-vinylcarbazole) (PNVCz) is distinct from other vinylaromatic polymers studied to date in that no emission from monomeric species is observed [80]. The fluorescence spectrum of PNVCz in fluid solution is supposed to result from radiative energy loss from two spectrally distinct excimers [80-83]. The lower energy excimer is generally accepted to be formed in a totally eclipsed conformation [80], whereas the higher energy excimer is assumed to involve a dimeric conformation in which partial overlap of the aromatic nuclei is attained [80,83]. Time-resolved fluorescence spectroscopic evidence has suggested that these excimers are interconvertible [81,82], although a recent study has suggested that such may not be the case [84].

The fluorescence decay characteristics of PNVCz have been shown to be complex and resolvable in terms of three components [85-88]. The decay data have been analysed in terms of kinetic schemes involving three excited state species which to varying degrees within each scheme, are interconvertible [85-88]. Ng and Guillet [88] and Tagawa et al [87] have analysed their data in terms of excited monomer and two excimers, whereas Roberts et al [85,86] favour an analysis involving three excimeric species. The situation may be summarized in terms of scheme (5) [89].

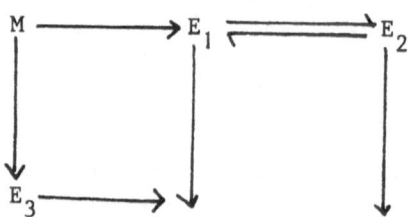

Scheme 5

The interpretations of Tagawa et al [87] and Roberts et al [86] vary in respect of the identity of species E_1. Tagawa et al [87] favour a 'relaxed monomer', whereas Roberts et al [86] favour a structure similar to the higher energy excimer E_3, but subjected to kinetic isolation due to its configuration. The higher energy excimer (E_1 and E_3) is supposed to constitute (by virtue of its extremely rapid formation; <10 ps) a spectral species corresponding to preformed excimer sites. The lower energy excimer, on the other hand, shows a 'growing in' on the nanosecond scale and is supposed to originate from either relaxed monomer [87] or from a fraction (E_1) of the higher energy species [85,89].

De Schryver et al [84] studying diastereoisomeric biscarbazolyl pentanes have produced convincing evidence implicating isotactic sequences in formation of the lower energy excimer and syndiotactic sequences for the higher energy excimer in poly(N-vinylcarbazole).

4.2. Derivation of Rate Coefficients

A prime objective in postulation of a kinetic scheme to describe a physical system is the use of that scheme to derive kinetic parameters which in turn allow comparison between one system and another. In polymer photophysics, the use of copolymer systems allows in principle the derivation of individual rate constants by a series of extrapolations and thereby removal of the effects of the fixed but unknown local chromophore concentration imposed on a system by study of the homopolymer alone.

The application of such an extrapolative procedure requires that a plausible kinetic mechanism is promoted and that terms descriptive of the intramolecular chromophore concentration are adopted. Subject to the constraints imposed by adoption of kinetic scheme 3 it is apparent that [17,25,26,71]

$$i_m(t) = A_1 \exp(-\lambda_1 t) + A_2 \exp(-\lambda_2 t) + A_3 \exp(-\lambda_3 t) \tag{6}$$

and

$$i_d(t) = A_4 \exp(-\lambda_1 t) + A_5 \exp(-\lambda_2 t) + A_6 \exp(-\lambda_3 t) \tag{7}$$

where

$$\lambda_1 + \lambda_2 = k_M + k_{DM}[M] + k_D + k_{MD} \tag{8}$$

and

$$\lambda_3 = k_t + k_M \tag{9}$$

Table (II) summarizes the procedures for deriving rate constants from the rate parameters λ_i resulting from analysis of $i_m(t)$ according to Equation (6). It should be noted that a measure of the self-consistency of the scheme, the extrapolative procedures and the intramolecular concentration functions adopted is afforded by the fact that each of the derived parameters may be checked through comparison of values obtained by alternative extrapolation approaches.

Table II. Procedures deriving rate constants [17]

Method	Procedure	Function derived	Derived parameter
(a) (b)	Measurement of unquenched monomer lifetime extrapolation of λ_1 to $[M] = 0$	k_M k_M^M	k_M k_M^M
(c)	$\dfrac{\partial(\lambda_1 + \lambda_2)}{\partial[M]}$	αk_{DM}	αk_{DM}
(d)	$\lambda_2 \to k_{DM}$ as $[M] \to \infty$	αk_{DM}	αkDM
(e)	$\lambda_1 + \lambda_2 \to k_M + k_D$ as $[M] \to 0$	$k_M + k_{MD} + k_D$	$k_{MD} + k_D$ [by combination with (a) or (b)]
(f)	$\lambda_1 \lambda_2 \to k_M(k_{MD} + k_D)$ as $[M] \to 0$	$k_M(k_{MD} + k_D)$	$k_{MD} + k_D$ [by combination with (a) or (b)]
(g)	$\dfrac{\partial(\lambda_1 \lambda_2)}{\partial[M]} = k_{DM} k_D$	$\alpha k_{DM} k_D$	k_D [by combination with (c) or (d)]
(h)	$\lambda_1 \to k_D$ as $[M] \to \infty$	k_D	k_D
(i)	(e) or (f) plus (g) or (h)	k_{MD}	k_{MD}
(j)	Extrapolation of λ_3 to $[M]' = 0$	k_M	k_M
(k)	$\dfrac{\partial \lambda_3}{\partial[M]'} = k_i'$	$\beta k_i'$	$\beta k_i'$

Implementation of this extrapolation method requires appropriate concentration terms to be adopted to describe the intramolecular concentration dependence of the rate parameters λ_i. In this context it has been assumed that the variation in aromatic content in the copolymers dominates the variations in photophysical behaviour across the the range of aromatic compositions and that microcompositional terms derived from steady-state analysis of I_D/I_M are sufficient to characterize the dependence of λ_1, λ_2 with aromatic content. These assumptions seem to be justified in terms of the adequate plots obtained for the copolymer systems studied to date. A typical example is shown in Fig. 2 for the copolymer system 1-vinylnaphthalene-co-methyl methacrylate.

The procedure has been applied to the copolymer series 1-vinylnaphthalene-methyl methacrylate [17], 1-vinylnaphthalene-methyl acrylate [71], acenaphthylene-methyl methacrylate [26] and styrene-methyl methacrylate [72]. The consistency of the data derived from alternative extrapolations are good as exemplified for the 1-vinylnaphthalene-methyl methacrylate copolymer series in Table (III)

Table III, Kinetic parameters for 1-vinylnaphthalene-methyl
          methacrylate copolymers [17]

Kinetic parameter	Value (10^7 s^{-1})	Method (see text)
k_M	1.72	(a)†
	2.15	(a)╪
	1.3	(b)
αk_{DM}	12.1	(c)
	12.0	(d)
$k_{MD} + k_D$	6.5	(e)
	6.2	(f)
k_D	2.2	(g)
	2.2	(h)
k_{MD}	4.2	(i)

† Determined using 2 x 10^{-5}M 1-methylnaphthalene.
╪ Determined using copolymer containing less than 0.5 mole % naphthalene chromophore.

In addition the agreement between rate constants such as k_D which are expected to be relatively invariant between copolymer series lends credibility to the application of this kinetic model to the chromophores considered.

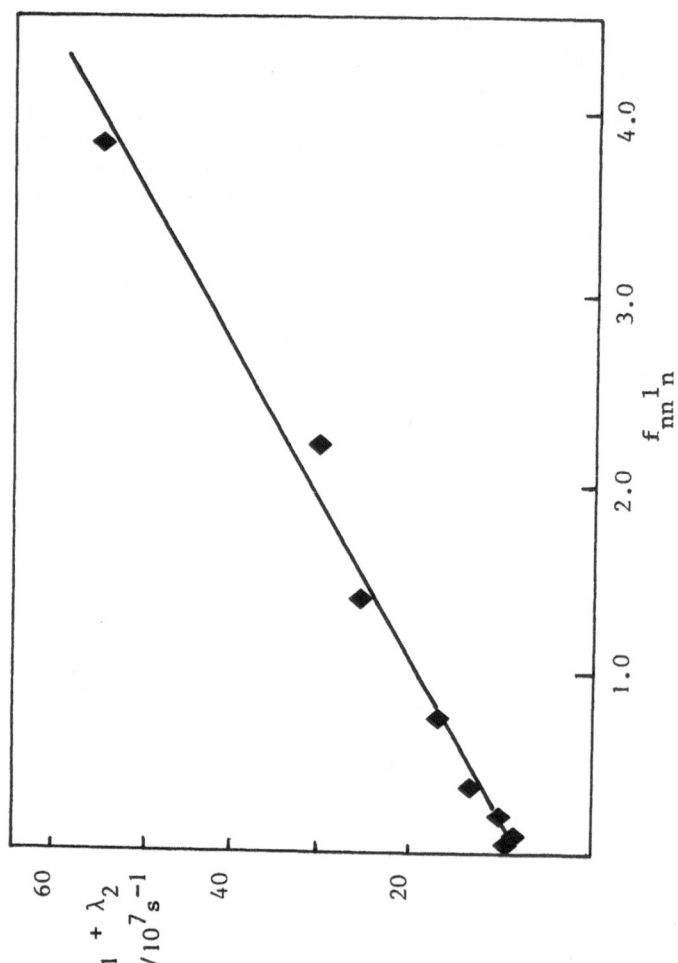

Figure 2. Sum of decay parameters $\lambda_1 + \lambda_2$ as a function of $f_{nn}\,^1_n$ for copolymers of 1-vinylnaphthalene and methyl methacrylate in tetrahydrofuran at 298K (After Phillips et al. [17])

4.3. Conclusions and Consolidation

Generation of mechanisms consistent with the photophysics of a given
series of macromolecules is a laudable exercise since a means is created
whereby the kinetic behaviour of the polymers is characterized. This
in turn will promote greater understanding of subsequent energy trapping
and photochemistry of macromolecules. Generally it has been assumed
that the multiexponential sum functions employed in decay curve fitting
have photophysical significance and that the "lifetimes" generated
thereby are related to the existence of kinetic entities representative
of groups of chromophores distinguishable within the overall distribution of photoactive species.

The models which have been generated (and discussed in sectiom 4.1)
may not be as conflictory as might at first appear. There does seem to
be an influence of tacticity upon polymer photophysical behaviour, supported by kinetic and spectroscopic evidence from low molar mass ana-
olgues [20] . These influences of tacticity may be responsible, for
example, for the differences in kinetic behaviour observed in homopolymers
and may be reflected in the 'kinetic isolation' of monomeric chromophores observed in some types of macromolecules [17,18].

Superimposed on these microcompositional effects are those of intramolecular concentration effects revealed in molecular weight effects
upon time-resolved emission data [73,90] and the influence of intramolecular chromophore concentration upon the luminescence characteristics of copolymers. The self-consistency of rate-parameter data
derived from time-resolved measurements on series of copolymers is good
evidence for the assignment (in the copolymers studied to date) of observed decay times as averages representative of chromophore distributions centred on species of the type M_1^*, M_2^* and D^* in kinetic scheme (3).

In some systems, such as those containing carbazoles and sterically hindered naphthalenes, there seems to be good spectroscopic evidence
for the involvement of more than one type of excimeric state. There are
without doubt systems in which, owing to the sheer complexity of the
processes involved, no average life-times will be adequate in description of the decay kinetics, i.e. average decay parameters will not be
characteristic of local chromophore environments within the total emitting
distribution. Examples of such situations may be apparent in consideration of polymers containing pyrene [91] or oxadiazole [92,93] fluors.

5. SINGLET MIGRATION IN POLYMERS

Whilst the models adopted for energy trapping at potential excimer sites
generally assume, and are consistent with, the occurrence of singlet
energy migration, the migrative capability of fluorescent chromophores
organized in macromolecular assemblages in solution, lacks definitive
empirical demonstration. Considerable effort has been devoted to
solution of this problem and is subject ot the briefest of reviews in
this article. Efforts to establish the existence of singlet energy
migration have centred upon the manifestation of emission depolariza-

tion [29, 45-47, 94-96   ], enhanced population of intramolecular
energy traps [95, 97-99] and promoted intermolecular fluorescence
quenching by 'external' quenchers of singlet states incorporated into
macromolecules in high local concentrations [90, 100-103]. Each experiment is beset by a number of problems concerned with the complexity of
macromolecular physics, polymer characterization and solubilization
characteristics.

5.1. Fluorescence Anisotropy

In principle, energy migration may be inferred from the depolarization
of emission concomitant upon energy migration. Fluorescence depolarization may be observed from chromophores emitting without energy trapping
or from energy traps such as excimers or species incorporated as intramolecular traps within the polymer chain. In practical terms, emission
characteristics must be sampled in general from chromophores dispersed
in solid matrices to obviate the depolarization which accompanied the
segmental diffusion experience in fluid solutions (as described by
Professor Monnerie elsewhere in this volume). The constraints imposed
by this sanction complicate the issue since the local strains induced in
glassy media have the potential to produce depolarizing effects in the
observed emissions. However, comparison of the anisotropy characteristics of the fluorescence of energy traps upon direct excitation with
those experienced following energy trapping should furnish evidence regarding prior energy migration which is subject to a lesser degree of influence of matrix induced effects, especially if the intrinsic polarization data are consistent with those obtained from independent experiments
in fluid media.

Studies of copolymer systems have shown that the extent of depolarization of fluorescence increases as the chromophore content of the
macromolecule is extended in glassy media [29,44,45,47,104-107]. Furthermore estimates of intrinsic polarization obtained in certain copolymer
systems are commensurate [29,44,45,47] with those evident in either
steady state [108] or transient excitation [109] conditions. Such
observations are consistent with the occurrence of intramolecular energy
migration. In these studies it is significant that excitation flux conditions were such as to obviate significant contribution from triplet
energy migration and resultant interference from delayed fluorescence.

Examination of the emission anisotropy of intramolecular energy
traps in copolymeric species following energy transfer from polymeric
donors can also yield important evidence regarding energy migration in
the polymeric donor. McInally et al [94] have observed significant depolarization of fluorescence from energy traps incorporated both as
terminal and intrachain acceptors within macromolecules and concluded
that energy migration occurred within those polymers examined. It is
important to note [110] that the data were corrected for the effects of
direct excitation of the acceptor species and the data correlated with
efficiencies of energy transfer from the polymeric donor. Supportive
evidence of singlet energy migration in several polymer systems incor-

porating intramolecular energy traps has been afforded by studies of
other antenna polymers' by Guillet et al [95,96,99,111

The primary dispute regarding energy migration in polymers as
evinced by anisotropy measurements seems to centre upon polystyrene.
MacCallum et al. [112-114] have reported that the anisotropy of the excimer emission both in fluid media and polymer films is indicative of a
high degree of retention of the polarized excitation characteristics.   In
contrast, Gupta et al [115] and Gardette and Phillips [116] have
disputed this observation. In the latter case, transient and steady
state anisotropies indicated that energy migration occurred. It is
significant to note that it is difficult to rationalize the observation of highly anisotropic emission of excimer emission in fluid media
with the occurrence of depolarization through the agency of facile
segmental mobility in such solutions as indicated by fluorescence depolarization [117] and e.s.r. measurements [118,119].

Lastly, it is pertinent to discuss recent results on a 'photozyme'
system consisting of sulphonated polyvinylnaphthalene in aqueous medium
[120]. The photozyme polymer was used to solubilize 9-phenylanthracene.
Emission polarization measurements indicated that the phenylanthracene
experienced a reasonably high microviscosity within the photozyme, yet,
upon excitation through transfer from the photozyme donor, exhibited
totally depolarized emission, consistent with singlet energy migration
within the polymeric donor. In this system, the constraints which might
be imposed by solid matrices are not encountered.

5.2. Intermolecular Quenching

In principle singlet energy migration may be detected through its enhancement of the degree of quenching of an excited state chromophore in
a macromolecule by a low molar mass external quencher. [90,100-103]
Generally, the treatment involves estimation of the migration diffusion
coefficient by reference of the efficiency of quenching of the polymer
emission to that for a small molecule model compound using the same
quencher. Frequently the Stern-Volmer equation is used to compare relative quenching rate coefficients.

There are several problems in adoption of the method:
(a) If the Stern-Volmer equation is applied directly it is assumed that
the fluorescence decays of both low molar mass model compound and polymer may be adequately described by a single exponential function. As
has been discussed, such is seldom the case in polymers. Attempts to
compensate for non-exponential decay in the polymer require adoption of
an adequate kinetic model for the polymer photophysics. Usually Birks
scheme has been adopted [102,121,122] the validity of which, as discussed
in section 4, is dubious. Holden et al. [103] have circumvented this
problem by studying the oxygen quenching of the fluorescence of poly(9-
phenanthrylmethyl methacrylate) in fluid solution. The decay of the unquenched polymer fluorescence is mono-expential [97] and it was inferred from the quenching data that singlet energy migration occurred
between the phenanthryl moieties.

(b) In comparison of the quenching efficiencies of polymer and small molecule model by a common quencher it is usually assumed that similar probabilities of quenching in the "encounter complex" are applicable regardless of steric constraints which might be imparted by the polymer chain.

(c) Closely related to problems regarding the influence of the polymer chain upon collisional deactivation are considerations of the partitioning of quencher within the coils of the macromolecule [123,124]. These considerations are especially pertinent in situations where the quencher employed is a non-solvent for the polymer [125,126]

5.3 Intramolecular Quenching

The photophysical characteristics of polymers incorporating varying amounts of intramolecular energy traps have furnished evidence of energy migration in polymers. The endeavours of Guillet et al [95,97-99, 127] bear particular mention in this context. Earlier reports concerned studies of energy trapping by anthracene acceptors (copolymerized generally as 9-anthrylmethyl methacrylate) from excited singlet naphthyl donors. The polymeric donors included poly(1-naphthylmethyl methacrylate) [127], poly[2-(1-naphthylethyl methacrylate)] [127] and poly(2-vinyl naphthalene) [98]. The energy transfer was studied under both steady state and transient excitation conditions. The data were interpreted as being consistent with the existence of energy migration via a moderate number of Forster transfers followed and accompanied by Forster energy transfer to the anthracene trap when the photoexcited donor lies within a sphere of radius R_0 characteristic of the dipolar interaction [128]. The concept is illustrated schematically in Figure 3.

Further evidence for these proposals has been obtained through the study of the intramolecular quenching of phenanthryl fluorescence in 9-phenanthrylmethyl methacrylate / 9-anthrylmethyl methacrylate copolymers [95,97]. The analysis is facilitated by the monoexponential decay of phenanthrene emission in the absence of intramolecular quenchers consequent upon the lack of excimer formation in this system. Under conditions of reasonably high trap concentrations, the decay curves became non-exponential (as is to be expected in the presence of energy migration and transfer of extents such that neither dominates the overall trapping kinetics).

These reports contain convincing evidence for the existence of energy migration within these polymer species. The kinetics were complicated in these antenna systems due to the relatively large Forster radii produced by the choice of donor-acceptor chromophores. The partial masking of the migrative influence by long-range transfers was avoided in an ingenious series of experiments [99] involving 2-vinylnaphthalene/phenylvinylketone copolymers. The forbidden nature of the S_0-S_1 transition of the ketone substantially decreases R_0 relative to systems studied earlier. In addition the complexities induced by excimer formation were avoided by studying the species in rigid glassy matrices. Steady state quenching efficiencies and non-exponentiality of donor decays were con-

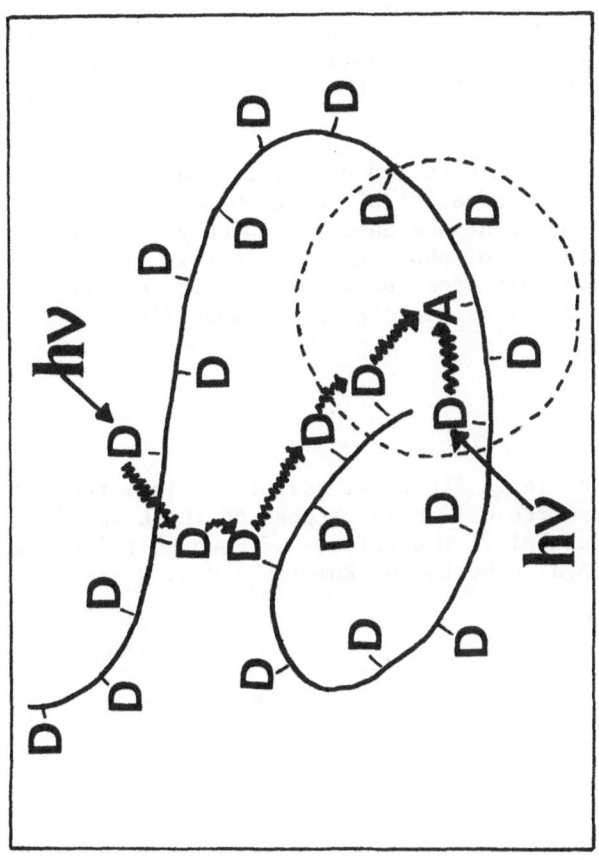

Figure 3. Conceptualization of energy transfer mechanisms between donor, D, and acceptor, A, species in an antenna polymer. (After Holden et al. [128])

sistent with migration of donor energy over a maximum of ca. 30 chromophores during the excited state lifetime.

Relevant to the dispute concerning energy migration in polystyrene
is a report on intramolecular energy trapping by the oxazole scintillation
solute, PPO, incorporated as an intramolecular trap into the parent macromolecule. The results were consistent with population of the oxazole
acceptors by migration and transfer. Further, it was shown that the excimer plays a negligible role as donor in the transfer process. (A
similar conclusion for the naphthalene excimer in naphthylalkyl methacrylate polymers was reached by Guillet et al. [98]). These transient
excitation data are supportive of conclusions reached following
anisotropy measurements on the styrene/PPO copolymer system [94,116].

Clearly the debates which have occurred over the nature of macromolecular photophysics are indicative of the complexities of the systems
concerned. These very complications have undoubtedly contributed in a
major fashion to the interest of photochemists in this field. Activity
will no doubt continue in this area for some years to come particularly
in developments of time resolved techniques and modelling.

ACKNOWLEDGEMENTS

The authors wish to acknowledge financial assistance from the SERC and
the US Army European Research Office in respect of these sections of our
work discussed in this article. The efforts of coworkers including
R.F. Reid, I.McInally, A.J. Roberts, G. Rumbles and K. Davidson are much
appreciated.

REFERENCES

1. FORSTER, Th. and KASPER, K. (1954), Z. Phys. Chem. NF 1, 275.
2. BIRKS, J.B. (1970), Photophysics of Aromatic Molecules, Wiley, London.
3. HIRAYAMA, F. (1965), J. Chem. Phys. 42, 3163.
4. STEVENS, B. and BAN, M.J. (1964), Trans. Faraday Soc. 60, 1515.
5. BIRKS, J.B., LUMB, M.D. and MUNRO, I.H. (1964), Proc. Roy. Soc. London. Ser. A 280, 289.
6. FOX, R.B., PRICE, T.R., COZZENS, R.F. and McDONALD, J.R. (1972), J. Chem. Phys. 57, 534.
7. WU, S.K., JIANG, Y.C. and RABEK, J.F. (1980), Polym. Bull. 3, 319.
8. BOKOBZA, L. and MONNERIE, L. (1981), Polymer 22, 235.
9. DAVID, C., PIENS, M. and GEUSKENS, G. (1972), Eur. Polym. J. 8, 1019.
10. HARRAH, L.A. (1972), J. Chem. Phys. 56, 385.
11. ITO, S., YAMAMOTO, M. and NISHIJIMA, Y. (1976), Rep. Prog. Polym. Phys. Japan 19, 421.
12. ASPLER, J.S. and GUILLET, J.E. (1979), Macromolecules 12, 1082.
13. NAKAHIRA, T., MARUYAMA, I., IWABUCHI, S. and KOJIMA, K. (1970), Makromol. Chem. 180, 1853.
14. FRANK, C.W. (1974), J. Chem. Phys. 61, 2015.
15. McINALLY, I., STEEDMAN, W. and SOUTAR, I. (1977), J. Polym. Sci. Polym. Chem. Ed. 15, 2511.
16. PHILLIPS, D., ROBERTS, A.J. and SOUTAR, I. (1980), J. Polym. Sci. Polym. Lett. Ed. 18, 123.
17. PHILLIPS, D., ROBERTS, A.J. and SOUTAR, I. (1980), J. Polym. Sci. Polym. Phys. Ed. 18, 2401.
18. HOLDEN, D.A., WANG, P.Y.-K. and GUILLET, J.E. (1980), Macromolecules 13, 295.
19. DEMEYER, K. VAN DER AUWERAER, M., AERTS, L. and DE SCHRYVER, F.C. (1980), J. Chim. Phys. 77, 493.
20. DE SCHRYVER, F.C., DEMEYER, K., VAN DER AUWERAER, M. and QUANTEN, E. (1981), Ann. N.Y. Acad. Sci. 366, 93.
21. NISHIJIMA, Y. (1973), Prog. Polym. Sci. Japan 6, 199.
22. DAVID, C., PIENS, M. and GEUSKENS, G. (1979), Eur. Polym. J. 15, 373.
23. HOLDEN, D.A. and GUILLET, J.E. (1981), in 'Developments in Polymer Photochemistry', vol. 1 (ed. by N.S. Allen) Applied Science Publishers, London).
24. GHIGGINO, K.P., WRIGHT, R.D. and PHILLIPS, D. (1978), J. Polym. Sci. Polym. Phys. Ed. 16, 1499.
25. PHILLIPS, D., ROBERTS, A.J. and SOUTAR, I. (1982), J. Polym. Sci. Polym. Phys. Ed. 20, 411.
26. PHILLIPS, D., ROBERTS, A.J. and SOUTAR, I. (1981), Eur. Polym. J. 17, 101.
27. WANG, Y.C. and MORAWETZ, H. (1975), Makromol. Chem., Suppl. 1, 283.
28. DAVID, C., LEMPEREUR, M. and GEUSKENS, G. (1972), Eur. Polym. J. 8, 417.
29. REID, R.F. and SOUTAR, I. (1980), J. Polym. Sci., Polym. Phys. Ed. 18, 457.

30. SOMERSALL, A.C. and GUILLET, J.E. (1973), Macromolecules 6, 218.
31. BOUDEVSKA, H. BRUTCHKOV, C. and ANSTRUG, A. (1979), Makromol. Chem. 180, 1113.
32. ABUIN, E.A., LISSI, E.A., GARGALLO, L. and RADIC, D. (1979), Eur. Polym. J. 15, 373.
33. REDPATH, A.E.C. and WINNIK, M.A. (1980), J. Amer. Chem. Soc. 102, 6869.
34. REDPATH, A.E.C. and WINNIK, M.A. (1981), Ann. N.Y. Acad. Sci. 366, 75.
35. CUNIBERTI, C. and PERICO, A. (1980), Eur. Polym. J. 16, 887.
36. CUNIBERTI, C. and PERICO, A. (1981), Ann. N.Y. Acad. Sci. 366, 35.
37. SOUTAR, I. (1981), Ann. N.Y. Acad. Sci. 366, 24.
38. FOX, R.B., PRICE, T.R., COZZENS, R.F. and ECHOLS, W.H. (1974), Macromolecules 7, 937.
39. YOKOYAMA, M., TAMAMURA, T., ATSUMI, M., YOSHIMURA, M., SHIROTA, Y. and MIKAWA, H. (1975), Macromolecules 8, 101.
40. LINDSELL, W.E., ROBERTSON, F.C. and SOUTAR, I. (1981), Eur. Polym. J. 17, 203.
41. TORKELSON, J.M. LIPSKY, S. and TIRRELL, M. (1981), Macromolecules 14, 1601.
42. NISHIJIMA, Y. (1970), J. Polym. Sci. C31, 353.
43. DAVID, C., LEMPEREUR, M. and GEUSKENS, G. (1973), Eur. Polym. J. 9, 1315.
44. REID, R.F. and SOUTAR, I. (1978), J. Polym. Sci. Polym. Phys. Ed. 16, 231.
45. REID, R.F. and SOUTAR, I. (1977), J. Polym. Sci. Polym. Lett. Ed. 15, 153.
46. ANDERSON, R.A., REID, R.F. and SOUTAR, I. (1979), Eur. Polym. J. 15, 925.
47. ANDERSON, R.A., REID, R.F. and SOUTAR, I. (1980), Eur. Polym. J. 16, 945.
48. ALEXANDRU, L. and SOMERSALL, A.C. (1977), Polym. Sci. Polym. Chem. Ed. 15, 2013.
49. HILL, D.J.T., LEWIS, D.A., O'DONNELL, J.H., O'SULLIVAN, P.W. and POMERY, P.J. (1982) Eur. Polym. J. 18, 75.
50. GARDETTE, J.L., PHILLIPS, D., and SOUTAR, I., to be published.
51. MAJUMDAR, R.N., CARLINI, C., ROSATO, N. and HOUBEN, J.L. (1980), Polymer 21, 941.
52. NAKAHIRA, T., MINAMI, C., IWABUCHI, S. and KOJIMA, K. (1979), Makromol. Chem. 180, 2245.
53. NAKAHIRA, T., SAKUMA, T., IWABUCHI, S. and KOJIMA, K. (1980), Makromol. Chem. Rapid Commun. 1, 413.
54. LONGWORTH, J.W. and BOVEY, F.A. (1966), Biopolymers 4, 1115.
55. BOKOBZA, L., JASSE, B. and MONNERIE, L. (1977), Eur. Polym. J., 13, 921.
56. FRANK, C.W. and HARRAH, L.A. (1974), J. Chem. Phys. 61, 1526.
57. DE SCHRYVER, F.C., BOENS, N. and PUT, J. (1977), Adv. Photochem. 10, 359.
58. BOKOBZA, L., JASSE, B. and MONNERIE, L. (1980), Eur. Polym. J. 16, 715.

59. GOLDENBERG, M., EMERT, J. and MORAWETZ, H. (1978), J. Amer. Chem. Soc. 100, 7171.
60. VAN DER AUWERAER, M., GILBERT, A. and DE SCHRYVER, F.C. (1980), J. Amer. Chem. Soc. 102, 6.107.
61. MEEUS, F., VAN DER AUWERAER, M. and DE SCHRYVER, F.C. (1980), Chem. Phys. Lett. 74, 218.
62. LONGWORTH, J.W. (1966), Biopolymers 4, 1131.
63. ISHII, T., MATSUSHITA, H. and HANDA, T. (1975), Kobunshi Ronbunshu 32, 211.
64. ISHII, T., MATSUNAGA, S. and HANDA, T. (1976), Makromol. Chem. 177, 283.
65. ISHII, T., HANDA, T. and MATSUNAGA, S. (1977), Makromol. Chem. 178, 2351.
66. MacCALLUM, J.R. (1981), Eur. Polym. J. 17, 797.
67. MacCALLUM, J.R. (1982), in Photophysics of Synthetic Polymers (ed. D. Phillips and A.J. Roberts) Science Reviews Ltd., Northwood).
68. PHILLIPS, D., ROBERTS, A.J. and SOUTAR, I. (1981), Polymer 22, 427.
69. FREDRICKSON, G.H. and FRANK, C.W. (1983), Macromolecules 16, 572.
70. ITAGAKI, H., HORIE, K. and MITA, I. (1983), Macromolecules 16, 1395.
71. PHILLIPS, D., ROBERTS, A.J. and SOUTAR, I. (1981), Polymer 22, 293.
72. SOUTAR, I., PHILLIPS, D., ROBERTS, A.J. and RUMBLES, G. (1982), J. Polym. Sci. Polym. Phys. Ed. 20, 1759.
73. PHILLIPS, D., ROBERTS, A.J., RUMBLES, G. and SOUTAR, I. (1983), Macromolecules 16, 1597.
74. GHIGGINO, K.P., ROBERTS, A.J. and PHILLIPS, D. (1978), J. Photochem. 9, 301.
75. PHILLIPS, D., ROBERTS, A.J. and SOUTAR, I. (1983), Macromolecules 16, 1593.
76. ITAGAKI, H., OKAMOTO, A., HORIE, K. and MITA, I. (1982), Eur. Polym. J. 18, 885.
77. ITAGAKI, H., OBUKATA, N., OKAMOTO, A., HORIE, K. and MITA, I. (1981), Chem. Phys. Lett. 78, 143.
78. ITAGAKI, H., OBUKATA, N., OKAMOTO, A., HORIE, K. and MITA, I. (1982), J. Amer. Chem. Soc. 104, 4469.
79. GUPTA, A., LIANG, R., MOACANIN, J., KLIGER, D., GOLDBECK, R., HORWITZ, J. and MISKOWSKI, V.M. (1981), Eur. Polym. J. 17, 485.
80. JOHNSON, G.E. (1975), J. Chem. Phys. 62, 4697.
81. GHIGGINO, K.P., WRIGHT, R.D. and PHILLIPS, D. (1978). Eur. Polym. J. 14, 567.
82. HOYLE, C.E., NEMZEK, T.L., MAR, A. and GUILLET, J.E. (1978), Macromolecules 11, 429.
83. ITAYA, A., OKAMOTO, K. and KUSABAYASHI, S. (1976), Bull. Chem. Soc. Japan 49, 2092.
84. DE SCHRYVER, F.C., VANDENDRIESSCHE, J., TOPPET, S., DEMEYER, K. and BOENS, N. (1982), Macromolecules 15, 406.
85. ROBERTS, A.J. CURETON, C.G. and PHILLIPS, D. (1980), Chem. Phys. Lett. 72, 554.
86. ROBERTS, A.J., PHILLIPS, D., ABDUL-RASOUL, F. and LEDWITH, A. (1981) J.C.S. Faraday I. 77, 2725.
87. TAGAWA, S., WASHIO, M. and TABATA, Y. (1979), Chem. Phys. Lett. 68, 276.

88. NG, D. and GUILLET, J.E. (1981), Macromolecules 14, 405.
89. GHIGGINO, K.P., ROBERTS, A.J. and PHILLIPS, D. (1981), Adv. Polym.
     Sci. 40, 71
90. ISHII, T., HANDA, T. and MATSUNAGA, S. (1978), Macromolecules, 11,
     40
91. NEVES, M.E.P.J.M. and SOUTAR, I. (1983) to be published.
92. SOUTAR, I. (1982), in Photophysics of Synthetic Polymers (ed.
     D. Phillips and A.J. Roberts), Science Reviews Ltd., Northwood.
93. ANDERSON, R.A., BIRCH, D.J.S., DAVIDSON, K., IMHOF, R.E. and
     SOUTAR, I. (1982), in Photophysics of Synthetic Polymers (ed.
     D. Phillips and A.J. Roberts), Science Reviews Ltd., Northwood.
94. McINALLY, I., REID, R.F., RUTHERFORD, H. and SOUTAR, I. (1983),
     Eur. Polym. J., 15, 723.
95. NG, D. and GUILLET, J.E., (1982), Macromolecules, 15, 724.
96. HOLDEN, D.A. and GUILLET, J.E. (1982), Macromolecules, 15, 1475.
97. NG, D. and GUILLET, J.E. (1982), Macromolecules, 15, 728.
98. NG, D., YOSHIKI, K. and GUILLET, J.E. (1983), Macromolecules, 16,
     568.
99. HOLDEN, D.A., REN X.-X. and GUILLET, J.E. (1984), Macromolecules,
     17, 1500.
100. UENO, A., OSA, T. and TODA, F. (1977), Macromolecules, 10, 130.
101. HARGREAVES, J.S. and WEBBER, S.E. (1984), Macromolecules, 17, 1741.
102. WEBBER, S.E., AVOTS-AVOTINS, P.E. and DEUMIE, M. (1981),
     Macromolecules, 14, 105.
103. HOLDEN, D.A., NG. D. and GUILLET, J.E. (1982), Brit. Polym. J.
     1982*, 159.
104. DAVID, C., BAEYENS-VOLANT, D. and GEUSKENS, G. (1974), Eur. Polym.
     J., 12, 71.
105. SCHNEIDER, F. (1969), Z. Naturforsch. 24a, 863.
106. JOHNSON, G.E. (1980), Macromolecules, 13, 145.
107. YOKOYAMA, M., TAMAMURA, T., ATSUMI, M., YOSHIMURA, Y., SHIROTA, Y.
     and MIKAWA, H. (1975), Macromolecules, 8, 101.
108. KETTLE, G.J. and SOUTAR, I. (1978), Eur. Polym. J. 14, 895.
109. DRAKE, R.G., PHILLIPS, D. and SOUTAR, I. to be published.
110. MacCALLUM, J.R. (1981), Eur. Polym. J. 17, 209.
111. HOLDEN, D.A., SHEPHARD, S.E. and GUILLET, J.E. (1982),
     Macromolecules, 15, 1481.
112. MacCALLUM, J.R. and RUDKIN, L. (1977), Nature, 266, 338.
113. MacCALLUM, J.R. and RUDKIN, L. (1981), Eur. Polym. J. 17, 953.
114. MacCALLUM, J.R. (1982), Polymer, 23, 175.
115. GUPTA, M.C., GUPTA, A., HORWITZ, J. and KLIGER, D. (1982),
     Macromolecules, 15, 1372.
116. GARDETTE, J.-L. and PHILLIPS, D. (1984), Polymer Commun. 25, 366.
117. BROWN, K. and SOUTAR, I. (1974), Eur. Polym. J., 10, 433.
118. BULLOCK, A.T., BUTTERWORTH, J.H. and CAMERON, G.G. (1971), Eur.
     Polym. J., 7, 445.
119. FRIEDRICH, C., NOEL, C., RAMMASSEUL, R. and RASSAT, A. (1980),
     Polymer, 21, 232.
120. GUILLET, J.E. and SOUTAR, I., unpublished data.

121. ISHII, T., HANDA, T. and MATSUNAGA, S. (1979), J. Polym. Sci.
      Polym. Phys. Ed., 17, 811.
122. PFISTER, G., WILLIAMS, G.J. and JOHNSON, G.E. (1974), J. Phys.
      Chem., 78, 2009.
123. DUPORTAIL, G., FROELICH, D. and WEILL, G. (1971), Eur. Polym. J.
      7, 977.
124. MOLDOVAN, L. and WEILL, G. (1971), Eur. Polym. J. 7, 1023.
125. TAHA, I. and MORAWETZ, H. (1971), J. Polym. Sci. A2, 9, 1669.
126. TAN, K.L. and TRELOAR, F.E. (1980), Chem. Phys. Lett., 73, 234.
127. HOLDEN, D.A. and GUILLET, J.E. (1980), Macromolecules, 13, 289.
128. HOLDEN, D.A., RENDALL, W.A. and GUILLET, J.E., (1981), Ann. N.Y.
      Acad. Sci., 366, 11.

ANALYSIS OF FLUORESCENCE DECAY DATA FROM SYNTHETIC POLYMERS: HETEROGENEITY, MOTION AND MIGRATION

David Phillips
The Royal Institution
21 Albemarle Street
London W1X 4BS

Ian Soutar
Department of Chemistry
Heriot-Watt University
Riccarton
Edinburgh

This article is concerned with the analysis of results obtained from experiments in which the time-dependence of fluorescence from synthetic polymers is analysed. Of the methods available to study such fluorescence, that utilising time-correlated single photon counting detection is probably the most widely used. The method will not be described in detail here, but readers may refer to a recent comprehensive volume on the subject [1]. It is worthwhile here however including some discussion on the analysis of data obtained with the method.

Convolution

If the flash of light that excites the sample were infinitely narrow, and if the response of the detection system were infinitely fast, the observed decay curve would represent the true decay, or δ-pulse response, of the sample $G(t)$.

The form of the observed decay, $I(t)$, when the excitation function, $E(t)$, is not a δ-function can be deduced from the theory of impulse functions and leads to the convolution concept. Convolution, or folding together, occurs because molecules excited by photons at early times are decaying while others are being excited by photon in the tail of the excitation pulse. A simple deduction of the convolution equation can be obtained from consideration of Figure 1. The pump pulse is assumed to be a sum of δ-pulses of amplitude $E(t')$ at any time t'. Since the number of sample molecules excited at time t' is proportional to $E(t')$ the number at any later time $x - t'$ is proportional to $E(t')G(x - t')$. The total number of excited state molecules at time x, written $[A*](x)$ is then a sum over all times t' preceding time x or, for an infinite sum,

M. A. Winnik (ed.), Photophysical and Photochemical Tools in Polymer Science, 129–150.
© *1986 by D. Reidel Publishing Company.*

$$[A*](x) \alpha \int_o^x E(t')G(x - t')dt'. \tag{1}$$

This treatment neglects distortions introduced by the detection system, considered below.

Suppose that $H(t)$ is the δ-pulse response of the detection system, and $P(t)$ the measured time profile of the pump pulse, i.e., the instrument response function. $P(t)$ is a convolution of $E(t)$ and $H(t)$.

$$P(t) = E(t) \otimes H(t). \tag{2}$$

Writing the Laplace transform of a function $X(t)$ as $x(s)$ with

$$x(s) = L\{X(t)\} = \int_o^\infty e^{-st}X(t)dt,$$

it follows that

$$p(s) = e(s).h(s). \tag{3}$$

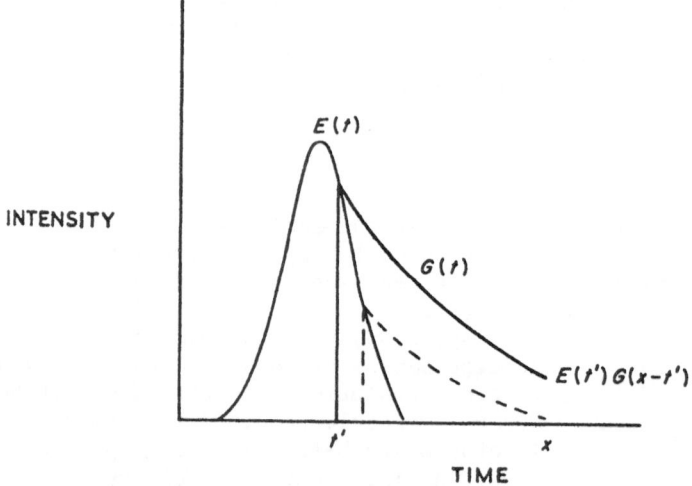

Figure 1.   Schematic representation of the effect of convolution. $E(t)$ idealized pump pulse profile; $G(t)$ decay law (assumed single exponential) of sample.

Similarly

$$a(s) = e(s).g(s) \tag{4}$$

and

$$i(s) = a(s).h(s). \tag{5}$$

Therefore

$$i(s) = e(s).g(s).h(s) = p(s).g(s) \tag{6}$$

and

$$I(t) = P(t) \textcircled{x} G(t) = \int_0^t P(t')G(t - t')dt', \tag{7}$$

where $I(t)$ represents the δ-function response of the sample distorted by convolution with both the pump pulse and the detector response, i.e., the measured decay curve. Thus the convolution integral can be solved for $G(t)$ if $I(t)$ and $P(t)$, measured under the same conditions of instrumental distortion, are known. There are many methods of achieving this, but non-linear least squares iterative reconvolution is that favoured by most.

Least squares fitting

With this technique the data points with the highest number of counts are more heavily weighted; moreover any section of the decay curve may be excluded from the analysis, a feature especially useful if distortions are present in the data. If the sample decay is a single exponential, $G(t)$ is given by

$$G(t) = a_1 \exp(-t/a_2) \tag{8}$$

and in order to linearize the fitting function Equation (8) is expanded to first order in a Taylor's expansion as a function of the parameters a_1 and a_2. A linear least-squares search is then carried out to find values of the parameter increments a_1 and a_2 that minimize the reduced X^2_v given by

$$X^2_v = \frac{\sum_{i=n_1}^{n_2} w_1 \{Y(t_i) - I(t_i)\}^2}{n_2 - n_1 + 1 - p} \tag{9}$$

where w_i, the weighting factor, is the reciprocal of the number of counts $Y(t_i)$ in channel i, n_1 and n_2 are the first and last channels of the section of the decay to be analysed, and p is the number of fitting parameters (two for a single exponential fit). The search for the minimum in X^2_v is performed according to Marquardt's technique.

When the minimum in X^2_v has been reached it is vitally important to have reliable criteria by which the fit can be judged. The actual

D. PHILLIPS AND I. SOUTAR

value should be close to 1; values of X^2_ν much less than 1 are symptomatic of poor statistics whereas values much in excess of 1 indicate a poor fit. If all distorted data are to be rejected we would accept results for which X^2_ν is less than 1.2, whereas if some level of distortion must be tolerated, fits with values of X^2_ν less than 1.4 may be acceptable if they are justified by some other criteria. Since acceptable values of X^2_ν are sometimes obtained for poor fits, it is usual to inspect a plot of the weighted residuals for nonrandom fluctuations. The weighted residual in channel i is given by

$$r_i = \sqrt{w_i}(Y(t_i) - I(t_i)) \tag{10}$$

It is generally less difficult to detect small deviations of the fitted from the observed curve in a plot of r_i vs. channel number rather than in the more traditional visual inspection of the two curves $Y(t_i)$ and $I(t_i)$. An even more sensitive plot is that of the autocorrelation function of the weighted residuals. The correlation of residual in channel i with the residual in channel i + j is summed over a number of channels, m, and normalized, i.e.,

$$C_{rj} = \frac{\dfrac{1}{m} \displaystyle\sum_{i = n_1}^{n_1 + m - 1} r_i \cdot r_{i + j}}{\dfrac{1}{n_3} \displaystyle\sum_{i = n_1}^{n_2} r^2_i} \tag{11}$$

In this expression $n_3 = n_2 - n_1 + 1$, the total number of channels in the section of the decay used in the fit. An upper limit, usually $n_{3/2}$, is put on j so that the number of terms, $m = n_3 - j$, summed in the numerator is sufficient to give proper averaging. According to Equation (11) $C_{r0} = 1$. In a successful fit C_{rj} for j = 0 is randomly scattered about zero although, because of the finite value of m, some high-frequency low-amplitude fluctuations are generally observed. These are clearly distinguishable from the type of correlation indicative of an incorrect fitting function or of distorted data.

Judgements based on inspection of the aforementioned plots are subject to the inevitable bias associated with subjective tests. Consequently we calculate the Durbin-Watson parameter DW, which is, in our opinion, more sensitive than X^2_ν to small nonrandom oscillations in the residuals. DW is calculated according to the equation

$$DW = \frac{\displaystyle\sum_{i = n_1 + 1}^{n_2} (r_i - r_{i - 1})^2}{\displaystyle\sum_{i = n_1}^{n_2} r^2_i} \tag{12}$$

Acceptable values for DW have been tabulated for up to 100 data points and five fittings parameters. Extrapolation of tables to more data points is quite straightforward. On the basis of our experience we conclude that single exponential fits yielding values of DW greater than 1.65 are generally successful. The corresponding values for double and triple exponential fits are 1.75 and 1.8, respectively. In addition we calculate a skewness factor, SK, given by

$$SK = \sqrt{\frac{n_3}{[\sum_{i=n_1}^{n_2} (r_i - r)^2]^3}} \sum_{i=n_1}^{n_2} (r_i - r)^3 \qquad (13)$$

and a kurtosis factor, K, given by

$$K = \frac{n_3 \sum_{i=n_1}^{n_2} (r_i - r)^4}{[\sum_{i=n_1}^{n_2} (r_i - r)^2]^2} \qquad (14)$$

In these equations r is the mean of the weighted residuals. For normally distributed residuals SK has a mean of zero and a standard deviation of $(6/n_3)^{1/2}$. Although we calculate these parameters routinely they are difficult to interpret and therefore we find them less useful than the Durbin-Watson parameter.

A very useful test, particularly when there is doubt about the suitability of a chosen fitting function is variation of the fitting range. Variation in the recovered parameters when channels representing earlier times are included in the fit is indicative of an incorrect fitting function. Usually, but not always, instrumental distortions affecting the early times data points lead to non-normally distributed residuals but the same values for the recovered parameters irrespective of the fitting range.

These tests are applied rigorously to all of the data obtained in our laboratories and, we believe, do permit some discrimination between alternative trial forms of G(t). (see below).

Expected form of G(t)

Single-exponential decay

It is perhaps worth stating at the outset the conditions under which single-exponential decay should be anticipated. Considering a single emitting component in the condensed phase, electronic excitation will be followed by rapid equilibrium of vibrational energy to produce the Boltzmann distribution of levels from which emission occurs. Since the equilibrium which maintains this distribution is usually

rapidly established compared with the timescale of electronic
relaxation processes, the depopulation of the excited state can be
represented by a single rate parameter, a pseudo-first order rate
constant multiplied by concentration of excited species.

$$\frac{d[M*]}{dt} = k'[^1M*]$$
(15)

We are, of course, familiar with the division of the pseudo-first
order decay constant k' into individual contributions, based upon a
simple scheme such as that below,

$$M + h\nu \rightarrow {}^1M* \qquad\qquad I_a$$
(16)

$${}^1M* \rightarrow M + h\nu_F \qquad\qquad k_R$$
(17)

$${}^1M* \rightarrow {}^3M* \qquad\qquad k_{ISC}$$
(18)

$${}^1M* + Q \rightarrow Quenching \quad k_Q[Q]$$
(19)

such that $k' = k_R + k_{ISC} + k_Q[Q]$, with the usual relationships below
holding.

$$\tau_F^{-1} = k_R + k_{ISC} + k_Q[Q]$$
(20)

$$\phi_F = \frac{k_R}{(k_R + k_{ISC} + k_Q[Q])}$$
(21)

It is important to remember that rate-constants relate to
bulk properties of molecular systems. Thus for example, a normal
thermal bimolecular rate-constant for the hypothetical reaction
(22) represents the rate observed in the bulk, and is thus averaged
over all initial energy distributions in A, B, angles and velocities

$$A + B \quad C + D$$
(22)

of approach, product internal energy distributions, trajectories and
kinetic energies. If experiments are performed in conditions such
that these parameters are specified, then the probability of reaction
observed will not relate to that observed in the bulk phase, and may
have different functional form.
  The common causes of deviations from single exponential decay of
fluorescence in molecular systems has been reviewed elsewhere [2], and
thus a digest only is given here. Briefly, these are:-

Heterogeneity

  For more than one simultaneously excited, non-interacting
species, the decay of total fluorescence will be described in

principle by (22). The situation with two non-interacting species is

$$I(t) = \sum_i A_i e^{-t/\tau_i} \qquad (23)$$

fairly commonly met, but as the number of species increases, interactions such as energy transfer are bound to become more probable, complicating the kinetics (or rather simplifying them in some concentration ranges).

In the extreme of a large number of non-interacting sites, such as molecules adsorbed on a solid surface, in defects in molecular crystals or in some polymeric species, the decay may be better described by a distribution of decay times, suitably weighted about some mean value.

A recent treatment by Albery et al [3] gives a rate-parameter k as a distribution represented by

$$k = \bar{K} \exp(\gamma x) \qquad (24)$$

Thus the decay of concentration C of a species from initial concentration C_o is given by

$$\frac{C}{C_o} = \frac{\int_{-\infty}^{+\infty} \exp(-x^2) \exp[-\tau \exp(\gamma x)]dx}{\int_{-\infty}^{+\infty} \exp(-x^2)dx} \qquad (25)$$

where

$$\tau = kt, \text{ and } \int_{-\infty}^{+\infty} (-x^2)dx = \pi^{\frac{1}{2}}$$

The width of the distribution can be obtained simply by using this analysis by measuring $t_{\frac{1}{2}}$, and $t_{\frac{7}{8}}$, the times taken for decay by $\frac{1}{2}$ and $\frac{7}{8}$ initial concentration respectively, and Equation (26).

$$\gamma = 0.92 \left[t_{\frac{7}{8}}/t_{\frac{1}{2}} - 3\right]^{\frac{1}{2}} \qquad (26)$$

In simple cases of two or three or perhaps four components, the derived A_i and resolved integrated areas under decay curves, $A_i\tau_i$, are respectively measures of the initial concentration of each species modified in one case by the radiative rate-constant, the second by the

$$A = k_R [^1M^*]_o \qquad (27)$$

$$A = I_a \phi_F [^1M^*_o] \qquad (28)$$

quantum yield of fluorescence. Without knowledge of respective values of k_R, or ϕ_F for each species, little can be said about initial concentrations.

It is possible in favourable cases to deconvolve successfully three components from a single experimental curve, although reliance on one such experimental curve would be fool-hardy. This is

illustrated in Figure 2 which displays analysis by this method of
simulated data obtained by convolving three components of respective A
and τ (shown in Table 1 as initial values), with a real instrument
response function using a cavity-dumped dye laser (see below) adding
random noise to simulate the experiment, and then analysing. Recovered
values of A and τ are very satisfactory for a three component fit (2b)
but a two-component, (four parameter) fit is seen to be unacceptable.
(Figure 2a) [4].

It is very important to stress here what these simulated fittings
mean. A triple-component, (six parameter) fit will certainly under
some circumstances simulate other, more complex forms of decay, and
thus great care must be taken in interpretation of data using such a
model. However, we have shown above that the technique can recover the
correct functional form, of the decay parameters i.e., while a triple
component fit is not automatically the correct functional form, it
certainly is not automatically inappropriate. Evaluation of the data
must rely on a range of experiments which test the model, and all

Table 1   Analysis of simulated three component decay curve

	Initial	Recovered
A_1	0.25	0.25
τ_1/ns	2.50	2.54
A_2	0.07	0.07
τ_2/ns	10.00	9.72
A_3	0.025	0.026
τ_3/ns	40.00	39.44

results must be compatible with this model. In many cases of multiple
species, more information can be gained by observing fluorescence in a
narrow wavelength range, thus reducing or even eliminating contributions
from one or more species. Sequential analysis of different wavelength
regions fixing decay times measured at wavelengths where kinetics are
simpler renders extraction of multiple components easier, through great
care must be taken in such a procedure. The technique of 'global'
analysis of data is particularly useful in such circumstances [5].

Relaxation processes

Of the many possibilities, that of reversible complex formation is
pertinent to polymers. The basic general scheme 1 leads to the

prediction that the decay of uncomplexed fluorophore, $^1A^*$, termed here monomer, and, complex, $^1AQ^*$, should follow the functional forms shown.

Scheme 1

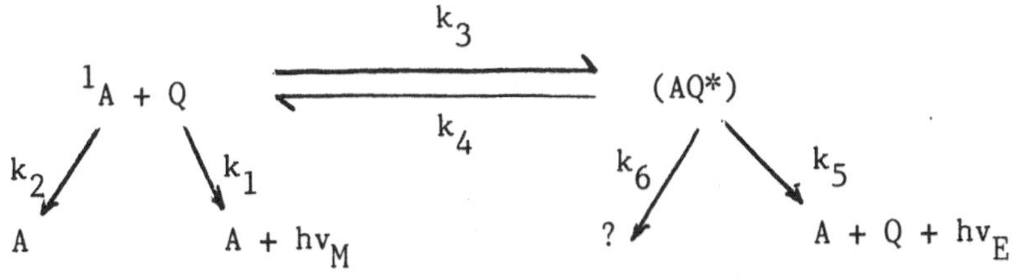

$$[^1A^*](t) = a_1 e^{-\lambda_1 t} + a_2 e^{-\lambda_2 t} \tag{29}$$

$$[^1AQ^*](t) = a_3 e^{-\lambda_1 t} + a_4 e^{-\lambda_2 t} \text{ where } a_3 = -a_4 \tag{30}$$

Table 2   Decay time data for CNN–TEA (+ 3 atm of Cyclohexane) at 188°C in the Gas Phase [6].

10^{-3} [TEA], M	monomer			exciplex			
	τ_1, ns	τ_2, ns	$a_1/(a_1 + a_2)$	τ_1, ns	τ_2, ns	a_3	a_4
0		24.1					
0.222	8.17	11.96	0.64	8.64	12.05	−0.46	0.46
0.530	5.41	10.13	0.82	5.80	11.60	−2.69	2.68
0.837	4.07	10.50	0.90	4.16	11.11	−3.68	3.70
2.13	2.00	11.23	0.98	1.96	10.97	−2.33	2.33
3.36	1.31	10.35	0.98	1.31	10.38	−1.93	1.92

That such kinetics can be observed in some systems is typified by the results shown in Table 2, for an exciplex-forming system α–cyano–naphthalene–triethylamine in the vapour phase, where it can be seen that the relationship $a_3 = -a_4$ is obeyed precisely [6]. Indeed, the precision of these measurements is such that deviations from expected values of a_3 and a_4 can be used as a monitor of <u>ground</u> state complex formation [6].

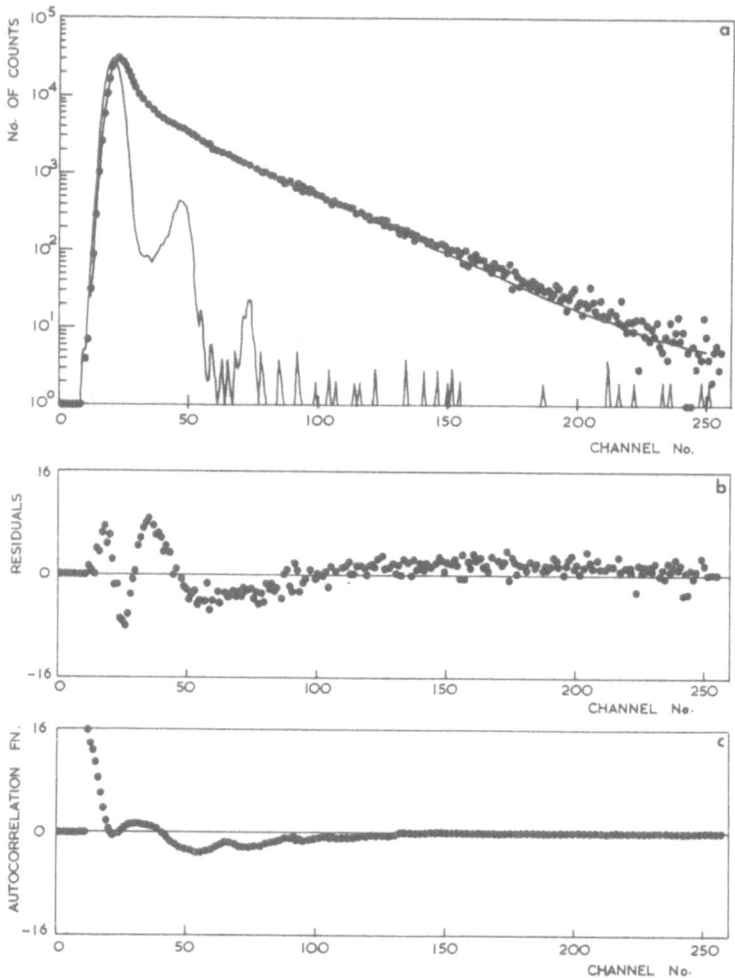

Figure 2a. Plots of (a) two-component fit to simulated three-component fluorescence decay (see text); (b) weighted residuals; (c) auto-correlation function.

Motion

The simplest correction to single-exponetial decay laws occurs as a result of transient effects in translational diffusional processes. An extensive review of the causes and consequences of these transient effects has been given elsewhere, and would be out of place here. In the diffusional quenching of molecule A* by B assuming the simple scheme below

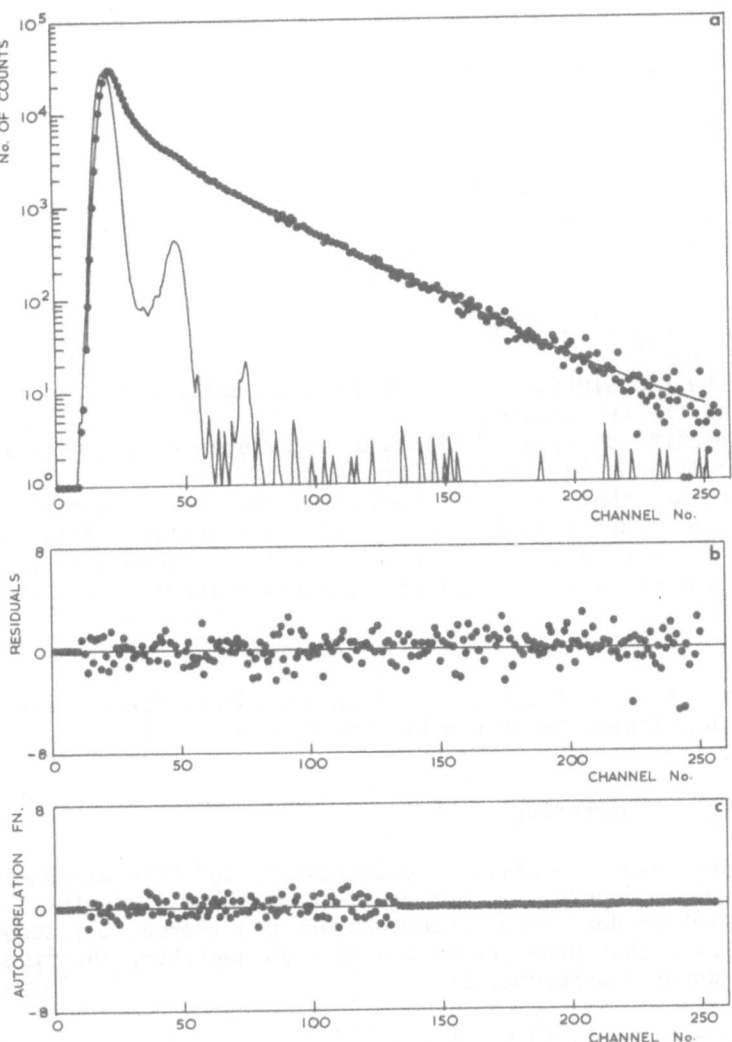

Figure 2b. Plots of (a) three-component fit to the same simulated three-component fluorescence decay as Figure 2a; (b) auto-correlation function.

$$A + h\nu \rightarrow {}^1A* \qquad (31)$$

$${}^1A* \rightarrow A \qquad (32)$$

$${}^1A* \rightarrow A + h\nu \qquad (33)$$

$$^1A* + B \xrightarrow{k(t)} \text{products} \tag{34}$$

the decay law of A*, monitored by fluorescence, has been shown to be

$$I(t) = \exp(-At - 2Bt^{\frac{1}{2}}) \tag{35}$$

where

$$A = \frac{1}{\tau_o} + 4\pi\sigma\, D_{AB}\, N'\, [B] \tag{36}$$

and

$$B = 4\pi\, D_{AB}\, \sigma^2\, N'\, [B] \tag{37}$$

where D_{AB} is the translational diffusion coefficient, and σ is the sum of the radii of the two species.

The exponential-$t^{\frac{1}{2}}$ term in (35) has been observed in quenching reactions in solution, but only on fast (< 1 ns) timescales [7]. This is an important point, that some deviations from simple exponential decay laws will only be observable if the appropriate experimental timescale is employed. For the restricted rotational motion, Itagaki et al [8] have proposed the decay law should obey a similar functional form,

$$I(t) = A\, \exp[-(at + bt^{\frac{1}{2}}] + B\, \exp(-ct) \tag{38}$$

although the theoretical basis for this is not clear. Decay laws for fluorescence anisotropy, which can be very complex, will not be discussed here.

Energy transfer and migration

This subject has been reviewed extensively, and this discussion will not be amplified here. In the simple case where randomly distributed immobile donors and acceptors are considered at a donor concentration such that donor-donor transfer is neglible, the time-dependence of donor fluorecence is

$$I(t) = \exp[-(t/\tau) - \gamma\, (t/\tau)^{3/s}] \tag{39}$$

where

$$\gamma = (4/3)\pi\, [1 - 3/s]N_A R_0^{\ 3} \tag{40}$$

with $S = 6$ for dipole-dipole transfers and N_A is the number of acceptor molecules / cm^3, R_0 is a constant proportional to the overlap of donor emission and acceptor absorption. In this case of dipole-dipole interaction formulated by Forster, the decay exhibits an exponential decay plus an additional $\exp(-t^{\frac{1}{2}})$ dependence. For exchange energy transfer, the decay rate has a different dependence.

$$I(t) = \exp[-t/\tau - \alpha g(\beta t)] \tag{41}$$

where

$$g(z) = 6z \sum_{M = 0}^{\infty} (-z)^M/[M!(M + 1)^4] \tag{42}$$

and the constants α, β can be related to macroscopic constants as with γ. In both cases the decay is represented by an initial non-exponential component followed at long times by the decay of the unquenched donor.

In cases where energy migration is a dominant feature of luminescence, as in molecular crystals, various forms of decay are expected depending upon circumstances, but relying upon solutions, usually complex, to the basic rate equations where $E(t)$ is the time-

$$\dot{E}(t) = -[k_E + k_L(t)]E(t) \tag{43}$$

$$\dot{T}(t) = -k_T T(t) + k_L(t)E(t) \tag{44}$$

dependent population of the initially excited (exciton) state, $T(t)$ the population of the trap state, k_E the decay rate constant for band states, k_T the decay rate constant for trap states, and $k_L(t)$ the time-dependent trapping rate functions, the form of which depends upon the effective transport topology [10]. For a strictly one-dimensional transport, Fayer has given the form of $k_L(t)$ as

$$k_L(t) = At^{-\frac{1}{2}} \tag{45}$$

For quasi-one-dimensional, two-dimensional, and three-dimensional diffusional processes, other forms are appropriate [10,11-14]. Thus very extensive theoretical and picosecond experimental work on electronic excited state transport in finite volumes of randomly distributed molecules has been reported, which shows that there are significant deviations in the behaviour of finite volume systems compared with the infinite volume systems considered above. The treatment is mathematically complex and the results will not be given here explicitly [12-14]. Frederickson and Franck [15] have used this treatment to suggest possible forms for the decay of monomer and growth and decay of excimer fluorescence in vinyl aromatic polymers where electronic energy migration might be a dominant process. This treatment is presented elsewhere in the volume, and will be discussed briefly below.

Heterogeneity in polymers

Even a polymer sample of narrow molecular weight distribution is just that, a distribution. Since the fluorescence process can be extremely sensitive to the environment of the fluorophore, in principle, a range of environments is being observed even in a non-interacting fluorophore, that is a molecule which does not interact with neighbouring chromophores through excimer formation or energy transfer or migration. This situation however, corresponds to that observed in a free chromophore in solution, and the decay should be modelled adequately by a rate constant. This is certainly not the case for interacting chromophores, where the local environment will be critical in determining the decay rate of any particular fluorophore. In a homopolymer, the principal cause of heterogeneity will be the tacticity of the polymer, isotactic, syndiotactic and atactic polymers being expected to behave very differently, as outlined elsewhere in this volume. In cases where nominally 'atactic' polymers consist of isotactic and syndiotactic sequences, the decay may in favourable simple cases be interpretable in terms of a summation of exponential decays of two kinetically distinct species. For a wide distribution of sites, a kinetic model recognising this heterogeneity may be more appropriate, although this yields information of limited usefulness.

    In copolymers, heterogeneity of environment of a chromophore by virtue of composition becomes of overriding concern.

Motion in polymers

    In a small molecule, rotational motion of a fluorophore in solution occurs on a ps timescale and the observed emission is thus averaged over all molecular orientations. In a synthetic polymer whole chain motion is in general very slow, but local motions may occur on the same timescale as fluorescence, and if these local motions, such as segmental motion, determine the rate of a process leading to emission, then the appropriate form of the rate expression for such motion must be employed, as suggested earlier [8].

Electronic energy transfer, migration in polymers

    Energy migration is thought to occur in many vinyl aromatic polymers, for example, and if this process is rate-determining for any observed fluorescence then the form of the expression used to model the fluorescence grow-in or decay must be correct, as described above.

    What prospect is there for analysing data, and recovering form appropriate for some of the processes identified above? In cases where the system is very carefully chosen to maximise the likelihood of observing a particular form of decay, these might be good. For example, in Figure 3 we show the decay of fluorescence from a crystal of a poly(diacetylene) which is highly ordered [16], leading to one-dimensional exciton diffusion, and with a very short trap decay time such that the form of the expression appropriate for the decay, an exponential $t^{-\frac{1}{3}}$ dependence, is clearly obeyed (Figure 4) [11].

    What of more complex systems, such as dilute solutions of flexible vinyl aromatic polymers?

    In our early work, it became clear that simple Birks kinetic schemes representing excimer formation and decay were inadequate for modelling fluorescence in excimer forming polymers [17-22]. The empirical observation that multiple (dual and triple) exponential terms could model successfully the decay of monomer and excimer fluorescence in poly(vinyl naphthalenes) led us to propose simple models which largely recognised the heterogeneity of the sample, and discriminated between two classes of monomer site, one which could rapidly lead to excimer formation through energy migration and rotational sampling, and one which could not, discussed in the previous article. These findings have been criticised by two groups, who derive, from entirely different reasoning, a monomer decay function of the form in Equation (46). (N.B. Equation (46) is the simplified form given by Frederickson and Franck) [15].

$$I(t) = A \exp[-(at + bt^{\frac{1}{2}}] + B \exp(-ct) \tag{46}$$

It has been shown that functions of the form of Equation (46) can in general simulate (through adjustment of the parameters) some curves of the form observed in empirical monomer and excimer decays. Alternatively, the functional type (47) has been shown to emulate dual

$$I(t) = \exp[-(at + bt^{\frac{1}{2}})] \tag{47}$$

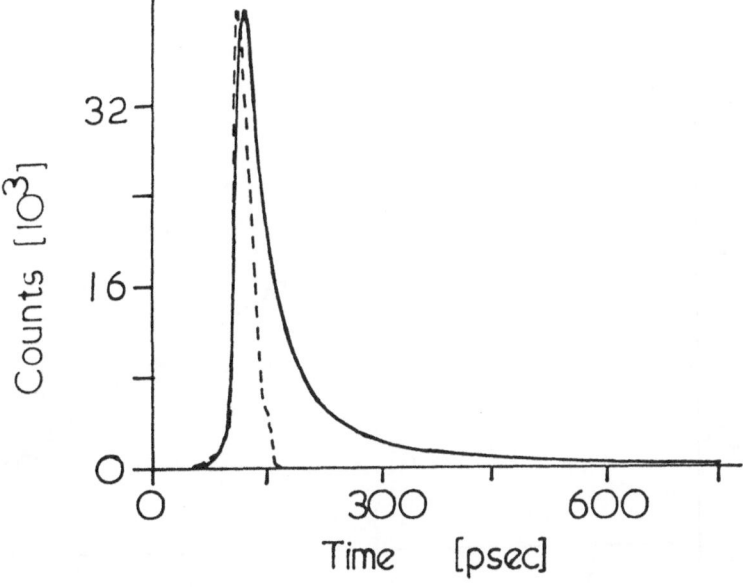

Figure 3. Luminescence decay of PDA-10H at 300 K (full curve). The laser excitation pulse is shown by the dashed curve [16].

exponential components of triple exponential fits, although the procedures adopted by these authors for testing acceptability if fitting are crude.

In the case of our own work on vinyl naphthalenes, the analysis was based upon monomer decay functions only, and is thus not wholly

secure. However, it is of relevance to note that in the copolymer
series studied to date [18,20,22] the two shorter decay components
obtained in triple exponential analyses increase as the basic
homopolymer repeat units are replaced by comonomers which will induce
chain flexibility. It is difficult to rationalise these results with
the trend expected in the presence of rotational diffusional as
described by Equation (47).

In the specific case of styrene polymers, we can test the simplest
alternative model, expression (47), against a multiple exponential
model, where we have shown, Figure 5, 6 that a dual component fit is
acceptable statistically for homopolymer monomer decay. Figure 7, 8
show that the simplified expression (47) is certainly unacceptable.

One clear deficiency in expression (47) is that reverse
dissociation of the excimer is omitted, although there is spectroscopic
evidence that it does occur slightly in poly(styrene), more strongly in
poly(vinyl naphthalenes). Inclusion of an additional exponential term
to account for this phenomenon may well improve the fitting of Equation
(47), but this function will only with difficulty be distinguished
from other simpler models.

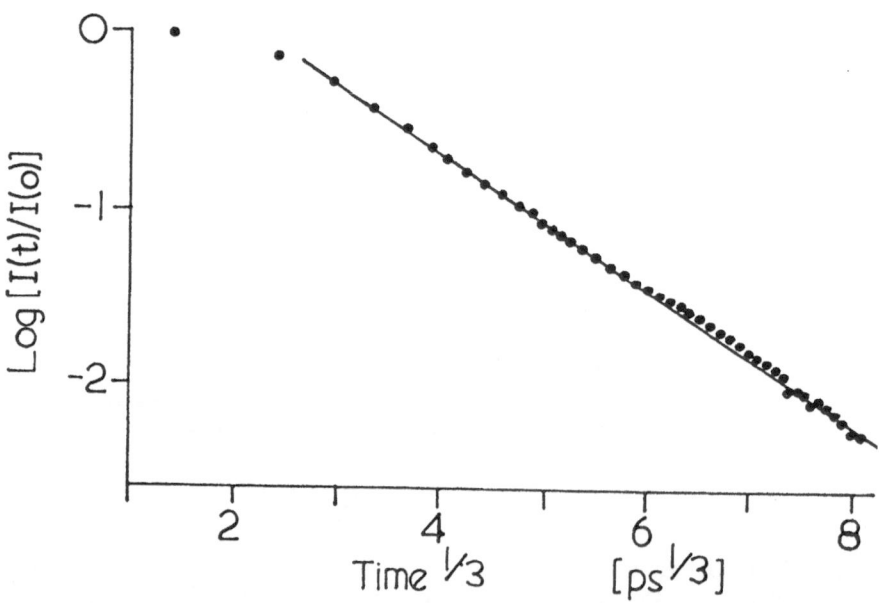

Figure 4. Luminescence decay of PDA-1OH showing $\exp[(\tau/t)^{1/3}]$
behaviour [16].

That simple models suffice in polystyrene polymers is further
illustrated by work we have carried out on polystyrene doped with a

small concentration of copolymerised phenyl oxazole POS [23]. In this work, careful study of the decay characteristics of emission at wavelengths selected to isolate the styrene monomer, styrene excimer, and POS trap, when fitted to an empirical three-component model without constraint, gave the same three decay constants, with of course, different weighing factors in different spectral regions, Figure 9. It seems to us to be inconceivable that the recovery of identical decay parameters could be coincidental. Moreover, the values recovered for decay of monomer correlate well with grow-in of trap. This would not be expected on the basis of a time-dependent trapping function. We feel that in the case of styrene polymers, the case for careful multiple-component fitting and subsequent physical interpretation is reasonable.

Conclusions

We have outlined above the complex nature of fluorescence decay signals expected in any study of the fluorescence of synthetic polymers, the techniques which may be used to analyse critically various models, and shown that under some circumstances, simple models may suffice. We should not underestimate the difficulty of the task

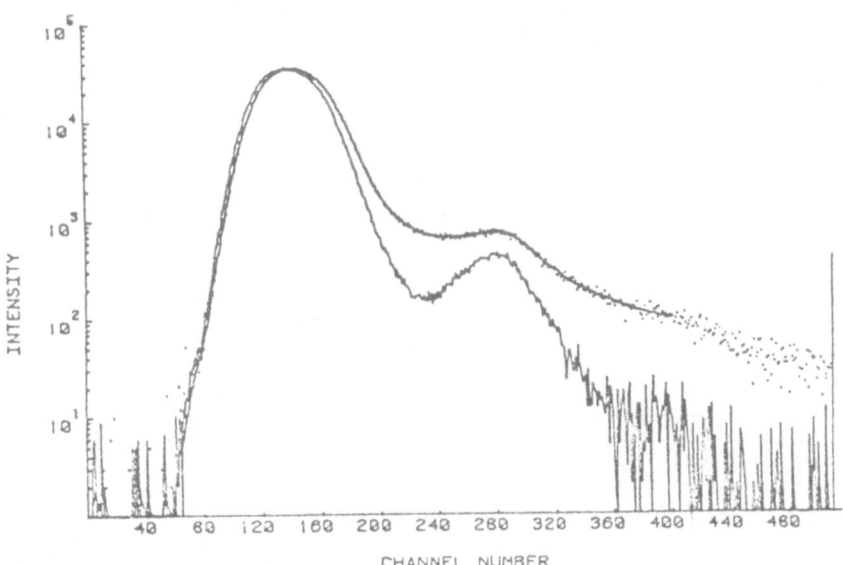

Figure 5. Double exponential fit to polystyrene decay measured at 270 nm with excitation at 257.25 nm.

facing us. One response in the face of this recognition is to abandon all hope of meaningful interpretation of results, a course we reject. The alternative is to attempt critical evaluation of models, and to

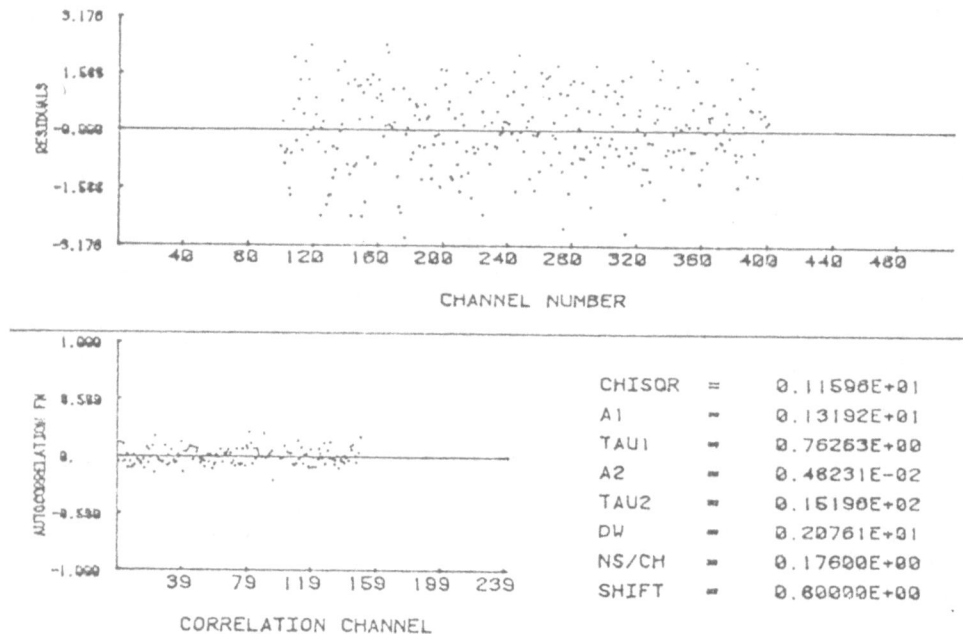

CHISQR   =   0.11596E+01
A1   =   0.13192E+01
TAU1   =   0.76263E+00
A2   =   0.46231E-02
TAU2   =   0.15196E+02
DW   =   0.20761E+01
NS/CH   =   0.17600E+00
SHIFT   =   0.60000E+00

Figure 6.   Residuals and autocorrelation for plots in Figure 5.

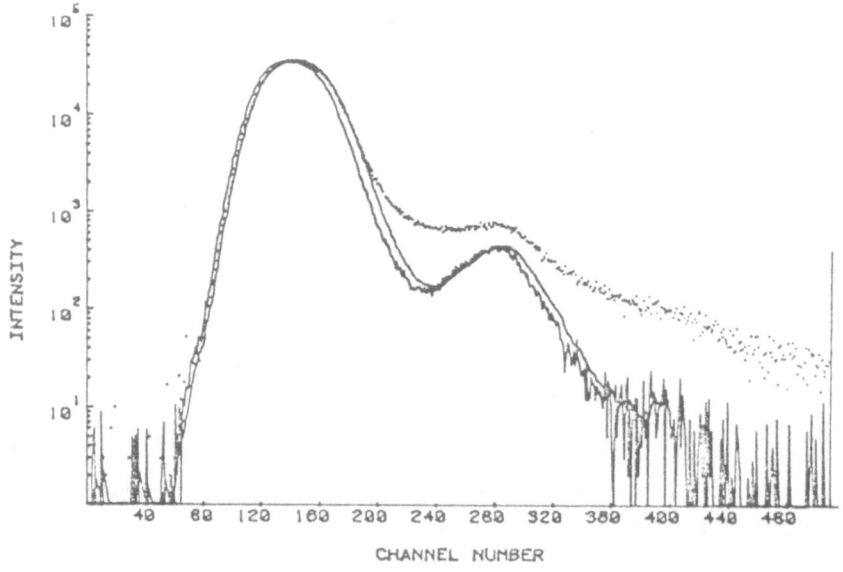

Figure 7.   Fluorescence decay of polystyrene measured at 270 nm fitted to the function: A exp-(bt + ct$^{1/2}$).

progress by devising experiments which will provide a means of
identifying the processes which are rate-determining in any system. It
is axiomatic in using such methods that the <u>simplest</u> model fully
compatible with all results must be selected, since one can <u>always</u>
replace this model by one which is more complex mathemtically. This
however, may defy interpretation in physical terms. A demonstration
that a very complex model is as compatible as a more simple model with
a particular set of data is <u>not</u> in itself cause for abandonment of the
simplest model. The way forwards as always, is to devise experiments
which further test the validity of the simpler model, and at the point
where this can be shown demonstrably, <u>experimentally</u> to be inadequate,
to abandon it in favour of the simplest <u>refined</u> model which is then
fully compatible with the results. This is the approach we have taken
and will continue to take.

It is clear that the origins of the photophysical behaviour of
polymers will continue to be the subject of lively debate for some
time. Activity in a field which has proved to be rapidly expanding and
stimulating during the past decade seems likely to maintain momentum in
the next few years.

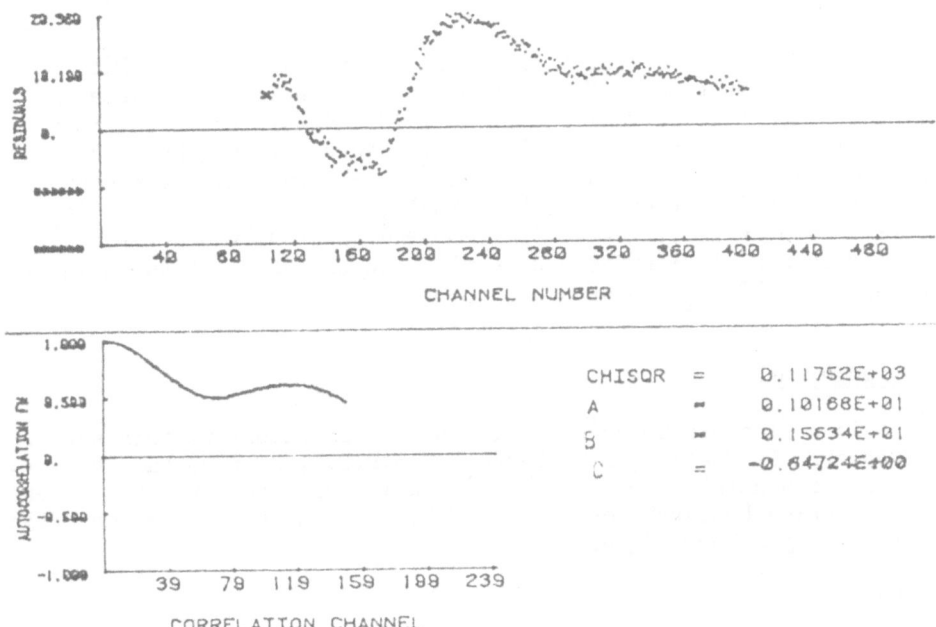

Figure 8. Residuals and autocorrelation function for plots in Figure 7.

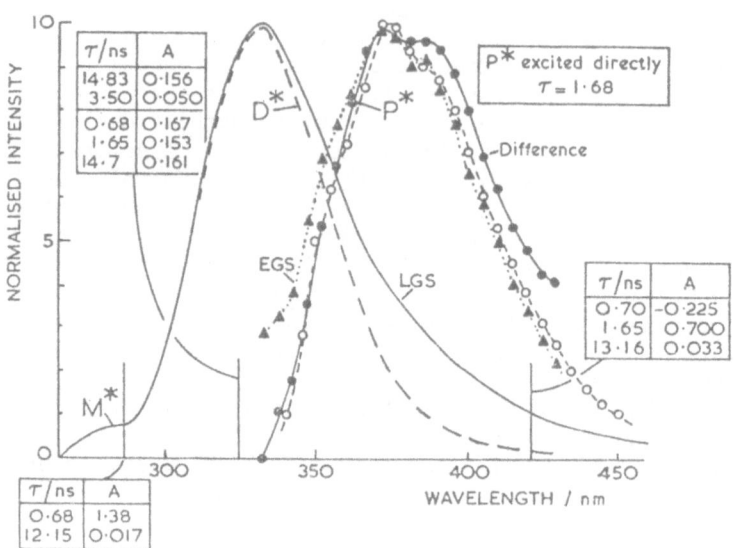

Figure 9. Fluorescence spectra and decay characteristics of POS
containing poly(styrene). M*, styrene monomer region, dual decay
kinetics. D*, styrene excimer region, triple decay characteristics
(double fit shown does not correlate with other wavelengths, thus
meaningless). P* is POS fluorescence, triple decay characteristics
when styrene excited (see box), but single, τ = 1.68 ns when excited
directly. EGS is early-gated time-resolved spectrum which matches
closely spectrum of D* excited directly, and difference between late-
gated spectrum LGS and known spectrum of D*.

Acknowledgements

It is a pleasure to acknowledge the contributions to this work of
A.J. Roberts, G. Rumbles, J-L. Gardette, S.D.D.V. Rughooputh and
R. Drake. Financial support for our work in this area from the Science
and Engineering Research Council and U.S. Army European Research Office
are gratefully acknowledged.

References

1.   'Time-correlated single photon counting', D.V. O'Connor and
     D. Phillips, Academic Press, 1984.

2.   'Non-exponential kinetics', D. Phillips in "Excited state probes
     in biochemistry and biology". Eds., A.G. Szabo and L. Masotti,
     Plenum Press, (in press).

3.   W.J. Albery, P.N. Barlett, C.P. Wilde and J.R. Darwent, J. Amer.
     Chem. Soc., 1985, 107, 1854.

4. A.J. Roberts, D.V. O'Connor and D. Phillips, Ann. N.Y. Acad. Sci., 1981, 366, 109.

5. J.R. Knutson, J.M. Beecham and L. Brand, Chem. Phys. Letters, 1983, 102, 501.

6. D.V. O'Connor, L. Chewter and D. Phillips, J. Phys. Chem., 1982, 86, 3400.

7. T.L. Nemzek and W.R. Ware, J. Chem. Phys., 1975, 62, 477.

8. H. Itagaki, K. Horie and I. Mita, Macromolecules, 1983, 16, 1395.

9. A. Kawski, Photochem. Photobiol., 1983, 38, 487.

10. 'Exciton coherence', M.D. Fayer in "Spectroscopy and excitation dynamics of condensed molecular systems". Eds., V.M. Agranovich and R.M. Hochstrasser, North-Holland, Amsterdam, 1983, p. 185.

11. I.G. Hunt, D. Bloor and B. Movaghar, J. Phys. C Solid State Phys., 1983, 16, L623.

12. C.R. Gochanour, H.C. Anderson and M.D. Fayer, J. Chem. Phys., 1979, 70, 4254.

13. M.D. Ediger and M.D. Fayer, J. Chem. Phys., 1983, 78, 2578.

14. R.J. Dwayne Miller, M. Pierre and M.D. Fayer, J. Chem. Phys., 1983, 78, 5138.

15. G.H. Frederickson and C.W. Frank, Macromolecules, 1983, 16, 572.

16. 'Luminescence of a fully polymerised poly(diacetylene)', D. Bloor, S.D.D.V. Rughooputh, D. Phillips, W. Hayes and K.S. Wong in "Proceedings of Winter School on electronic properties of polymers and related compounds". Eds., H. Kuzmany and S. Roth, Springer Verlag, (in press).

17. A.J. Roberts, I. Soutar and D. Phillips, J. Polym. Sci., (Polymer Letters, 1980, 18, 123.

18. D. Phillips, A.J. Roberts and I. Soutar, J. Polym. Sci. Polymer Physics, 1980, 18, 2401.

19. D. Phillips, A.J. Roberts and I. Soutar, Polymer, 1981, 22, 293.

20. D. Phillips, A.J. Roberts and I. Soutar, Eur. Polym. J., 1981, 17, 101.

21. D. Phillips, A.J. Roberts and I. Soutar, Polymer, 1981, 22, 427.

22. D. Phillips, A.J. Roberts and I. Soutar, J. Polym. Sci., (Polymer Physics), 1982, 20, 411.

23. D. Phillips, A.J. Roberts and I. Soutar, Macromolecules, 1983, 16, 1593.

POLYMER THEORY: A CONSPECTUS OF SEVERAL RECENT ADVANCES

Eugene Helfand
AT&T Bell Laboratories
Murray Hill, NJ 07922
USA

ABSTRACT. Some of the fundamental polymer theory underlying many phenomena currently being investigated by photophysical and photochemical techniques is reviewed.

1. **Equilibrium Properties of Dilute and Semi-dilute Solutions.** Excluded volume causes a polymer to favor expanded states. Large fluctuations, analogous to critical fluctuations, lead to non-ideal power law relations between properties. In the semi-dilute regime excluded volume correlations are screened. Scaling laws relate dilute and semi-dilute exponents.

2. **Relaxation of Single and of Entangled Macromolecules.** In the absence of hydrodynamic interactions (HI) the normal modes of a polymer are Rouse modes, which act as overdamped harmonic oscillators. With HI the Rouse modes are still nearly normal modes, but the relaxation spectrum is modified. The HI are screened in semidilute solutions. At higher concentrations and in bulk disentanglement by reptation and tube renewal dominate slow viscoelastic processes.

3. **Dynamics of Conformational Transitions.** The fastest backbone relaxations in polymers are conformational transitions. These must proceed along a reaction coordinate which is a localized mode. This frequently leads to cooperative, cranklike pairs of transitions and nonexponential correlation functions. Simulations of chains undergoing conformational transitions will be described.

1. EQUILIBRIUM PROPERTIES OF DILUTE AND SEMI-DILUTE SOLUTIONS

1.1 Questions to be Addressed

Some of the questions to be addressed in Section 1 are: i) what are excluded volume and the Θ point; ii) why are polymer coils expanded due to excluded volume, and what limits that swelling; iii) what is meant by a semi-dilute solution; iv) what is screening; v) how do screening distance, radius of gyration, and osmotic pressure vary with concentration in the semi-dilute regime; vi) what are scaling laws and how are they used; and vii) what are blobs?

1.2 Excluded Volume

A polymer molecule may be regarded as a chain of N beads. Each represents, let us say, a monomer, in which case N would be the degree of polymerization. We will restrict our attention to linear molecules. In the absence of interactions between the beads, other than short range interactions between those beads that are nearby in the chain sequence, the root-mean-squared end-to-end distance, R, goes like that of a random walk

M. A. Winnik (ed.), Photophysical and Photochemical Tools in Polymer Science, 151–191.
© 1986 by D. Reidel Publishing Company.

$$R \equiv \langle |\mathbf{r}_N - \mathbf{r}_1|^2 \rangle^{1/2} = aN^{1/2}. \tag{1.1}$$

This defines an effective length a for a monomer. In general we will pay little attention to dimensionless constants of order one, so that a may be taken as a characteristic of all monomeric distances: monomer length $\approx a$, concentration of monomer in the bulk $\approx a^{-3}$, range of interactions $\approx a$ (away from Θ conditions), etc. Thus, the radius of gyration is known to be $(1/6)\, aN^{1/2}$, but we will usually write $R_G \approx aN^{1/2}$.

However, a polymer molecule generally exhibits "excluded volume" interactions between monomers remote along the chain, when the monomers come close to each other. A simple model illustrating such forces is a chain of hard spheres. More realistically, the effective interactions of the monomers take into account a variety of effects, not literally "exclusions." The monomers exhibit both repulsive and attractive forces at different distances, as most chemical entities do. Even more significantly, if two monomers move apart the space tends to be filled by solvent, so what is important is the polymer-solvent interactions relative to those of monomer-monomer and solvent-solvent. This difference is embodied in the mixing parameter χ. Even in the absence of potential energy differences between monomers and solvents, the solvent creates an effective repulsion between monomers as it attempts to make its density, and that of the monomers, uniform over all space. This may be termed an osmotic force. Although the underlying excluded volume potential is this complex of phenomena, qualitatively the effect can be parameterized by a single variable, which plays a role similar to the second virial coefficient in a gas. This parameter is frequently called v, with the dimensions of a volume (like a gas second virial coefficient). In terms of the more familiar parameters

$$v \approx a^3 (\tfrac{1}{2} - \chi). \tag{1.2}$$

The contribution to the chemical potential of a monomer (i.e. the work of adding a monomer at a point $\mathbf{r}$), related to the interactions, may be taken as

$$kT\, vc\,(\mathbf{r}), \tag{1.3}$$

where $c(\mathbf{r})$ is the concentration (number of monomers/unit volume). This neglects higher virial coefficients and powers of concentration. (There are also terms proportional to concentration gradients arising from the fact that the monomer to be placed at $\mathbf{r}$ interacts with entities not at the same, but at neighboring points. These terms are minor and will be ignored.)

1.3 Distance Scales for a Dilute Polymer in Good Solvent

In the presence of excluded volume (net repulsion, i.e. $\chi < \tfrac{1}{2}$) a polymer coil will favor those configurations that are more diffuse (expanded, swollen). From the point of view of eq. 1.3, the effective energy will be lower if the concentration is lower. However, what is usually described in polymer science as elastic forces oppose the swelling, limiting it. The number of configurational states available to a swollen polymer is lower than in an unswollen state, giving rise to the entropic force responsible for rubber elasticity, and to resistance to swelling.

Flory used these ideas to obtain a prediction for the excluded volume swelling. He wrote an estimate, as follows, for the free energy as a function of the characteristic polymer size, say the radius of gyration, R_G. Let the concentration in the coil be estimated as N/R_G^3 out to the radius R_G, and zero beyond. This produces an excluded volume (free) energy for the whole molecule of $NkTv(N/R_G^3)$, where a multiplicative constant of O(1) has been neglected. The elastic free energy is taken as that required to stretch the ends to a distance of $O(R_G)$, namely, $kT R_G^2/(Na^2)$. The size R_G that minimizes the free energy (constants ignored)

$$kT \left[\frac{vN^2}{R_G^3} + \frac{R_G^2}{Na^2} \right] \tag{1.4}$$

is

$$R_G \approx a \left(\frac{v}{a^3} \right)^{1/5} N^{3/5} \tag{1.5}$$

— the famous 3/5 law of Flory. Except near the Θ point one has $v \approx a^3$ so we will usually write

$$R_G \approx aN^{3/5}. \tag{1.6}$$

This law is not precisely right, and the mean field type of arguments employed are not at all valid.[2] One now knows that the polymer molecule has a structure with large fluctuations, whose most important characteristic is their self similarity under changes of scale as the polymer grows very large. This is the type of structure, appearing near critical points, that we have learned to handle by renormalization group techniques. In general one can write

$$R_G \approx aN^\nu, \tag{1.7}$$

with the best estimates for ν being $\nu = 0.588 \pm 0.0015$.[3] We follow common practice, henceforth, and use $\nu = 3/5$, although more precise exponents for all properties to be discussed can be derived easily in terms of the proper ν. Without going into fundamental renormalization group arguments in this lecture, we will explore correct and useful ways of thinking about excluded volume effects. We follow lines suggested and developed primarily by de Gennes,[4] Edwards,[5] and their coworkers. Our ambition will be considerably less than that of Flory or those doing renormalization group calculations, i.e., we will not be attempting to derive a value of ν. Rather we will be pursuing the consequences of the modification of chain statistics caused by the excluded volume, as embodied in eq. 1.7.

1.4 Density Correlations

One useful means of gaining insight into the structure of a polymer with excluded volume is to consider the average density pattern about a given monomer, m. Consider a sphere of radius r about m. Within this sphere one expects to find an average $n_r \approx (r/a)^{1/\nu}$ monomers. The concentration at a distance r from the given monomer, which we will call $c_m(r)$, is the number of monomers in a spherical shell about the central monomer,

$$dn_r \equiv c_m(r)\,(4\pi r^2\,dr), \tag{1.8}$$

so that for $a \ll r \ll R_G$

$$c_m(r) \approx \frac{1}{a^3}\,\left(\frac{a}{r}\right)^{3-(1/\nu)} \approx \frac{1}{a^3}\,\left(\frac{a}{r}\right)^{4/3}. \tag{1.9}$$

[Exercise: Confirm that this $c_m(r)$ implies a scattering factor (Fourier transform) $S(q) \sim q^{5/3}$ for $qR_G \gg 1$.] In Fig. 1 this result is schematically plotted. The more rapid $1/r^{4/3}$ falloff in good solvent, compared to the Debye $1/r$ distribution for ideal chains ($\nu = \frac{1}{2}$ in eq 1.9), is a manifestation of swelling occurring on all scales of $n \gg 1$ in good solvents.

For $r \geq R_G$ the above argument breaks down because it would not make sense to use a $n_r \geq N$ in the above argument. One expects that $c_m(r)$ will fall off much more rapidly for $r \gg R_G$ as Fig. 1 shows. R_G may be regarded as a limiting distance scale for excluded volume behavior.

We have seen that at a distance r from a monomer the monomers within a range $n \approx (a/r)^{1/\nu}$ of m along the chain make the major contribution to c_m. On occasion, especially for photophysical processes where two monomers must come to proximity, one needs to know the behavior of a more detailed correlation function, $p(r,n)$, the probability that two monomer units n monomers apart on the chain are separated in space by a vector $\mathbf{r}$. For an ideal chain this is the well-known Gaussian distribution

$$p_{ideal}(r,n) = \left[\frac{3}{2\pi na^2}\right]^{3/2}\,\exp\left[-\frac{3r^2}{2na^2}\right]. \tag{1.10}$$

In good solvent we expect r to enter on a scale of an^{ν}; i.e. the properly scaled r variable is r/an^{ν}, so

$$p(r,n) = \left[\frac{1}{an^{\nu}}\right]^3\,f_p\left[\frac{r}{an^{\nu}}\right]. \tag{1.11}$$

The prefactor of the function $f_p(x)$ is necessary for normalization on integration over all space. The derivation of the functional form of f_p requires critical phenomena arguments beyond the scope of this lecture, but the result is[7]

$$f_p(x) \approx \begin{cases} x^{\theta} & x \ll 1, \\[2mm] x^{\sigma}\exp(-\kappa x^{\tau}), & x \gg 1, \end{cases} \tag{1.12}$$

with $\theta \approx 0.27$, $\sigma \approx 1.03$, and $\tau \approx 2.42$. Besides being related to ν, these exponents depend on a second exponent. It is important to note that $p(r,n) \to 0$ as $r \to 0$, unlike the ideal case where $p_{ideal}(r,n)$ goes to a nonzero function of n. This (related to the appearance of a second exponent) is a manifestation of excluded volume effects going beyond the picture of a mere coil swelling. [Justify the formula

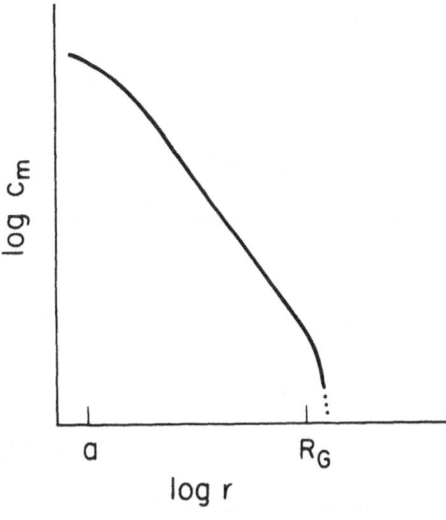

Fig. 1. Average concentration distribution, $c_m(r)$, about a point where one monomer is known to be located (schematic plot on a log-log scale). The $(a/r)^{4/3}$ law in good solvent applies out to a distance $\approx R_G$. There is a faster fall off thereafter due to an exhaustion of monomers from the same molecule.

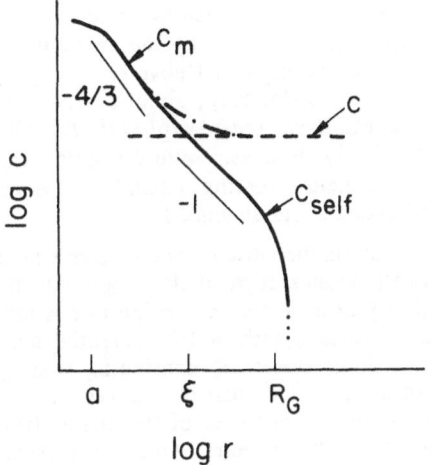

Fig. 2. Schematic plot of $c_m(r)$ in semidilute solution. For $r < \xi$ monomers of the same molecule outnumber monomers of other molecules, and the excluded volume correlations are as in dilute solution (unscreened). For $r > \xi$ other molecules' monomers are dominant and screening occurs. The total concentration adjusts rapidly to the background value. The concentration from the same molecule, $c_{self}(r)$, is higher than in dilute solution, as in an ideal chain of blobs.

$$c_m(r) \approx \int_1^N dn\, p(r,n)$$

and use this formula to derive eq. 1.9. Consider the results for $r \approx a$ and $r \geq R_G$.]

1.5 Semi-dilute Solutions

So far we have been considering the properties of individual polymer molecules. As the concentration increases, well above a value called c^*, the molecules begin to overlap significantly. The value of c^* is such that there is one polymer molecule (N monomers) for a volume equal to the radius of gyration cubed

$$c^* \approx \frac{N}{R_G^3} \approx \frac{1}{a^3 N^{3\nu-1}} \approx \frac{1}{a^3 N^{4/5}}. \tag{1.13}$$

In terms of volume fraction polymer, approximated by $\phi \approx ca^3$, one can write

$$\phi^* \approx 1/N^{4/5}. \tag{1.14}$$

By this criterion a polymer with degree of polymerization 5000 has $\phi^* \approx 0.001$. With $\phi \gg \phi^*$ the system is concentrated (highly overlapping) in molecules; but at the same time, with $\phi \ll 1$, highly dilute in monomers. The solution under such conditions is called semi-dilute.

1.6 Screening

Under conditions of significant molecular overlap an important screening effect occurs. Screening is the mitigation of the effect of an interaction brought about by the medium (in the present case, by a correlated adjustment). A classical example is Debye screening in electrolytic solutions, where correlations reflect not the bare field, ze/r, about an ion, but the field of the ion and the distorted cloud of counterions, $(ze/r)\exp(-r/\xi_D)$. The screening distance, ξ_D, depends on concentration. Similarly, in a semi-dilute solution the medium of other polymer molecules can screen the field creating excluded volume correlations, the most significant screening occurring beyond some distance ξ.[5]

To achieve a deeper understanding of screening let us further discuss the factors which enter into the creation of the pattern of density about the monomer m at the origin. In the presence of the excluded volume potential the probability of a monomer j being at $\mathbf{r}$ is less than in the absence of excluded volume not because j interacts with m (the potential may be of shorter range than r), but because a monomer k of the chain continuing from j might overlap m. Furthermore the density pattern of k around j must be like the pattern of j around m. These patterns must take into account the connectivity of the chains (the elastic forces), and allow for the entropy associated with fluctuations. One has learned from the study of critical phenomena that the resulting states have, statistically, certain invariance properties with respect to changes of distance scale, but we do not want to delve into renormalization group techniques here.

Now, let us see what happens in the presence of other molecules at a concentration of monomer $c \gg c^*$. In Fig. 2 we show the $1/r^{4/3}$ pattern $c_m(r)$ of an isolated molecule and the asymptotic concentration c. There is a characteristic distance scale ξ for which the two are about equal: $c \approx (1/a)^3 (a/\xi)^{4/3}$ or

$$\xi \approx a\phi^{-3/4}. \tag{1.15}$$

For distance $r \ll \xi$ most of the monomers contributing to $c_m(r)$ are from the same molecule as the one at the origin, and these adjust their density pattern as in dilute solution. Conversely, for $r \gg \xi$, the monomer j at $\mathbf{r}$ is more likely to belong to another molecule. Therefore it is more important that these molecules move away, which they do with relative ease, than that the central molecule undergoes a long range swelling. The adjustment of the medium is done in such a way as to quickly create an overall pattern $c_m(r) \approx c$ for $r > \xi$. (Show that for $c \approx c^*$, one has $\xi \approx R_G$.)

The value of $c_m(r)$ is composed of two parts:

$$c_m(r) = c_{self}(r) + c_{other}(r) \approx c, \qquad r \gg \xi; \tag{1.16}$$

c_{self} is the concentration from monomers from the same polymer; c_{other} from other molecules. For $c_{self}(r)$ one expects to have the Debye $1/r$ form for $r \gg \xi$ to minimize elastic effects now that excluded volume is screened; but it must also reflect, self-consistently, the short-range, unscreened structure of the molecule, as we will next see.

1.7 Blobs

A useful qualitative way of looking at the question of what c_{self} is, as well as addressing many other problems in semi-dilute systems, is in terms of blobs, a term introduced by de Gennes.[4] Let us regard the bonded units, expected to be within distance ξ of each other, as a blob. The number of monomers in a blob, g, is given by the distance-number relation of excluded volume theory, since these units are not shielded from excluded volume forces:

$$g \approx (\xi/a)^{5/3} \approx \phi^{5/4}. \tag{1.17}$$

(Show that the average distance between blobs $\approx$ their size ξ, so that they are nearly close-packed.) The polymer molecules, under semi-dilute conditions, may be considered as a freely-jointed chain of noninteracting blobs. Thus the radius of gyration of a molecule is the length of a blob, times the square root of the number of blobs in a chain:

$$R_G \approx \xi (N/g)^{1/2} \approx aN^{1/2}/\phi^{1/8}. \tag{1.18}$$

The c_{self} follows the Debye law, being the number of monomers per blob times the self-correlation of blobs

$$c_{self}(r) \approx g \, \frac{1}{\xi^3} \, (\frac{\xi}{r}) \approx \frac{\phi^{1/4}}{a^3} \, (\frac{a}{r}), \qquad r \gg \xi. \tag{1.19}$$

The difference, $c_{other}(r) \approx c - c_{self}(r)$, can be measured in a neutron scattering experiment on molecules partially labelled in a proper fashion. Its deviation from c is a manifestation of the so-called correlation hole, usually discussed for bulk polymers.

1.8 Scaling Laws

We turn now to the technique of scaling laws, which is so very powerful in predicting the properties of polymer systems.[4] As an illustration, let us consider once more the dependence of scaling distance on concentration in a semi-dilute solution with good solvent.

We noted that in a dilute system the correlations of concentration fall off rapidly beyond a distance of $O(R_G)$, so R_G may be regarded as a screening length. In semi-dilute solutions the screening length is a function of concentration, with the concentration scale being set by c^*, the overlap concentration. This is because $c \approx c^*$ is when we expect behavior to change. Thus, we may write ξ as the dilute value times a function of ϕ / ϕ^*:

$$\xi = aN^{3/5} f_\xi(\phi/\phi^*), \qquad (1.20)$$

$$= aN^{3/5} f_\xi(\phi N^{4/5}). \qquad (1.21)$$

For small x one takes $f_\xi(x) \approx 1$. For $\phi \gg \phi^*$ (but still dilute in monomer, $\phi \ll 1$) we hypothesize that $f_\xi(x) \sim x^{p_\xi}$ (the power law form is another consequence of renormalization group invariance). Thus, for semi-dilute concentrations,

$$\xi \approx 2\phi^{p_\xi} N^{(3/5) + (4/5) p_\xi}. \qquad (1.22)$$

But for $\phi \gg \phi^*$, the screening distance should be independent of N, so $p_\xi = -3/4$, and

$$\xi \approx a\phi^{-3/4} \qquad (1.23)$$

as in eq 1.15.

The reader will learn best about the application of scaling laws by using them, and so we pose some exercises:

i) From the fact that $R_G \approx aN^{3/5}$ for $c \ll c^*$ and $R_G \propto N^{1/2}$ for $c \gg c^*$ rederive eq. 1.18.

ii) From $g \approx N$ for $c \ll c^*$ derive eq. 1.17 directly.

iii) From $c_{self} \approx a^{-3}(a/r)^{4/3}$ for $c \ll c^*$, and $c_{self} \propto r^{-1}$ for $c \gg c^*$ rederive eq. 1.19.

Scaling laws serve as a means of relating behavior in one range of parameters to behavior in another. Examples are $c \ll c^*$ going to $c \gg c^*$; $r \ll \xi$ going to $r \gg \xi$; $v \gg N^{-1/2}$ going to $v \ll N^{-1/2}$; or $v \gg \phi$ going to $v \ll \phi$. (Show that the third of these follows from eq. 1.5, and the fact that $R_G \approx aN^{1/2} v^0$ for small v.) The full laws governing behavior in the transition region are embodied in the functions, $f(x)$, for $x \approx 1$, not just the asymptotic behaviors of f. These functions are not as easily obtained from theory, although attention is being devoted to the subject. We emphasize that the asymptotic nature of most scaling laws (power laws) must be kept in mind, and the applicability of asymptotic forms vigilantly checked.

1.9 Osmotic Pressure

As a final topic we shall use scaling law arguments to derive the concentration dependence of the osmotic pressure for polymer in a good solvent. The virial expansion for the osmotic pressure is a power series in polymer concentration

$$\frac{\pi}{kT} = \frac{c}{N} + A_2 c^2 + \cdots, \tag{1.24}$$

$$= c_P + A_{2P} c_P^2 + \cdots. \tag{1.25}$$

The second form, in terms of polymer molecule concentration, $c_P = c/N$, is less conventional but physically revealing. The first (ideal) term arises from the number of ways of placing the molecules, while the second is the leading contribution from interactions. Since c^* serves as a measure of where intermolecular interactions crossover from being unimportant to being important, we write, more generally, that the osmotic pressure is the dilute limit times a function of c/c^*,

$$\frac{\pi}{kT} = \frac{c}{N} f_\pi (\frac{c}{c^*}) = \frac{c}{N} f_\pi (ca^3 \, N^{4/5}), \tag{1.26}$$

valid away from Θ conditions. To apply a scaling argument we assume that for c/c^* large, $f_\pi(x) \sim x^{p_\pi}$. In the semi-dilute regime π should be independent of N, so $p_\pi = 5/4$ and

$$\pi a^3 / kT \approx \phi^{9/4} \tag{1.27}$$

as was shown by des Cloiseaux.[8]

In the dilute regime we may assume that $f_\pi(x)$ is expandable so that

$$\frac{\pi}{kT} = \frac{c}{N} + \frac{\alpha_2 c^2 a^3}{N^{1/5}} + ..., \tag{1.28}$$

$$= c_P + \alpha_2 (N^{3/5} a)^3 c_P^2 + ... \tag{1.29}$$

The second form indicates that in dilute solutions (we repeat, away from Θ conditions) the molecules interact strongly with each other (interaction energy $\geqslant kT$) for separations of order R_G. Thus the second virial coefficient goes like R_G^3 [and α_2 is a dimensionless constant of O(1)].

1.10 Concluding Remarks

Time precludes a discussion of the very interesting, but more complex, behavior of polymer systems as Θ conditions are approached. This is achieved either by varying solvent quality (e.g. by varying temperature or solvent composition), or by increasing concentration to the point where the screening distance loses concentration dependence as it approaches the characteristic monomer distance, a.

Indeed, there is a myriad of important and fascinating equilibrium polymer topics for which significant insight has been gained recently — phase behavior, blends, block copolymers, mesomorphic phases, surfaces and interphases, to name just a few. This brief lecture can serve, at most, as a springboard to wider ranging and deeper study, such as with de Gennes' book,[4] various reviews,[5,6] and the literature.

In this lecture we have examined how polymer molecules assume a swollen state to avoid monomer-monomer interactions. Another way of looking at this is to consider an ensemble of every polymer configuration and eliminate (or give lower Boltzmann weighting to) those with monomer-monomer overlap. The remaining conformations will be the more expanded ones. One manifestation is that the concentration of a chain about a given monomer falls off like $c_m \propto 1/r^{4/3}$, rather than the $1/r$ of an ideal polymer. When molecules overlap significantly, for volume fraction $\phi^* \ll \phi \ll 1$, with $\phi^* \approx N^{-4/5}$, the system is semi-dilute. In a semi-dilute solution excluded volume correlations are screened beyond the distance where $c_m \approx c$. This screening distance is given by $\xi \propto \phi^{-3/4}$. A useful description of semi-dilute solutions may be achieved by picturing the molecules as ideal chains of blobs. Scaling laws are valuable for taking properties known for one range of parameters and extending them to another range. The technique is illustrated by deriving several semi-dilute behavior laws from the dilute behaviors. In particular the semi-dilute osmotic pressure equation $\pi/kT \approx a^{-3}\phi^{9/4}$ is derived in this way.

We have provided answers to the questions posed in Section 1.1. We hope that this will enable the reader to understand other explanations of polymer solution properties. As always, the ultimate test is whether the reader can pose new questions.

2. RELAXATION OF SINGLE AND OF ENTANGLED MACROMOLECULES

2.1 Introduction

Polymeric systems exhibit relaxations which have an extremely broad range of characteristic times, from nanoseconds to seconds, and in some cases hours. The most significant fast relaxation of the backbone is related to conformational transitions, which will be discussed in Section 3. Slower relaxations involve a distance scale where the motion of individual atoms and bonds is not evident.

Such slow dynamics is important for rheological properties. While coarse-scaled variables are changing little the finer-scaled variables come to a constrained, or local equilibrium. The resulting free energy, a function of the coarse variables' values, serves as a potential for motion. In ideal polymers with no excluded volume effects (all time permits us to treat here) the forces are linear in the stretch of the chain. These are the usual elastic forces, entropic in origin. The fluctuations in the rapidly relaxing degrees of freedom, including the solvent, create friction and random forces, so that the slow variables undergo stochastic motion. In this section we will examine the resulting slow relaxations in dilute solution, and also at higher concentrations where entanglement effects dominate.

2.2 Rouse Model

Rouse provided a pictorial representation of the above concepts with the bead-spring model.[9,10] Consider the polymer as a sequence of beads, $j = 1, \cdots, N$, connected by harmonic bonds of characteristic (root-mean-squared) size a_B. The beads can be thought of as collections of monomers, the position variables, r_j, representing the center of the jth group of monomers. One of the questions we will later address is the conditions under which properties are independent of the size of this grouping. (For many purposes it is best to assume immediately this independence and go to the model of a polymer as an elastic string,[4] but we will retain the discrete interpretation for a while.) As a potential between neighboring beads we employ

$$\tfrac{1}{2} \, \kappa \, |\mathbf{r}_{j+1} - \mathbf{r}_j|^2 \, ,$$

with $\kappa = 3 \, kT/a_B^2$ the elastic force constant for ideal chain statistics.

The dynamics is governed by Langevin's equations of motion

$$m_B \frac{d^2 \mathbf{r}_j}{dt^2} = - \zeta_B \frac{d \mathbf{r}_j}{dt} + \kappa \, [(\mathbf{r}_{j+1} - \mathbf{r}_j) - (\mathbf{r}_j - \mathbf{r}_{j-1})] + \mathbf{F}_j^{(R)}(t) \, , \tag{2.1}$$

which says the acceleration equals the sum of the frictional, spring, and random forces. (For the end beads one spring force term is to be omitted.) In the high friction limit, appropriate to polymers, the momentum rapidly relaxes, so one can neglect the acceleration and write this equation as a force balance (divide by ζ_B)

$$\frac{d \mathbf{r}_j}{dt} = - \frac{\kappa}{\zeta_B} \sum_{\ell} A_{j\ell} \mathbf{r}_{\ell} + \mathbf{B}_j(t) \, . \tag{2.2}$$

where $\mathbf{B}_j(t)$ is a random velocity resulting from the random force. The matrix

$$\mathbf{A} = \begin{pmatrix} 1 & -1 & 0 & & & \\ -1 & 2 & -1 & & & \\ 0 & -1 & 2 & & & \\ & & & \ddots & & \\ & & & & 2 & -1 \\ & & & & -1 & 1 \end{pmatrix} \tag{2.3}$$

describes the connectivity of the chain. The eigenvectors of $\mathbf{A}$, labeled with $p = 0, 1, \ldots, N-1$, are

$$\phi_{jp} = \left(\frac{2}{N}\right)^{\frac{1}{2}} \cos \frac{p(j - \frac{1}{2})\pi}{N} \, , \quad p \neq 0,$$

$$\phi_{jo} = (1/N)^{\frac{1}{2}} \, , \tag{2.4}$$

corresponding to eigenvalues

$$\lambda_p = 4 \sin^2 \frac{p\pi}{2N} \, , \tag{2.5}$$

$$\simeq \frac{p^2 \pi^2}{N^2} \, , \quad \text{for } \frac{p}{N} \ll 1 \, . \tag{2.6}$$

The coupled equations of motion can be decoupled by going to normal modes. These variables, the Rouse modes, are defined by

$$\tilde{r}_{p\alpha}(t) = \sum_{j=1}^{N} r_{j\alpha}(t)\, \phi_{jp}\,, \quad \alpha = x,y,z. \tag{2.7}$$

One can also say that the motion of any bead can be decomposed into a linear combination of Rouse mode motions

$$\mathbf{r}_j(t) = \sum_{p=1}^{N-1} \phi_{jp}\, \tilde{\mathbf{r}}_p(t)\,. \tag{2.8}$$

The Rouse coordinates are essentially the Fourier components of the spatial position of the beads as a function of position, j, along the chain (what we will call chain space). The pth mode has a "wavelength" of $2N/p$ beads in chain space. [Exercises: 1) An 8 bead polymer has beads at the points $r_{jx} = -R\cos\left[\left(j-\frac{1}{2}\right)\pi/8\right]$, $r_{jy} = R\sin\left[\left(j-\frac{1}{2}\right)\pi/8\right]$, $r_{jz} = 0$. Draw the molecule and determine the normal coordinates. 2) The normal coordinate $\tilde{r}_{5x}$ is increased by Δ in problem 1. Resketch the molecule. 3) A polymer molecule has normal coordinates $\tilde{\mathbf{r}}_p$. The qth normal coordinate is increased by $\Delta\tilde{\mathbf{r}}_q$. Prove that the change of end to end distance is zero for q even, and $-2\Delta\tilde{\mathbf{r}}_q\cos q\pi/2N$ for q odd.] From eq. 2.2 (multiply eq. 2.2 by 2.4 and sum on j) it follows that the Rouse coordinates act as independent Brownian oscillators; i.e., their dynamics is governed by

$$\frac{d\tilde{\mathbf{r}}_p}{dt} = -\frac{1}{\tau_p}\tilde{\mathbf{r}}_p + \mathbf{B}_p\,, \tag{2.9}$$

with $\mathbf{B}_p$ a noise term, and

$$\tau_p = \frac{\zeta_B}{\kappa\lambda_p}\,, \tag{2.10}$$

$$\simeq \frac{\zeta_B N^2}{\pi^2 \kappa p^2}\,. \tag{2.11}$$

We skip the steps whereby the stress modulus (that part attributable to the polymer) is written in terms of the normal modes and averaged, and content ourselves with the resulting formula

$$G(t) \simeq kT\, c_P \sum_{p=1}^{N-1} \exp(-p^2 t/\tau_R)\,, \tag{2.12}$$

with c_P the (number) concentration of polymer molecules, and

$$\tau_B \equiv \zeta_B/\pi^2\kappa = \zeta_B a_B^2/3\pi^2 kT \approx \frac{a_B^2}{(kT/\zeta_B)} ,\qquad (2.13)$$

$$\tau_R = N^2\tau_B = 2(N_0\zeta) R_G^2/\pi^2 kT .\qquad (2.14)$$

N_0 is the degree of polymerization and ζ is the monomer friction constant, so $N_0\zeta$ is the total friction of the polymer. Note that τ_R is independent of the definition of the Rouse bead size.

We next wish to examine the qualitative implications of the modulus eq. 2.12. For $t=0$ we get $G(0) = kT \, c_p N$. The quantity $c_p N$ is the concentration of Rouse bonds in the system, and $G(0)$ may be related to the rubber elasticity modulus result; viz., each bond contributes about kT to the modulus.[1] Of course, eq. 2.12 cannot be taken literally back to $t=0$, but represents a law appropriate after initial fast conformational relaxation processes.

For some purposes it is better to write the number of Rouse bonds as the degree of polymerization, N_0, divided by N_B, the number of monomers in a Rouse bead: $N = N_0/N_B$. Then

$$G(0) = \frac{kTc}{N_B} = \frac{kTca^2}{a_B^2} ,\qquad (2.15)$$

where c is the monomer (number) concentration and a is the effective length of a monomer.

For a particular $t \geqslant \tau_B$ we can regard the series in eq. 2.12 as composed of two types of terms. For $p > p_t = N(\tau_B/t)^{1/2}$, we can say the terms $\exp(-p^2 t/N^2\tau_B)$ are $\simeq 0$, while for $p < p_t$ the terms are $\simeq 1$. Thus, as time evolves, the details of the high p relaxation spectrum become unimportant. Indeed even the number of modes N, related to the size of the beads, becomes unimportant. The system has relaxed out to wavelengths, in beads, of $(t/\tau_B)^{1/2}$. The relaxed segments contain $N_t = N_B(t/\tau_B)^{1/2}$ monomers. The concentration of unrelaxed chains is $(c/N_B)(\tau_B/t)^{1/2} = c/N_t$, and kT times that is indeed an estimate of the modulus at time t (the result can be derived more rigorously by changing sum to integral on p in eq. 2.12)

$$G(t) = \frac{kTc}{N_B} \left[\frac{\tau_B}{t}\right]^{1/2} ,\qquad (2.16)$$

$$= \frac{kTc}{N_t} .\qquad (2.17)$$

To reiterate, the fine scale detail of the orientational anisotropy, manifest in stress, is being lost for distance scales in chain space less than N_t monomers.

Note the analogy between eq. 2.17 and eq. 2.15; i.e., the modulus is still kT (times a constant) for each unrelaxed segment, but these are taken as having N_t monomers at time t. N_B is, thus, the number of monomers in a relaxed segment at $t=0$, in the sense of extrapolating back the full eq. 2.12.

Chain wavelengths N/p correspond to real space separations $a_B(N/p)^{1/2}$ or real space wave vectors

$$q \approx \left[\frac{p}{N}\right]^{1/2} \frac{1}{a_B} \approx p^{1/2}/R_G . \qquad (2.18)$$

Thus we expect the dynamical scattering intensity $S(q,\omega)$ to have a maximum at the relaxation rate characteristic of waves of wavevector, q:

$$\omega_{max} \approx 1/\tau_p \approx N^2[p(q)]^2/\tau_R \approx q^4 R_G^4/\tau_R . \qquad (2.19)$$

The modulus law,

$$G(t) \approx \frac{kTc}{N_B} \left[\frac{\tau_B}{t}\right]^{1/2} , \qquad (2.16)$$

holds for $\tau_B \leq t \leq \tau_R$. For $t \geq \tau_R$ only one term of the modulus eq. 2.12 remains important, namely that for $p = 1$. Thus,

$$G(t) \approx \frac{kTc}{N_0} \exp(-t/\tau_R) \qquad (2.20)$$

for $t \geq \tau_R$. Note that by the time $t \approx \tau_R$ the modulus value has fallen to only $O(N_B/N_0)$ of its $t=0$ value. Nevertheless, in the viscosity, given by

$$\eta = \int_0^\infty dt\ G(t), \qquad (2.21)$$

the time spans $t < \tau_R$ and $t > \tau_R$ each contribute the same order of magnitude. One finds for the viscosity attributable to the polymer

$$\eta_P = \frac{\pi^2}{6} \frac{kTc}{N_0} \tau_R , \qquad (2.22)$$

$$= c\ \zeta\ R_G^2/3 \approx N_0\ c\ a^2\zeta . \qquad (2.23)$$

Thus η_P is proportional to the molecular weight. It is also seen, in actuality, to be independent of number of monomers in a Rouse bead.

2.3 Hydrodynamic Interactions and the Rouse-Zimm Theory

The above discussion has neglected hydrodynamic interactions, which arise as follows. In a fluid of viscosity η_0, if force $\mathbf{F}$ is exerted at a point $\mathbf{r}_1$, then the fluid develops an extra field of motion. The induced velocity at point $\mathbf{r}_2$ is

$$\mathbf{u}(\mathbf{r}_2) = \mathbf{T}(\mathbf{r}_{12}) \cdot \mathbf{F}(\mathbf{r}_1) , \qquad (2.24)$$

where $\mathbf{T}(\mathbf{r})$ is the Oseen tensor:

$$\mathbf{T(r)} = \frac{1}{8\pi\eta_0 r}\left[1 + \frac{\mathbf{rr}}{r^2}\right].\tag{2.24}$$

Look again at eq. 2.2, which says that the velocity at bead j is balanced by $1/\zeta_B$ times the bond force (and the random term). To this must be added the hydrodynamic interactions, the velocity fields created by the spring forces on the other beads and transmitted through the fluid. Thus

$$\frac{d\mathbf{r}_j}{dt} = -\kappa \sum_m \left[\frac{1}{\zeta_B}\delta_{jm}\mathbf{1} + \mathbf{T(r}_j - \mathbf{r}_m)\right] \cdot \sum_\ell A_{m\ell}\mathbf{r}_\ell + \mathbf{B}_j .\tag{2.25}$$

This is now a nonlinear equation in positions. It cannot be analyzed exactly, but a qualitative and semiquantitative treatment can be given by preaveraging the Oseen tensor. Essentially one replaces the Oseen tensor's $a_B/|\mathbf{r}_j - \mathbf{r}_m|$ by $1/|j - m|^{1/2}$, in keeping with the ideal relation between real and chain space. An influence $1/n^{1/2}$ (with $n = |j - m|$) in chain space is long range, since

$$\int^{n_c} dn\,(1/n^{1/2}) \approx n_c^{1/2} ;\tag{2.26}$$

i.e., the integral diverges unless cut off at some n_c. For $\mathbf{r}_j$ related as in a pure Rouse mode of index p, the integral to consider is more like

$$\int^N dn\,\frac{1}{n^{1/2}}\cos\frac{\pi np}{N} \approx \left(\frac{N}{p}\right)^{1/2},\tag{2.27}$$

with the chain space wavelength N/p serving as cutoff. Thus the normal mode equation is more like

$$\frac{d\tilde{\mathbf{r}}_p}{dt} = -\frac{1}{\tau_B'}\left(\frac{N}{p}\right)^{1/2}\left(\frac{p}{N}\right)^2 \tilde{\mathbf{r}}_p + B_p\tag{2.28}$$

(cf. eq. 2.9). The relaxation rate spectrum is[11]

$$\tau_p \approx \tau_B'(N/p)^{3/2}, \quad \text{for } p \gg 1.\tag{2.29}$$

Using the relation between chain and real space wavenumbers, $p \approx (q\,R_G)^2$, developed above, a maximum in the dynamical scattering is expected at $\omega_{max} \approx (q\,R_G)^3$. Modifying other arguments above one gets

$$G(t) \approx G(\tau_B')(\tau_B'/t)^{2/3}, \quad t \ll \tau_{RZ},\tag{2.30}$$

a terminal relaxation time

$$\tau_{RZ} \approx N^{3/2}\tau_B',\tag{2.31}$$

and a viscosity proportional to $M^{1/2}$.

Clearly, we are just scratching the surface of the Rouse, Rouse-Zimm and related models here. One additional point we wish to make is that, as Freed and Edwards[12] showed, in the presence of a background of other molecules the hydrodynamics interactions are screened, with a real space screening distance ξ_H. This introduces another effective cutoff in eq. 2.26. We do not have time to trace the consequences of this. We will, however, raise the question, which has received much attention in the literature, of whether ξ_H and the static screening distance scales are the same (have the same power law dependence on concentration). Also we will point out that with a p-independent cutoff, the spectrum of relaxation, and consequently various properties, becomes Rouse-like.

A number of scaling arguments relevant to the dynamics of polymer solutions have been given by de Gennes.[4]

We hope the discussion here has again stimulated the reader to go further. This is besides our goal of providing a basis for understanding some aspects of the dynamics in the presence of entanglements.

2.4 Introduction to Entanglement Effects

At high concentration, or in bulk, special consideration must be given to the effects of entanglement constraints, which are not easily expressed as a force. Indeed, disentanglement becomes the dominant relaxation process. Enormous progress has been made in understanding and modeling disentanglement based on the reptation model proposed by de Gennes,[4,13] the primitive path models of Doi and Edwards,[14] and the tube concept of Edwards.[15] We will begin by introducing the simplest concepts; and later try to indicate what ideas are being employed today to achieve a more complete, accurate, and extended picture.

One of the clearest manifestations of the onset of entanglement dominated relaxation is the observation of viscosity as a function of molecular weight, M. The data indicate a Rouse-like $\eta \propto M$ for low molecular weight, but this changes to a higher power law, apparently $\eta \propto M^{-3.4}$ for higher molecular weights. The change occurs at a characteristic molecular weight proportional to what is called the molecular weight between entanglements, M_e. The term is not to be taken literally since, as we will see, one can not say here there is an entanglement and here there is not. One of our goals will be to attach a more physical meaning to M_e. With M_e is associated a degree of polymerization between entanglements $N_e = M_e/M_{monomer}$, and a characteristic distance between entanglements, $a_e = N_e^{1/2}a$ (for ideal statistics, as is appropriate for high concentration).

2.5 Reptation in an Effective Tube

There are two fundamental ideas on which a theory of relaxation of entanglements is based. The first is that the effect of entanglements on a molecule (we will call it the central molecule) is something like confining that molecule to a tube.[14,15] Motion lateral to the tube is strongly inhibited. A linear molecule can, however, move along the tube, a motion de Gennes termed reptation.[13] This motion can be Brownian motion of the molecule as a whole or Brownian motion of internal modes. The former is more important, so for a while we will consider it alone.

2.6 Disengagement (Reptation) Time and Diffusion

The tube may itself be regarded as a random walk made up of independent segments. The characteristic size of each segment (length and diameter) will be taken as a_e, and the tube segment is spanned by N_e monomers of the central polymer. The number of segments constituting the tube is thus

$$N = N_0/N_e . \tag{2.32}$$

The tube end-to-end distance, $a_e N^{1/2}$, and that of the polymer, $a N_0^{1/2}$, must be identical. This is consistent with all the above definitions. The length of the tube is $L = Na_e$.

The friction constant for the Brownian motion of the molecule as a whole along the tube is $\zeta_P = N_0\zeta = NN_e\zeta$, which corresponds to a diffusion constant along the tube of $D_P = kT/NN_e\zeta$. (It has been assumed that, in bulk, hydrodynamic interactions are well screened.) Thus the characteristic time for the molecule to move one tube length, the "disengagement" or "reptation" time, is

$$\tau_d \approx L^2/D_P ,$$

$$= \frac{N^3 N_e^2 a^2 \zeta}{\pi^2 kT} = \frac{N_0^3 a^2 \zeta}{\pi^2 N_e kT} , \tag{2.33}$$

where a π^2 has been inserted for later convenience. Note τ_d varies as the cube of the molecular weight.

The diffusion constant of the molecules can next be estimated. In the time τ_d the molecule is displaced a distance R in real space (as opposed to moving a distance L along the coiled tube). Thus the molecules' diffusion constant is

$$D \approx R^2/\tau_d ,$$

$$\approx \frac{N_e kT}{N_0^2 \zeta} . \tag{2.34}$$

(Cf. to the diffusion constant $kT/N_0\zeta$ in the absence of entanglements and hydrodynamic interactions, and note that the two become equal for $N_0 \approx N_e$.)

2.7 Viscoelasticity

Assume that upon straining a system the molecules are affinely deformed; i.e., the shapes of the molecules are distorted similarly to the distortion of the sample (on all but the shortest distance scales). Rapidly some relaxation occurs by conformational transitions, followed by Rouse relaxation of short wavelength modes (with enough nodes so that even severe entanglements hardly interfere). We can assume that after a period of time, τ_e, short compared with τ_d, modes out to chain wavelengths $N_e a$ relax in Rouse-like fashion (defining N_e). The time $\tau_e \approx N_e^2 \tau_B/N_B^2$ is the Rouse relaxation time for chains of N_e monomers. According to eq. 2.17 the modulus at time τ_e is

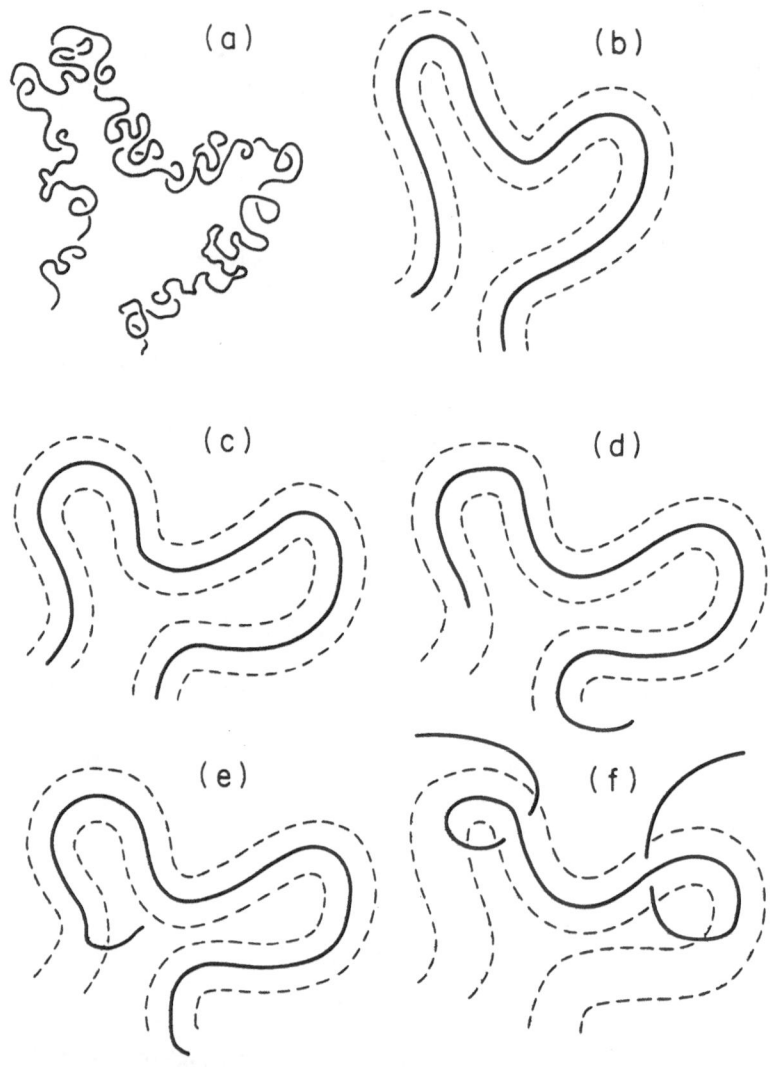

Fig. 3.   (a) Polymer molecule. (b) A representation of the molecule, shown as the solid line, in which wiggles on a distance scale $< a_e$ are suppressed (the primitive path). The confining tube is shown dashed. (c) Affine deformation of the tube and enclosed molecule. (d) Early motion in which molecule emerges isotropically from the tube at one end. (Emerged segment is confined by a new, isotropic tube, not shown.) Other end is pulled up the tube destroying the constraining influence of a portion of the tube (shaded). This is because as the molecule moves back (e) the other end emerges isotropically. (f) After some time, t, a portion of the time-zero deformed tube has been destroyed as a constraint. It is the portion that has been visited by a chain end between time 0 and t. The modulus is proportional to the fraction of undestroyed tube, $\mu(t)$.

$$G_N^0 \approx kTc/N_e , \tag{2.35}$$

independent of molecular weight. Thereafter relaxation is much slower, being inhibited by entanglements. It proceeds by reptation, rather than by the free chain Brownian motion of Rouse theory. G_N^0 is known as the plateau modulus because as a manifestation of the paucity of relaxation processes with characteristic rates between τ_e^{-1} and τ_d^{-1} the storage modulus plateaus in this frequency range. (There are a few processes in this range which we will discuss later.) Plateau modulus measurements enable us to estimate N_e as $N_e \approx kTc/G_N^0$, and to take as the "distance between entanglements" $a_e \approx aN_e^{1/2}$.

As shown in the appendix polymeric stress is related to orientational anisotropy. After the time τ_e one can assume that the confinement of the molecule to a tube prevents reorientation of the polymer segments of length N_e except by escape from the tube. (For the moment we regard the tube as permanent.) Since such escape occurs on the time scale τ_d we can estimate the viscosity as

$$\eta \approx G_N^0 \tau_d , \tag{2.36}$$

$$\approx N_0^3 \, \frac{ca^2\zeta}{N_e^2} \tag{2.37}$$

(note $\eta_{reptation} \approx \eta_{Rouse}$ for $N_0 \approx N_e$). The viscosity is predicted to vary as the molecular weight cubed. This is close to, but not, $M^{3.4}$. The difference remains one of the compelling mysteries of polymer science.

An examination of the time-dependent relaxation modulus, $G(t)$, in the terminal region, gives much greater insight into the essence of the reptation process. We have assumed that the tube is capable of holding segments of the polymer (or modes of wavelength longer than N_e) in a fixed orientation, affinely related to the sample's deformation. Relaxation can occur only by reptative escape from the tube. Let us look at the escape process in greater detail. The molecule moves back and forth along the tube. Consider an early stage (Fig. 3), where a motion in one direction has been made, let us say a distance ξ along the tube to the right. A portion of molecule, the end, emerges from the tube. As it emerges it is free to orient itself in any direction (Fig. 3d). Hence the stress is only a fraction $(L-\xi)/L$ of its pre-reptation, plateau value. Next, perhaps the molecule moves back towards the left, as in Fig. 3e. The left end will then emerge with equal probability in any direction. Once vacated - i.e., the first time the molecule's end has moved inward through it - a section of tube is no longer effective as an orientational constraint. One says that that section of tube has been destroyed. At time t the modulus is equal to the product of: (i) the plateau value, and (ii) the fraction, $\mu(t)$, of tube which has not been visited by either of the molecule's ends (cf. Fig. 3f). For Brownian motion of the molecule as a whole, the calculation of $\mu(t)$ is an exercise in diffusion theory. One finds

$$G(t) = G_N^0 \mu(t) , \tag{2.38}$$

$$= G_N^0 \left[\frac{8}{\pi^2} \sum_{q \, odd} \frac{1}{q^2} \exp\left(-\frac{q^2 t}{\tau_d} \right) \right] . \tag{2.39}$$

Note a crucial difference between this series and the one for the Rouse modulus, eq. 2.12. The weighting $1/q^2$ in the summand means that even as $t \rightarrow 0$ the first term represents 81% of the series (as opposed to $1/N$ of the series for the Rouse series) so the relaxation looks, for practical purposes, like a single exponential.

2.8 Entanglement Net, Unentangled Loops, and Primitive Path

In the above description there is much that has been skipped over or defined rather loosely. Let us reexamine and refine the tube picture. Consider a molecule entangled with a "net." We mean this net to represent the other molecules. For the moment though, for conceptual and calculational purposes, we will take the net as regular. This is schematically illustrated in Fig. 4, where a molecule is rendered as a two-dimensional unrestricted walk on a lattice. The lattice is permeated with a regular array of crosses representing the net. The changes of state of the molecule will be such that the molecule does not pass through a cross. In three dimensions, picture the net as composed of lines in various directions. The net divides space into regions, which we will call cages.

A walk, or piece of walk, is an unentangled loop if it starts and ends in the same cage, and in between is topologically unentangled with the net (loosely speaking, does not encircle a cross or line). This means that one can grab the ends of the unentangled loop and pull the polymer totally into the cage without cutting through the net.

In general a polymer molecule consists of a sequence of unentangled loops and other steps that go from cage to cage. Imagine that we take some walk and pull out the unentangled loops. The sequence of cages, through which the remaining steps pass, may be regarded as the constraining tube; and a path from remaining cage center to cage center will be called the primitive path (although we will frequently use the terms tube and primitive path interchangeably).

Recently Helfand, Pearson, and Rubinstein[16,17] have been able to make quantitative statements about various probability distributions associated with entanglement of a walk with a net. We have determined the probability that an N_0 step walk forms an unentangled loop; and the probability that an N_0 step walk has a K cage primitive path. The average number of segments (cages) in a tube, $N = \bar{K}$, was determined; and thus $N_e = N_0/K$. An entropy, hence free energy, for deviations from the average was determined. Its use will be discussed later.

In a system of polymer molecules, what should be viewed as the net? Clearly the net is composed of the other molecules. However on a time scale τ_e the unentangled loops of the other molecules go through many conformations, allowing molecules to pass. Thus they are not entanglements. We believe, physically, that only the primitive paths of the other molecules are to be regarded as constituting the entanglement net of a given molecule. Unfortunately, this seems to contradict some experimental data. We will have to return to this matter later.

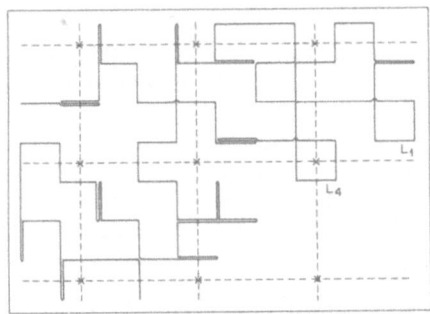

Fig. 4. (a) A typical random walk on a lattice, representing a polymer. The entanglement net, represented by crosses in two dimensions, is here placed on a lattice too, although it would not in general be. The net spacing is larger than the polymer step length. Dashed lines divide the space into cages. Some closed loops, such as L_1, are unentangled, while congruent ones like L_4 are entangled. L_1 is part of a large unentangled loop beginning and ending in the cage to the left. The polymer shown has a tube of 10 cages, 2 of which are visited twice. (b) A molecule, represented by a solid line, has its motion obstructed by other molecules, represented by circles. However, some of these encounters are with unentangled loop portions of other molecules (open circles) which move out of the way on the relatively rapid time scale, τ_e. These are not part of the entanglement net. The dashed line indicates the primitive path. Loops L_1 and L_2 are evidently unentangled. Loop L_3 is unentangled when the constraint of the open circles is removed.

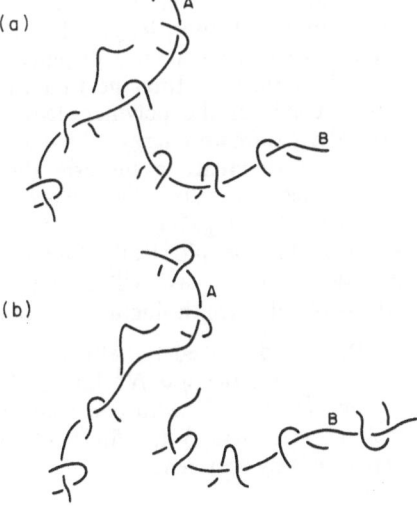

(a)

(b)

Fig. 5. (a) Two molecules, A and B, are shown as represented by their primitive paths. (b) An end of B has reptated past A destroying a portion of B's tube. When that end moves back it can do so in any direction, loosing all memory of its original orientation. B was also part of the tube of A so that the illustrated motion represents a tube renewal process. A segment of A can make a limited step of distance $\approx a_e$ until the next entanglement is encountered. The attachment of the released segment of A at both ends to the rest of the molecule allows only limited randomization of the orientation of this segment.

2.9 Early Stages of Relaxation

With this as background we can discuss some of the relaxation processes taking place on time scales $\ll \tau_d$. At any moment a polymer molecule can be described by the following variables: the sequence of cages forming the primitive path; the number of monomers associated with each cage, N_{e1}, N_{e2}, N_{eK}; and the particular conformation of each unentangled loop. These particular conformations relax, but to do so they must move through a fluid of other unentangled loops, which are relaxing at the same rate. Basically, though, this can be regarded as a relaxation of Rouse-like modes up to chain distance scale N_e, in time τ_e.

The fluctuations in the number of monomers in each unentangled loop also relax. For N_{ei} to go up by one, either neighbor must go down one, making the average leveling of these instantaneous fluctuations a diffusion problem along the tube. If the time for transfer of a monomer is τ_e, then the terminal relaxation time is $N^2\tau_e$, and there is a Rouse-like spectrum of shorter relaxation times down to τ_e.

2.10 Relaxation by Tube Contraction

We spoke previously of reptation as Brownian motion of the molecule as a whole, and how that destroyed tube constraints as the molecule's ends passed inward to parts of the tube not previously reached. Superimposed on this motion are all the (odd) internal modes. By contraction these may pull the ends to unvisited tube points, destroying tube constraint.[18]

As we saw earlier, such contractions are opposed by entropy barriers associated with the improbability of short tubes. The thermally probable fluctuations, those with a free energy increase $\leq kT$, are fluctuations of fractional length $\Delta L/L \leq N^{-\frac{1}{2}}$, where L is the average tube length.

{Exercise: Prove this result using the fact that the probability of a long unentangled loop (zero tube length) goes like $\exp(-\nu' N_0/N_e)$ [with a prefactor $\approx (N_0/N_e)^{-3/2}$ which we ignore here[16]]; and that a reasonable fit for the probability of a tube length $L - \Delta L$ is proportional to $\exp[-\nu'(N_0/N_e)(\Delta L/L)^2]$. The constant ν' depends weakly on the details of the net and has been measured as $\nu' \simeq 0.6$ for several bulk polymers.[19]}

There are two manifestations of relaxation by tube fluctuations. For short times the upper limits of tube length fluctuations are being established. We must calculate the typical (root-mean-square) displacement of the end, λ, of a polymer molecule *along* a tube in a short time t. Imagine that this displacement is created by uniformly contracting n Rouse bonds of the polymer, leaving the rest unperturbed. Each bond is stretched λ/n inward creating an energy $\approx \kappa n(\lambda/n)^2$. For this to be $\approx kT$, one must take $n \approx (\lambda/a_B)^2$. The time to create a uniform disturbance of n Rouse beads (or for the effect of a disturbance of the end bead to diffuse to the nth bead) is (cf. eq. 2.14) $t \approx n^2\tau_B \approx (\lambda/a_B)^4\tau_B$. Thus the typical magnitude of end displacement grows like $\lambda \propto t^{\frac{1}{4}}$. This is physically manifest also as an $\omega^{-\frac{1}{4}}$ decrease of the dynamically loss modulus. Such a law will hold until $n = N$, i.e., until $t \approx N^2\tau_B \approx \tau_R$. Thereafter the effects of tube length fluctuation are to make the chain look shorter by a distance $\approx L/N^{\frac{1}{2}}$.

Doi has discussed the effect of this on the molecular weight dependence of viscosity.[18] He finds that the $\eta \propto N^3$ law of DE theory is still correct but slowly approached like $\eta \propto N^3(1-\alpha N^{-\frac{1}{2}})$, with α a constant. This approach makes the slope of log η *vs.* log N look greater than 3 for finite N, which Doi feels may make the observation of a 3.4 exponent only apparent.

Another case where tube contraction is important is the relaxation of branched polymers,[19,20] as epitomized by star-polymers. Because the branch point is highly immobile, reptation by Brownian motion of an arm as a whole is strongly inhibited. Escape from the tube can only occur by contraction of the primitive path. Any significant contraction has a high free energy (discussed in the exercise above). The time for a fractional contraction ξ, i.e. $\Delta L = \xi L$, is

$$t \approx \tau_R \left\{ \left[\frac{N_e}{N_a} \right]^{1/2} \frac{1}{\xi} \right\} \exp \left[\nu' \left[\frac{N_a}{N_e} \right] \xi^2 \right], \qquad (2.40)$$

where N_a is the degree of polymerization of the arm, and the relatively unimportant prefactor in { } is discussed elsewhere.[19] Solving this equation for ξ as a function of t, and again taking the modulus as proportional to the plateau modulus and the fraction of undestroyed tube, one finds

$$G(t) = G_N^0 [1 - \xi(t)], \qquad (2.41)$$

which fits the data very well. The viscosity

$$\eta = \int_0^\infty G(t) dt$$

can be thought of as the sum over segments of the modulus associated with each segment, (G_N^0/N), times the time at which that segment is destroyed by tube contraction. Eq. 2.40 results in the time to destroy the last segment of tube overwhelming this sum, so

$$\eta \approx G_N^0 \tau_R \left(\frac{N_e}{N_a} \right)^{3/2} \exp \left[\nu' \frac{N_a}{N_e} \right]. \qquad (2.42)$$

Star polymers, indeed, have enormous viscosities and very low diffusion coefficients, reflecting the exponential dependence of the time for complete contraction on molecular weight. A time of 1600s, characterizing complete escape of a polyisoprene arm of molecular weight 100,000, has been observed.[19]

2.11 Origins of Nonlinearity

The viscoelastic properties of polymers show many nonlinear manifestations: e.g., flow-rate-dependent viscosities, nonlinear stress relaxation as a function of strain, and normal stresses. Doi and Edwards[14] attribute nonlinearity to lack of symmetry between tube stretching and contraction for a tube undergoing affine deformation. On average the whole tube will always be lengthened by a deformation, though the effect starts with terms second order in the strain. In the early stages after deformation, up to τ_e, the polymer will fill the elongated tube; i.e., it will have a longer primitive path and shorter unentangled loops. In a time τ_R it will contract back into a tube of length $L = N a_e$, relaxing some stress. The more stretched sections of tube will constrain more polymer. All of this produces a plateau stress nonlinear in the strain. In the terminal stage of relaxation, reorientation still occurs by escape from the tube, so the fractional stress reduction is still given by the same $\mu(t)$ discussed above. Thus, after a single step strain the stress is a product of a function of time and a function of strain. The behavior can be described for other strain histories, such as double step or oscillatory strain, although it is not as easy. These measurements are sensitive probes of the physics.

2.12 Tube Renewal

To this point the tube has been regarded as being effectively permanent – long lived on the time scale for reptation. (We did address the question of relaxation of obstruction by unentangled loops, but ended up totally disregarding them as part of the entanglement net.) A simple argument shows that the tube must itself be relaxing, undergoing what is called tube renewal. Picture the primitive paths of two molecules, A and B. Primitive path B contributes to forming the tube of A, and vice versa, as in Fig. 5. Consider now that B's end passes A, the type of process we spoke of earlier that destroyed A's constraint on B. This also frees A to move a segment, N_e monomers, a distance a_e, until the next entanglement (primitive path) is encountered.

Every reptation process (new release of an end segment) of one polymer is a tube renewal process of another. Although their numbers are equal, their effects are not. The reptative release of an end allows that segment to completely reorient. The reorientation which occurs when a middle segment is released by a tube renewal process is limited by the attachment of that segment at both ends to the remainder of the polymer. There is a sequence of jumps of the central molecule, each occurring on a time scale τ_d, allowed by tube renewal processes. This constitutes a Brownian motion, but the chain attachments make it of the Rouse type. What is the effective friction constant of a segment? The diffusion constant for a segment can be written both as kT/ζ_e and as a_e^2/τ_d, and ζ_e is determined by equating the two expressions.

The result of tube renewal is a fractional relaxation of stress, by a factor we will call $R(t)$. The functional form of $R(t)$ is like that predicted by Rouse theory, eq. 2.12. Beware, because the theory is being applied for time of the order of the fundamental bead relaxation time, which it should not have been, but that is a deeper level of detail.

It is appropriate to write for the stress relaxation[21]

$$G(t) = G_N^0 R(t)\mu(t) , \qquad (2.43)$$

when $\mu(t)$ describes the average fraction of polymer remaining in anisotropic tube, and $R(t)$ tells what fraction of the stress (anisotropy) of the chains in these tube portions has not relaxed by tube renewal.

Each of the processes for moving ends past entanglements – reptation as a whole, primitive path length fluctuations, contraction in stretched tubes – contributes to tube renewal. The different time scales for each, as well as the distribution of time scales arising in a polydispersed sample, introduce new features in the relaxation modulus.

Although reptation and tube renewal contribute equal magnitudes to viscoelastic relaxation and viscosity,[21] reptation dominates diffusion. Consider a time τ_d when reptation moves the molecule's center of mass a distance of order of the radius of gyration, i.e., a squared distance $a_e^2 N$. In τ_d there have also been about N tube renewal processes, each of which moves the center of mass a_e/N. These add incoherently to produce a squared displacement $N(a_e/N)^2 = a_e^2/N$. The contribution of tube renewal to the diffusion constant is thus only $1/N^2$ that of reptation.

2.13 Distance Between Entanglements

It is useful to focus on issues where anomolies exist, in order to discover whether some key effect has been omitted. Thus we conclude with a discussion of the concentration dependence of the distance between entanglements.

The "number of segments between entanglements," N_e, was identified above as the boundary between the wavelength of Rouse modes which could and could not rapidly relax orientation in the presence of entanglements. In real space the associated distance between entanglements is $a_e \approx a\,N_e^{1/2}$. The Rouse theory, shown above to be consistent with rubber elasticity ideas, relates N_e to the plateau modulus by $G_N^0 \approx c\,kT/N_e$.

Edwards and Evans[22] argument for the concentration dependence of a_e essentially comes down to the following (I hope they agree). The entanglement constraints arise from the polymers as lines. Therefore, they say, if a_e is to depend on concentration it must be in units of line length per unit volume, ca. This is a quantity with dimension (length)$^{-2}$, so to make a length it must be raised to the $-1/2$ power. The length a_e should go like the "only" physical length that determines it, so

$$a_e \approx (ca)^{-1/2} \approx a/\phi^{1/2} . \qquad (2.44)$$

(A physical, rather than dimensional, argument is given in ref. 22).

However, Rubinstein and Helfand[17] have argued that the concentration which determines the spacing between entanglements is the length of primitive path per unit volume, ca^2/a_e, again a (length)$^{-2}$. Thus

$$a_e \approx (ca^2/a_e)^{-1/2} , \qquad (2.45)$$

or

$$a_e \approx a/\phi \qquad (2.46)$$

(with a large constant[12]). In good solvents this is modified to $a_e \approx a/\phi^{3/4}$, just like screening length.

Experiments[23,24] indicate that in Θ solvents $a_e \approx a/\phi^{0.6}$. The whole issue requires more study.

2.14 Summary

The coupled Brownian motion of polymer segments can be resolved into a linear combination of the motion of Rouse modes, essentially a Fourier transformation in position along the chain. The Rouse modes behave as independent, overdamped, harmonic oscillators. Hydrodynamic interactions arise from the flow field accompanying one segment's motion creating flow at other segments. Increase of concentration leads to a screening of hydrodynamic interactions.

At high concentration, and in bulk, viscoelasticity and diffusion are dominated by relaxation of constraints imposed by entanglements. A tube-like condition is created about each molecule, inhibiting lateral motion, but allowing reptative motion along the tube. Brownian motion, in this reptative fashion, leads to escape of a molecule from its original tube in a time $\tau_d \propto M^3$. In describing reptation and the tube, chain lengths (or Rouse wavelengths) less than N_e and spatial dimensions less than a_e are unimportant, since relaxation on smaller scales can occur without major inhibition by entanglements. The entanglement net picture provides a more physical description of the tubes' properties. Can the net be regarded as the other molecules' primitive paths?

Strain affinely deforms the tube. This creates a directional anisotropy of the enclosed molecule, and a resulting stress. The stress relaxes by reptative escape from the distorted tube. The viscosity is the product of the plateau modulus and the disengagement time τ_d, so $\eta \propto M^3$.

The tube itself is relaxing (renewing) as surrounding molecules reptate away releasing constraints. This allows the central molecule to undergo a series of small motions, each on a time scale of τ_d and distance scale of a_e. The result is a Rouse-like spectrum of relaxation. Other relaxation processes, such as fluctuations of primitive path length and equilibration of monomer concentration per length of tube, contribute to the early stage of relaxation and to nonlinearity.

The reptation theory is powerful in its predictive capabilities, but nagging irreconcilabilities with observations ($\eta \propto M^{3.4}$ and $a_e \propto \phi^{-0.6}$) may indicate that something fundamental has been omitted.

3. DYNAMICS OF CONFORMATIONAL TRANSITIONS

3.1 Introduction

The type of motion we shall be considering in this Section is the rotation of a bond producing a conformational transition. Rotation (or torsional) motion about one bond of a chain takes place in a potential such as that illustrated in Fig. 6. There are potential wells corresponding to a trans state or conformation (t), and two gauche states (g^+, g^-). The barriers separating the wells are high enough that equilibrium within a well is essentially established in a time short compared with the transition time between wells. For a polymer in solution the lifetime in a well is of the order of nanoseconds.

Again we see the idea of a hierarchy of well-separated relaxation times in the macromolecule: relaxations in a single conformational state, conformational transitions, long wavelength collective modes, and reptation. These processes are manifest in physical phenomena with matching time scales. Each may be thought of as fixed when considering faster time scale processes. For slower time scale processes each may be regarded as at equilibrium, and fluctuations about that equilibrium contribute noise and frictional loss.

Conformational transitions are probed by dielectric relaxation, NMR, dynamical light scattering, ESR, fluorescence depolarization, intramolecular excimer formation, ultrasonics, and other experiments. Some of these will be discussed in great detail by other authors, so we will not expound on them further.

The theory of transitions from well to well over a potential barrier, such as in Fig. 6, is actually quite a simple theoretical problem for one free bond, as in butane. Unfortunately, Fig. 6 is only a part of the picture needed to understand conformational transitions in polymers. In the insert to Fig. 6 we show a part of a polymer molecule. One might rotate any bond to create the type of potential changes shown, and to produce conformational transitions. However, the molecule is not likely to follow this course spontaneously in a conformational transition. Merely rotating any one bond would cause the long, attached chains (much longer than shown) to swing in enormous arcs. Such a large scale motion

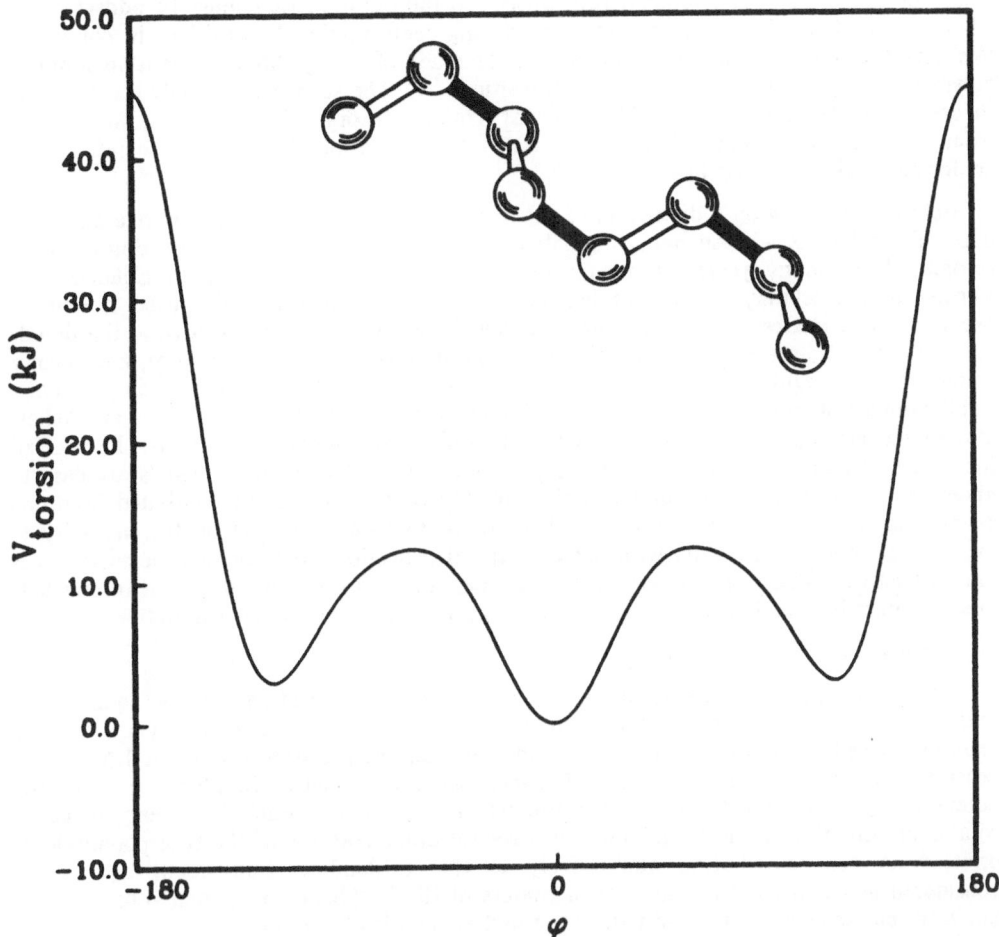

Fig. 6.   Potential for rotation of a single bond in a chain such as polyethylene. (The cis barrier is artifically high for simulation purposes.) Interaction between remote carbons is not included. If the bonds of the molecular model inset are labelled from left to right, then the black bonds 2, 4, and 6 are g^+, g^-, and g^+, respectively, while bonds 3 and 5 are t. One can see that a single bond rotation produces a swinging of the tails. For example, consider rotating white bond 3, holding the third and fourth carbons stationary, and picture the swinging of the tails (especially more remote polymer not shown).

would be met by strong frictional resistance. So the question that must be addressed is: what are the dynamical pathways followed during conformational transitions in polymers? We have taken a three-pronged approach to the study of this question. One is to simulate polymers and characterize the observed transitions.[25,26] The second is to apply reaction rate theory[27,28] in the form of a multidimensional Kramers theory.[29] The third is to determine, from theory, the functional form of the time-correlation functions associated with conformational transitions in polymers.[30,26]

Before embarking on a discussion of the results of these studies let us add one historical note. The difficulty with swinging the polymer tails in a conformational transition has been recognized for many years. A means of circumventing was proposed by Schatzki.[31,28] Verdier and Stockmayer[32] had earlier invoked a similar principle but used it only to produce Rouse modes. We know now that slow Rouse modes are insensitive to the details of the faster time-scale dynamics. The proposed motions are completely local, and involve going from one equilibrium rotational isomeric state to another by moving only a finite, small number of atoms. Mechanisms of this class have come to be known as crankshaft motions (a term applicable in the strictest sense only to the Schatzki proposal). Because of the limited amount of motion and the simplicity of the dynamics these models are easy to understand, analyze, and simulate. This probably contributes to the continued attention devoted to them. The crankshaft idea has helped to focus attention on the necessity to localize the motion associated with conformational transitions, but complete localization is too restrictive. There are theoretical objections that can be raised to the crankshaft mechanism,[25] but the bottom line is that no signs of it are found in our simulations.

3.2 Simulations

An assumed potential energy function must both force a collection of atoms to assume a realistic polymer backbone structure, and create reasonable conformational dynamics. We have performed simulations wherein the bonds are kept at a distance of about 1.53A by a potential quadratic in the separation between successive bonds. Similarly, a quadratic potential keeps the bond angles near the tetrahedral value. Each bond experiences a rotational potential as shown in Figure 6. Several other features of the true potential are omitted as not contributing qualitatively to the mechanics. Substituent groups are considered as collapsed onto the carbon centers of the backbone. Excluded volume forces between remote carbons are omitted, as are hydrodynamic interactions.

The equations governing the motion of each carbon are taken to be high friction Langevin equations analogous to eq. 2.2. The full nonlinearity of the potential forces is retained, though, to correspond to an energy with multiple wells, between which transitions could occur. The equations of motion had to be solved numerically. Long, representative trajectories were generated. We used a chain of 200 carbon centers, and carried out the integration until 40,000 transitions had occurred at each temperature. By using periodic boundary conditions in chain space the polymer ends are eliminated, making all bonds equivalent. Transition rates (for single and cooperative processes) are extracted by a fairly sophisticated (hazard) analysis of the intervals between transitions. The reader interested in further detail is referred to the original publications.[25,26,33] We concentrate here on results.

An Arrhenius plot of the logarithm of the transition rate *vs.* reciprocal temperature reveals that the activation energy is about equal to the barrier height.[25] This is in accord with experimental observations. If there is cooperativity then it is not being reflected in the activation energy.

A more direct search for cooperativity of transitions was more revealing. Immediately following any transition of the conformational state of a bond we looked at the rate of transition of neighboring bonds. For a short period of time after the first transition one could clearly see an enhancement of the transition rate of the second neighbor bond. Other neighbors showed little or no such correlation. In order to characterize this phenomenon more fully we took advantage of the tremendous detail available in a computer simulation to create Table I, as follows. Every pair of transitions of second neighbor bonds can be written as a chemical-like equation with initial and final state: e.g., $ttg^+ \rightarrow g^+tt$. All 48 possible transitions are listed in Table I, with some symmetry-related ones grouped together, as are forward and reverse reactions. A pair of transitions of second neighbor bonds was categorized and counted if the two transitions occurred within a short time of each other during which the rate enhancement was significant. A small number of such events can be expected on a statistical basis even for independent transitions, but the Table clearly reveals that two types of pair transitions are occurring with much greater than random frequency. What are the distinguishing features of these cooperative transitions that dominate the list?

The two types are illustrated in Figure 7. They are

$$\cdots g^{\pm}tt \cdots \; \rightleftarrows \; \cdots ttg^{\pm} \cdots \, , \tag{3.1}$$

$$\cdots ttt \cdots \; \rightleftarrows \; \cdots g^{\pm}tg^{\mp} \cdots \, . \tag{3.2}$$

Actually these two are quite similar, and quite different from every other transition in Table I. Note that in both cases the bond between is trans, and the transforming bonds (black in Fig. 7) undergo a counterrotation with respect to each other. If the rotational angles of the sequence are originally $\phi_1, 0, \phi_3$, then finally they are $\phi_1 \pm 120, 0, \phi_3 \mp 120$. Specifically eq. 3.1 may be written $\pm 120, 0, 0 \rightleftarrows 0, 0, \pm 120$, while eq. 3.2 is $0, 0, 0 \rightleftarrows \pm 120, 0, \mp 120$. This has an important geometric consequence.[28] As a result of the middle bond being trans the two outer bonds are parallel. When the two outer bonds counterrotate a crank-like motion occurs. The material attached to the two outer carbons (carbons 2 and 5 of the figures, numbering left to right) will translate as a result of the counterrotational motion. This, too, is quite evident in the figures. Note that the first and fifth bonds have translated in the final state relative to the initial state. This is true of the whole of the tail, not fully shown. For all pair transitions in Table I, except eqs. 3.1 and 3.2, this is not true; i.e., the tails undergo a rotation in going from initial to final state. Thus the two observed cooperative transitions greatly reduce the tail motion and the accompanying frictional resistance.

This is enlightening but still a number of questions are left to be answered. A translation of the tails reduces, but does not eliminate, the enormous frictional resistance of the tails. The motion as illustrated is still not local. How does full localization occur? Furthermore, while many cooperative crank-like pairs of transitions were observed, individual conformational transitions were also observed. The implication is that swinging of the tails between initial and final states can occur. How? Finally, why is the activation energy one barrier height even when two transitions occur?

3.3 Kinetic Theory

A kinetic theory analysis sheds a great deal of light on the questions just posed.[27,28] One can, indeed, use such a theory to predict, with fair accuracy, the rates of the

Table I. Number of each type of pair transition of second neighbor bonds occurring within a short time interval, and at various temperatures

temperature (°K)	330	372	425
% cooperative	29.3	29.2	29.4
$g^{\pm}tt \rightleftharpoons ttg^{\pm}$	5513	5394	5188
$ttt \rightleftharpoons g^{\pm}tg^{\mp}$	4151	4162	4163
$ttt \rightleftharpoons g^{\pm}tg^{\pm}$	438	389	330
$g^{\pm}tt \rightleftharpoons ttg^{\mp}$	411	344	303
$t^{\mp}g^{\pm}t \rightleftharpoons tg^{\pm}g^{\pm}$	394	428	569
$tg^{\pm}t \rightleftharpoons g^{\pm}g^{\pm}g^{\pm}$	352	376	442
$\left.\begin{array}{l} tg^{\pm}_{\pm}t \rightleftharpoons g^{\pm}_{\mp}g^{\pm}_{\pm}g^{\mp}_{\pm} \\ tg^{\pm}t \rightleftharpoons g^{\pm}g^{\pm}g^{\pm} \end{array}\right\}$	192	248	313
$g^{\mp}g^{\pm}t \rightleftharpoons tg^{\pm}g^{\mp}$	179	241	331
$\left.\begin{array}{l} g^{\pm}g^{\pm}t \rightleftharpoons tg^{\pm}_{\pm}g^{\pm}_{\pm} \\ tg^{\pm}g^{\pm} \rightleftharpoons g^{\pm}g^{\pm}t \end{array}\right\}$	76	79	90
$tg^{\pm}t \rightleftharpoons g^{\mp}g^{\pm}g^{\mp}$	40	41	39

Table II. Localized reaction mode for the central (16th) bond transforming in a 32 carbon chain all other bonds of which are trans. Carbon 16 follows bond 16.

l	$b_0\lvert\rho_l^0\rvert$ $\times 10^2$	$D_{\phi l}$	$D_{\theta l}$	D_{bl} $\times 10^2$
16	7.41	1.54	−0.128	1.82
17	3.60	0.17	0.048	1.07
18	1.80	−0.32	−0.061	1.53
19	4.53	0.02	−0.005	0.66
20	1.14	−0.15	−0.019	0.98
21	3.29	−0.06	−0.011	0.53
22	1.02	−0.06	−0.005	0.58
23	2.01	−0.07	−0.011	0.41
...				

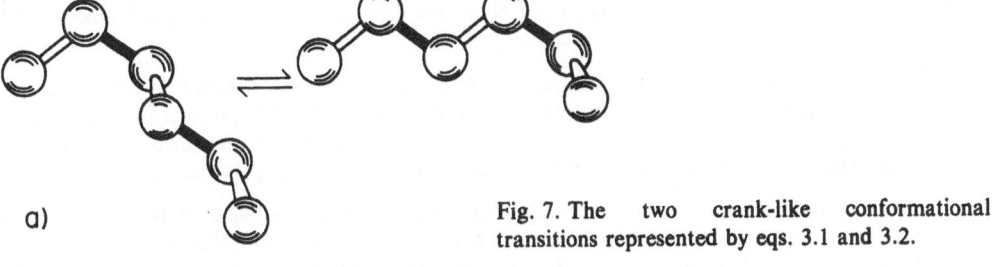

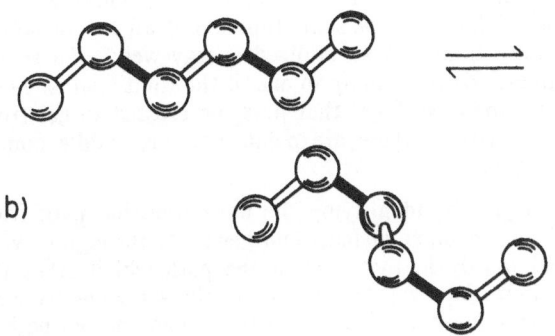

a)

Fig. 7. The two crank-like conformational transitions represented by eqs. 3.1 and 3.2.

b)

Fig. 8. The potential energy is a function of all the degrees of freedom but is here represented schematically by equipotential curves as a function of two variables. Wells correspond to stable rotational isomeric states. Conformational transitions take place by passage through a transition plane. The reaction coordinate represents the most likely path. It usually passes through or very near the saddle point. For all friction constants equal it is a path of steepest descent. A vector in the direction of the reaction coordinates has projection on the axis representing any degree of freedom proportional to the amount of motion of that degree of freedom in the reaction coordinate path (the D's of Table II).

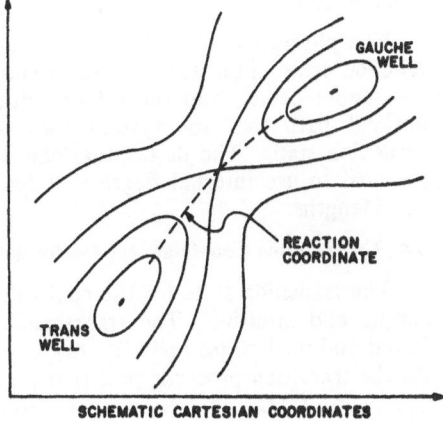

conformational transitions. Here we are more interested in descriptive aspects of the process. The reader who is curious about how rates can be calculated is referred to the original literature.[27]

The conformational transition consists of the passage of some bond angle from one energy well to another over a reaction barrier. We mentioned earlier that it is necessary to have more than one bond rotate, because a tremendous amount of motion of the attached tails during the transition would create frictional resistance, slowing the reaction.

The rates are molecular weight independent, confirming our belief that the transition is taking place in some localized fashion, rather than involving the whole chain (and very slow Rouse modes). This we will clarify in a moment. In order to have the transforming bond twist, but for material remote from this bond to remain essentially stationary, there must be some distortion of neighboring degrees of freedom. A proper analysis of the problem requires consideration of all the degrees of freedom of the molecule. The transformation can be viewed as follows. The energy is some function of all the variables. In the multidimensional space of variables there are generalized energy wells. These are regions containing a single energy minimum corresponding to one of the rotational isomeric states. The wells are separated by generalized surfaces that must be crossed to get from one conformation to another. On these surfaces there are points that are saddle points. This is all illustrated schematically in Fig. 8.

A multidimensional kinetic theory begins by identifying the most probable path from one well to a neighboring well through the reaction surface, and generally through or very near a saddle point. One is interested mostly in that part of the path which takes the system from the saddle point energy minus about kT on one side to the same energy level on the other side. This is the rate determining part of the transition. The optimal path is called the reaction coordinate. For a system undergoing Brownian motion, and with the friction constant of every atom equal, the reaction coordinate can be shown to be a path of steepest descent on both sides of the saddle point. Having identified the reaction coordinate one can essentially apply the usual one-dimensional kinetic theory along it. The only difference is that the energy along the reaction coordinate should be supplemented with an entropy term, which is a measure of the freedom to follow paths deviating from the reaction coordinate.

The physical information we seek about the pathways of transitions is contained in the reaction coordinate. One wishes to characterize the transition state (state corresponding to the saddle point), and the relative displacement of all the degrees of freedom along the optimal path that the system uses to approach and (symmetrical) descend from the transition state. The description can be in terms of Cartesian coordinates, but it is more physical to use internal degrees of freedom – rotation (torsion) angles, bond angles, and bond lengths.

3.4 The Reaction Coordinate for Conformational Transitions in Polymers

The transition state for the conformational change of one bond (not a pair transition) is simple and intuitive. The transforming bond, call it bond j, is at the barrier between initial and final states and all other internal degrees of freedom are at their energy minima. As the transition proceeds past that point the jth torsion angle will go from $\phi_j = \phi_{barrier}$ to $\phi_{barrier} + \alpha(t)D_{\phi j}$. The other torsional angles will go from $\phi_i = \phi_{i,\,min}$ to $\phi_{i,\,min} + \alpha(t)D_{\phi i}$. The bond angles will go from $\theta_i = \theta_{tetrah}$ to $\theta_{tetrah} + \alpha(t)D_{\theta i}$. The bond lengths will go from $b_i = b_{min}$ to $b_{min} + \alpha(t)D_{bi}$. We see that $\alpha(t)$ measures how far along the reaction coordinate the system has moved, with zero corresponding to the transition state, positive values to progress into the final well, and negative values to states before the saddle point is reached. The D's (with some arbitrary normalization) indicate the relative amount of motion of the different degrees of freedom. If the ratio $D_{\phi i}/D_{\phi j}$ is small in magnitude then the ith bond rotates very little relative to the rotation of the transforming bond. The above description is strictly valid only for small displacements from the transition state, but is generally valid for energy descents up to kT. However, cooperativity can not be handled this way.

Without doing any mathematics one can state, on physical grounds, what one expects the reaction coordinate to look like. (Working with a stick model will help the reader follow this argument.) The transforming bond is at its barrier and all other degrees of freedom are at their stable energy minima when the system is in its transition state. The transforming bond rotates some to get the system into the final state well. If all the other internal degrees of freedom stayed unchanged (at their energy minima) the tails would swing, and the frictional resistance would enormously slow the transition. This can not be the fastest path. Neighboring degrees of freedom must distort (move off their minima) in order to localize the total motion. If the distortion is confined to only a few near degrees of freedom the energy of the distortion will be high and the total energy will not be descending rapidly, or the distortion energy may even be greater than the decreasing energy of the transforming bond. If the distortion is widely spread the distortion energy decreases (inversely with the number of degrees of freedom it spreads over) but the friction of the greater total motion slows the transition. The reaction coordinate represents the optimal compromise.

Since the aim is to decrease the total system energy as rapidly as possible the system will favor distortion of soft (low force constant) degrees of freedom. The softest degrees of freedom are other bond torsions, so the reaction coordinate will invoke these rotations as much as possible. Some bond angle bending, and even a bit of bond stretching, helps the system localize the reaction pathway.

Of all the distortions of neighboring degrees of freedom that localize the reaction coordinate one is most effective. As seen earlier, counterrotation of a neighbor bond parallel to the transforming bond will convert a wide swing of the tail into a much smaller translation. Then other distortions can be invoked to eliminate asymptotically even that translation. When a first neighbor is trans the second neighbor bond is parallel, and it is found to counterrotate.

These principles are quantitatively illustrated in Table II, where the reaction coordinate is given for the $t \rightarrow g^+$ transition of bond 16 (the central bond, going from the 15th to the 16th carbon) of a C_{32} chain, with all other bonds trans. Column 1 shows the magnitude of the displacements of the various carbons for the reaction coordinate. The normalization is arbitrary; only ratios of values are significant. One sees that rather than a swinging tail, with displacement increasing away from the transforming bond, the mode is localized with motion falling off. In column 2 one sees the prominent counterrotation of the second neighbor (18th) bond and to a lesser extent of further even neighbors. The odd neighbor bonds also rotate somewhat to contribute to localization, helping to remove the tail translation. There is some bond angle bending and even a bit of bond stretching. (If the bond angle force constant is tightened, inhibiting its participation, the transition rate decreases.[34])

This discussion pertains strictly to the small motions associated with getting through the transition state. However, we see that the transition of one bond frequently induces significant counterrotation of a parallel second neighbor bond. That can continue. It greatly assists the second bond in undergoing a transition of its own, as is observed in the simulations. Sometimes the second bond is not brought all the way to the point of its own transition (or it may be up against the cis barrier). Then the second neighbor falls back to its old well, and a single transition is observed overall.

The distorted degrees of freedom will eventually fall back into equilibrium. The swinging of the tail for a single transition will eventually occur, but not in the rate determining part of the reaction coordinate. The tiny tendency toward tail swinging will be nearly lost in the host of other transitions and internal degree-of-freedom fluctuations.

The activation energy is one barrier height because cooperative pair transitions occur successively rather than simultaneously.

The above analysis suggested other implications, which have been confirmed by simulations. Since the reaction mode is localized one can expect similar effects even in smaller molecules. Indeed we do find enhanced cooperative pair transitions for middle bonds of octane, and slower rates for the middle as opposed to the end bonds.[27] In the polymer we expected that a bond, both of whose neighbors are gauche, will have a slower rate of transition. This, too, is observed.[25]

3.5 Time Correlation Functions

Experimentalists do not have as easy a time as computer simulators examining the complex details of processes such as conformational transitions. They can vary conditions and see if the changes in quantities like transition rates are as expected. Generally the most detailed measurements of a linear dynamical process are determinations of time correlation functions. Simple processes frequently exhibit a correlation function consisting of one, or a few, exponential decays. However with many degrees of freedom nonexponential relaxation often occurs. For this reason it is worthwhile to examine several time correlation functions related to conformational transitions to see if the mechanics is reflected in some characteristic way. The study also serves as a guideline to the experimentalist and simulator as to how to convert time correlation data into transition rates for the fundamental processes.

As a guideline to the type of result to be expected we consider the following model.[30] There is a chain of elements each of which is in one of two states, $+1$ or -1 (frequently denoted $+$ or $-$). We can let $\mu_i(t)$ denote the state of element i at time t. There are two ways that the state of an element can change. It can undergo the forward or back transitions

$$+ \rightleftarrows - \qquad\qquad (3.3)$$

at a rate λ_0. Also a pair of nearest neighbor elements can simultaneously undergo the compensating transitions

$$+- \rightleftarrows -+ \qquad\qquad (3.4)$$

at a rate λ_1.

The analogy to conformational transitions is clear, though imperfect. The states are the bond conformations, here two (of equal weight) rather than three (of unequal weight). The single transitions are the uncorrelated single conformational transitions. The pair transitions represent the crank-like counterrotations of neighboring states (first rather than second neighbors).

The object is to calculate the time correlation function between the state of element i at time zero and the state of element $i + l$ at time t:

$$c_l(t) = \langle \mu_{i+l}(t)\mu_i(o)\rangle . \tag{3.5}$$

It is exactly calculable. We will discuss here only the autocorrelation function $c_0(t)$ given by

$$c_0(t) = \exp(-2\lambda_0 t)\,\exp(-2\lambda_1 t)I_0(2\lambda_1 t) , \tag{3.6}$$

where I_0 is a modified Bessel function. For short times this looks like

$$c_0(t) \simeq \exp[-2(\lambda_0+\lambda_1)t], \qquad \lambda_1 t \ll 1; \tag{3.7}$$

while for long times

$$c_0(t) \sim \exp(-2\lambda_0 t)\,\frac{1}{(4\pi\lambda_1 t)^{\frac{1}{2}}} , \qquad \lambda_1 t \gg 1 . \tag{3.8}$$

Physically one can understand these results.

At time zero the state of element i is specified and all other elements are random. A pair process exchanges that information to a neighboring element and transfers the ignorance of state to i. The information continues to diffuse up and down the chain. The state of element i is coherent with its initial state only if at time t the information has diffused back to the origin. In eq. 3.8 the factor with $t^{-\frac{1}{2}}$ will be recognized as the probability of a return to the origin of a one-dimensional walk. The factor $\exp(-2\lambda_0 t)$ represents the superposed continuous loss of coherence by single transitions.

For a chain of elements with three states, closer to the polymer model, the correlation functions for an analogous dynamic model can not be exactly determined.[34] One can develop a theory which ignores all the terms that do not have a diffusional character. One must take into account that there are two types of deviations from equilibrium that can diffuse. We will not present more detail but merely mention that the resulting formulas can be found elsewhere, and shown to fit well to various conformational state time correlation functions determined from the simulations.[26]

The experimentalist does not ordinarily measure a conformational state time correlation function. Usually a technique is sensitive to some first or second order tensor, an orientational quantity such as a dipole moment along or perpendicular to the chain, the direction of a C-H bond, a polarizability, or a transition moment of a chromophore. It is even more difficult to analyze these objects' orientational time correlation functions. Furthermore they may exhibit both rapid relaxation due to conformational transitions and slow relaxation due to coupling to long wavelength modes. We have suggested fitting the short time part to the simplest function which accounts for single and pair transitions,[26] eq. 3.6. The results are quite satisfactory for fitting data from the simulations. The fit is good, too, for fluorescence depolarization[35] and NMR[36] data, although the values of the single and pair rates in the former are a bit surprising.

Other cooperative transition models, crankshaft models being an example, have been suggested. These, too, can give rise to diffusion of conformational states, and hence to functional forms similar to eq. 3.6 and especially the asymptotic behavior of eq. 3.8. Monnerie will discuss these models elsewhere in this volume.[37]

3.6 Experimental Evidence

While there have been numerous measurements of polymer fast relaxation rates which can be attributed to conformational transitions, few have contributed directly to a determination of the transition mechanism. Let us cite several of the results which do provide some indirect evidence.

Rates appear to be inversely proportional to viscosity. Activation energies are close to one barrier height, after an allowance is made for the temperature dependence of viscosity.[38,39]

An indication that the conformational transition can be brought about in a localized fashion comes from the work of Liang, Dickerson, and Miller.[40] They have measured transition rates in small polymer loops that are formed when polymer molecules absorb from solution onto a substrate. The adsorbed monomers serve as pinning points, between which short polymer loops penetrate into the solution. Incorporated nitroxide labels, in loops larger than about four to six bonds, produce ESR line shapes similar to those for molecules in solution. The ESR line shape is an indicator of reorientational freedom associated with backbone transitions.

Jelinski, Dumais, and Engel[41] have performed a revealing experiment involving solid state NMR of polybutylene terephthalate,

$$\left[\overset{\overset{\displaystyle O}{\parallel}}{C} - \bigcirc - \overset{\overset{\displaystyle O}{\parallel}}{C} - O - CH_2CH_2CH_2CH_2 - O\right]$$

The high friction associated with the large aryl groups keeps them fairly stationary during a conformational change in the butylene segment. This enhances the need for transitions to occur in a very localized fashion. The NMR looks at orientational relaxation of the C-H bonds in the $-CH_2-CH_2-CH_2-CH_2-$ sequence. By deuteration of the central two carbons the authors show that the central C-H bonds reorient more rapidly than the outer two. A crank-like counterrotational transition of the first and third C-C bonds does, indeed, reorient the two central C-H bonds without reorienting the two outer ones. The outer ones translate in such a motion, just like the tails.

A high degree of creativity will be called for in devising experiments that truly reveal what motion is taking place during a conformational transition. Perhaps some pathways will have to be blocked. (It would be valuable to know, in any event, what changes in the molecule block or open transition pathways.) Perhaps multiple quantum NMR will be capable of measuring the motion of whole sequences of atoms.[42]

3.7 Summary

The rate determining part of a conformational transition process is the part as one bond moves past the barrier separating the two rotational isomeric states (similar considerations apply to other transitions, such as trans-cis isomerization of a double bond). During this process the attached polymeric tails can not swing. In order to localize the motion neighboring degrees of freedom must temporarily distort, and relax after the rate determining part of the barrier to transition has been passed. One extremely effective means of localizing tail motion created by rotation of one bond is counterrotation of another parallel bond. This converts the large scale swinging of a tail into a much less severe translational motion. When two second neighbor bonds are separated by an intermediary trans bond they are parallel. The rotation of one past its transition barrier can be aided by counterrotation of the other. Frequently the second bond is thus brought all the way to the point of its own transition. Such crank-like pairs of cooperative, counterrotational transitions of second neighbor bonds separated by a trans are frequently observed in simulations. They are an important feature of the dynamics of conformational transitions in polymers. The functional forms of time correlation functions reflect the cooperativity associated with the counterrotation mechanism.

Appendix: A Microscopic Description of Stress

In these notes we shall discuss a microscopic picture of stress, especially in a polymeric system.

Stress enters in a development of hydrodynamics when one considers the equation of conservation of momentum. The rate of change of momentum in some volume element at point r is written as the acceleration produced by external forces on that element and a (negative) flux of momentum across the surface. The flux of momentum has two parts. The first is the momentum associated with the average velocity, $u(r)$, of the fluid at r. Thus momentum density in the α direction (with $\alpha = x, y$, or z) is $\rho(r)u_\alpha(r)$, where $\rho(r)$ is the mass density at r. This momentum is transported in the β direction at a rate $u_\beta(r)$. Therefore this contribution to the flux of α momentum in the β direction is $\rho(r)u_\alpha(r)u_\beta(r)$. Additional observed momentum transfer is called minus the stress tensor. The stress tensor can be separated into contributions from two molecular sources. One is also kinetic, and arises from the fact that the particles have a distribution of velocities about the average fluid flow velocity. We can write this term as a statistical average

$$\sigma_{\alpha\beta,K}(r) = -\left\langle \sum_i m_i (u_{i\alpha}-u_\alpha)(u_{i\beta}-u_\beta)\, \delta(r_i-r) \right\rangle, \tag{A.1}$$

where the sum is over all the particles in the system, although the δ-function picks out only the ones actually at the point of interest. Here u_i is the actual velocity of particle i. This term is of little concern in condensed media because the particle velocity deviations from the average relax more rapidly than the typical observation times.

Of greater interest in polymer science is the momentum transported from one side of a surface to the other when material on one side exerts forces on, thus accelerating, material on the other side. For simplicity we will discuss only pair forces. If particle i is on one side of a surface while j is on the other, i's force on j accelerates j, changing the momentum on j's side of the surface. This potential energy (force) contribution can be

FORCES
ON UPPER
HALF SPACE

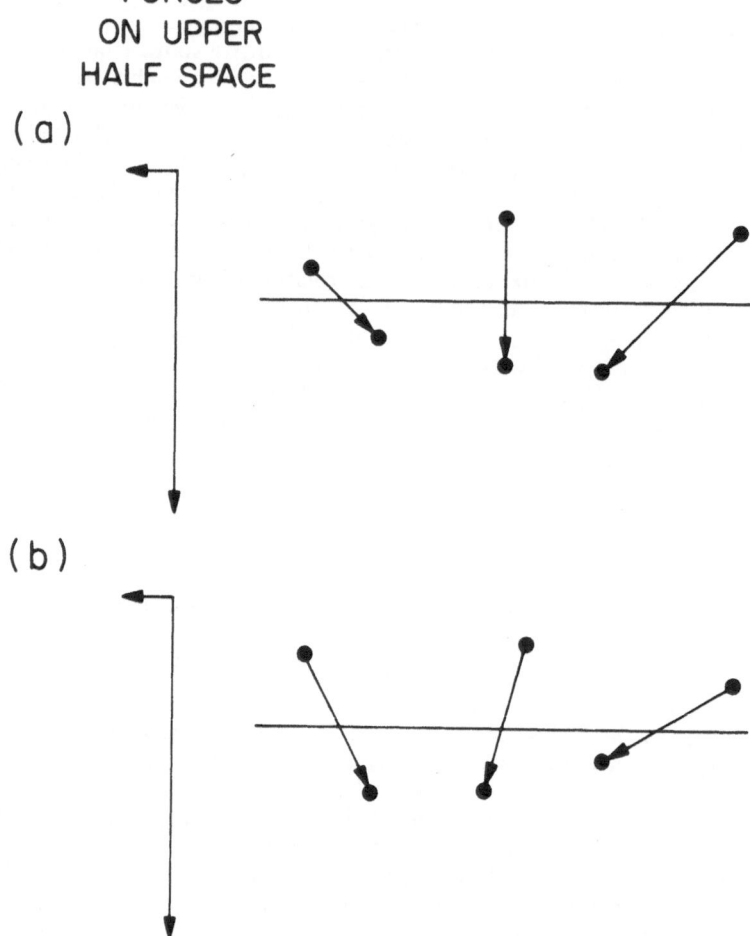

Fig. 9.   Bonds crossing a surface element. (a) The bond directions are isotropic but the
length distribution is anisotropic. (b) The bond length distribution does not
depend on direction, but the orientational distribution is anisotropic. In both
cases there is a net force on material in the upper half-space exerted by
material in the lower half-space, hence a momentum flux across the surface.

written

$$\sigma_{\alpha\beta,V} dS = -\left\langle \sum_{i,j} F_{ji,\alpha} \right\rangle . \tag{A.2}$$

The sum is over all pairs, of particles in the system which satisfy the condition that the vector $r_j - r_i$ from particle i to j penetrates the surface element dS in the positive direction. The normal to the surface element is in the β direction.

Consider a pair of particles separated by $r_j - r_i$. The probability of this vector penetrating dS is proportional to $(r_j - r_i) \cdot e_S \equiv r_{ji,\beta}$. One can replace the constraint on the summation by the probability of the constraint being satisfied,[43] and sum over all pairs to obtain

$$\sigma_{\alpha\beta,V} = -\frac{1}{2}\left\langle \sum_{ij} r_{ji,\beta} F_{ji,\alpha} \delta(r - r_i) \right\rangle \tag{A.3}$$

the usual virial expression. The factor of ½ comes from removing the condition of $(r_j - r_i) \cdot e_S$ being only positive.

In a polymer system the forces due to the polymer backbone give rise to the long-lived stress, so that frequently all other forces are ignored. Indeed only the coarse-grained (or long-wavelength) modes are persistent. For these one can employ the Rouse bead picture and an elastic force $-\kappa(r_j - r_i)$, where i and j are neighbors on the chain and κ is defined in Section 2.2. Thus we can write for the long-lived polymer contribution

$$\sigma_{\alpha\beta} \simeq \frac{3kTc_P}{a_B^2} \sum_{j=1}^{N-1} \langle b_{j,\alpha} b_{j,\beta} \rangle , \tag{A.4}$$

where b_j is the bond vector $r_{j+1} - r_j$, c_P is the concentration of polymer molecules, and the sum runs over the "beads" of one molecule. (Note the root-mean-squared value of $b_{j\alpha}$ is $a_B / 3^{1/2}$).

We can express eq. A.4 (and eq. A.3) pictorially as follows. If the bonds are isotropically distributed in direction but have an anisotropic distribution of lengths there will be a stress, as in Figure 9a. On the other hand, even if the distribution of lengths is at equilibrium, if the orientational distribution is anisotropic then there is a stress, as illustrated in Figure 9b.

In a nonglassy, amorphous polymer any stretch of the real bonds is rapidly relaxed. The whole stress is attributable to the bond orientation and the forces required to maintain that orientation against Brownian motion. Stress relaxation can be paralleled by the relaxation of other measures of real bond orientation, such as optical anisotropy. The Rouse or reptative description is a convenient means of describing the slow relaxation of the bond orientations. If one focuses on a sequence of N_B unstretched real bonds with a preference for one direction, and calls that a Rouse bond, that bond will appear to be oriented and stretched (in that its root-mean-squared end-to-end distance will be greater than $N_B^{1/2} a$).

REFERENCES

1. P. J. Flory, *Principles of Polymer Chemistry* (Cornell Univ. Press, Ithaca, NY, 1953).

2. J. des Cloiseaux, *J. Physique* **31**, 715 (1970); *ibid* **37**, 431 (1976).

3. J. C. Guillou and J. Zinn-Justin, *Phys. Rev.* **B21**, 3976 (1980).

4. P. G. de Gennes, *Scaling Concepts in Polymer Physic* (Cornell University Press, Ithaca, NY, 1979).

5. S. F. Edwards, in ref. 6; S. F. Edwards, in *Molecular Fluids*, ed. by R. Balian and G. Weill (Gordon & Breach, London, 1976) and references therein; S. F. Edwards, *Proc. Phys. Soc.* **88**, 265 (1966).

6. R. A. Pethrick and R. W. Richards, *Static and Dynamic Properties of the Polymeric Solid State* (D. Reidel Publ. Co., Dordrecht, Holland, 1982).

7. J. des Cloiseaux, *Phys. Rev.* **A10**, 1665 (1974).

8. J. des Cloiseaux, *J. Physique* **36**, 281 (1975).

9. P. E. Rouse, Jr., *J. Chem. Phys,.* **21**, 1272 (1953).

10. H. Yamakawa, *Modern Theory of Polymer Solutions* (Harper and Row, New York, 1971).

11. B. H. Zimm, *J. Chem. Phys.* **24**, 269 (1956).

12. K. F. Freed and S. F. Edwards, *J. Chem. Phys.* **61**, 3626 (1974); K. F. Freed and A. Perico, *Faraday Symp. Chem. Soc.* **18**, 29 (1983).

13. P. G. de Gennes, *J. Chem. Phys.* **55**, 572 (1971).

14. M. Doi and S. F. Edwards, *J. Chem. Soc., Faraday Trans. 2,* **74**, 1789, 1802, 1818 (1978); M. Doi, *J. Chem. Soc. Faraday Trans. 2* **75**, 38 (1979).

15. S. F. Edwards, *Proc. Phys. Soc.* **92**, 9 (1967).

16. E. Helfand and D. S. Pearson, *J. Chem. Phys.* **79**, 2054 (1983).

17. M. Rubinstein and E. Helfand, *J. Chem. Phys.* **82**, 2477 (1985).

18. M. Doi, *J. Polym. Sci. Polym. Lett. Ed.* **19**, 265 (1981).

19. D. S. Pearson and E. Helfand, *Macromolecules* **17**, 888 (1984).

20. M. Doi and N. Y. Kuzuu, *J. Polym. Sci., Polym. Phys. Ed.,* **18**, 775 (1980).

21. W. W. Graessley, *Adv. Polym. Sci.* **47**, 67 (1982).

22. S. F. Edwards and K. E. Evans, *J. Chem. Soc., Faraday Trans. 2*, **77**, 1913 (1981).

23. V. R. Raju, E. V. Menezes, G. Marin, W. W. Graessley, and L. J. Fetters, *Macromolecules* **14**, 1668 (1981).

24. M. J. Struglinski and W. W. Graessley, submitted for publication.

25. E. Helfand, Z. R. Wasserman, T. A. Weber, *Macromolecules* **13**, 526 (1980).

26. T. A. Weber and E. Helfand, *J. Phys. Chem.* **87**, 2881 (1983).

27. J. Skolnick and E. Helfand, *J. Chem. Phys.* **72**, 5489 (1980); E. Helfand and J. Skolnick, *ibid.* **77**, 5714 (1982).

28. E. Helfand, *J. Chem. Phys.* **54**, 4651 (1971).

29. H. A. Kramers, *Physica* **7**, 284 (1940).

30. C. K. Hall and E. Helfand, *J. Chem. Phys.* **77**, 3275 (1982).

31. T. F. Schatzki, *J. Polym. Sci.* **57**, 496 (1962); *Polym. Preprints* **6**, 646 (1965).

32. P. H. Verdier and W. H. Stockmayer, *J. Chem. Phys.* **36**, 227 (1962).

33. E. Helfand, *J. Chem. Phys.* **68**, 1010 (1978).

34. E. Helfand, Z. R. Wasserman, and T. W. Weber, J. Skolnick, and J. H. Runnels, *J. Chem. Phys.* **75**, 4441 (1981).

35. J. L. Viovy, L. Monnerie, and R. Brochon, *Macromolecules* **16**, 1845 (1983); J. L. Viovy, L. Monnerie, J. C. Morola, to be published.

36. J. J. Connolly, E. Gordon, A. A. Jones, *Macromolecules* **17**, 722 (1984).

37. L. Monnerie, this volume.

38. A. A. Jones, K. Matsuo, K. F. Kuhlmann, F. Gény, and W. H. Stockmayer, *Polym. Preprints* **16**, 578 (1975); B. Baysal, B. A. Lowry, H. Yu, and W. H. Stockmayer, *Dielectric Properties of Polymers*, ed. by F. E. Karasz (Plenum, New York, 1972).

39. H. Morawetz, *Science* **203**, 405 (1979); T.-P. Liao and H. Morawetz, *Macromolecules* **13**, 1228 (1980).

40. T. M. Liang, P. N. Dickenson, and W. G. Miller, in *Polymer Characterization by ESR and NMR*, ed. by A. E. Woodward and F. A. Bovey, (ACS Symposium Series 142, American Chemical Society, Washington, DC, 1980).

41. L. W. Jelinski, J. J. Dumais, and A. K. Engel, *Macromolecules* **16**, 492 (1983).

42. W. S. Warren, D. P. Weitekamp, and A. Pines, *J. Chem. Phys.* **73**, 2084 (1980); W. S. Warren and A. Pines, *ibid.* **74**, 2808 (1981).

43. There are corrections in the case of large density gradients; J. H. Irving and J. G. Kirkwood, *J. Chem. Phys.* **18**, 817 (1950).

LOCAL MOLECULAR DYNAMICS STUDIES OF POLYMER CHAINS - IN SOLUTION AND
IN BULK - USING THE FLUORESCENCE ANISOTROPY DECAY TECHNIQUE

Lucien MONNERIE, Jean-Louis VIOVY
Ecole Supérieure de Physique et de Chimie Industrielles de
Paris
10, rue Vauquelin
75231 Paris Cedex 05 - France

ABSTRACT - The fluorescence anisotropy decay (FAD) technique is first
described, then the different expressions which have been proposed for
the orientation autocorrelation function (OACF) of polymer chains are
presented. Typical FAD curves of dilute and concentrated solutions of
polystyrene labelled with an anthracene group in the middle of the chain
are compared to the various OACF expressions and discussed. In the case
of bulk polybutadiene, FAD results obtained either on anthracene labelled chains or on 9,10 dialkylanthracene probes free in the polymer
matrix, show that the same type of OACF as for polymer solutions can
account for the experimental data. Besides, the temperature dependence
of the correlation time of the labelled polybutadiene appears to agree
with the WLF equation derived from macroscopic viscoelastic measurements,
proving that the segmental motions of about 20 bonds which lead to the
FAD of labelled polybutadiene participate in the glass transition processes of this polymer.

I. INTRODUCTION

Molecular motions in polymer solutions or bulk polymers have been inves·
tigated by various spectroscopic techniques such as dielectric relaxation, ^{1}H and ^{13}C NMR, ESR, quasi-elastic Light Scattering or Neutron
Scattering. However, there are still many unanswered questions concerning the type of orientation autocorrelation function able to describe
local polymer motions in solution or in bulk and the size of these
motions. Furthermore, for bulk polymers, the relation of these local
motions to the glass transition phenomena observed in macroscopic mechanical measurements is controvertible
    The Fluorescence Anistorpy Decay technique, performed either on
polymers labelled with fluorescent groups or on fluorescent probes
dissolved in a bulk polymer matrix, can yield unique information on
the type of orientation auto-correlation function able to describe the
segmental motion of the polymer chain as well as on the nature of the
motion performed by the probe molecule.

M. A. Winnik (ed.), Photophysical and Photochemical Tools in Polymer Science, 193–224.
© 1986 by D. Reidel Publishing Company.

In this paper we first describe briefly the fluorescence anisotropy decay method and the equipment necessary and we emphasize the
advantages of synchrotron sources. We will indicate several chemical
reactions which can be used to label a polymer. Then, we will present
and discuss typical examples of fluorescence anisotropy decay studies
on dilute and concentrated polymer solutions and on bulk polymers.

II. FLUORESCENCE ANISOTROPY DECAY TECHNIQUE

II.1 - Principle

When absorbing a light of suitable wavelength, a molecule behaves like
an electric dipole oscillator, called the "absorption transition moment
M_o", with a fixed orientation with respect to the geometry of the molecule. In the same way, an "emission transition moment M" is associated
with the fluorescence emission. Thus, as far as fluorescence polarization is concerned, the fluorescent molecules can be represented by their
absorption and emission transition moments. When polarized light is used,

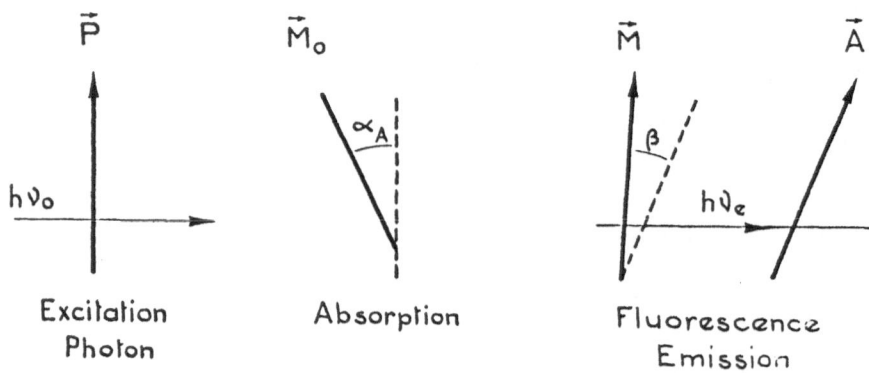

Fig. 1 - Polarized absorption and fluorescence emission. $\vec{P}$ = polarizer
$\vec{A}$ = analyzer

(Fig. 1), the probability of absorption is proportional to $\cos^2 \alpha_A$. In
the the same way, the fluorescence intensity measured through an analyzer A is proportional to $\cos^2 \beta$. Thus, for P and A direction of polarizer and analyzer, the observed luminescence intensity is proportional
to $\cos^2 \alpha_A \cdot \cos^2 \beta$.
    Due to the lack of phase correlation between excitation and emis-
sion light, fluorescence emission can be described as resulting from
three independent radiations, respectively, polarized along OX, OY, OZ
with intensities I_X, I_Y, I_Z (Fig. 2). The total fluorescence intensity
I is equal to :

$$I = I_X + I_Y + I_Z = I_{\parallel} + 2 I_{\perp}$$

One normally measured fluorescence emission either at right angles to
the excitation direction or along this direction. The emission polarization is usually characterized by the emission anisotropy, r :

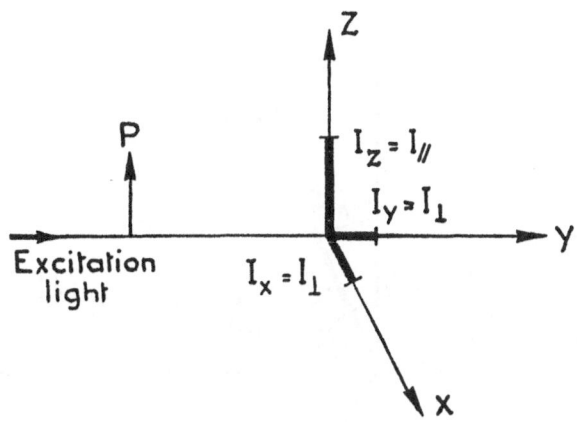

Fig. 2 - State of polarization of fluorescence emission for a vertically
polarized excitation light.

$$r = (I_\| - I_\perp)/(I_\| + 2 I_\perp)$$

When a set of molecules, isotropically distributed, is excited at time
t = 0 by an infinitely short light pulse, vertically polarized, the
absorption creates a temporary anisotropic population of excited molecules. If molecular motions of the fluorescent species occur during the
fluorescence lifetime, the anisotropy of the fluorescence emitted at time
t will be affected. Thus, the measurement of fluorescence anisotropy
yields information on the dynamics of fluorescent molecules.

More quantitatively, let us consider the emission transition moment
$\vec{M}_E(t)$ of a molecule i which emits a fluorescence light at time t
(Fig. 3).

Taking into account that the excited molecules are symmetrically
distributed around the polarization direction OZ, the emission anisotropy
r(t) of the fluorescence light emitted at time t by the set of molecules
is (1) :

$$r(t) = \overline{(3 \cos^2 \alpha_E(t) - 1)/2}$$

where the bar means an average over the molecules emitting at time t.
This quantity can be expressed as :

$$\overline{(3 \cos^2 \alpha_E(t) - 1)/2} = (\overline{(3 \cos^2 \alpha_A(0) - 1)/2})(3 \cos^2 \lambda - 1)/2)$$

$$\overline{((3 \cos^2 \Theta(t) - 1)/2)}$$

where λ is the mean angle between absorption and emission transition
moments, $\Theta(t)$ the angle through which the emission transition moment
of a molecule i rotates during time t between absorption and emission.

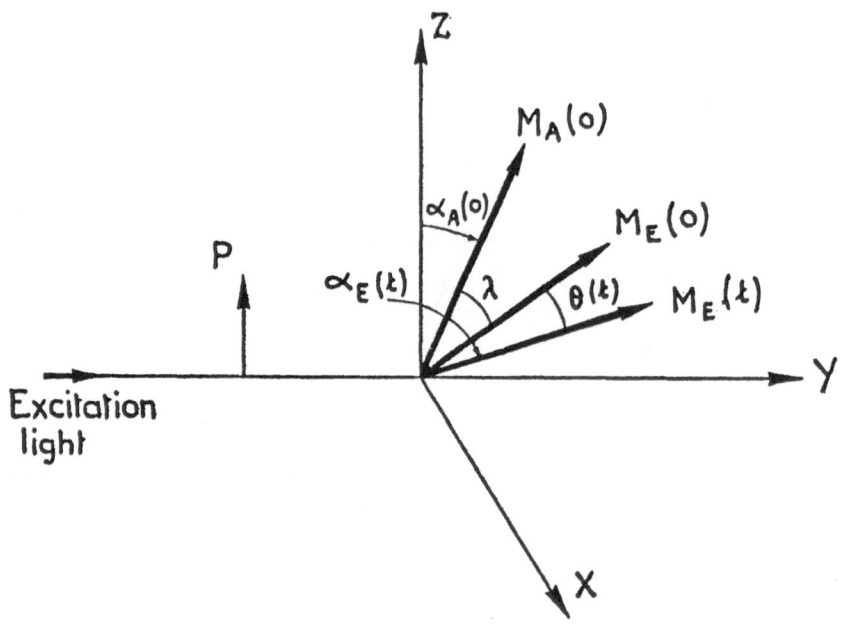

Fig. 3 - Orientation of transition moments in the reference frame
          for mobile molecules

For motionless molecules isotropically distributed, r does not depend
on time and its value, r_o, called the "fundamental emission anisotropy"
is characteristic of the fluorescent molecule considered :

$$r_o = 0.4(3 \cos^2 \lambda - 1)/2$$

For mobile molecules isotropically distributed, this leads to

$$r(t) = \overline{r_o(3 \cos^2 \Theta(t) - 1)/2}$$

The quantity $\overline{(3 \cos^2 \Theta(t) - 1)/2}$ corresponds to the orientation autocorrelation function (OACF) of the transition moment M of the fluorescent molecule and reflects the type of motion:

$$r(t) = P_2(\vec{M}_{(0)} \cdot \vec{M}(t))$$

where P_2 is the second Legendre polynomial.

II.2 - Equipment

The fluorescence anisotropy decay (FAD) technique is now a well-known
technique, and the general features of the experiment have been explained in full details by several authors (2, 3). We briefly present here
the apparatus used in our studies, which has already been described

elsewhere (4). Its block diagram is given in Fig. 4. The excitation
wavelength is selected from synchrotron radiation by a double holographic grating monochromator. It is vertically polarized. The emission
wavelength is selected by a single holographic grating monochromator.

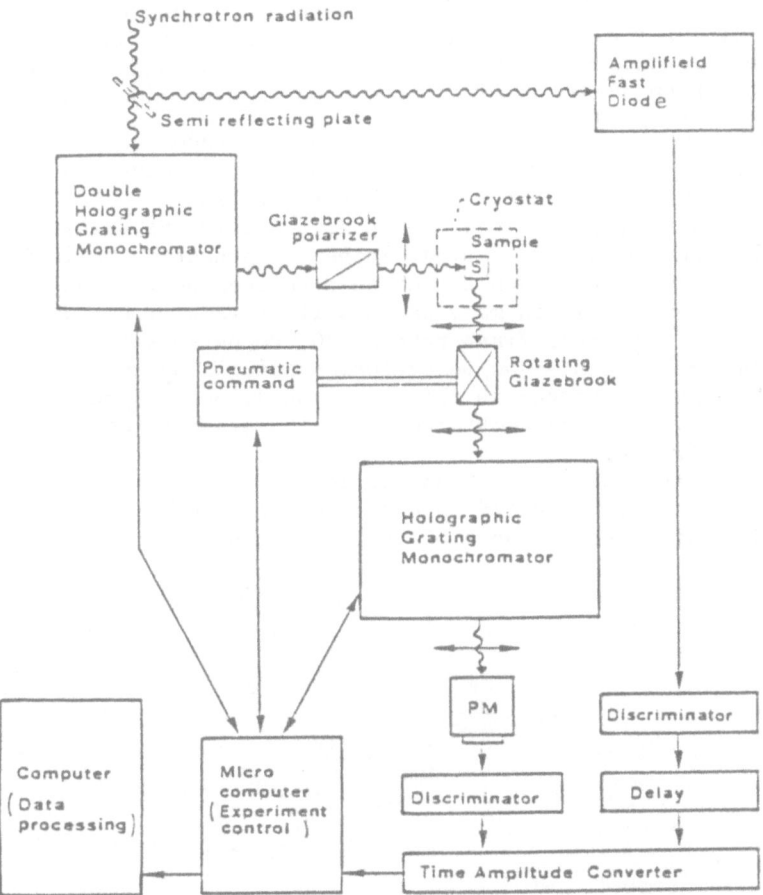

Fig. 4 - Block diagram of the Fluorescence Anisotropy Decay experiment

The single photon counting method is used (2, 5), i.e. no more than one
fluorescence photon is received by the photomultiplier for one excitation pulse. On receiving this photon, the cathode of the PM ejects a
photoelectron which is amplified in order to produce an electric pulse
at the anode. This pulse is sent to the "start" inject of a time to
amplitude converter (TAC) at a time t_1, which is compared with the time t_2
at which the delayed fluorescence pulse is received by the "stop" imput.
At its output, the TAC sends a pulse whose amplitude is proportional to
t_2-t_1 to an analog -to- digital converter and then to a computer, which

stores one count in the channel corresponding to the time $t_0 = t_2 - t_1$.

After many excitation cycles, the number of counts accumulated in one channel is proportional to the probability of emission at a given time after excitation. The errors on these numbers of counts are random and follow the well known Poisson distribution. The emission decay is recorded alternatively in vertical (Parallel) and horizontal (Perpendicular) polarization using a computer-driven rotating Polarizer

Finally, in the data analysis, one uses an iterative non linear least square reconvolution procedure to account for the finite width of the excitation pulse. One can achieve good precision in the determination of r(t) if :
(i) The counting rate is high, to minimize the random relative error proportional to : $n_i^{-1/2}$, where n_i is the number of counts in channel i;
(ii) One remains in the single photoelectron regime;
(iii) The excitation pulse is as short as possible, as stable in time as possible, and correctly sampled. Indeed, synchrotron radiation fulfills these requirements much better than conventional light sources.

The light source is in many cases a flash lamp with a pulse width around 2 ns and a typical flash rate of 20,000 pulse/sec. However, much more accurate experiments can be performed using a synchrotron radiation. The results presented in this paper on dilute polymer solutions and bulk polymers have been obtained with the synchrotron source LURE-ACO at Orsay (France).

In a synchrotron, an electron (or positron) bunch issued from a linear accelerator is injected into a vacuum ring chamber (Fig. 5) and maintained colinear to the axis of the chamber by intense magnetic

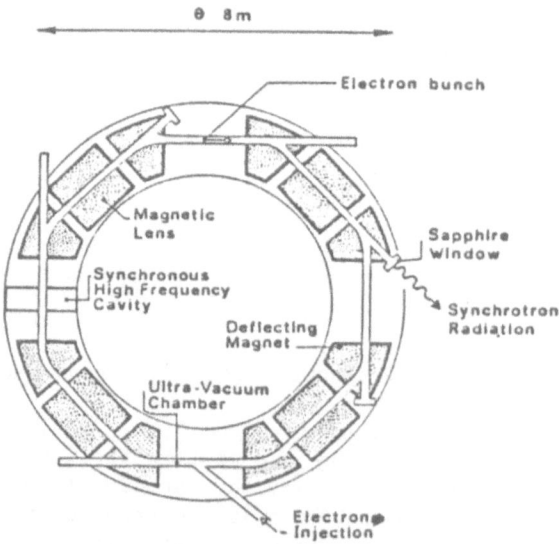

Fig. 5 - Schematic view of the cyclosynchrotron LURE-ACO (Orsay, France)

fields. In the curved parts of the ring, the relativistic electrons
emit pulsed light with a continuous spectrum ranging from X-ray to
infra-red. The width and shape of the light pulses depend on the size
of the electron bunch but not on the wavelength. It can range from
50 picoseconds to several nanoseconds depending on the characteristics
of the synchrotron, and it is very stable in time. In ACO, the pulse
width is about 1 ns, and the intensity decays exponentially with a
typical lifetime of 10 hours or more. These characteristics are compared with those of a typical flash lamp in Table 1.

Characteristics	Flash lamp	Synchrotron Radiation ACO-LURE(Orsay - France)
Pulse width	$\simeq$ 2 ns	$\sim$ 1 ns
Pulse shape	Asymmetric	gaussian
Pulse shape stability	$<$0.1 ns/h	Long time evolution (0.02 ns/h)
Intensity variation	Random fluctuation (several percent)	Exponential decay (Cte $\sim$ 10 h)
Max. Counting rate in single photon regime	400 c.s^{-1}	15 000-500 000 c.s^{-1}
Required counting time	1-10 h	10 min
Time interval between pulses	$>$10 μs.	75 ns
Excitation wavelengths	Discrete set or continuous spectrum	Continuous spectrum (X ray - IR)
Correction of P.M. charac- teristics with wavelength	Approximate	Direct calibration from scattering at emission wavelength
Intensity calibration for I_V, I_H	Accuracy $\approx$ 10^{-2}	Accuracy $\approx$ 10^{-3}
Absolute accuracy on r	$|\Delta r|$ = 0.01	$|\Delta r|$ = 0.01 with emission mono- chromator $|\Delta r|$ = 0.004 with emission optical filter
Correlation times available minimum	$\approx$ 0.5 ns	$\approx$ 0.1 ns
maximum	$\approx$ 50-400 ns depending on fluorescence lifetime	$\approx$ 200 ns
Accessibility	Easy	Difficult (yearly schedule, propo- sals)

Table 1 - Comparison between the performances of LURE-ACO and those
of a classical flash lamp

The light flux is at least two orders of magnitude greater than
that of conventional pulsed flash sources and the very high repeti-
tion rate $v_e \simeq$ 13 MHz enables the use of high fluorescence counting
rates v_f in the single photon regime ($v_f < 10^{-2} v_e$). Indeed, even
with sharp monochromation ($\Delta\lambda$ = 4 nm) both in excitation and emission,
v_f is mainly limited by the response of the electronics. For decays
longer than the period of ACO (73 ns), the decays following successive
pulses partly overlap. This overlap can be accounted for in the reconvolution procedure, and a satisfying determination of correlation times up
to about 200 ns can be achieved.
The continuous spectrum of the synchrotron allows a direct sampling of the excitation pulse at the emission wavelength, avoiding the
perturbations due to the wavelength-dependence of the P.M. response,
which affects the precision of the deconvolution.

All these characteristics allows one to measure FAD with a statistic quality and a reliability unattainable with flash sources. Moreover, in the case of polymers, it is generally not possible to purify samples as much as one would wish to perform a fluorescence experiment in confortable conditions, and the free choice of wavelength permitted by the continuous spectrum of the synchrotron source is essential.

II.3 - Quantitative criteria for data interpretation

The purpose of the studies reported in this paper is to compare the experimental anisotropy curves to the different expressions of orientation autocorrelation function (OACF) which have been proposed for simple or polymer molecules.The knowledge of the statistical distribution of errors on each channel is an essential tool for this comparison. It provides objective criteria to decide whether a discrepancy between a model and a set of data is significant or not, and to compare different models. Among these criteria, the best known is the reduced χ^2 (6), which should be 1 for purely statistical deviations but which increases with systematic deviations from the model. However, several other statistical criteria can be used (6, 7, 8). We recently developed (9) a systematic method for discussing theoretical expressions, using statistical criteria, together with physical criteria, aimed at discriminating expressions which may fit the data correctly but lead to parameters which behave in a non-physical way. For instance, if a model is supposed to reflect the true OACF of the chain backbone, the best fit parameters should not vary significantly when this OACF is sampled with a different experimental window.

Also, the fundamental anisotropy r_o cannot exceed 0.4, for theoretical reasons.

Of course, the comparison with experiments by a single technique cannot prove that a model is the unique realistic representation of molecular motions.

III - FLUORESCENT LABELING OF POLYMER CHAINS

Various types of labeling can be performed : at the chain end, along the main chain or on side-chain, depending on the chemical reaction used and on the chemical structure of the polymer chain. One important feature concerns the position of the transition moment relative to the chain backbone. Indeed, it is necessary to avoid any rotation of the transition moment independently of the chain backbone.

One of the most appropriate label is the anthracene group. Its optical properties are very convenient and many reactive derivatives are available for labeling. A review about anthracene labeling can be found in Ref. (10).

A very interesting labeling to investigate dynamics of polymer chains is the main chain labeling. In the case of polymerizations, the best way, when it is possible, is to perform an anionic polymerization and to terminate the living chains with 9,10 bis bromomethyl anthracene (Fig. 6, scheme 1) (11). Polystyrene, polyisoprene and other polydienes

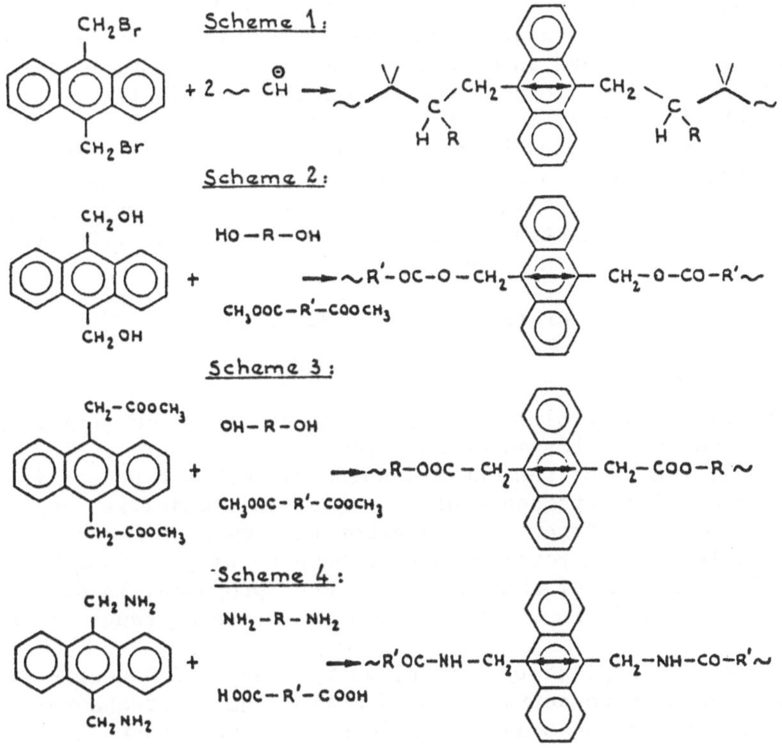

Scheme 1:

Scheme 2:

Scheme 3:

Scheme 4:

Fig. 6 - Different types of main-chain labeling with anthracene derivatives. Double arrows indicate the direction of transition
moment

have been labelled by this way. When dealing with condensation polymers,
such as polyesters, polyamides, some anthracene derivatives bearing ester,
alcohol or amine groups in positions 9 and 10 can be incorporated by
copolycondensation.

IV - MODELS FOR THE ORIENTATION AUTOCORRELATION FUNCTION OF A POLYMER CHAIN

It has been recognized for a long time that the orientation dependence
of a vector fixed to a polymer chain cannot be represented by a simple isotropic rotational diffusion model. In such a model (12) the
orientation is assumed to follow a vector joining the center of a sphere
to a point performing a random brownian diffusion on the surface of that
sphere. According to this model which describes well the orientation of
spherical objects or infinitely thin rigid rods, the OACF is an exponential function (13).

Generalized rotational diffusion models were built up later to
account for the orientational relaxation of anisotropic rigid bodies
or for the orientational relaxation in an anisotropic medium (19, 20).

But these models are not very realistic for actual flexible polymer
chains, and one cannot expect to understand polymer dynamics without tur-
ning to models which take into account the molecular nature of polymers.
In polymers, each bond is subjected to particular anisotropic constraints
due to neighboring bonds. Rouse (21) proposed to model the chain by a
sequence of beads separated by springs. The random forces exerted by the
viscous environment are localized on the beads. In spite of its crude-
ness, this early model contains the two essential features of polymer
dynamics, i.e. the connectivity and the flexibility. It leads to a master equation for the orientation probability :

$$\frac{dP_\Theta(x,\ t)}{dt} = W\ \frac{d^2 P_\Theta(x,\ t)}{d^2 x}$$

Where $P_\Theta(x,\ t)$ is the probability that a bond with a curvilinear coordinate x along the chain has an orientation Θ at time t.

This equation is of the one-dimension-diffusion type and it leads
to a $t^{-1/2}$ long time dependence of the OACF. This model is supposed to
be valid only on a distance scale greater than that of the "sub-chain"
i.e. the smallest chain portion large enough to be gaussian. Thus, more
realistic descriptions of the chain have been proposed, in order to get
OACF expressions valid in the whole time and distance range of experiments.

In the model proposed by Valeur, Jarry, Geny and Monnerie (VJGM),
the chain is assumed to perform 3-bond motions on a tetrahedral lattice,
to account for the fixed bond angles imposed to real backbone (22). This
model leads to a one-dimension diffusion equation for the orientation
probability, with the following expression for the OACF :

$$M_2(t) = \exp(-\ t/\tau_2)\ \exp(t/\tau_1)\mathrm{erfc}\ (t/\tau_1)^{1/2}$$

τ_1 is related to the inverse of the rate of 3-bond motions, and the τ_2
exponential term is introduced to account for out-of-lattice motions.

This relation presents an unrealistic infinite first derivative
at t = 0, due to the continuous approximation made in the analytical
treatment, and further refinements were proposed to avoid this defect.
For example, under the continuous approximation, Bendler and Yaris have
performed an arbitrary truncation in the mode analysis (23).

The expression for the OACF is then :

$$M_5(t) = \frac{1}{2}(\frac{\pi}{2})^{1/2}(\tau_1^{-1/2} - \tau_2^{-1/2})\{\mathrm{erfc}[\ (t/\tau_2)^{1/2}] - \mathrm{erfc}[\ (t/\tau_1)^{1/2}]\}$$

This expression presents a satisfactory behavior at short times.
However, it seems difficult to correlate the parameters τ_1 and τ_2,
which are the inverses of the arbitrary cutoff frequencies, with molecular quantities.

Another solution is to use the master equation in its discrete
from and to perform an exact mode analysis on the resulting Hückel
matrix arbitrarily truncated. In this case the truncation is directly
associated with the finite length of the chain which is taken into

account in the calculation. In fact, this procedure, proposed by Jones and Stockmayer (24) does not lead to a closed expression for the OACF, but to an infinite series of expressions corresponding to different truncations. This makes the comparison of the J S model with experiments rather lengthy. Since similar ideas can now be accounted for by closed expressions, we will not present here the detailed discussion of the J S model (for a more complete discussion, see Ref. (25)).

In contrast with these models, which start from a rather crude representation of the chain, but allow a direct computation of the OACF, Hall and Helfand (26) proposed recently a model able to predict conformational correlation functions (CCF) for rather realistic molecular potentials. The OACF cannot be derived from these CCF, at least at the present time. However, Hall and Helfand suggested that the CCF for a chain of two-state elements :

$$C_{ii}(t) = \exp(-t/\tau_2)\exp(-t/\tau_1)I_0(t/\tau_1)$$

may be a good approximation for the OACF (I_o represents the modified Bessel function of order 0).In this case, τ_2 should be associated with isolated conformational jumps, and τ_1 to correlated ones.

Several refinements of the HH model have been proposed recently (9, 25, 27). For the sake of simplicity, we do not present here these models, nor their detailed discussion using FAD, since they rely on the same essential physical assumptions, and lead to the same physical conclusions as the original HH model. For a complete discussion, the reader is invited to refer to Ref. (25).

The different models outlined above are not based exactly on the same representation of the polymer chain, and do not involve the same analytical treatment. However, it is worth noting that they all agree with the prediction of an OACF which contains the two following features (see Table 2).

Abbreviation	Expression	Ref
WW	$r(t) = r_o \exp(-(t/\tau_1)^{\beta})$	28
VJGM	$r(t) = r_o \exp((\frac{1}{\tau_1} - \frac{1}{\tau_2})t)\text{erfc}((t/\tau_1)^{1/2})$	22
BY	$r(t) = 0.5\, r_o(\pi/t)^{1/2}(\tau_2^{-1/2} - \tau_1^{-1/2}).(\text{erfc}((t/\tau_1)^{1/2}) - \text{erfc}((t/\tau_2)^{1/2}))$	23
HH	$r(t) = r_o \exp(-t/\tau_2) \exp(-t/\tau_1)I_0(t/\tau_1)$	26
GDL	$r(t) = r_o \exp(-t/\tau_2) \exp(-t/\tau_1)[I_0(t/\tau_1) + I_1(t/\tau_1)]$	9
IR	$r(t) = r_o \exp(-t/\tau_1)$	13
RR	$r(t) = r_o((1 - \beta)\exp(-t/\tau_1) + \beta)$	19

Table 2 - Expressions for the anisotropy time dependence used through the paper

(i) A non exponential short time term, characteristic of a one-dimension diffusion. This term, with characteristic time τ_1, is associated with

some "elementary motion" which must diffuse along the chain because of
the connectivity.
(ii) An exponential term, with characteristic time $\tau_2 > \tau_1$ which
reflects some finite damping or truncation of this diffusion.

In the case of bulk polymers it is questionable if the above described molecular models would be able to account for the experimental
results. This will be discussed later in this paper.

On the other hand, some phenomenological distributions of relaxation times, such as the well known Williams-Watts distribution (28)
(see Table 2, WW) provide a rather good description of dielectric relaxation experiments in polymer melts, but they are not of much
help in understanding molecular phenomena since they are not associated
with a molecular model. In the same way, the glass transition theories
account well for macroscopic properties such as viscosity, but they are
based on general thermodynamic concepts such as the free volume (29, 30)
or the configurational entropy (31, 32), and they completely ignore the
nature of molecular motions.

V - DILUTE SOLUTIONS OF LABELLED CHAINS

We present here the results obtained for a well known polymer : Polystyrene labelled with anthracene in the middle of the chain (25) PSAPS,
Fig. 7. This polymer was studied at 25°C in dilute solution in different

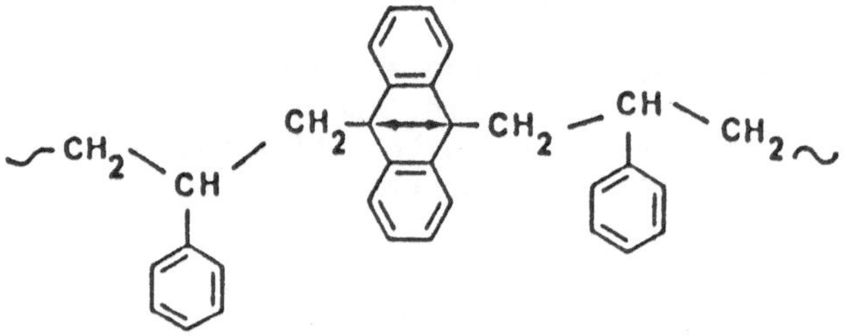

Fig. 7 - Polystyrene labelled with anthracene in the middle of the
chain (PSAPS)

mixtures of ethylacetate and tripropionin (glyceryl tripropionate) in
order to vary the viscosity in a wide range of values (from 0.4 to 8 cp).
For each solution, the different models were fitted to the data. This
comparison once again shows the definitely non exponential character of
the OACF. The monoexponential best fit to one of the experimental anisotropies (recorded at a viscosity $\eta = 5.4$ cp), presented in Fig. 8,
shows a systematic deviation clearly apparent in the weighted residuals
(upper curve). The reduced χ^2 are also much larger than 1, as can be
checked on the examples given in Table 3. On the contrary, the 3 expressions WW, BY and HH lead in all cases to a fit which is satisfying from
a curve-fitting point of view (see Fig. 9 and 10). In these cases the
weighted residuals are randomly distributed. Only the VJGM model always
leads to high χ^2 and poor fit.

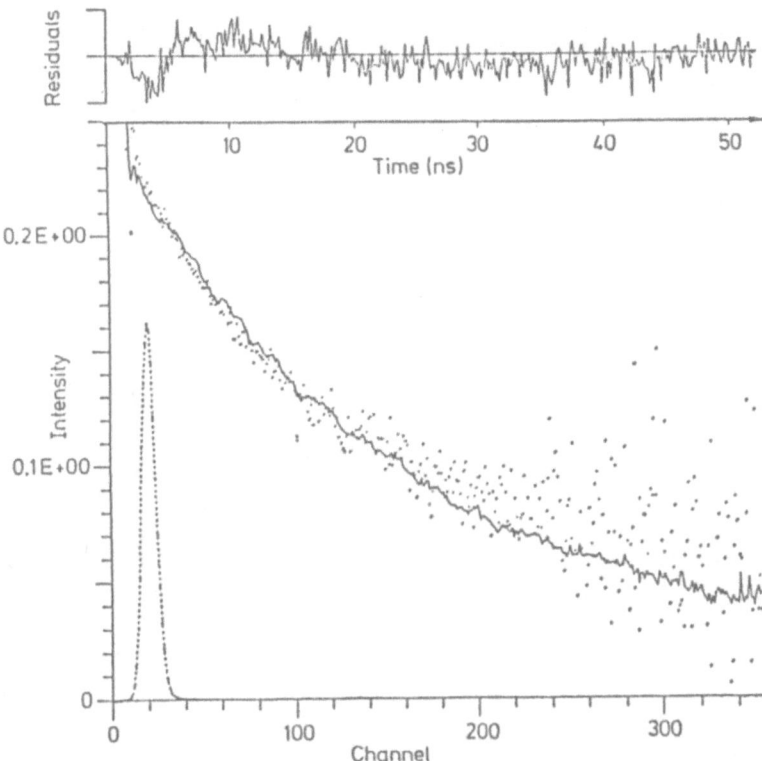

Fig. 8 - Comparison of the isotropic rotation model to the experimental anisotropy of PSAPS in solution (η = 5.4 cp). Dots are experimental data. The continuous line is the best fit OACF reconvoluted by the measured instrumental function (exciting pulse). This pulse is plotted as a dash-dot line (arbitrarily scaled). The upper graph represents the weighted residuals

Model	Viscosity cp		
	0.43	2.28	5.40
IR	2.56	3.22	1.83
RR	1.33	1.21	1.13
WW	0.97	1.09	0.98
VJGM	1.44	1.25	1.05
BY	1.06	1.11	1.00
HH	1.06	1.14	1.02

Table 3 - Reduced χ^2 for different samples and different models

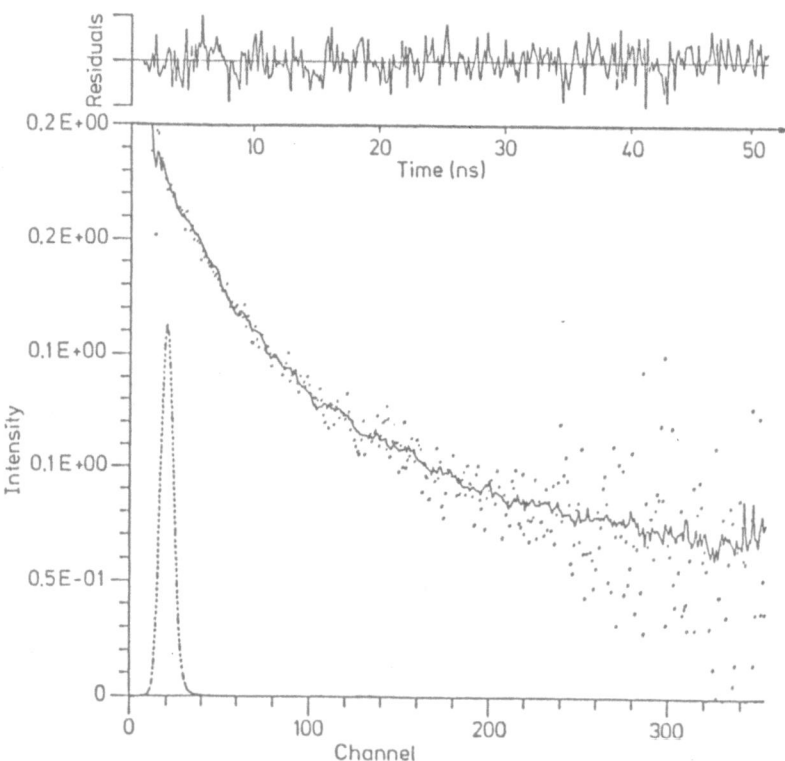

Fig. 9 - Comparison of the WW expression to the experimental aniso-
         tropy of PSAPS (same representation as Fig. 8).

This result is not suprising, since this model predicts an unrealistic
infinite slope at time zero. Nevertheless, to our knowledge, it is the
first time that this theoretical weakness is demonstrated by experimental results.

This discussion of models can be improved by considering the physical meaning of the obtained values of the parameters. For instance,
we have shown that the satisfying χ^2 values obtained with the WW model
correspond to parameters which are not physically reasonable and stable
as regards to variations in the experimental window (see Table 4).

Similar conclusions can be drawn when one uses models describing
the anisotropic motion of rigid bodies : good curve fitting may be
obtained fortuitously, but, in contrast with the case of the BY or HH
expressions, this fitting does not lead to stable and significant parameters.

These results were corroborated by studies of labelled polystyrene
in other solvents such as toluene, styrene and pure tripropionin, at
different temperatures. The main conclusions of these studies of polymer

dynamics in dilute solution are :

- the connectivity and flexibility of the chains lead to theoretical OACF which present unique features. Indeed, these features are observed experimentally. Only specific models taking into account the molecular structure of polymer chains are able to account for the experimental OACF.
- among the different closed expressions derived from specific polymer models, further distinctions can be made, and only the BY and HH models remain acceptable at the present level of experimental precision. But, the arbitrary cutoff frequencies in the BY model are difficult to interpret on a molecular scale.
- one can notice that the HH expression is simply the product of a diffusive term with characteristic time τ_1 and of a damping term with characteristic time τ_2. Since these two processes seem to be a general feature of polymer dynamic models, the agreement between the HH model and experiment is a satisfying observation. Moreover, it suggests that this expression can be used rather extensively to analyse experimental data, even if the molecular origin of the two processes remains to be determined.
- FAD is able to measure precisely the characteristic time of the diffusive process, τ_1 and, in some cases, the caracteristic time τ_2 of the damping process (with a lower precision). The ratio τ_2/τ_1 seems to depend on the solvent, but not on the temperature. In the different solvents that we used, the ratio τ_2/τ_1 for labelled polystyrene varies in the range : $3 < \tau_2/\tau_1 < 30$.

    Thus, in the case of polymer solutions, Synchrotron excited FAD leads to results consistent with previous experimental data and with the most recent theoretical expectations. Moreover, this technique is able to measure the OACF with a precision unavailable previously, and gives access to new information on polymer dynamics.

VI - CONCENTRATED SOLUTIONS OF LABELLED CHAINS

The dynamics of moderately concentrated and entangled polymer solutions has been widely investigated using various techniques, a review can be found in Ref. (33). The most important features of these semi-dilute solutions can be described within a model based on the "reptation" idea (34).

The lowest critical concentration, C^*, of polymer solutions is associated with the overlap threshold of the coils :

$$C^* \simeq N/(\overline{R^2})^{3/2}$$

where N is the number of statistical units and $\overline{R^2}$ the mean square end to end distance. From a dynamic point of view, C should correspond to the appearance of a deviation from the behavior of diluted chains.

A second critical concentration, C^{**}, has been introduced (35) which corresponds to the onset of a dominant pseudo-gel behavior :

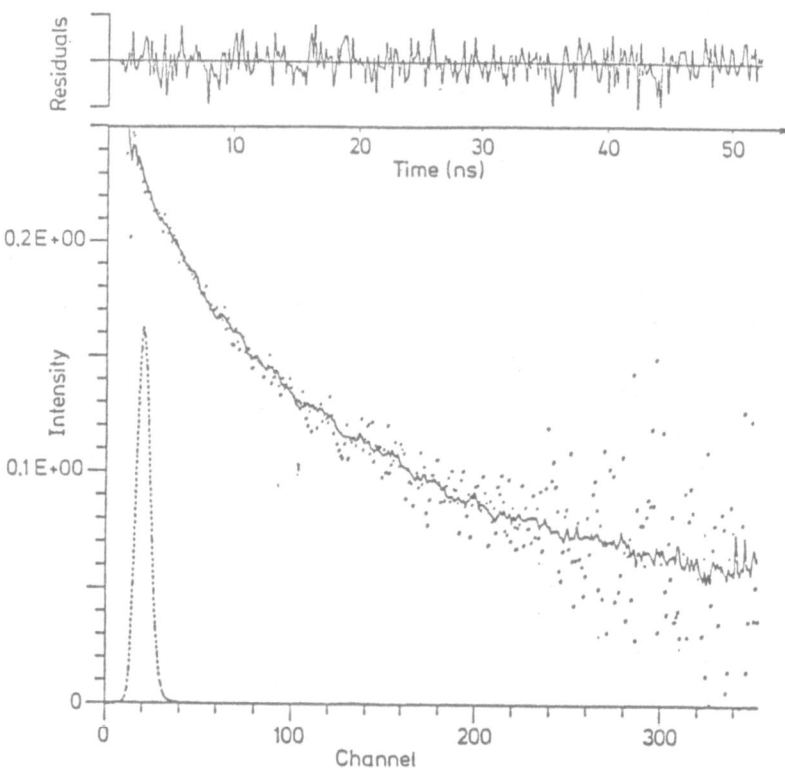

Fig. 10 - Comparaison of the HH expression to the experimental anisotropy of PSAPS (same representation as Fig. 8)

Model	r_o	τ_1 ns	τ_2 ns	τ_2/τ_1 (or β)	χ^2
RR	0.166	2.76		β = 0.018	1.337
WW	0.725	1.81		β = 0.712	0.972
VJGM	0.171	147	3.26	0.002	1.436
BY	0.200	0.857	8.40	9.8	1.057
HH	0.199	2.55	6.71	2.62	1.062

Table 4 - One example of the best-fit parameters obtained when different models are fitted to the experimental anisotropy of
PSAPS (viscosity 0.43 cp)

$$C^{\star\star} = (18 \ \pi^2)^{1/2} \ N^{-4/5}/A$$

where A is a constant which depends on the local structure. $C^{\star\star}$ is larger than $C^{\star}$ but no definite ratio can be assessed and values ranging from 2 to 15 have been found, depending on the type of experiment.

In both cases, it is assumed that the friction coefficient does not depend on the concentration. In fact, it appears that such an assumption is no longer valid for concentrations over 0.1 weight by weight.

We carried out an investigation of the dynamics of polymer chains in concentrated solutions of polystyrene, using FAD (36). A small amount (< 0.01 w/w) of polystyrene labelled with anthracene in the middle of the chain (PSAPS) was added to a toluene solution of unlabelled polystyrene (PS) with molecular weight ranging from 23,000 to 1,300,000. The experiments were carried out at 25°.

It appears that there is no effect of the polymer concentration on the shape of the anisotropy decay curves and the same conclusions as on dilute solutions are obtained about the most appropiate expression of the orientation autocorrelation function.

The concentration dependence of the correlation time τ_1, characteristic of the diffusive process along the chain sequence and related to the correlation time of the elementary motions, is shown in Fig. 11. τ_1 has been calculated using the GDL expression given in Table 2. First of all, there is no molecular weight the labelled or unlabelled polymer influence for either. This point has been carefully checked at a concentration of 0.28 w/w with the labelled polystyrene of M.W. 300,000 and two unlabelled polystyrenes with M.W. 420,000 and 23,000, respectively. Secondly, the critical concentration $C^{\star\star}$, determined from viscosity measurements and indicated as circled numbers in Fig. 11, does not lead to any change in the dynamic behavior investigated by FAD. Indeed, we will see from FAD studies on anthracene probes with alkyl tails in a bulk polymer that the FAD technique is sensitive to dynamic processes occuring over a range of about 30 bonds. Thus, they are not affected by the entanglements which influence the viscosity behavior. Finally, the increase of τ_1 with concentration reflects the increase of friction coefficient.

VII - BULK POLYMER WITH LABELLED CHAINS

FAD studies on concentrated solutions have shown that the same orientation auto correlation functions which are used in dilute solutions can be satisfactorily applied. So, it is interesting to investigate bulk polymers to check if the constraints arising from the surrounding in a dense medium would yield a different type of OACF. Furthermore, it is questionable if the segmental motions observed by FAD can be related to the molecular processes involved in the glass-rubber transition.

For these reasons, we have recently performed (9) FAD experiments on polybutadiene chains labelled with anthracene in the middle of the

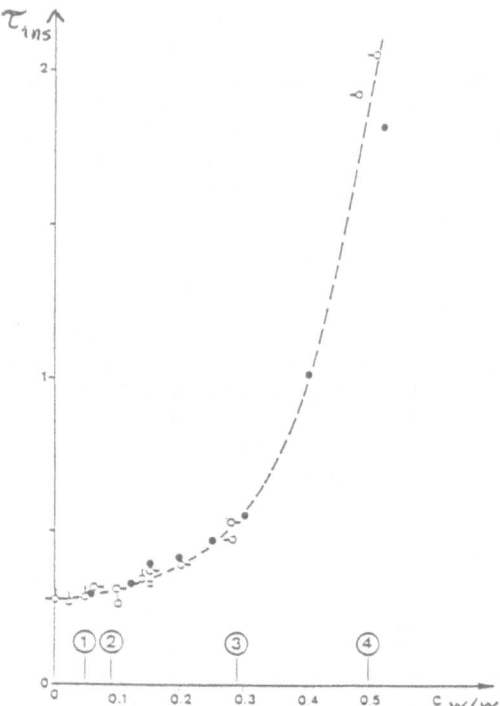

Fig. 11 - Concentration dependence of τ_1. The symbols are the following:

$(M.W.\times 10^{-3})$	$(M.W.\times 10^{-3})$	τ_1	$c_v^{\star\star}$
300	1,300	♂	①
300	420	○-	②
300	23	-○	④
52	68	●	③
30	23	☐	④

$c^{\star\star}$ corresponds to viscosity measurements.

chain (Fig. 12), embedded in a matrix of homopolymer unlabelled chains. The concentration of labelled chains was about 0.01 w/w.

The evolution of the experimental anisotropy as a function of the temperature is shown in Fig. 13. As expected, the decay rate increases as the temperature increases.

It has to be pointed out that the calorimetry glass-transition temperature of polybutadiene is - 90°C.

For the highest temperature (t > 50°C), it can be noticed that the anisotropy decays from a value close to the fundamental anisotropy of 9,10 dimethylanthracene to almost zero in the time window of the experiment (about 60 ns). This means that the initial orientation of a backbone

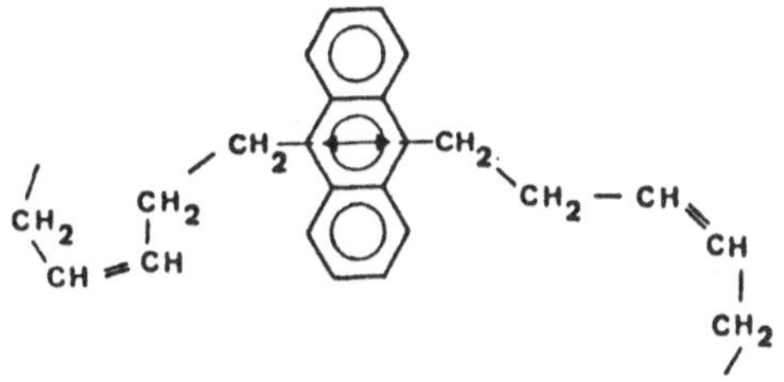

Fig. 12 - Polybutadiene labelled with anthracene in the middle of the
         chain (PBAPB)

segment is almost completely lost within this time. This possibility to
check directly the efficiency of motions associated with the involved
relaxation is a very useful advantage of FAD. In particular, it indicates that in the temperature range 50°C ~ 80°C, we sample continuously
and almost completely the elementary brownian motion in polymer melts.
Processes too fast to be observed by this technique involve only very
small angles of rotation and cannot be associated with backbone rearrangements. On the other hand, the processes too slow to be sampled
involve motions with only a very low residual orientational correlation,
i.e. they are important only on a scale much larger than the size
of conformational jumps.

The different expressions available for the OACF (Table 2) were
tested from the data. As in the case of polymers in solution, the non-
polymer models (rotational diffusion or restricted rotation (RR)) give
a poor fit. This is shown in Fig. 14 where the best fit OACF for the RR
model is compared with the experimental data (Θ = 62.7°C). Thus, the
very specific character of main-chain orientation relaxation is as apparent in melt polymers as in was in solution.

Furthermore we have also observed that the WW expression leads to
very unstable best fit parameters, and non-realistic values for r_o. Even
from a mere curve-fitting point of view, this expression is not as
good as the others (compare Figs. 15 and 16). As an example the best
fit parameters obtained for the different models at 62.7°C are given in
Table 5.

Thus, the WW expression, which appeared to account well for the distribution of relaxation times observed in multimolecular experiments
(37, 38), does not correspond to the shape of the OACF for one given
vector of the chain. This observation may be surprising at first ;
But, multimolecular experiments involve the time evolution of many
intrachain and interchain correlations. Thus, one should expect significant differences from the OACF probed in FAD.

On the other hand, the recent models which accounted well for the
dynamic behavior in solution (HH model and related) seem to account
also for the OACF in melt polymers (see Table 5). This rather good
fit can also be checked by direct visual observation of Fig. 16 (the
small discrepancy observed in channels 20 to 40 corresponds to the
exciting pulse. It may betray a residual scattered stray light, and the

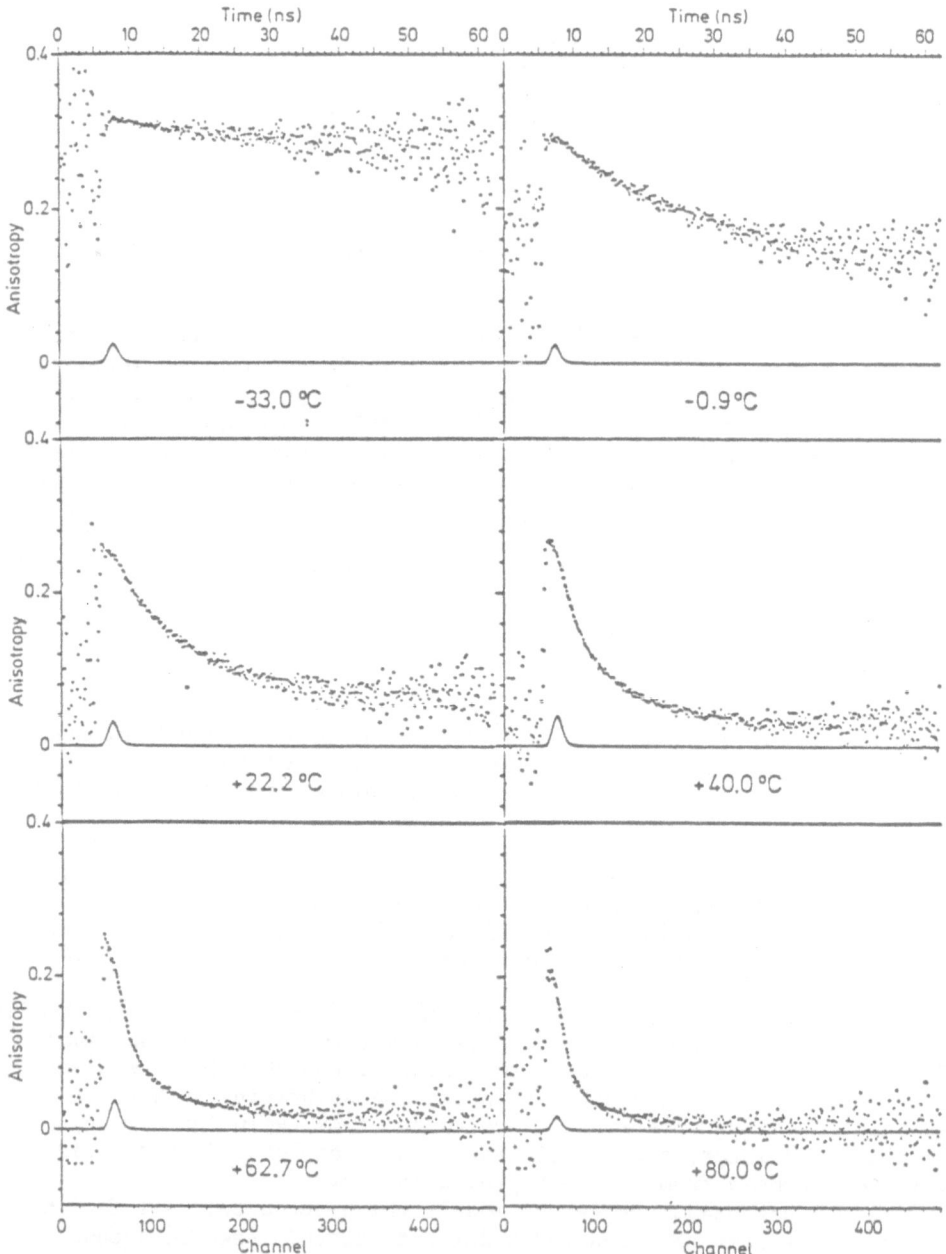

Fig. 13 - Evolution of the FAD of PBAPB in bulk PB as a function of
         temperature.

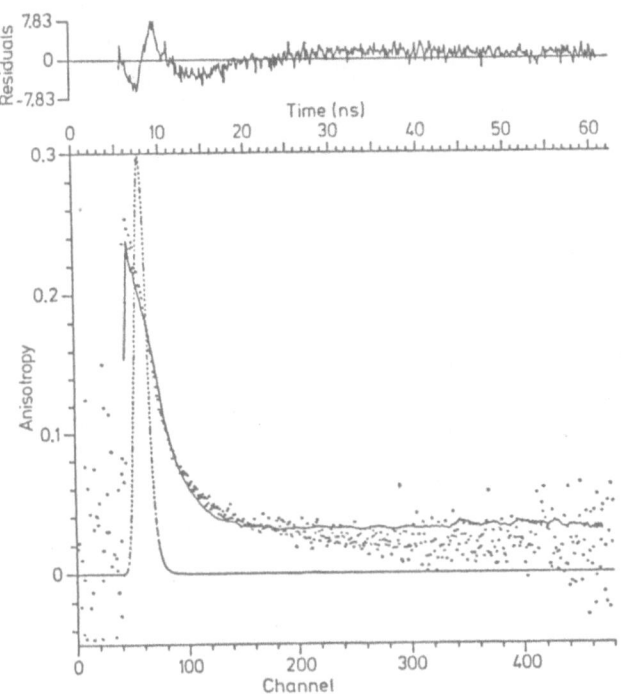

Fig. 14 - Comparison of the RR model to the experimental anisotropy of PBAPB at 62.7°C (same representation as Fig. 8).

Model	Truncation	x^2	r_o	$\tau_1$ns	$\tau_2$ns (or β)	τ_2/τ_1
VJGM	1	1.9709	0.336	1.07	30.6	28.6
	2	1.774	0.345	0.99	31.5	31.8
	3	1.134	0.424	0.60	29.1	48.5
	4	2.416	0.322	1.13	28.9	25.6
	5	2.2754	0.311	1.61	19.5	11.7
WW	1	1.8650	0.333	1.568	0.394	
	2	1.7213	0.347	1.40	0.381	
	3	1.299	0.596	0.28	0.275	
	4	2.200	0.330	1.61	0.400	
	5	1.0017	0.307	1.98	0.446	
HH	1	1.2477	0.234	1.222	96.8	43.6
	2	1.233	0.234	2.22	101.1	45.5
	3	1.1053	0.212	2.73	89.0	32.6
	4	1.434	0.234	2.23	95.2	42.6
	5	1.8238	0.234	2.26	83.8	37.1
BY	1	1.1917	0.242	0.641	151	235
	2	1.1773	0.242	0.638	152	238
	3	1.168	0.230	0.718	139	193
	4	1.327	0.242	0.647	144	222
	5	1.4664	0.240	0.681	112	164

Table 5. Best fit parameters obtained when different models are fitted to the anisotropy of PBAPB at 62.7°C using different experimental windows 1:0 ns-55 ns;2:1.3 ns-55 ns;3:3.2 ns-55 ns;4:0 ns-37 ns;5:0 ns-17 ns

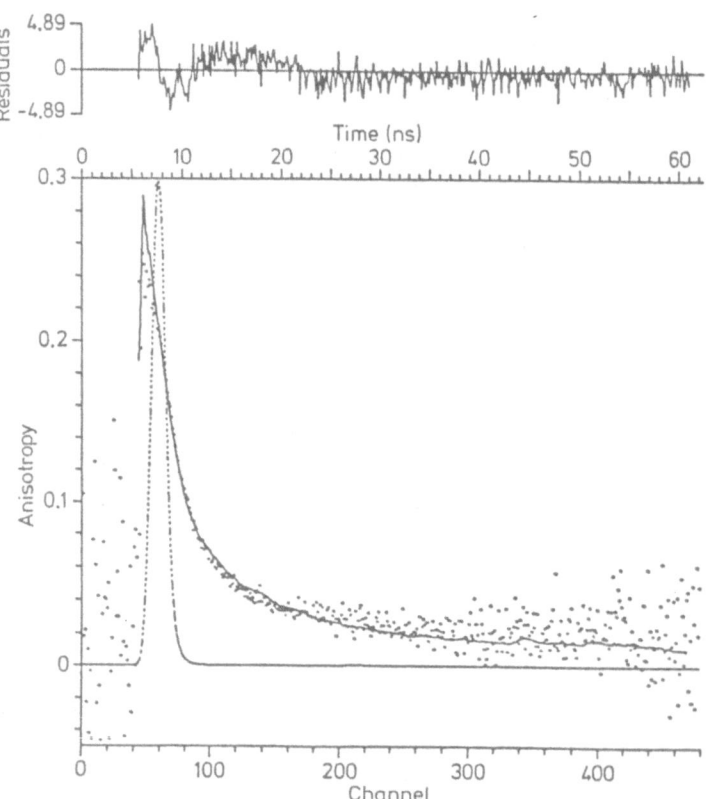

Fig. 15 - Comparison of the WW expression to the experimental anisotropy of PBAPB at 62.7°C (same representation as Fig. 8).

difficulty to perform precise optical experiments in bulk polymers).

Moreover, the ratios τ_2/τ_1 obtained at different temperatures are
similar to the values obtained for labelled polybutadiene in dilute
solution in toluene ($\simeq$ 30). Thus, the surrounding chains seem to affect
local dynamics only at the level of the friction coefficient. Intrachain connectivity remains the essential non isotropic constraint on
local dynamics in bulk polymers and models for the isolated chain are
applicable. This is consistent with the idea that topological interchain
effects only act on a large scale (entanglement distance)and do not
change local dynamic processes.

Now let us look at the temperature effects. The macroscopic dynamic properties of polymers have a well known but rather poorly understood temperature dependence, and it is very important to study the
corresponding temperature evolution of molecular dynamics.

The HH model fits correctly experimental anisotropies in all the
temperature range explored (- 55°C/80°C). But the long time loss term

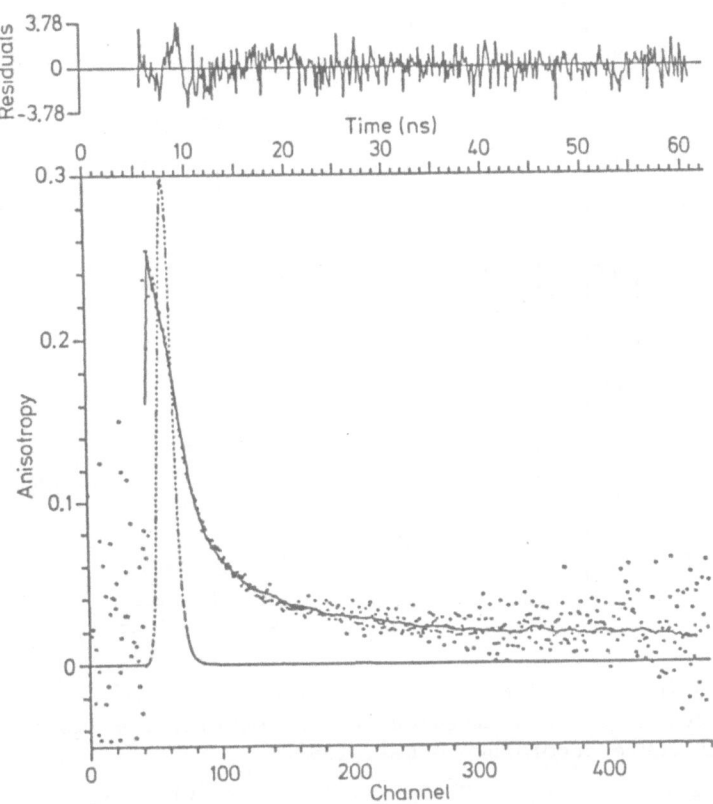

Fig. 16 - Comparison of the HH expression to the experimental anisotropy of PBAPB at 62.7°C (same representations as Fig. 8).

rapidly falls far off the experimental window when the temperature is
decreased, and its characteristic time τ_2 could be determined with a
reasonable precision only above 40°C. In the range 40°C/80°C, the
ratio τ_2/τ_1 does not seem to vary significantly, and remains rather
high ($\simeq 30$). If one accepts the separation of the OACF into a one-dimension diffusion term and a damping term, this observation implies that
the processes responsible for the damping (whatever their molecular
nature is) are slow compared to the diffusive ones. Indeed, when the
HH expression is used without damping ($\tau_2 = \infty$), fitting remains rather
satisfying and χ^2 increases only slighly.

The values of τ_1, plotted on Fig.17 lead to an apparent activation
energy of 40 ± 5 KJ/mole. To compare these results with macroscopic
relaxation, we used the well-known WLF (39) time-temperature superposition equation. According to this equation, the value of the principal or
"glass - transition" relaxation time τ_a at a temperature T_a can be deduced from its value τ_b at T_b as follows :

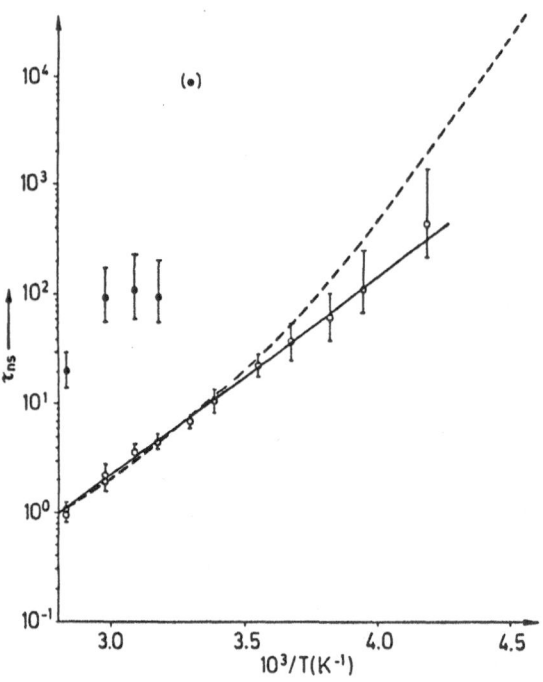

Fig.17-Arrhenius plot of the best fit parameters τ_1 (o) and τ_2 (●)
(HH model) for PBAPB embedded in a PB matrix. The dashed line is
the WLF curve according to Ref. (39)

$$\log\left(\frac{\tau_a}{\tau_b}\right) = \frac{c_1^o(T_b - Tg)}{c_2^o + T_b - Tg} - \frac{c_1^o(T_a - Tg)}{c_2^o + T_a - Tg}$$

Where Tg is the glass transition temperature and c_1^o and c_2^o are
phenomenological paramaters. In figure 17, we have arbitrarily chosen
for τ_h the value of τ_1 at 40.9°C, i.e, 4.54 ns, and applied this equation together with the parameters c_1^o and c_2^o given in Ref (39) from low
frequency mechanical measurements. The corresponding "WLF curve" (dotted
line) fits well with the other measured values of τ_1. Of course, a
similar procedure can be applied to τ_2, but the poorer precision on
this parameter makes it less significant.
        The local reorientation processes observed in FAD and the macros-
copic relaxation have the same temperature behavior. Similar observa-
tinuous excitation (40). However, in this latter technique, the slopes
of the curves depend on the choice of the model, so that the confidence one could put in the agreement (or discrepancy) between spectroscopic and mechanical results relies directly on the confidence one
put in the model arbitrarily chosen to treat the fluorescence polarization data.
        In the present set of experiments, we have first demonstrated that
the model used to interpret the data accounted precisely for the orien-

tation relaxation in a rather wide range of temperatures. Thus, our results strongly support the idea that the local reorientation processes observed in FAD are indeed the elementary processes of macroscopic viscoelasticity.

VIII - BULK POLYMER WITH FREE PROBES

In the studies described above on FAD of labelled polybutadiene, the type of orientation autocorrelation function able to account for the experimental results has been defined. However, no information has been obtained about the size of the chain sequence involved in the motion observed by FAD of labelled polybutadiene. In order to obtain an estimate of this sequence length, FAD measurements have been performed (41) on various 9,10 dialkylanthracene probes (Fig. 18), denoted C_n, with n varying.

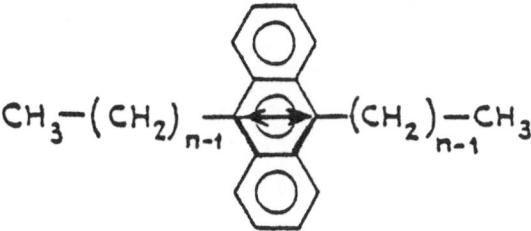

$$CH_3-(CH_2)_{n-1}--(CH_2)_{n-1}-CH_3$$

Fig. 18 - General formula of 9,10 dialkylanthracene. The double arrow represents the transition moment direction

from 1 to 16. A small amount of probe (a few 10^{-6} w/w) is dissolved in a polybutadiene matrix through chloroform solution and solvent removal in vacuum.

The FAD curves at 29.8°C for C_6 to C_{14} probes are shown in Fig.19. The data show quite clearly that the orientation decorrelation rate decreases when the size of the alkyl substituents is increased. During the time of the experiment, the anisotropy decreases to zero in the case of C_6, but it remains significantly higher than zero for C_{14} and even more for C_{16} which is not represented. Furthermore, the shape of the OACF also varies with the length of the tails : the longer the alkyl tail, the more pronounced is the non-exponential character.

As the alkyl substituents are similar to very short polyethylene chains, it seemed interesting to check if their motions could be described by models characteristic of polymer dynamics. Among those, the Jones-Stockmayer (JS) expression (24) is of particular interest since it describes a finite chain with an odd (and generally moderate) number of bonds performing "3-bond" motions in a tetrahedral lattice as in the VJGM model (22). In the JS model the damping of the orientational memory along the polymer chain is replaced by a finite orientation diffusion involving U kinetic units on each side of the central bond whose orientation has been changed by a "3-bond" motion. The resulting expression for the anisotropy time dependence is

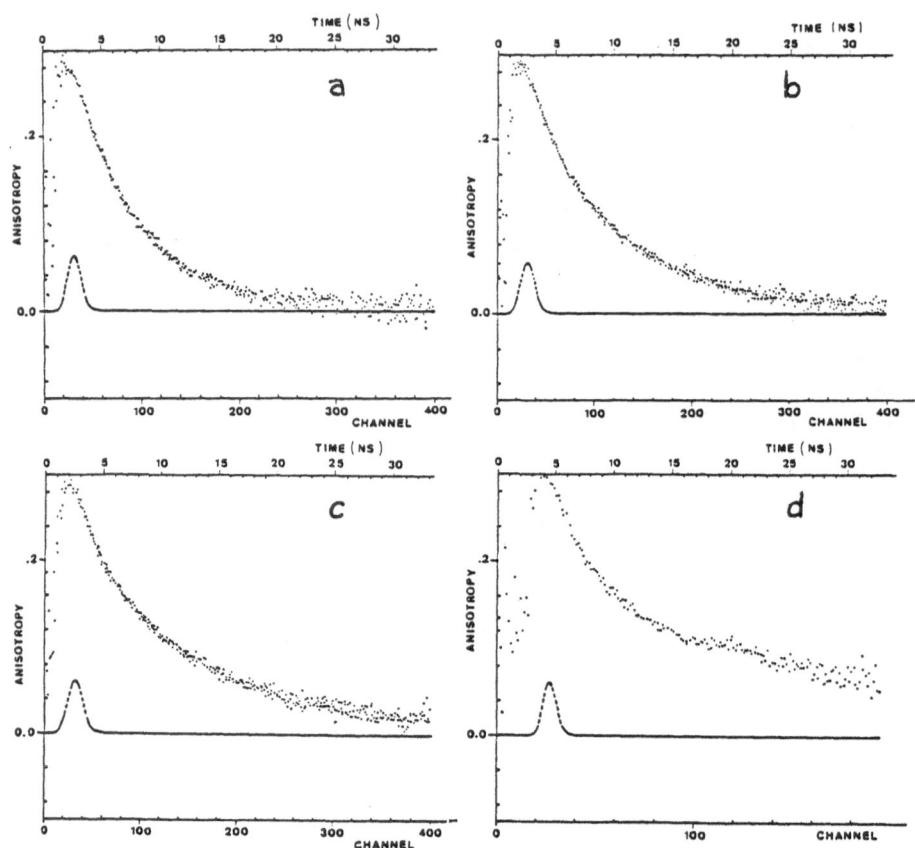

Fig. 19 - Experimental Fluorescence Anisotropy Decay for different C_n probes in polybutadiene. a/ n = 6; b/ n = 8; c/ n = 10; d/ n = 14

JS $r(t) = r_o \sum_{i=1}^{U} a_i \exp(- t/\tau_i)$ where the a_i coefficients are given in Ref (24). The dependence of the number U of kinetic units yielding the best fit with the number n of CH_2 groups in the probe tails is shown in Fig. 20. The slope indicates that a kinetic unit in the JS model would correspond approximately to 3 carbon atoms in aliphatic chains in a polybutadiene matrix.

Besides, it is interesting to examine how the H-H or GDL expressions, given in Table 2 and used for polymer chains in solution or in bulk, can fit the FAD curves of the various C_n probes. It appears that both functions yield a very good fit for C_{14} and C_{16} and in this latter probe the characteristic times τ_1 and τ_2 are the same as in labelled polybutadiene. This gives an estimate of the size of the sequence whose motion is reflected through the FAD of the anthracene label : about 30 bonds. It is worth noting that for shorter tails the quality of the fit rapidly decreases because of a higher contribution of the overall motions compared to the internal (polymer-like) motions. Thus,

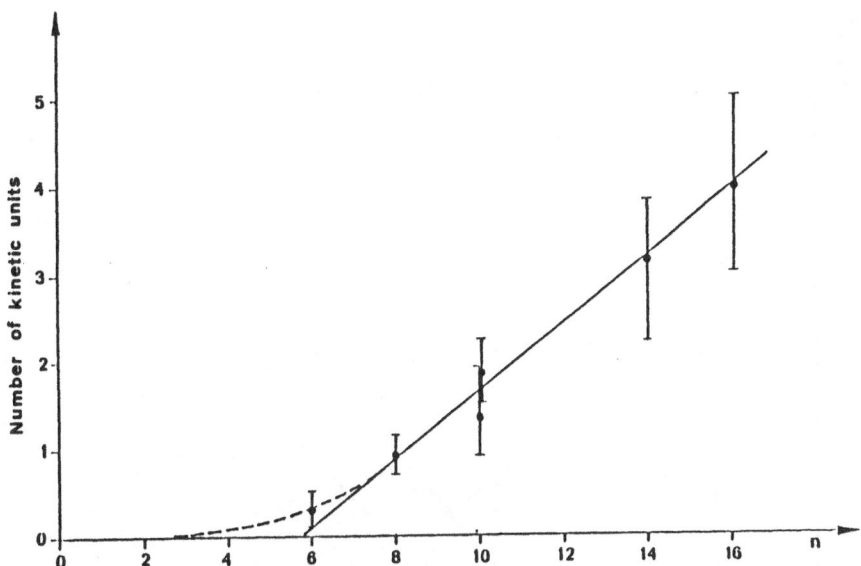

Fig. 20 - Number of kinetics units leading to the best fit for each C_n probe versus the number n of carbon atoms in the alkyl tail.

the C_6 probe behaves roughly like a rigid object whereas C_8 and C_{10} probes are in the crossover region between overall and polymer-like motions.

The temperature dependence of the correlation times, is shown in Fig. 21 for a few C_n probes. The behaviour of τ_1 for C_{14} and C_{16} is identical to the one observed for labelled polybutadiene and it agrees with the W.L.F. equation, indicating that the motions of these probes are coupled to the segmental motions of the polybutadiene matrix which are involved in the glass transition phenomenon. A more surprising result involves the temperature dependence of the overall motion correlation time of C_6. Indeed, it also agrees with the same WLF evolution, showing that the overall motion of probe molecules as small as C_6 is coupled to local motions of the polybutadiene matrix which are involved in the glass transition.

On the contrary, a different behaviour is observed with C_1 probe (Dimethylanthracene). Indeed, due to the geometry of the molecule, depicted in Fig. 22, the C_1 molecules behaves in FAD as an oblate spheroid, the motion of which is characterized by two relaxation times : $\tau_\parallel$ around the y-axis, $\tau_\perp$ around the x-axis (in the spheroidal approximation x and z axes are equivalent). The temperature dependence of $\tau_\parallel$, $\tau_\perp$ as well as that of the mean relaxation time $\bar\tau$, are shown in Fig. 23. The temperature dependence seems Arrhenian but the apparent activation energy is much smaller than the W.L.F. one. Thus the local chain motions associated with the motion of the C_1 probe are not involved in the glass transition of polybutadiene; they could be related to the β-transition of this polymer.

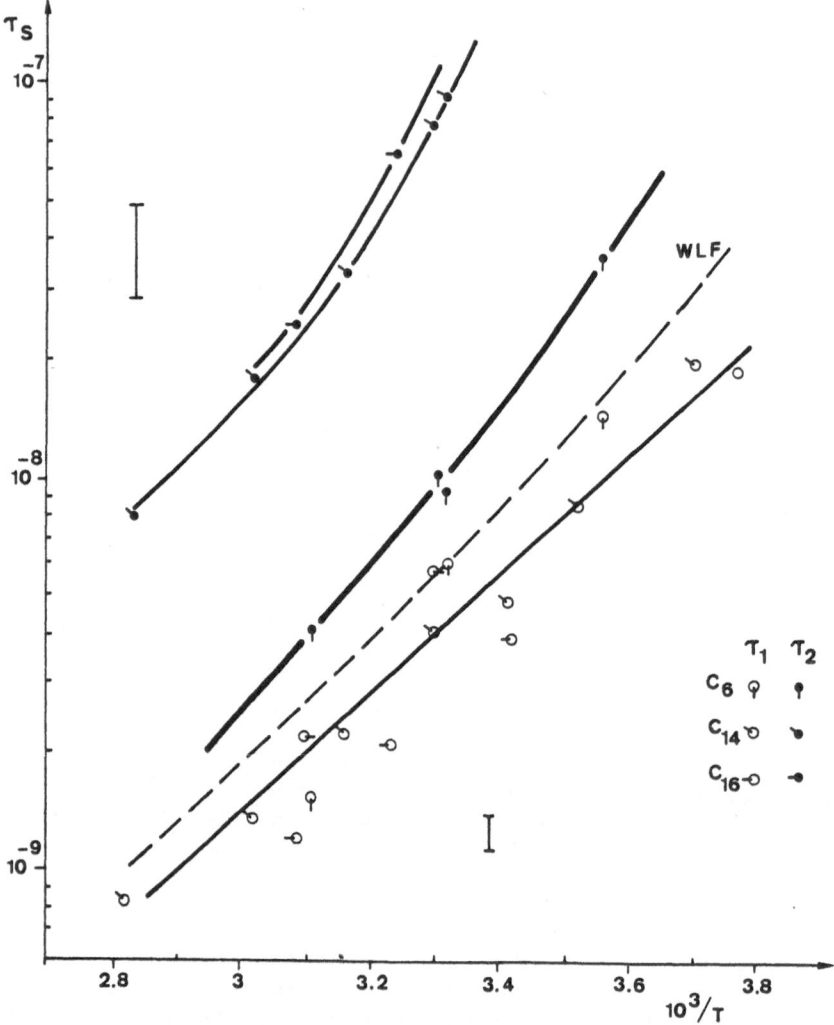

Fig. 21 - Arrhenius plot of the correlation times of various C_n probes
in bulk polybutadiene. The dashed line corresponds to the
WLF expression.

IX - CONCLUSION

In this paper we have shown that the fluorescence anisotropy decay
technique is a powerful tool to examine the orientation autocorrelation
function of labelled chains in solution or in bulk polymers, as well
as at of free probes in a polymer matrix. It clearly appears that
OACF of polymer chains has a specific nature due to the chain connectivity. This implies a non exponential short time term characteristic

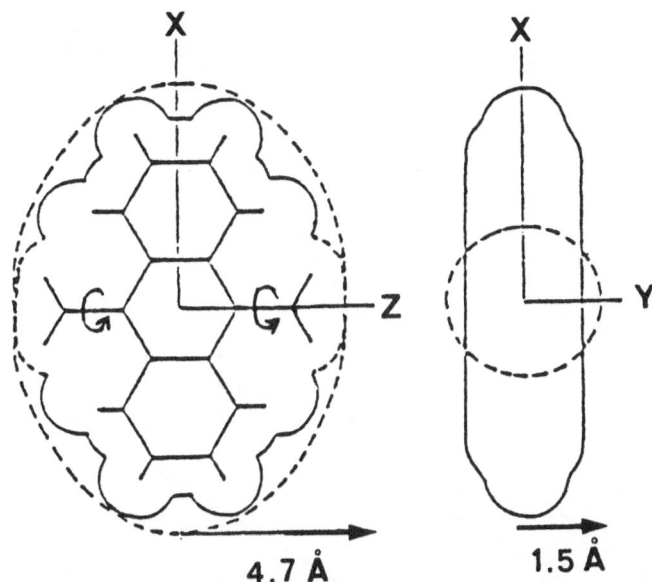

Fig. 22 - Molecular dimensions of 9,10 dimethylanthracene and molecular symmetry axes

of a one-dimension diffusion of bond orientation along the chain sequence. The expression proposed by Hall and Helfand accounts fairly well for the experimental results obtained with labelled chains either in solution or in bulk at a temperature at least 80°C above the glass transition temperature. This unambiguously proves that the polymer chains undergo the same types of segmental motions in the two cases, independently of the constraints arising from the surrounding segments in the bulk state. It is worth noting that this knowledge of the OACF of a polymer chain is very important for it allows to use the H-H expressions to treat NMR relaxation data and to get further details on molecular mechanisms involved in polymer segment motions. Studies on this topic are in progress in our laboratory.

It has also been shown that the segmental motions which are responsible for the orientation decorrelation observed by FAD of labelled polybutadiene, involve about 30 bonds and these motions participate to the glass transition phenomenon. On the contrary, the overall motion of 9,10 dimethylanthracene probe in a polybutadiene matrix is associated with local chain motions which are not involved in the glass transition processes of this polymer.

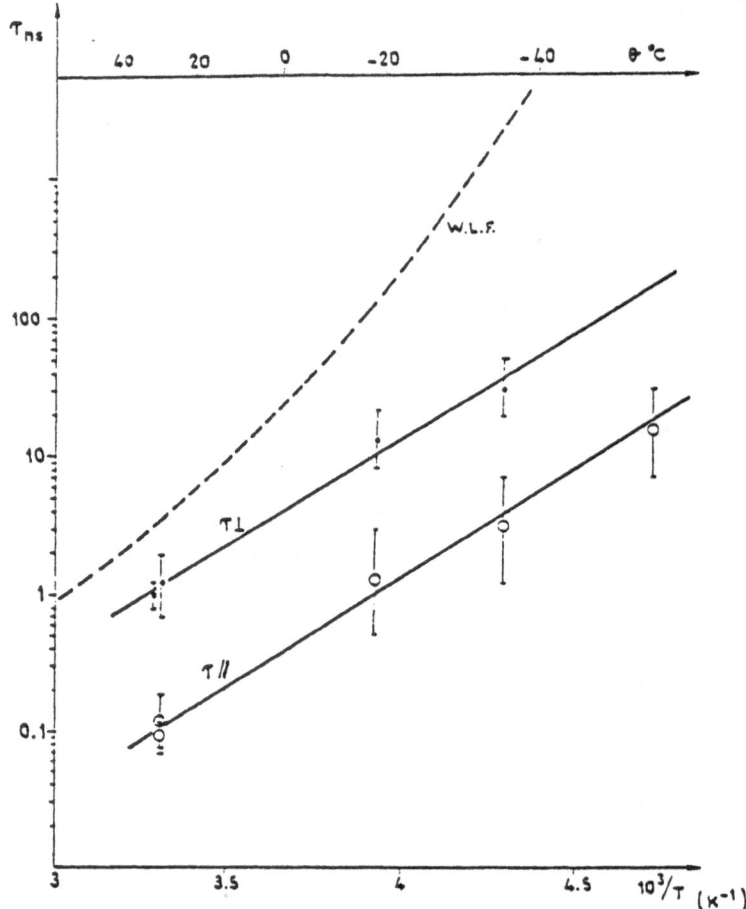

Fig. 23 - Arrhenius plot of the correlation times $\tau_{\parallel}$ and $\tau_{\perp}$ of 9,10
dimethylanthracene in bulk polybutadiene. The dashed line
corresponds to the WLF expression

References

1. Pesce, A.J., Rosen, C.G., Pasby, T.L.: 1971, "Fluorescence Spectroscopy", Marcel Dekker.

2. Wahl, P.:1975, New Tech. Biophys. Cell. Biol. 2, pp.233-183.

3. Wahl, P.: 1975, "Decay of Fluorescence Anisotropy" in "Concepts in
Biochemical Fluorescence" ed. by R.F. Chen and M. Edelhoch, Marcel
Dekker, pp. 1.

4. Brochon, J.C.: 1980, in "Protein Dynamics and Energy Transduction"
ed. by Shin'ichi Ishiwata, Taniguchi Foundation, pp. 163-189.

5. Herons, R.W., Mc Whirter, P., Rhoderick, E.H.: 1956, Proc. Roy. Soc. A 234, pp. 565-583.

6. Bevington, P.R.: 1969, "Data Reduction and Error Analysis for the Physical Sciences", Mc Graw Hill.

7. Rayner, D.M., Mc Kinnon, A.E., Szabo, A.G. : 1976, Can. J. Chem. 54, pp. 3246-3259.

8. Durbin, J., Watson, G.S., : 1951, Biometrika, 38, pp. 159.

9. Viovy, J.L., Monnerie, L., Merola, F.: Macromolecules, in press.

10. Krakoviak, M.G.; in Anufrieva, E.V., Gotlib, Yu. Yu.: 1981, Adv. Polym. Sci., 40, pp. 1-68.

11. Valeur, B., Monnerie, L.: 1976, J. Polym. Sci., Polym. Phys. Ed., 14, pp. 11-27.

12. Debye, P.: 1929, "Polar Molecules", Chemical Catalog Co.

13. Perrin, F.: 1929, Ann. Phys. 12, pp. 169-275.

14. Perrin, F.: 1934, J. Phys. Le Radium, 5, pp. 497-511.

15. Perrin, F.: 1936, J. Phys. Le Radium, 7, pp. 1-11.

16. Favro, L.D.: 1960, Phys. Rev., 119, pp. 53-62.

17. Hu, C.M., Zwanzig, R.: 1974, J. Chem. Phys. 60, pp. 4354-4357.

18. Youngren, G.K., Acrivos, A.: 1975, J. Chem. Phys., 63, pp. 3846-3848.

19. Kinosita, K., Kawaro, S., Ikegami, A.: 1977, Biophysical Journal, 20, pp. 289-305 .

20. Wahl, P.: 1975, Chem. Phys. 7, pp. 210-219.

21. Rouse, P.E.: 1953, J. Chem. Phys., 21, pp. 1272-1280.

22. Valeur, B., Jarry, J.P., Gény, F., Monnerie, L.: 1975, J. Polym. Sci. , Polym. Phys. Ed., 13, pp. 667-674, pp. 675-682, pp. 2251.

23. Bendler, J.T., Yarris, R.: 1978, Macromolecules, 11, pp. 650-655.

24. Jones, A.A., Stockmayer, W.H.: 1975, J. Polym. Sci., Polym. Phys. Ed., 15, pp. 847-861.

25. Viovy, J.L., Monnerie, L., Brochon, J.C.: 1983, Macromolecules, 16, pp. 1845-1852.

26. Hall, C.K., Helfand, E.: 1982, J. Chem. Phys., 77, pp. 3275-3282.

27. Weber, T.A., Helfand, E.: 1983, J. Phys. Chem., 87, pp. 2881-2889.

28. Williams, G., Watts, D.C.: 1970, Trans. Farad. Soc., 66, pp. 80-85.

29. Cohen, M.H., Turnbull, D.: 1959, J. Chem. Phys., 31, pp. 1164-1169; 1961, ibidem, 34, pp. 120-125.

30. Cohen, M.H., Grest, G.S.: 1979 Phys. Rev. B., 20, pp. 1077-1098.

31. Gibbs, J.H., Di Marzio, E.A. 1958, J. Chem. Phys., 28, pp. 373-383; and pp. 807-813.

32. Adams, G., Gibbs, J.H.: 1965, J. Chem. Phys., 43, pp. 139-146.

33. de Gennes, P.G., Léger, L. : 1982, Annual. Rev. Phys. Chem., 33, pp. 49-61.

34. de Gennes, P.G.: 1971, J. Chem. Phys., 55, pp. 572-579.

35. Klein, J.: 1978, Macromolecules, 11, pp. 852-858.

36. Viovy, J.L., Monnerie, L.: Macromolecules, in press.

37. Williams, G.: 1979, Adv. Pol. Sci., 33, pp. 59-92.

38. Patterson, G.D.; 1982, Adv. Pol. Sci., 48, pp. 125-159.

39. Ferry, J.D.: 1980 "Viscoelastic Properties of Polymers" 3 rd ed., Wiley.

40. Jarry, J.P., Monnerie, L.: 1979, Macromolecules, 12, pp. 927-932.

41. Viovy, J.L., Frank, C.W., Monnerie, L.: Macromolecules, in press.

FLUORESCENCE PROBING OF THE LOCAL DYNAMICS OF POLYMERS : A MODEL APPROACH

J. Vandendriessche, R. Goedeweeck, P. Collart and F.C. De Schryver
KULeuven, Department of Chemistry,
Celestijnenlaan 200 F
B-3030 Heverlee (Leuven), Belgium

I.   INTRODUCTION

The importance of the ground state configurational distribution of a polymer chain in solution in relation to its physical properties has been well recognized. Differences in the photophysical properties of polymers with different tacticity have been noted in the past. However, an in depth study of the relation between the configurational distribution and polymer photophysics is still, in view of the complexity of the possible relaxation pathways in a polymer system, a difficult task.

It has been our aim over the last years to tackle part of this problem by studying different model systems (dyads) mimicking the local configurational and conformational distribution of polymer systems. Using the proper fluorescence of chromophores covalently linked to a chain unit reflecting either vinyl polymers or polypeptides, the differences in the photophysical behaviour were correlated to information of the ground state distribution obtained by other techniques. Factors influencing the ground state distribution were found to affect the excited state properties. Such an analysis permits the acquisition of data related to chain dynamics of these systems. It is however clear that increasing the number of repeating units will influence the dynamic behaviour of the base unit and that furthermore excitation transfer will influence the (re)distribution of the excitation energy over the chromophores present. To gain some insight into this problem a triad model system was investigated but further work along this line is necessary to understand the different parameters that are of importance.

It is furthermore clear that a study of the spectral properties of a polymer system as such is an insufficient base of information and that a study of the time dependence of the emission intensity is a prerequisite.

M. A. Winnik (ed.), Photophysical and Photochemical Tools in Polymer Science, 225–261.
© *1986 by D. Reidel Publishing Company.*

II. LOCAL DYNAMICS OF POLYVINYL AROMATIC SYSTEMS

II.1. 2,4-diarylpentanes and 1,1'-diaryldiethylethers

An important contribution to the understanding of the spectral and dynamical properties of polymers in solution is made by the study of the ground and excited state behaviour of 2,4-diarylpentanes and 1,1'-diaryl diethyl ethers. These compounds do exist in two diastereoisomeric forms. The properties of the meso isomer can be correlated with the properties of an (isolated) isotactic dyad in the polymer. The racemic form is used as model for the processes of a syndiotactic dyad in the polymer.

$$CH_3 \diagdown \underset{\underset{R}{|}}{CH} \diagup X \diagdown \underset{\underset{R}{|}}{CH} \diagup CH_3$$

X : A = $-CH_2-$ R : 1 : phenyl
        B = O   2 : 1-naphthyl
                                         3 : 2-naphthyl
                                         4 : 9-anthryl
                                         5 : 1-pyrenyl
                                         6 : 9-carbazolyl

The ground and excited state properties of A1, A3, A5, A6, B2, B3, B4 and B5 will be discussed.

II.1.1. 2,4-diphenylpentanes (A1).

The assignment of the configuration and the calculation of the conformational distribution within a configuration in the ground state could be obtained from ^{1}H-NMR measurements (Bovey et al., 1965). In meso-2,4-disubstituted pentanes the methylene protons H_B and H_C are heterosteric (fig.1). H_B and H_C are in different environments in all conformations and do not exchange environments upon conformational equilibrium between mirror image forms. H_B and H_C will thus always have different chemical shifts. Figure 1 represents the staggered conformations of meso 2,4-diphenylpentane. Not shown in figure 1 are the mirror image forms of TG, $T\bar{G}$ and $G\bar{G}$ (i.e. GT, $\bar{G}T$ and $\bar{G}G$). In the nomenclature of the staggered conformations capital letters are used. This is done to distinguish these conformations from the propane conformations as referred to in articles concerned with excimer formation in the latter compounds. Note that T is an alkyl group between H_B and H_C; G is an aryl group between H_B and H_C; $\bar{G}$ means an H between H_B and H_C.

In racemic 2,4-disubstituted pentanes, the methylene protons H_B and $H_{B'}$ are homosteric. On the average H_B and $H_{B'}$ are in the same environment. This can be derived from figure 2 when the identomers of TG, $T\bar{G}$ and $G\bar{G}$ (i.e. GT, $\bar{G}T$ and $\bar{G}G$) are drawn. The presence or absence of a difference in chemical shift of the methylene protons makes the assignment of the configurations possible.

Information about the conformer populations is derived from the vicinal coupling constants. The observed vicinal couplings in a substi-

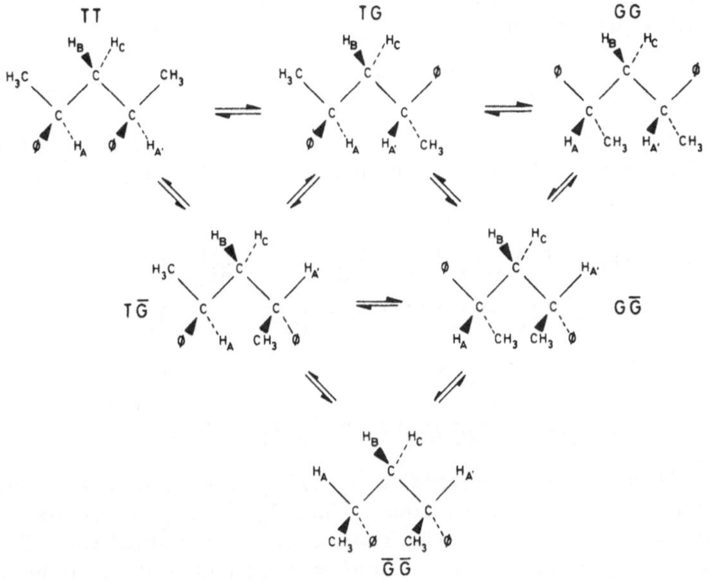

Figure 1.   Staggered conformations of meso-2,4-diphenylpentane.

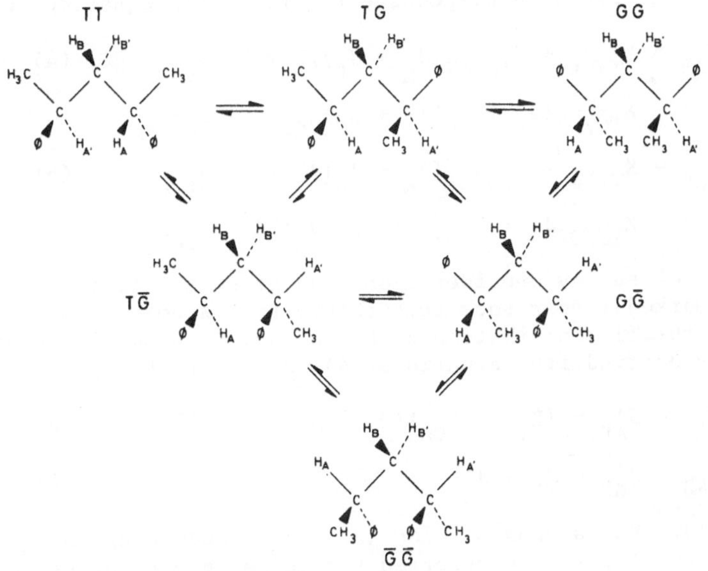

Figure 2.   Staggered conformations of racemic-2,4-diphenylpen-
tane.

stituted ethane are the average over the conformations present (rapid conformational equilibrium) :

$$J_{obs} = \sum_i X_i \, J_i \qquad (1)$$

where X_i and J_i represent the mole fraction and vicinal coupling of each conformer. For meso-2,4-diphenylpentane the following equations can be written

$$J_{AB} = X_{TT}J_t + X_{TG/GT}(J_t + J_g)/2 + X_{GG}J_g + \qquad (2)$$
$$X_{T\bar{G}/\bar{G}T}(J_t + J_g)/2 + X_{G\bar{G}/\bar{G}G}J_g + X_{\bar{G}\bar{G}}J_g$$

$$J_{AC} = X_{TT}J_g + X_{TG/GT}(J_g + J_t)/2 + X_{GG}J_t + \qquad (3)$$
$$X_{T\bar{G}/\bar{G}T}J_g + X_{G\bar{G}/\bar{G}G}(J_t + J_g)/_2 + X_{\bar{G}\bar{G}}J_g$$

In equation (2) and (3) an equality of all gauche and trans couplings is assumed. Experimentally it is found that $J_{AB} \approx J_{AC} = 7.4$ Hz. This equality excludes the possibility that either conformation TT or conformation GG is strongly preferred. Conformation $\overline{GG}$ can also be excluded since the couplings should be equal and in the order of 2-3 Hz. The same conclusions are obtained when steric considerations are made. Steric considerations, supported by nonbonded interaction energy calculations (McMahon and Tincher, 1965) indicate that in general the TG/GT conformation will be very strongly preferred. This is in agreement with the observed couplings (Bovey et al., 1965).

For racemic 2,4-diphenylpentane the following equations can be written :

$$J_{AB} = X_{TT}J_t + X_{TG/GT}(J_g + J_t)/2 + X_{GG}J_g + \qquad (4)$$
$$X_{T\bar{G}/\bar{G}T}(J_t + J_g)/2 + X_{G\bar{G}/\bar{G}G}J_g + X_{\bar{G}\bar{G}}J_g$$

$$J'_{AB} = X_{TT}J_g + X_{TG/GT}(J_g + J_t)/2 + X_{GG}J_t + \qquad (5)$$
$$X_{T\bar{G}/\bar{G}T}J_g + X_{G\bar{G}/\bar{G}G}(J_t + J_g)/2 + X_{\bar{G}\bar{G}}J_g$$

On the basis of energy considerations and given the fact that J_{AB} and J'_{AB} differ markedly near room temperature for racemic A1 one can state that either the TT conformation or GG conformation must be preferred. It is easily derived from equations (4) and (5) that

$$J_{AB} - J'_{AB} = (X_{TT} - X_{GG})(J_t - J_g) \qquad (6)$$

$$J_{AB} + J'_{AB} = J_t + J_g \qquad (7)$$

A value of 10.0 Hz is chosen   for $(J_t - J_g)$, according to J_t and J_g values from the literature reported for large numbers of ethanes and cyclic compounds. The question remains :which conformation is the most stable one. For racemic A1 no real proof is found, but in analogy with dichloro, dibromo, dicyanide and diol pentanes, for which energy calculations were performed (McMahon and Tincher, 1965), it is assumed that the TT conformation is the more stable. The conformational distribution is independent of the solvents reported (CS_2, CCl_2 and o-dichlorobenzene) but changes with temperature. $\Delta H^\circ_{rac} = 6.9$ kJmol^{-1}; $\Delta S^\circ_{rac} = 13.4$ Jmol^{-1}K^{-1} (Bovey et al., 1965).

Another technique used in the analysis of the configuration and conformations of the 2,4-diphenylpentanes is I.R. spectroscopy (Jasse et al., 1973). The influence of the configuration and conformation of the different isomers in solution was shown by the alteration of the shape of the absorption bands in the range 500-600 cm^{-1}.

For meso Al only one absorption band is observed (550 cm^{-1}). This band is assigned to the TG/GT conformation. Upon temperature variation only thermic broadening of the band could be observed. For racemic Al two absorption bands are observed (544 and 553 cm^{-1}). The temperature dependence of the two bands was studied and a ΔH°_{rac} of $\pm$ 8.4 kJmol^{-1} was found.

Theoretical calculations (Gorin and Monnerie, 1970) of the conformational energy of racemic Al reveal that the TT conformation is indeed the most stable. These calculations take into account Van der Waals interactions, electronic interactions and electrostatic interactions. For meso Al the TG/GT conformation is indeed an energetic minimum but according to the calculations the TT conformation should even be more stable (9.6 kJmol^{-1}) than the TG/GT conformation. Gorin and Monnerie (1970) explain the absence of experimental observations of the TT conformation as due to the shape of the minimum. Small variations in angles of the chain and/or of the phenyl moiety do increase the energy of the system rapidly. This corresponds to an important entropic decrease when the TT conformation is reached.

In a study of the conformational characteristics of polystyrene Yoon et al. (1975) calculated the conformational energy contours for a meso and a racemic dyad, taking into account the interactions of the chain with solvent. The energy content at the TT minimum in meso Al is about 3.8 kJmol^{-1} higher than the energy content of the TG/GT minima. The energy barrier between the TT and TG/GT conformations lies about 10.4 kJmol^{-1} higher than the TG/GT conformation. In racemic Al the energy of TT is 10.0 kJmol^{-1} lower than the GG minimum. TT can be converted to GG via the TG/GT conformation. The latter lies 14.2 kJmol^{-1} higher than TT. The barrier between TT and TG/GT equals about 16-21 kJmol^{-1}. Between TG/GT and TT the barrier is about 8.4 kJmol^{-1}. A direct equilibrium between TT and GG can be excluded since the barrier for this process is larger than 25 kJmol^{-1}.

A fourth contribution to the knowledge of ground state conformations and equilibria was obtained from ultrasonic relaxation measurements on the 2,4-diphenylpentanes in solution (Froelich et al., 1976). A single relaxation is observed for meso Al as well as for racemic Al. For meso Al the relaxation was assigned to the conformational equilibrium between TG/GT and TT. Their difference in enthalpy and entropy was estimated at 5.8 $\pm$ 2.9 kJmol^{-1} and -5.4 $\pm$ 4.2 Jmol^{-1}K^{-1} respectively. It was argued that the TG/GT state is the lower energy conformation since a negative ΔS is in agreement with a degenerate lower state and hindered rotation of the phenyl rings in the TT conformation will also reduce ΔS. The equilibrium in racemic A1 has an enthalpy difference of 5.8 + 0.8 kJmol^{-1} and an entropy difference of 13.0 + 1.3 Jmol^{-1}K^{-1}. Froelich et al. (1976) stipulated that these latter values agree with the values found by Bovey et al. (1965) but that the assignment of the equilibrium TT $\leftrightarrow$ GG cannot be certain because of the earlier reported presence of the TG/GT state (Yoon et al., 1975). The positive ΔS value, however, is consistent with a GG upper state where the phenyl rings are less restricted to rotation than in the TT conformation.

With all this information about ground state conformations and equilibria it is interesting to look at the photophysical behavior of the meso and racemic 2,4-diphenylpentane in solution.

Stationary state measurements (Longworth and Bovey, 1966, Bokobza et al., 1977) as well as measurements of the transient behavior of the excited states (De Schryver et al., 1982a) were reported. Longworth and Bovey (1966) published the uncorrected fluorescence spectra of meso A1 and racemic A1 at 298 and 77 K. At 77 K the spectra are, within experimental error, identical and resemble the spectrum of toluene at 77 K. At 298 K excimer emission is detected for meso A1 and racemic A1. The ratio of the intensity of the excimer emission (I_D, at 313 nm) over the intensity of the locally excited state emission (I_M, at 283 nm) equals 2.45 for meso A1 and 0.74 for racemic A1. The difference in excimer forming capacity is explained based on the NMR data reported above (Bovey et al., 1965). The meso isomer starting from the TG/GT form reaches the TT conformation via an easy bond rotation, leading to a full overlap of the phenyl chromophores. The racemic isomer is predominantly TT (75%), with about 25% of GG present at room temperature. The rotations to TG/$\overline{G}$T (also full overlap of the phenyl chromophores) are less favourable leading to less excimer species within the radiative lifetime of the excited phenyl groups. At 77 K, however, the time required for bond rotation to the appropriate conformation exceeds the radiative lifetime and therefore no excimer emission is observed.

Bokobza et al. (1977) measured the corrected fluorescence spectra of meso and racemic A1 in function of temperature. Applying the equations derived for the monomolecular analogue of the scheme of Birks (Scheme I) the slope of a plot of ln ϕ_D/ϕ_M as a function of the reci-

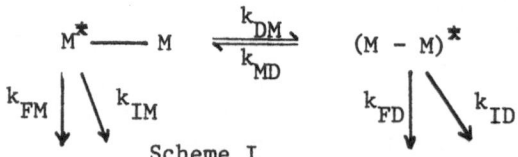

Scheme I

procal temperature yields the activation energy of excimer formation. In this scheme a direct equilibrium between one locally excited state and one excimer is proposed. The values of 8.4 and 19.2 kJmol^{-1} respectively for the activation energy of excimer formation for meso A1 and racemic A1 support the conclusion that meso A1 forms more easily an excimer conformation than racemic A1. No justification was given for the use of the kinetic scheme in view of the different conformations present in racemic A1.

Using a time-correlated single-photon counting technique, De Schryver et al. (1982a) measured the time dependence of the fluorescence intensity of meso Al and racemic Al at 282 nm (locally excited state emission) and at 335 nm (excimer emission). Both compounds can be analyzed within scheme I. The activation energy of excimer formation in meso Al equals 9.6 kJmol^{-1}, this value is, within experimental error, the same as found from stationary state measurements and suggests that meso Al forms the excimer TT starting from the locally excited TG/GT state.

The activation energy of the excimer formation in racemic Al equals 18.8 kJmol^{-1}. This barrier is lower than the minimal barrier of 25 kJmol^{-1} calculated for the direct equilibrium between TT and TG (Yoon et al., 1975). Furthermore no explanation was given for the fact that two ground state species are excited (TT and GG) but only one locally excited state species is observed. Probably the excimer TG is formed via the TG state (a common intermediate for TT and GG as can be seen from figure 2), the latter step being the rate determining (Moens,1985). Kinetic, thermodynamic and spectral data are reported in table I.

Reports on ground or excited state properties of 1,1'-diphenyldiethylether are not known to the authors.

II.1.2. 2,4-dinaphthylpentanes (A3) and 1,1'-dinaphthyldiethyl ethers (B2, B3).

II.1.2.1. 1,1'-di(1-naphthyl)diethyl ether (B2). To our knowledgde no information about the ground state conformations and equilibria in B2 is available. A comparison of the relative position of methine and methyl protons in the ^{1}H-NMR spectra of the two diastereoisomers of B2 with the relative position of methine and methyl protons in Al, A3, A6 and B4 made the assignment of "meso" and "racemic" possible. This assignment is confirmed when the relative positions of the ^{13}C-NMR signals of methine and methyl carbons are compared in the same series of compounds.

Data on the photophysical properties of B2 in solution are scarce. The fluorescence spectrum of meso B2 in isooctane measured as a function of temperature indicates the presence of two emitting excimers (De Schryver et al., 1981). More information on the excited state properties of meso and racemic B2 are reported by Demeyer (1982). The ratio I_D/I_M (400nm/330nm) at 293 K in isooctane equals 1.76 for meso B2 and 0.89 for racemic B2 indicating easier excimer formation in meso B2. The decay curves obtained for meso B2 as well as for racemic B2 at 330 nm (locally excited state emission) and 400 nm (excimer emission region) cannot be analyzed with two exponentially decaying terms. This indicates a very complex excited state behaviour. Furthermore, upon irradiation of B2 with $\lambda> 280$ nm (isooctane, 10^{-4} M) the formation of cyclomers is observed. The structure of these cyclomers is obtained from ^{1}H-NMR measurements and represented in figure 3.

The cyclomer formed in meso B2 ($\bar{\alpha}$ = 1.010^{-2}) is an exocyclomer which at 77 K in a rigid matrix upon irradiation with λ =250 nm can be cleaved . A broad structureless emission band with maximum at 390 nm is observed. This maximum corresponds to the emission of the hypsochromic excimer of meso B2 in solution. The cyclomer formed in racemic B2 ($\bar{\alpha}$ = 0.410^{-2}) is also an exocyclomer which at 77 K in a rigid matrix upon irradiation (λ =250 nm) cleaves and an emission band with a maximum at 410

Table I. Spectral, kinetic and thermodynamic properties of
2,4-diarylpentanes and 1,1'-diaryldiethyl ethers

a. Spectral properties

Compound	Solvent (b)	Excimer Max. (nm)	$\nu_{oo} - \nu_{excim}$ (cm^{-1})	FWHM (cm^{-1})
Racemic A1	IO	330	5000	5900
Meso A1	IO	330	5000	5900
Racemic B2	IO	405	6500	6500
Meso B2	IO	385/420	5200/7300	4600
Racemic A3	THF	405	6500	--
Meso A3	THF	405	6500	4100
Racemic B3	IO	390	5500	4200
Meso B3	IO	375/390	4500/5500	4300
Racemic B4	CHX	480	5100	4000
Racemic A5	IO	490	6100	3600
Meso A5	IO	450/490	4300/6100	3600/4000
Racemic B5	IO	505	6700	3700
Meso B5	IO	450/490	4300/6100	3600/4600
Racemic A6	IO	370	2200	2600
Meso A6	IO	420	5400	3400

b. Kinetic and thermodynamic properties[c]

Compound	Solvent[b]	E_{DM} kJ mol^{-1}	k'_{DM} 10^{10} s^{-1}	E_{MD} kJ mol^{-1}	k'_{MD} 10^{12} s^{-1}	E_{ID} kJ mol^{-1}	k'_{ID} 10^{7} s^{-1}	$\Delta H°$ kJ mol^{-1}	$\Delta S°$ Jmol^{-1} K^{-1}
Racemic A1	IO	18.8	--	37.7	--	--	--	-18.8	-46.0
Meso A1	IO	9.6	10	--	--	0.4	55	--	--
Racemic B2	IO	9.6	--	--	--	--	--	--	--
Meso B2	IO	9.6	--	--	--	--	--	--	--
Racemic A3	THF	23.0	--	--	--	--	--	--	--
Meso A3	THF	20.1	--	--	--	--	--	--	--
Meso A5	IO	11.7 20.8	1.6 150	--	--	--	--	--	--
Racemic B5	IO	16.0 16.0	33 260	--	--	--	--	--	--
Meso B5	IO	8.3 18.7	6.5 330	--	--	--	--	--	--
Meso A6	IO	16.0	83	35.8	14	1.6	2.2	-19.8	-23.5

(a) The data in this table are taken from the references cited
in the text; (b) IO = isooctane, THF = tetrahydrofuran, CHX =
cyclohexane; (c) E_{DM}, E_{MD} = activation energy of excimer formation and dissociation respectively, k'_{DM}, k'_{MD} : preexponential factors of the rate constants of excimer formation and
dissociation respectively, E_{ID}, k'_{ID} activation energy and pre-
exponential factor of the rate constant of non radiative decay
of the excimer, $\Delta H°$, $\Delta S°$ enthalpy and entropy differences on
the excimer formation process.

nm, corresponding to the excimer emission observed in racemic B2 in solution, is observed.

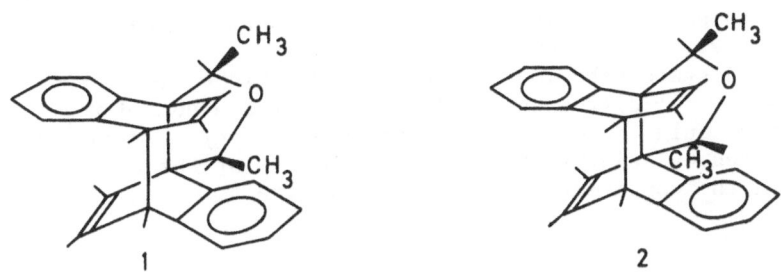

Figure 3.   Cyclomers of meso B2 (1) and racemic B2 (2).

Until now no information is available on the ground or excited
state properties of 2,4-di(1-naphthyl)pentane.

II.1.2.2. 2,4-di(2-naphthyl)pentane (A3) and 1,1'-di(2-naphthyl)diethyl
ethers (B3). Information about the conformational distribution in the ground state of A3 is obtained from conformational energy
calculations (Ito et al., 1982). The results for meso A3 and racemic
A3 are comparable with the results described earlier (meso A1 and racemic A1). Meso A3 has one predominant conformation TG/GT. The TT excimer conformation is 19.2 kJmol^{-1} higher in energy, only 1-2 % populated
at room temperature. Ito et al. (1982) assume that only the full overlap of the chromophores is possible in this TT excimer conformation.
No evidence or argument is given to support this assumption. Racemic
A3 has two important ground state conformations : TT and GG. TT is the
more stable conformation (82 % at room temperature). The excimer con-
formation T$\bar{\text{G}}$ in racemic A3 is reached via one rotation (TT → T$\bar{\text{G}}$,barrier
27 kJmol^{-1}) or two rotations (TT → TG → T$\bar{\text{G}}$, barrier 13-17 kJmol^{-1}). GG
reaches the excimer conformation via the same intermediate TG. Also
for racemic A3 the excimer conformation is suggested to have a full
overlap of the chromophores.
      Ito et al.(1981, 1982) reported also on the photophysical proper-
ties of meso and racemic A3. The same observation as before is made
when the yield of excimer formation of meso A3 is compared with the
yield in racemic A3. Meso attains the excimer geometry more readily, assuming that the radiative rate constant of the excimer and the stabili-
sation of the excimer are of comparable magnitude. Ito et al. (1981)
were able to analyse the fluorescence spectra and decay curves of meso
A3 and racemic A3 within scheme I. The activation energy for excimer
formation equals 20.1 and 23.0 kJmol^{-1} for meso A3 and racemic A3 respectively. Large differences in excimer formation rate constants were
reported (7.9 10^8 s^{-1} ↔ 8.7 10^7 s^{-1} at 298 K meso A3 ↔ racemic A3).

These differences were correlated with the conformational changes required for excimer formation (Ito et al. 1982). In meso A3 the excimer
conformation is formed via a one rotation process. In racemic A3 more
than one rotation takes place before the excimer geometry is reached.
For both compounds the excimer geometry is supposed to be the full overlap geometry. A more complete analysis of ground state conformations
and excited state behaviour of meso and racemic A3 seems necessary when
the data on 1,1'-di(2-naphthyl)diethyl ether are considered.

The calculation of the conformational energy of meso and racemic B3
was reported by Pajot (1983). The definition of the four angles who
determine the conformation is given in figure 4.

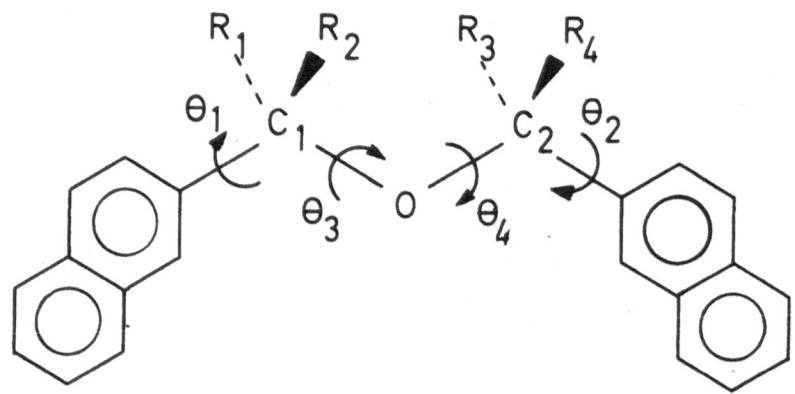

Figure 4. Definition of configurations and conformations in
B3 R_1 = R_3 = CH_3, R_2 = R_4 = H, θ_1 - θ_4 = o →
meso B3 in GG conformation; R_1 = R_4 = CH_3, R_2 = R_3
= H, θ_1 - θ_4 = o → racemic B3 in GG conformation.

A conformation is represented with the matrix (θ_1, θ_3, θ_4, θ_2). The
excimer geometries are defined in figure 5.
Furthermore a distinction is made whether the methyl group is equatorial
(e) or axial (a) versus the naphthalene moiety (fig. 6)
The energy of the e conformations is too high due to steric crowding and
will not be considered further.

Meso B3 has three ground state conformations of comparable energy.
M_1 is defined by (0, +60, -20, 0). This is an intermediate conformation
between GG and TG. This conformation is chosen as reference and its
energy is put equal to 0 kJmol^{-1}. A second minimum is situated at
(+140, +100, -20, 0). This is the TG conformation and lies about 1.3

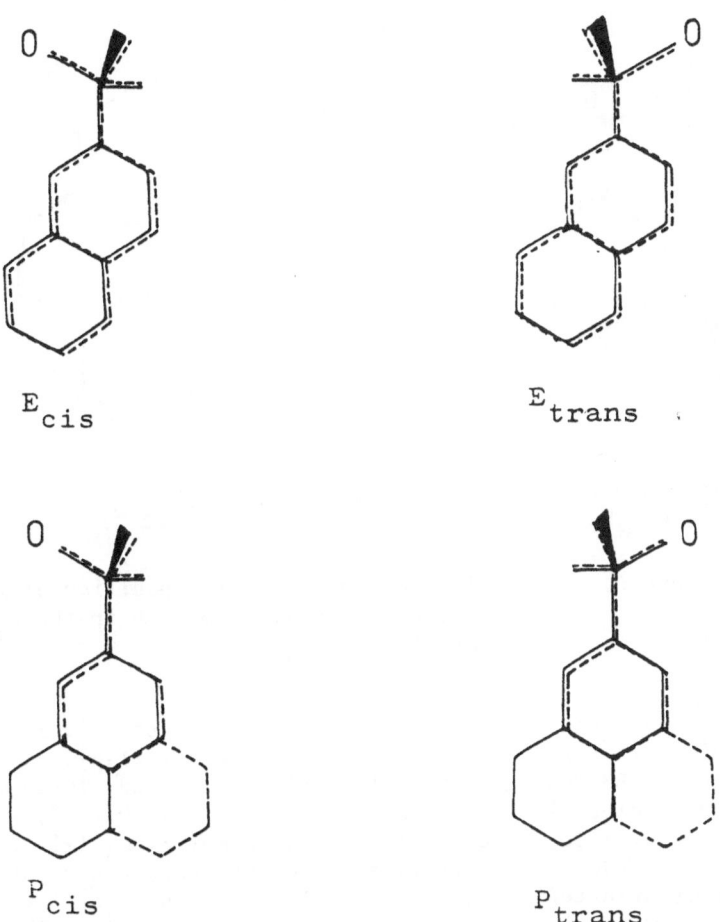

Figure 5.   Excimer geometries in B3.   P = partial overlap,
            E = eclipsed overlap.

kJmol^{-1} above M_1.   A third minimum is situated at (+120, +120, −110,−60).
This is the TT conformation with partial overlap of the chromophores
$P_{trans\ a}$ (see fig.5). $P_{trans\ a}$ is 7.5 kJmol^{-1} more stable than M_1.
When solvatation is accounted for this $P_{trans\ a}$ will be higher in energy, but the conclusion remains that $P_{trans\ a}$ has a comparable energy
with the other ground state conformations.
A summary of the results on meso B3 is given in figure 7.
In figure 7 the excited state energies were obtained by adding the a-
mount of energy for the (0, 0) transition and a stabilization energy
appropriate for the excimer geometries which was taken from a theoretical study (Padmar Malar and Chandra, 1980).   From figure 7 it is clear
that one can expect very efficient excimer formation and direct excitation of an excimeric conformation $P_{trans\ a}$.   It is evident that a com-

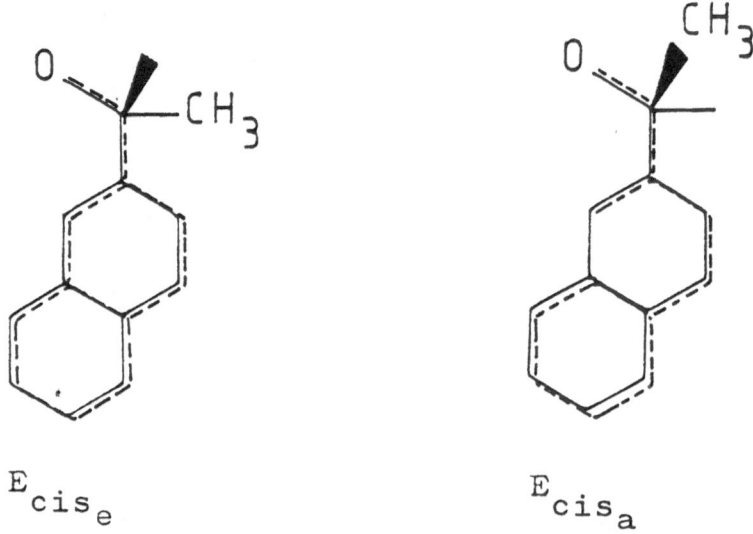

E_{cis_e} E_{cis_a}

Figure 6.   Equatorial (e) and axial (a) position of the methyl
            groups versus naphthalene moiety in the E_{cis} exci-
            mer geometry.

plex photophysical behaviour will be observed in meso B3 : more than one
type of excimer is formed (full overlap, partial overlap).
        In racemic B3 four ground state minima are found. Three of these
minima are very close in energy and geometry. M_1 (-10, +40, +100, +150)
and M_2 (-10, +80, +100, +150) differ by 2.1 kJmol^{-1}, M_3 (+150, 100, 100,
+150) can be regarded as an arylrotamer of M_2 (only the naphthyl moiety
is differently oriented). M_3 is destabilized 2.5 k J mol^{-1} versus M_2.
These three conformations are situated between TT and TG. The fourth
minimum is the more stable conformation : M_4 (0, 20, 20, 0) (-8.8 k J
mol^{-1} versus M_1). M_4 is the GG conformation and is separated by an
energy barrier of 9.6 kJmol^{-1} from the M_1 conformation. In none of these
ground state conformations a partial or full overlap of the chromophores
is present. The formation of the TG excimer geometries ($P_{cis\ a}$, $P_{trans\ a}$
$E_{cis\ a}$, $E_{trans\ a}$) is an energetically difficult process as can be seen
in figure 8. In figure 8 the results of the calculation are summarized.
The two partial overlap excimer conformations are in the ground state
more stabilized than the full overlap excimer conformations. This does
not mean that they will be formed in the excited state since no information of the barrier in the excited state is available.
        De Schryver et al. (1981) and Demeyer (1982) described the photo-
physical properties of meso and racemic B3. No cyclomers were found
upon irradiation. Again the ratio I_D/I_M differs markedly : 40.0 for meso
B3 and 0.3 for racemic B3. At 77 K no difference in emission spectra
between meso and racemic B3 is detected. The excitation spectra of meso
B3 in isooctane at room temperature however differ upon wavelength of
analysis. A broadening of the ^1L_a absorption band is found when analyzing at λ< 380 nm. This effect is lowered when the temperature is de-

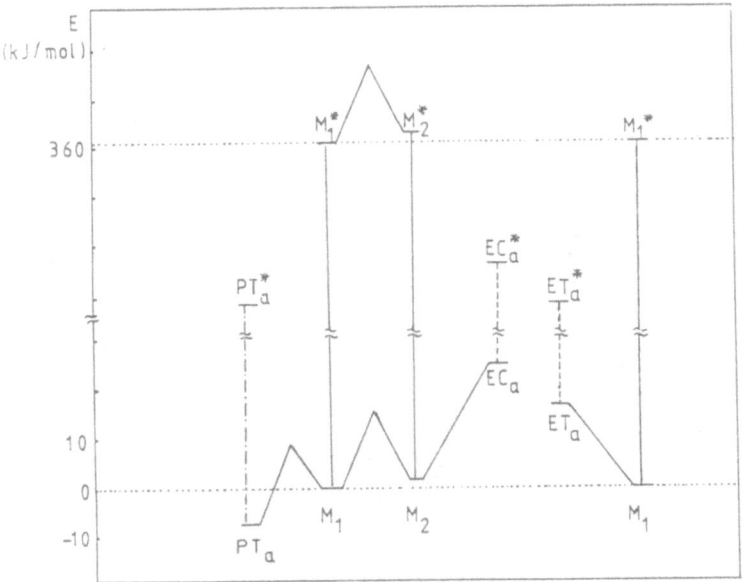

Figure 7. Conformation energy profile of meso B3.

creased. The wavelength dependence of the excitation spectra is due to the presence of the $P_{trans\ a}$ conformation in the ground state. Furthermore, the I_D/I_M ratio of meso B3 but not of racemic B3 is excitation wavelength dependent. A slight shoulder at 375 nm in the excimer emission band of meso B3 is assigned to the excimer emission from the $P_{trans\ a}$ conformation. The high I_D/I_M ratio in meso B3 once again is correlated with the accessibility of excimer conformations. In racemic B3 the excimer conformations can only be reached via large energy barriers, this results in a low excimer emission yield(assuming again comparable radiative rate constants and stabilization of the excimers).

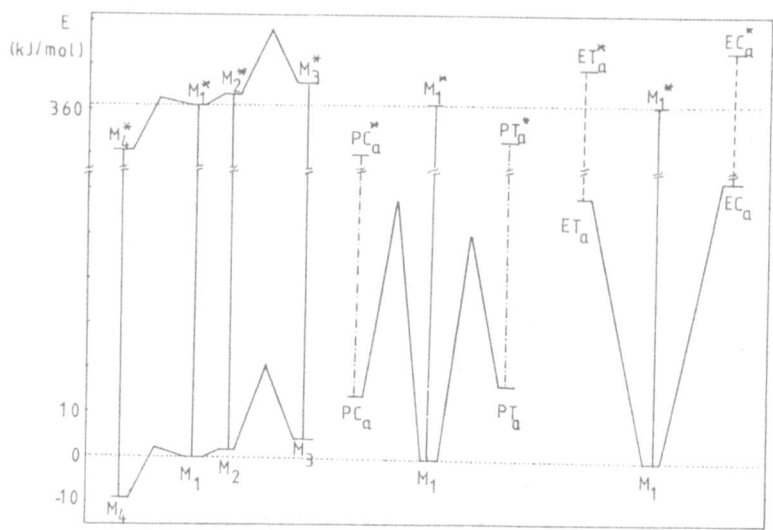

Figure 8.   Conformation energy profile of racemic B3.

    The time dependent profile of the locally excited state emission of
meso B3 in isooctane cannot be described with a two exponential decay
function. The analysis is probably complicated by the large spectral
overlap of excimer emission and the low quantum yield of locally excited
state emission. The decay curves in the excimer region (430 nm) are
described as the sum of three exponents. One of the exponents is very
short (2.1 ns at 178 K) and has a negative preexponential factor. ·This
complex behaviour of excited meso B3 is understood when ground state
conformations are taken into account. The TTP_{trans} a conformation will
rapidly upon excitation form a partial overlap excimer, no rotation
around a C−naphthyl bond is possible (to form a full overlap excimer).

M_1 and M_2 are in fact two arylrotamers (see conformational energy calculations) and can form the partial overlap as well as the full overlap TT excimers upon excitation.

The decay curves of racemic B3 in the locally excited state emission region (330 nm) are described by single exponential between 200 and 240 K, where nearly no excimer is formed, and a sum of two exponentials from 240 till 350 K. The decay curves in the excimer region (420 nm) were initially described (De Schryver et al. 1981) as a difference of two exponentials, but improvement of experimental conditions lead to a more accurate description of the decays with a three exponential decay function (Demeyer 1982). These data indicate that in the case of racemic B3 more than one excimer is formed. A more detailed study of the wavelength dependence of the decay curves and the use of global analysis should reveal more information about the photophysical behaviour of meso and racemic B3.

II.1.3. __1,1'-di(9-anthryl)diethyl ether (B4).__ Becker and Andersson (1982) reported on the ground and excited state properties of B4. A first important result is that on ^{1}H-NMR time scale the two anthryl chromophores in meso as well as in racemic B4 nearly have come to a standstill at -30°C. This effect is due to the steric hindrance of the CH_3 groups upon the anthryl proton 1 and 8.

The presence of six signals in the aromatic region in the ^{1}H-NMR spectrum of meso B4 suggests an orientation of the anthryl chromophores such that proton 1 and 8 have a different chemical shift. This condition is met in the TT conformation (fig.9) where H-1 but not H-8 will be affected by the proximity of the ether oxygen. In the ^{1}H-NMR spectrum of racemic B4 at -30°C nine signals are detected in the aromatic region. This is explained when the TT conformation (fig.9) is assumed as the predominant present conformation.

The assignment of the conformations is confirmed when the absorption spectra of meso and racemic B4 are compared.In meso B4 the extinction coefficient of the S_0-S_1 transition is lower than in racemic B4.Furthermore the S_0-S_1 spectrum of meso B4 is broadened. This indicates the presence of a geometry where intramolecular π-orbital interaction is feasible. As can be seen from figure 9 π-orbital interaction is large in the TT conformation of meso B4. Both diastereoisomers undergo intramolecular photochemical symmetrical ($4\pi + 4\pi$) cycloaddition with about equal quantum efficiency ($\Phi = 0.25$). However, the radiative deactivation is different for meso and racemic B4. Meso B4 is virtually non-fluorescent ($\Phi = 410^{-3}$). Racemic B4 fluoresces ($\Phi = 0.014$) from both a locally excited state and more prominently from an excimer state (λ_{max} 480 nm). The authors argue, taking in mind the starting geometry, that the geometry of this emitting excimer of racemic B4 deviates from the perfect sandwich structure and may be one in which the two anthracene moieties are only partially overlapping.

II.1.4. __2,4-di(1-pyrenyl)pentanes (A5) and 1,1'-di(1-pyrenyl)diethyl__
       __ethers (B5).__ ^{1}H-NMR data on ground state conformations and populations of A5 are reported by Collart et al.(1985). Meso A5 is predominantly TG at room temperature (99 % in chloroform, 94 % in cyclohexane.

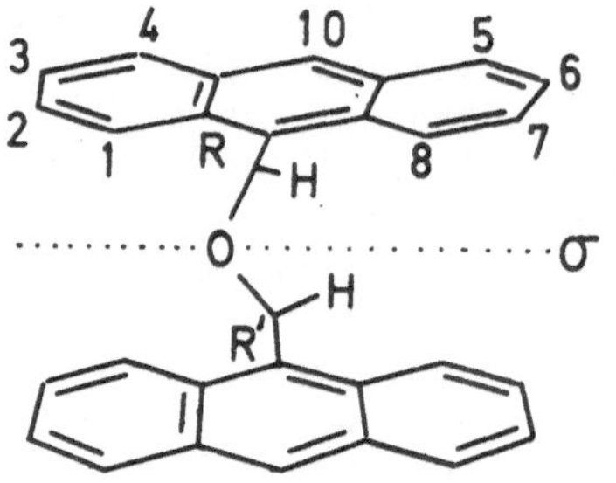

Figure 9. Ground state conformations of meso B4 (1) and racemic B4 (2) (Becker and Andersson, 1982).

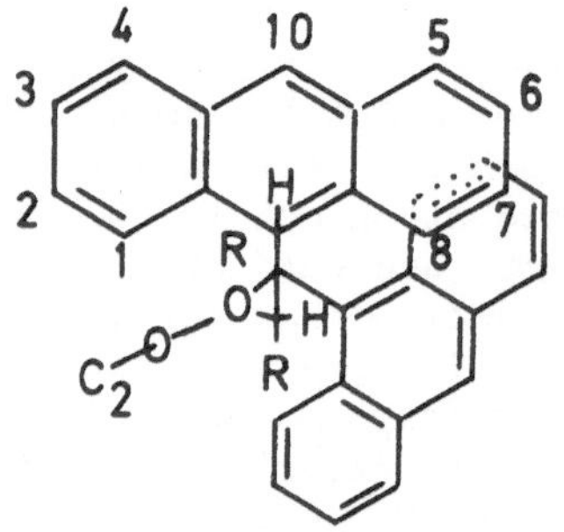

Racemic A5 has two predominant ground state conformations : TT and
GG. An important change in the population of the conformations is observed in racemic A5 which is dependent upon the nature of the solvent.
In alkane solvents mainly the TT conformation is present (75 %); in
chloroform and tetrahydrofuran 50 % TT is present at room temperature.
In the TT conformation of racemic A5 the influence of the ring current
of one chromophore upon the resonance signals of the protons of the
second chromophore is observed. This effect is similar to the effect
described earlier for racemic B4.
       Only information about the photophysical properties of meso A5 has
been reported. The fluorescence spectrum of meso A5 in isooctane at
room temperature shows predominant broad excimer emission with maximum
at 480 nm (I_D/I_M=23). Lowering the temperature decreases the excimer
emission (I_D/I_M = 0.08 at 183 K). The fluorescence decay curves of
meso A5 in the monomer region (377 nm) are analyzed as the sum of a two
exponential function. The decay curves in the excimer region are fitted
to a triple exponential function. One of the exponents has a negative
preexponential factor. The complex photophysical behaviour of meso A5
cannot be interpreted with the one ground state conformation present if
one does not include the possibility of aryl rotamers and consequently
the existence of partial overlap excimers. The existence of more than
one excimer is confirmed when fluorescence decay curves in function of
wavelength of detection in the excimer region were analyzed. The shorter

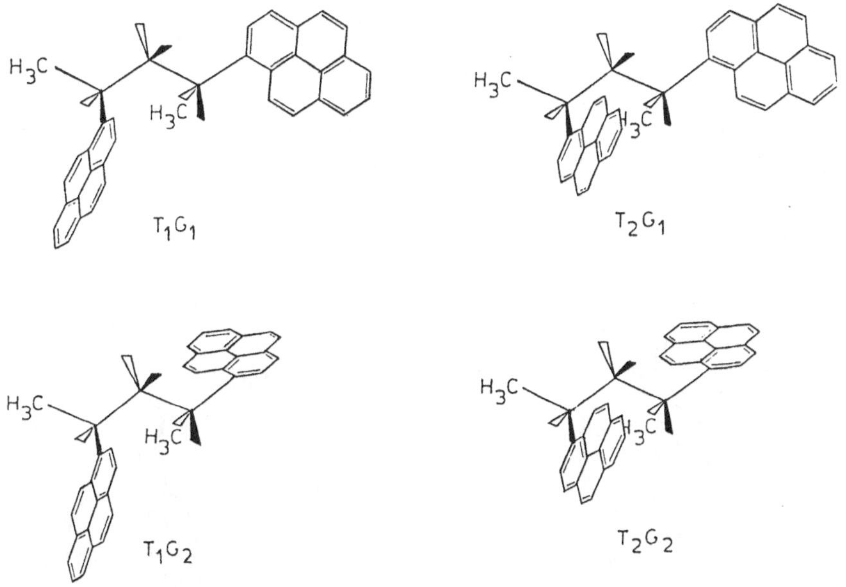

Figure 10.   Ground state rotamers of the TG conformation of
             meso A5.

decaying component contributes more to the decays measured at shorter
wavelengths.   Four different ground state TG conformations are drawn
in figure 10.   The resulting excimer geometries are drawn in figure 11.
The T_2T_2 geometry is less likely due to steric hindrance of the methylene bridge upon the excimer.

Information about the ground state conformational distribution in
meso and racemic B5 is obtained through comparison of the ^{1}H-NMR spectra
of A5 and B5 (Collart et al., 1985).   It is concluded that only the TG
conformation is present in meso B5 at room temperature in all solvents
investigated.   An indication of the presence of aryl rotamers in meso B5
TG is given by the strong broadening of the aromatic resonance signals
at 180 K.   In racemic B5 a comparable effect of spreading of aromatic
signals as in racemic A5 is observed.   This leads to the conclusion that
the TT conformation is present.

At low temperature (180 K) three separate signals were observed for
the methine and methyl hydrogen resonance signals, again indicating the
presence of rotamers in racemic B5.

The fluorescence spectra of meso and racemic B5 show large differences (Collart et al., 1983).   Both compounds emit excimer fluorescence
but the ratio I_D/I_M at room temperature equals 18 and 1.15 respectively
for meso B5 and racemic B5.   Only the fluorescence decay curves of meso
B5 are reported (Collart et al., 1985).   The photophysical behaviour of

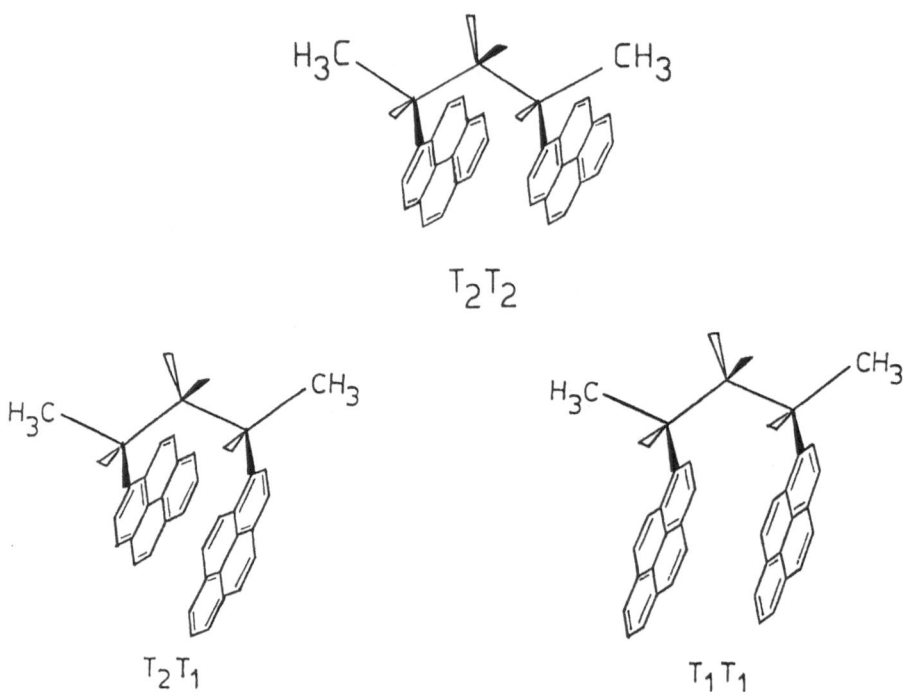

Figure 11.   Excimer geometries in the TT conformation of
             meso A5.

meso B5 is similar to the behaviour of meso A5. The same conclusions
are drawn from the analysis.

II.1.5.   2,4-di(N-carbazolyl)pentane (A6).   The assignment of the configuration and the calculation of the conformational distribution within
the configuration in the ground state was done from ^{1}H-NMR measurements
in analogy with the method used by Bovey et al.(1965) for the 2,4-di-
phenylpentanes (De Schryver et al. 1982; Vandendriessche et al., 1984).
In meso A6 J_{AB} = 7.98 Hz and J_{AC} = 8.08 Hz. It is thus concluded that
within experimental error, in meso A6 100 % of the molecules are present
in the TG/GT conformation at room temperature. At higher temperatures
(393 K) a small deviation from the equality $J_{AB}=J_{AC}$ is observed. This
could be an indication of the presence of the ground state conformation
(≃ 6 % TT ?). At lower temperatures a splitting of the aromatic part
of the ^{1}H-NMR spectrum is observed. This splitting is comparable to the
one reported for N-isopropylcarbazole (NIPCz) and is due to the fact
that the rotation of the carbazole chromophores is slow on ^{1}H-NMR scale.
The main contribution to the ^{1}H-NMR spectrum of racemic A6 at room
temperature are the TT (88 %) and the GG (12 %) conformations. At low
temperatures a splitting of the aromatic part of the ^{1}H-NMR spectrum occurs which is much more important than for meso A6 and NIPCz : eight
signals are observed. This effect is explained by the influence of the

ring current of the first chromophore on the protons of a part of the second chromophore and vice versa when the molecule is in the TT conformation. This behaviour is comparable to the behaviour of racemic B4 (Becker and Andersson 1982) and racemic A5 and racemic B5 (Collart et al., 1985) described above.

Sundararajan (1980) reported the conformational energy maps of the meso and racemic dyads in poly(N-vinylcarbazole). For the meso dyad two possible candidates for ground state population are found. The TT minimum is about 9.6 kJmol^{-1} lower in energy than the TG/GT minimum. The author, however, mentions that the TT meso state energy rises drastically when the value of the interaction parameter which takes into account the solvent-chain interactions is changed. In the racemic dyad the TT conformation is 6.7 kJmol^{-1} lower in energy than the TG/GT conformation and 10.4 kJmol^{-1} lower in energy than the GG conformation.

Data on the excited state properties of meso and racemic A6 are reported by De Schryver et al. (1982b), Masuhara et al. (1983), Evers et al. (1983) and Vandendriessche et al. (1984). The S_0 -S_n absorption spectra of meso A6, racemic A6 and NIPCz are identical (De Schryver et al., 1982b). This suggests that in meso A6 the population of TT is low and that in the TT conformation in racemic A6 the partial overlap of the chromophores induces a little or no interaction.

The fluorescence spectra of meso and racemic A6 differ largely (De Schryver et al., 1982b). The fluorescence spectrum of meso A6 in solution at room temperature shows, besides alkyl carbazole emission, an excimer band with maximum at 420 nm. Lowering the temperature does increase the alkyl carbazole emission and decrease the excimer emission. At low temperatures nearly no excimer fluorescence is observed, indicating that the TG conformation is the more stable. In analogy with the excimer emission observed in 1,3-di(N-carbazolyl)propane, the 420 nm excimer band in meso A6 is assigned to the full overlap excimer (TT). The fluorescence spectrum of racemic A6 in solution at room temperature shows alkyl carbazole emission and an excimer emission band with maximum at 370 nm. When the sample is cooled to 133 K in isopentane (still liquid) the latter emission is almost the sole contributor to the spectrum (De Schryver et al., 1982b). Cooling the sample below 133 K (methylcyclohexane/isopentane, 1 : 1, glass formation) the contribution of the alkyl carbazole emission increases again and becomes the sole contribution to the spectrum at 84 K (Evers et al., 1983). This behaviour is explained by Vandendriessche et al. (1981). The most stable conformation in racemic A6 is the TT conformation. This conformation can form, upon excitation, a partial onverlap excimer via minor rotations of the chain and the chromophores (<20 deg).No real energy barrier is present between the TT locally excited state and the TT excimer state. At room temperature the 88 % TT will form efficiently the TT* partial overlap excimer. The 12 % GG do not form the partial overlap excimer nor the full overlap excimer (no indication in fluorescence spectra and decay curves) and emits alkyl carbazole fluorescence. Upon cooling, the GG population decreases and more and more (TT)*excimer will be observed. When the molecule is inbedded in a glass (below 133 K) the small rotations to form the partial overlap excimer are hindered and more and more TT locally excited state (alkyl carbazole fluorecence) will be observed at lower temperatures.

Due to the large spectral overlap of the carbazole and partial overlap excimer emission it is not possible to analyse the decay curves of
racemic A6 accurately. A complicating factor on top of this spectral
overlap is the lifetime of the partial overlap excimer : 17 ns at 300 K
(compare with 15 ns for NIPCz). No growing in of the partial overlap
excimer was detected (De Schryver et al., 1982b).

The decay curves obtained for meso A6 can be analyzed within scheme
I (Table I). This can be understood when one assigns the locally excited
state to the TG/GT conformation and the excimer state to the TT conformation. These two conformations are separated through a one rotation process. A full kinetical and thermodynamical analysis is reported for
meso-A6 in isooctane (Vandendriessche et al., 1984).

S_1-S_n absorption spectra of meso and racemic A6 were reported by
Masuhara et al. (1983). The growing in of a new absorption band at 880
nm in the time-resolved absorption spectra of meso A6 in THF corresponds
to the growing in of the excimer emission at 460 nm. No large changes
were observed in the time-resolved absorption spectra of racemic A6.
This indicates that the interaction of the two carbazole chromophores in
the partial overlap excimer is weak, and that the interaction of the two
carbazole chromophores in the full overlap excimer is strong.

Differences in excited state properties of meso and racemic A6 were
also reported on quenching experiments. The existence of emitting triple
complexes is reported (Masuhara et al., 1982) as well as the absorption
spectra of the resulting ionic dicarbazolyl compounds when these triple
complexes are brought in a very polar medium (Masuhara et al., 1983).

II.2. 2,4,6-triphenylheptane (TPH)

A second step in the understanding of the polymer photophysics,
after having investigated the dimeric compounds, is the study of well
characterized configurational and conformational higher homologues.
Until now such model systems are known only for polystyrene.

Three diastereoisomeric forms are separated : syndiotactic (s-TPH),
isotactic (i-TPH) and heterotactic (h-TPH) 2,4,6-triphenylheptane and
the configuration is assigned using ^{1}H-NMR (Lin et al. 1966). The ground
state conformations and their population are investigated with ^{1}H-NMR
(Pivcova et al., 1969), IR-spectroscopie (Jasse et al., 1973) and depolarized Rayleigh scattering (Fourche et al., 1977). Energy calculations
were performed by Gorin (1970). The results of all these measurements
and calculations are summarized in table II.

Bokobza et al.(1977) reported the fluorescence spectra of i-TPH and
s-TPH. The quantum efficiency of excimer formation (Φ_{EF}) at 25° in
s-TPH equals the Φ_{EF} of racemic A1. The quantum efficiency of excimer
formation in i-TPH is larger than in meso A1. This was explained as
the result of a high delocalized excitation energy in i-TPH. Fluorescence spectra as well as fluorescence decay measurements of all three
isomers were reported recently (Moens 1985). The fluorescence spectra
of s-TPH, i-TPH and h-TPH in isooctane at room temperature are represented in figure 12. The fluorescence intensities of locally excited
state and excimer emission of h-TPH are not the average of the intensities in the spectra of i-TPH and s-TPH.

Table II. Stable conformations in 2,4,6-triphenylheptanes

Isomer	Chain conform.	T (K)	Equilibrium[1] composition	ΔH° kJmol^{-1}	ΔS° Jmol^{-1} K^{-1}	E[2] kJmol^{-1}
i-TPH	TGTG/GTGT	343	0.78			267.1
		413	0.73			
	GTTG	343	0.22	4.4	10.5	267.5
		413	0.27			
s-TPH	TTTT	293	0.52			247.0
		413	0.38			
	GGTT/TTGG	293	0.48	4.6	9.9	263.3
		413	0.62			
h-TPH	TTTG	293	0.41			258.3
		393	0.34			
	TTGT	293	0.44	1.0	4.0	250.8
		393	0.41			
	GGTG	293	0.15	6.4	13.4	273.4
		393	0.24			

1. Pivcova et al. 1969
2. Gorin, 1970

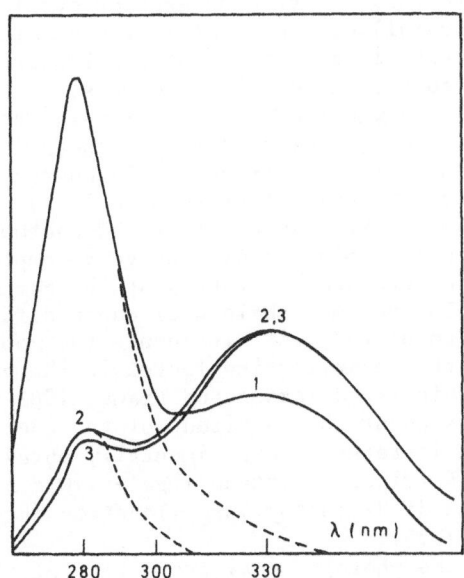

Figure 12. Fluorescence spectra of s-TPH (1), i-TPH (2) and
            h-TPH (3) in isooctane at room temperature. The
            samples have identical optical densities at wave-
            length of excitation.

The decay curves of the three diastereoisomers are analyzable within scheme I. The rate parameters and thermodynamic data obtained for s-TPH and i-TPH correspond to the rate parameters and thermodynamic data of racemic Al and meso Al respectively. It is concluded that the excimer formation in a dyad is not influenced by the neighbouring configurational identical dyad. In h-TPH the same rate parameters and thermodynamic properties of the excimer formation process are found as in i-TPH and meso Al. This fact together with the very similar fluorescence spectra (fig. 12) are interpreted as follows : h-TPH consists of an iso- and a syndiotactic dyad. The two chromophores belonging to the iso-tactic dyads have the same excimer forming properties as i-TPH and meso Al. The remaining chromophore will upon excitation transfer its energy to the neighbouring chromophore. This latter process is very fast compared to the other deactivating processes so that only the excimer formation in the isotactic dyad will be observed.

II.3. Polymer Photophysics

This compilation of data proves the importance of configuration and population of conformations in the intramolecular excimer formation in model systems of polyvinylaromatics. Besides the above mentioned parameters another important aspect in the photophysical behaviour of the compounds is the role of the substitution pattern on the chromophore. When a C_2 symmetry axis of the chromophore is in direct line with the bond which links the chromophore to the chain, the photophysical behaviour is simplified. When this situation is not met, different chromophore orientations lead to more than one type of excimer overlap. Bulky chromophores (carbazolyl, anthracyl) may interact in the excited state in the conformation which is very close to the conformation present in the ground state, producing also partial overlap excimers.

The general photophysical behaviour of the 2,4-diarylpentanes is comparable to the behaviour of the 1,1'-diaryldiethyl ethers. The introduction of the oxygen causes only slight differences in energy barriers and thus in rate parameters and activation energies (table I).

From the preceeding discussion it is clear that the photophysical behaviour of the polyvinyl aromatic polymers will be very complex. Polymer photophysics will even be more complex than just the sum of the properties of the dyads. The polymer chain will cause different starting conformations to be populated, will influence the mobility of the rotations. A possible further complicating factor is the energy migration along the polymer chain (Fredrickson and Frank, 1983 ; Fredrickson et al.,1984). An idea about the magnitude of the energy migration rate constant in solution is recently experimentally obtained from the study on heterotactic 2,4,6-triphenylheptane ($k_{ET} \gg 10^9 s^{-1}$) (Moens 1985) and from the study on 1,3-di(2-naphthyl)propylacetate ($k_{ET} \sim 2.10^9 s^{-1}$) (Vandendriessche et al., 1985).

A detailed discussion of the photophysical properties of the polyvinyl aromatic polymers is not given. At this moment for none of these polymers a full, complete analysis is reported and generally accepted, altough many attempts were and are made (Semerak and Frank, 1984; Kauffmann et al., 1985; Soutar and Phillips, this volume).

III. LOCAL DYNAMICS OF POLYPEPTIDES

III.1. Nature of the polypeptide conformation

One of the most challenging problems of biophysics is that of the physical basis for the unique and specific structure of proteins. Whereas the conformation of synthetic polymers such as polyvinylaromatic compounds is more or less determined by statistical (entropic) factors, the overall conformation of polypeptides is mainly the result of specific interactions between amino acid residues.

Hydrogen bonding, dipolar interactions, hydrophobic interactions, salt bridges and disulfide bonds, together with the rigidity of the amide bond which is generally confined to the planar trans configuration, eliminate a large fraction of possible conformations, leading, in the ideal case, to a "native" conformation. The combined effect of these interactions cause the spontaneous folding of the chain to a specific secondary (e.g. α-helix or β-pleated sheet) and tertiary structure (mainly thightly folded globular arrangements dominated by the tendency of apolar groups to be internal and of polar residues to be external).

The resulting conformation is totally coded in the primary structure of the chain (the amino acid sequence). This is expressed by the "one sequence, one conformation" rule, which is dramatically confirmed by the observation that denatured proteins regain their biological activity once the appropriate environment is restored (Anfinsen et al., 1961 and 1975).

III.2. Some basic strategies of conformational studies on biopolymers

All the stabilizing phenomena mentioned above are short range interactions. As a conclusion, the behaviour of the peptide chain can be understood by a systematic examination of the individual parameters affecting the local molecular dynamics of low molecular weight model compounds.

Only when the molecule is small enough that the contributions of every residue can be monitored separately, spectroscopic techniques offer their full power as a tool for obtaining quantitative information about the conformations and conformational transitions of the system.

Another approach is to analyze the global structure under study. However, the most serious obstacle is the large amount of variables present in a macromolecular system. A protein structure may entail hundreds of significant variables and few people can retain this much for immediate recall and correlation.

Therefore, the system must be simplified for detailed biophysical studies. Some approaches are : observing only one part of the system (e.g. transition metal in the active site or a probe attached by the investigator), comparing two systems that are almost identical (e.g. mutant proteins compared to the native proteins) or isolating discrete states of the system (e.g. changing the temperature, pH or salt concentration up to a level that allows only one secondary structure to exist).

However, only qualitative information can be obtained which does not allow a complete understanding of the link between sequence and conformation. It is hoped that the knowledge of this relation would satis-

fy the ultimate outlook : to predict the polypeptide conformation from
its primary structure (Review by Maser et al. 1984).

III.3.   Some results obtained by spectroscopy of the ground state and
         conformational calculations.

     The conformational preference of model peptides in solution are
usually determined by various spectral measurements. The X-ray structure analysis, which is the absolute method, can only be applied to cases where the substance can be obtained in the crystalline state.
     Besides the theoretical calculations, three spectroscopic techni-
ques proved to be very useful : circular dichroism (C.D.), I.R.- and
N.M.R.-spectroscopy.

III.3.1.   Theoretical calculations.
III.3.1.1.   Ramachandran maps. Figure 13 displays the polypeptide chain
in its all trans (planar zig-zag) form.

Figure 13.   The α-L-polypeptide chain in the all-trans form.

Due to the amide resonance, the peptide function is kept in the planar
trans conformation. As a consequence, the overall conformation can be
specified by the values of the torsion angles ϕ_i and ψ_i of each amino
acid residue. The ϕ,ψ coordinates of a variety of ordered forms of
polypeptides are given in table III.
G.N. Ramachandran and colleagues were the first to investigate the exclusion of many hypothetical conformations on the basis of unfavourable
steric overlaps (hard sphere model). They constructed steric contour
diagrams which graphically represent the allowed ϕ,ψ values (figure 14)
(Ramachandran et al., 1963 and 1968).

III.3.1.2.   Conformational energy calculations. These steric diagrams
do not show gradations in energy within the sterically permissible regions. In order to obtain the contribution of non-bonded interactions,
hydrogen bonding, intrinsic energy barriers of internal rotations and
electrostatic interactions, it is necessary to compute the potential
energy as a function of ϕ and ψ. Since all terms are assumed to be independent, they are treated by means of separate semi-empirical functions (Sheraga et al., 1971).
     The computer program ECEPP (Empirical Conformational Energy Program

Table III. Approximate torsion angles for some regular structures (IUPAC-IUB, 1970).

Structure	ϕ	ψ
Right-handed α helix [α-poly(L-alanine)]	$-57°$	$-47°$
Left-handed α helix	$+57°$	$+47°$
Parallel-chain pleated sheet	$-119°$	$+113°$
Antiparallel-chain pleated sheet [β-poly(L-alanine)]	$-139°$	$+135°$
Polyglycine II	$-80°$	$+150°$
Collagen (triple helix)	$-51°,-76°,-45°$	$+153°,+127°,+148°$
Poly(L-proline) I	$-83°$	$+158°$
Poly(L-proline) II	$-78°$	$+149°$

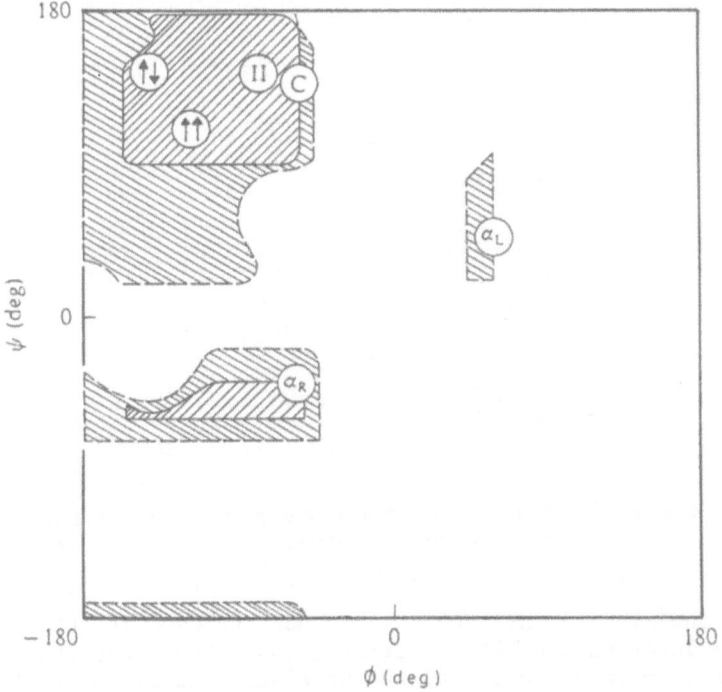

Figure 14. Steric contour diagram for an L-alanyl residue in a polypeptide chain. /// zones show "normal" and \\\ zones show "outer-limit" contours. Coordinates of right- and left-handed α-helices (α), parallel (↑↑) and antiparallel (↑↓) pleated sheets, polyglycine II (II) and collagen (C) are denoted (see also table III) (Ramachandran et al. 1963).

for Peptides) is most frequently applied to these conformational energy
calculations (Momany et al., 1974). The inability of ECEPP to predict
energy minima in regions of φ-ψ space where experimental β-turns occur
has been overcome by the improved program ECCEP83 (Chumam et al., 1984).
    Application of these techniques to the glycyl residue leads to an
energy contour diagram comparable to the sterically allowed regions.
The alanyl residue, however, has three distinct low energy regions, close to the right-handed α-helix (region I), the left handed α-helix (region II) and the β-pleated sheet (extended chain, region III) (compare
figure 15 with figure 14), the energy of region I being much lower than
that of region II (Brant et al., 1967). This provides a rational basis
for the well-known preference of L-polypeptides for the right-handed α-
helical conformation.

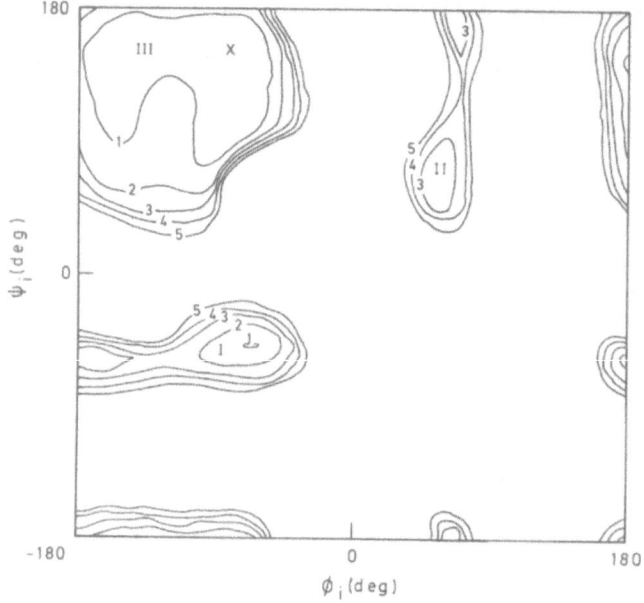

Figure 15.   Energy contour diagram for an L-alanyl residue in
             a polypeptide chain .

    When the dipolar interactions are not included in the calculations,
region III loses much of its stability, indicating that the extended peptide chain is mainly stabilized by the electrostatic interactions between the antiparallel amide dipoles.

III.3.2. Circular dichroism. In a C.D. experiment, the variation of
the ellipticity, obtained by measuring the difference in the absorbance
of left-handed and right-handed circularly polarized light, is monitored
as a function of wavelength.
    The C.D. of small chiral molecules is often very weak. Since

the chirality of a large secondary structure induces intense Cotton
effects,   especially polymers are studied. Figure 16 shows
some typical C.D. curves of polypeptides.

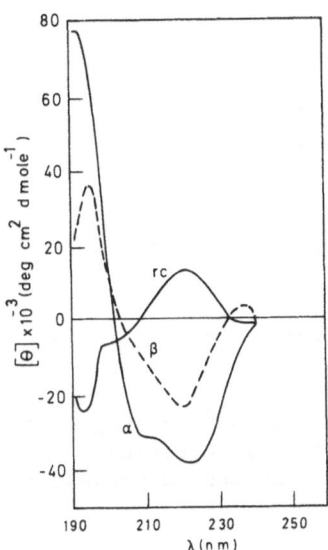

Figure 16.   C.D. spectra for α-helix, β-sheet and random coil
conformation (Saxena et al., 1981).

By using this technique, conformational changes of polypeptides as
a function of environment (solvent composition, temperature, ligand
binding etc.) are easily detected. By a C.D. study on some oligomeric
methionines with various chain lengths (Toniolo et al., 1979), it is
shown that up to a chain length of n=6 the peptides adopt an essentially
unordered conformation. At the level of the heptamer the onset of the
α-helix can be observed. The helix contents grows clearly as the chain
length increases (Rinaudo et al., 1976).

III.3.3. Infrared spectroscopy. Solvents relevant to peptide studies
such as methylene chloride and dimethylformamide cannot be used for C.D.
spectroscopy. Additional problems with regard to the interpretation of
C.D. spectra is generated by overlapping curves arising from aromatic
side chains.
      I.R. studies are generally not affected by these problems. Espe-
cially the spectral regions of the conformationally sensitive amide absorption bands are studied by I.R. spectroscopy. In practice, the best
information is obtained from spectra of a diluted solution of small peptides. Spectra in the solid state or from polypeptides show low resolution of interesting absorption bands.
      The first I.R. experiments on dipeptides were carried out by Tsuboi
et al. (1959). Many efforts have been done by other investigators such
as Avignon et al. (1969), Maxfield et al. (1979), Cung et al. (review,
1972) and Rao et al. (1983). A general conclusion is that dipeptides of

the formula $CH_3-CONH-CHR-CONH-R'$ adopt two sets of conformations :
1) a C_5 conformation, stabilized by a weak hydrogen bond between the
$C=O$ and NH functions of the same amino acid residue, thereby creating
a five-membered ring and leading to an extended peptide chain with alternating side chains.

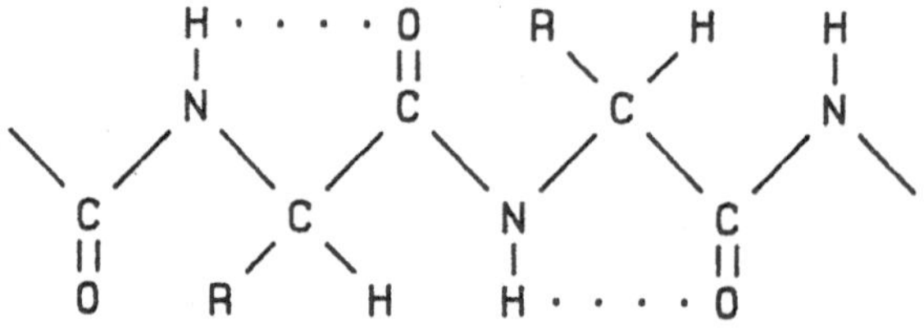

Figure 17. C_5 conformation ($\nu_{NH} = 3420$ cm^{-1}).

2) a folded C_7 conformation, with an hydrogen bond between the $C=O$ and
NH functions of vicinal residues, creating a seven-membered ring. The
side group of the first residue can be present in an axial (C_7^{ax}) or an
equatorial (C_7^{eq}) configuration.

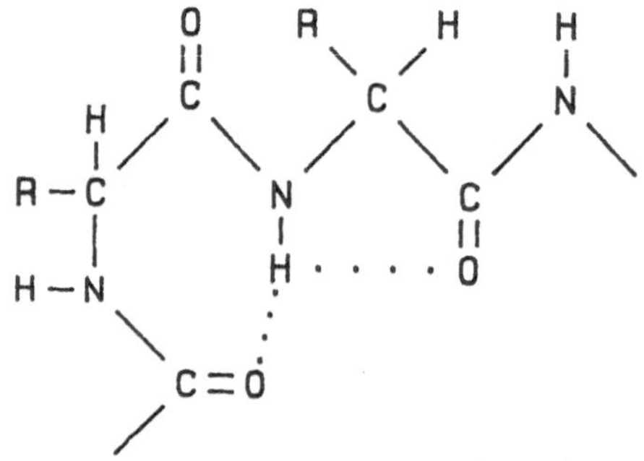

Figure 18. C_7 conformation ($\nu_{NH} = 3335$ cm^{-1}).

A tripeptide which has an additional third amide function can also
adopt the C_{10} conformation (often called β-turn)(Figure 19).
These conformations were identified by their different stretching
frequencies of the NH function involved in the intramolecular hydrogen
bond (C_7 and C_{10}, however, are difficult to distinguish). At concentrations above 10^{-2} M, intermolecular aggregation becomes possible, having
also a typical NH absorption frequency.
From the relative absorbances some information can be obtained about
the equilibrium constants of conformational transitions (Boussard et al.,

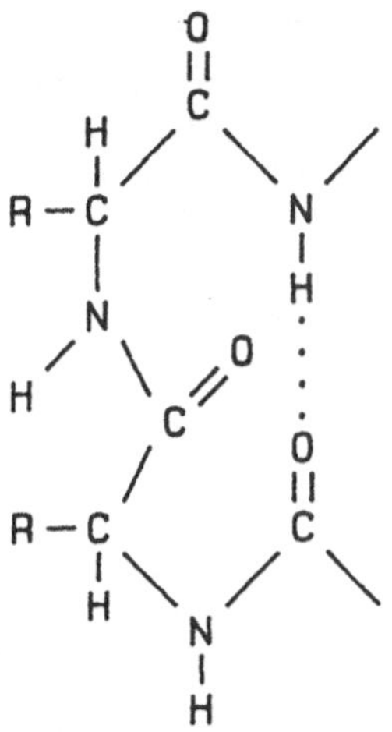

Figure 19. C_{10} conformation (ν_{NH} = 3350 cm^{-1}).

1985) but no I.R. data are available on transition rates or average conformational lifetimes. In addition, only the infrared transparant solvents can be used, thereby limiting the information about the solvent influence on the conformational equilibria.

III.3.4. N.M.R. The most frequently used approach in ^{1}H-N.M.R. experiments has been the investigation of the change of the chemical shift δ
of the amide protons as a function of temperature, concentration and
solvent composition. A general conclusion is that the temperature dependence dδ (NH)/dT is much smaller for an intramolecular hydrogen bond
than for hydrogen bonding with the solvent (e.g. DMSO-d6) (Goodman et
al., 1981 and 1975, Ribeiro et al., 1980 and 1982, Stevens et al., 1980,
Naider et al., 1980).
        So called titration experiments also are able to distinguish between
amide protons exposed to and shielded from the solvent. In agreement
with the C.D. experiment mentioned above, Goodman et al. (1975) observed
by means of ^{1}H-N.M.R. that oligoalanines can adopt a secondary structure
from a chain length of n=5.
        A number of more advanced methods were developed for conformational
studies on peptides, the most powerful being the determination of the
correlation between coupling constants and the dihedral angle of the
bond involved (Bystrov plot). This technique will be discussed in some
more detail now (review by Bystrov, 1976).

III.3.4.1. Torsional angle ϕ. For this angle, the use of the vicinal
^{1}H-NC$^{\alpha}$-^{1}H coupling has now become a routine procedure. However, the
Karplus-type angular dependence presents four angles to select from for
most values of the coupling constant (figure 20). In order to remove
this ambiguity, determination of the vicinal ^{13}C'-NC$^{\alpha}$-^{1}H coupling constant is recommended (figure 20). The combined usage of both constants
often permits discrimination between the positive and negative region
of the ϕ angle.

III.3.4.2. Torsional angle ψ. The uncertainty in determining ψ from
N.M.R. experiments is very high since only one coupling constant, the
vicinal $^1H-C^\alpha C'-^{15}N$ coupling, is calibrated experimentally. Mostly,
for a complete characterization of the spatial structure of the peptide
backbone, it is important to compare the observed spin-spin couplings
with the φ,ψ energy map (figure 20).

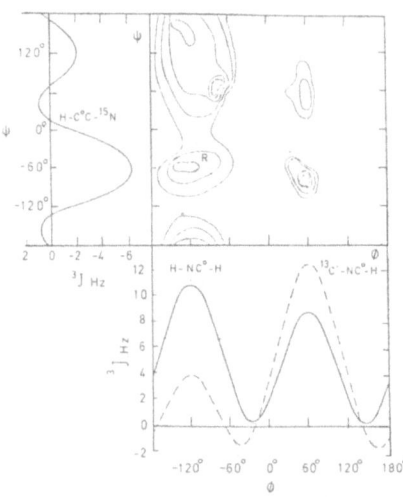

Figure 20. Comparison of the angular dependences of the vici-
            nal $^{13}C-NC^\alpha-^1H$ and $^1H-NC^\alpha-^1H$ (angle φ) and
            $^1H-C\alpha C'-^{15}N$ (angle Ψ) constants with the φ,Ψ
            conformational energy maps of a peptide fragment.

III.3.4.3. Torsional angle χ₁. This angle determines the rotation of
the side chain of the peptide fragment about the $C^\alpha-$ C^β bond. Three
vicinal coupling constants $^1H-C^\alpha C^\beta-^1H$, $^1H-C^\beta C^\alpha-^{13}C'$ and $^{15}N-C^\alpha C^\beta-^1H$ are
available for its determination. The joint use of two different types
of constants makes it possible to determine the population of two staggered rotamers, X_1 and X_2, and thereby from the relation $X_1+X_2+X_3 = 1$
also that of the third. The rotamer populations determined in this manner are very crude because of the number of simplifications made in correlating the constants with the rotamer distribution function and the
assumption of equivalency of all the constants in the coupling between
gauche oriented nuclei.
     It should be stressed that these N.M.R. methods are unreliable in
the case of aromatic side chains for which no experimental calibration
of the φ,ψ dependence of the coupling constants is available.

III.4. Fluorescence of biopolymers

III.4.1. Intrinsic protein fluorescence (review by Eftink, 1981)
     Though all three natural amino acids with aromatic side chains ex-
hibit fluorescence, in most cases only the tryptophan residues are detected due to the efficient electronic energy transfer occuring in the
sequence phenylalanine → tyrosine → tryptophane.
     It is assumed that only the tryptophane residue emits at wavelengths
above 350 nm, but Rayner et al. (1978a and 1978b) suggested the possibility that excited state ionization of the tyrosine phenolic group may
occur in proteins under neutral pH conditions, leading to a bathochromic
emission from this residue.
     In addition, some other phenomena can complicate the interpretation
of protein fluorescence. The most important problem is obviously the
non-exponential decay of tryptophane emission, which is initially ascribed to the different environments of this residue in proteins (Burstein
et al., 1973 and Weber, 1960), later to emission from solvent equilibra-
ted ^1L_a and ^1L_b states of tryptophane (Rayner et al., 1978b) and eventually to separate conformational forms of the amino acid (Szabo et al.,
1980 and Petrich et al., 1983).
     Further, it has been shown that fluctuations in the tertiary and
secundary structure of biopolymers occur on a time scale comparable to
the fluorescence process, leading to a complex decay profile (Weber,
1976 and Grinwald et al., 1977). Finally, the low emission quantum yield
and short fluorescence lifetimes of proteins leads to experimental difficulties for which a solution has become available only the last few years.

III.4.2. Extrinsic fluorescence of probes (review by Galley et al.,1979)
     The covalent binding to a residue with a functional side group,or
the adsorption at the biopolymer surface of fluorescent probe molecules,
is often accompagnied by changes in decay time and emission characteristics, providing information about the environment of the probe.
     The most common probes used are 1-anilino-8-naphtalene-sulphonate
(ANS), 2-p-toluidinyl-6-naphtalene-sulphonate (TNS) and, more recently,
1-pyrenyl-butyric acid and N(1-pyrenyl)-maleimide. The fluorescence of
the latter compound (Betcher-Lange et al., 1978, Graceffa et al., 1978,
Lehrer et al., 1981 and Lee et al., 1976) is totally quenched when solu-
bilized in the water phase, but after a reaction with the thiol group of
cysteine, the probe becomes highly emissive. It has been shown however
that this pyrene derivative also strongly adsorps on the hydrophobic
parts of the protein and that an aminolysis (Wu et al., 1976) or hydroly-
sis (Lux et al., 1981) of the probe can occur within the protein.
     Other difficulties in the application of this technique are e.g. the
sensitivity of the probe to environmental relaxation processes or conformational transition of the chain under study within the probe's lifetime
(e.g. helix-coil transition).

III.4.3. Fluorescence studies on model peptides. The above mentioned
difficulties encountered in the study of the emission of biopolymers have
pointed the research interests towards model systems.
     First, the homopolymers of synthetic amino acids with aromatic side

groups have been studied : polyphenylalanine (Leroy et al., 1974 and
Lapp et al., 1971), aromatic esters of polyglutamates (Ueno et al.,1974
and 1976) and -aspartates (Ueno et al., 1975 and 1977), poly(1- and 2-
naphtyl)alanine (Sisido et al., 1983a and 1983b), poly(1-pyrenyl)alani-
ne (Egusa et al., 1983) and poly-N_ϵ-(9-carbazolyl)carbonyl-L-lysine
(Biddle et al., 1984). These compounds all form an α-helical structure
in solution but only in the latter case and of poly(1-pyrenyl)alanine
(PPA) a significant excimer emission has been observed.

   Measurements of the circularly polarized fluorescence (CPF) of this
compound(PPA) as a function of temperature indicate the existence of two
kinds of excimers; one with a negative CPF dissymmetry with an emission
maximum around 460 nm and predominant at higher temperatures, the other
with a positive CPF dissymmetry emitting at wavelengths longer than 500
nm and mainly formed at lower temperatures (Egusa et al., 1985). In
the emission spectrum of poly-N_ϵ-(9-carbazolyl)-carbonyl-L-lysine, even
four distinct species are observed besides the emission of locally excited carbazole.

   In our own study, the emission properties of both diastereoisomers
of N-acetyl-bis(1-pyrenyl)alanine-methylester (figure 21) have been studied (Goedeweeck et al., 1984 and 1985).

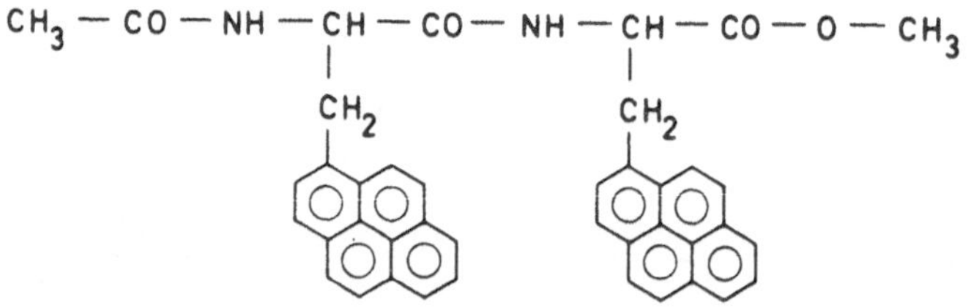

Figure 21.   N-acetyl-bis(1-pyrenyl)alanine-methylester.
             D,D and L,L : threo (t-ABPE);   D,L and L,D :
             erythro (e-ABPE).

   In contradiction to the dimeric model compounds of polyvinylaromatics
the fluorescence spectrum of these dipeptides is very dependent on solvent properties :
- in inert solvents, unable to form hydrogen bonds, a high value for the
ratio of the quantum yields of emission from the excimer and from the
locally excited state ϕ_{exc}/ϕ_{py} is observed (about 3.0 for e-ABPE and
1.0 for t-ABPE).
- in hydrogen accepting solvents, this value is considerably smaller and
the decrease is proportional to the hydrogen accepting properties of the
solvent (in terms of the Taft basicity parameter β (Mortimer et al.,
1983). In dimethylformamide, practically no excimer emission is observed.
- in solvents with a low hydrogen donating tendency, also lower values
of ϕ_{exc}/ϕ_{py} are observed, compared to the inert solvents, but in very
acid solvents (in terms of the acidity Taft parameter α) such as tri-
fluoroethanol very high values are obtained (about 8.0 for e-ABPE and
3.0 for t-ABPE).
   These solvent influences and the observation that the erythro di-

astereoisomer has a larger efficiency of excimer formation, in all solvents studied, than the threo diastereoisomer were explained by means
of a consecutive kinetic scheme (Scheme II) :

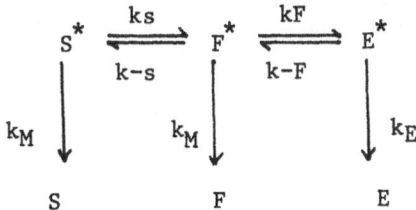

Scheme II

These dipeptides are present in two sets of ground state conformations :
$C_5(S)$ which does not allow a rotation of the side chains to an excimer
geometry within the lifetime of excited pyrene and $C_7(F)$ in which the side
chains are at the same side of the molecule, enabling excimer (E) formation.
  In inert solvents, the molecule stabilizes itself by folding. Consequently, a high C_7 population is present and accordingly, intense excimer emission is observed. Addition of hydrogen bonding solvents destroys the intramolecular hydrogen bond of C_7, the conformational equilibrium shifts towards C_5 and less excimer is formed. The same arguments are valid in the case of alcohols with low acidity. In very acid
solvents, C_5 loses all its rigidity because the free electron pair of
the NH function is captured in an hydrogen bond with the acid solvent
proton, thereby prohibiting the amide resonance. In such solvents, the
peptide chain behaves as an alkane chain, so ABPE should be compared
with the corresponding bis(1-pyrenyl)hexane. The corresponding high
value for $\emptyset_{exc}/\emptyset_{py}$ is observed.
  When compared in the same solvent, the threo diastereoisomer has a
lower folded population than the erythro diastereoisomer due to a large
steric hindrance between the side groups which is significantly less
in the erythro configuration. Consequently, less excimer is observed
for t-ABPE (Zachariasse et al., 1976).
  This interpretation is confirmed by means of the time correlated
single photon counting technique : at low temperatures, the decay of
the locally excited state can be described by a two exponential function.
Since the excimer dissociation to the locally excited state is only com-
petitive at temperatures above -20°C, these two decay parameters are to
be ascribed to two ground state conformations. From both lifetimes, the
ratio of their preexponentials and the lifetime of the unquenched chromophore all the rate constants present in the kinetic scheme and the
ratio of the populations C_7/C_5 could be calculated.
  The solvent induced conformational equilibrium shift is confirmed
by these measurements : in toluene a ratio of 3.0 and 0.75 is calculated for e- and t-ABPE respectively. In ethylacetate, these values decrease to 0.8 and 0.4 respectively (values at -20°C). By a temperature
study, the activation energies on the rate constants and the enthalpy

and entropy change on folding from C_5 to C_7 can be calculated. These values are given in table IV.

Table IV. Preexponentials k_i^o and activation energies E_i^a of k_S, k_{-S} and k_F and enthalpy and entropy change on folding for e- and t-ABPE in toluene and ethylacetate.

	e-ABPE (EtOAc)	t-ABPE (EtOAc)	e-ABPE (toluene)	t-ABPE (toluene)
k^o_S, s^{-1}	6.3×10^9	1.9×10^{11}	1.7×10^{11}	4.3×10^{11}
E^a_S, k Jmol^{-1}	14.6	24.7	20.1	23.8
k^o_{-S}, s^{-1}	3.5×10^9	7.1×10^9	1.5×10^{10}	1.8×10^{10}
E^a_{-S}, kJ mol^{-1}	13.0	15.9	17.6	16.7
k^o_F, s^{-1}	2.5×10^{10}	6.1×10^9	1.5×10^{10}	6.1×10^{10}
E^a_F, kJ mol^{-1}	14.2	8.8	13.0	12.5
ΔH_{fold}, kJmol^{-1}	1.7	8.8	2.5	7.1
ΔS_{fold}, kJmol^{-1}	4.2	25.1	20.9	29.3

The steric hindrance between the side chains in the folded conformations of the threo diastereomer is depicted by the larger activation energy of folding, the lower activation energy of excimer formation and the larger enthalpy change on folding. The folding process is entropy controlled indicating that the extended C_5 conformation has less entropic freedom than the folded C_7 conformation. Some phenomena responsible for the rigidity of C_5 have been mentioned above. In addition the following aspects having a similar effect should be considered :
- The solvent molecules are organized around the dipeptide by hydrogen bonds with the peptide function and C_5 should be more apt to perform this solvent structuring than C_7.
- In the C_5 conformation the pyrene side chains are immobilized by a dipole-induced dipole interaction between the NH function and the system of the aromate. In C_7, the NH is involved in an intramolecular hydrogen bond, resulting in a freely rotating pyrene.
As a conclusion, it can be stated that the use of adequate model peptides allows one to obtain information on the molecular dynamics of the peptide chain by means of fluorescence techniques. Both conformational and configurational effects have to be taken into account when analyzing the photophysics of these aromatic peptides. Moreover, rate constants and thermodynamic parameters of conformational transitions can be obtained. Until now, no other technique offers the possibility to obtain quantitative information on the molecular dynamics of peptides and, to our knowledge, no other reports on the rate of conformational transitions of peptides can be found in literature.

ACKNOWLEDGEMENTS

The authors thank the University Research Fund, the Fonds voor Fundamen-
teel Kollektief Onderzoek, Agfa-Gevaert, IWONL, NFWO and ERO for financial
support of part of this research either by financially supporting the
laboratory or by providing a scholarship to one of them.

REFERENCES

Anfinsen, C.B., Haber, E., Sela, M., White, Jr., H.F., *Proc.Natl.Acad.
Sci. USA*, 47, 1309 (1961) ·
Anfinsen, C.B., Sheraga, H.A., *Adv.Prot.Chem.*, 29, 205 (1975).
Avignon, M., Huong, P.V., Lascombe, J., *Biopolymers*, 8, 69 (1969).
Becker, H.-D., Andersson, K., *J.Org.Chem.*, 47, 354 (1982).
Betcher-Lange, S.L., Lehrer, S.S., *J.Biol.Chem.*, 253, 3757 (1978).
Biddle, D., Chapoy, L.L., *Macromolecules*, 17, 1751 (1984)
Bokobza, L., Jasse, B., Monnerie, L., *Europ.Polym.J.*, 13, 921 (1977).
Boussard, G., Marraud, M., *J.Am.Chem.Soc.*, 107, 1825 (1985).
Bovey, F.A., Hood, F.P., Anderson, E.W., Snyder, L.C., *J.Chem.Phys.*, 42,
3900 (1965).
Brant, D.A., Flory, P.J., *J.Mol.Biol.*, 23, 47 (1967).
Burstein, E.A., Vedenkin, N.S., Ivkova, M.N., *Photochem.Photobiol.*, 18,
263 (1973).
Bystrov, V.F., *Progr. NMR Spectr.*, 10, 41 (1976).
Chuman, H., Momany, F.A., Schafer, L., *Int.J.Pept.Prot.Res.*, 24, 233
(1984).
Collart, P., Demeyer, K., Toppet, S., De Schryver, F.C., *Macromolecules*,
16, 1390 (1983).
Collart, P., Toppet, S., Zhou, Q.F., Boens, N., De Schryver, F.C., *Macro-
molecules*, 18, 1026 (1985).
Cung, M.T., Marraud, M., Néel, J., *J.Ann.Chim.*, 7, 183 (1972).
Demeyer, K., Ph.D. Thesis, Catholic University of Leuven, Belgium (1982).
De Schryver, F.C., Demeyer, K., Van der Auweraer, M., Quanten, E., *Ann.
N.Y. Acad.Sci.*, 366, 93 (1981).
De Schryver, F.C., Moens, L., Van der Auweraer, M., Boens, N., Monnerie,
L., Bokobza, L., *Macromolecules*, 15, 64 (1982a).
De Schryver, F.C., Vandendriessche, J., Toppet, S., Demeyer, K., Boens,
N., *Macromolecules*, 15, 406 (1982b).
Eftink, M.R., Ghiron, C.A., *Anal. Biochem.*, 114, 199 (1981)
Egusa, S., Sisido, M., Imanishi, Y., *Chem.Lett.*, 1307 (1983).
Egusa, S., Sisido, M., Imanishi, Y., *Macromolecules*, 18, 882 (1985).
Evers, F., Kobs, K., Memming, R., Terrell, D.R., *J.Am.Chem.Soc.*, 105,
5988 (1983).
Fourche, G., Jasse, B., Maelstaf, P., *Polym.J.*, 9, 537 (1977).
Fredrickson, G.H., Frank, C.W., *Macromolecules*, 16, 572 (1983).
Fredrickson, G.H., Andersen, H.C., Frank, C.W., *Macromolecules*, 17, 54
(1984).
Froelich, B., Noel, C., Jasse, B., Monnerie, L., *Chem.Phys.Lett.*, 44,
159 (1976).
Galley, W.C., Milton, J.G., *Photochem.Photobiol.*, 29, 179 (1979).

Goedeweeck, R., De Schryver, F.C., *Photochem.Photobiol.*, 39, 515 (1984).
Goedeweeck, R., Van der Auweraer, M., De Schryver, F.C., *J.Am.Chem.Soc.*, 107, 2334 (1985).
Goodman, M., Ueyama, N., Naider, F., *Biopolymers*, 14, 901 (1975).
Goodman, M., Saltman, R.P., *Biopolymers*, 20, 1929 (1981).
Gorin, S., *J.Chim.Phys.*, 67, 885 (1970).
Gorin, S., Monnerie, L., *J.Chim.Phys. Physicochim.Biol.*, 67, 869 (1970).
Graceffa, P., Lehrer, S.S., *J.Biol.Chem.*, 255, 11296 (1978).
Grinwald, A., Steinberg, I.Z., *Biophys.J.*, 19, 74 (1977).
Ito, S., Yamamoto, M., Nishijima, Y., *Bull.Chem.Soc.Jpn.*, 54, 35 (1981).
Ito, S., Yamamoto, M., Nishijima, Y., *Bull.Chem.Soc.Jpn.*, 55, 363 (1982).
IUPAC-IUB Commission on Biochemical Nomenclature, *Biochem.*, 9, 3471 (1970).
Jasse, B., Lety, A., Monnerie, L., *J.Mol.Struct.*, 18, 413 (1973).
Kaufmann, H.F.,Weixelbaumer, W.-D., Buerbaumer, J., Schmoltner, A.-M., Olaj, O.F., *Macromolecules*, 18, 104 (1985).
Lapp, C.F., Laustriat, G., *J.Chim.Phys.*, 68, 1333 (1971).
Lee, T., Heintz, R.L., *Eur.J.Biochem.*, 66, 105 (1976).
Lehrer, S.S., Graceffa, P., Betteridge, D., *Ann. N.Y. Acad.Sci.*, 366, 285 (1981).
Leroy, E., Lapp, C.F., Laustriat, G., *Biopolymers*, 13, 507 (1974).
Lim, D., Kolinsky, M., Petranek, J., Doskocilova, D., Schneider, B., *Polym.Lett.*, 4, 645 (1966).
Longworth, J.W., Bovey, F.A., *Biopolymers*, 4, 1115 (1966).
Lux, B., Gérard, D., *J.Biol.Chem.*, 256, 1767 (1981).
McMahon, P.E., Tincher, W.C., *J.Mol.Spectrosc.*, 15, 180 (1965).
Maser, F., Bode, K., Rajasekharan Pillai, Mutter, M., *Adv.Pol.Sci.*, 65, 177 (1984).
Masuhara, H., Vandendriessche, J., Demeyer, K., Boens, N., De Schryver, F.C., *Macromolecules*, 15, 1471 (1982).
Masuhara, H., Tamai, N., Mataga, N., De Schryver, F.C., Vandendriessche, J., *J.Am.Chem.Soc.*, 105, 7256 (1983).
Masuhara, H., Tamai, N., Mataga, N., De Schryver, F.C., Vandendriessche, J., Boens, N., *Chem.Phys.Lett.*, 95, 471 (1983).
Maxfield, F.R., Leach, S.J., Stimpson, E.R., Powers, S.P., Sheraga, H.A., *Biopolymers*, 18, 2507 (1970).
Moens, L., Ph.D. Thesis, Catholic University of Leuven, Belgium (1985).
Momany, F.A., Carruthers, L.M., McGuire, R.F., Sheraga, H.A., *J.Phys. Chem.*, 78, 1595 (1974).
Mortimer, J.K., Abbaud, J.L., Abraham, M.H., Taft, R.W., *J.Org.Chem.*, 48, 2877 (1983).
Naider, F., Ribeiro, A., Goodman, M., *Biopolymers*, 19, 1791 (1980).
Padma Malar, E.J., Chandra, A.K., *Theor.Chim.Acta*, 55, 153 (1980).
Pajot, E., Ph.D. Thesis, l'Université Pierre et Marie Curie, Paris, France (1983).
Petrich, J.W., Chang, M.C., McDonald, D.B., Fleming, G.R., *J.Am.Chem. Soc.*, 105, 3824 (1983).
Pivcova, H., Kolinsky, M., Lim, D., Schneider, B., *J.Polym.Sci. C*, 22, 1093 (1969).
Ramachandran, G.N., Ramakrishnan, C., Sasisekharan, V., *J.Mol.Biol.*, 7, 95 (1963).

Ramachandran, G.N., Sasisekharan, V., *Adv.Prot.Chem.*, 23, 283 (1968).
Rao, C., Balaram, P., Rao, C.N.R., *Biopolymers*, 22. 2091 (1983).
Rayner, D.M., Szabo, A.G., *Canad.J.Chem.*, 56, 743 (1978a).
Rayner, D.M., Krajcarski, D.T., Szabo, A.G., *Canad.J.Chem.*, 56, 1238 (1978b).
Ribeiro, A., Saltman, R.P., Goodman, M., *Biopolymers*, 19, 1771 (1980).
Ribeiro, A., Saltman, R.P., Goodman, M., Mutter, M., *Biopolymers*, 21, 2225 (1982).
Rinaudo, M., Domard, A., *J.Am.Chem.Soc.*, 98, 6360 (1976).
Saxena, V.P., Wetlaufer, D.B., *Proc.Natl.Acad.Sci. USA*, 66, 969 (1971).
Semerak, S.N., Frank, C.W., *Adv.Polym.Sci.*, 54, 32 (1984).
Sheraga, H.A., *Chem.Rev.*, 71, 195 (1971).
Sisido, M., Egusa, S., Imanishi, Y., *J.Am.Chem.Soc.*, 105, 1041 (1983a).
Sisido, M., Egusa, S., Imanishi, Y., *J.Am.Chem.Soc.*, 105, 4077 (1983b).
Stevens, E.S., Sugawara, N., Bonora, G.M., Toniolo, C., *J.Am.Chem.Soc.*, 102, 7048 (1980).
Sundararajan, P.R., *Macromolecules*, 13, 512 (1980).
Szabo, A.G., Rayner, D.M., *Biochem.Biophys.Res.Comm.*, 94, 909 (1980).
Toniolo, C., Bonora, G.M., Salardi, S., Mutter, M., *Macromolecules*, 12, 620 (1979).
Tsuboi, M., Shimanouchi, T., Mizushima, S., *J.Am.Chem.Soc.*, 81, 1406 (1959).
Ueno, A., Toda, F., Iwakura, Y., *Biopolymers*, 13, 1213 (1974).
Ueno, A., Ishigura, F., Toda, F., Uno, K., Iwakura, Y., *Biopolymers*, 14, 353 (1975).
Ueno, A., Osa, T., Toda, F., *J.Polym.Sci., Polym.Lett.Ed.*, 14, 521 (1976).
Ueno, A., Osa, T., Toda, F., *Macromolecules*, 10, 130 (1977).
Vandendriessche, J., Palmans, P., Toppet, S., Boens, N., De Schryver, F.C., Masuhara, H., *J.Am.Chem.Soc.*, 106, 8057 (1984).
Vandendriessche, J., Collart, P., De Schryver, F.C., Zhou, Q.F., Xu, H. J., *Macromolecules*, accepted for publication, November 1985.
Weber, G., *Biochem.J.*, 75, 335 (1960).
Weber, G., in *Excited States of Biological Molecules*, Birks, J.B., (Ed), 363, London – New York – Toronto – Sidney, Wiley 1976.
Wu, C.W., Yarbrough, L.R., Wu, F.Y.-H., *Biochem.*, 15, 2863 (1976).
Yoon, D.Y., Sundararajan, P.R., Flory, P.J., *Macromolecules*, 8, 776 (1975).
Zachariasse, K., Kühnle, W., *Z. Phys. Chem.* (Wiesbaden), 101, 267 (1976).

SPECTROSCOPIC STUDIES OF RATES OF HINDERED ROTATION IN THE BACKBONES OF HIGH POLYMERS AND THEIR ANALOGS.

H. Morawetz
Department of Chemistry
Polytechnic Institute of New York
333 Jay Street
Brooklyn, NY U.S.A.

ABSTRACT. The concept of "crankshaft-like motion" in the backbones of chain molecules implies much slower conformational transitions than in small molecules if two energy barriers have to be surmounted simultaneously. Studies of the rates of hindered rotation around the amide bonds in solutions of piperazine polyamides by NMR and of the photochemical and dark isomerization of polyamides with azobenzene residues in the chain backbone revealed no difference between these rates in the polymers and their analogs. Polyoxyethylene with a dibenzylacetamide residue in the middle of the chain exhibited similar excimer emission as N,N'-di(p-methylbenzyl)acetamide. These observations imply that only one energy barrier is surmounted in conformational transitions of polymer backbones.

1. Introduction.

Kuhn and Kuhn (1) first pointed out that rotations in the backbone of chain molecules with the displacement of part of the chain (Fig. 1a) is strongly resisted by the viscous medium. They suggested that this could be avoided by two correlated rotations, so that only a short segment of the chain has to move (Fig. 1b). This concept was later introduced by Schatzki (2) and Stockmayer (3) (who were unaware of the suggestion of Kuhn and Kuhn) and became known as a "crankshaft-like motion". It seems to have been generally taken for granted that the two potential energy barriers which have to be surmounted in such a process are passed simultaneously (4-6) and this assumption implied that the doubling of the activation energy for hindred rotation in the backbone of polymers would make it slower by orders of magnitude in comparison with hindered rotations in analogous small molecules.

2. Experimental results.

2.1. NMR studies.

In a program designed to check on the validity of the "crankshaft-like

M. A. Winnik (ed.), Photophysical and Photochemical Tools in Polymer Science, 263–267.
© *1986 by D. Reidel Publishing Company.*

(a)

(b)

Figure 1. Schematic representation of a conformational transition in a flexible chain molecule. (a) Rotation around a single bond. (b) A correlated rotation around two bonds in a "crankshaft-like motion".

motion" concept, we studied first piperazine polyamides and analogous small molecules by NMR spectroscopy (7). In a molecule such as diacetyl piperazine

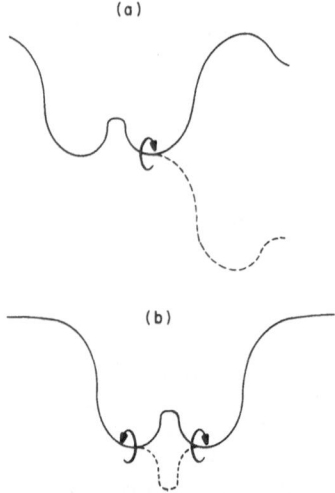

the methylene groups are nonequivalent and exhibit separate NMR absorption peaks at ambient temperatures, where rotation around the N-C partial double bond is slow. Recording spectra in the temperature range of peak coalescence can be interpreted in terms of the activation parameters for hindered rotation and, using this approach, no significant difference was found in the conformational mobility of polymers such as poly(succinyl piperazine) and diacetylpiperazine. It was realized that in this polyamide rotation around the bond connecting the two methylenes of the succinyl residues would involve a much smaller energy barrier than rotation around the N-C amide bond, so that this system would be relatively insensitive in distinguishing between a single and two correlated rotations, but poly(terephthaloyl piperazine)

which contains only "stiff" bonds also behaved in an essentially identical manner as its analog, dibenzoylpiperazine.

2.2 Isomerization of azobenzene residues.

The photoisomerization of trans-azobenzene to the cis form and the thermally activated back reaction can be followed conveniently by UV spectroscopy. We used these processes to compare the behavior of azobenzene
residues in the backbones of polymer chains and their low molecular
weight analogs. This approach has the advantage over the NMR technique
in that we are not limited to solutions of high fluidity, but can study
also very viscous and even glassy systems. Data obtained on the dark reaction indicated again no difference between the rates of azobenzene residues in polyamide backbones and in small molecules (8). The isomerization rate of the polymer remained unchanged even when the solution
concentration was raised to the point where the molecular chains were
heavily intertwined.
     Although these results seemed incompatible with the concept of
two simultaneous hindered rotations, they suffered again from the fact
that isomerization with a high activation energy was observed in chain
molecules with many bonds characterized by a high conformational mobility. To avoid this problem, we carried out a similar study of the photochemical reaction. This seemed attractive since a study of the photoisomerization of azobenzene (9) had indicated that the excited trans
and cis isomers are distinct species separated by an energy barrier of
about 3 kcal/mole, similar to the barrier for conformational transition
in a paraffinic chain. The results (10) are represented in Figure 2.

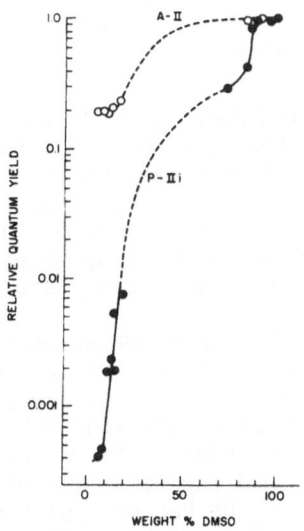

Figure 2. Relative quantum yields in the trans-cis photoisomerization
of azobenzene residues in the backbone of a polyamide (●) and its low
molecular weight analog (o) in mixtures of unlabeled polyamide and dimethyl sulfoxide.

Here the quantum yield of a small number of azobenzene residues built
into the backbone of a polyamide was compared with the quantum yield of
an analogous low molecular weight azobenzene derivative in a series of
polyamide - DMSO mixtures. As expected, the quantum yield for the azobenzene residues in the polymer backbones was very small in systems
containing a high polymer concentration, since here conformational
transitions of the polymer chains would necessarily be cooperative,
while the photoisomerization of the small molecule was relatively insensitive to the polymer concentration. However, in highly dilute solution there was absolutely no difference between the photoisomerization
efficiency of azobenzene residues built into the polymer chains and the
small azobenzene derivatives. This identity in the behavior of the polymer and its analog in dilute solution shows that the isomerization of
azobenzenes in the chain backbone can be accomplished without a simultaneous hindered rotation in another portion of the chain.

An interesting result was also obtained (10) in a study of polyamides with stilbene residues in the chain backbone. In that case, an
increasing polymer concentration led to a decreasing fluorescence intensity. This effect can be understood since the trans-cis isomerization
and the emission from the excited stilbene moiety compete with each
other. Thus, an increasing concentration of chain molecules which hinders the photoisomerization favors the fluorescence. This effect is observed long before the systems becomes glassy and it is, therefore, distinct from the enhancement of fluorescence in rigid media in dyes
whose emission is quenched by internal motions of the excited molecules (11).

2.3. Intramolecular excimer emission.

The lifetime of many aromatic fluorophores is of the order of 10^{-8}s,
similar to the relaxation times for conformational transitions in paraffinic chains. Thus, the ratio of excimer and "monomer" emission intensities can be used to estimate the rate constant for the hindered
rotation required to form the excimer conformation in a molecule containing two aromatic residues. A detailed study based on this principle
has been carried out in De Schryver's laboratory and is reported elsewhere in this volume.

We have used intramolecular excimer formation to compare rates of
conformational transitions in the backbone of a long chain molecule and
in a small analogous compound (12). To this end, we prepared N,N-diben-
zylacetamides carrying in the two para positions substituents of increasing length. We found that Arrhenius plots of the ratio of excimer
and "monomer" emission intensity had identical slopes with the dibenzylacetamide built into the center of the backbone of a polyoxyethylene
with a polymerization degree of 136 and a small dibenzylacetamide derivative. In fact, this ratio decreased by only 27% when para methyl
substituents were replaced by the long polyoxyethylene chains. This was
taken as conclusive proof that only a single hindered rotation is involved in conformational transitions of long chain molecules in dilute
solution.

3. Concluding remarks.

Although the experiments detailed above show that a single hindered rotation can take place in the backbone of a long chain molecule in dilute
solution, this does of course not mean that this process can be accompanied by the large displacement of a substantial portion of the chain as
represented schematically in Figure 1a. How then is this contradiction
to be resolved ? We suggested (12) that the stress introduced into the
chain by one hindered rotation is relieved by many small distortions of
the internal angles of rotation which would require very little energy.
As explained by Helfand elsewhere in this volume, the results of computer simulations of the dynamics of conformational transition in long polymer chains agrees with our conclusion that the activation energy of
the process corresponds to the passage over a single energy barrier. The
simulation method makes it possible to give a detailed description of
the nature of the transition which could hardly be obtained by exper-
imentation. In particular, it indicates (13) that one transition increases the probability of a later compensating transition. Another computer simulation study by Perchak and Weiner (14) led to similar results
showing that two near-simultaneous barrier crossings need not involve a
significantly higher activation energy than a single transition.

4. Acknowledgement.

We are indebted to the National Science Foundation for their support of
this study by Grants GH-33134 and DMR-77-05210.

5. References.

(1) W. Kuhn and H. Kuhn, Helv.Chim. Acta,29,857 (1946)
(2) T.F. Schatzki, J.Polym.Sci.,Pt.C,14,139 (1966)
(3) W.H. Stockmayer, Pure Appl.Chem.,15,539 (1967)
(4) R.F. Boyer, Rubber Chem Tech.,34,1303 (1963)
(5) G. Allegra, J.Chem.Phys.,61,4910 (1974)
(6) S. Mashimo, Macromolecules,9,91 (1976)
(7) Y. Miron,B.R. McGarvey and H. Morawetz, Macromolecules,2,154 (1969)
(8) D. Tabak and H. Morawetz, Macromolecules,3,403 (1970)
(9) S. Malkin and E. Fischer, J.Phys.Chem.,66,2482 (1962)
(10) D. T.-L. Chen and H. Morawetz, Macromolecules,9,465 (1976)
(11) G. Oster and Y. Nishijima, J.Am.Chem.Soc.,78,1581 (1956)
(12) T.P. Liao and H. Morawetz, Macromolecules,13,1228 (1980)
(13) E. Helfand, Z.R. Wasserman and T.A. Weber, Macromolecules,13,526
     (1980)
(14) D. Perchak and J.H. Weiner, Macromolecules ,14,585 (1981)

LIGHT-INDUCED CONFORMATIONAL CHANGES OF SYNTHETIC POLYMERS

Masahiro IRIE
The Institute of Scientific and Industrial
Research, Osaka University
Ibaraki, Osaka 567
Japan

ABSTRACT. Several proposals to construct artificial photoresponsive
polymer systems have been reviewed. Photoisomerizable chromophores
are incorporated into the polymer systems and used to actuate the
polymers to control their properties. Cis-trans isomerization of
unsaturated linkages in the polymer backbone, for example, directly
reflects a conformational change. A change in the intramolecular interactions due to generation of charges or strong dipoles in the polymer
chain also induces the conformational change. Using a laser flash
photolysis combined with a light scattering detection method, the
dynamics of the conformational change of the polymers has been followed
in the microsecond range. Applications of the systems to photostimulated shape changes of polymer gels and sol-gel transitions are described.

1. INTRODUCTION

Biological systems have developed various kinds of photoactive organs
to adopt themselves to the environmental electromagnetic radiation,
sunlight. In plants, for example, photosynthesis systems have been
evolved to utilize light as an energy source. At the same time,
organisms have developed other systems that measure and respond to the
light intensity or duration to find favorable conditions for survival.
In the latter systems, the light is used as information, such as
phototropism, phototaxis or vision. They are named "photoresponsive
systems".[1]
      In a manner similar to biological systems, light can be used in
organic photochemistry not only as an energy source for chemical
synthesis but also as an information source or a trigger for subsequent
events. Recently, such applications to construct "artificial photoresponsive systems" have been undertaken by several research groups,
where light is used as a trigger or a signal for the reversible control
of functions of polymers[2] as well as low molecular weight compounds,
such as crown ethers,[3] cyclodextrins,[4] or microcapsules.[5]
      The photoresponsive polymer systems are among the earliest, in
which the chain conformation is reversibly controlled by photo-

M. A. Winnik (ed.), Photophysical and Photochemical Tools in Polymer Science, 269–291.

irradiation. The conformational change should produce a concomitant
change in physical and chemical properties of polymer solutions and
solids. In this chapter, several attempts to enforce reversible
conformational changes of polymer chains by photoirradiation and recent
progress in their applications are reviewed.

2. PHOTOCONTROL OF POLYMER CONFORMATIONS

It is well known that the chain conformation of polymers depends on
solvent quality and temperature. In good solvents, polymers have an
expanded conformation, while they shrink in bad solvents at low
temperature. Polyelectrolytes change their conformation depending on pH
and salt concentration. It is obviously a tedious method to change
these conditions to regulate the chain conformation. If we could
control the conformation reversibly by photoirradiation outside the
sample, the method would be superior to the above method in the response
time, reversibility and easy procedure.

Figure 1 illustrates several proposals to use photoisomerizable
chromophores as a tool to enforce conformational changes. Many photo-
sensitive molecules are known to be transformed under photoirradiation
into other isomers, which can return to the initial state either
thermally or photochemically,

$$A \underset{h\nu,\Delta}{\overset{h\nu}{\rightleftharpoons}} B$$

The reactions are named photoisomerizations. The isomerizations are
always accompanied by certain physical and chemical property changes.
The property changes of the chromophores, such as dipole moment and/or
geometrical structural changes, have been successfully utilized as a
driving force to induce a conformational change of the polymer chain,
when the chromophores are incorporated into the polymer as "photoreceptor molecules".

(1) and (2) are the mechanisms proposed for the first time by

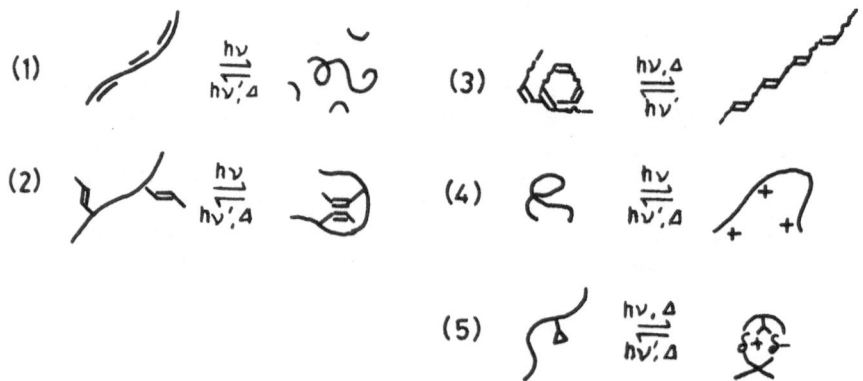

Figure 1. Schematic illustrations of photostimulated conformational
changes of polymer chains.

by Lovrien in 1967.[6] If a polymer is involved in equilibrium interaction with some photoisomerizable low molecular weight chromophores in
solution, a different conformation may result under the influence of
light due to the change of interactions between the polymer and the
low molecular weight chromophores. The system described by Lovrien is
the mixture of poly(methacrylic acid) and chrysophenine G (1, CHP).

$$C_2H_5O - \langle \bigcirc \rangle - N{\overset{.}{=}}N - \langle \bigcirc \rangle - CH{\overset{CH}{=}}CH - \langle \bigcirc \rangle - N{\overset{.}{=}}N - \langle \bigcirc \rangle - OC_2H_5$$

with SO_3Na above and SO_3Na below.

1

Upon ultraviolet irradiation, trans CHP isomerized to the cis form
(around 10%) and the aqueous solution viscosity decreased as much as 80
%. It was suggested that the anionic linear and planar all-trans CHP
would attach itself to the hydrophobic poly(methacrylic acid) backbone,
leading to an extended polymer conformation. In the cis form, the azo-
dyes are much more water-soluble. Consequently, the cis form was
envisaged as binding less strongly so that the polymer chain would be
less extended.
       van der Veen and Prins carefully re-examined the above system,[7]
but could not obtained the change of viscosity in the same order of
magnitude. Under the conditions of 43 % cis CHP, the reduced viscosity
was found to be only 5 % lower than the value after complete reversion
of CHP to the trans form. They concluded that the decrease of viscosity
reported by Lovrien is incorrect probably due to some impurity of the
dye and/or polymer. The photoregulation mechanism postulated by Lovrien,
however, is a probable one, which can expain the small change in the
viscosity.
       A polymer - low molecular weight chromophore system has also been
investigated by Negishi et. al..[8] They observed 13 % reduction of the

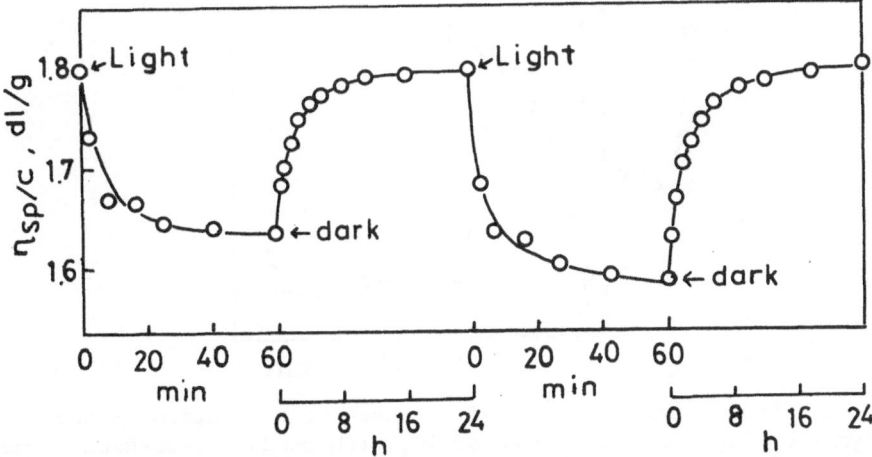

Figure 2. Photostimulated reversible viscosity change of HEMA - VPy
copolymer (HEMA; 62 mole%) - CHP complex in water at 30°C.

reduced viscosity in the aqueous solution of 2-hydroxyethylmethacrylate
(HEMA) - N-vinylpyrrolidone (VPy) copolymer and CHP as shown in Figure
2.

The second system described by Lovrien, which belongs to mechanism
(2), is poly(methacrylic acid) with pendant azobenzene groups.[6] In
aqueous solution, the viscosity was found to increase by ultraviolet
irradiation. Matějka et. al. have developed the system to styrene -
maleic anhydride copolymer with pendant azobenzene groups.[9] The
copolymer exhibited a pronounced photodecrease of the viscosity in 1,4-
dioxane solution, i.e. a reversible decrease by 24 - 30 % in the reduced
viscosity of the solution after ultraviolet irradiation. In THF, the
viscosity decreased by 1 - 8 %. The contraction of the dimensions of
the copolymer coil is explained as follows: trans to cis isomerization
induces a strong dipole in the azo bond. These dipoles become
mutually oriented and attract each other so that compact coil
conformations are preferred. In the dark, the viscosity of the copolymer solution returned to the original value as shown in Figure 3. The
increase rate of the viscosity, however, was much slower, by a factor
of 1/2.5 to 1/7, than the rate of cis to trans isomerization of the
pendant azobenzene chromophores measured by optical absorption in the
dark. The discrepancy requires further examination of the postulated
mechanism of the conformational change.

The mechanism (3) is the most simple one. Trans-cis isomerization
of organic molecules about an unsaturated linkage is a well known
photochromic phenomenon. When we incorporate the trans-cis photoisomerizable chromophores into the backbone of a polymer chain, photoinduced isomerization of the chromophores is expected to induce a
conformational change of the polymer chain. Table I summarizes polymers
having photoisomerizable unsaturated linkages in the polymer backbones

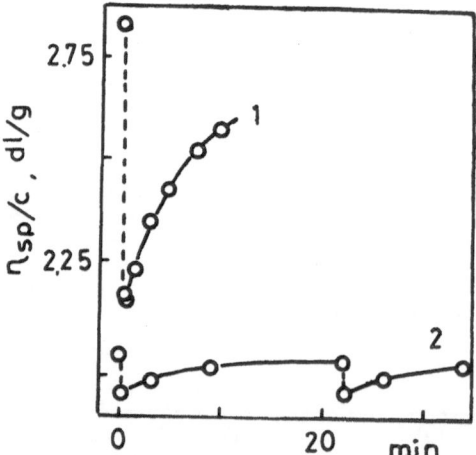

Figure 3. Photostimulated viscosity change of a solution of the
copolymer of styrene - maleic anhydride with pendant azobenzene groups:
1: 1,4-dioxane solution (polymer conc, 0.052 g/dl); 2: THF solution (
polymer conc, 0.061 g/dl); irradiation (----); dark (——).

TABLE I. Photoresponsive polymers with photoisomerizable unsaturated linkages in the polymer backbone

Structure		Reference

so far examined, mostly containing azobenzene groups.[10-14]

Azobenzene is a well known photochromic molecule, which undergoes isomerization from trans to cis form under ultraviolet irradiation, while the cis form can return to the trans form either thermally or photochemically. During the course of the isomerization, azobenzene

shows a large structural change. The distance between 4 and 4' carbons decreases from 9.0 to 5.5 Å.

Polyamides with azobenzene groups in the polymer backbone described by Irie et. al. are among the earliest in which trans-cis isomerizable chromophores are used to regulate the polymer conformation.[10,11] The intrinsic viscosity,[η] , of polyamide (2) in N,N-dimethylacetamide was found to decrease from 1.22 to 0.5 dl/g on ultraviolet irradiation (410 >λ> 350 nm) and to return to the initial value in 30 h. in the dark at 20°C (Figure 4). The slow recovery of the viscosity in the dark was accelerated by visible light irradiation (λ> 470 nm). On alternate irradiation of ultraviolet and visible light, the viscosity reversibly changed as much as 60 % (Figure 5).

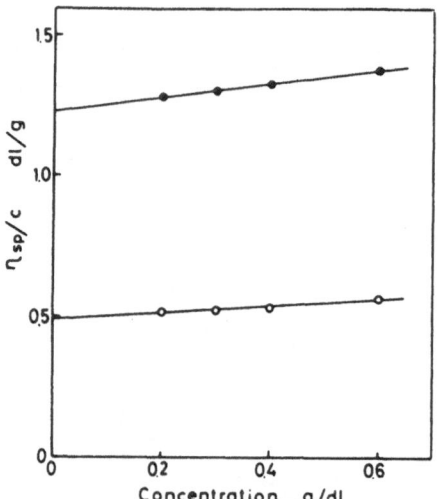

Figure 4. Viscosity of polyamide (2) in N,N-dimethylacetamide at 20°C
(●) in the dark before photoirradiation and (o) under irradiation
with ultraviolet light (410 >λ> 350 nm).

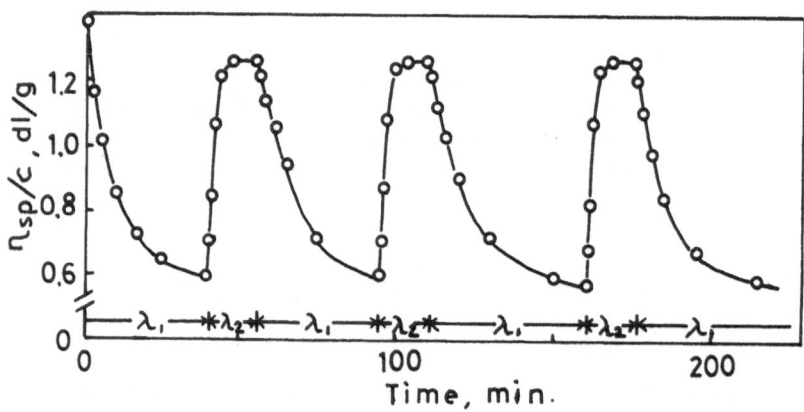

Figure 5. Photostimulated viscosity change of N,N-dimethylacetamide
solution of polyamide (2) on alternate irradiation with ultraviolet
light (410 >λ_1> 350 nm) and visible (λ_2> 470 nm) light at 20°C.
Concentration of the polymer was 0.9 g/dl.

Before irradiation, the polymer has a rod like conformation. The isomerization from trans to cis form kinks the polymer chain, resulting in a compact conformation and a decrease in the viscosity. The compact conformation returns to the initial extended conformation either thermally or by visible light irradiation, causing an increase in the viscosity.

According to the above illustration, rigidity of the polymer chain is expected to alter the amount of photodecrease of the solution viscosity. When the azobenzene residues are connected by rigid phenylene groups, the resulting viscosity change should be large, while it should become small when the connecting groups are flexible, as long methylene chains. The effect of backbone structure on the photodecrease of solution viscosity is summarized in Table II. As expected, the amount of photodecrease of the viscosity decreased with increasing number of methylene groups in the polymer backbone. The stiffest polymer, that having phenylene residues, gives a large photodecrease, while the viscosity of polymers having flexible long methylene chains is hardly reduced by photoirradiation. The absence of photodecrease in the polymer with 12 methylene groups suggests that flexible methylene chains act as a strain absorber. The conformational change induced by the isomerization of the azobenzene residues is relaxed in the connecting flexible methylene chains, resulting in no change of the shape of the polymer.

TABLE II. Effect of backbone structure of the polymers on the photodecrease of the solution viscosity.

Polymer	R	$\eta_{sp}(UV)/\eta_{sp}(dark)$[a]
Polyamide	$+p\text{-}C_6H_4+$ [b]	0.37
Polyamide (3)	$+(CH_2)_4+$ [b]	0.59
Polyamide (3)	$+(CH_2)_8+$ [b]	0.80
Polyamide (3)	$+(CH_2)_{12}+$ [b]	0.96

a) $\eta_{sp}(UV)$ and $\eta_{sp}(dark)$ are specific viscosities under irradiation with ultraviolet light ($410 > \lambda > 350$ nm) and in the dark before irradiation in N-methyl-2-pyrrolidone in the presence of LiCl (1.3 M), respectively.

b) R group in $+NH\text{-}C_6H_4\text{-}N{=}N\text{-}C_6H_4\text{-}NHCO\text{-}R\text{-}CO+_n$

Similar experiments have been carried out for polyamides(4) and (5) by Blair et. al..[12] They could not, however, observe any decrease in the intrinsic viscosity under ultraviolet irradiation in contrast with our result, though a small decrease was detected in the reduced viscosity at high polymer concentration. The absence of the photo-decrease in the viscosity is possibly due to inclusion of flexible piperazine segments in the polymer chain.

Recently, Neckers et. al. demonstrated that polyureas with back-bone azobenzene groups also showed photoviscosity effects under ultra-violet irradiation.[13] Stille et. al. reported that the viscosity of polyquinoline (7) with backbone stilbene groups in di-m-cresyl phosphate /m-cresol decreased by as much as 24 % in the intrinsic viscosity value under the influence of ultraviolet light.[14] The decrease is ascribed to the trans to cis isomerization of the stilbene groups. Because of the simplicity of the mechanism, the concept to introduce photorespon-siveness to polymers may be applied to other poly-condensation or poly-addition polymers.

The type (4) employs an electrostatic force of repulsion between photogenerated charges as a driving force of a conformational change. Triphenylmethane leucoderivatives have been used as photoreceptor molecules in order to produce positive charges in the pendant groups of polymers.[15] The chromophore dissociates into an ion-pair under ultraviolet irradiation with production of an intensely colored triphenylmethyl cation. The cation thermally recombines with the counter anion as follows.[16,17]

Triphenylmethane leucohydroxide residues were introduced into the pendant groups by copolymerizing the vinyl derivative (8, R=CH=CH$_2$, X=OH) with N,N-dimethylacrylamide. In the dark before irradiation, a methanol solution containing the copolymer is pale green. Upon ultra-violet irradiation (λ> 270 nm), the solution becomes deep green. The appearance of the deep green color means that the pendant triphenyl-methane leucohydroxide residues dissociate into triphenylmethyl cations and hydroxide ions. Concurrently with the coloration, the reduced viscosity of the solution showed a remarkable increase from 0.55 to 1.6 dl/g as depicted in Figure 6. After removal of the light, the viscosity returned to the initial value with a half-life of 3.1 min. The viscos-ity change indicates that the polymer chain expands on ultraviolet light irradiation and shrinks in the dark. The good correlation between the viscosity change and the absorption intensity at 620 nm implies that the expansion of the polymer conformation is induced by photodissocia-tion of the pendant triphenylmethane leucohydroxide residues and that the disappearance of the positive charges causes the shrinkage of the chain conformation.

To check the validity of the above expansion mechanism, the

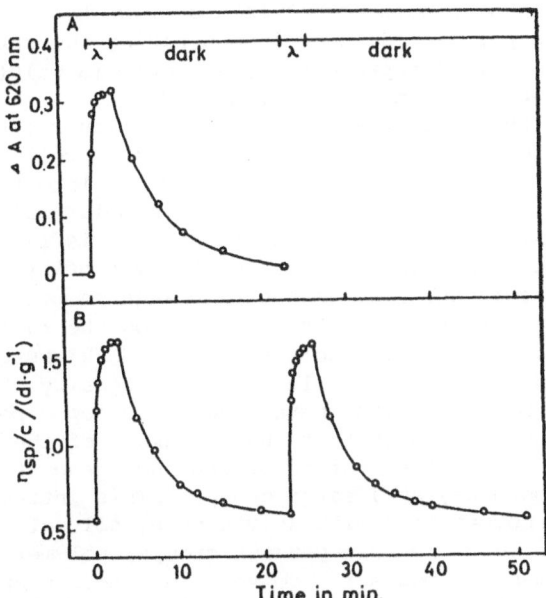

Figure 6. Photostimulated (A) color change and (B) viscosity change of poly(N,N-dimethylacrylamide) having pendant triphenylmethane leuco-hydroxide groups (9.1 mole%) in methanol at 30°C. Polymer concentration was 0.06 g/dl. λ>270 nm.

concentration dependence of the viscosity was examined. In the dark before photoirradiation, the dependence was linear; the reduced viscosity decreased with decreasing the polymer concentration. During ultraviolet irradiation, the reduced viscosity steeply increased at low polymer concentration. The viscosity during photoirradiation was 4 times larger than the viscosity in the dark at 0.04 g/dl. The concentration dependence is characteristic of polyelectrolytes and is clear evidence that electrostatic force of repulsion between positive charges expands the polymer chain. In addition, the photostimulated increase of the viscosity was suppressed in the presence of salt (10^{-3} M LiBr). The salt effect is definite evidence of the above-mentioned mechanism.

Formation of strong dipoles along the polymer chain would also be expected to change the conformation, as depicted by the mechanism (5). This approach has been taken using spirobenzopyran or azobenzene groups as photoreceptor molecules.[18 - 21] Spirobenzopyrans are well known to undergo under ultraviolet light irradiation ring opening with production of merocyanines having strong dipoles; the merocyanines can return to the initial spiropyrans either thermally or photochemically.

We can use the change of dipole moment as a driving force of conformational change of the polymer chain by incorporating the chromophores
into the pendant groups. A representative example is poly(methyl methacrylate) with pendant spirobenzopyran groups.[18]
     Figure 7 shows the viscosity of poly(methyl methacrylate) with
spirobenzopyran (13 mole%) in benzene in the dark as well as during
photoirradiation ($\lambda > 310$ nm). The viscosity in dichloroethane is also
shown as an example of the viscosity behavior in polar solvents. In
benzene, the intrinsic viscosity during irradiation is 17 % lower than
the viscosity in the dark. The decrease of the viscosity depends on
the polarity of the solvent. The viscosity change in polar dichloroethane is only 1 %. The effect of the polarity on the ratio of
viscosity during photoirradiation to that in the dark was examined in
several solvents and is shown in Table 3. The viscosity change
decreased almost in parallel with increasing microscopic polarity, E_{T30}.
     The solvent effect suggests that the shrinkage of the polymer chain
is mainly caused by specific solvation of the photogenerated merocyanines
by the poly(methyl methacrylate) ester groups. This intramolecular
solvation occurs in competition with solvation by solvent.   The intramolecular attractive force between pendant groups, polymer - polymer
interaction, overcomes the polymer - solvent interaction and the polymer
takes a contracted conformation. The possibility of the decrease being
caused by intramolecular dipole-dipole interaction between merocyanine
side groups in a polymer chain can be rejected by the following results:
(1) the viscosity of benzene solution of polystyrene having spirobenzopyran side groups does not show any response to the photoirradiation,

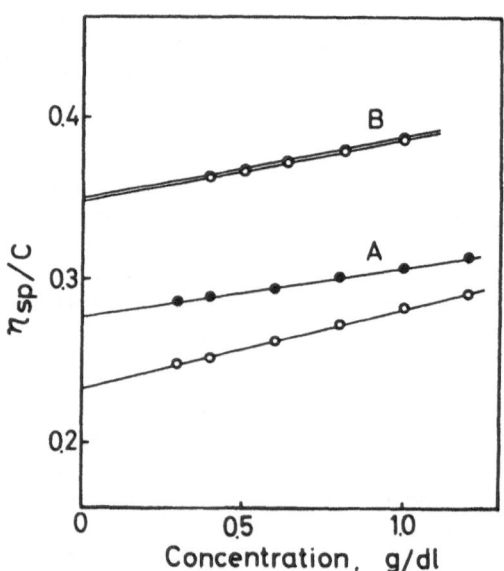

Figure 7. Viscosity of poly(methyl methacrylate) having spirobenzopyran
residues (13 mole%) at 30°C: (A) in benzene (●) in the dark (o) under
irradiation ($\lambda > 310$ nm); (B) in dichloroethane.

TABLE III. Solvent effect on the photostimulated viscosity change.

Solvent	$[\eta]_p / [\eta]_d$	E_{T30} [a]
benzene	0.83	34.1
dioxane	0.87	36.1
THF	0.95	37.4
ethylacetate	0.93	38.1
chloroform	0.97	39.1
dichloroethane	0.99	(41.1)[b]

a) E_{T30} value of Dimroth at 25°C. b) Value of dichloromethane

and (2) the photodecrease of the viscosity of benzene solution was highest at a rather low spirobenzopyran content of 17 mole %.

Poly(methacrylic acid) with pendant spirobenzopyran groups also showed photostimulated conformational changes in methanol.[19] Visible light irradiation of the solution increased the viscosity, while ultraviolet light irradiation caused a decrease in the solution viscosity. The photostimulated conformational change is interpreted by the change of balance between polymer - polymer and polymer - solvent interactions due to the formation and disappearance of polar merocyanine groups. Alternate irradiation of visible and ultraviolet light brings about reversible viscosity changes of as much as 40 % and the increase/decrease cycles of the viscosity can be repeated many times without any noticeable fatigue.

Photostimulated conformational changes observed for polypeptides with pendant azobenzene residues,[22,23] and for polystyrene with pendant azobenzene residues in cyclohexane[21] may also be classified into the category (5).

Several proposals to induce conformational changes of polymers by photoirradiation have been summarized from the view point of the mechanisms. Although polymer systems so far examined are limited, at the present stage of research, it seems possible to regulate the chain conformation of almost all kinds of polymers by applying one of these five mechanisms.

3. DYNAMICS OF CONFORMATIONAL CHANGES

It is of particular interest to know how fast a long polymer chain changes its conformation in response to a short laser pulse. The dynamics of the conformational changes, accompanied by large size change of the total polymer chains, can be appropriately studied with photo-responsive polymer systems undergoing conformational changes as a consequence of photoisomerization such as trans-cis isomerization of azobenzene, because the latter reaction proceeds very fast, less than 10^{-7} s, under the influence of a laser pulse. Using a laser flash photolysis method combined with a light scattering detection system, direct measurement of the conformational change has been carried out.[24,25,26]

As can be seen from the Debye equation (1),

$$\frac{K c}{R_\theta} = \frac{1}{\bar{M}_w} + \frac{16\pi^2 \langle s^2 \rangle}{3\lambda_0^2 \bar{M}_w} \sin^2(\theta/2) + 2A_2 \qquad (1)$$

the light scattering intensity R_θ is correlated to the weight average molecular weight $\bar{M}_w$, the mean square radius of gyration $\langle s^2 \rangle$ and the second virial coefficient A_2. Here, $K = (2\pi^2 n_0^2/N_A \lambda_0^4)(dn/dc)^2$, c is the polymer concentration, n_0 is the refractive index of the solvent, dn/dc is the specific refractive index increment, λ_0 is the wavelength of the incident light and N_A is Avogadro's number. Expansion of the polymer chain leads to an increase in $\langle s^2 \rangle$, which causes a decrease of the light scattering intensity (R_θ, LSI), whereas upon shrinkage of the chain, which leads to a decrease of $\langle s^2 \rangle$, the LSI increases.

Polyamide (2) was irradiated with a single 20 ns flash (530 nm) in N,N-dimethylacetamide.[24] Kinetics of cis to trans isomerization of backbone azobenzene residues was followed by time-resolved optical absorption measurement and a subsequent conformational change of the total polymer chain was followed by time resolved light scattering measurement. Before each laser experiment, the polymer was initially converted to a compact conformation by continuous ultraviolet irradiation and then the unfolding process was followed by the laser flash photolysis method.

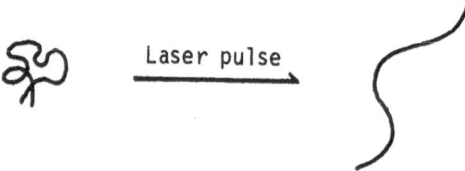

Laser pulse

Figure 8 shows oscillograms demonstrating the decrease and increase of optical absorption at 310 and 410 nm of polyamide (2) and model compound (10) during and after flash.

It is interesting to note that in the case of the polymer, the rapid spectral change during the flash is followed by a somewhat slower process (Figure 8b). While the total change of the absorption occurs during the flash in the case of low molecular weight compound (Figure 8d), about 10 % of the total change is completed, in the case of the polymer, during a period about 100 ns after the flash, corresponding to a relaxation time of $10^{-8} - 10^{-7}$ s. Obviously a small fraction of the electronically excited azobenzene groups cannot convert to the stable trans configuration without restraint, which might be caused by relatively slow motion of chain segments. Even if the relatively slow relaxation process is included, it is safe to say that the isomerization from cis to trans of both the polyamide and the model compound completes in 100 ns.

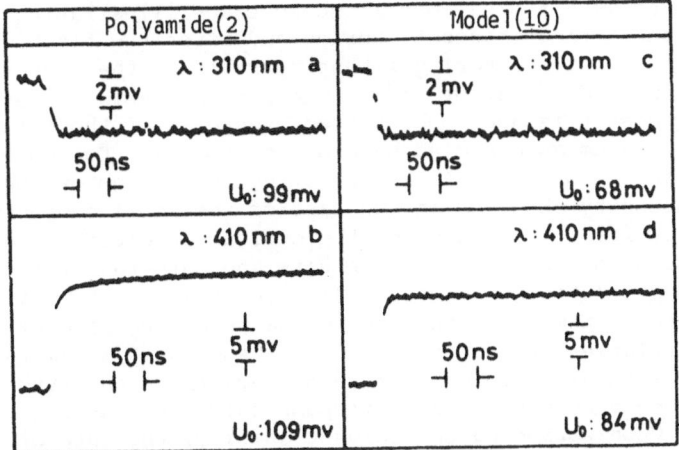

Figure 8. Cis to trans isomerization of polyamide (2) and model compound (10) at 22°C in N,N-dimethylacetamide (1.1 x 10^{-2} g/l) measured by optical aborption. The oscillograms illustrate the decrease and increase of optical absorption at 310 (cis form) and 410 nm (trans form) during and after irradiation with 20 ns flash of 530 nm.

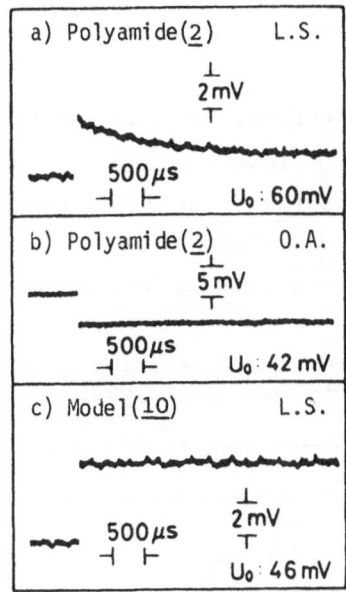

Figure 9. Chain unfolding and cis to trans isomerization of polyamide (2) in N,N-dimethylacetamide solution (0.31 g/l). The oscillograms illustrate changes of light scattering intensity (a, c) at 514 nm and optical absorption at 514 nm (b) during and after a 20 ns flash of 530 nm light. Traces a and b, polyamide (2); trace c, model compound (10).

Figure 9 shows a typical oscilloscope trace illustrating the change of the light scattering intensity during and after the flash. The decrease of the light scattering intensity reflects the conformational change involving a decrease of $<s^2>$. The initial rapid increase of the scattering intensity is due to concurrent decrease of the optical absorption at 514 nm as depicted by oscillogram (b). For comparison, Figure 9 c shows an oscillogram illustrating the change of the light scattering intensity observed in the case of the model compound (10). This trace only reflects changes of the optical absorption and specific refractive index increment. These results indicate that the conformational change of the total polymer chain occurs in around 1 ms.

The nature of the solvents changed the unfolding rate of the polymer chain (Table IV). In good solvents, such as N,N-dimethylacetamide or N,N-dimethylformamide, the unfolding proceeded rather quickly, whereas the unfolding was retarded in poor solvents containing miscible non-solvent. Upon worsening the solvent quality, the rate of unfolding was retarded. Although solvent quality influences the relaxation time, it is allowed to say that the unfolding time of the total polymer chain fall within the range of 0.5 to 1.1 ms.

The large difference in the response time of the optical absorption and the light scattering intensity suggests a two step mechanism for the photostimulated unfolding process. During the isomerization of the backbone azobenzene residues, the total chain conformation remains in the initial compact conformation. After the isomerization is completed, the conformation slowly relaxes to the more stable extended conformation in 1 ms.

Compact conformation (cis form)

10^{-7} s $\downarrow$ isomerization

Compact conformation (trans form)

10^{-3} s $\downarrow$ unfolding

Extended conformation (trans form)

TABLE IV. Conformational relaxation of polyamide chains subsequent to cis → trans isomerization of azobenzene groups in the backbone.

Solvent	$\tau_{1/2}$, s [a]
N,N-dimethylacetamide	5.8×10^{-4}
N,N-dimethylformamide (DMF)	4.7×10^{-4}
4 : 1 DMF - ethanol[b]	7.2×10^{-4}
3 : 2 DMF - ethanol[b]	11.0×10^{-4}
4 : 1 DMF - water[b]	8.8×10^{-4}

a) Halflives of the decrease of the light scattering intensity at 514 nm at 22°C. b) Volume ratio.

The scheme implies that the compact conformation having trans azobenzene residues is a constrained form, which stores a certain strain energy in the chain conformation. The strain energy is a driving force of expansion and it is released during the unfolding process.

Polystyrene with azobenzene pendant groups changes its solubility in cyclohexane upon irradiation with light of a specific wavelength; ultraviolet light caused precipitation of the polymer, while visible light resolubilized it.[21] Structural change of the pendant azobenzene groups from trans to cis form altered the balance of polymer – polymer and polymer – solvent interactions, resulting in contraction of the polymer chain and finally in its precipitation. The dynamics of the coil contraction and the subsequent precipitation can be appropriately studied by time-resolved light scattering measurement. Kinetics of trans to cis isomerization was followed by time resolved optical absorption measurement.[25]

Figure 10 shows oscillograms of the polystyrene with pendant azo-benzene groups (4.3 mole%) demonstrating the decrease at 360 nm and the increase at 440 nm during and after the flash. As can been seen from the oscillograms, the change of the optical absorption occurs during the flash in less than 10^{-8} s. The change of the absorption is ascribed to trans → cis isomerization of the azobenzene groups. From the very fast trans → cis isomerization, it was inferred that neighbour phenyl groups did not interact strongly with trans azobenzene groups.

In the microsecond time range after the flash, an increase of the light scattering intensity was observed. A typical oscillogram demonstrating the intensity change is shown in Figure 11. It can be seen that the intensity decreased initially. This decrease is due to the concurrent increase of the optical absorption at 514 nm (b) and of the decrease of dn/dc as a consequence of trans → cis isomerization. The relatively slow increase of the light scattering intensity is considered to reflect the conformational change involving a decrease of both the radius of gyration of the coil and the second virial coefficient. The halflives of the intensity increase for the various

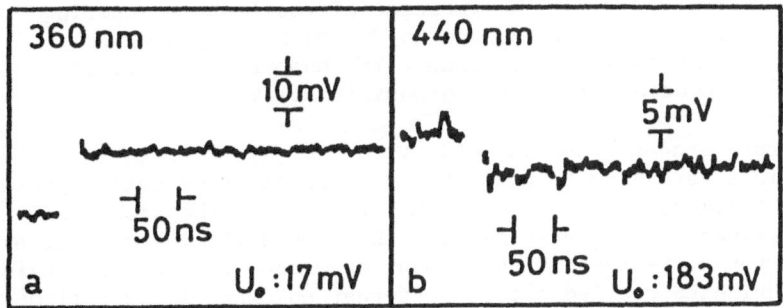

Figure 10. Trans to cis isomerization of pendant azobenzene groups after irradiation of cyclohexane solution of polystyrene with azobenzene groups (4.3 mole%). Oscilloscope traces depicting the change of the optical absorption at (a) 360 and (b) 440 nm.

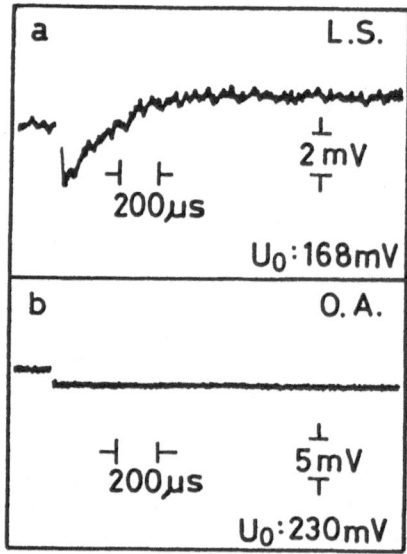

Figure 11. Oscillograms illustrating the increase of the light scattering intensity (a) and of the optical absorption (b), both at 514 nm, after irradiation of polystyrene with pendant azobenzene groups in cyclohexane solution (0.13 g/l) with a laser flash of 347 nm light at 25°C.

TABLE V. Conformational relaxation of polystyrene with pendant azobenzene groups.

Sample	$M_w \times 10^{-4}$	$\tau_{1/2}$ s [a]
PS-A-4.3 [b]	2.7	2.1×10^{-4}
PS-A-5.6 [b]	2.0	4.0×10^{-4}
PS-A-6.5 [b]	1.8	8.5×10^{-4}

a) Halflives of the increase of the light scattering intensity at 514 nm and 25°C. b) Polystyrene with pendant azobenzene groups. The number indicates the content of azobenzene groups in mole%.

copolymers are summarized in Table V.

It is interesting to note that the halflives of coil contraction of the copolymers decrease with the decrease in the azobenzene content and the values are slightly smaller than the halflives of coil expansion measured for polyamide having azobenzene groups in the main chain.

Experiments carried out at different polymer concentrations yielded the same halflife of the intensity change. This result is clear evidence that the process observed in the range of several hundred microseconds is an intramolecular reaction. The rate of coil contraction depended on temperature: $\tau_{1/2}= 4 \times 10^{-4}$s at 25°C and $\tau_{1/2}= 3 \times 10^{-4}$

s at 35°C. Since the coil contraction depends on the mobility of both
the segments and solvent molecules, it is expected to occur faster at
higher temperature.

The fact that the isomerization of only a few azobenzene groups per
chain induces significant conformational changes is attributed to a
severe perturbation of the balance of polymer - solvent and polymer -
polymer interactions by trans → cis isomerization of pendant azobenzene
groups. Concurrent with the trans → cis isomerization, the dipole
moment of the molecule increases from 0.5 to 3.1 D. It is feasible,
therefore, that cis azobenzene groups are more prone to interact with
styrene base units than trans azobenzene groups because of dipole -
induced dipole interactions. The phenomena observed with the copolymer
dissolved in cyclohexane can be, then, interpreted as follows: trans
azobenzene groups interact more favorably with solvent molecules than
with styrene base units, whereas the reverse situation is true for cis
azobenzene groups, which interact less strongly with solvent molecules
than with styrene base units. The enhanced capability of cis azobenzene
groups of interacting with other segments of the chain is considered to
give rise to a shrinkage of the polymer chain. The process is schematically illustrated in Figure 12. The trans to cis isomerization of the
pendant azobenzene groups completes during the laser flash in less than
10^{-8} s. During the isomerization, the total chain conformation remains
in the initial random conformation. The interactions between the cis
form azobenzene groups and styrene base units slowly change the total
chain conformation to more stable contracted form in 0.1 - 0.8 ms.
When different polymers collide with each other at a later stage, intermolecular interaction will also become operative. As a consequence,
the polymer chains will aggregate and precipitation will be observed.

4. APPLICATIONS

4.1. Photostimulated Deformation of Polymer Gels

It seems possible to amplify the photostimulated conformational change
of polymer chains in solution at the molecular level into the shape
change of polymer gels or solids at the macrosize level. A proposal to

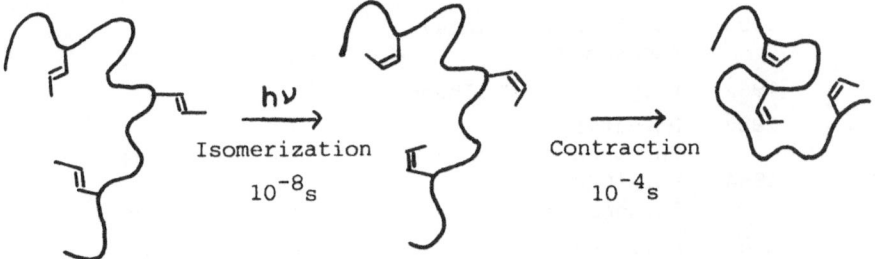

Figure 12. Schematic illustration of a conformational change of
polystyrene with pendant azobenzene groups.

use structural changes of isomerizable chromophores at the molecular
level for direct conversion of photon energy into mechanical work has
been made for the first time by Merian.[27] The system described by
Merian was nylon filament fabric, 6 cm wide and 30 cm long, containing
15 mg/g of azo-dye. After exposure to a xenon lamp at a distance of 30
cm, the dyed fabric was found to be 0.33 mm shorter. Since then, many
materials were reported to show photostimulated deformation.[28-35]
Table VI summarizes the results. The deformation so far reported,
however, is limited to less than 10 %. One exceptional material is
poly-4-(N,N-dimethylamino)-N-γ-D-glutamanilide, which displayed up to
35 % dilation in N,N-dimethylformamide when exposed to light in the
presence of CBr_4, though the phenomenon was _irreversible_.[36]

Recently, polyacrylamide gels having triphenylmethane leuco
derivatives were reported to show large _reversible_ photostimulated
dilation by Irie.[37] Triphenylmethane leuco derivatives dissociate into
ion pairs under ultraviolet irradiation as described in section 2. The
cations thermally recombine with counter anions. By incorporating the
chromophores into the pendant groups of polymers, it becomes possible
to produce reversibly positive charges in polymer gels.

Figure 13 shows photoresponsive behavior of the gel having triphenylmethane leucohydroxide groups as measured by the change of weight.
Upon ultraviolet light irradiation ($\lambda > 270$ nm), the gel swelled as much
as 3 times in weight in around 1 h and the dilated gel again deswelled
in the dark to the initial weight in 20 h. The cycles of dilation and
contraction of the gel by photoirradiation could be repeated several
times.

TABLE VI. Photostimulated deformation of photoresponsive
polymers.

Year	System	$\Delta L/L_0$	Reference
1966	Nylon - Azo dye	0.1 %	27
1970	Polyimide - Azobenzene	0.6 %	28
1971	Poly(HEMA)gel - Azo dye	1.2 %	29
1973	Poly(ethylacrylate) - spirobenzopyran	2-3 %	30
1979	MAH-co-Styrene gel - Azobenzene	-	31
1980	Poly(ethylacrylate) - Azobenzene	0.25 %	32
1982	Polyamide - Stilbene	-	33
1982	Polystyrene - Spirobenzopyran	-	34
1984	Poly(HEMA)gel - Azobenzene	2 %	35
1984	Polyacrylamide - Triphenylmethane	120 %	37
1985	Polyquinoline - Stilbene	1-5 %	14

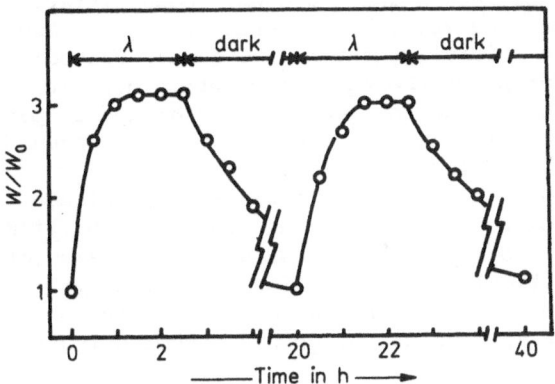

Figure 13. Photostimulated dilation and contraction of polyacrylamide gel having pendant triphenylmethane leucohydroxide groups with light of wavelength longer than 270 nm at 25°C. The content of leucohydroxide groups was 3.7 mole%. Initial pH of external water phase was 6.6.

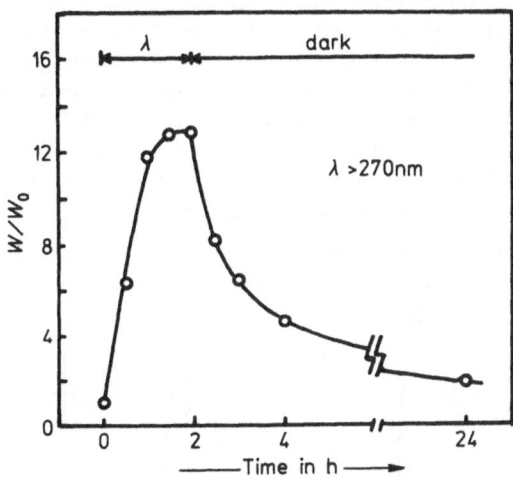

Figure 14. Photostimulated dilation and contraction of polyacrylamide gel having pendant triphenylmethane leucocyanide groups with light of wavelength longer than 270 nm at 25°C. The content of leucocyanide groups was 3.1 mole%.

The gel having triphenylmethane leucohydroxide groups swelled even in the dark when the aqueous solution became acidic. At pH of 4.0, the gel weight increased as much as 20 times in comparison with its weight at pH 8.0. To minimize the pH-sensitivity, the hydroxide was replaced by cyanide. Figure 14 shows the photoresponsive behavior of the gel having triphenylmethane leucocyanide groups. Upon ultraviolet irradiation, the weight increased as much as 13 times, 1200 %, and the dilated gel contracted to the initial weight in the dark. Photodilation was observed regardless of pH of the aqueous solution down to pH 4.0. The large expansion of the gel having leucocyanide groups is partly due to the high ionic dissociation quantum yield of the leucocyanide as compared with that of the leucohydroxide.

The gel film tended to curl upon irradiation, because the boundaries have more freedom to deform than the center of the film. Although this situation made it difficult to measure the dimension of the film precisely, the photostimulated deformation was around 2.2 times in each dimension. This value agreed with a deformation value of 2.35 times as estimated from the 13-fold weight change assuming isotropic expansion.

The photo-deformable gels have potential applications not only to photo-mechanical conversion but also to photo-driven molecular valve and photo-controlled slow release of drugs.[38]

4.2. Sol - Gel transition

Three dimensional infinite network formation in a polymer by chemical or physical process leads to gels. The gels are classified into two types, irreversible and reversible gels. In the latter case, the gels are cross-linked by physical interaction between certain points on different polymer chains and sol-gel transition is induced by changing the temperature (heat mode). At temperature lower than T_{gel}, the temperature of gelation, the polymer solution stops flowing, while above T_{gel} the gel again melts to flow. The sol-gel transition can be

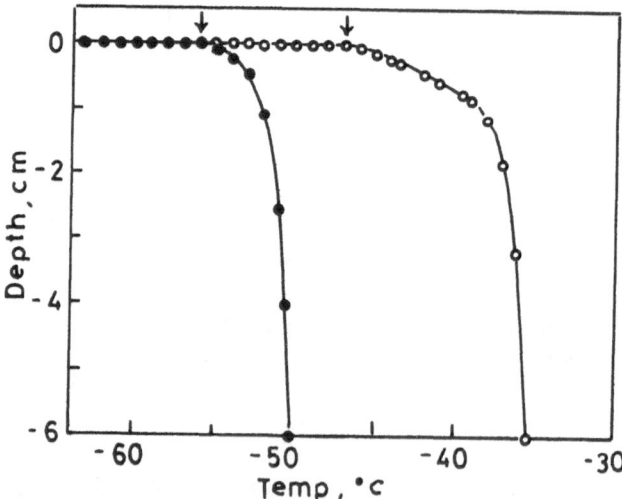

Figure 15. Gel-sol transition of polystyrene with pendant all trans (●) and 62 % cis azobenzene groups (o) in CS_2, measured by the ball-drop method. Arrows indicate melt temperatures of the gels. The content of azobenzene groups is 10.5 mole%. Polymer concentration was 200 g/l.

induced reversibly by photoirradiation (photon mode) by using a CS_2 solution of polystyrene with pendant azobenzene groups.[39]

The sol - gel transition temperature was determined by the two methods: test tube tilting and a ball drop method. In the second method, a steel ball was placed on the top of the gel and the point at which the depth-temperature curve deviated from the horizontal was taken as T_{gel}. Figure 15 shows the temperature dependence of ball-sinking into the CS_2 solution of polystyrene with all pendant trans- and partially photoisomerized cis-azobenzene groups. The cis fraction is 62 mole%. A stable gel was obtained from a CS_2 solution containing the copolymer (azobenzene content, 10.5 mole%), when the solution was cooled at -78°C for 5 h. As the temperature was raised slowly at the rate of 0.5°C/min, the ball on the top of the gel of polystyrene with all trans-azobenzene pendant groups began to sink at -56°C, while the ball on the gel with partially cis-azobenzene groups still remained on the top at this temperature. The ball began to sink at -47°C. The gel melting temperature was 9°C higher than that of the polystyrene with trans azobenzene groups. The gel - sol transition temperature, -47°C, of polystyrene with cis azobenzene groups reverted to -56°C, when the gel was irradiated for 20 min with visible light ($\lambda >$ 450 nm) at -78°C, prior to raising the temperature. The visible light converts the pendant azobenzene groups from cis to trans form. This result indicates that the sol - gel transition can be regulated by photoirradiation (photon mode). The difference in the gel melting temperature between the polymers with trans and cis azobenzene groups increases with the increase in the content of azobenzene in the pendant groups and with the amount of cis form azobenzene groups. These results suggest that cis azobenzene moieties in the pendant groups make inter-polymer cross-linking junctions, which are more stable than the original inter-polymer interactions between polystyrene chains. Although the trans azobenzene groups also have some interaction among them , the attractive force is considered to be weaker than the interaction among cis azobenzene groups, because of the difference in the dipole moment of the two iso-mers. This is, so far as the author knows, the first example of photo-controlled phase transition in a polymer system.

4.3. Other Applications

Because of limitation of this chapter length, only two applications are

TABLE VII. Photoregulation of physical and chemical properties of polymer solutions and solids.

Solution	Solid
Viscosity 2,6-15,18-20	Membrane potential [42]
pH - value [11,17]	Wettability [43]
Solubility [21,40]	Shape 27-38
Metal ion capture [13,41]	Sol - gel transition [39]
	T_g [44]

described. Table VII includes several other properties so far reported,
which are regulated by photoirradiation. In these applications, photoisomerization reactions are used to trigger the polymers to control
their properties in aiming at accomplishing certain works. Photophysical and photochemical reactions can serve as tools not only for
analyzing the polymer properties but also for regulating the properties.

5. REFERENCES

1. "Molecular Models of Photoresponsiveness", G. Montagnoli, B. F.
        Erlanger ed., Plenum, New York (1983)
2. M. Irie, ibid., p 291
3. a) S. Shinkai, ibid., p 325
    b) M. Irie, M. Kato, J. Am. Chem. Soc., 107, 1024 (1985)
4. A. Ueno, K. Takahashi, T. Osa, J. Chem. Soc. Chem. Commun., 1981, 94
5. Y. Okahata, H-J. Lim, S. Hachiya, Makromol. Chem. Rapid Commun., 4
        303 (1983)
6. R. Lovrien, Proc. Nat. Acad. Sci., 57, 236 (1967)
7. G. van der Veen, W. Prins, Photochem. Photobiol., 19, 191 (1974)
8. N. Negishi, M. Takahashi, A. Iwazawa, K. Matsuyama, I. Shinohara,
        Nippon Kagaku Kaishi., 1977, 1035
9. L. Matějka, K. Dušek, Makromol. Chem., 182, 3223 (1981)
10. M. Irie, K. Hayashi, J. Macromol. Sci. Chem., A13, 511 (1979)
11. M. Irie, K. Hirano, S. Hashimoto, K. Hayashi, Macromolecules, 14,
        262 (1981)
12. H. S. Blair, H. I. Pogue, J. E. Riordan, Polymer, 21, 1195 (1980)
13. G. S. Kumar, P. DePra, D. C. Neckers, Macromolecules, 17, 1912, 2463,
        (1984)
14. E. K. Zimmermann, J. K. Stille, ibid., 18, 321 (1985)
15. M. Irie, H. Hosoda, Makromol. Chem. Rapid Commun., 6, 533 (1985)
16. M. L. Herz, J. Am. Chem. Soc., 97, 6777 (1975)
17. M. Irie, ibid., 105, 2078 (1983)
18. M. Irie, A. Menju, K. Hayashi, Macromolecules, 12, 1176 (1979)
19. A. Menju, K. Hayashi, M. Irie, ibid., 14, 755 (1980)
20. M. Irie, K. Hayashi, A. Menju, Polymer Photochem.,1, 233 (1981)
21. M. Irie, H. Tanaka, Macromolecules, 16, 210 (1983)
22. F. Ciardelli, O. Pieroni, A. Fissi, J. Houben, Biopolymers, 23,
        1423 (1984)
23. A. Ueno, K. Takahashi, J. Anzai, T. Osa, J. Am. Chem. Soc., 103,
        6410 (1981)
24. M. Irie, W. Schnabel, Macromolecules, 14, 1246 (1981)
25. M. Irie, W. Schnabel, ibid., 18, 394 (1985)
26. M. Irie, W. Schnabel, Makromol. Chem. Rapid Commun., 5, 413 (1984)
27. E. Merian, Text. Res. J., 36, 612 (1966)
28. F. Agolini, F. P. Gay, Macromolecules, 3, 349 (1970)
29. G. van der Veen, W. Prins, Nature, Phys. Sci., 230, 70 (1971)
30. G. Smets, G. Evans, Pure Appl. Chem. Suppl. Macromol. Chem., 8, 357
        (1973)
31. a) L. Matějka, K. Dušek, M. Ilavský, Polymer Bull., 1, 659 (1979)
    b) L. Matějka, M. Ilavsky, K. Dušek, O. Wichterle, Polymer, 22,
        1511 (1981)

32. C. D. Eisenbach, Polymer, **21,** 1175 (1980)
33. Y. Osada, E. Katsumura, K. Inoue, Makromol. Chem. Rapid Commun., **2,** 411 (1981)
34. H. S. Blair, H. I. Pogue, Polymer, **23,** 779 (1982)
35. K. Ishihara, N. Hamada, S. Kato, I. Shinohara, J. Polym. Sci. Chem., **22,** 121 (1984)
36. A. Aviram, Macromolecules, **11,** 1275 (1978)
37. M. Irie, D. Kungwatchakun, Makromol. Chem. Rapid Commun., **5,** 829 (1984)
38. M. Irie, D. Kungwatchakun, K. Hayashi, Polym. Prep. Jpn, **34,** 458 (1985)
39. M. Irie, R. Iga, Makromol. Chem. Rapid Commun., **6,** 403 (1985)
40. M. Irie, T. Iwayanagi, Y. Taniguchi, Macromolecules, **18** (1985) in press
41. S. Shinkai, H. Kinda, M. Ishihara, O. Manabe, J. Polym. Sci. Chem., **21,** 1551 (1983)
42. M. Irie, A. Menju, K. Hayashi, Nippon Kagaku Kaishi, **1984,** 227
43. K. Ishihara, N. Hamada, S. Kato, I. Shinohara, J. Polym. Sci. Chem., **21,** 1551 (1983)
44. M. Irie, M. Mohri, K. Hayashi, Polym. Prep. Jpn., **34,** 716 (1985)

The Cyclization of Polymer Chains in Solution

Mitchell A. Winnik
Department of Chemistry and Erindale College
University of Toronto
Ontario, Canada M5S 1A1

ABSTRACT. The cyclization properties of polymer chains serve as a useful
vehicle for examining various theoretical predictions about the polymer
conformation and dynamics. Cyclization phenomena are particularly sensitive to excluded volume phenomena. This chapter provides a review of
pertinent theoretical concepts and a description of experiments, based
upon luminescence spectroscopy, which allow one to examine the
predictions of the theory.

1. INTRODUCTION

The cyclization properties of flexible chain molecules have been of
interest to chemists since the early days of this century (1).
Originally the concern focussed on what kinds of local factors (torsional
strain, steric effects) affected the cyclization of flexible organic
molecules. Then emphasis shifted to the entropic factors affecting
polymer cyclization (2). More recently one has begun to appreciate that
many topical issues in polymer theory, particularly excluded volume
effects and models for chain dynamics, find rather vivid expression in
cyclization-related phenomena. Thus studies of cyclization equilibria
and dynamics provides new insights into polymer behavior and new
approaches to testing predictions of contemporary theory.

Experimental methods which permit precise measurements to obtain
cyclization equilibrium constants K_{cy} and cyclization rate constants k_{cy}
all depend upon spectroscopic studies of polymers covalently labelled
with appropriate dye molecules (3). The various spectroscopic methods

M. A. Winnik (ed.), Photophysical and Photochemical Tools in Polymer Science, 293–324.

for examining cyclization all operate on the same principle. Imagine a polymer molecule A~~~~Q containing a group A at one end and a group Q at the other, chosen so that A and Q react to give an observable event. This could be formation of a donor–acceptor complex, an acid–base interaction or a chemical reaction. Measurement of the extent or rate of that reaction would provide a measure proportional to the probability W(0) of cyclization or the dynamics of cyclization. The meaningful experimental quantities are K_{cy} and k_{cy}. The experiments must be carried out in such a way that either of these values can be obtained.

Information on cyclization dynamics is obtained when the reaction between A and Q is so fast that it occurs on every encounter: the reaction becomes diffusion controlled. The most convenient means of carrying out such experiments is via fluorescence or phosphorescence quenching (4) where A is excited electronically to A* with light of an appropriate wavelength and Q is chosen to be a diffusion-controlled quencher of A*. One has to establish that Q and A* indeed react at the diffusion controlled rate.

This chapter will review experiments and theory pertinent to the use of excited state quenching processes to study cyclization phenomena in polymers. Most of the emphasis will be on cyclization dynamics, but this will be examined in comparison with results and expectations for corresponding cyclization equilibria.

2.   THEORETICAL OVERVIEW

Any polymer can be represented as a series of n bonds of length l. Each conformation has a characteristic end-to-end length r. A population of chains can be described by various distribution functions and ensemble averages. Equilibrium chain properties are usually described in terms of the end-to-end distance distribution function W(r), as well as the mean-squared end-to-end distance R_F^2 and the mean-squared radius of gyration R_G^2.

Chain dynamics are described in terms of time autocorrelation functions. These describe the rate of decay of the probability that a chain with a specific configuration at time t has that same configuration at a later time ($t + \Delta t$). The two important correlation functions in the theory of cyclization dynamics are $\rho(t)$ describing the relaxation of the end-to-end distance, and s(t) describing the re-encounter probability of two chain ends in proximity (5,6). $\rho(t)$ describes the decay of the probability that the two chain ends separated by a distance r at some specified time will have the same separation at some later time.

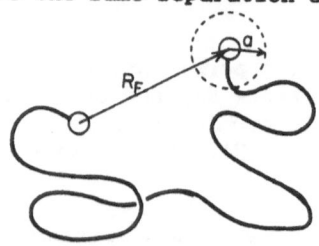

To understand s(t) imagine a volume centered at one chain end in which a chemical reaction between the chain ends would occur. One could call this volume the reaction sink $\underline{S}$ and for simplicity refer to it as a sphere of radius $\underline{a}$. The sink autocorrelation function s(t) describes the decay of the probability in time that a chain with both ends in $\underline{S}$ at time $\underline{t}$ will still have both ends in $\underline{S}$ at a later time ($\underline{t}$ + Δt). Obviously when Δt is very small this probability will be large, and at long times the probability will decay to zero as the positions of the chain ends become uncorrelated.

2.1. Cyclization Equilibria

The theoretical model of random flight chains (7) generates a gaussian form for W(r). This form applies to any polymer model, including the rather realistic rotational isomeric state model (7), for the case of sufficiently long chains where excluded volume interactions are neglected. There is evidence from neutron and light scattering that real chains of sufficient length in theta–solvents and at modest extensions are reasonably well described in terms of a gaussian W(r) (8).

The cyclization probability W(0) can be obtained by taking the limit of the radial distribution function $4\pi r^2 W(r)$ as $\underline{r} \rightarrow 0$, obtaining for gaussian chains the expression

$$W(0) = (\frac{3}{2})^{3/2} R_F^{-3} \tag{1}$$

Since R_F here is proportional to $N^{1/2}$, one obtains the prediction that $W(0) \sim N^{-3/2}$. Experimental evidence for cyclization of poly(dimethyl siloxane) [PDMS] is in accord with this prediction (9).

Even in the absence of excluded volume W(r) is not necessarily gaussian in shape. W(r) is non–gaussian for short chains because of bond angle correlations. The gaussian description also breaks down at the extremities of W(r) because, obviously, for real chains $\underline{r}$ cannot exceed $\underline{r}_{max}$; and, even in theta–solvents, $\underline{r}$ cannot equal exactly zero because of the steric requirements of the end groups.

In good solvents excluded volume becomes important. R_F and R_G increase, and their dependence upon chain length increases. Both vary as N^ν where ν takes a value very close to 3/5. If one takes the form of eq. 1 as an appropriate description of the cyclization of chains in the presence of excluded volume, one would predict that $W(0) \sim R_G^{-3} \sim N^{-9/5}$. This expression is incorrect because it overlooks a second effect of excluded volume on W(r).

When excluded volume effects are important, there is a severe depression in W(r) at small values of $\underline{r}$ (10). This arises from correlations in pair repulsions between elements of the chain, and these act to lower the probability of the two chain ends being in proximity. If one could create a population of chains with their ends in proximity, the gradient of W(r) with respect to $\underline{r}$ would create an entropic force to drive the separation of chain ends. Theoretical treatments which take into account this second factor suggest that $W(0) \sim N^{-1.92}$.

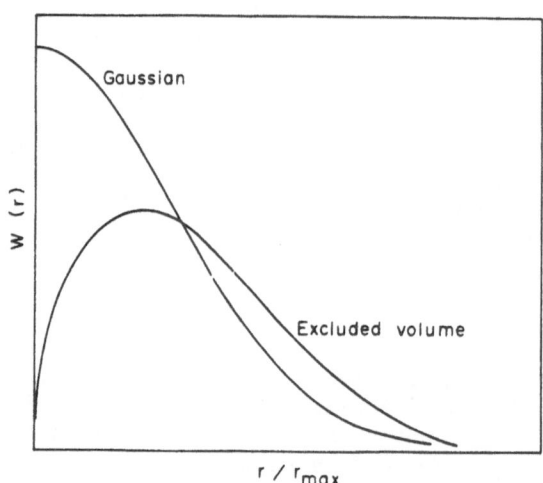

Figure 1. A plot of W(r) vs r/r_max for gaussian chains and for
self-avoiding (excluded volume) chains.

2.2. Cyclization Dynamics

The theory of cyclization dynamics was first presented by Wilemski and
Fixman [WF] (5). A number of curious features of the theory prompted
detailed attention by Doi (11), by Perico and Cuniberti (12), and by
others (13). The theory is developed in terms of the bead-and-spring
Rouse-Zimm [RZ] model (14). Unrealistic in detail, this model is quite
useful for describing low frequency, large amplitude chain motions. The
RZ model, figure 2, treats the chain as a series of $\underline{n}$ beads connected by
($\underline{n}$-1)harmonic springs of root-mean-squared length $\underline{b}$.
      The coupled harmonic springs lead to a description of chain motion
as a hierachy of orthogonal modes, each with its characteristic
relaxation rate λ_i. Mathematically one writes down a matrix equation
whose eigenfunctions Q_{il} describe these motions and with the λ_i as the
corresponding eigenvalues. The lowest frequency motion is associated
with translational diffusion of the center of mass, and of the higher
frequency modes, only the odd ones (i = 1, 3, ...) lead to a change in
the end-to-end displacement. In the Rouse model, the friction
coefficients (ξ) of each bead add: the chains are said to be free
draining. Real polymers experience hydrodynamic coupling between chain
elements close in space although remote along the chain contour. The
motion of one bead carries along solvent, which in turn drags along
nearby elements of the chain. The Zimm model accommodates these
interactions by assuming that solvent is trapped inside the coil during
the motion (i.e., chains are non-draining) and the total friction is
reduced. The Zimm model provides a much better description of polymers
in dilute solution, particularly those of very high molecular weight,
than the Rouse model.

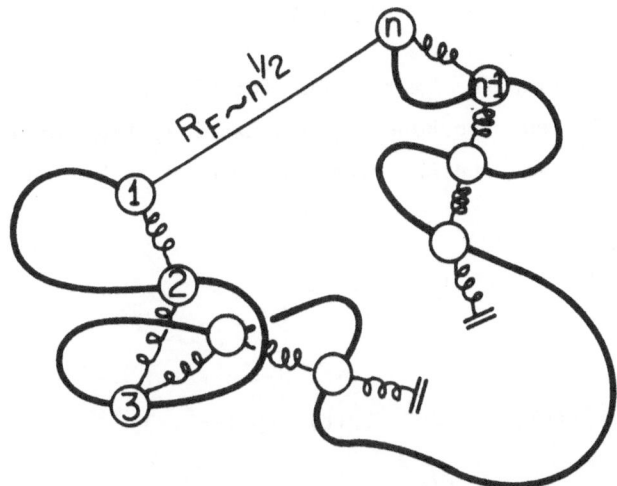

Figure 2. A depiction of the bead-and-spring (Rouse Zimm) model
for a polymer chain. The chain is represented by $\underline{n}$ beads and
$(\underline{n-1})$ springs of length $\underline{b}$.

Polymers of intermediate and low molecular weight experience partial
draining. This leads to a reduction of the overall friction experienced
by the chain compared to the Rouse model, but not to the extent of full
non-draining conditions. Calculations of chain dynamics based upon the
RZ model for partial draining conditions frequently introduce a reduced
bead friction coefficient, (ξ_r), a dimensionless quantity which measures
the strength of the hydrodynamic coupling (15).

WF theory, which neglects excluded volume, predicts that the
diffusion controlled cyclization rate constant $k_{cy}{}^D$ decreased as $\underline{n}^{-2}$ (or
N^{-2}) for free-draining chains, or, more realistically, as $N^{-3/2}$ for
non-draining chains (5).

The controversial aspect of the original WF theory which elicited
such attention was the way the sink term denoting cyclization was
introduced into the model. This aspect and the theory in general are
described in detail in a marvelous review by Cuniberti and Perico (6).
These authors have taken great pains to make the WF theory accessible to
experimentalists (12). They extended the theory to the case of finite
chains, under partial draining conditions, and showed how one could use
intrinsic viscosity data to obtain the parameters $\underline{b}$ and ξ_r necessary to
calculate theoretical values of $k_{cy}{}^D$, the diffusion controlled
cyclization rate constant, for specific polymers. Their equations are
summarized in Table 1.

In these expressions $k_{cy}{}^D$ is obtained from a rather complicated
integral equation which depends upon the end-distance autocorrelation
function (t) contained in the z(t) term, and a sink term $\underline{Y}$ which treats
cyclization in terms of a spherical reaction volume with a capture radius
$\underline{a}$. Note that $\underline{Y}$ depends upon $\underline{a}/R_F$, so that at a constant value of $\underline{a}$, $\underline{Y}$
depends on chain length.

Table I

Cuniberti and Perico (6) Expressions for k_{cy}^D in Terms of the Rouse
Eigenfunctions $Q_{i\ell}^R$ and the Rouse λ_ℓ^R and Exact λ_ℓ Eigenvalues for Finite
Chains, and the Empirical Parameters $\underline{a}$, $\underline{b}$, and ξ_r.

$$k_{cy}^D = \left[\int_o^\infty (\frac{K(t)}{K(\infty)} - 1) \exp (k_1 t) \, dt \right]^{-1} \qquad \text{I.1}$$

$$K(t) = \text{erf } [z(t)] - (2/\pi^{1/2})z(t) \exp (-z^2(t)) \qquad \text{I.2}$$

$$K(\gamma) = \text{erf } (\gamma) - (2/\pi^{1/2})\gamma \exp (-\gamma^2) \qquad \text{I.3}$$

$$z(t) = \gamma[1 - \rho^2(t)]^{-1/2} \qquad \text{I.4}$$

$$\gamma = (3/2)^{1/2} a/R_F \qquad \text{I.5}$$

$$\rho(t) = \frac{1}{n-1} \sum_{\ell=1}^{n-1} \frac{(Q_{n,\ell}^R - Q_{o,\ell}^R)^2}{\lambda_\ell^R} \exp (-(\frac{k_B T}{2b^3 \xi_r \eta_o})\lambda_\ell t) \qquad \text{I.6}$$

$$Q_{i,\ell}^R = \frac{(2-\delta_{\ell,0})^{1/2}}{n} \cos \frac{\pi\ell}{n} [i+\tfrac{1}{2}] \qquad \text{I.7}$$

$$\lambda_\ell^R = 4 \sin^2 \frac{\pi\ell}{2n} \qquad \text{I.8}$$

$$\lambda_\ell \approx \frac{\pi^2\ell^2}{n^2} \left| 1 + 2\xi_r (\frac{3(n-1)}{\ell\pi}) \right|^{1/2} \qquad \text{I.9}$$

In the limit of long chains (D) the non-draining translational
diffusion constant):

$$k_{cy}^D = 1.98 \, D/R_F^2 \qquad \text{I.10}$$

The term (t) contains information on chain length. CP (6) have shown that it suffices in these calculations to use eigen functions Q_i from the simple Rouse model, whereas one needs λ_ℓ from the full solution of the matrix equations for partially draining chains. For large enough $\underline{n}$, the expression in Table 1 provides useable values of λ_ℓ.

In the first expression, k_1 is the experimental cyclization rate constant. If the reaction between $\underline{A}^*$ and $\underline{Q}$ is less than diffusion controlled, $k_1 < k_{cy}{}^D$, and these terms are related by the expression

$$\frac{1}{k_1} = \frac{1}{k_{cy}{}^D} + \frac{1}{k\,W(0)} \tag{2}$$

where k is the rate constant describing the reaction rate for an $\underline{A}^*/\underline{Q}$ pair inside the reactive volume $\underline{S}$. If $k \rightarrow \infty$, $k_1 = k_{cy}{}^D$. When k is very small, $k_1 = k\,W(0)$. Operationally, the term $\exp(k_1 t)$ in the $k_{cy}{}^D$ expression has little effect because $K(t)$ decays so fast that $\exp(k_1 t)$ is essentially equal to unity over the entire time scale for which the value of the integral is significant.

These equations have been described in detail for two reasons. First they allow one to predict values for $k_{cy}{}^D$ for specific polymers, as PC did for polystyrene and PDMS (12). More importantly, one sees that the relationship between the size of the capture radius $\underline{a}$ and the value of $k_{cy}{}^D$ is by no means obvious, and furthermore, that the $\underline{\gamma}$ term can have an influence on the chain length dependence of the cyclization rate.

These equations apply only to experiments in theta-solvents since the theory neglects excluded volume. Calculations based upon the equations in Table 1 predict that $k_{cy}{}^D \sim N^{-1.44}$ for polystyrene and PDMS of molecular weights of 10^4 to 10^5.

We have recently recognized that by changing end groups on a polymer, one could effectively vary the capture radius in the $\underline{A}^*/\underline{Q}$ interaction (16). This would allow us to test how the sink term enters into the cyclization expression. Stephen Maj in my group wrote a computer program to solve the equations in Table 1 using parameters appropriate to polystyrene. If $\underline{a}$ is varied with other factors kept constant, the cyclization rate changes as shown in Figure 3: he finds that $k_{cy}{}^D$ increases in proportion to $\underline{a}$, but that the sensitivity to $\underline{a}$ falls off for longer chains (larger $\underline{n}$).

3.   EXPERIMENTAL METHODS

Three types of experiments have been applied to the study of diffusion controlled cyclization of polymers. These are excimer formation in polymers containing pyrene groups at both chain ends [Py-polymer-Py*[1]], exciplex formation in polymers containing a pyrene at one end and a dimethylaminophenyl group at the other [DMA-polymer-Py*[1]], and triplet-triplet [TT] annihilation between anthracene groups at both ends of a polymer [A*[3]-polymer-A*[3]]. A fourth obvious approach, as yet to be reported, is intramolecular phosphorescence quenching in a polymer

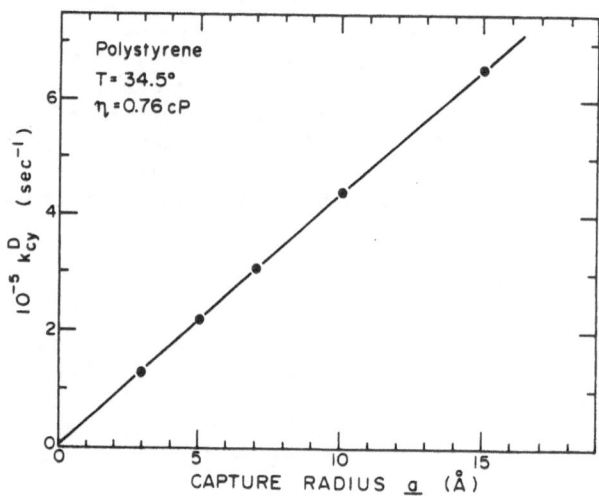

Figure 3. Plot of $k_{cy}{}^D$ calculated for polystyrene of M = 37500 in cyclohexane at 35°C using the equations in Table 1 and appropriate parameters from CP, ref (6), as a function of the size of the capture radius $\underline{a}$.

containing an α–diketone such as benzil at one chain end and an appropriate quencher group at the other chain end [Bz*³–polymer–Q].

Py-polymer-Py

Py-polymer-DMA

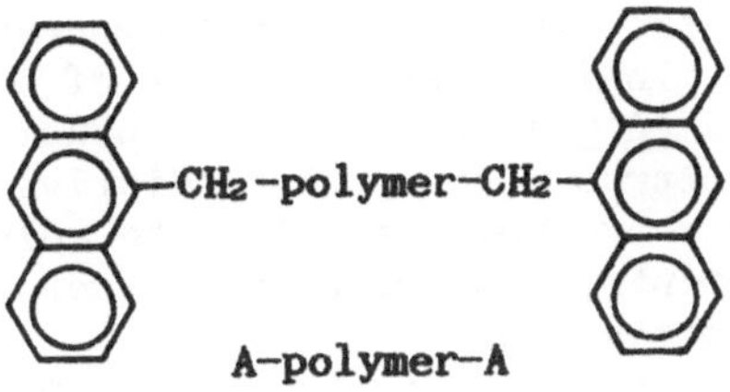

-CH₂-polymer-CH₂-

A-polymer-A

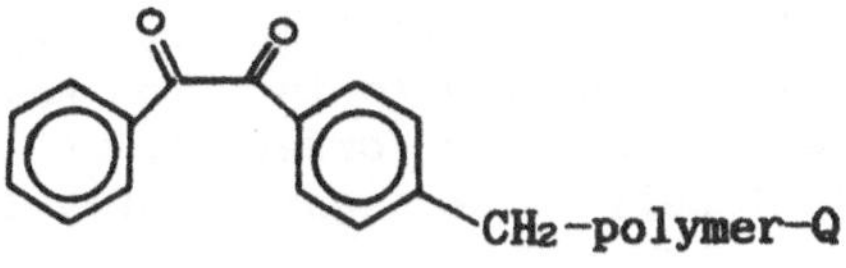

CH₂-polymer-Q

Bz-polymer-Q

Although synthesis will not be discussed in this chapter, this is the proper point to emphasize that the synthesis and characterization of end-functionalized macromolecules with narrow molecular weight distribution [MWD] are often the most difficult part of the entire problem. Several different types of polymers of the form Py-polymer-Py have been studied because of the ready availability of the corresponding polymers with -OH groups on both ends. The techniques of polymerization via living polymers is best understood for polystyrene. TT annihilation studies have been reported only for A-polystyrene-A, and exciplex measurements are known only for are known only for DMA-polystyrene-Py. The data available for testing the theory of cyclization dynamics derive almost exclusively from experiments on polystyrene.

Time scale is an important concern in these experiments. To test the magnitude of the exponent in the expression $k_{cy} \sim N^{-\alpha}$ one needs to examine a wide range of chain lengths, corresponding to a large change in k_{cy}. Pyrene derivatives have fluorescence lifetimes of ca. 200 ns. Pyrene exciplex methods allow k_{cy} values to be determined over the range of 10^8 s^{-1} to 10^5 s^{-1}. Accurate measurement of weak pyrene excimer emission is somewhat easier; k_{cy} values down to ca. 3×10^4 s^{-1} can be determined. Triplet methods are needed to determine cyclization rates in the millisecond domain.

3.1. Excimer and Exciplex Spectroscopy

Excimers and exciplexes are bound binary complexes in an electronically excited state, formed from a ground state species (Q) interacting with a molecule or groups (A*) in its excited state (4). In excimers $\underline{A} = \underline{Q}$ and the (AA)* pair is held together largely by an exciton resonance interaction. The spectroscopic properties of the (PyPy)* excimer are essentially unaffected by solvent polarity.

Exciplexes have large dipole moments, comparable to that of a contact ion-pair. The binding energy is that of a donor-acceptor interaction, [$(\underline{AQ})^*$ <-> $(\underline{A}^-\underline{Q}^+)$ or $(\underline{A}^+\underline{Q}^-)$], depending upon which species, $\underline{A}$ or $\underline{Q}$, acts as the electron donor. The radiative and non-radiative properties of exciplexes are very sensitive to solvent polarity. In sufficiently polar solvents, dissociation to free ions is the major pathway of exciplex deactivation.

Both exciplexes and excimers form at the diffusion-controlled rate. Dissociation back to $\underline{A}^*$ and $\underline{Q}$ can be important and must be taken into account in the analysis of the experimental data. One important distinction between the two processes is that when exciplex formation is sufficiently exergonic, it is known to be preceded by electron transfer to form a solvent separated ion pair (17), e.g., $\underline{A}^-//\underline{Q}^+$, which in a second step collapses to the exciplex-contact ion pair $(\underline{AQ})^*$ <-> $(\underline{A}^-\underline{Q}^+)$. The interaction distance for exciplex formation, at least in polar and polarizable solvents, is likely to be significantly larger than that for excimer formation.

3.2. Excimer and Exciplex Kinetics

According to the photochemists jargon, the locally excited state of an excimer- or exciplex-forming species is called the monomer, whose emission intensity is denoted I_M, with an intensity decay profile $I_M(t)$. Similarly the excimer or exciplex intensity is described by I_E and $I_E(t)$. Many of the problems of multiple excimers and different classes of conformational populations discussed in the chapters by Soutar and Phillips (18) and by DeSchryver et al. (19) do not occur here. The polymers here are long and flexible enough to permit the excited complexes to reach the geometry of the energy minimum dictated by the chromophore interaction. Furthermore, individual local conformational changes occur on a much faster time scale than that sampled in these experiments. Consequently, the kinetic mechanism in Scheme I is adequate to accommodate all of the experimental data. It is a two-state model (one monomer population, one excimer or exciplex), the intramolecular analog of the scheme popularized by Birks (4). Scheme I neglects transient effects (19) on the diffusion controlled cyclization.

Scheme I

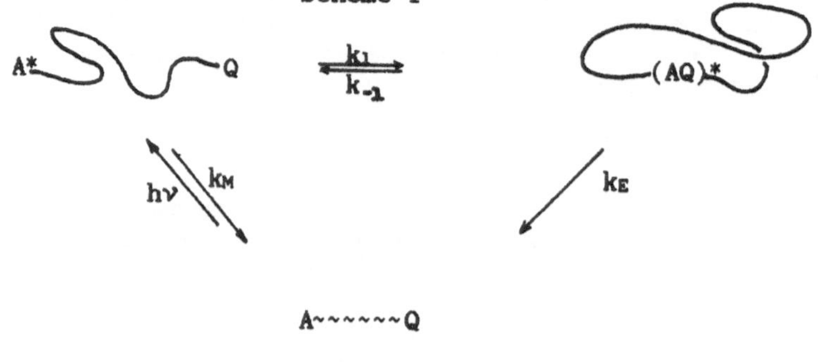

In Scheme I, k_1 describes the cyclization rate; and k_{-1}, the ring opening dissociation rate. In polymer systems where the molecules have a finite polydispersity, the angle brackets on $\langle k_1 \rangle$ represent an ensemble average. The terms k_E and k_M represent, respectively, the reciprocal lifetimes of the excimer (or exciplex) and the locally excited (monomer) state.

Two coupled differential equations describe the excited state populations:

$$d[Py^*]/dt = (k_1+k_M)[Py^*] - k_{-1}[(PyPy)^*] \tag{3}$$

$$d[(PyPy)^*]/dt = (k_E+k_{-1})[(PyPy)^*] - k_1[Py^*] \tag{4}$$

Solution of these equations gives the expressions to which data are fit.

$$I_M(t) = A\exp(-\lambda_1 t) + \exp(-\lambda_2 t) \tag{5}$$

$$I_E(t) = B[\exp(-\lambda_1 t) - \exp(-\lambda_2 t)] \tag{6}$$

$I_M(t)$ decays as a sum of two exponential terms; $I_E(t)$ grows in and decays as a difference of two exponential terms, with the same short and long values in each experiment. The λ's are related to the rate constants in Scheme I by the expressions

$$2\lambda_1, \lambda_2 = (k_1+k_{-1}+k_M+k_E) \mp [\{(k_1+k_M) - (k_{-1}+k_E)\}^2 + 4k_1 k_{-1}]^{1/2} \tag{7}$$

$$A = \frac{(k_1+k_M) - \lambda_1}{\lambda_2 - (k_1+k_M)} \tag{8}$$

To determine k_1, one needs to measure $I_M(t)$ and $I_E(t)$ in order to verify that similar λ's are obtained in each experiment, and then substitute the values of λ_1, λ_2 and $\underline{A}$ into eqs. 7 and 8, with k_M known independently from measurements made on a model polymer containing pyrene at one end only. At room temperature and below, the dissociation rate constant k_{-1} is quite small. When $4k_1 k_{-1}$ is small compared to the squared term in the brackets of eq. 8, $I_M(t)$ becomes exponential, and k_1 can be determined from

$$k_1 = \lambda_1 - k_M \tag{9}$$

Alternatively one can use steady state fluorescence intensities to obtain information about the cyclization process. For example, the ratio of excimer to pyrene fluorescence quantum yields is

$$\frac{\phi_E}{\phi_M} = \frac{\phi_E^o}{\phi_M^o} \frac{k_E}{k_M} \left(\frac{k_1}{k_E + k_{-1}}\right) \tag{10}$$

where ϕ_E^o and ϕ_M^o are the quantum efficiencies of the pure components. Perhaps the most useful application of these measurements is in the study of polymers of different lengths in one solvent at a single temperature where the term $(k_E + k_{-1})$ does not vary. Then the ratio of I_E/I_M values for two chain lengths is equal to the ratio of their cyclization rate constants

$$\frac{(I_E/I_M)_1}{(I_E/I_M)_2} = \frac{(k_1)_1}{(k_1)_2} \tag{11}$$

3.3. Kinetics of TT Annihilation

In order to observe intramolecular TT annihilation both chromophores at the chain ends must be excited to their triplet states (20). Since the probability of double excitation is low, the presence of singly excited species must be taken into account.

SCHEME II

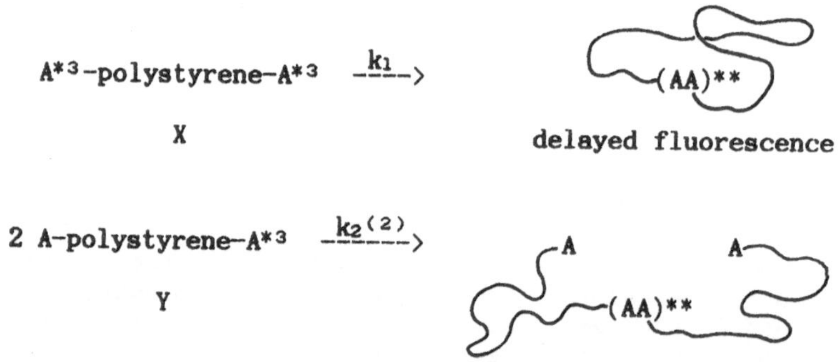

$$A^{*3}\text{-polystyrene-}A^{*3} \xrightarrow{\ k_1\ } \quad (AA)^{**}$$

X   delayed fluorescence

$$2\ A\text{-polystyrene-}A^{*3} \xrightarrow{\ k_2^{(2)}\ } \quad A \quad A \quad (AA)^{**}$$

Y

For the mechanism in Scheme II, k_T describes the decay rate of A^{*3}; $k_2^{(2)}$, the rate of bimolecular TT annihilation, and k_1, the cyclization rate. Mita (20) has shown that under these experimental conditions, the reaction rates are adequately described by the equations

$$-\frac{d[X]}{dt} = (k_1 + 2k_T)[X] \tag{12}$$

$$- \frac{d[Y]}{dt} = k_T[Y] + 2k_2^{(2)}[Y]^2 - 2k_T[X] \tag{13}$$

where [X] and [Y] are the concentrations of doubly and singly excited polymers. Note that TT annihilation is neglected as an additional source of A^{*3}. Experimentally one measures either the decay of the intensity of delayed fluorescence $I_{DF}(t)$ or the decay of the absorption $A_T(t)$ of light due to A^{*3}.

$$I_{DF}(t) \sim k_1[X] + k_2^{(2)}[Y]^2 \tag{14}$$

$$A_T(t)_T \sim 2[X] + [Y] \tag{15}$$

Solving eqs. 12 and 13 and substituting the results into the expressions 14 and 15, one obtains for $I_{DF}(t)$ and $A_T(t)$

$$I_{DF}(t) \sim Z_1 k_1 X_0 \ \exp[-(k_1+2k_T)t] + Z_2^2 k_2^{(2)} Y_0^2 \ \exp(-2k_T t) \tag{16}$$

$$A_T(t) \sim 2Z_1 X_0 \ \exp[-(k_1+2k_T)t] + Z_2 Y_0 \ \exp(-k_T t) \tag{17}$$

where the pre-exponential terms unfortunately also depend upon time (6,20), and X_0 and Y_0 are the initial concentrations of X and Y produced by flash photolysis. The measured signals are not simply a sum of two exponential terms but have a more complicated time dependence. At very long times, the k_T term dominates the $A_T(t)$ measurement, and only bimolecular processes contribute to the measurement. If k_1 is very much larger than k_T and one examines the system in a time domain where Z_1 and Z_2 remain reasonably constant in time, the two rate constants can be evaluated from examination of the short-time and long-time portions of the decay curves (6,20).

4. EXPERIMENTAL RESULTS: POLYSTYRENE

Fluorescence spectra of pyrene-end-capped polystyrene are shown in figure 9 for two different chain lengths, M_n = 2900 and M_n = 15500. In both cyclohexane at 35.4° (figure 4a) and in toluene at 22° (figure 4b) the shorter chain gives a more intense excimer emission than the longer chain. These results are in accord with the WF theory, which predicts that cyclization rates decrease with increasing chain length (21). Comparing figures 4a and 4b one also notices that there is significantly more excimer emission in cyclohexane at 34.5°, a theta-solvent for polystyrene, than in toluene, a good solvent for PS. One sees that chain length and solvent quality are important parameters which affect polymer cyclization.

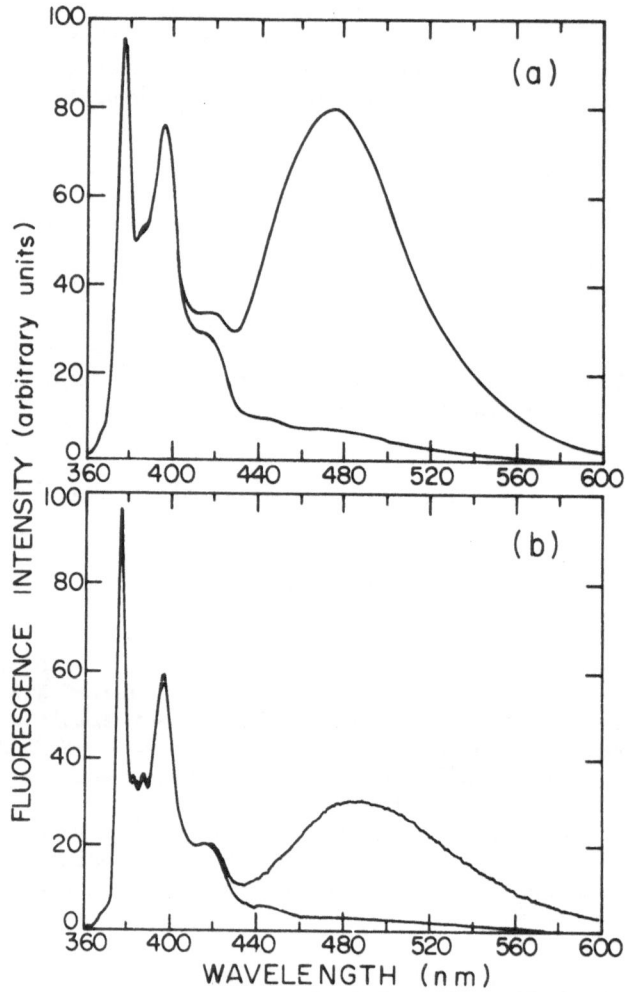

Figure 4. Fluorescence spectra of Py-polystyrene-Py, 2 x 10⁻⁶ M. (a) In cyclohexane at 34.5°, M_n 15600 (lower curve). (b) In toluene at 22°, M_n 2900 (upper curve) and M_n 15600 (lower curve). Both pairs of spectra are normalized at 378 nm.

Fluorescence decay experiments on these molecules also indicate that cyclization rates decrease strongly with increasing chain length. A similar conclusion can be drawn from TT annihilation experiments on A-polystyrene-A.

4.1. Results in Theta Solvents

For Py–polystyrene–Py and DMA–polystyrene–Py samples of M < 30000 in cyclohexane at 34.5°, fluorescence decay and steady–state studies of pyrene excimer formation give essentially identical values of the cyclization rate constant $\langle k_1 \rangle$. For higher molecular weights, λ_1 is nearly equal to k_M; eq. 9 gives $\langle k_1 \rangle$ values with a large uncertainty. Under these circumstances, eq. 11 provides more reliable data. Measurements of I_E/I_M become progressively more difficult for samples for which the $\langle k_1 \rangle$ values are less than 2 or 3 x 10^4 sec^{-1}.

For short chains, particularly in cyclopentane, where the $4k_1k_{-1}$ term in eq. 7 is significant, $I_M(t)$ decays as a sum of two exponential terms. Under these circumstances, values of k_E and k_{-1} can also be determined. For Py–polystyrene–Py at 34.5° these values are 3.5 x 10^6 sec^{-1} and 1.8 x 10^7 sec^{-1}, respectively, independent of chain length, within experimental error. The former is very sensitive to temperature (E_a = 11 kcal/mol) whereas k_E depends only weakly upon temperature (E_a = 1.5 kcal/mol). From this data, an apparent binding energy of $-\Delta H_0$ = 8 kcal/mol can be calculated for the intramolecular pyrene excimer. This compares to a value of 8 kcal/mol reported by Birks (22) for bimolecular excimer formation from pyrene itself in cyclohexane.

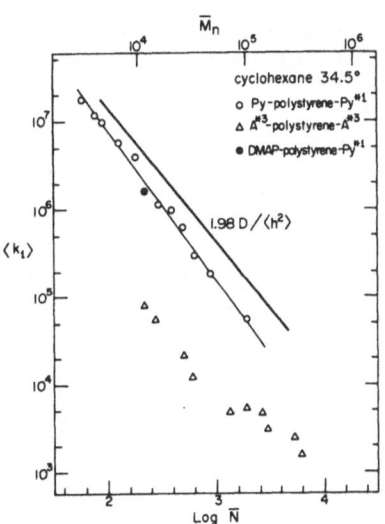

Figure 5. Log–log plot of $\langle k_1 \rangle$ vs N for polystyrene in dilute solution at the theta temperature in cyclohexane. (o) Values determined from pyrene excimer formation in Py*1–polystyrene–Py, taken from Ref. (21). (△) Values determined from A(t) measurements on A*3–polystyrene–A*3, taken from Ref. (20). The solid line is calculated from the expression $k_{cy}^D = 1.98D/R_F^2$, the asymptotic limit for long chains using data for PS in cyclohexane at 34.5°.

The chain length dependence of $\langle k_1 \rangle$ is shown in Figure 5. Data from excimer formation give a linear log–log plot, with values bracketed by those predicted by eq. I.1 (Table 1) as calculated by CP (6), using the parameters a = 60 Å and r = 0.23, and those predicted from eq. I.0 using the literature values (23) of D = 1.21 x 10^{-4} M$^{-0.49}$ (cm^{2}sec^{-1}) and $R_F{}^2$ = 4.9 x 10^{-17} M (cm^2) for polystyrene in cyclohexane at 34.5°. Eq. I.1 generates a slope of –1.43 over the molecular weight range indicated in figure 5, whereas the D/$R_F{}^2$ plot has a slope of –1.49. If a first order correction is applied to the experimental data to accommodate the fact that the polydispersity of the two highest molecular weight samples is somewhat higher than that of the other samples, the least squares slope is –1.68 ± 0.10.

Also shown in Figure 5 are data obtained from $A_T(t)$ measurements on TT annihilation experiments on A–PS–A in cyclohexane. Here, too, $\langle k_1 \rangle$ decreases with increasing N, but the magnitude of these values is ca. 30 times smaller than that obtained from excimer studies. A line of slope –3/2 passes reasonably through the data for M < 10^5; including all the data, however, gives a smaller slope. The origin of the discrepancy between the two sets of data is not clear at this time.

One criticism is often directed to these kinds of experiments: The frictional resistance to motion of the chromophore, because of its size, may affect the cyclization rates. DMA–polystyrene–Py has a group on one end only slightly larger (the Me$_2$N–substituent) than the polymer pendant group. As shown in figure 6, these polymers cyclize at the same rate in cyclohexane at 34.5° as Py–polystyrene–Py (24). Figure 6 also provides an interesting insight into the accuracy and precision of the individual measurements.

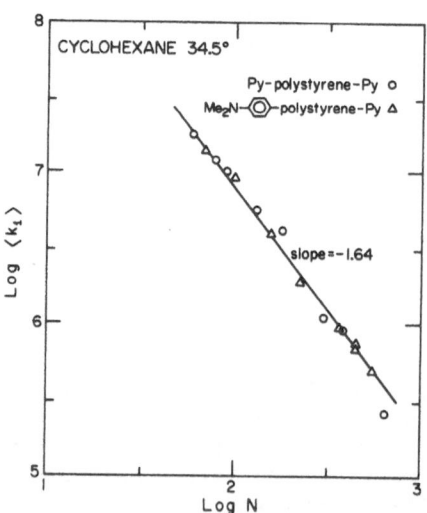

Figure 6. Log–log plot of $\langle k_1 \rangle$ vs N for Py–polystyrene–Py (o) and for DMA–polystyrene–Py in cyclohexane at 34.5°.

4.2. Results in Good Solvents

Excluded volume effects should contribute to polymer cyclization in good
solvents. TT annihilation experiments have been carried out in benzene,
at 22° for $A_T(t)$ measurements and at 30° for delayed fluorescence
measurements (20). Excimer and exciplex formation has been studied
primarily in toluene at 22° (22,25-27). These various results can be
compared by normalizing them to a common η_0/T value, here that of toluene
at 22°. Here η_0 is the solvent viscosity. The validity of this correction is substantiated by the finding that $\eta_0\langle k_1\rangle$ values for CMAP-poly-
styrene-Py in benzene, toluene and tetrahydrofuran are within a few
percent of one another (27).

The $\langle k_1\rangle$ values obtained from intramolecular excimer formation in
toluene are substantially smaller than corresponding values obtained in
cyclohexane (21,25). The data appear to have a slight downward curvature
in the log-log plot, Figure 7. $\langle k_1\rangle$ values of 2×10^{-4} sec^{-1} are
obtained from measurements at the very limit of meaningful I_E/I_M values.
The $\langle k_1\rangle$ values for the sample of M = 41,000 has a significant
uncertainty, and the M = 10^5 sample has been deleted from the plot.

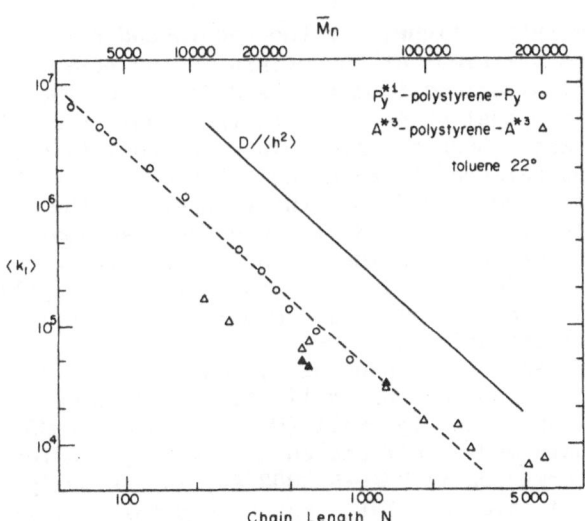

Figure 7. Log-log plot of $\langle k_1\rangle$ vs N for polystyrene in toluene
at 22°. (o) Values determined from pyrene excimer formation in
Py*¹-polystyrene-Py, taken from Ref. (21). Values, (benzene,
30°) corrected to toluene at 22° determined from ( ) A(t) and
( ) I_{DF} measurements on A*³-polystyrene-A*³, taken from
Ref. (20). The solid line is calculated from $k_{cy}^D = 1.98$
D/R_F^2, using data for PS in toluene at 22°.

The normalized $\langle k_1 \rangle$ values from both transient triplet absorption and delayed fluorescence studies on A–polystyrene–A are comparable in magnitude to those obtained from excimer experiments on Py–polystyrene–Py; and over a small range of M, the data coincide. Treated alone, the $A_T(t)$ data show a significantly smaller slope.

According to the literature values (29) of D and R_F for PS in toluene [D = 3.92 x 10^{-4} $M^{-0.574}$ (cm^2/sec); $R_F{}^2$ = 9.36 x 10^{-18} $M^{1.19}$ (cm)] $k_1 \sim D/R_F{}^2$ should decrease as $N^{-1.76}$. A line of this slope passes nicely through the excimer data and some of the TT annihilation data. This treatment assumes that the only consequence of excluded volume is expansion of the mean dimensions of the chain, so that $\langle k_1 \rangle$ is affected only by factors that affect D and $R_F{}^2$. With this assumption, $k_1{}^D$ values can be calculated from eq. I.1, since even in good solvents $\langle h^2 \rangle$ = 6 $\langle R_G{}^2 \rangle$ to within a few percent.

Experimental values of $\langle k_1 \rangle$ in toluene are significantly smaller than values calculated (3,6,25) from $k_{cy}{}^D$ = 1.98 $(D/R_F{}^2)$. The difference is about a factor of six. This difference suggests that another factor besides the change in chain dimensions is responsible for the decreases in cyclization rate in good solvents.

4.3. Solvent Effects

According to theory(cf. figure 1), the end–to–end distance distribution function W(r) has a minimum at the origin in the presence of excluded volume. The minimum arises from the tendency of polymer segments at or near the origin to repel other segments from their vicinity, including the other chain end. Such a minimum in a probability distribution function is often referred to as a "correlation hole." The depth and breadth of the minimum depend upon the strength of the excluded volume interaction (solvent quality) (30). In a theta–solvent, the minimum should disappear entirely. Evidence for the existence of the correlation hole would take the form of large solvent effects on cyclization properties, particularly cyclization equilibria, unanticipated from corresponding solvent effects on R_G.

Two sets of experiments provide evidence for the role of the correlation hole in suppressing cyclization in good solvents. Both involve polymers sufficiently small (M_n = 11000 and 2900) that a change in solvent has only modest effects on R_G. First, studies of intramolecular exciplex formation in DMA–PS13000–Py, in various good solvents for PS (e.g, methylene chloride, THF, benzene and toluene) give virtually identical values of $\eta_0 \langle k_1 \rangle$. Values determined in cyclopentane (θ = 22°) and in cyclohexane (θ = 34.5°), are more than a factor of two greater (27). These differences are too large to be explained only by changes in mean polymer dimensions.

A more compelling experiment involves solvent effects on room temperature cyclization of Py–PS2900–Py. In this case data are available in sufficient detail to permit both $\langle k_1 \rangle$ and k_{-1} to be evaluated (31). It is unfortunate that the data are not as simple as one would like: $I_M(t)$ and $I_E(t)$ decays, particularly in poor solvents, give λ_1, λ_2 values that differ by 20%, so that the actual $\langle k_1 \rangle$ and k_{-1} values one calculates depend upon which set of data one uses. Nevertheless one finds from either set that not only is $\langle k_1 \rangle$ diminished in good vs. poor solvents, but k_{-1} is enhanced. Thus their ratio $\langle K_{cy} \rangle = \langle k_1 \rangle / k_{-1}$ is an order of magnitude larger in cyclopentane and acetone than in benzene, toluene or THF. Such results are consistent only with some special factor, the correlation hole in $W(r)$, suppressing cyclization in good solvents.

4.4. Upper and Lower Theta Temperatures

Cyclopentane is an attractive solvent in which to study polystyrene properties because it has two accessible theta temperatures. The normal or lower theta temperature occurs at 22°. Below that temperature, high molecular weight polymer would precipitate because of unfavorable energies of mixing between solvent and polymer. At much higher temperatures, precipitation would also occur, but here the driving force is entropic and derives from a mismatch in the free volumes associated with the polymer and solvent. Such experiments have to be carried out in sealed tubes, at temperatures well above the normal boiling point of the solvent. For PS in cyclopentane the upper theta temperature occurs at 150° (32). At both critical temperatures the mean polymer dimensions are predicted to be compressed to their unperturbed values commensurate with a gaussian form for $W(r)$.

This theoretical picture leads to the prediction that excluded volume interactions would become of increasing and then decreasing importance as a solution of PS in cyclopentane is heated from 22° to 150°. By inference, cyclization would be favored at the extremes of temperature, and furthermore the magnitude of the critical cyclization exponent in the relationship $k_{cy} \sim N^{-\alpha}$ would go through a maximum as the temperature was raised.

Andrew Sinclair (23) in my group set out to test this prediction by measuring cyclization rates for eight samples of DMA-polystyrene-Py in cyclopentane over a temperature range of −10° to 130°C. These are the eight samples previously shown in figure 6. They range in molecular weight from 3000 to 28000. Fluorescence decay measurements were all consistent with the kinetic equations derived from Scheme I and in accord with steady state I_E/I_M determinations. The great risk in these experiments is that the change in the magnitude exponent may not be large enough to fall outside experimental error.

Sinclair's results (33) are shown in figure 8 below. It is quite clear that the exponent changes with temperature: it passes through a maximum at approximately 70° and decreases thereafter as the temperature is raised. In these experiments it takes a maximum absolute value less than that found for these same molecules in toluene (see below) where the magnitude of the excluded volume interaction is larger. Curiously, the magnitude of α seems to level off at the extremes of temperature, achieving near −10° (rather than 22°) the value predicted by CP for cyclization in a theta-solvent. It may not be accidental that this temperature corresponds to the upper critical solution temperature for low molecular weight polystyrene in cyclopentane [e.g., for M = 37000, UCST = −5° (32)].

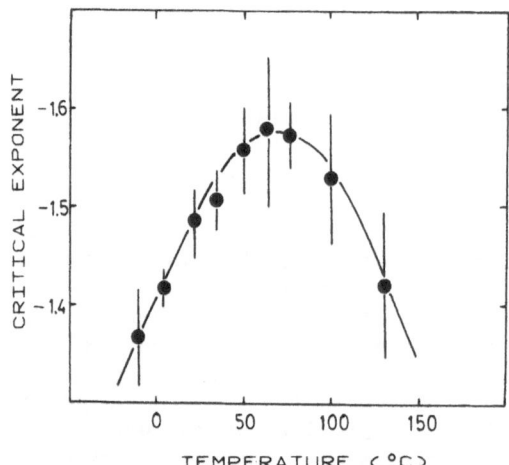

Figure 8. A plot of the critical exponent from the expression
$\langle k_1 \rangle \sim N^{-\alpha}$ for DMA–polystyrene–Py samples of M = 3000 to 28000
measured in cyclopentane at various temperatures. The error
bars represent one standard deviation.

4.5. The Effect of the Capture Radius

WF theory as formulated by CP for finite chains, Table 1, predicts that
an increase in the capture radius $\underline{a}$ will lead to an increase in k_{cy}, but
that the magnitude of the increase will diminish with increasing chain
length. Calculations based on this model, figure 2, suggest a linear
increase in cyclization rate with an increase in capture rate. The
proportionality constant gets smaller for longer chains. These predictions make an inviting target for experimental assessment.
    Chemical reactions, particularly photochemical interactions, are
characterized by distance–dependent rate constants rather than sharp
spherical reaction boundaries. Furthermore, the detailed shapes of the
potential energy surfaces for the kinds of interactions discussed in this
chapter are largely unknown. Nevertheless if one is interested in a more
qualitative assessment of the capture radius prediction, evidence should
be available from a comparison of the cyclization rates of Py–polystyrene–Py and DMA–polystyrene–Py. In polar and polarizable solvents,
exciplex formation in the latter polymer should be preceded by rate-
limiting electron transfer to generate the solvent separated ion pair.
Estimates for this interaction distance are typically 7 Å to 10 Å (16),
larger than the 5 Å normally invoked for pyrene excimer formation (6).
    Indeed in toluene at 22°, the cyclization rate constant for the
exciplex forming polymer is substantially larger than that for the
excimer forming polymer, figure 9. I find it remarkable that such a
subtle feature of the experiment is effectively predicted by the WF
model. While the difference in the slopes of the two lines in the log-

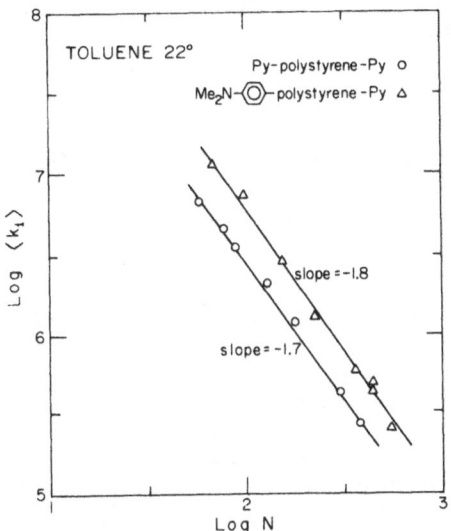

Figure 9. Log-log plot of ⟨k_1⟩ vs N for Py-polystyrene-Py and for DMA-polystyrene-Py in toluene at 22°.

log plot of figure 9 is not significant, there is every indication that the ratio of rate constants for the two polymers diminishes as the chains become longer. We find for the shortest polymers in figure 9 that ⟨k_1⟩ (exciplex) is 2.5 times that of ⟨k_1⟩ (excimer), whereas for the longest chains, this ratio has dropped to 1.5. Presumably for long enough chains the data would fall on the same line.

4.6. Other Polymers

Studies of end-to-end cyclization rates have been reported for poly-(ethylene oxide) [PEO] (34,35), poly(dimethylsiloxane) [PDMS] (36) and poly(tetramethylene oxide) [PTHF] (37). These experiments have been largely preliminary, and the data are not anywhere near as detailed or complete as for polystyrene. The fundamental scientifc issue is the relationship between cyclization rates and the detailed chemical struct-ure of the polymer.

This issue can take many forms. On one level one might be interested in what parameters could usefully be introduced into the equations of Table I to predict $k_{cy}D$ values corresponding to experimental cyclization rates. Beyond that, one might be interested in the relation-ship between chain stiffness and $k_{cy}D$. Ultimately one would like to be able to relate dynamic relaxation rates to chemical structure.

There are other reasons, as well, to study cyclization rates of polymers other than polystyrene. Polymers such as PDMS, PTHF and PEO undergo faster motion for a given chain length than polystyrene. Faster cyclization means that a given set of spectroscopic probes can be used to study a wider range of chain lengths, giving greater confidence and precision in the determination of critical exponents such as that in the $k_{cy}{}^D$ vs N relationship.

$$Py(CH_2)_3 \overset{\overset{O}{\|}}{C}OCH_2CH_2-(OCH_2CH_2)_{n-1}-O\overset{\overset{O}{\|}}{C}(CH_2)_3Py$$

(Py-PEO-Py)

The first experiments on the diffusion-controlled end-to-end cyclization of polymers were reported by Cuniberti and Perico (34) in 1977. These polymers were studied by steady-state fluorescence spectroscopy in air-saturated solution, and relative $\langle k_1 \rangle$ values were obtained from measurements of I_E/I_M. Absolute values of $\langle k_1 \rangle$ were obtained by Cheung et al. (35), for a single sample of Py-PEO-Py of M = 9000 in a variety of solvents, using a combination of steady-state and fluorescence decay techniques. With their value of $\langle k_1 \rangle$ in THF solution, the I_E/I_M results of Cuniberti and Perico (34) can be calibrated. It is curious, figure 10, that the plot of log (I_E/I_M) vs log N for these polymers has a slope significantly less steep than $-3/2$.

Also shown in figure 10 are results for cyclization of Py-PDMS-Py in toluene, obtained by Winnik and co-workers (36). A single sample of the polymer 1-pyrene-$(CH_2)_5Si(Me_2)$-PDMS-Si$(Me)_2(CH_2)_5$-1-pyrene of broad molecular weight distribution was synthesized and then fractionated using

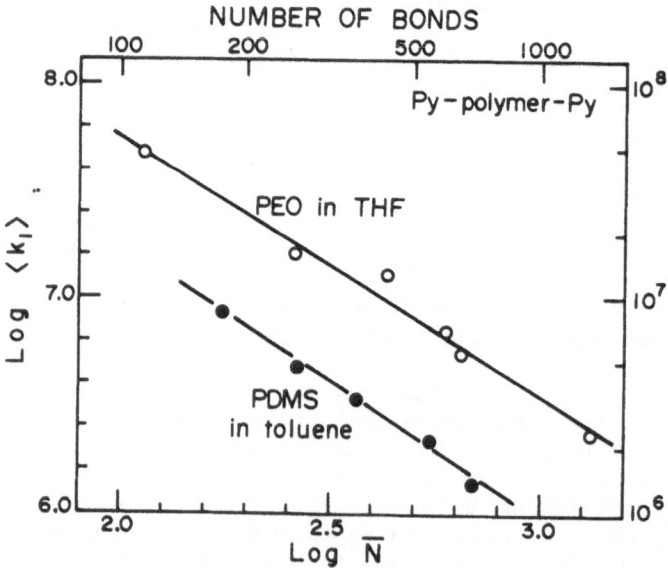

Figure 10. Log-log plot of $\langle k_1 \rangle$ vs N for Py-PEO-Py in tetrahydrofuran (o) and Py-PDMS-Py in toluene (●), both at 22°. A line of slope -1.3 is drawn through the Py-PDMS-Py data.

analytical gpc columns to obtain the samples shown in figure 10. A line
of slope -3/2 is drawn through these data and fits reasonably well, even
though toluene is not a theta solvent for PDMS [the Mark-Houwink exponent
is 0.68 (23).]

   According to the PC model and the equations in Table 1, PDMS chains
are predicted to cyclize by a factor of two faster than polystyrene
chains of comparable length. To make this comparison, one would need
experimental results for both polymers carried out in their respective
theta-solvents. Unfortunately, appropriate data for PDMS and other
polymers are not yet available. One can compare $\langle k_1 \rangle$ values for various
polymers if one keeps in mind that three factors contribute to the
differences observed. These are intrinsic chain dynamics, differences in
mean dimensions, and differences in the strength of the excluded volume
interactions. In this qualitative spirit, values of $\langle k_1 \rangle$ are compared
in Table II for various polymers having ca. 660 bonds between their chain
ends.

TABLE II

Comparison of $\langle k_1 \rangle$ Values for Different Polymers in Toluene at 22°

	M_n	N	$\langle k_1 \rangle (sec^{-1})$
Py-polydimethylsiloxane-Py	25800	692	1.4×10^6
Py-poly(ethylene oxide)-Py	9600	655	8.3×10^5
Py-polystyrene-Py	33600	650	9.0×10^4

4.7. Cyclization of Groups Internal to the Chain

Interesting information about internal chain dynamics can be obtained by
examining cyclization in polymers which have a small number of excimer
forming pendant groups attached to the chain interior. Such polymers
with a statistical distribution of pyrene groups are not difficult to
prepare. The group of Cuniberti and Perico (38) has devoted considerable
attention to the study of poly(vinyl acetate) [PVAc] containing 1% Py
groups.

   They observed that I_E/I_M increased with M, reaching an asymptotic
level at $M = 10^5$. I_E/I_M increased linearly with η_o^{-1} in methanol-
ethylene glycol mixtures – both poor solvents for the polymer – whereas
at low viscosities, this plot was curved for ethyl acetate-glycerol
triacetate mixtures. To the extent that cyclization depends upon coil
expansion, as reflected in the intrinsic viscosity $[\eta]$ of the polymer,
these differences should disappear in a plot of $\eta_o [\eta] I_E/I_M$ vs $[\eta]$
or η_o. While these plots show some scatter, the data are best fit by
a line of zero slope. This result suggests that in this system, solvency
factors which affect $[\eta]$ through coil expansion have the same effect on
the average (internal) cyclization rate of the polymer.

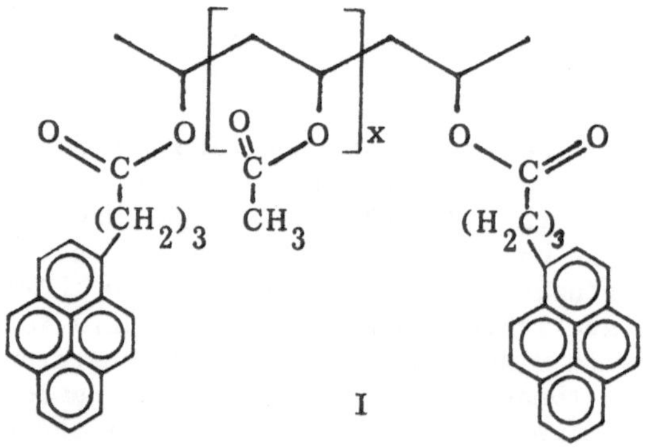

I

Winnik and co-workers (39) have prepared the polystyrene derivatives II shown below with evenly-spaced pyrene pendant groups. They too
observed that I_E/I_M was very sensitive to solvent quality and that this
sensitivity was more pronounced in the higher molecular weight polymers.
The most important results reported so far on both sets of polymers
containing pendant internal pyrene groups concern their behavior in the
presence of elevated concentrations of unlabelled polymer. These results
will be examined in the following section.

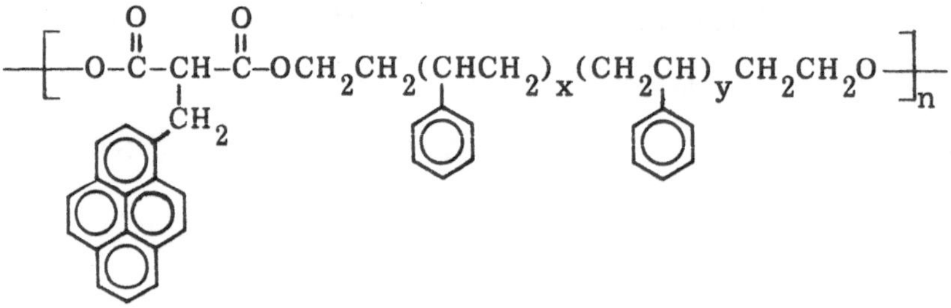

II

5. EXPERIMENTS IN CONCENTRATED SOLUTION

One of the great virtues of labelled-chain experiments is that one can
study the behaviour of a marked chain in the presence of a vast excess
of unlabelled chains. In consequence one can study the properties of
individual polymer molecules over the entire range of very dilute to very
concentrated polymer solutions without interference from intermolecular
interactions. Many phenomena are open to study in this way. These

include screening of excluded volume effects in experiments sensitive to chain conformation, as well as screening of hydrodynamic interactions and entanglement effects in measurements of chain dynamics.

To date, only three studies have appeared which examine the consequences of macromolecule concentration on the cyclization of labelled polymers. These have involved pyrene excimer formation: end-to-end cyclization for polymers with Py- groups on the chain ends, and internal cyclization for polymers containing Py- groups along the chain backbone. In none of these experiments are the chains long enough, or the solutions sufficiently concentrated, for entanglement effects to be important.

5.1. End-to-End Cyclization

Cyclization rate constants have been obtained (40) for Py-polystyrene-Py samples of M_n = 9200 and 25000 as a function of unlabelled polystyrene concentration for weight fractions up to 0.6. $\langle k_1 \rangle$ values decrease in the presence of added unlabelled polymer, but the concentration sensitivity of $\langle k_1 \rangle$ depends on the quality of the solvent, figure 11. In cyclohexane just above the theta temperature (35°), one observes a monotonic decrease in $\langle k_1 \rangle$ by a factor of ca. 4 for Py-PS9200-Py, and by a factor of ca. 8 for Py-PS25000-Py. In toluene, a good solvent for polystyrene, $\langle k_1 \rangle$ is substantially slower at infinite dilution than in a theta-solvent, a fact which was attributed to the effect of the correlation hole on cyclization. At increased polystyrene concentrations, $\langle k_1 \rangle$ decreases only modestly, so that above c = 200 mg/ml, $\langle k_1 \rangle$ values in good and poor solvents are comparable. As the various symbols in figure 11 indicate, the values of $\langle k_1 \rangle$ obtained depend upon the weight concentration c of the matrix chains but not on their molecular weight.

Two factors operate to affect $\langle k_1 \rangle$ when the concentration of matrix polymer is increased. Polymer-polymer interactions first act to screen hydrodynamic coupling during chain motion (41). This has the effect of increasing the net friction each chain experiences and acts to retard chain motion, which becomes more Rouse-like at large c. This is the factor responsible for the decrease in $\langle k_1 \rangle$ in cyclohexane at 35°.

Polymer-polymer interactions also act to suppress excluded volume effects for chains in good solvents. Mean coil dimensions shrink, and the minimum at low r disappears in the distribution function W(r), figure 1. These circumstances lead to an _increase_ in $\langle k_1 \rangle$ as well as in W(0). In other words a labelled polymer in dilute solution in a good solvent would experience opposing effects on cyclization as the concentration of matrix chains is increased: Excluded volume screening would lead to an enhancement of $\langle k_1 \rangle$, whereas hydrodynamic screening would lead to retardation. The data in figure 11 for Py-polystyrene-Py suggest that these factors almost exactly cancel.

5.2. Internal Cyclization

CP have studied the influence of matrix chains concentration on the extent of pyrene excimer formation in their labelled poly(vinyl acetate) derivative I.

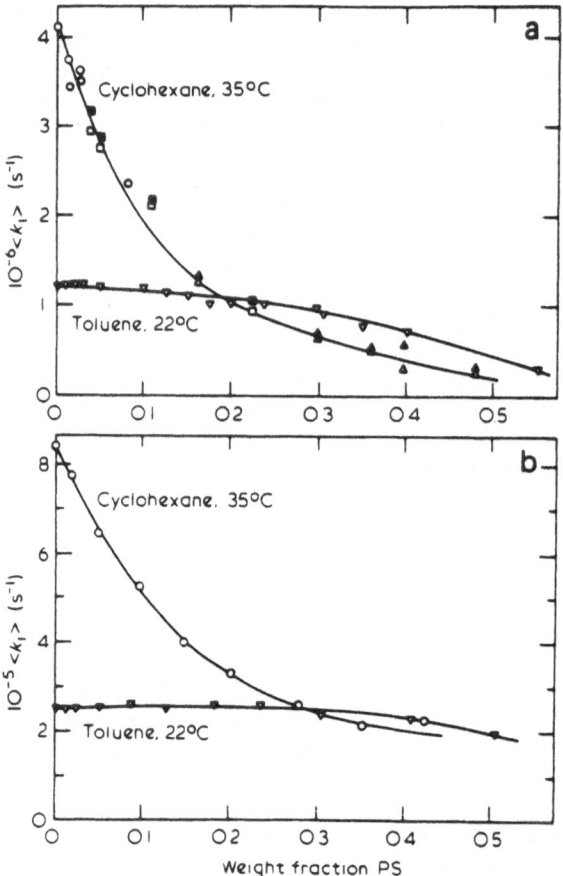

Figure 11. A plot of $\langle k_1 \rangle$ vs the weight fraction of polystyrene in the solution. (a) Test chain of $M_n = 9200$ in the presence of unlabelled polystyrene of various molecular weights, cf. Ref. (40). (b) Test chain $M_n = 25000$ in the presence of polystyrene of $M_n = 17500$. Open points refer to steady-state measurements; closed points to fluorescence decay measurements.

Addition of unlabelled PVAc to dilute solutions of the labelled polymer causes changes in I_E/I_M (6,41). The nature of these changes differs in poor solvents (methanol, 20°) and good solvents (THF, 20°). As the lower line in figure 12 indicates, there is an <u>increase</u> in I_E/I_M for I in THF for PVAc concentrations above $c = 0.2$ g/ml which then levels off at higher concentrations. In the poor solvent, one observes first a <u>decrease</u> in I_E/I_M in this PVAc concentration range followed by an increase and a subsequent decrease at higher concentrations. The initial changes occur at concentrations approximately equal to the concentration

c^* when the coils of unlabelled polymer overlap. Here c^* is taken to be
equal to $M/N_A(2R_G)^3$, where N_A is Avogadro's number.

In both poor and good solvents, one expects the screening of
hydrodynamic interactions to increase the net friction felt by the chain,
causing the mean cyclization rate $\langle k_1 \rangle$ to decrease. In good solvents,
this retardation is compensated by excluded volume screening. From the
net increase in I_E/I_M for I in THF, it appears that the latter effects
predominate. In contrast to the results for Py-polystyrene-Py, here the
cyclization rate is quite sensitive to the molecular weight of matrix
chains. In addition, there seems to be no simple explanation at this
time for the more complicated behavior of I + PVAc in a poor solvent.

The results for the polystyrene sample containing evenly spaced
pyrene groups are shown in figure 13 (42). Here, for a sample with
M_n = 14000 (n ~ 4), the dependence of I_E/I_M is presented as a function of

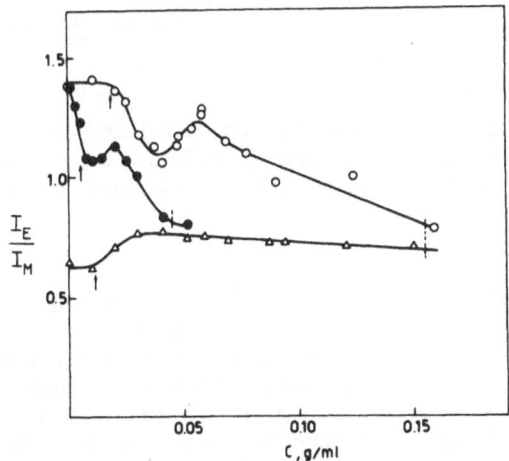

Figure 12. A plot of I_E/I_M for the poly(vinyl acetate) derivative I as a function of the concentration of unlabelled
polymer of M = 1.8 x 10⁵ (o) and M = 2 x 10⁶ (●) in methanol;
and M = 1.8 x 10⁵ (△) in THF.

$\underline{c}$ for experiments carried out in a good solvent (toluene), a modest
solvent (2-butanone), and a theta solvent (cyclohexane, 35°), for
unlabelled polystyrene concentrations up to a weight fraction of 0.7.
Even for these low molecular weights [M(PS) = 17500] the samples of
highest concentrations are rubbery solids that barely flow on the time
scale of weeks. In these experiments II is present at a concentration of
10 ppm.

One sees in figure 13 that in a theta-solvent, I_E/I_M for II
decreases monotonically with increasing c, mimicking the behavior of end-
labelled polymers. In the good solvent toluene, I_E/I_M at [PS] = 0 is
substantially reduced over that in cyclohexane, but is relatively
uninfluenced by even very large amounts of added polystyrene. Increases

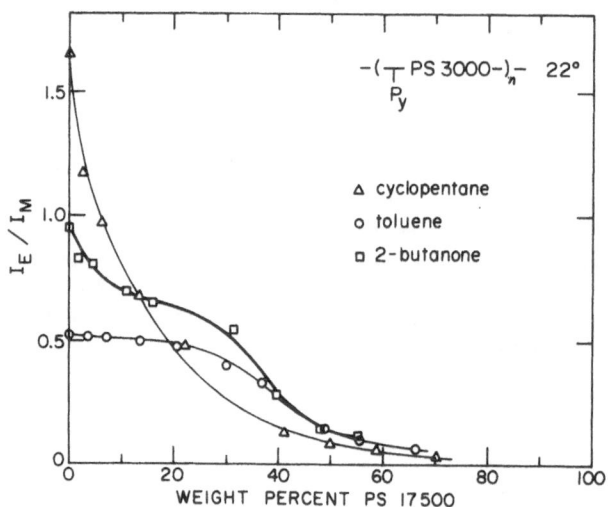

Figure 13. A plot of I_E/I_M for the labelled polymer II as a
function of the concentration of unlabelled polystyrene of
M = 17500 in cyclohexane at 35°, 2-butanone at 22° and toluene
at 22°.

in bulk solution viscosity of several orders of magnitude reduce I_E/I_M
for the labelled polymer in a theta-solvent by only a factor of ca. 3
and barely affect its internal cyclization rate in a good solvent.
    In 2-butanone, an intermediate behavior is observed. There is a
pronounced decrease in I_E/I_M at low $\underline{c}$, followed by a plateau region at
intermediate $\underline{c}$. At high polystyrene concentrations I_E/I_M decreases once
more, with values superimposing on those in toluene at similar
polystyrene concentrations. One has rather detailed insights into the
interplay of excluded volume and hydrodynamic interactions on internal
chain dynamics, and the screening effects of added unlabelled polymer.

6. CONCLUSION

Fluorescence quenching techniques provide powerful methods for studying
cyclization processes. One requires polymers of the form $\underline{A}$~~~~$\underline{Q}$, where
proximity of $\underline{A}^*$ and $\underline{Q}$ gives rise to an observable event. Many such
pairs of groups exist; their choice represents one of the most delicate
aspects of the experimental approach. One must know the mechanistic
details of the $\underline{A}^*/\underline{Q}$ interaction. Interpretation of the experimental
results requires the data from the intramolecular reaction to be compared
with corresponding data for the bimolecular reaction of suitable ~~$\underline{A}$ and
~~$\underline{Q}$ molecules.

The power of these techniques derives from their sensitivity and versatility. One can study fast processes which are diffusion controlled and slower processes which depend upon a conformation equilibrium or pre-equilibrium. Molecules can be examined at very low concentrations. Perhaps the most important application, and the least explored, is the study of traces of labelled chains in semidilute and concentrated polymer solutions. These experiments give one the power to examine the effect of the environment on the cyclization of individual polymer chains.

6. REFERENCES

1. (a) H. Freundlich and A. Krestovnikoff, Z. Phys. Chem. 1911, 76, 26.

    (b) For an early review, see G.M. Bennet, Trans. Faraday Soc., 1941, 37, 794.

2. (a) W.H. Kuhn, Kolloid Z., 1934, 68, 2.

    (b) H. Jacobson and W.H. Stockmayer, J. Chem. Phys., 1950, 18, 1600.

    (c) P.J. Flory and J.A. Semlyen, J. Am. Chem. Soc., 1966, 18, 3209.

3. S. Tazuke and M.A. Winnik, chapter 2 in this volume.

4. J.B. Birks, 'Photophysics of Aromatic Molecules,' Wiley-Interscience, New York, 1971.

5. G. Wilemski and M. Fixman, J. Chem. Phys., 1974, 60, 866.

6. C. Cuniberti and A. Perico, Prog. Polym. Sci., 1984, 10, 271.

7. P.J. Flory, 'Statistical Mechanics of Chain Molecules,' Wiley-Interscience, New York, 1969.

8. L.H. Sperling, Polym. Eng. Sci., 1984, 24, 1.

9. J.F. Brown and G.M. Slusarczuk, J. Am. Chem. Soc., 1966, 88, 3209.

10. (a) P.G. deGennes, 'Scaling Concepts in Polymer Physics,' Cornell University Press, Ithaca, N.Y., 1979.

    (b) J. des Cloizeaux, J. Phys. (Paris), Colloq., 1978, C2, 135.

11. M. Doi, Chem. Phys., 1975, 9, 455; S. Sunagawa and M. Doi, Polym. J., 1975, 7, 604; ibid. 1976, 8, 239.

12. A. Perico and C. Cuniberti, J. Polym. Sci. Polym. Phys. Ed., 1977, 15, 1435.

13. (a) A. Perico and M. Battezzati, J. Chem. Phys., 1981, 75, 4430.

    (b) M. Battezzati and A. Perico, J. Chem. Phys., 1981, 74, 4527.

    (c) M. Battezzati and A. Perico, J. Chem. Phys., 1981, 75, 886.

    (d) A. Szabo, K. Schulten and Z. Schulten, J. Chem. Phys., 1980, 72, 4350.

14. H. Yamakawa, 'Modern Theory of Polymer Solutions,' Harper and Row, New York, 1971.

15. A. Perico, P. Piaggio and C. Cuniberti, J. Chem. Phys., 1975, 62, 4911.

16. (a) J.D. Simon and K.S. Peters, J. Am. Chem. Soc., 1981, 103, 6403.
    (b) J.D. Simon and K.S. Peters, J. Am. Chem. Soc.,1982, 104, 6542.

    (c) W. Hub, S. Schneider,F. Dörr, J.D. Oxman and F.D. Lewis, J. Am. Chem. Soc., 1984, 106, 701.

    (c) W. Hub, S. Schneider,F. Dörr, J.D. Oxman and F.D. Lewis, J. Am. Chem. Soc., 1984, 106, 708.

    (d) A. Weller, Z. Phys. Chem. (Wiesbaden), 1976, 101, 371.

    (e) A. Weller, Z. Phys. Chem. (Wiesbaden), 1982, 130, 129.

    (f) A. Weller, Z. Phys. Chem. (Wiesbaden), 1982, 133, 93.

17. I. Soutar and D. Philips, chapter 6 in this volume.

18. J. Vandendriessche, R. Goedeweek, P. Collart and F. DeSchryver, chapter 10 in this volume.

19. R.M. Noyes, Prog. Reaction Kinetics, 1961, 1, 129.

20. (a) H. Ushiki, K. Horie, A. Okamoto and I. Mita, Polym. J., 1981, 13, 191.

    (b) K. Horie, W. Schnabel, I. Mita and H. Ushiki, Macromolecules, 1981, 14, 1422.

21. (a) M.A. Winnik, A.E.C. Redpath, K. Paton and J. Danhelka, Polymer, 1984, 25, 91.

    (b) A.E.C. Redpath and M.A. Winnik, Ann. N.Y. Acad. Sci., 1981, 10, 359.

22. J.B. Birks, Rep. Prog. Phys., 1975, 38, 903.

23. J. Brandrup and E.H. Immergut, editors, 'Polymer Handbook,' 2nd ed., section IV, J. Wiley, New York, 1975.

24. M.A. Winnik, A.M. Sinclair and G. Beinert, Macromolecules, 1985, 18, 1517.

25. M.A. Winnik, Accounts Chem. Res., 1985, 18, 73.

26. A.E.C. Redpath and M.A. Winnik, J. Am. Chem. Soc., 1984

27. M.A. Winnik, A.M. Sinclair and G. Beinert, Can. J. Chem., 1985, 63, 1300.

28. A.M. Sinclair, M.A. Winnik and G. Beinert, J. Am. Chem. Soc., 1985, in press.

29. (a) J. Raczek and G. Meyerhoff, Proceedings of the 27th International Symposium on Macromolelcules, Strasbourg, France, July 1981, pp. 700-703.

    (b) J. Raczek, Eur. Polym. J., 1983, 19, 607.

30. Y. Oono and K.F. Freed, J. Phys. A. Math. Gen.,1982, 15, 1931.

31. M.A. Winnik, X.B. Li and J.E. Guillet, J. Polym. Sci. Polym. Symp., in press, 1985.

32. S. Saeki, N. Kuwahara, S. Konno and M. Kaneko, Macromolecules, 1973, 6, 589.

33. A.M. Sinclair, Ph.D. Thesis, University of Toronto, 1986.

34. C. Cuniberti and A. Perico, Eur. Polym. J., 1977, 13, 369.

35. S.T. Cheung, A.E.C. Redpath and M.A. Winnik, Makromol. Chem., 1982, 183, 1815.

36. A.E..C. Redpath, P. Svirskaya, J. Danhelka and M.A. Winnik, Polymer, 1983, 24, 319.

37. S. Slomkowski and M.A. Winnik, submitted for publication (1985).

38. (a) C. Cuniberti and A. Perico, Eur. Polym. J., 1980, 16, 887.

    (b) C. Cuniberti and A. Perico, Ann. N.Y. Acad. Sci., 1981, 366, 35.

39. M.A. Winnik, X.B. Li and J.E. Guillet, Macromolecules, 1983, 16, 992.

40. A.E.C. Redpath and M.A. Winnik, Polymers, 1983, 24, 1286.

41. C. Cuniberti, L. Musi and A. Perico, J. Poly. Sci. Poly. Lett.,
     1982, 20, 265.

42. M.A. Winnik, X.B. Li and J.E. Guillet, Macromolecules, 1984, 17,
     699.

FOLDING AND DYNAMICS OF PROTEINS STUDIED BY NON-RADIATIVE ENERGY TRANSFER MEASUREMENTS

Elisha Haas
Department of Life Sciences
Bar-Ilan University
Ramat-Gan 52100
Israel

ABSTRACT. The mechanism of protein folding, the transition of an unordered chain to a well-defined three-dimensional structure, is investigated by means of site specific fluorescence labeling and nanosecond fluorescence decay measurements of nonradiative excitation energy transfer. Donor and acceptor chromophores are attached specifically to well-defined sites on the polypeptide chain and the labeled derivatives are purified by high performance liquid chromatography. The intramolecular distance distributions between two labeled sites in each derivative is deduced from donor fluorescence decay curves. This information is used to determine the average distance and degree of order in the labeled segments throughout the folding pathway. Local dynamics of the protein is determined from the effect of viscosity on the transfer efficiency and, hence, on donor fluorescence decay kinetics. Side chain and backbone thermal motions on the nanosecond time scale are observed, mainly in the unordered denatured state. The theory for another approach to monitor slower motions in donor and acceptor labeled chain molecules was developed, based on autocorrelation analysis of the donor fluorescence emission intensity. The time dependence of conformational fluctuations in small ensembles of chain molecules can be detected over a wide range of time scales. The methods used for the study of protein folding and of flexible or unordered states of proteins are applicable for the study of any other polymeric chain molecules in solution of in the bulk.

INTRODUCTION

Living organisms derive their wealth of characteristic structures and activities from their constituent macromolecules. Proteins are the major group of chain molecules which mediate the self assembly and development of all known organisms as well as most of their complex activities directed towards their sustainment and self-multiplication (metabolism and reproduction). The recognition that proteins are a group of heteropolymers with well-defined sequences and three-dimensional structures has led to the development of protein

325

M. A. Winnik (ed.), Photophysical and Photochemical Tools in Polymer Science, 325–350.
© 1986 by D. Reidel Publishing Company.

crystallography and structural analysis. The three-dimensional structure of many proteins in their crystalline state is known to 2Å or 3Å
resolution, and invaluable insight into the structural basis of function and mode of action has been gained (1). A second aspect of protein structure and function is their dynamic flexibility revealed even
in the crystalline state. In solution, protein molecules continuously
fluctuate over a wide range of frequencies and amplitudes from side
chain rotations to segmental and domain motions and even transient
unfolding/refolding transitions (2,3). These spontaneous fluctuations
of protein structure enable all dynamic aspects of life, including
catalysis, regulation, development, adaptation and locomotion. Little
is known about this essential characteristic of proteins mainly due to
limitations of experimental methods. Investigation of the conformational dynamics of protein molecules depends on the development of new
methods with appropriate spatial and temporal resolution. The folded
states of proteins are only marginally stable under physiological conditions and may be disrupted by changes of solution conditions, such
as temperature, pressure, pH or addition of a variety of denaturants.
When denatured (unfolded) protein molecules are placed under conditions where the folded state is stable, they spontaneously refold
from the open random coil conformations back to the compact precisely
ordered native conformations. All the information required for an
unfolded protein to refold to its unique native conformation is
available in its amino acid sequence (4), so that it should be
possible to predict the folded conformation from the sequence alone.
This, however, is not yet possible, due to our ignorance of the exact
nature of the wealth of interactions stabilizing the native conformation. Moreover, most globular proteins in a denatured state complete
the attainment of the unique native structure within seconds after
transfer to folding conditions, while a random search through the
enormous space of possible conformations would require a much longer
time (5). Hence, it seems that the amino acid sequence does not
simply code for a single final conformation. Instead, it probably
codes for stepwise formation of intermediate structures of some
nature, which thus direct the pathway to the native conformation (6).

Experimental study of the mechanism of folding depends on development of methods which can detect local, partially ordered structures. This, in turn, leads to the regime of statistical approaches,
searching for distributions of chain configurations and dynamics of
segmental motions. At this point, concepts and methodologies used by
polymer chemists are useful for the protein chemists and vice versa.

Investigation of short-lived transient structures and their
dynamics requires a means of monitoring the relative position of each
residue relative to the molecular framework with appropriate time
resolution. This seems to be unattainable by present diffraction and
spectroscopic methods due to the dynamic and statistical nature of
the conformations studied. In the present study, this is accomplished
by measuring the transfer of electronic excitation energy between
chromophores attached to well-defined sites along the chain molecule,
one chromophore serving as an energy donor and the other serving as an
acceptor. As will be shown in the following sections, energy transfer

measurements yield detailed intramolecular distances distribution functions and rates of conformational transitions over the time scale of nanoseconds and longer. The application of these methods to the study of folding and dynamics of globular proteins will be shown.

2. THEORETICAL BACKGROUND

2.1. Non-radiative Excitation Energy Transfer (Förster Theory)

A molecule in an electronically excited state (donor) can transfer its excitation energy to another molecule (acceptor) (7,8) provided that a number of conditions are fulfilled by the participants (9): (a) the energy donor must be luminescent; (b) the emission spectrum of the donor should have some overlap with the absorption spectrum of the acceptor; (c) the distance between the donor and the acceptor should not exceed the limit of 50Å or 60Å. The theoretical basis for the mechanism of long-range nonradiative energy transfer has been worked out with both classical and quantum mechanical approaches. In both treatments the interactions between energy donor and acceptor are assumed to be of the dipole-dipole type and therefore this type of energy transfer is sometimes referred to as transfer by dipole-dipole interactions. The theoretical basis will not be described here; some basic considerations were described in an earlier review (10).

For an isolated pair of chromophores which fulfill the requirements for energy transfer by dipole-dipole interactions, the transfer efficiency is a well-defined function of the dis- tance, r, between the donor and the acceptor. It may thus serve to monitor this distance. The probability, $n_{D \to A}$, of energy transfer from a donor, D, to an acceptor, A, per unit time, is given by (7):

$$n_{D \to A} = \frac{9,000(\ln 10)\kappa^2 \eta_0}{128\pi^5 n^4 N r^6 \tau} \int_0^\infty \frac{f(\bar{\nu})\varepsilon(\bar{\nu})}{\bar{\nu}^4} d\bar{\nu} = \frac{1}{\tau} \cdot \frac{R_0^6}{r^6} \tag{1}$$

where η_0 is the quantum yield of the donor fluorescence in the absence of an acceptor, n is the refractive index of the medium, N is Avogadro's number, r is the distance between the donor and the acceptor chromophores, τ is the lifetime of the donor in the absence of the acceptor, $f(\bar{\nu})d\bar{\nu}$ is the normalized fluorescence intensity of the donor in the wavenumber range ν to $\nu + d\nu$, $\varepsilon(\nu)$ is the absorption coefficient of the acceptor at the wavenumber $\bar{\nu}$, R_0 is defined by Eq. 1 and is given by:

$$R_0^6 = 8.8 \times 10^{-25} \eta_0 \kappa^2 n^{-4} J \tag{2}$$

J being the integral in Eq. 1. The energy transfer process competes with spontaneous decay of the donor, characterized by the rate constant $1/\tau$. Thus, the probability ρ that the donor will not be deexcited within the time, t, following excitation is given by:

$$-(1/\rho)(d\rho/dt)=(1/\tau)+(1/\tau)(R_0/r)^6 \qquad (3)$$

and the efficiency, E, for energy transfer is expressed by:

$$E = R_0^6/(R_0^6+r^6) \qquad (4)$$

Thus, the efficiency of transfer is 50% when $r=R_0$. The orientational dependence of the probability of energy transfer is contained in a factor referred to as κ^2, where $\kappa^2 = \cos\theta_{DA} - 3\cos\theta_D \cos\theta_A$ (θ_{DA} is the angle between the donor and acceptor dipoles, and θ_D and θ_A are the angles between the donor and acceptor dipoles, respectively, and the line joining them (Figure 1)). It has been shown (11) that by proper choice of donor and acceptor chromophores exhibiting relatively low limiting polarization of fluorescence for the electronic transitions involved in the transfer process, the orientational dependence of the transfer probability can be made small or even insignificant. This has been the case for the experiments to be described and will be assumed to be the case in the following theoretical calculations.

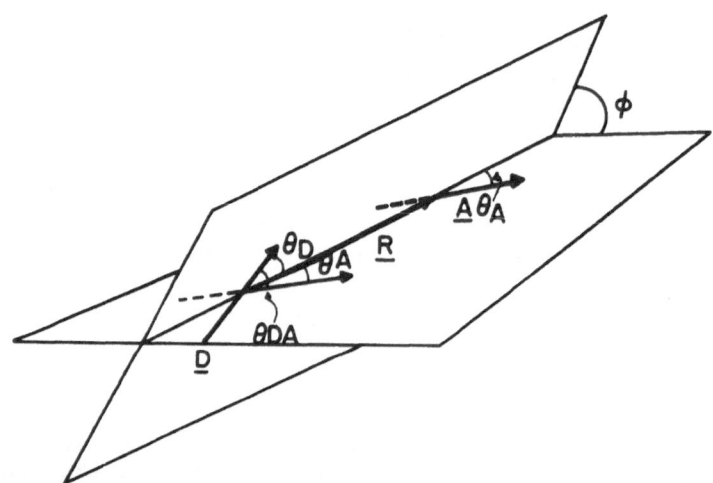

Figure 1. Geometrical representation of the two dipoles D (donor) and A (acceptor). R, the vector connecting the centers of the two dipoles, makes an angle θ_A with A and θ_D with D. The angle between the two dipoles is θ_{DA} and the angle between their planes is ϕ.

The range of R_0 values usually encountered for pairs of chromophores (up to about 50A), is of suitable magnitude for the study of the dimensions of most biopolymers, such as proteins and supramolecular assemblies, e.g., ribosomes (12,13). In the following it is shown that the time dependence of the fluorescence decay of the donor contains information on the <u>distributions</u> of distances between the two

probes. A method for determination of intramolecular distances distributions is presented. It is further shown that the dynamics of the
labeled molecules are expected to affect the probability of nonradiative energy transfer and should therefore be amenable to study by
energy transfer. Two methods for determination of intramolecular
dynamics by energy transfer are described. One is based on analysis of
donor fluorescence decay kinetics and is sensitive to conformational
changes occurring on the time-scale of the fluorescence lifetime,
i.e., 10^{-9}-10^{-8} sec. The second method involves analysis of
fluorescence intensity fluctuations upon illumination of a small
number of labeled molecules by a light beam of constant intensity.
This method allows detection of conformational transitions which are
slower than the lifetime of the excited state. The two methods are
thus complementary in terms of time scales.

2.2. Intramolecular Distributions of Distances between Labeled Sites: Computation from Donor Fluorescence Decay Kinetics

The efficiency, E, of energy transfer between a donor-acceptor pair
separated by a distance r, is given by Eq. 4. For flexible labeled
chain molecules the interchromophoric distance is not unique and the
efficiency of energy transfer from donor to acceptor measured for an
ensemble of molecules is an average quantity, E , given by:

$$E = \int_0^\infty f(r) \ [R_0^6/(R_0^6 + r^6)]dr \tag{5}$$

where f(r)dr expresses the fraction of donor-acceptor pairs with a
separation distance in the range r to r+dr. It is obviously not
possible to evaluate f(r) from a single measurement of E . It is,
however, possible to reconstruct f(r) by measuring the fluorescence
decay kinetics of the donor (14). If the donor-acceptor distances do
not change during the lifetime of the excited state, the donor
fluorescence decays monoexponentially for each subpopulation of
molecules (characterized by a donor-acceptor separation r) with a
rate constant $(1/\tau)$ $\{1- (R_0/r)^6\}$. The decay kinetics, I(t), for
the total population of donor-acceptor labeled molecules in response
to an infinitely short excitation pulse is thus given by (15):

$$I(t) = k \int_0^{r_m} f(r)\exp[-(t/\tau)-(t/\tau)(R_0/r)^6]dr \tag{6}$$

where k is a proportionality factor and r_m is the largest possible
donor-acceptor separation (fully extended conformation).

2.3. Intramolecular Dynamics of Labeled Chain Molecules: Investigation by Fluorescence Decay Measurements

Let us consider a population of flexible labeled molecules with an
equilibrium interchromophoric distance distribution of f(r) (as can be
determined by the method described in the previous section) and with

free movement of labeled segments under the influence of Brownian motions. Consider the changes which occur with time in a sample of this population which has been excited at time t=o by an infinitely short excitation light pulse. The number, N*(r,o) of excited molecules in the sample for which the donor-acceptor separation is r at t=o, is proportional to f(r), because all molecules in the ground state have equal probability to be excited. N*(r,t), the number of molecules with an excited donor which have an interchromophoric distance r at time t after excitation decreases with time, as a result of both spontaneous decay of the excited state and transfer of excitation energy.

The probability of non-radiative energy transfer to the acceptor is considerably higher for molecules with short interchromophoric distances. The sample of excited labeled molecules will rapidly be depleted of conformers with short interchromophoric distance. This is equivalent to a perturbation of the equilibrium distribution of conformations. N*(r,t) the distribution of intramolecular distances of the molecules with an excited donor will thus, within the lifetime of the excited state of the donor, depart from the equilibrium distribution. This is equivalent to generation of a concentration "gradient" within the sample of excited molecules. Brownian motion, if permitted, will cause a net transfer of excited molecules from the well populated long interchromophoric distance fractions to the depleted, short interchromophoric distance fractions. The transferred molecules will, in turn, rapidly transfer their excitation energy and disappear from the sample. Thus, the gradient is sustained throughout the lifetime of the excited state and a net acceleration of energy transfer takes place; this will, of course, affect the transfer efficiency.

Analysis of the time course of fluorescence decay can yield information about the conformational dynamics of the labeled molecules. Obviously, the motion of labeled chain segments is not a free diffusion, since conformational forces direct the motions towards the equilibrium distance distributions. It has been shown in detail (10, 16) (and therefore will not be repeated here) that, based on the above consideration, one can derive the differential equation describing the changes in N*(r,t) with time:

$$\frac{\partial \overline{N}(r,t)}{\partial t} = -\frac{1}{\tau}\overline{N}(r,t) - \frac{1}{\tau}(\frac{R_o}{r})^6\overline{N}(r,t) + \frac{1}{N_o(r)}\frac{\partial}{\partial r}[N_o(r)D\frac{\partial \overline{N}(r,t)}{\partial r}] \qquad (7)$$

where $N_o(r) = N*(r,o)$, $\overline{N}(r,t) = N*(r,t)/N_o(r)$ and D is the intramolecular segmental diffusion coefficient defined as RT/f, where f is a combination of conformational and frictional forces affecting the motion of chain segments.

The equation above has to be solved subject to the initial and boundary conditions N(r,o)=1 and $\overline{N}$(a,t)=o, the latter condition expressing the assumption that an acceptor approaching a donor to within some limiting distance a (a being much smaller than R_o) causes immediate quenching of the donor excited state. In Eq. 7, the first and second terms express the decrease in N*(r,t) due to spontaneous decay and energy transfer, respectively, whereas the third term expresses the changes in the interchromophoric distances due to Brownian motion. D refers to the diffusional flux of one labeled site of the chain through the surface of a sphere of radius r whose center is the origin of the coordinate system; the other labeled site being conceptually assumed to be at the center of the sphere. This treatment applies to conformational transitions which are stochastic in nature. This seems to be the case for the elementary steps of most transitions. The fluorescence decay curve of the donor is readily obtained from $N^*(r,t)$ or $N(r,t)$ by the relationship:

$$I_c(t) = k \int_a^{r_m} N^*(r,t)dr = k \int_a^{r_m} N_0(r)\overline{N}(r,t)dr \qquad (8)$$

where k is a proportionality factor and r_m is the interchromophoric distance of the fully extended conformation. τ and R_0 are obtained from independent measurements and calculation. $N_0(r)$ is proportional to $f(r)$ obtained independently from measurements of donor fluorescence decay kinetics in highly viscous medium. This leaves D as the only unknown parameter in the analysis of experiments in which significant Brownian motions are allowed (e.g., low viscosity). It is determined fitting $I_c(t)$ (Eqs. 7, 8) to the experimental decay curve by a reconvolution procedure. It should be noted that this method is useful in monitoring conformational changes occurring within the lifetime of the donor excited state. Proper choice of the donor and manipulation of the solution environment allow measurements in the range of 1 to 100 nsec. Slower motions should be monitored by a different approach, as outlined in the next section.

2.4. Intramolecular Dynamics of Chain Molecules Monitored by Fluorescence Intensity Fluctuations

The method described above is effective only if there are appreciable conformational transitions in the molecule during the lifetime of the donor excited state. Thus, many problems of interest, such as chain dynamics in highly viscous solvents or in polymer melts, as well as slow structural fluctuations in folded biopolymers are not approachable by the above method. In order to extend the applicability of energy transfer measurements to the study of slow molecular movements, a different approach has been taken. This involves the analysis of intensity fluctuations of fluorescence emitted from an ensemble of a small number of chain molecules that carry donor-acceptor pairs. The idea underlying this method is as follows. When a collection of labeled chain molecules is excited by light of steady intensity, the contribution of each molecule is proportional to its instantaneous interchromophoric distance. The overall instantaneous fluorescence intensity emitted by the donor chromophores is the sum of all individual contributions. For an infinitely large ensemble of molecules, the observed instantaneous fluorescence intensity is the value expected for summation over all fractional contributions interchromophoric distances and does not vary with time. When a small (constant) number of molecules whose conformation is frozen is illuminated under the same conditions, a constant emission at low intensity will be observed. However, when the molecules in the small illuminated ensemble are free to undergo conformation changes, their instantaneous interchromophoric distances distribution function may deviate from equilibrium, with concomitant fluctuations in donor fluorescence intensity. The magnitude of these fluctuations relative to the time average intensity would obviously become larger as the sample size becomes smaller. The duration of each intensity fluctuation reflects the time scale on which the conformational changes in the labeled molecules take place. Hence, the temporal behavior of the intensity fluctuations contains the information about the kinetics of intramolecular dynamics of macromolecules.

The time scale of intensity fluctuations is determined by the time dependence of the autocorrelation function, $AC(\tau)$, of the fluorescence intensity defined by:

$$AC(\tau) = (1/T) \int_{O}^{T} I(t) \cdot I(t + \tau)dt \qquad (9)$$

where $I(t)$ and $I(t + \tau)$ are the instantaneous donor fluorescence intensities at time t and $t + \tau$, respectively. For long integration intervals T, $AC(\tau)$ averages out intensity fluctuations that are completely random, but retains information about processes that change systematically with time. This information may be obtained from experiments, provided that $AC(\tau)$ can be related to $I(t)$ and that $I(t)$ can be related to molecular parameters. This is done in the following pages in which an analysis of the fluctuations in donor fluorescence intensity is presented for the case of intramolecular energy transfer, followed by numerical calculations of the magnitude of fluctuations expected for realistic cases. As will be shown, the intensity fluctuations are expected to be large enough to render the present approach experimentally feasible.

It is convenient to calculate the autocorrelation function of fluorescence intensity for a <u>single</u> molecule. A study of the emission of a single molecule is, of course, experimentally not feasible. Nevertheless, there is a simple relationship between the autocorrelation function $AC(\tau)(n)$ of the intensity of $I(t)(n)$ emitted by n molecules and the autocorrelation $AC(\tau)$ of the intensity $I(t)$ emitted by a single molecule, provided that the intensity fluctuations of each molecule are not correlated with those of other molecules. The relationship is derived as follows (17):

$$I(t)(n) = \sum_{i=1}^{n} I_i(t), \qquad (10)$$

where $I_i(t)$ is the intensity emitted at time t by the <u>i</u>th molecule. $Ac(\tau)(n)$ is thus given by

$$AC(\tau)(n) = \frac{1}{T}\int_0^T I(t)(n) \cdot I(t + \tau)(n)dt$$

$$= \frac{1}{T}\int_0^T \sum_{i=1}^n I_i(t) \sum_{j=1}^n I_j(t + \tau)dt$$

$$= \frac{1}{T}\int_0^T \sum_{i=1}^n I_i(t) \cdot I_i(t + \tau)dt$$

$$+ \frac{1}{T}\int_0^T \sum_{\substack{i=1 \\ i \neq j}}^n \sum_{j=1}^n I_i(t) \cdot (t + \tau)dt. \tag{11}$$

For a collection of n identical molecules the first term in Eq. 11 equals $n \cdot AC(\tau)$. The second term in this equation equals $n(n-1)<I>^2$ ($<I>$ being the average intensity of a single molecule) because it is assumed that the intensities of the different molecules are not mutually correlated. Equation (11) therefore reduces to

$$AC(\tau)(n) = n \ AC(\tau) + n(n - 1)<I>^2. \tag{12}$$

Equation (12) thus permits us to deduce the autocorrelation function of the intensity emitted by a collection of n molecules if the auto-correlation function for a single molecule is known. The subsequent calculations will aim at the analysis of the fluorescence intensity fluctuations of a single molecule. $AC(\tau)$ will be analyzed for the case involving intramolecular energy transfer in terms of macromole-cular dynamics. The analysis given (17) is carried out as follows. Let us start with a molecule whose end-to-end distance is ρ at $t = o$. (For simplicity, we assume that the donor and the acceptor chromo-phores are attached to the ends of a segment within a chain of mole-cules.) Let $N(\rho,r,\tau)dr$ denote the probability of finding this mole-cule at time τ with an end-to-end distance in the range r to r+dr. $N(\rho,r,\tau)$ is obtained by solving the equation that governs the beha-vior of $N(r,\tau)$, subject to the initial conditions $N(r,o) = \delta(\rho-r)$, where $N(r,\tau)dr$ is the fraction of molecules in the ensemble of poly-mer chains that have an end-to-end distance r at time τ. Following the same arguments given in the derivation of Eq. 7 (16), a similar equation is obtained for the change of $N(r,\tau)$ on the correlation time scale (17),

$$\frac{\partial \overline{N}(r,\tau)}{\partial \tau} = D \ \frac{1}{N_o(r)} \cdot \frac{\partial}{\partial r} \left| N_o(r) \cdot \frac{\partial \overline{N}(r,\tau)}{\partial r} \right| \tag{13}$$

where $N_o(r)$ is the equilibrium distances distribution in the sample, $\overline{N}(r,\tau) = N(r,\tau)/N_o(r)$ and D is defined above (Eq. 7) (16). $N(\rho,r,\tau)$ is obtained from solution of Eq. 13 subject to the initial conditions, $N(r,o) = \delta(\rho-r)$.

Let us consider a molecule which starts at a conformation having an end-to-end distance ρ at $t=0$, its probability of occurrence is proportional to $N_0(\rho)d\rho$ and its intensity of donor emission is proportional to $\rho^6/(\rho^6+R_0^6)$. At time τ, its contributions will be a summation of the probability of its transition into other conformations having an end-to-end distance r, $N(\rho,r,\tau)$ given by the solution of Eq. 13, multiplied by the relative quantum yield of the corresponding conformers, which is a function of their end-to-end distance, r: $I(\rho,r) = r^6/(r^6+R_0^6)$. If we define $AC_\rho(\tau)d\rho$ as that molecule's contribution to the autocorrelation function, we obtain:

$$AC_\rho(\tau)d\rho = N_0(\rho)d\rho\frac{\rho^6}{\rho^6+R_0^6}\int_0^{r_m} N(\rho,r,\tau)\frac{r^6}{r^6+R_0^6}\,dr \qquad (14)$$

Finally, integration over all initial conformations yields $AC(\tau)$;

$$AC(\tau) = \int_0^{\rho_m} AC_\rho(\tau)d\rho = \int_0^{\rho_m} N_0(\rho)\frac{\rho^6}{\rho^6+R_0^6}\int_0^{r_m} N(\rho,r,\tau)\frac{r^6}{r^6+R_0^6}\,drd\rho \qquad (15)$$

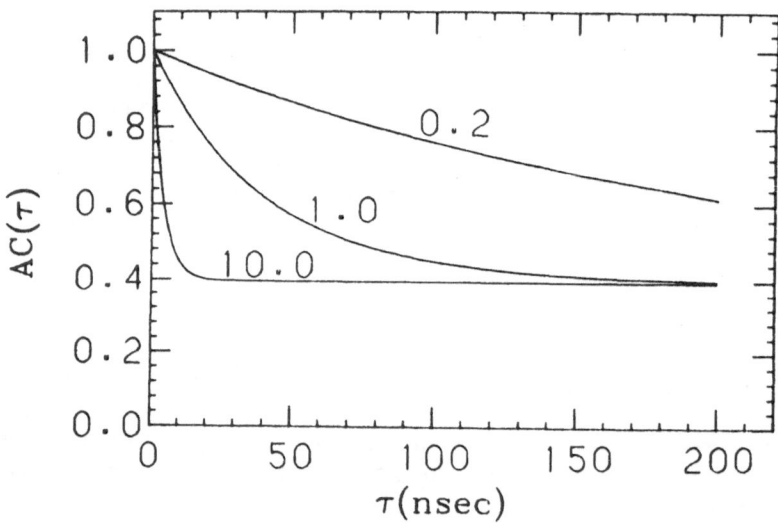

Figure 2. A simulation of the autocorrelation function, $AC(\tau)$, of the donor fluorescence calculated for different diffusion coefficients of one molecular end relative to the other end. The calculations were performed for a model oligopeptide whose end-to-end distance distribution function is given in Fig. 5, Ref. 15 (curve 8). R_0 was assumed to be 25Å. The time scale is given in units of 10^{-6} sec and each curve is marked by the value assumed for the intramolecular diffusion coefficients in units of 10^{-10} cm^2/sec. $AC(\tau)$ is given in arbitrary units. (Reprinted with permission from Ref. 17).

Equation 15 gives $AC(\tau)$ in terms of macromolecular dynamics and thus sets a basis for application of this approach. In terms of the dynamics of the chain molecule, the interpretation of $AC(\tau)$ (Eq. 15) should be straightforward. This is because: (a) the dynamics has been characterized by a single adjustable parameter, D, (b) the fluorescence intensities are well-defined functions of distance between the chromophores, and (c) the equilibrium distribution function, $N_0(r)$, can be obtained by independent experiments, as has been shown above.

To illustrate the expected time course of the autocorrelation function, $AC(\tau)$, computer simulations were carried out based on parameters obtained experimentally for a series of oligomer chain molecules (15,16). By use of Eq. 15, $AC(\tau)$ was calculated for oligomer chain molecules having the distribution function $f(r)$ (depicted in ref. 15). Fig. 2 illustrates the effect of varying D on the shape of the autocorrelation function. Note that changing the value of D affects the time course of $AC(\tau)$ only by stretching or contracting the function along the time-coordinate; otherwise, the function remains the same. Figure 2 further shows the ratio of $AC(o)$ to $AC(\infty)$; the ratio is reasonably high and hence this simulation shows that the experimentally recorded time course of $AC(\tau)$ could be analyzed and interpreted in terms of macromolecular dynamics.

The two methods described above for the study of intramolecular dynamics of chain molecules, i.e., the method of fluorescence decay kinetics and the method of intensity fluctuations, are complementary. The two methods cover different time domains of molecular movements, the first being applicable to the study of changes occurring in the nanosecond time scale, and the second extending to movements slower than the time scale of the lifetime of excited state of the energy donor.

3. Applications

The applicability of the above approach to the study of intramolecular distances and dynamics has first been demonstrated for a series of flexible oligopeptides (10,15,16). Based on the results obtained for model compounds, the experimental and theoretical methods were applied in a study of protein folding and dynamics. In the following we shall review experiments done towards the investigation of protein denaturation and folding by fluorescence labeling and energy transfer measurements.

3.1. Determination of intramolecular distance distributions in a globular protein

The problem of understanding the mechanism of protein folding, briefly described above, is essentially a problem of decoding a code. This code governs the translation of the unidimensional information contained in the sequence of amino acids to the three-dimensional information expressed in the native structure. This code is undoubtedly very complex and many experiments (1,6,18) point to the importance of

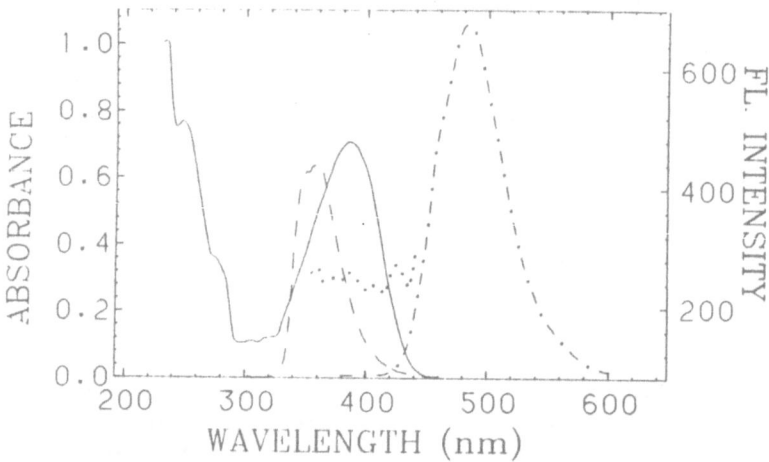

Figure   3.   Fluorescent
probes used for labeling
BPTI and RNase.

the pathway of folding in formation of the native structure within
surprisingly short times.

The theoretical and experimental approach outlined in the previous sections are suitable for investigation of the mechanism of protein folding. The rationale is as follows: first, a series of globular
protein derivatives, each carrying a donor and an acceptor at well-
defined sites, are prepared. Each derivative is useful for monitoring
the distance between a pair of labeled residues at any time.   The
larger the number of derivatives, the more complete is the information.   The synopsis of data obtained by following the folding

Figure 4.   Spectral characteristics of the donor-acceptor pair of
chromophores used for labeling BPTI.   (----) Emission spectrum of MNA
(centered at 360 nm) overlaps the absorption spectrum of DA-coum
(recorded for N-ε-DA-coum-Lys[26]-BPTI) (————).   Emission of DA-
coum (centered at cf. 480 nm) was recorded for the same derivative
using excitation at 386 nm (-·-·-·-) (band width 2 nm), and
corrected using quinine sulfate in 1N H_2SO_4 as a reference.
Excitation polarization spectrum of the DA-coum fluorescence is shown
(····).   Note the spectral overlap and the large Stokes shift of
the acceptor.

transition of many derivatives can yield a detailed anatomy of the structure of the molecule along the pathway.

An essential element in this program is the preparation of pure protein derivatives site specifically labeled by a donor and acceptor and their characterization. Several methods were developed for specific labeling of globular proteins and applied in the preparation of fluorescent derivatives of bovine pancreatic trypsin inhibitor (BPTI) and ribonuclease (RNase). Methoxy-naphthyl derivatives are used as donor chromophores, in particular 2-methoxy-1-naphthyl-methylenyl (MNA) and 2-naphthoxy-acetyl (NAA) were used as donor chromophores (Fig. 3). These two probes are characterized by spectral characteristics suitable for measurements by the present methods (Fig. 4). In particular, they both show monoexponential decay when attached to the protein in the absence of an acceptor under all experimental conditions tested so far. This is a key element in the analysis of fluorescence decay curves of donor-acceptor labeled derivatives, in which all deviations from monoexponential decay are interpreted as distributions of distance. 7-dimethylamino-coumarine-4-acetyl (DA-coum) or 2,4-dinitrophenyl residues were used as acceptors.

3.2. BPTI

The pancreatic trypsin inhibitor, BPTI, is a small single domain globular protein with elements of ß sheet and α helix and three disulfide bonds (Fig. 4). It is a reasonable model for investigation of the mechanism of folding of a globular protein. Procedures for the attachment of donor and acceptor probes to any one of the five amino groups of BPTI (four ε-amino groups of lysine side chains and the α-amino group of arginine-1 at the N-terminus) were developed (19-21). Purification methods based on affinity chromatography and high performance liquid chromatography (HPLC) were developed, as well as HPLC-tryptic peptide mapping for unequivocal determination of the labeled sites and degree of specificity of the labeling procedures (19,20,22). In the following we describe results obtained using one series of labeled BPTI: $N-\alpha-MNA-arg^1-N-\epsilon-DA-coum-lys^n-BPTI$ (n=15, 26,41,46) (abbreviated (1-n) BPTI) (23), i.e., each derivative carrying a donor label at the N- terminus and an acceptor at one of its lysine residues. All four BPTI derivatives retained full inhibitory capacity towards trypsin (except (1-15)BPTI which has a reduced inhibitory activity (20)). Samples of the three derivatives (1-n)BPTI (n=26,41,46) were denatured and reduced in 6M guanidinium hydrochloride (GuHCl) and 0.1M DTT. Each derivative restored its original stochiometric inhibition capacity towards trypsin after removal of the denaturant and air oxidation. This indicates that the probes attached to the amino group do not perturb the capacity of the labeled derivatives to refold to the active structure and, hence, they are suitable models for investigation of protein folding. The absorption spectra of all four derivatives were almost identical, showing the absorption bands corresponding to the protein, the donor and the acceptor. However, the excitation and emission spectra of the four derivatives showed wide variation. The emission spectra of (1-n)BPTI

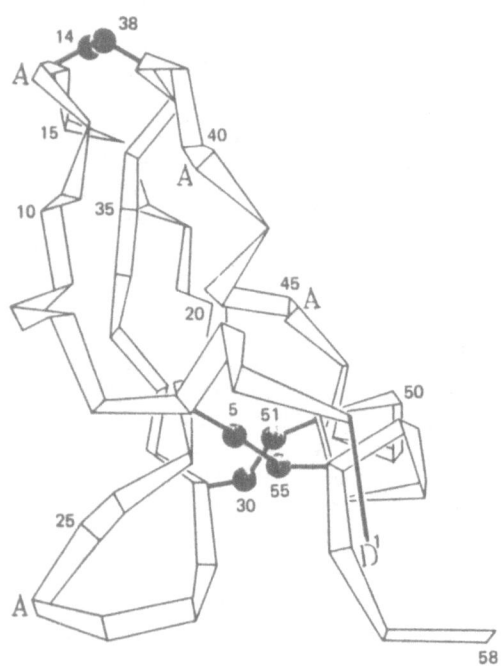

Figure 5. The folded structure of BPTI represented by
ribbon diagram of the polypeptide backbone. Sites
which were labeled by the
donor (D) and by the acceptor (A) are marked (after
Creighton (1).).

(n=15,26,41,46) are shown in Fig. 6. The two well-resolved emission
bands, at 360nm and 480nm, correspond to the emission of the excitation energy donor (MNA) and the acceptor (DA-coum), respectively. The
ratio of donor emission to acceptor emission found for (1-15)BPTI is
larger than that observed for (1-26)BPTI. The high transfer efficiency of the latter is expected from the known interresidue distance
calculated from x-ray crystallographic data (Table I).

Two additional spectra are shown in Fig. 6. The lowest curve
(broken line) is the emission spectrum of BPTI labeled by the acceptor, without a donor, in which DA-coum was attached to the amino group
of lysine residue 41. No emission is observed at the 360 nm band.
The spectrum marked S is the emission spectrum of MNA-DA-coum-lys
(see below).

A quantitative estimation of average transfer efficiencies is
obtained from acceptor-excitation spectra. We have selected for this
study a pair of probes which is characterized by well-resolved
emission bands and, hence, one can follow the emission of the acceptor
in a spectral range in which donor emission is negligible (> 480nm,
see Fig. 4). The large Stokes shift of the acceptor emission allows
direct determination of transfer efficiencies from acceptor excitation
spectra. It should be noted that in this mode of transfer efficiency
measurements one relies on analysis of the shape of spectrum alone
independent of any additional input, such as concentration or absorbances required for analysis of transfer efficiencies from donor
emission spectra. Thus, the shape of the excitation spectrum is an
instrinsic property of the system dependent of variation of transfer

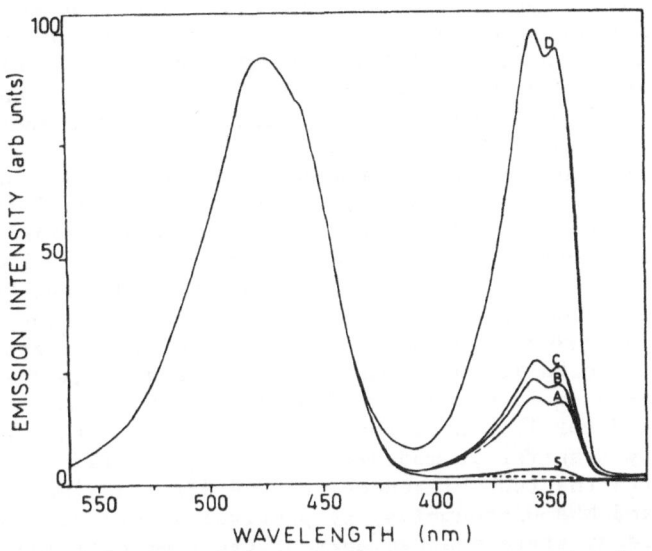

Figure 6. Fluorescence emission spectra (uncorrected) of N-ε-lys^{41}-
DA-coum-BPTI (-----), a derivative labeled by the acceptor only, and
(1-n)BPTI (n=41,26,46 and 15) labeled by a donor at the N-terminal
amino group and the acceptor at ε-amino groups of lysine residues 41,
26,46 and 15, curves A through D respectively. Excitation wavelength
295nm, excitation bandwidth 12nm, emission bandwidth 2nm. All spectra
were recorded on a Perkin-Elmer MPF-44 spectrofluorometer at room temperature and normalized to the same intensity at 480nm. The emission
at 360nm is the donor fluorescence; it is considerably higher for
(1-15)BPTI which has a larger interchromophoric distance and, hence,
lower transfer efficiency. The curve marked S is the emission
spectrum of a model compound MNA-DA-coum-lys, which is characterized
by 97% transfer efficiency.

efficiency only. This is so since all measurements are made under the
same setting of the emission monochromator; no wavelength dependent
variation in the detector sensitivity can affect the results and
quenching effects also do not affect the spectrum unless affecting the
transfer efficiency as in the case of quenching of donor emission and
thus, changing R_o. Quantitative analysis of excitation spectra is
simpler when spectra are compared to those of a derivative characterized by zero transfer efficiency and a derivative known to have a
constant high transfer efficiency. The first derivative is readily
available, any DA-coum-BPTI derivative supplies the "baseline" of
direct acceptor excitation in the spectral range of donor absorption.
As shown in Fig. 4 the pair of probes used here are ideal from this
point of view since in the range of 290-340 nm, which is between the
main absorption bands of the protein and the acceptor, the acceptor
absorption is relatively low, with a "window" free for detection of
donor excitation. The second derivative was prepared by attaching

both probes to each one of the two amino groups of lysine. Curve S in Fig. 6 shows that only very low intensity of emission is monitored in the donor emission spectral band for this compound. Fluorescence decay measurements show that this residual emission is characterized by a lifetime of 0.2 nsec. Thus, transfer efficiency in this reference compound is 97±2 %.

An essential control should be done in order to make sure that in the concentration range of dilute solutions of BPTI used in the present study, no intermolecular excitation energy transfer interferes with our measurements. This is done simply by recording acceptor excitation spectrum for a solution containing $(DA-coum)_2$-BPTI and $(MNA)_2$-BPTI at concentrations $10^{-6}M$ to $10^{-5}M$ with an excess of the donor labeled protein (five-fold relative to the acceptor labeled protein). The excitation spectrum of the acceptor recorded in the above experiment was identical to that obtained for DA-coum-BPTI in the absence of MNA-BPTI. Hence, this experiment shows that only intramolecular energy transfer is measured for the (1-n)BPTI derivatives.

Equipped with the two references, 0% and 97% transfer efficiencies (DA-coum BPTI and MNA-DA-coum-1ys, respectively), one can approach the quantitative steady state measurements of transfer efficiencies. This is achieved by recording the acceptor excitation spectra. In Fig. 7 are shown the excitation spectra of (1-15)BPTI (B), (1-41)BPTI (A), along with the excitation spectra of the two references, all recorded at an emission wavelength of 480 nm under identical conditions, though at different concentrations. Comparison of excitation spectrum of MNA-DA-coum-lysine with that of DA-coum-BPTI clearly shows the contribution of the excitation of donor to the emission of the acceptor. The excitation bands at 285, 295, 323 and 338 nm are characteristic of MNA excitation or absorption spectra. All spectra were normalized to equal emission intensity when excited at 382 nm (at this wavelength, there is no absorption by the donor). Once normalized, one can measure the difference between the excitation spectrum of any doubly labeled derivative, and the DA-coum-BPTI (zero transfer). The ratio of integrated area under this difference spectrum in the range of 285 to 338 nm to the difference spectrum calculated by the same procedure for MNA-DA coum-lysine (multiplied by 100/97) gives quantitatively the excitation energy transfer efficiency.

The same clear differences in transfer efficiencies between (1-41) BPTI and (1-15)BPTI found in the emission spectra appear in the excitation spectra. In (1-41)BPTI (Fig. 7), a more efficient excitation energy transfer takes place; this is interpreted as a shorter average interchromophoric distance. Fig. 8 shows the excitation spectra of (1-46)BPTI (A) and (1-26)BPTI (B) in the native state in aqueous neutral buffer.

Intramolecular distances determined from steady state measurements are average distances. These are simple average quantities only when all molecules have exactly the same distance. A more quantitative analysis of excitation energy transfer in flexible molecules can be achieved by analyzing donor fluorescence decay kinetics and the deconvolution procedure described above.

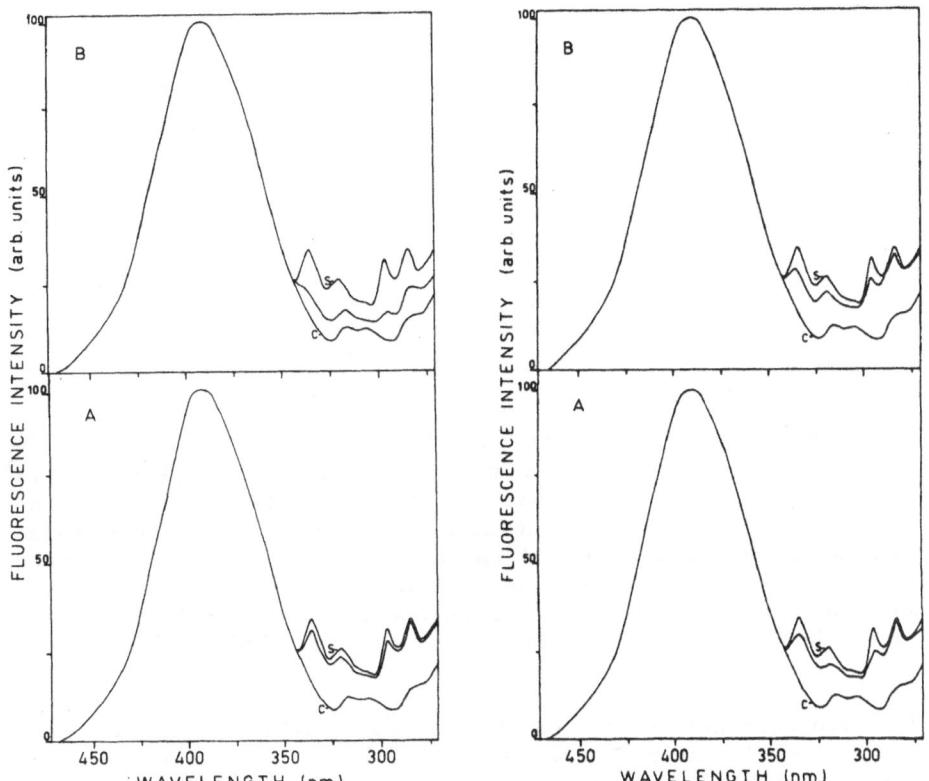

Figure 7 (left). Acceptor fluorescence excitation spectra (uncorrected) (emission wavelength 480nm) of BPTI derivatives and reference
derivatives of 97% and 0% excitation transfer. The spectrum marked C
is that of N- ε -DA-coum-lys[41]-BPTI, a derivative without donor and
hence 0% transfer. The spectrum marked S is that of MNA-DA-coum-lys,
a reference compound with 97% transfer efficiency. The unmarked
traces are those recorded for (1-41)BPTI in A and (1-15)BPTI in B
under identical conditions. All spectra were recorded at room temperature and normalized to constant intensity at 380nm, where no
absorption by the donor takes place. The derivatives (concentration
less than 2×10^{-6}M) were dissolved in 0.05M Bicine buffer pH 7.5
with 0.1M NaAc. Excitation bandwidth 2nm, emission bandwidth: 12nm.
The excitation bands at 285, 295, 323 and 338 are characteristic for
MNA, the donor. These spectra give a direct demonstration of the
acceptor fluorescence contributed by donor absorption and excitation
energy transfer.

FIGURE 8 (right). Acceptor excitation spectra of (1-26)BPTI (A) and
(1-46) BPTI (B). Same conditions and designations as in Fig. 7.

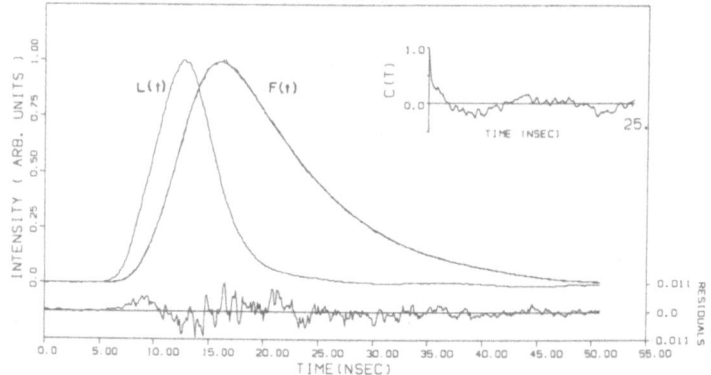

Figure 9. Fluorescence decay of the donor, MNA, in MNA-BPTI in 50% glycerol in 0.05M Bicine buffer, pH 7.5 at -30°C. The experimental curve was fitted to a monoexponential decay function. The shortest pulse, L(t), is the trace of the excitation pulse. The second pulse, F(t), shows the experimental and calculated donor fluorescence decay curves. The lower inset shows the deviation between the two curves, and the upper right inset shows the autocorrelation of the residuals. These show that the decay rate is monoexponential (RMS of fit 0.0030).

The fluorescence decay of the labeled BPTI derivatives were measured using the sampling method (24,25). In the absence of an acceptor, MNA-BPTI has very close to monoexponential decay with a lifetime of 6.8 ± 0.1 nsec and 8.1 ± 0.15 nsec in the same buffer but with 50% glycerol at -30°C (Figure 9). Only a very small, almost random deviation between the calculated and experimental traces is observed (less than 2%). This deviation sets the limits of significance of analysis of energy transfer measurements in terms of distribution of distances. The four double-labeled derivatives (1-n)BPTI (n=15,26,41, 46) had a shorter lifetime, with a relatively small deviation from monoexponential decay. Figure 10A shows the fluorescence decay of the donor chromophore, MNA, in (1-15)BPTI in aqueous solution at neutral pH. The experimental curve was fitted to a monoexponential decay function with an average lifetime of 3.9 ± 0.1 nsec. The deviation from monoexponential decay is relatively small; the deviations between the calculated and experimental traces are small and the root mean square (RMS) deviation is 0.0050. This small deviation means that a relatively narrow distribution of distances between the labels exists in the protein. The donor fluorescence decay curves for each derivative were analyzed with a least squares procedure (Fig. 10B) using Eq. 6 for calculation of $I_c(t)$, and a skewed Gaussian expression for f(r).

$$f(r) = 4 \pi r^2 \exp[-a(r-b)^2].$$

This expression was derived for describing the end-to-end distance distributions of flexible chain molecules (26). This function is

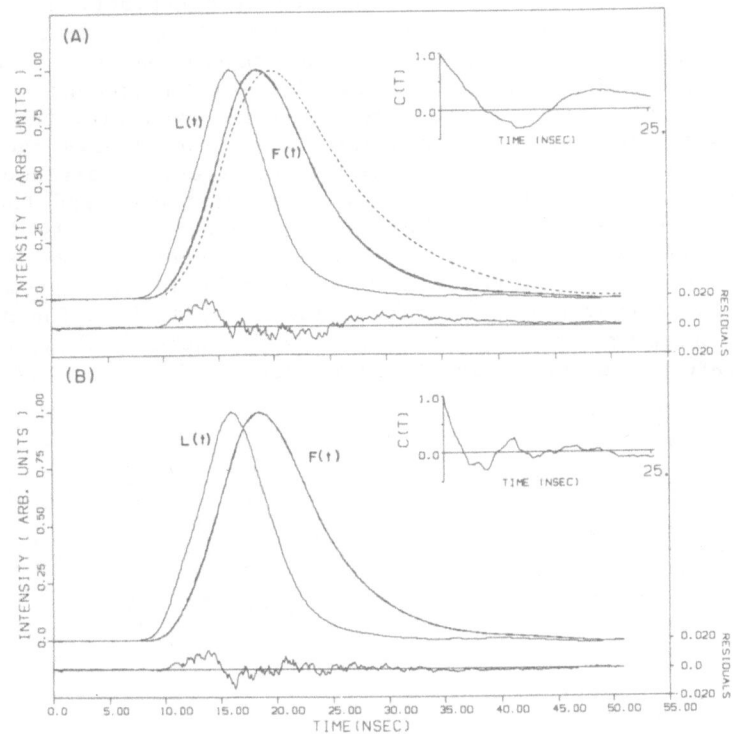

Figure 10. Fluorescence decay of the donor, MNA, in mono- and double-labeled BPTI. $\underline{A}$. The fluorescence decay of the donor in (1-15)BPTI fitted to a monoexponential decay function. The shortest pulse (L(t)) is the trace of the excitation pulse. The intermediate pulse, (F(t)), shows the experimental and calculated donor fluorescence decay curves. The lower inset shows the deviation between the two curves, and the upper right inset shows the autocorrelation of the residuals (RMS of fit 0.0050). These show that the decay rate is not monoexponential, but that deviation is relatively small. The fluorescence decay of MNA in the absence of an acceptor (monoexponential), with a lifetime of 6.8 ± 0.1 sec) is traced (----), showing the extent of decrease of donor lifetime by energy transfer. $\underline{B}$. The same decay curve shown in A was reanalyzed using Eq. 6. The best fit obtained with the parameters a = 0.043 and b = 30.5 is improved compared to the monoexponential fit shown in $\underline{A}$. RMS of fit equals 0.0035. The corresponding distribution function is shown in Fig. 11.

flexible enough to adopt most experimental distributions, within the experimental noise, by variation of the two free parameters a and b. R_o, the Förster critical distance, was calculated from the spectral characteristics of the probes (using Eq. 2); in aqueous buffers this is 29.5±1 Å (23). In Fig. 10B, note the improved fit, the randomness of the deviations, low autocorrelation of the residuals and the lower RMS, which was 0.0035. The resulting interresidue distances distributions are shown in Fig. 11; the average distances and the width at half height of the distributions are presented in Table I. Table I also gives the distances between the amino groups which are carrying the two labels in each derivative as calculated from atomic coordinates of the nitrogen atoms derived from x-ray diffraction data (27).

The agreement between transfer efficiencies calculated by the steady state measurements and the fluorescence kinetics measurements

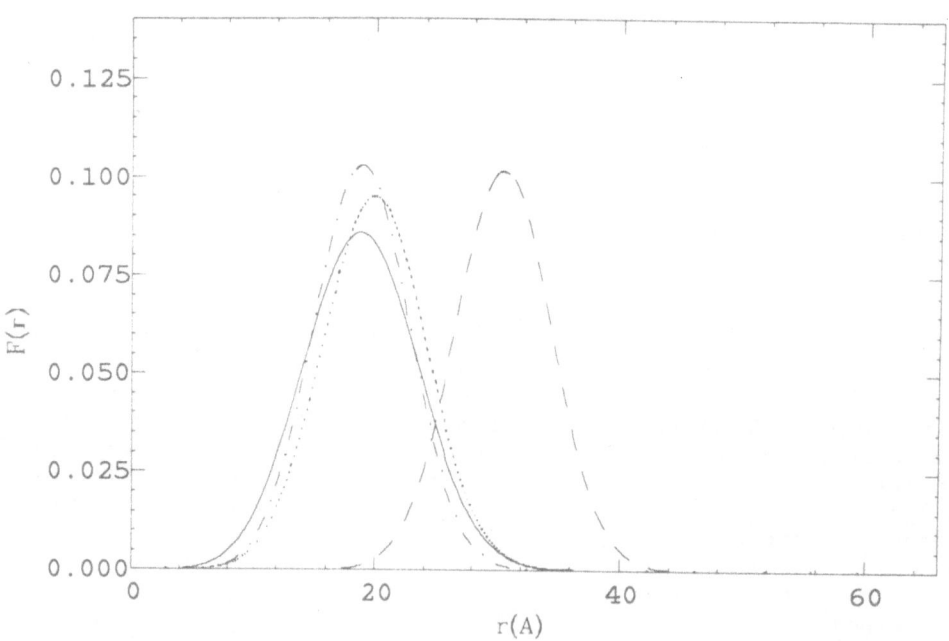

Figure 11. Normalized interchromophoric distances distributions calculated directly from donor fluorescence decay curves of the four derivatives: (1-n)BPTI (n=15,26,41,46) (- - -) (1-15)BPTI, (·—·—·—) (1-26)BPTI, (————) (1-41)BPTI and (····) (1-46)BPTI.

TABLE 1

Intramolecular distances of (1-n)BPTI (n=15,26,41,46) determined by nonradiative energy transfer

Derivative[a]	E[b]	Wi[c](A)	Rm[d](A)	Rx[e](A)
1 - 15	0.45 ± 0.05	8 ± 2	31 ± 1	31.7
1 - 26	0.78 ± 0.05	9 ± 2	20 ± 1.5	16.8
1 - 41	0.80 ± 0.05	10 ± 2	19 ± 1.5	19.1
1 - 46	0.75 ± 0.05	9 ± 2	21 ± 1.5	21.7

(a) Each BPTI derivative was labeled by a donor on the α-amino group of Arg^1 and an acceptor at the lysine residue indicated.
(b) Transfer efficiency determined from excitation spectra;
(c) Width at half height of the interchromophoric distances distributions shown in Fig. 11.
(d) Average distance, $<R^2>^{\frac{1}{2}}$, for each distribution shown in Fig. 11.
(e) Average distance determined from x-ray crystallography data (27).

is an important control that supports the reliability of the interpretation of the decay curves in terms of distances distributions. The striking feature emerging from the comparison given in Table I is that the average distances determined by our method are very close to those determined from the crystal structure by x-ray crystallography. The relative differences between the distances between the four pairs of labeled sites are reproduced, the distance between residue 1 and 15 being by far the largest one. The other three distances are quite similar, with the distance between residues 1 and 46 being slightly 'larger than the rest, and the distance between the probe attached to lys-41 and arg-1 the shortest, as found in the crystalline state. Indeed, the centers of the probes, the distance beween which is measured by the transfer efficiency measurements, are removed from the labeled amino groups. However, the probes' rotational freedom compensates for this extension and, hence, the comparison given in Table I shows that the spectroscopic measurements reproduce quite closely the crystallographic distances.

The width of the calculated interchromophoric distance distributions is obtained from the extent of deviation from monoexponential decay of the donor. This, in turn, includes contributions from experimental noise; segmental flexibility of the protein and mostly from the rotational freedom of probe, the side chain and the "arm" connecting it to the main chain.

The results shown in Fig. 11 show an interchromophoric distribution narrower than the actual equilibrium distribution functions due to the effect of the dynamic rotation of the probes which enhance transfer efficiency, as discussed above. In high viscosity solution, a slightly wider distribution is observed. The width of the distributions observed in the native state are the lower limit of resolution of the present method in terms of measurements of conformational flexibility. Yet, the uncertainty in the determination of the average of each intramolecular distance distribution is considerably smaller than its width, since Eq. 6 properly weighs the contribution of each fraction of the distance distribution. The results presented above show that we have a reliable method for measuring intramolecular distances and set the limits of the resolution in monitoring the static flexibility of the protein chain. Based on these results, one can proceed to study the unfolding of the protein.

The five derivatives, MNA-BPTI and (1-n)BPTI (n=15,26,41,46) were dissolved in 6M guanidine hydrochloride in 50% glycerol solution in 0.1M bicine buffer pH 7.5 and their fluorescence decay measured at low temperature (-32°). MNA-BPTI gave again a monoexponential decay. Only small changes in average distances or width of distributions were observed for the double-labeled derivatives. When ß-mercaptoethanol was added, a dramatic change in excitation and emission spectra and fluorescence decay kinetics was observed. Fig. 12 shows the denaturation of (1-15)BPTI based on analysis of donor fluorescence decay curves measured in the above solvents at -32°C. Clearly, the average distance is decreased, showing that in the native state this segment of BPTI is quite stretched comared to a random coil conformation. The width of the distribution is increased two-fold (from 14Å to 27Å) a direct manifestation of the transition to an unordered conformation. Moreover, in the reduced state a skewed Gaussian distribution is observed. This shows that even under conditions considered fully denaturing, a considerable fraction of the molecules retain the stretched

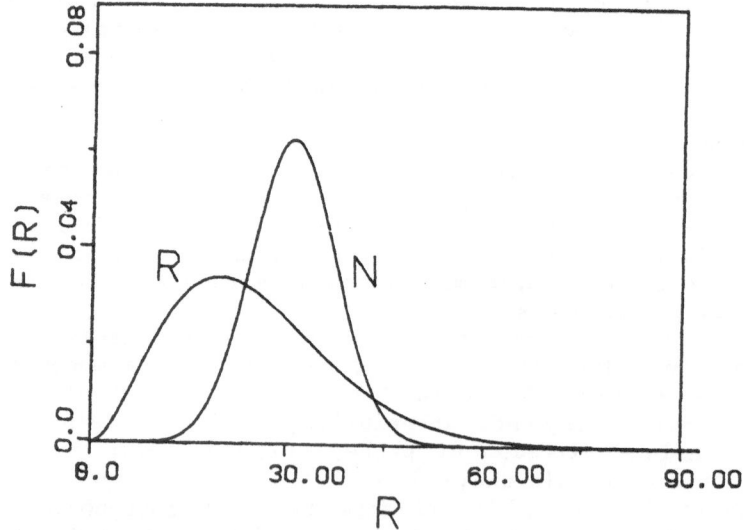

Figure 12. Distribution of distances betweeen residues 1 and 15 in BPTI in viscous solution (50% glycerol, -32°C) (N) and with added guanidine HCl and ß mercaptoethanol (R).

conformation in the N-terminal segment characteristic for the native conformation. On the contrary, the distance between residues 1 and 26 or 1 and 46 is increased upon reduction, with a transition to a wider distribution. Thus, the measurements described above yield very detailed information on the local structure of segments of the protein, in the denatured state. It seems that some residual structures exist in the N-terminal segment even in the denatured state.

3.3. Ribonuclease

Models proposed to account for the fast folding of this enzyme assume a sequential mechanism for the folding transition. Kinetic experiments (18) reveal at least two intermediates. Recent investigations (28,29) have suggested that the guanidine induced unfolding of RNase-A proceeds sequentially with the N-terminal region unfolding before the rate-limiting step. In order to obtain further information about the unfolding in this localized region of the molecule, and hence about its sequential unfolding, we have first labeled the N-terminal region of the enzyme. A non-fluorescent acceptor, 2,4-dinitrophenyl (DNP) was covalently attached to the N-terminal α-amino group, and then a fluorescent donor, ethylenediamine-monoamide of 2-naphthoxyacetic acid (ENA) was covalently linked in the vicinity of residue 50 (75% at Glu 49 and 25% at Asp 53) (22). The calculated R_0 (Förster critical distance) for this pair of chromophores in 50% glycerol at neutral pH is 36Å. The sites of labeling were determined by peptide mapping (22). The derivatives possessed full enzymatic activity and underwent reversible thermal unfolding transitions. The fluorescence decay of the donor, ENA, attached to RNase-A in the absence of an acceptor, is monoexponential under all solvent compositions and temperatures studied. The fluorescence decay of DNP-ENA-RNase (donor- and acceptor-labeled proteins) was measured in 50% glycerol in 0.1M Hepes pH 7.5 (native state) and in the same solvent with guanidine hydrochloride added to a final concentration of 6M (denatured state). A third measurement was carried out under the latter conditions with the addition of dithiothriethol (DTT) to a final concentration of 0.1M (reduced denatured state). All three states of the protein were measured at −32°C, a temperature at which these solutions become highly viscous. The intramolecular distance distributions calculated from the donor fluorescence decay kinetics (Eq. 6) are shown in Fig. 13 and Table II. The native state is characterized by a relatively narrow distribution, with width at half height of 13±2Å, very similar to the width found for the labeled BPTI derivatives in this solvent and temperature. This is mainly due to the rotational freedom of the probes. The denatured state is characcterized by an increased average lifetime. The corresponding calculated intramolecular distances distribution functions (Fig. 13) has an average of 56±3Å and width of 23±3Å in the denatured state which are further increased upon reduction to 60±4Å and 34±6Å, respectively, in the reduced denatured state. Clearly, these experiments monitor a transition to an unordered state of the N-terminal segment, with a more compact structure retained in the non-reduced state. It is of interest to note that Eq. 20 of

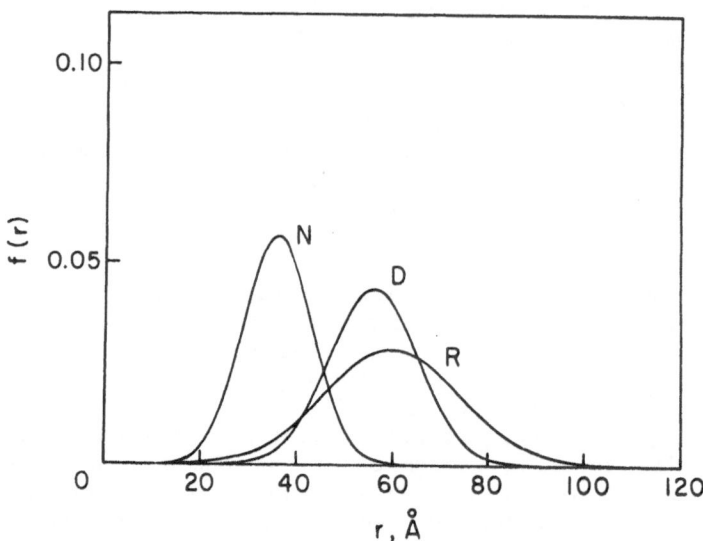

Figure 13. Donor-acceptor distance distribution functions in DNP-ENA-RNase calculated from donor fluorescence decay curves using Eq. 6 in three states: Native (N), denatured by guanidine HCl (D), and denatured and reduced by 0.1 M DTT (R).

TABLE II

Analysis of fluorescence decay curves of DNP-ENA-RNase[a]

State	τ (nsec)[b]	τ_{DA}(nsec)[c]	$R^d(\overset{\circ}{A})$	$W^e(\overset{\circ}{A})$
Native[f]	13.2 ± 0.1	7.3 ± 0.1	36 ± 2	13 ± 2
Denatured[g]	10.8 ± 0.1	9.8 ± 0.1	56 ± 2	23 ± 3
Reduced/denatured[h]	9.9 ± 0.1	9.1 ± 0.2	60 ± 4	34 ± 6

(a) All measurements were carried out at a low temperature, $-32°C$, i.e., high viscosity.
(b) Donor fluorescence lifetime in the absence of an acceptor.
(c) Donor fluorescence in the presence of an acceptor on the molecule analyzed in a model assuming monoexponential decay. τ_{DA} is reduced compared to τ.
(d) RMS distance between the labels.
(e) Width at half height of the interchromophoric distribution.
(f) In 50% glycerol in Hepes buffer 1 M pH 7.5.
(g) Same as (f) with added guanidine HCl to final concentration of 6 M
(h) Same as (g) reduced by 0.1 M DTT.

Tanford (30), which pertains to the unperturbed end-to-end distance of a polypeptide chain gives a value of $59\pm7A$ for a polypeptide of 50 residues, which is quite close to the above result. However, the latter is not the unperturbed distance (since guanidine hydrochloride is not a θ solvent) and the actual distance would be expected to be larger than the unperturbed distance. It should be kept in mind that in the reduced state, transfer efficiency is very low and the uncertainty in the distances distribution function is increased. Hence, this result should be discussed with caution.

The methods described above for the determination of intramolecular distance distributions are unique in their ability to monitor local conformations in chain molecules in exact structural terms. The measurements are carried out in dilute solutions over a wide range of temperatures and solvents (it might be possible to make these measurements in the solid state). The time resolution is limited by mixing methods, since any conformational transition slower than nanoseconds should be detectable. The shape of intramolecular distances distribution functions obtained by the analysis of fluorescence decay curves contains important information on the conformational state of the molecule. The experiments described above show that the methods developed are very useful in the study of protein folding, the conformation of flexible oligopeptides and should prove very useful in the study of conformational properties of polymers in solution and in melts. The two approaches developed for the study of conformational dynamics of chain molecules cover a very wide range of time scales. It has been shown experimentally that intramolecular diffusion is measured in flexible peptides and unfolded proteins. The same approach, using the same experimental setups and analytical tools, are useful for evaluation of static and dynamic flexibility of polymers.

ACKNOWLEDGEMENT. The work reviewed in this paper was promoted by indispensable contributions from my teachers, colleagues and students. In particular, I am grateful to Prof. E. Katchalski-Katzir and Prof. I.Z. Steinberg, who introduced me to the fascinating field of macromolecular structure and laid down the theoretical foundations from which this work has evolved. I am also grateful to my collaborators at Cornell University, Prof. H.A. Scheraga and Dr. C.M. McWehrther, with whom I have a continuous stimulating collaboration studying RNase folding. I am grateful to M. Wilchek, A. Tishbi, D. Amir, D. Levy, J. Levine, L. Varshavsky and R. August for their important contributions to this work. P.S. Kim has carefully read the manuscript and his comments helped clarify many points. Special thanks to Mrs. B. Lederhendler for the artful preparation of the layout of this paper. The gift of BPTI (Trasylol ®) by Bayer AG is gratefully acknowledged. The work reviewed here was supported by the U.S.-Israel Binational Science Foundation. An EMBO travel grant is gratefully acknowledged.

4. REFERENCES

1. Creighton, T.E. (1984) "Proteins," Freeman, San Francisco.
2. Gurd, F.R.N., Rothgeb, T.M. (1979) Adv. Prot. Chem. 33, 73.
3. Lakowicz, J.R. and Weber, G. (1973) Biochemistry 12, 4171.
4. Anfinsen, C.B. (1967) Harvey Lectures 61, 95.
5. Levinthal, C. (1968) J. Chim. Phys. 65, 44.
6. Creighton, T.E. (1984) Adv. Biophys. 18, 1-20.
7. Förster, Th. (1948) Ann. Phys. 2, 55.
8. Förster, Th. (1951) "Fluoreszenz Orgnicher Verbindungen," Vandenhoeck and Ruprecht, Götingen.
9. Förster, Th. (1959) Disc. Farady Soc. 27, 7.
10. Steinberg, I.Z., Haas, E., Katchalski-Katzir, E. (1983) In: "Time Resolved Fluorescence Spectroscopy in Biochemistry and Biology", R. B. Cundal and R.B. Dale (eds.), Plenum, p. 411.
11. Haas, E., Katchalski-Katzir, E., Steinberg, I.Z. (1978) Biochemistry 17, 5064.
12. Steinberg, I.Z. (1971) Ann. Rev. Biochem. 40, 83.
13. Stryer, L. (1978) Ann. Rev. Biochem. 47, 819.
14. Grinvald, A., Haas, E., Steinberg, I.Z. (1972) Proc. Natl. Acad. Sci. USA 69, 2273.
15. Haas, E., Wilchek, M., Katchalski-Katzir, E., Steinberg, I.Z. (1978) Proc. Natl. Acad. Sci. USA 72, 1807.
16. Haas, E., Katchalski-Katzir, E., Steinberg, I.Z. (1978) Biopolymers 17,
17. Haas, E., Steinberg, I.Z. (1984) Biophys. J. 46, 429.
18. Kim, P., Baldwin, R.L. (1982) Ann. Rev. Biochem. 51, 459.
19. Amir, D., Varshavski, L., Haas, E. (1985) Biopolymers 24, 623.
20. Amir, D., Haas, E. (1985) Int. J. Pept. Prot. Res. (in press).
21. Amir, D., Levy, D.P., Levin, J., Haas, E. (1985) (submitted).
22. McWerther, C.A., Haas, E., Leed, A.R., Scheraga, H.A. (1985) (submitted).
23. Amir, D., Haas, E. (1985) Biopolymers (in press).
24. Hundley, L., Coburn, T., Garwin, E., Stryer, L. (1967) Rev. Sci. Instrum. 38, 488.
25. Hazan, G., Grinvald, A., Maytal, M., Steinberg, I.Z. (1974) Rev. Sci. Instrum. 45, 1602.
26. Edwards, C.F. (1965) Proc. Phys. Soc. London 85, 613.
27. Wlodawer, A., Walter, J., Huber, R. and Sjolin, L. (1984) J. Mol. Biol. 180, 301.
28. Lin, S.H., Konishi, Y., Denton, M.E., Scheraga, H.A. (1984) Biochemistry 23, 5504.
29. Lin, S.H., Konishi, Y., Nall, B.T., Scheraga, H.A. (1985) Biochemistry (in press).
30. Tanford, C. (1968) Adv. Prot. Chem. 23, 121.

ROTATIONAL DYNAMICS OF BIOLOGICAL MACROMOLECULES

Edward Blatt and Thomas M. Jovin
Department of Molecular Biology
Max Planck Institute for Biophysical Chemistry
D-3400 Göttingen
Federal Republic of Germany

ABSTRACT. The rotational diffusion of biological macromolecules is very sensitive to shape, size, and environmental constraints. Large scale and global motions occur in the μs-ms time domain, as in the case of intrinsic membrane proteins. We describe methods for measuring the rotational motion of macromolecules in free, liganded, and cell-surface associated states. They are based on long-lived photophysical processes and the detection of a polarized emission process: phosphorescence, delayed fluorescence, ground state depletion (monitored by fluorescence), and fluorescence from transient intermediates. In every case, the signals derived from different combinations of excitation and emission polarization states are used to construct a time-resolved polarization (anisotropy) function in addition to the primary decay curve. Thus, lifetimes and rotational correlation times can be determined, generally by a multi-component exponential analysis.

1. INTRODUCTION

Biological macromolecules-proteins, nucleic acids, carbohydrates and lipids-exhibit two prominent characteristics: they are large and are generally associated into complexes. It is the latter property which often confers a high degree of specificity upon biological systems. Functional molecules which exhibit catalytic activity or otherwise participate in signaling and metabolic cellular processes must engage in some form of motion. Due to different levels of structural organization-primary, secondary, tertiary, and quaternary-the dynamics of macromolecular assemblies occur in different spatio-temporal domains. Thus, motions about single bonds of intrinsic or extrinsic chromophores characteristically

351

M. A. Winnik (ed.), Photophysical and Photochemical Tools in Polymer Science, 351–370.
© *1986 by D. Reidel Publishing Company.*

take place in the ps-ns domain, whereas segmental motions
may occur in nanoseconds. Global dynamics, especially of
long linear molecules such as DNA and membrane-associated
proteins extend into the μs-ms region. In general, one can
anticipate mixed behavior with contributions from concerted
and coupled torsional, bending, and tumbling modes. Their
relative magnitudes will depend upon the degree to which
the molecule in question senses a predominantly aqueous
environment or a highly retarding medium. For example,
biological membranes feature a lipid bilayer structure with
an operative *microviscosity* about 100 times the viscosity
of water (1). A further characteristic of biological
membranes is a pronounced degree of structural anisotropy.
Thus, non-spherical molecules respond to constraints
imposed by such surroundings by showing strong
orientational preferences.

The above properties and phenomena can be assessed with
great sensitivity and precision by the measurement of
rotational diffusion, usually based upon the combined use
of polarized excitation and deactivation processes. The
faster motions alluded to above are particularly well
adapted to the techniques of nuclear magnetic relaxation
and fluorescence depolarization, the formalisms for which
are extensively documented (references 2 and 3,
respectively, and citations therein; other chapters in this
volume). Optical anisotropy decay measurements with longer
time resolution have been very effective in studies of
biological and model membrane systems (reviewed in 4-6).
The requirement for a μs-ms spectroscopic "clock" implies
that suitable photophysical processes must be used.
Furthermore, in the case of inherently dilute systems such
as suspensions of living cells, sensitivity of detection is
a major consideration. In this respect, light emission
processes offer significant advantages compared to
measurements based upon light absorption. They include
delayed luminescence (phosphorescence, fluorescence)
involving the long-lived triplet state (7-12) and other
two-laser techniques to be described below.

Our aims in this chapter are to introduce the
principles underlying the methods based on delayed
luminescence and other photophysical mechanisms, to
describe the experimental designs enabling the measurement
of rotational diffusion, and to illustrate applications in
studies of the rotational properties of biological
macromolecules.

2. PHOTOPHYSICAL PROCESSES

2.1 Delayed Luminescence Anisotropy (**DLA**)

The delayed luminescence of chromophores originates from

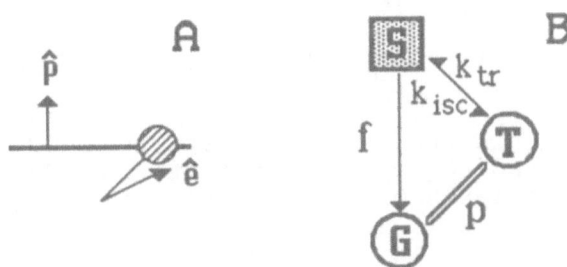

Figure 1. Delayed Luminescence Anisotropy (**DLA**). A) The
sample (lined sphere) is excited with polarized light
(vector $\hat{p}$). The emission is observed at 90° through an
analyzer set at an angle corresponding to vector $\hat{e}$). B) The
ground state **G** is excited to the singlet state **S** (square)
which decays directly (fluorescence **f**) or indirectly
(phosphorescence **p** or delayed fluorescence by thermal
reactivation to **S**) through the triplet state **T**. The circled
states are long-lived and the hollow line connecting **T** and
G denotes a slow decay process. Non-radiative modes are
omitted (see text).

the first excited triplet state T_1 after absorption of
light into the excited singlet state S_1 and intersystem
crossing to T_1 (Figure 1). Relaxation to the ground state
can proceed either by thermal reactivation to S_1 followed
by emission (delayed fluorescence) or by direct emission of
a photon (phosphorescence). Because the S_1- T_1 transition
is spin forbidden, triplet state emission processes are
long-lived in comparison with direct fluorescence, and
lifetimes range from μsec to seconds, depending on
experimental conditions such as temperature (13), solvent
composition and concentration of quenching molecules (e.g.,
oxygen). Triplet quantum yields and lifetimes are
particularly sensitive to the presence of external or
internal heavy atoms, accounting for the extensive use of
halogenated chromophores. Since T_1 is lower in energy than
S_1, phosphorescence spectra are red-shifted relative to the
corresponding fluorescence spectra. (The properties of the
triplet state are reviewed in 14).
   From the various deactivation pathways, the following
expressions may be defined for the prompt fluorescence and
delayed excited-state lifetimes, τ_1 and τ_2, respectively
(see also 8,9):

$$\tau_1 = (k_f + k_{nr}{}^S)^{-1}$$

$$\tau_2 = (k_p + k_{nr}{}^T + k_{tr}[1 - \Phi_{isc}])^{-1} \tag{1}$$

where k_f, k_p are the radiative rate constants and $k_{nr}{}^S$, $k_{nr}{}^T$ are the non-radiative rate constants for the deactivation pathways associated with the singlet and triplet states, respectively; k_{tr} is the rate constant for thermal reactivation of T_1 and intersystem crossing. Φ_{isc} is defined in Equation 4. At all times t,

$$\text{fluorescence intensity} \quad \propto \quad k_f S_1(t)$$

$$\text{phosphorescence intensity} \quad \propto \quad k_p T_1(t) \tag{2}$$

in which $S_1(t)$ and $T_1(t)$ are the fractional concentrations of excited state species. In experiments using pulse excitation of short duration (e.g. 10 ns) relative to the period of data collection, it is not necessary to deconvolute the delayed emission response function. Under these conditions, the luminescence decay for a single species is monoexponential:

$$S_1(t) = (k_{tr}/k_{isc})\Phi_{isc}{}^2 exp(-t/\tau_2)$$

$$T_1(t) = \Phi_{isc} exp(-t/\tau_2) \tag{3}$$

where k_{isc} is the rate constant for singlet-triplet intersystem crossing (Figure 1). The quantum yields associated with the various excited states and deactivation pathways may be defined as

$$\Phi_f = k_f \tau_1$$

$$\Phi_{isc} = k_{isc} \tau_1 \tag{4}$$

$$\Phi_p = k_p \tau_2$$

From the above, it can be seen that even in the simplest case, the time constants, amplitudes and yields are determined by an array of rate constants, the values of which are dependent on a number of parameters including solvent composition, temperature, and the nature of the

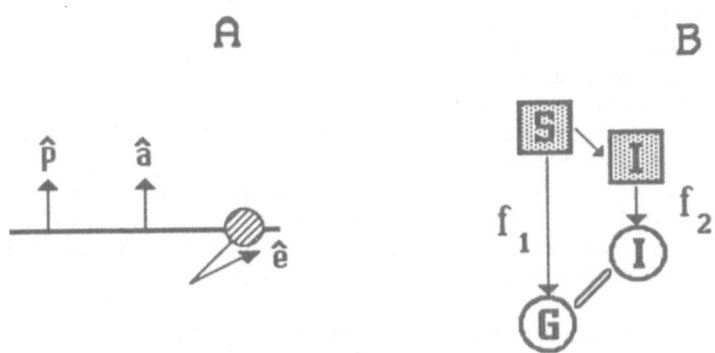

Figure 2. Fluorescence Recovery Anisotropy (**FRA**) and
Transient Fluorescence Anisotropy (**TFA**). A) To the scheme
in Figure 1 has been added a second monitoring light source
(vector **â**) oriented at an arbitrary angle to the
propagation direction. The steady-state fluorescence it
evokes is measured through the analyzer (**ê**). B) Expanded
photophysical scheme corresponding to the **TFA** measurement.
The excited single state **S** decays by fluorescence (f_1) or
by chemical conversion to an excited state intermediate
(stippled **I**) which rapidly deactivates to a long-lived
ground state species (circled **I**), giving off fluorescence
(f_2). The wavelength of the monitoring source **â** (panel A)
is chosen so as to excite the steady-state fluorescence of
intermediate **I**.

chromophore. The last factor is of particular importance in
delayed luminescence techniques and may be exploited to
enhance or diminish quantum yields of interest (see Section
4.1).

2.2 Fluorescence Recovery Anisotropy (**FRA**)

In the scheme of Figure 1, it can be perceived that the
population of the triplet state is necessarily accompanied
by a reciprocal depletion of the ground state. For certain
purposes, it is more convenient and sensitive to exploit
the latter process by monitoring the steady-state
fluorescence using a second continuous light source (Figure
2). As the triplet state decays, the ground state
"recovers" and the time course of its fluorescence

intensity and polarization provides information similar to
that derived from **DLA** but with two significant advantages:
(i) the sensitivity of the measurement can be greatly
enhanced due to the fact that excitation photons are
continuously pumping the system; and (ii) many common
fluorophores can be utilized as probes. This technique,
also known as *fluorescence depletion* (11,15,16) or
fluorescence recovery spectroscopy (17), has been applied
to observations of single or multiple cells in the
microscope by Garland and colleagues (11,15,16) and to
solutions and suspensions of cells (18).

2.3 Transient Fluorescence Anisotropy (**TFA**)

In addition to excited states, other photochemical
intermediates can be exploited for the purpose of
establishing long-lived photoselective probes of molecular
motion. In the case of suitable compounds, chemical
reactions proceed during the excited singlet state, thereby
establishing photochemical intermediates with desirable
properties. A classical example is intramolecular proton
transfer which proceeds in the ps time scale following the
dramatic change in pK accompanying the changed electronic
structure. Thus the photochromism of salicylideneaniline
(SAN), shown schematically in Figure 3, is characterized by
a *tran-cis* isomerization (ps), a large decrease in phenolic
pK, a fast (ps) tautomeric proton transfer (enol→quinoid),
and a slow (µs) ground state proton translocation (19,20).
In this case, the system is pumped in the near u.v. and the
intermediate monitored in the blue part of the spectrum
(see Figure 2).

2.4 Comparative Features of **DLA**, **FRA**, **TFA**

The above techniques differ greatly in important features
such as sensitivity and the potential for exploiting
intrinsic chromophores. A comparative analysis is given in
Table I (see also ref. 11). One should note at this point
that other methods for studying slow rotation of biological
macromolecules are available but will not be treated in
this chapter: i) linear dichroism using the triplet state
and intrinsic probes (4,21,22); ii) fluorescence recovery
after photobleaching (FRAP, 23); and iii) saturation
transfer electron paramagnetic resonance (ST-EPR, 24).

3. ROTATIONAL DIFFUSION

3.1 Photoselection and Anisotropy

The measurement of rotational diffusion is based upon the
creation of an oriented population of excited molecules by

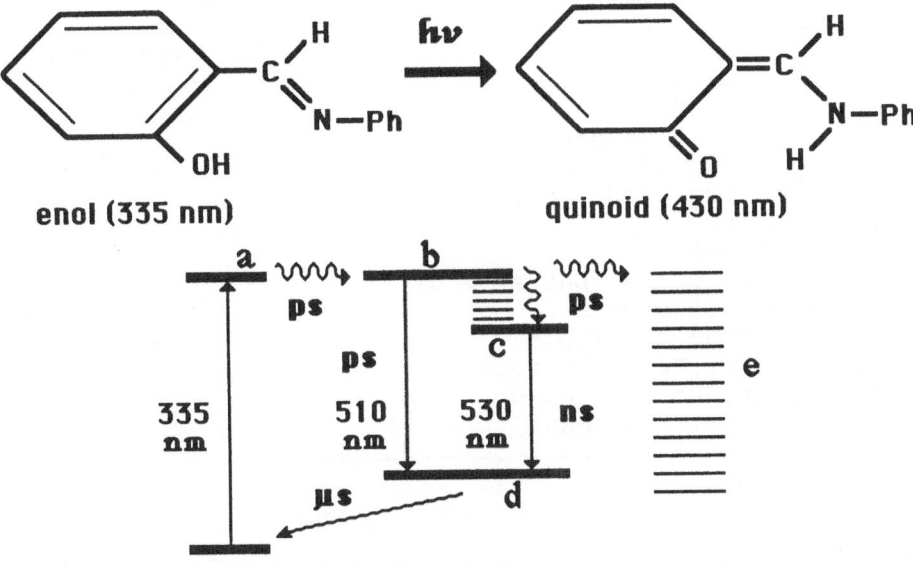

Figure 3. Photochromism of salicylideneaniline (SAN). Adapted from references 19,20. A number of photochemical species can form after the initial proton transfer reaction, the distribution of which depends upon solvent conditions and temperature: (**a**) excited enol; (**b**) excited quinoid; (**c**) excited state after solvent relaxation. The ns decay process shown is not observed at room temperature. Instead, species (**c**) is reactivated to (**b**) followed by emission to (**d**), the ground state quinoid; (**e**) other photochemical products. The intermediate (**d**) can be exploited in rotational studies because it has a lifetime in the μs range, one which is also very dependent upon environmental conditions (e.g. polarity, temperature).

optical selection from an initial distribution. The latter may be isotropic, as in the case of molecules or cells randomly oriented in solution or suspension. Alternatively, biological structures may be oriented on surfaces, as in typical microscope preparations. We will deal in this chapter exclusively with a priori globally random samples. Experimentally, photoselective excitation is achieved with plane-polarized light so that molecules with transition moments for absorption lying parallel or at a small angle to the electric vector of the incident light will be preferentially excited (the probability varies with the square of the difference cosine). Following excitation, the polarization of the emission or absorption decays as a

Table I. Comparative features of **DLA, FRA, TFA**

Feature	DLA	FRA	TFA
low quantum yield (phosphorescence)	•		
high sensitivity (fluorescence)	•	•	•
high sensitivity (2nd light source)		•	•
large anisotropy	(•)	•	•
long lifetime(s)		•	
dark reference state (better SNR)	•		•
insensitivity to oxygen			•
intrinsic probes available	•	(•)	•
interference of "prompt" emission	•	•	(•)
susceptibility to bleaching		•	•
critical optical alignment		•	

result of Brownian rotation and other motions. The emission anisotropy as a function of time may be defined in terms of the two components polarized parallel ($I_\parallel$) and perpendicular ($I_\perp$) to the excitation vector, according to the relation:

$$r(t) = [I_\parallel(t) - I_\perp(t)]/[I_\parallel(t) + 2I_\perp(t)] \qquad (5)$$

in which the denominator represents the total emission, a function we denote S(t). In the general case of an irregular rigid body (ellipsoid) rotating in an isotropic medium, the anisotropy decays as a sum of five exponential terms. The number of components corresponding to an ellipsoid of revolution rotating freely, to the same body undergoing uniaxial rotation (the usual model for intrinsic membrane proteins), and to a sphere reduce to 3, 2, and 1, respectively. Thus, the anisotropy decay for the sphere is given by

$$r(t) = r_o exp(-t/\phi) \qquad (6)$$

in which ϕ, the rotational correlation time, is equal to $V\eta/kT$ (V is the molecular volume, η the solution viscosity, k Boltzmann's constant, and T the absolute temperature). r_o is the fundamental anisotropy in the absence of rotation and is related to the angle δ between the absorption and emission transition moments by the equation:

$$r_o = 2/5 \; P_2(\cos\delta) \tag{7}$$

where $P_2(x) = (3x^2-1)/2$ is the second Legendre polynomial. Values of r_o are dependent on the nature of the chromophore and the photoselection regime used to elicit fluorescence or phosphorescence.

Biological macromolecules in general are not absolutely rigid structures. The formalism for treating librational motion (3,5), segmental flexibility (25,26), and combinations of torsional and bending motions (27) is available. Furthermore, one has to consider the orientational constraints mentioned earlier which lead to an observed limiting anisotropy r_∞, expressions for which can be derived in a model-independent manner (3,28), as is shown schematically for a membrane protein in Figure 4. The equilibrium distribution is related to r_∞ according to the expression (28):

$$r_\infty = 0.4 \; P_2(\hat{\mu}_e \cdot \hat{\mu}) \, P_2(\hat{\mu}_p \cdot \hat{\mu}) \, [<P_2(\hat{\mu} \cdot \hat{d})>]^2 \tag{8}$$

in which the correlation function is equivalent to the familiar order parameter (28,29).

3.2 Generalized Formalism for **DLA, FRA, TFA**

The construction of the S(t) and r(t) functions depends upon specification of the experimental configuration, in particular the polarization states of the pump source and emission analyzer in the case of **DLA** (Figure 1), and the corresponding additional information for the monitoring beam in the case of **FRA** and **TFA** (Figure 2). Thus, the evaluation of Equation 5 may require the selection of particular geometries (17,30). The expressions for arbitrary orientations and for arbitrary light intensities are available but complex (30).

4. EXPERIMENTAL DESIGN

4.1 Choice of Probes

Two interrelated factors are of consequence, the type of chromophore and the particular photochemical process being monitored. In general, the triplet quantum yields of most aromatic molecules are very low so that observation of phosphorescence is only possible at extremely low temperatures (e.g., 77 K). However, substitution with heavy

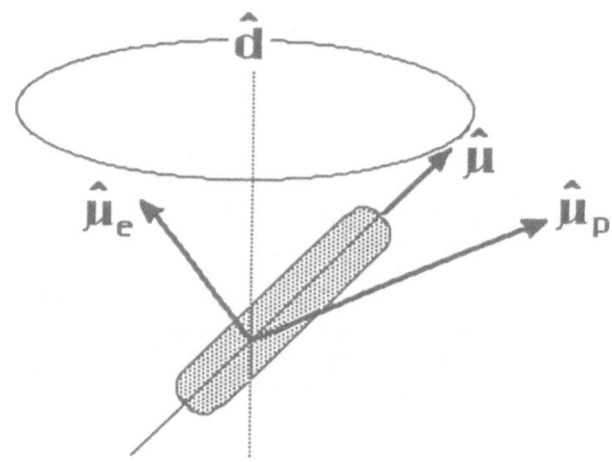

Figure 4. Model for a membrane protein undergoing
rotational relaxation. The protein is represented as a
stippled cylinder.The cylindrical axis of symmetry (vector
$\hat{\mu}$) moves about the normal (vector $\hat{d}$) to the plane of the
lipid bilayer membrane (not shown). The transition moments
for absorption ($\hat{\mu}_p$) and emission ($\hat{\mu}_e$) are at fixed angles
to the molecular axis.

atoms usually enhances k_{isc}, thereby increasing Φ_{isc} and
Φ_p. Table II lists some excited state parameters for three
commonly used chromophores (see also 8,16,21). Erythrosin
(tetraiodofluorescein) has a high Φ_{isc} and thereby
constitutes a very useful phosphorescence and delayed
fluorescence probe. Eosin (tetrabromofluorescein), with
$\Phi_{isc}= 0.71$ and $\Phi_f = 0.19$, may be used both as a
phosphorescence or depletion probe (the suitability of the
latter can be assessed by the ratio Φ_f/Φ_{isc}; ref. 16).
Finally fluorescein, with a very low Φ_{isc} but high
Φ_f/Φ_{isc}, should be appropriate for depletion experiments.
Other considerations include the required laser power, the
sensitivity, and the temporal region of interest.

4.2 Instrumentation

Figure 5 is a simplified diagram of the triplet state

Table II. Lifetimes and quantum yields of
fluorescein and derivatives in free solution.

Chromophore	τ_f (ns)	τ_p (ms)	Φ_{isc}	Φ_f	Φ_f/Φ_{isc}
fluorescein	4	–	0.05	0.92	18.4
eosin	0.9	1	0.71	0.19	0.23
erythrosin	0.1	0.5	0.98	0.02	0.02

τ_f and τ_p are the fluorescence and phosphorescence
lifetimes, respectively.

spectrometer previously described (9; for earlier versions
see 7,8,31). The configuration is such that the triplet
state can be monitored by phosphorescence or delayed
fluorescence, triplet-triplet absorption or fluorescence
depletion. The same configuration can also be used for **TFA**
measurements. The pump (actinic) laser consists of a Lambda
Physik (Göttingen, FRG) Model EMG50 excimer laser operating
with XeCl at 308 nm coupled to a dye laser yielding 10 ns
pulses of tunable, u.v. or visible, polarized light. Up to
10 mJ/pulse of visible light between 500 nm and 520 nm is
obtained with a Coumarin 307 (Lambda Physik) laser dye. The
emission is collected at right angles by a photomultiplier
which is protected from prompt fluorescence by electronic
gating (8). Appropriate cutoff or bandpass filters are
placed between the sample compartment and the photomultiplier; alternatively a monochromator can be used for
spectral scans. The two polarized components of emission
are obtained by rotating a sheet polarizer 90° after a
preselected number of laser flashes. The data are time
averaged by strobing the contents of a Biomation 8100
(California, USA) transient recorder into a microprocessor.
The anisotropy and total intensity profiles (Equation 5)
are then generated and analysed by a computer.
    In the case of **FRA** experiments, a steady-state
population of chromophores in S_1 is first obtained by use
of a continuous wave argon laser (model 90 Innova,
Coherent, California, USA). The light is made to pass
through the sample compartment in spatial coincidence with
the pump beam by the use a Melles Griot (California, USA)
pellicle beam splitter. The uncoated, <5 µ thick pellicle
beam splitter transmits 92% of the excimer laser light and
8% of the argon laser intensity. The polarization state of
the argon beam is established by interposing a polarizer
between the mirror and the beam splitter. In order to
obtain the true fluorescence recovery response of the probe

molecule, two blank records, corresponding to the
individual use of excimer or argon lasers, are collected,
summed and subtracted from each sample data file (18). Such
a sequence, obtained for erythrosin-5'-isothiocyanate
immobilized in a solid matrix of polymethylmethacrylate
(PMMA) is shown in Figure 6. Comparable data for an eosin
probe in PMMA are given elsewhere (18).

    TFA experiments with SAN were performed using the same
configuration as for **FRA** but with a laser dye
(p-terphenyl) tuned to 341 nm and the monitoring argon
laser set to 488 nm.

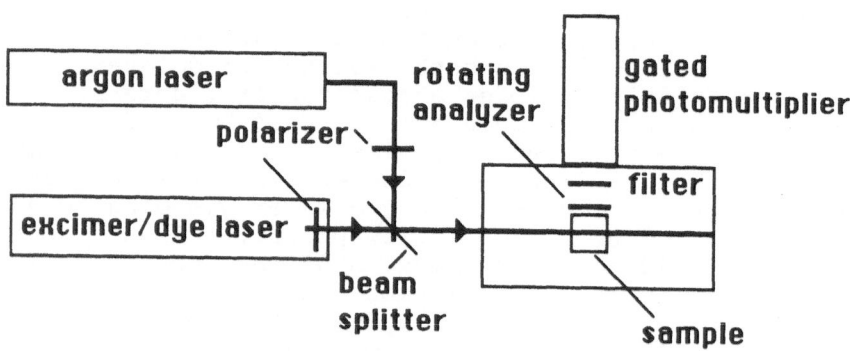

Figure 5. Schematic diagram of triplet state spectrometer.
Not shown are the active electro-optical elements, the
components required for absorption dichroism measurements,
nor the microprocessor control units. The same
configuration functions for **TFA** measurements.

5. APPLICATIONS

Two general strategies using extrinsic probes may be
employed in the investigation of macromolecular rotation.
The first is to allow the chromophore to simply partition
within the host structure, while the second is to attach
the chromophore covalently in a random position or to
particular sites of the macromolecule. In this section, we
discuss some examples of experiments in our laboratory
utilizing these two approaches according to the three
methods described above (**DLA**, **FRA**, **TFA**).

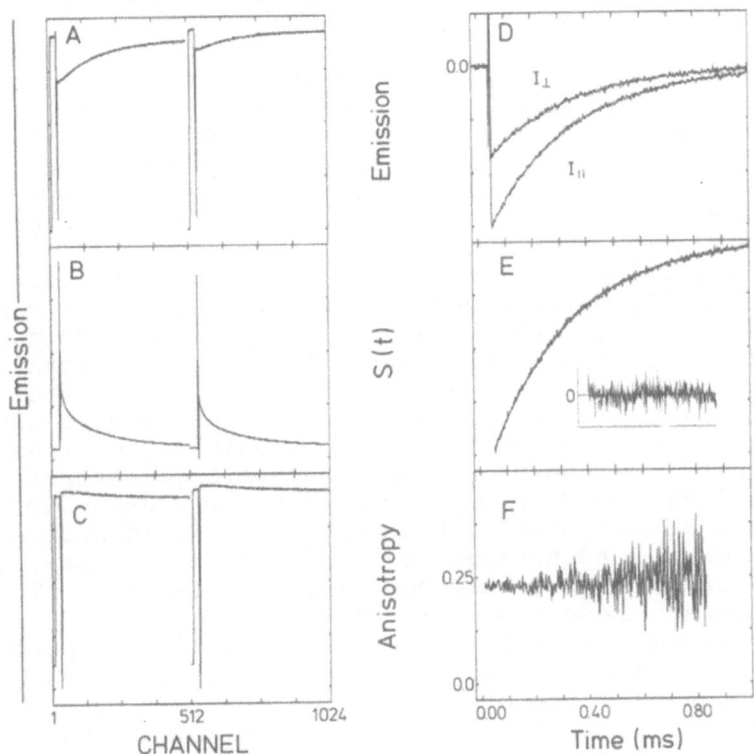

Figure 6. Sequence of measurements required for the
fluorescence depletion technique (**FRA**). Shown is the
complete data set for erythrosin-5'-isothiocyanate in a
solid matrix of polymethyl- methacrylate. In traces (A) -
(C) the parallel (to the actinic laser) [channels 1-512]
and perpendicular [channels 513-1024] emission components
are shown for: (A) depletion signals (both lasers active);
(B) excimer (actinic) laser only; (C) steady-state
fluorescence (i.e., with only the continuous wave argon
laser); (D) polarized emission components after subtraction
of curves (B) and (C) from (A). (E) S(t) curve and
residuals plot for a single exponential fit with τ_t = 301
μs. The same sample, measured by **DLA** (phosphorescence)
gave a τ_p = 308 μs. The excellent agreement is in
accordance with the photophysical scheme of Figure 1.(F)
corresponding anisotropy time course, showing a constant
r = 0.22, a value lower than expected for reasons discussed
elsewhere (18). The actinic and monitoring beams were
polarized vertically and horizontally, respectively.

5.1 **DLA**: Phosphorescence of Eosin-labeled Fatty Acids in Phospholipid Bilayers

The structural and dynamic properties of polymerized surfactant aggregates such as detergent micelles, vesicles and bilayers have been studied extensively (32). From a biological aspect, it is of interest to determine in which way these structures mimic the properties of natural membranes (33). Most luminescent anisotropy studies of lipid rotation have employed the fluorescence characteristics of incorporated probes (34,35), as the time scales of lipid rotation are usually in the ns regime. However, a recent electron spin study using incorporated phospholipid spin-labels (36), indicated that rotation about the long axis of dimyristoyl-phosphatidylcholine (DMPC) lipids below the phase transition occurrs with time constants of about 60-100 μs. Such values lie within the time domain of phosphorescence anisotropy measurements.

We have investigated further the rotation of DMPC lipids by incorporating two eosin fatty acid probes (dodecanoyl- and hexadecanoyl-amidoeosin) and measuring the time-dependent phosphorescent anisotropies (37). The eosin moieties of these reporter molecules are located close to the membrane surface. Figure 7 shows typical experimental results at two temperatures. A number of features serve to illustrate the type of information provided by such studies. The phosphorescence emission at both temperatures displayed a time-dependent anisotropy which could be fit to an equation of the form

$$r(t) = (r_0 - r_\infty)\sum_{i=1}^{j}\alpha_i\exp(-t/\phi_i) + r_\infty \tag{9}$$

where ϕ_i are the rotational correlation times, α_i are the corresponding fractional amplitudes and r_∞ was defined earlier (Equation 8). It can be seen from Figure 7 that r_∞ is higher at 10 °C than at 30 °C. This result is typical in studies of bilayer structures (38), and reflects the increase in lipid chain ordering as the temperature is lowered below the phase transition of DMPC ($T_t \sim 23$ °C). Another interesting feature evident in Figure 7 is the finding that below the T_t two exponentials were required to fit the data (j = 2 in Equation 9), whereas above T_t one exponential was adequate. This phenomenon is often observed in luminescence studies of phospholipids (35,39), and can be attributed to solubilization of the probe in more than one site below T_t. An alternative explanation, however, has been advanced for the data shown in Figure 7 (37). The single correlation time above T_t and the longer correlation time below T_t (ϕ_2) were assumed to be reporting eosin/lipid head-group interactions (discrete jumps on the surface of the bilayer), whereas the shorter correlation time below T_t (ϕ_1) was assigned to the above-mentioned long-axis rotation (36). The latter motion is not observed under our experimental conditions (above T_t) since it has a rotational correlation time in the ns time range (36).

Erythrosin has also been used as a phosphorescence probe to assess the effects of photopolymerization on surfactant vesicle surface morphology (40).

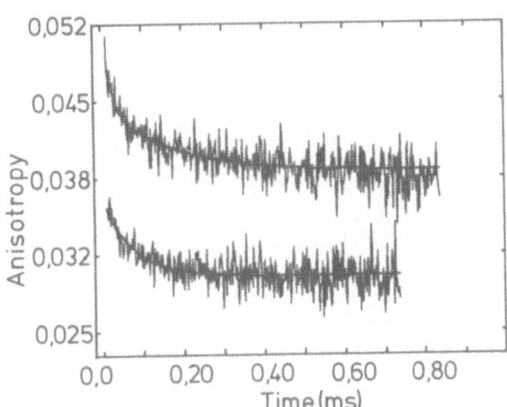

Figure 7. Phosphorescence anisotropy decay of the dodecanoyl-amidoeosin fatty acid probe in DMPC vesicles at 30 °C (lower curve) and 10 °C (upper curve). The solid lines represent the best fit analyses according to Equation 9. At T = 30 °C: j = 1, ϕ = 62 µs, r_∞ = 0.030; at T = 10 °C: j = 2, ϕ_1 = 17 µs, ϕ_2 = 265 µs, α_1 = 0.55, α_2 = 0.45, r_∞ = 0.039. Data of Blatt and Corin (37).

5.2 FRA: Labeled Protein (Concanavalin A)

As indicated earlier, only probes with relatively high values of Φ_{isc} are suitable for phosphorescence emission measurements. The triplet state of fluorescein (see Table II) is not detectable by phosphorescence emission techniques but should provide a suitable probe for the fluorescence depletion technique (11,15,16). Figure 8 shows **FRA** data for the rotational diffusion at 6 °C and in 98%

glycerol, 5 mM phosphate buffer, pH 7.0, of the lectin
Concavalin A (ConA) labelled with fluorescein-5'-
isothiocyanate (41). Fitting the decay to a single
exponential function yielded a rotational correlation time
of 169 μs. This value can be compared to a φ = 198 μs
expected for the hydrated tetrameric form of ConA (102 kD)
in 98% glycerol at 6 °C if sphericity is assumed. In
buffer, the anisotropy was centered around zero indicating
a complete depolarization in the sub-μs time domain (7).

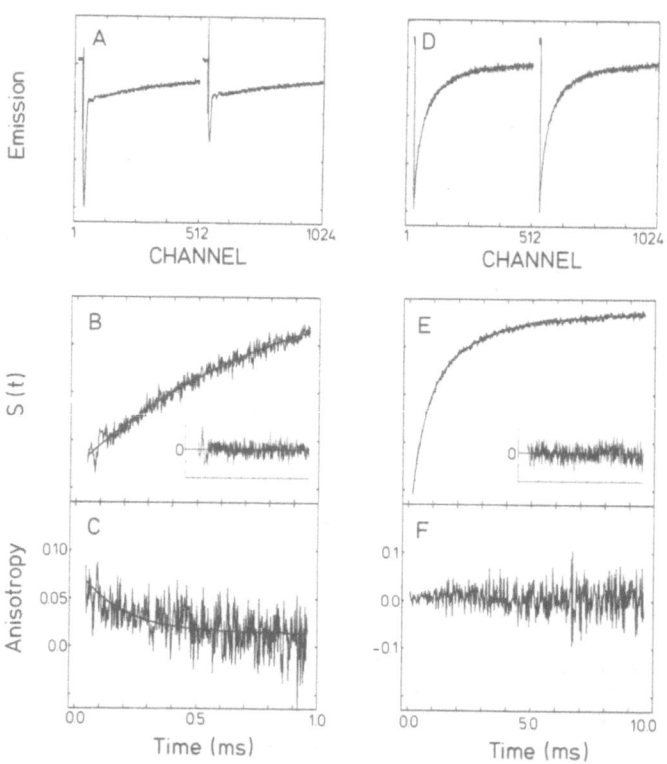

Figure 8. Fluorescence recovery anisotropy (**FRA**) of ConA
labelled with fluorescein-5'-isothiocyanate. Conditions:
6 °C in 98% glycerol [(A) - (C)] and in 5 mM phosphate
buffer, pH 7.0 [(D) - (E)]. Panels A and D represent the
initially acquired data, as in Figure 6A. Analyses of S(t)
curves gave a lifetime τ of 850 μs and 1270 μs in glycerol
and buffer, respectively. In glycerol, a rotational
correlation time of φ = 169 μs was obtained, a value in
agreement with that expected for tetrameric ConA. Data of
Blatt, Corin, & Jovin (41).

5.3 **TFA**: Salicylideneaniline in EtOH.

The photophysical scheme of Figure 3 was investigated by
monitoring the time course of fluorescence from the quinoid
ground state intermediate. A multicomponent decay process
was observed in EtOH at 25 °C (Figure 9). The apparent
steady-state anisotropy was high (ca. 0.25) and constant,
reflecting the very short fluorescence lifetime. The
significance of the complex decay kinetics is unclear but
clearly depends upon the nature of the solvent. For
example, in cyclohexane the corresponding lifetime was 1/6
that obtained in EtOH; the fluorescence anisotropy was also
higher.

Similar data have been obtained for SAN and related
compounds in other solvents (42). Of particular interest is
the enzyme cofactor pyridoxal 5'-phosphate (PLP) which also
displays proton transfer and tautomerism (43). In the form
of a Schiff base adduct to proteins, including glycogen
phosphorylase with which PLP is normally associated, it can
be used to measure rotational diffusion by following the
transient dichroism of the photophysical intermediate(s)
(44). In preliminary experiments, the **TFA** method has
yielded anisotropy decay curves derived from the transient
fluorescence (42).

5.4 Other Studies and Perspectives

The determination of rotational diffusion by the various
methods based on the triplet state have been applied in our
laboratory to numerous systems involving membrane, cell
surface receptors and antigens, and complexes of proteins
and nucleic acid components (see tabulated data in 9,31).
For example, the dynamics of the polypeptide growth hormone
epidermal growth factor (EGF) bound to its specific
receptors on cells (45) and on plasma membrane fragments
(46) have been determined. The hormone in this and other
systems appears to induce the association (microclustering)
of the intrinsic membrane proteins constituting the
receptors, a possibly crucial initial event in triggering
the complex chain of biological responses.

Clearly, the techniques discussed in this chapter are
most useful in combination with other complementary
approaches to the study of molecular motion and
interactions. Examples other than those already alluded to
are the chemical kinetics related to drug-DNA binding (47),
and the accessibility of probes associated with proteins
and membranes to diffusable quenchers (18,37). The search
will continue for more sensitive modes of measurement and
for reporter groups which exert a minimal perturbation of
the target molecules.

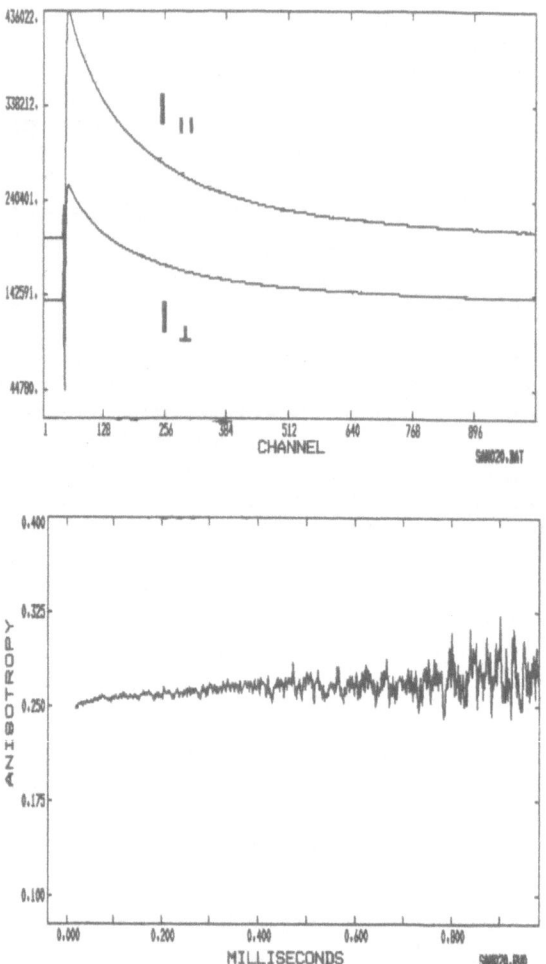

Figure 9. Transient fluorescence anisotropy (**TFA**) of salicylidenaniline in EtOH. The solution was 10 μM. The upper panel depicts the two polarized components of emission at >520 nm. The lower panel is the corresponding anisotropy. The decay of intensity was resolved into 3 components: 32 μs, 112 μs, and 297 μs, with fractional amplitudes of 0.14, 0.43, and 0,43, respectively. Data of Matko and Jovin (42).

6. REFERENCES

1. Shinitzky, M., Ed. *Physiology of Membrane Fluidity, Vol. 1,* (CRC Press, Boca Raton, FL) (1984).
2. Lipari, G. & Szabo, A. *J. Am. Chem. Soc.* **104**: 4546 (1982).
3. Szabo, A. *J. Chem. Phys.* **81**:150 (1984).
4. Cherry, R.J. *Biochim. Biophys. Acta* **559**: 289 (1979).
5. Kinosita, K. Jr., Kawato, S. & Ikegami, A. *Adv. Biophys.* **17**: 147 (1984).
6. Bayley, P.M. & Dale, R.E., Eds. *Spectroscopy and the Dynamics of Molecular Biological Systems,* (Academic, Orlando, FL) (1985).
7. Austin, R.H., Chan, S.S. & Jovin, T.M. *Proc. Natl. Acad. Sci. U.S.A.* **76**: 5650 (1979).
8. Jovin, T.M., Bartholdi, M., Vaz, W.L.C. & Austin, R.H. *Ann. N.Y. Acad. Sci.* **366**: 176 (1981).
9. Corin, A.F., Matayoshi, E.D. & Jovin, T.M., p. 53 in ref. 6.
10. Moore, C. Boxer, D. & Garland, P. *FEBS Lett.* **108**: 161 (1979).
11. Garland, P.B. & Johnson, P., p. 95 in ref. 6.
12. Murray, E.K., Restall, C.J. & Chapman, D., p. 119 in ref. 6.
13. Turro, N.J., Liu, K.C., Chow, M.F. & Lee, P. *Photochem. Photobiol.* **27**: 523 (1978).
14. McGlynn, S.P., Azumi, T. & Kinosita, M. *Molecular Spectroscopy of the Triplet State* (Prentice-Hall, Englewood Cliffs, NJ) (1969).
15. Johnson, P. & Garland, P.B. *FEBS Lett.* **132**: 252 (1981).
16. Johnson, P. & Garland, P.B. *Biochem. J.* **203**: 313 (1982).
17. Wegener, W.A. *Biophys. J.* **46**: 795 (1984).
18. Corin, A.F., Blatt, E. & Jovin, T.M. , submitted for publication.
19. Barbara, P.F., Rentzepis, P.M. & Brus, L.E. *J. Am. Chem. Soc.* **102**: 2786 (1980).
20. Ottolenghi, M. & McClure, D.S. *J. Chem. Phys.* **46**: 4613 (1967).
21. Cherry, R.J. *Methods Enzymol.* **88**: 248 (1982).
22. Cherry, R.J. p. 79 in ref. 6.
23. Smith, B.A. & McConnell, H.M. *Proc. Natl. Acad. Sci. U.S.A.* **75**: 2759 (1978).
24. Thomas, D.D., Eads, T.M., Barnett, V.A., Lindahl, K.M., Momont, D.A. & Squier, T.C., p. 239 in ref. 6.
25. Wegener, W.A. *J. Chem. Phys.* **76**: 6425 (1982).
26. Wegener, W.A. *Biopolymers* **21**: 1049 (1982).
27. Hogan, M., LeGrange, J. & Austin, R. *Nature* **304**: 752 (1983).
28. Lipari, G. & Szabo, A. *Biophys. J.* **30**: 489 (1980).

29. Jähnig, F. *Proc. Natl. Acad. Sci. U.S.A.* **76**: 6361 (1979).
30. Szabo, A., Clegg, R.M. & Jovin, T.M., in preparation.
31. Matayoshi, E.D., Corin, A.F., Zidovetzki, R., Sawyer, W.H. & Jovin, T.M. in *Mobility and Recognition in Cell Biology*, Sund, H. & Veeger, C., Eds. (de Gruyter, Berlin), 119 (1983).
32. Fendler, J.H. & Tundo, P. *Acc. Chem. Res.* **17**: 3 (1984).
33. Radhakrisnan, R., Gupta, C.M., Erni, B., Robson, R.J., Curatolo, W., Majumdar, A., Ross, A.H., Takagaki, Y. & Khorana, H.G. *Ann. N.Y. Acad. Sci.* **346**: 165 (1980).
34. Chen, L.A., Dale, R.E., Roth, S. & Brand, L. *J. Biol. Chem.* **252**: 2163 (1977).
35. Blatt, E., Sawyer, W.H. & Ghiggino, K.P. *Aust. J. Chem.* **36**: 1079 (1983).
36. Marsh, D. *Biochemistry* **19**: 1632 (1980).
37. Blatt, E. & Corin, A.F., submitted for publication.
38. Seelig, J. & Seelig, A. *Quart. Rev. Biophys.* **13**: 19 (1980).
39. Vincent, M., Gallay, J., de Bony, J. & Tocanne, J.F. *Eur. J. Biochem.* **150**: 341 (1985).
40. Reed, W., Lasic, D., Hauser, H. & Fendler, J.H. *Macromolecules* **18**: 2005 (1985).
41. Blatt, E. Corin,, A.F. & Jovin, T.M., unpublished data.
42. Matko, J & Jovin, T.M., in preparation.
43. Walters, L.S., Cornish, T.J., Askins, H.W. & Ledbetter, J.W. *Anal. Biochem.* **127**: 361 (1982).
44. Cornish, T.J. & Ledbetter, J.W. *FEBS Lett.* **154**: 378 (1983).
45. Zidovetzki, R., Yarden, Y., Schlessinger, J. & Jovin, T.M. *Proc. Natl. Acad. Sci. U.S.A.* **78**: 6981 (1981).
46. Zidovetzki, R., Yarden, Y., Schlessinger, J. & Jovin, T.M. *The EMBO J.*, in press.
47. Corin, A.F. & Jovin, T.M., submitted for publication.

THE FLUORESCENCE POLARIZATION TECHNIQUE AS A TOOL TO INVESTIGATE CHAIN
ORIENTATION AND RELAXATION IN BULK POLYMERS

Lucien MONNERIE
Ecole Supérieure de Physique et de Chimie Industrielles de
Paris
10, rue Vauquelin
75231 Paris Cedex 05 - France

ABSTRACT - Fluorescence polarization and its application to orientation
measurements of uniaxially stretched polymers are described. Results
obtained on anthracene labelled polystyrene chains embedded in normal
polystyrene and stretched at various temperatures and strain rates are
presented and discussed. An improvement of the Doi-Edwards slip-link
model is required to account for the results and a modified model is
proposed. Finally, an illustration of fluorescence polarization measurements on mobile and oriented polyisoprene networks is shown.

I. INTRODUCTION

A recent renewal of interest in the viscoelasticity of entangled bulk
polymers has arisen from the theoretical approach of chain relaxation
developed by de Gennes (1), Edwards and Doi (2, 3). These theories are
based on a molecular description of chain motions which involves reptation of the chain along the tube formed by neighbouring chains.
    Although such theories can be tested through their predictions of
macroscopic viscoelastic properties of bulk materials, such as viscosity and elastic and loss moduli, it is very tempting to obtain direct
information on chain dynamics on a molecular scale. For this purpose
spectroscopic techniques such as small-angle neutron scattering, infra-
red dichroism and fluorescence polarization (F.P.) are particularly
suitable. Small angle neutron scattering yields information either on
the molecular dimensions of the whole chain or on the structure of part
of the chain, whereas infrared dichroism and fluorescence polarization
applied to stretched samples lead to information about the chain segment
orientation.
    In this paper we will deal exclusively with fluorescence polariza-
tion but a presentation of the two other techniques and their application to the molecular viscoelasticity of bulk polymers can be found in
Ref. (4).
    After presenting the fluorescence polarization technique,
we will consider its application to bulk polystyrene stretched above its
glass transition temperature, looking at the influence of the molecular
weight of the matrix or of the labelled chain.

M. A. Winnik (ed.), Photophysical and Photochemical Tools in Polymer Science, 371–396.

II. ORIENTATION DISTRIBUTION FUNCTION

Before considering the fluorescence polarization technique which leads
to a measurement of the orientation distribution of characteristic vectors, we introduce some convenient quantities to describe the orientation of uniaxially symmetric systems.

Each vector is characterized by the angle Θ that its direction
makes with the symmetry axis. The orientation distribution is represented by a function $f(\Theta)$ which can be expanded in terms of Legendre poly-
nomials in $\cos \Theta$ as follows :

$$f(\Theta) = \sum_{\ell} b_{\ell} P_{\ell}(\cos \Theta)$$

with

$$b_{\ell} = (1/2 \, \pi)(2 \, \ell + 1)/2 < P_{\ell}(\cos \theta) >$$

where $< P_{\ell}(\cos \Theta) >$ is the value of $P_{\ell}(\cos \Theta)$ averaged over the distribution :

$$< P_2(\cos \Theta) > = (1/2) < 3 \cos^2 \Theta - 1 >$$

$$< P_4(\cos \Theta) > = (1/8) < 35 \cos^4 \Theta - 30 \cos^2 \Theta + 3 >$$

$$< \cos^n \Theta > = \int_0^{\pi} f(\Theta) \cos^n \Theta \; \sin \Theta d\Theta.$$

It has recently been shown (5) that for an uniaxial distribution
the determination of $< P_2 >$ and $< P_4 >$ is sufficient in most cases.

III. FLUORESCENCE POLARIZATION

III.1. Principles

Fluorescent molecules have the property of re-emitting in the form of
visible light part of the energy acquired by the absorption of luminous
radiation. After illumination by a very short pulse at time t_o, the
fluorescent light emitted at time $t_o + u$ is proportional to $\exp(-u/\tau)$,
where τ is the mean lifetime of the excited state (usually called the
fluorescence lifetime). The most frequent τ values range from 1 to
100 ns. When absorbing light of a suitable wavelength a molecule behaves as an electric dipole oscillator with a fixed orientation with respect to the geometry of the molecule. Such an equivalent oscillator is
termed an absorption transition moment, M_o. In the same way, for the
fluorescence emission we have an emission transition moment, M. When
such a molecule receives an incident beam polarized along the P direction (fig. 1), the absorption probability is proportional to $\cos^2 \alpha_o$.
In the same way, the fluorescence intensity measured through an analyser, A, is proportional to $\cos^2 \beta$. Thus for the P and A directions of
polarizer and analyser the observed luminescence intensity is proportional to $\cos^2 \alpha_o \cos^2 \beta$. Owing to the lack of phase correlation between excitation and emission lights, fluorescence emission can be des-

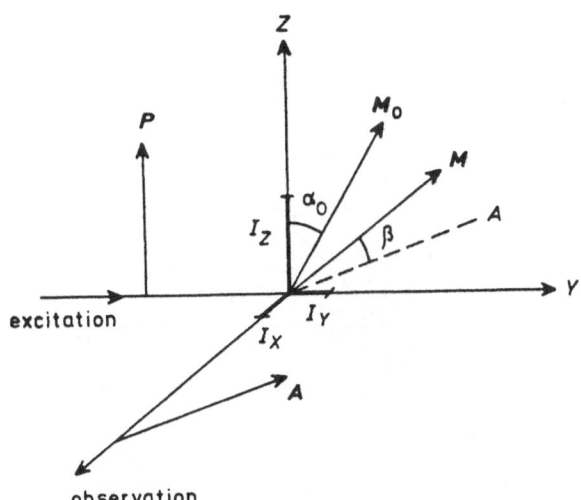

Fig. 1 - Polarized absorption and fluorescence emission : P, polarizer,
A, analyser

cribed as resulting from three independent radiations respectively polarized along the X, Y and Z axes with intensities I_x, I_y and I_z. The
Curie symmetry principle, applied to excitation light polarized along Z,
leads to $I_x = I_y$. The fluorescence polarization is characterized by the
emission anisotropy :

$$r = (I_{\parallel} - I_{\perp})/(I_{\parallel} + 2I_{\perp})$$

where $I_{\parallel}$ and $I_{\perp}$ correspond to the fluorescence intensity obtained with
an analyser direction parallel and perpendicular, respectively, to that
of the polarizer.

III.2. Orientation of uniaxially symmetric systems

For our present purpose we are mainly interested in the use of fluorescence polarization to look at the orientation distribution of fluorescent molecules. The main results which can be derived are presented
below; more details can be found in a recent review (6) and in the original paper. (7).

In the following we will assume that the transition moments in both
absorption and emission coincide with a molecular axis M of the molecule,
whose direction is specified by the spherical polar angle $\Omega = (\alpha, \beta)$ in
the reference frame (fig. 2). Let us introduce the angular functions
$N(\Omega_o, t_o)$, the orientation distribution of M at time t_o (M_o in fig. 2),
and $P(\Omega, t | \Omega_o, t_o)$, the conditional probability density of finding at
position Ω at time t a vector M which was at position Ω_o at time t_o.
After illuminating the sample by a linearly polarized short pulse
of light at t_o, the intensity emitted at time (t_o+u) for the P and A

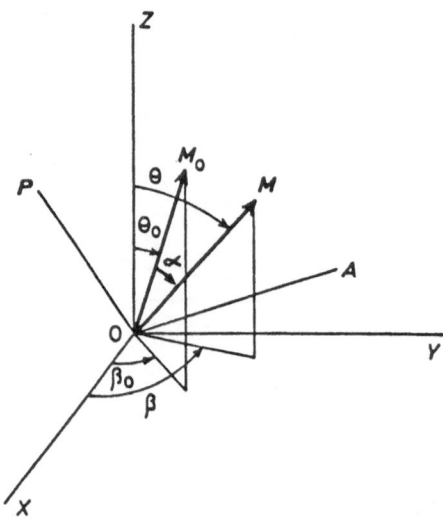

Fig. 2 - Illustration of the angles which define the orientation of
molecular axis M_o at time t_o and M at time t_o+u with respect
to the fixed frame OXYZ.

directions of polarizer and analyser is given by :

$$i(P,A,t_o+u) = K \int\int N(\Omega_o,t_o)P(\Omega,t_o + u|\Omega_o,t_o)$$

$$\times \cos^2(P,M_o)\cos^2(A,M)\exp(- u/\tau)d\Omega_o d\Omega$$

where K is an instrumental constant. In this expression t_o corresponds
to the macroscopic evolution of the sample, for example in a rheological experiment, whereas u corresponds to a microscopic reorientational
motion on the scale of the fluorescence lifetime τ. In most cases the
t_o dependence of N and P can be ignored within the time τ(ca.10^{-8} s),
and the fluorescence intensity emitted under continuous excitation is
given by :

$$i(P,A,t_o\tau) = \int_o^\infty i(P,A,t_o + u)du$$

In order to derive convenient expressions, it is interesting to do an
expansion of the angular functions of Ω_o and Ω in a series of spherical
harmonics $Y_\ell^m(\Omega_o)$ and $Y_\ell^m(\Omega)$. Taking into account the properties of spherical harmonics, we obtain :

$$\cos^2(\vec{P},\vec{M}_o) = \sum_{k=0,2} \sum_{m=-k}^{+k} P_k^m(\gamma_o, \varphi_o)\overline{Y_k^m}(\Omega_o)$$

$$\cos^2(\vec{A},\vec{M}) = \sum_{\ell=0,2} \sum_{n=-\ell}^{\ell} a_\ell^n(\gamma_o, \varphi_o)Y_\ell^n(\Omega)$$

$$K \, N(\Omega_o,t_o)P(\Omega_o t_o + u | \Omega_o,t_o) = \sum_{k=0}^{\infty} \sum_{\ell=0}^{\infty} \sum_{m=-k}^{+k} \sum_{n=-\ell}^{+\ell} f_{k\ell}^{mn}(t_o,u)$$

$$x \; Y_k^m(\Omega_o)\overline{Y_\ell^n(\Omega)}$$

$$i(\vec{P},\vec{A},t_o,\omega) = \sum_{k=0,2} \sum_{\ell=0,2} \sum_{m=-k} \sum_{n=-\ell} P_k^m \, a_\ell^n \, f_k^{mn}(t_o,u)$$

$$x \; e^{-u/\tau}$$

with $\quad f_{k\ell}^{mn}(t_o,u) = K < Y_k^m(\Omega_o)\overline{Y_\ell^n(\Omega)} >$

(the angular brackets denote an ensemble average)

$$i(\vec{P},\vec{A},t_o,\tau) = \sum_{k=0,2} \sum_{\ell=0,2} \sum_{m=-k}^{+k} \sum_{n=-\ell}^{+\ell} P_k^m \, a_\ell^n \, f_{k\ell}^{mn} \, t_o, \tau)$$

where $f_{k\ell}^{mn}(t_o,\tau) = \int_o^\infty f_{k\ell}^{mn}(t_o,u)e^{-u/\tau} \, du$

Thus, any fluorescence intensity is expressed as a linear combination of 36 $f_{k\ell}^{mn}$ coefficients, in agreement with Frehland's result (8). All $f_{k\ell}^{mn}$ values are proportional to K, and f_{00}^o is simply

$$f_{00}^o = K/4 \; \pi$$

When $\vec{P}$ or $\vec{A}$ lies along the fixed frame axes, some coefficients vanish, leading to the following relations :

$$i(Z,Z) = (4 \, \pi/9)[\, f_{00}^o + 2(5)^{-1/2}f_{20}^o + 2(5)^{-1/2}f_{02}^o + 4(5)^{-1} \, f_{22}^o]$$

$$i(Z,X) = (4 \, \pi/9)[\, f_{00}^o + 2(5)^{-1/2}f_{20}^o - (5)^{-1/2}f_{02}^o - 2(5)^{-1} \, f_{22}^o]$$

$$i(X,Z) = (4 \, \pi/9)[\, f_{00}^o - (5)^{-1/2}f_{20}^o + 2(5)^{-1/2}f_{02}^o - 2(5)^{-1} \, f_{22}^o]$$

$$i(X,X) = (4 \, \pi/9)[\, f_{00}^o - (5)^{-1/2}f_{20}^o - (5)^{-1/2}f_{02}^o +$$
$$(5)^{-1}f_{22}^o + 3(5)^{-1} \, f_{22}^2]$$

$$i(X,Y) = (4 \, \pi/9)[\, f_{00}^o - (5)^{-1/2}f_{20}^o - (5)^{-1/2} \, f_{02}^o +$$
$$(5)^{-1}f_{22}^o - 8(5)^{-1} \, f_{22}^2]$$

In the case of an uniaxial distribution of the molecular axes M, simplification occurs in the intensity expressions and only six independent $f_{k\ell}^{mn}$ quantities remain :

$$f_{00}^o; \; f_{20}^o; \; f_{02}^o; \; f_{22}^o; \; f_{22}^1; \; f_{22}^o$$

All of them are proportional to the instrumental constant K and in addition, one gets :

$$f^o_{00} = K/4\,\pi$$

It is convenient to introduce the following quantities :

$$G^{(0)}_{20} = \frac{1}{2} < 3\,\cos^2\Theta_o - 1 >$$

$$G^{(0)}_{02} = \frac{1}{2} < 3\,\cos^2\Theta - 1 >$$

$$G^{(0)}_{22} = \frac{1}{4} < 3\,\cos^2\Theta_o - 1)(3\,\cos^2\Theta - 1) >$$

$$G^{(1)}_{22} = \frac{9}{16} < \sin\Theta_o\,\cos\Theta_o\,\sin\Theta\,\cos\Theta\,\cos(\beta-\beta_o) >$$

$$G^{(2)}_{22} = \frac{9}{64} < \sin^2\Theta_o\,\sin^2\Theta\,\cos 2(\beta-\beta_o) >.$$

the various involved angles are those defined in Fig. 2 and the angular brackets denote an ensemble average. The quantities G^o_{20} and G^o_{02} represent the second moments of the distribution of vectors $\vec{M}$ and $\vec{M}$ respectively and they describe the molecular orientation independently of molecular mobility. The functions $G^{(m)}_{22}$ depend on both orientation and mobility.

Further investigation is required to decide if, during the fluorescence lifetime, motions can occur or not. We will hereafter refer to these situations as "mobile" or "frozen" systems, respectively.

   i/ Uniaxial frozen systems.

Assume that absorption and emission transition moments are parallel (the effect of electronic delocalization will be treated later). In the above G expressions $\Theta = \Theta_o$, $\beta = \beta_o$ and the quantities $G^{(0)}_{20}$ ($=G^{(0)}_{02}$) and $G^{(m)}_{224}$ can be rewritten with only two independent quantities, $\cos^2\Theta$ and $\cos^4\Theta$. All the information on the fluorescence intensities may be displayed in a 3 x 3 tensor I :

$$I = K
\begin{vmatrix}
\frac{3}{8} < \sin^4\Theta > & \frac{1}{8} < \sin^4\Theta > & \frac{1}{2} < \sin^2\Theta\,\cos^2\Theta > \\
\frac{1}{8} < \sin^4\Theta > & \frac{3}{8} < \sin^4\Theta > & \frac{1}{2} < \sin^2\Theta\,\cos^2\Theta > \\
\frac{1}{2} < \sin^2\Theta\,\cos^2\Theta > & \frac{1}{2} < \sin^2\Theta\,\cos^2\Theta > & < \cos^4\Theta >
\end{vmatrix}$$

whose elements I_{ij} can be easily obtained by reporting the G expressions in the general intensity expressions.

This is identical to the result given in (9). The apparatus constant can be obtained from :

$$K = \sum_i \sum_j I_{ij} = (8/3)I_{XX} + 4\,I_{XZ} + I_{ZZ}$$

in such a way that the second and fourth moments of the orientation distribution can be derived from fluorescence intensity measurements by the relations :

$$< \cos^2 \alpha > = (I_{ZZ} + 2\, I_{XZ})/((8/3)I_{XX} + 4\, I_{XZ} + I_{ZZ})$$

$$< \cos^4 \alpha > = I_{ZZ}/((8/3)I_{XX} + 4\, I_{XZ} + I_{ZZ})$$

ii/ Uniaxial mobile systems

In this case both orientation and mobility contribute to the fluorescence polarization. Nevertheless such systems can be treated (7). Indeed, if we assume that during the fluorescence lifetime (10^{-10} to 10^{-7} s) the orientation distribution does not change, we obtain from the general expressions of G :

$$G_{20}^o = G_{02}^o = (1/2) < 3 \cos^2 \Theta_o - 1 > = < P_2(\cos \Theta_o) >$$

On the other hand, for mobile systems, G_{22}^o depends on both orientation and mobility, in such a way that $< P_4(\cos^2\Theta_o) >$ cannot be obtained. Furthermore, intensity measurements corresponding to polarizer and analyser directions along the X and Z axis are not sufficient for deriving G_{20}^o. The set of quantities G_{20}^o and G_{22}^m must be obtained from the measurement of five intensities, i(P, A), corresponding to P and A orientations which are not contained in the same plane, excluding the use of only a straight-through optical arrangement.

Concerning the mobility, from the available G_{20}^o and G_{22}^m quantities, one can derive :

a/ The mean mobility amplitude M

$$M = < (3 \cos^2 \alpha - 1)/2 > = G_{22}^o + (16/3)G_{22}^1 + (16/3)G_{22}^2$$

describing the motion performed during the considered time interval. Thus, for a Dirac-pulse excitation :

$$M(t) = < (3 \cos^2 \alpha(t) - 1)/2 >$$

whereas for a continuous excitation, one gets :

$$M(\tau) = \int_o^\infty M(t) \exp(- t/\tau)dt$$

where τ is the fluorescence lifetime.

b/ Three angular correlation functions which contain information on the anisotropy of the motion (7).

iii/ Effect of electronic delocalization

The effect of non-parallel absorption and emission moments on orientation measurements have been considered by several authors (9, 10). For uniaxial systems, an easy correction has been derived (7). The delocalization results in the fact that the measured intensities i($\vec{P},\vec{A}$) do not lead to the G quantities above defined, but to γ coefficients which are related to them by the relations :

$$\gamma_{20}^{o} = (5 \ r_{o}/2)^{1/2} \ G_{20}^{o}$$

$$\gamma_{22}^{m} = (5 \ r_{o}/2) \ G_{22}^{m} \quad \text{with } m = 0,1,2$$

where r_{o} is the fundamental emission anisotropy, which must be determined on an unoriented sample from an additional measurement of emission anisotropy.

iiii/ Birefringence and light-scattering corrections

The birefringence effect occurs when the direction of either the polarizer or the analyser does not coincide with a principal direction of the refractive index tensor. It has been shown (7) that only the quantity γ_{22} is modified and becomes :

$$\gamma_{22}^{1} = (5 \ r_{o}/2) \ b \ G_{22}^{1}$$

where b is a correction factor which can be experimentally determined (11).

When dealing with crystalline polymers, light-scattering occurs and modifies the state of polarization of both the excitation and emitted light. The principle of a method of correction has been given in (10) and applied to straight-through measurements (12).

III.3. Instrumentation for measuring orientation in uniaxial systems

For systems in which the orientation does not change in time, successive fluorescence intensity measurements can be performed for the required polarizer and analyser directions, leading to a determination of $< P_{2}(\cos \alpha) >$ and $P_{4}(\cos \alpha) >$.

Measurements are performed under continuous excitation. For frozen systems, a straight-through optical arrangement perpendicular to the sample is very convenient. In-front illumination has been proposed (13) but the effect of refraction of the light inside the sample must be considered (11). Measurements under optical microscope have been performed either on stretched samples (10) or during stretching (12).

When dealing with mobile systems a more elaborated equipment is required, and it has been developed in our laboratory (11). This apparatus permits simultaneous measurements of the intensities required for determining orientation and mobility even during stretching. The optical system is represented in Fig. 3.

III.4. Chain labelling

In order to use fluorescence polarization for studying orientation of polymers, it is necessary to get labelled chains. Among the various fluorescent molecules which can be used, anthracene is one of the most interesting ones.

To get information on the chain orientation, it is necessary to bind covalently the anthracene group to the polymer and preferentially in the middle of the chain. Such a labelling can be easily achieved with

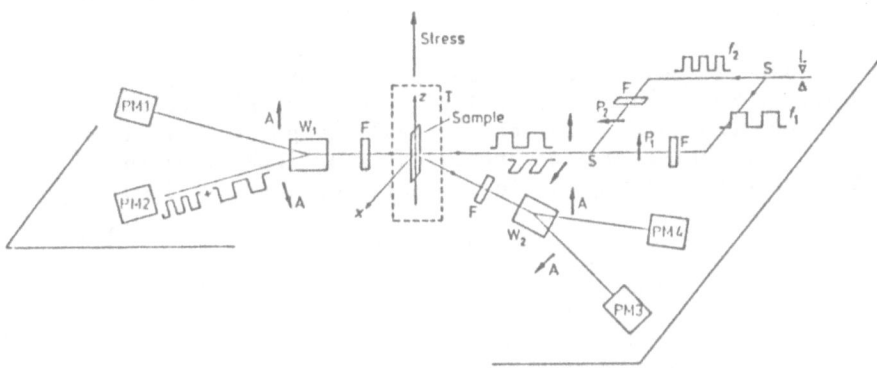

Fig. 3 - Optical equipment : L, mercury lamp (HBO 200 W);S, beam
        splitter; F_1, F_2 modulation frequencies of the mechanical
        choppers; F, optical filter; P_1, P_2 polarizers; W_1, W_2,
        Wollaston prisms, A, analysing direction; PM, photomultiplier;
        T, temperature chamber.

polymers which can be polymerized by anionic polymerization, as for
instance polystyrene, polyisoprene... In this case, living anionic polymers are prepared and deactivated by 9,10-bis bromomethylanthracene,
resulting in the following labelled species :

- labelled polystyrene :

$$\sim CH-CH_2 \longrightarrow CH_2-CH \sim$$

- labelled polyisoprene :

$$\sim CH_2-\underset{CH_3}{C} = CH-CH_2-CH_2 \longrightarrow CH_2-CH_2-CH = \underset{CH_3}{C}-CH_2 \sim$$

The double arrow represents the direction of the transition
moment. It is worth noting that its orientation is fixed relative to
the local chain axis.
    A small amount (~ 0.1 to 1 % w/w) of labelled   polymer is mixed
in solution with normal polymer, in such a way that the final concentration of anthracene in the polymer sample is around 10^{-6} w/w. After
solvent drying, the samples are pressure moulded.

IV. ORIENTATION AND CHAIN RELAXATION IN POLYSTYRENE UNIAXIALLY STRET-
    CHED ABOVE Tg

A series of experiments on polystyrene has recently been performed in
our laboratory. The fluorescence polarization measurements were carried
out during the stretching using the optical equipment shown in Fig. 3,
adapted to a stretching machine specifically designed (14) for this purpose and represented in Fig. 4. This machine operates at constant strain

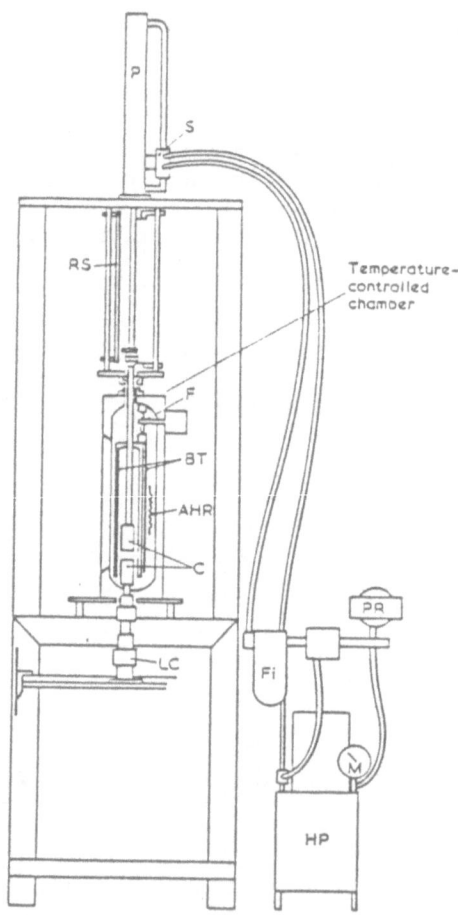

Fig. 4 - Stretching machine and temperature controlled chamber; P, dou-
         ble hydraulic plunger; S, servovalve; RS, resistance sensor;
         C, clamps; PR, pressure reservoir; M, manometer; HP, hydraulic
         pump; LC, load cell; F, fan; BT, blowing tubes; AHR, additio-
         nal heating resistances; F_i, filter

rate, $\dot{\varepsilon}$ (in a range from 2.10^{-3} s^{-1} to 2.10^{-1} s^{-1}) up to a 600 %

deformation for a sample initially 6 cm in length between the jaws. The
stress is simultaneously recorded. The temperature controlled chamber
can be regulated from room temperature up to 150°C. The temperature is
controlled within $\pm$ 0.002°C and is homogeneous along the stretching axis
to at least 0.033°C.

In the following, the draw ratio is defined as $\lambda = l/l_o$ (l_o is the
initial length of the sample between jaws, l is the length after drawing).
The stress has to be understood as the nominal stress defined by
$\sigma = F/S_o$ where F is the tensile strength and S_o the initial section of
the sample.

IV.1. Effects of temperature and strain rate

Anthracene labelled polystyrene chains (called PAP) with a number average molecular weight, $\overline{M}_n$ = 287,000, have been incorporated (at 1 % w/w
concentration) in a polystyrene matrix (PS 200) with a narrow M.W. distribution ($\overline{M}_n$ = 191,000, $\overline{M}_w$ = 207,000). The samples moulded under pressure and annealed to obtain transparent bubbleless samples and to avoid
any birefringence. Sample size was 8 cm long, 2 cm wide and 0.2 thick.
Tg of the resultant polymer blend measured by differential scanning
calorimeter at a heating rate 10°C min^{-1} was 107.5°C.

A typical example of stress-strain and orientation-strain curves is
given in Fig. 5. In this temperature range, polystyrene samples undergo
a homogeneous deformation. The stress-strain curves show two main
regions. The first, at small extension ratio, is characterized by a rapid
increase of stress and is attributed to the glassy deformation. The
second, at higher extension ratio, in which the stress increases slowly
with the deformation, reflects a rubber-like deformation, with a possible contribution from flow. In contrast, orientation curves, in which
the orientation is plotted against the extension ratio λ, show a regular increase of orientation during stretching.

The influence of temperature appears clearly on these two plots. It
can be seen that, at a given strain rate, the first part of the stress-
strain curve is very sensitive to temperature whereas the slope of the
second part decreases as temperature rises. It can also be noted that
the orientation falls continuously and at 135°C is too low to be measurable.

Fig. 6 shows the behaviour of stress and orientation at Tg + 9°C
for various strain rates. It is clearly seen that only the first part
of the stress-strain curve is affected by the strain rate, but the orientation is not modified within the accuracy of our experiments. Comparison with Fig. 7 shows that at a higher temperature at Tg + 14.5°C, both
the stress reflecting the rubber-like behaviour and the orientation are
functions of the strain rate.

It must be noted that the orientation given by our fluorescence
polarization measurements refers to the anthracene group, the transition
moment of which lies along the chain axis and is located in the middle
of the chain. This particular position implies that the measured orientation reflects the orientational behaviour of statistical chain segments of the central part of the chain. It seems difficult to state the
size of this central part precisely. Owing to the molecular weight of

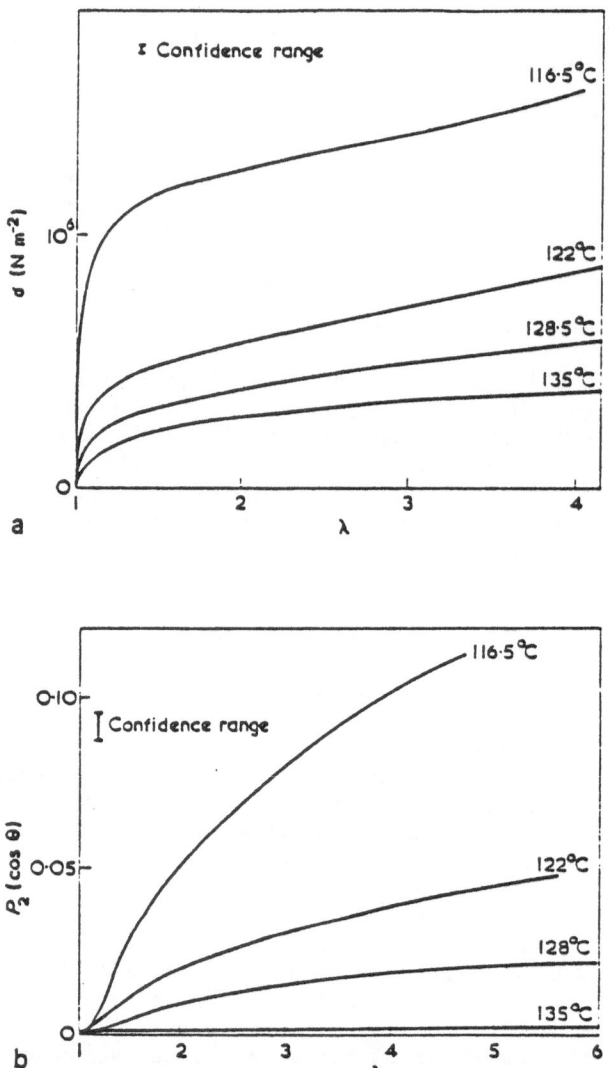

Fig. 5 - a/ Stress and b/ orientation function versus draw ratio as a
function of temperature . Strain rate, $\dot{\varepsilon}$ = 0.115 s^{-1} .

the labelled chain (M_n = 287,000), it is reasonable to assume that the
orientation of the chain-end segments is not taken into account by this
method.

Comparison between Fig. 5a and 5b seems to indicate that orientation does not originate from the glassy part of the stress but is related to the rubber-like component. At a given temperature, the slope of
the rubbery part of the stress-strain curve and its dependence on the

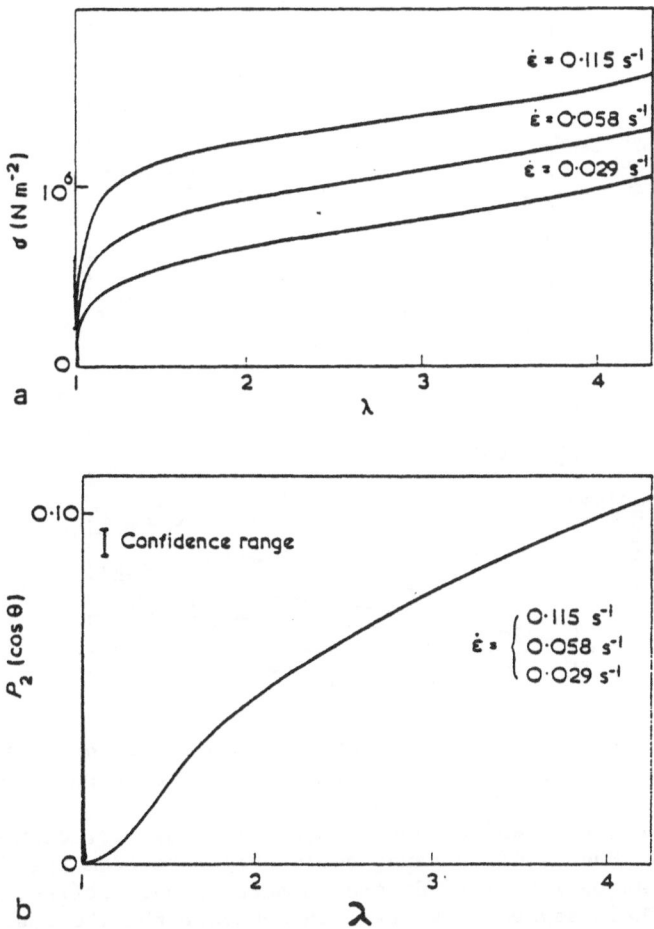

Fig. 6 -a/ Stress and b/ orientation function versus draw ratio as a function of strain rate. Stretching temperature, T = 116°5 C.

strain rate governs the orientation behaviour. The shrinking experiments performed on deformed samples at a temperature of 120°C for 24 h, show that, within our experimental conditions (strain rate, temperature, molecular weight), the contribution of the terminal flow zone is negligible.

Thus, in our experiments, the orientation observed through fluorescence polarization is not directly related to the complete true stress, but depends on the deformation in the rubbery plateau zone. This result is rather different from that which is obtained with birefringence measurements in which stress and orientation are related through a constant coefficient. Also, infra-red dichroism measurements performed in our laboratory with films issued from the same polystyrene batch lead to the

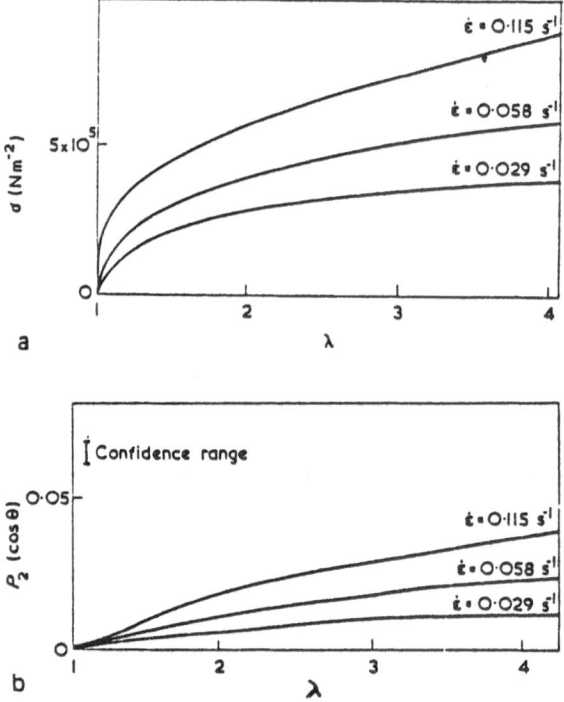

Fig. 7 - a/ Stress and b/ orientation function versus draw ratio as a
function of strain rate. Stretching temperature, T = 122°C.

conclusion that stress and orientation obtained by i.r. dichroism are
identical (16). Such a discrepancy could originate from the fact that
birefringence as well as i.r. dichroism measurements reflect an average
over all the chain segments, whereas in our case the fluorescence polarization measurements are only sensitive to the central part of the
chain.

From results presented in Fig. 5 b and 7 b it can also be noted
that an increase in temperature may have the same effect on orientation
as a decrease of strain rate, so that time and temperature seem to
affect the system in the same way. Such a behaviour must be attributed
at a molecular level to the participation of relaxation phenomena during
stretching.

IV.2. Influence of the molecular weight of the polymer matrix

In order to pursue the investigation of chain relaxation
during stretching, a set of experiments has been performed on the
influence of the molecular weight of the polystyrene matrix on the
orientation of the labelled polystyrene PAP 287 ($\overline{M}_n$ = 287,000). The
stretching was performed at 128.5°C for several constant strain rates $\dot{\epsilon}$.

The M.W. characteristics of the matrix chains are reported in Table I.

Polymer	$\overline{M}_n$	$\overline{M}_w/\overline{M}_n$
PS 200	190 000	1.17
PS 400	420 000	1.24
PS 600	660 000	1.15
PS 900	855 000	1.19
PS 1300	1 300 000	1.28
PAP 300	287 000	

Table I - Characteristics of the polystyrene
matrices

The orientation function of the PAP chain in the various matrices is shown as a function of the draw ratio in Fig. 8.

It can be seen from these plots that the orientation of the PAP chains is the same in PS 900 and PS 1300 matrices at a given strain rate, and quantitative comparison of these curves leads to the conclusion that the orientation of the labelled chain is independent of the strain rate for these two matrices. This is not the case for the other matrices. For $\dot{\varepsilon}$ = 0.029 or 0.059 s^{-1} the orientation of the PAP decreases as the MW of the matrix decreases. This effect tends to disappear when stretching is performed at higher strain rate (i.e., 0.115 s^{-1}) where the behavior of the fluorescent chain is the same within the accuracy of our experiments in the matrices from PS 400 to PS 1300, but the orientation is still significantly lower in PS 200 ($M_{matrix} < M_{PAP}$).
The influence of the strain rate and MW of the matrix is better demonstrated in Figure 9. The orientation function $< P_2(\cos\theta) >$ measured at a draw ratio λ_o = 4 is plotted against the ratio of the molecular weights of the matrix and the labelled chains for different strain rates. This variable affords a comparison between the relative length of the PAP chains and the matrix chains.
At this point, it is of interest to note that the measured orientation is related to the behavior of the central part of the chain and does not basically represent an average over all chain units. During stretching, polymer chains may experience relaxation. The characteristic time of an experiment at a given strain rate $\dot{\varepsilon}$ is on the order of $\dot{\varepsilon}^{-1}$, so that only relaxation mechanisms with times in the range of $\dot{\varepsilon}^{-1}$ can contribute.
From this point of view, some features appearing in Fig. 9 can be explained qualitatively. If the length of the chains of the matrix is greater than that of the PAP chains, at a given $\dot{\varepsilon}$ the relaxation of the labelled chain becomes slower, leading to higher orientation. On the other hand, if the surrounding chains are shorter than the PAP chains, the relaxation of these species is more important and lower orientation is observed. Such an experimental fact suggests a strong coupling between the relaxation of the labelled chain and the relaxation of the

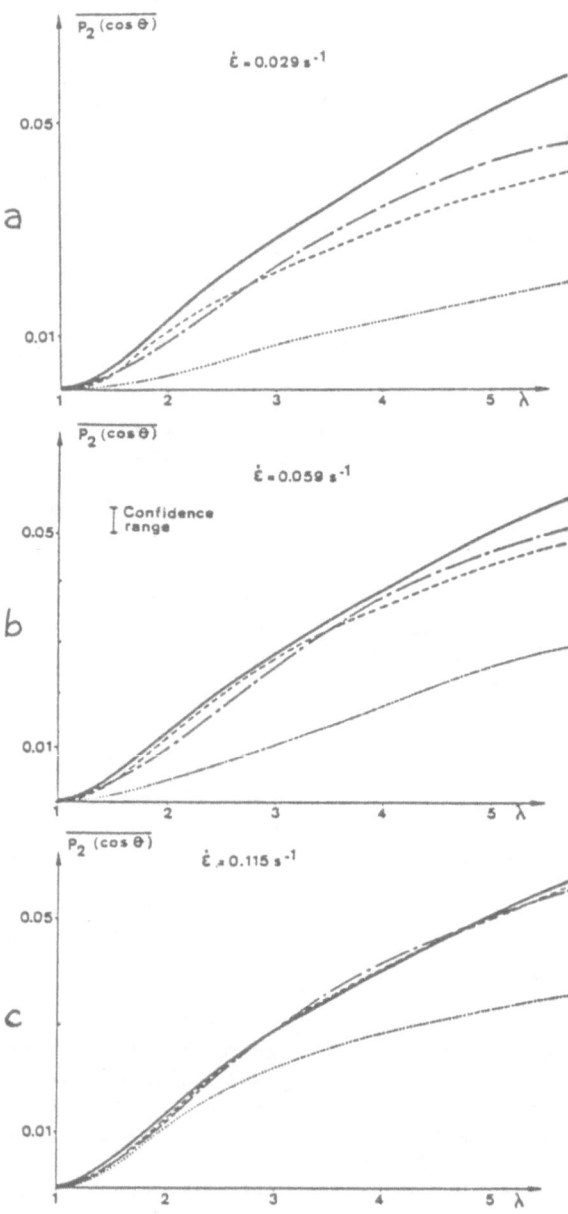

Fig. 8 - Orientation function versus draw ratio at various strain rates
a/ $\dot{\varepsilon}$ = 0.029 s^{-1}, b/ $\dot{\varepsilon}$ = 0.059 s^{-1}, c/ $\dot{\varepsilon}$ = 0.115 s^{-1}. The curves correspond to (•••••) PS 200, (− −) PS 400, (− · −) PS 600,
(——) PS 900 and PS 1300

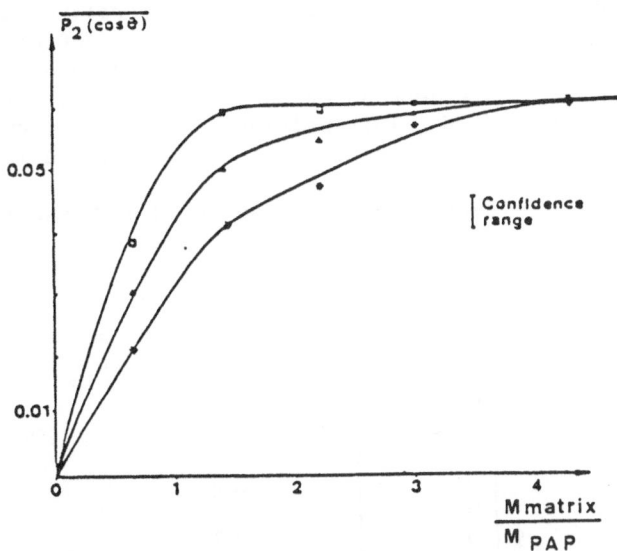

Fig. 9 - Orientation function measured at a draw ratio, $\lambda_o = 6$, as a function of the strain rate versus the ratio $M_{(matrix)}/M_{PAP}$ (□) $\dot{\varepsilon} = 0.115$ s^{-1}, (▲) $\dot{\varepsilon} = 0.059$ s^{-1}, (*) $\dot{\varepsilon} = 0.029$ s^{-1}.

matrix chains. The first macroscopic effect of this interaction consists in an apparent modification of the relaxation times of the labelled chain depending on the length of the surrounding chains. These conclusions are valid for the strain rates used here.

When the MW of the matrix increases, the apparent relaxation times are shifted toward larger values in such a way that no further influence of the strain rate can be discerned, under the experimental conditions considered here, and $P_2(\cos \Theta)$ approaches a limiting value. When the strain rate is lower, the time domain involved is larger and a higher matrix MW is required to reach this limit of the orientation function.

A deeper insight can be reached by considering the molecular models of polymer viscoelasticity which have recently been developed by Doi and Edwards (2, 3) based on the tube model. In this model, a chain is constrained within a hypothetical tube representing the highly entangled surrounding chains. Such a tube is fixed relative to the motion of the individual chain. The tube constraints have been formalized by Doi and Edwards through the slip-link model, in which a chain is trapped by small friction-less rings, through which it can pass freely and which represent the topological constraints. The spatial position of the slip-links is also fixed relative to the motion of the chain.

The relaxation of a suddenly- deformed polymer melt, the strain then being kept constant, is described in terms of three different relaxation processes occuring on various time scales.

We start from an affinely deformed chain and consider the relaxation of this chain in fixed surroundings.

   At short times, the slip-links act as fixed crosslinks and the
chain can relax between these entanglement points like a Rouse chain
with fixed ends. This first relaxation time τ_A, is independent of the
length of the chain, and only depends on the average number of monomers
between entanglement points, N_e. This relaxation process is quite fast
at a temperature above the glass transition temperature and it will
not be considered later.
   During the second relaxation process, which is pictured in Fig. 10.

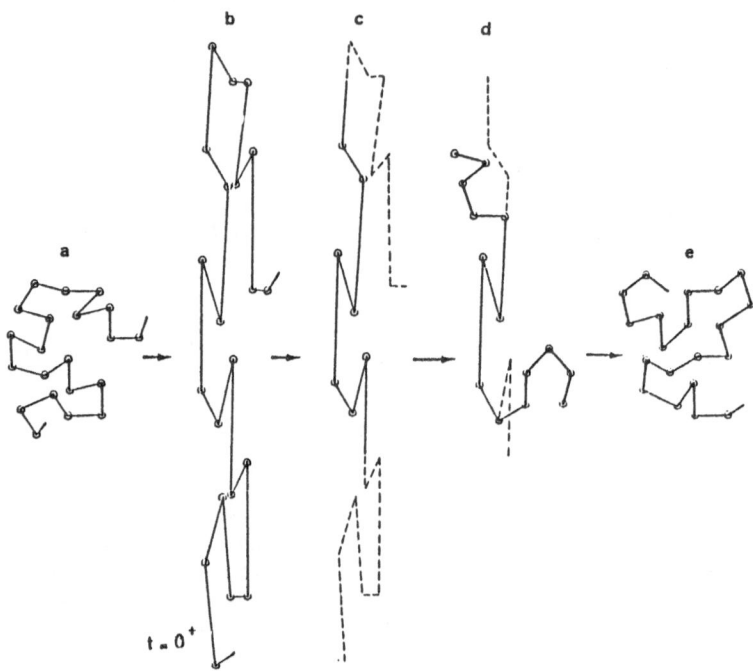

Fig. 10 - A given chain embedded in a network of chemically identical
         chains. Different relaxation stages of the primitive chain
         after a step uniaxial stretching : (a) initial isotropic
         state; (b) step-strained primitive chain at time 0^+; (c) pri-
         mitive chain at the end of the self-retraction process; (d) pri-
         mitive chain during reptation; (e) primitive chain after com-
         plete disengagement. Slip-links are represented by small cir-
         cles. For easier comparison, in each figure the state of the
         primitive chain at the end of the previous stage is represen-
         ted by a dotted line. Newly created parts of the tube are
         drawn in heavy line.

the chain shrinks into the deformed tube in order to recover the equilibrium monomer density per unit arc length. This process is characterized by a relaxation time τ_B, related to τ_A by :

$$\tau_B = 2(N_o/N_e)^2 \, \tau_A$$

where N_o is the number of monomers per chain.

In the last stage, shown in Fig. 10, the chain disengages itself from the original deformed tube by the reptation process proposed by de Gennes (1). This occurs in a time τ_C such that

$$\tau_C = 6(N_o/N_e)^3 \, \tau_A = 3(N_o/N_e) \, \tau_B$$

Analytical expressions for these three relaxation times show that if $N_o \gg N_e$ (which is a basic condition for these models of entangled chains) the processes are well separated in time.

Coming back to our study, we should obtain at a given strain rate, according to this model, the same orientation of the labelled chain, independently of the MW of the matrix when $M_{matrix} > M_{PAP}$, i.e., if the environment is fixed relative to the motion of the labelled chain. Such behaviour is observed qualitatively at the highest strain rate (0.115 s^{-1}) but for lower rates, the orientation still increases with increasing MW of the matrix. Before trying to improve on this model, it is useful to determine the relaxation motions involved in our experiments. The first relaxation process occurs at very short times τ_A and does not create a change in the orientation since the distribution of the end-to-end vector of the labelled chains is not affected, nor is any vector joining two consecutive entanglement points. As concerns the third relaxation process, the fact that the deformation was recoverable by annealing above Tg proves that no flow occured and thus the reptation time τ_C was not reached under the experimental time and temperature conditions. Hence, in our experiments the orientation is mostly affected by the shrinking of the chain into its deformed tube.

During stretching both PAP chains and matrix chains experience retraction occuring respectively in times τ_B^{PAP} and τ_B^{matrix}.

Since the labelled chain is the same in the different matrices, we can check the retraction process by plotting the orientation at a given draw ratio as a function of the dimensionless parameter $\dot{\varepsilon} \, \tau_B^{matrix}$, since in a stretching experiment we are dealing with a time range of the order of $\dot{\varepsilon}^{-1}$. In fact, as τ_B^{matrix} scales as M^2, it is equivalent to plot the orientation function versus the quantity $\dot{\varepsilon} \, M_{(matrix)}^2$. As shown in Fig. 11 all the points fall fairly well on a single curve, proving that the orientation function under our experimental conditions is completely controlled by the chain retraction process.

Such an effect of the molecular weight of the polymer matrix on the retraction time of a labelled chain is not predicted for the long chains considered in this study. Indeed, if we consider that N_e for polystyrene is around 160, the labelled chain participates to about 20 entanglements and even the smallest matrix molecular weight corresponds to more than 10 entanglements per chain. Under such conditions, the Doi-Edwards theory does not predict any coupling between the behaviour of one given chain and that of the surrounding chains. The Doi-Edwards model would correspond to a free chain in a polymer network because the surrounding chains are not considered as able to retract or reptate. For matrix chains of a finite length, after a step deformation, the retraction pro-

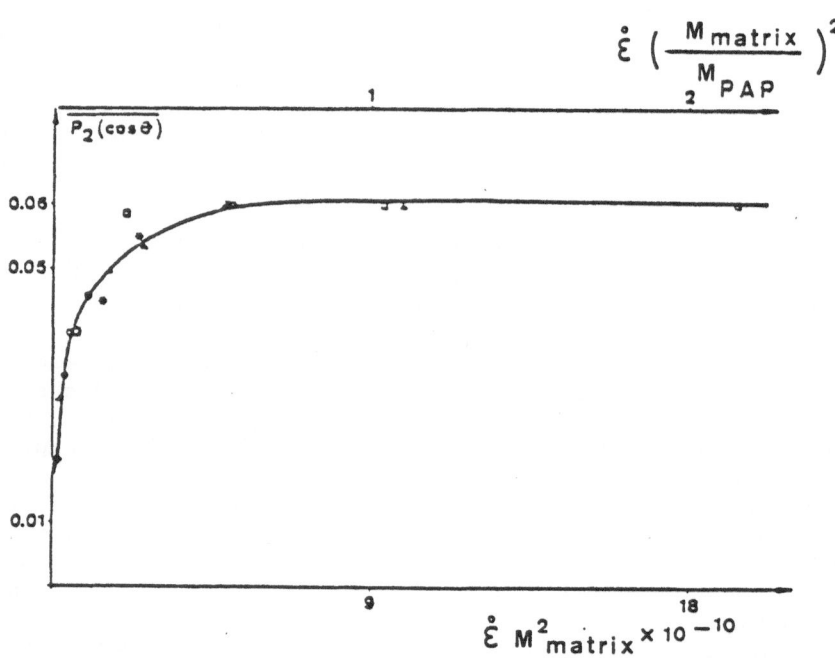

Fig. 11 - Orientation function measured at $\lambda_0 = 6$, versus the parameter $\dot{\varepsilon} \, M^2_{(matrix)}$: data obtained from ($\square$) $\dot{\varepsilon} = 0.115$ s^{-1}, ($\blacktriangle$) $\dot{\varepsilon} = 0.059$ s^{-1}, ($\ast$) $\dot{\varepsilon} = 0.029$ s^{-1}; ($\bigcirc$) data obtained for $M_{(matrix)} = M_{PAP}$

cess inside their own tubes and the reptation process out of the original tubes occur. During these relaxation processes, the constraints on the labelled chain arising from the matrix chains will be changed. For example, when a matrix chain end reptates away from the neighborhood of the labelled chain, the latter has an opportunity to escape out of its original tube. During this time another matrix chain (or the same one) may reptate across the old tube so that the tube of the labelled chain is permanently modified. In this way, the labelled chain will relax faster. This coupling mechanism between the relaxation of a labelled chain and the reptation of the surrounding chains is called the tube renewal. It occurs both for deformed or undeformed chains. The same type of mechanism has been applied to deformed chains considering the retraction processes (17). The effect of the retraction processes of the surrounding chains on the retraction of the labelled chain is depicted in Fig. 12. As the retraction mechanism only occurs in deformed systems, the coupling shown in Fig. 12 and called tube relaxation, is specific for these systems. Of course, the relaxation of the labelled chain is enhanced by the tube relaxation mechanism and the shorter are the matrix chains, the faster the labelled chain relaxes. As the retraction characteristic time scales as M^2, the effect of the matrix molecular weight on the contribution of the tube relaxation to the relaxation of the labelled chain should be proportional to M^2_{matrix}. This matrix

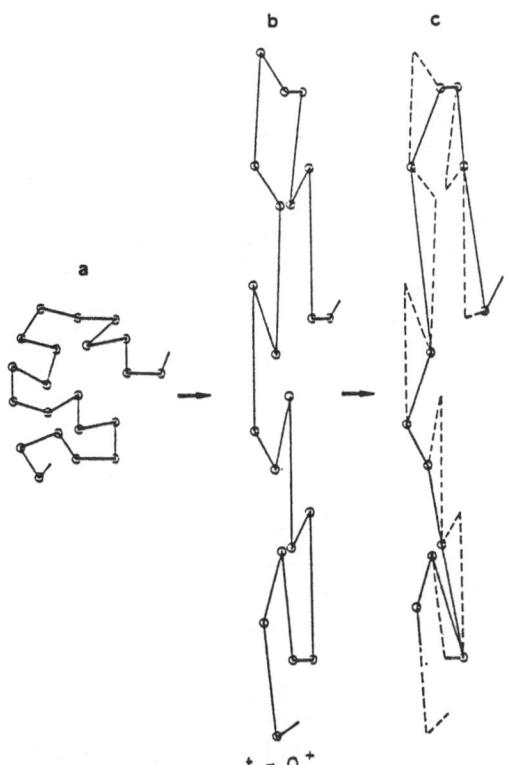

$$t \simeq 0^+$$

Fig. 12 - Modification of the path of a labelled chain due to the retraction of the matrix chains: (a) isotropic initial stage;
(b) step-stretched primitive chain of the labelled chain at
time 0^+; (c) path of the primitive chain of the labelled
chain after retraction of the surrounding chains, showing
the local loosening due to random loss of slip-links.

M.W. dependence is actually observed, as shown in Fig. 11. A comparison
between the observed and calculated effects of the matrix M.W. on the
orientation of a labelled chain is shown in Fig. 13.

A satisfactory agreement is obtained, proving the validity of the
proposed improvement of the Doi-Edwards model by this self-consistent
approach.

IV.3 - Influence of the molecular weight of labelled chains

Another interesting feature is the influence of the molecular weight
of the labelled chain when that of the polymer matrix is kept constant.
Such experiments have been recently performed in our laboratory (18)
at 128.5°C with a PS 160 matrix. Results are presented in Fig. 14 at a
strain-rate value $\dot{\varepsilon} = 0.115$ s^{-1}; similar behaviour is observed at other
strain rates.

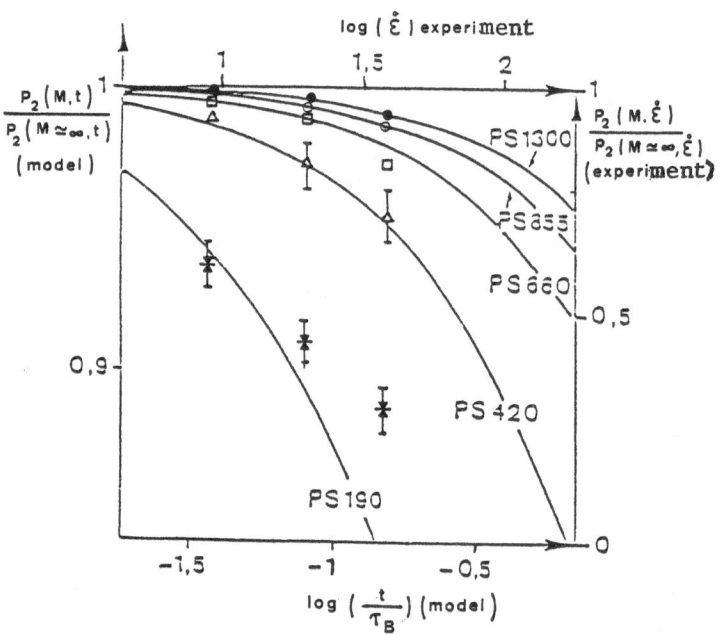

Fig. 13 - Comparison of the observed and calculated matrix molecular
         weight effect on the orientation of an anthracene-labelled
         polystyrene chain. The full curves represent the calculated
         behaviour

      First, there is an increase in orientation with the molecular
weight of the labelled chain.
      However, a more surprising result is the rather high orientation
obtained for PAP 17 chains. Indeed, as the molecular weight of this
labelled polymer is around the mean molecular weight between entanglements
for PS ($M_e \simeq 16000$), one would have expected a rather low orientation
or even no orientation if only topological constraints were considered.
Such chains are too short to be oriented efficiently by the deformation
of the physical network. On the other hand, it seems that their orientation comes from the anisotropy of the surrounding medium. Note that similar orientation effects have been observed for pendant polyisoprene
chain in a chemically crosslinked network (19). The pendant chains
labelled inside the chain or at the end of the chain exhibit an orientation which increases with λ, although always remaining lower than the
orientation of the labelled chain involved in the permanent network.
Furthermore, free fluorescent probes, made up of 9,10-dialkylanthracene
with 16 CH_2 groups, are oriented at $\lambda > 3$. Thus from these results it
appears that, in addition to orientation arising from topological effects
(a physical entanglement network) and which could be described by mole-
cular-viscoelasticity theories based on the tube concept, there is
another contribution arising from the interactions with the surroundings
anisotropic medium.

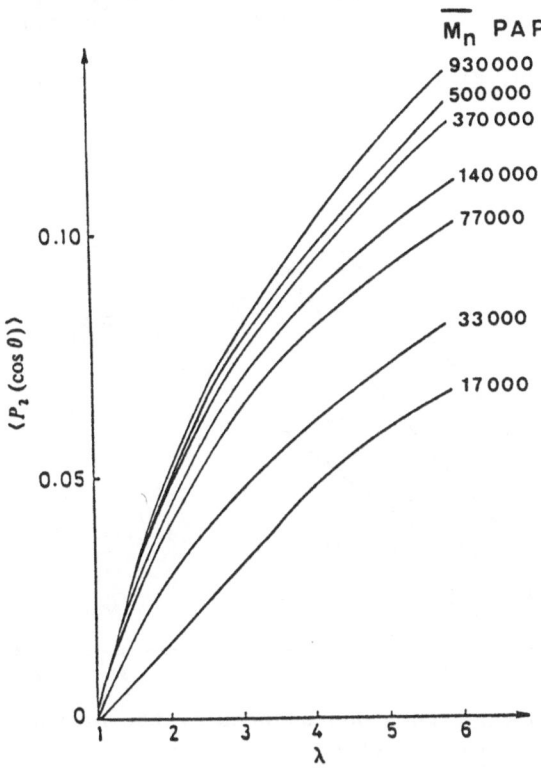

Fig. 14 - Orientation function plotted against draw ratio for various
          molecular weights of the PAP-labelled chains in a PS 160
          matrix. Stretching temperature, T = 128°5 S.

Although it could be argued that this effect is specific to fluorescence polarization and originates from the perturbation introduced
by the fluorescent label, the fact that the fluorescence polarization
behaviour observed at 128.5°C for PAP 287 in a matrix of $\overline{M}_n$ = 200,000,
is similar to that found from infra-red dichroism measurements for the
matrix itself indicates that the contribution from the anisotropic
medium is also involved to some extent in infra-red dichroism. Indeed
the statistical unit of a polymer chain corresponds to an anisotropic
object which can also be oriented by interaction with the strained
surroundings.

V. ORIENTATION AND MOBILITY IN MOBILE, UNIAXIALLY STRETCHED POLYISO-
   -PRENE NETWORKS

As described here above, fluorescence polarization can be applied to
uniaxial mobile systems. Such measurements have been performed in our
laboratory during stretching of polyisoprene networks (20), using the
optical equipment shown in Fig. 3 adapted to a stretching machine operating at constant cross-head displacement.

Labelled polyisoprene chains with an anthracene group in the middle were incorporated in polyisoprene Shell IR 307 before crosslinking by dicumyl peroxide.

As an illustration, Fig. 15 shows the strain and temperature depen-

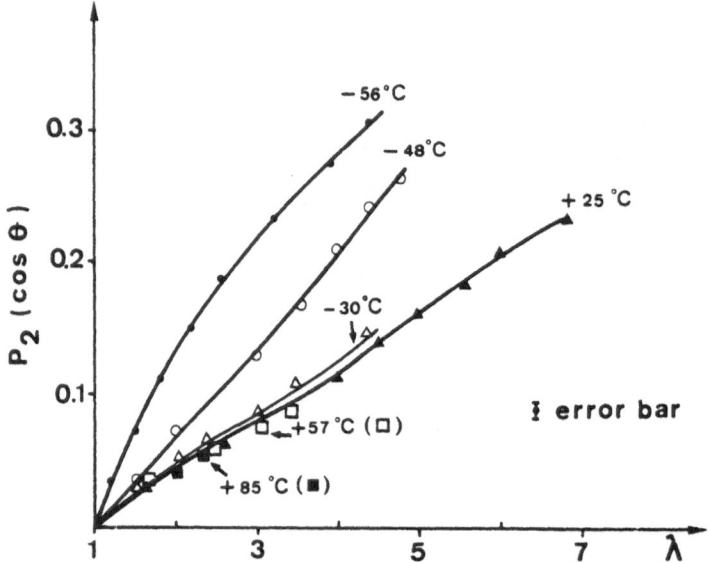

Fig. 15 - Strain and temperature dependences of the orientation func-
          tion versus the draw ratio

dence of the orientation function.

Concerning the mean mobility amplitude, M, above defined, its strain dependence is reported in Fig. 16 for various crosslinking densities.

VI. CONCLUSION

The studies reported above on the orientation of uniaxially stretched polymer melts show that fluorescence polarization is a very powerful technique to investigate the molecular behaviour.

The measurements can be performed either on stretched samples below Tg or during stretching above Tg. As only the orientation behaviour of the labelled species is observed, it allows one to look at the molecular weight dependence of the chain relaxation processes of the labelled chain as well as the effect of the molecular weight of the matrix. Experiments carried out on polystyrene have clearly shown that the Doi-Edwards treatment of the shrinking of the chain in its tube has to be improved and a modified model has been proposed.

Fluorescence polarization should lead in the future to a new insight in molecular viscoelasticity of polymers.

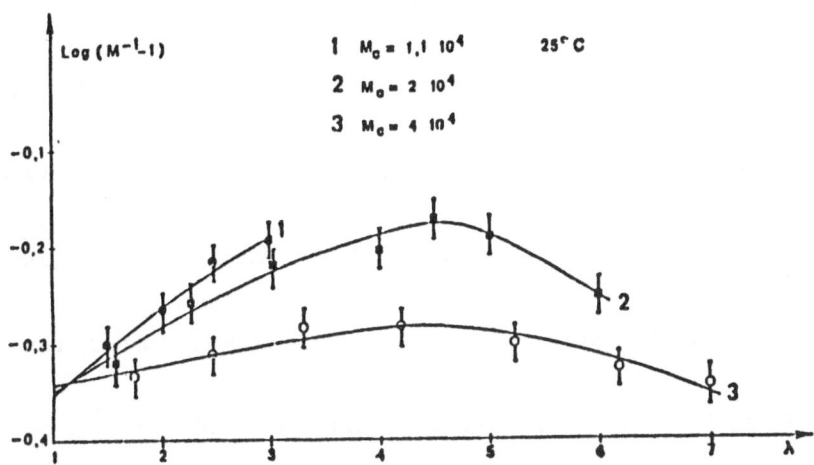

Fig. 16 - Mean mobility amplitude at 25°C for anthracene labelled polyisoprene networks versus the draw ratio. M_c denotes the average molecular weight of chain strands between adjacent crosslinks.

References

1. de Gennes, P.G. : 1971, J. Chem. Phys., 55, pp 572-579

2. Doi, M., Edwards, S.F. : 1978, J. Chem. Soc., Faraday Trans 2, 74,
pp 1789-1801, pp 1802-1817, pp 1818-1832

3. Doi, M. : 1980, J. Polym. Sci., Polym. Phys. Ed., 18, pp 1005-1020

4. Monnerie, L.: 1983, Faraday Symp. Chem. Soc., 1983, pp. 57-81

5. Bower, D.I. : 1981, J. Polym. Sci., Polym. Phys. Ed., 19, pp. 93-107

6. Monnerie, L. : 1982, in "Static and Dynamic Properties of the Polymeric
Solid State" ed. by R.A. Pethrick and R.W. Richards (N.A.T.O. Series)
D. Reidel, Dordrecht, pp. 383-413.

7. Jarry, J.P., Monnerie, L.: 1978, J. Polym. Sci., Polym. Phys. Ed.,
16, pp. 443-445

8. Frehland, E. : 1975, Z. Naturforsch., 30a, pp. 1241-1246

9. Despers, C.R., Kimura, I.: 1967, J. Appl. Phys., 38, pp. 4225-4233

10. Nobbs, J.H., Bower, D.I., Ward, I.M., Patterson, D. : 1974, Polymer,
15, pp. 287-300

11. Jarry, J.P., Sergot, P., Pambrun, C., Monnerie, L.: 1978, J. Phys.E: Sci. Instrum., pp. 702-706

12. Pinaud, F., Jarry, J.P., Monnerie, L.: 1982, Polymer, 23, pp. 1575-1580

13. Nishijima, Y., Onogi, Y., Asai, T.: 1966, J. Polym. Sci., Part C, 15, pp. 237-250

14. Fajolle, R., Tassin, J.F., Sergot, P., Pambrun, C., Monnerie, L.: 1983, Polymer, 24, pp. 379-382

15. Tassin, J.F., Monnerie, L.: 1983, J. Polym. Sci., Polym. Phys. Ed., 21, pp. 1981-1992

16. Lefebvre, D., Jasse, B., Monnerie, L.: 1983, Polymer, 24, pp. 1240-1244

17. Viovy, J.L., Monnerie, L., Tassin, J.F. : 1983, J. Polym. Sci., Polym. Phys. Ed., 21, pp. 2427-2444

18. Tassin, J.F., Ayrault, C. : private communication

19. Queslel, J.P.: 1982, Thèse de Dr-Ingénieur, Paris

20. Jarry, J.P., Monnerie, L.: 1980, J. Polym. Sci., Polym. Phys. Ed., 18, pp. 1879-1890.

PHOTOCHEMICAL METHODS FOR MEASURING POLYMER DIFFUSION

Barton A. Smith
Department K91
IBM Almaden Research Center
650 Harry Road
San Jose, CA 95120-6099

ABSTRACT. Over the past 15 years, a number of transient optical grating techniques have been developed for measurements of the transport properties of materials. Such methods have been used to measure the tracer diffusion coefficients of polymer molecules which have been labeled with photochromic or fluorescent dyes. The present paper describes the common features of these techniques, and gives an example of how Fluorescence Redistribution After Pattern Photobleaching has been used to study the diffusion of polymer molecules in the melt.

1. PRINCIPLES

1.1. Optical Interference

When two monochromatic, mutually coherent beams of light intersect, a standing wave interference pattern of light intensity is established in the region of intersection. The spatial period p, or peak-to-peak distance, of the standing wave pattern is determined by the wavelength λ of the light and the angle between the two beams, Θ.

$$p = \frac{\lambda}{2 \sin(\Theta/2)} \qquad 1$$

For the special case in which the two light beams enter a sample through a single plane surface, and the normal to that surface lies in the plane of the beams and bisects the angle between them, p in the sample will be independent of the index of refraction of the material, and can be calculated from the external wavelength and angle. The light intensity I as a function of position x will be:

$$I(x) = A + B \sin ax \qquad 2$$

where A if the average light intensity, B is the amplitude of the intensity modulation, and a is the spatial frequency: $a = 2\pi/p$. If the two light beams are of equal intensity, $A = B$; otherwise $A > B$.

M. A. Winnik (ed.), Photophysical and Photochemical Tools in Polymer Science, 397–406.
© *1986 by D. Reidel Publishing Company.*

1.2. Creation of Grating

There are many ways in which the optical properties of a material can be modified by the interaction of that material with light. For example, light can be absorbed and converted into heat, which raises the temperature of the material and thus alters its refractive index. In the present study, we will be interested in photochemical reactions which produce detectable changes in the optical properties of molecules of interest. Whatever the type of interaction, the periodic pattern of intensity (equation 2) produced by the intersecting coherent beams of light will induce in the sample a periodic pattern in the optical properties. This pattern can be detected in various ways, for instance by its ability to diffract light. The creation of a diffraction grating by optical interference forms the basis for optical holography.

In general, the response of the sample to light may not be linear in intensity, and thus higher order spatial frequency terms may be present in the induced grating. For our case in which a particular molecular species is created by the light, the concentration C as a function of position in the sample can be represented by:

$$C(x) = C_{ave} + \sum_{i=1}^{\infty} H_i \sin iax \qquad\qquad 3$$

where C_{ave} is the average concentration of the optically induced species, and H_i is the amplitude of each Fourier component present in the pattern.

1.3. Decay of Grating

After the initial exposure to light, the induced pattern may not be permanent, but may disappear in time after the light source is removed. We will be interested in the case in which the decay of the pattern is caused by the diffusion of optically modified molecules. Assuming that the concentration of the molecules of interest as a function of time and position is changed only by diffusion, and not by another process such as chemical reaction, the concentration must obey the diffusion equation:

$$\frac{\partial C(x,t)}{\partial t} = D\frac{\partial^2 C(x,t)}{\partial x^2} \qquad\qquad 4$$

in which t is the time, and D is the diffusion coefficient of the optically labeled molecules. The solution to Equation 4 with initial conditions given by Equation 3 is:

$$C(x,t) = C_{ave} + \sum_{i=1}^{\infty} \exp(-D(ia)^2 t)H_i \sin iax \qquad\qquad 5$$

We see from Equation 5 that the form of the pattern remains the same as diffusion progresses, and that the pattern amplitude simply decays with time. We also see that the higher spatial frequency terms decay more rapidly.

1.4. Detection of Grating

If the optically induced species alter the refractive index or absorbance of the sample, a phase or amplitude diffraction grating is produced. If a (non-perturbing) light beam is directed through the sample, a diffracted beam will be produced for each spatial frequency present. The intensity of each diffracted beam will be proportional to the amplitude of its particular Fourier component in the pattern, and each beam will decay exponentially with time.

$$I_i(t) = I_i(0)\exp(-D(ia)^2 t)\qquad\qquad 6$$

I_i is the intensity of the i'th diffracted beam. Thus the diffusion coefficient can be calculated from the measured rate of the decay and the known spatial frequency.

In Fluorescence Redistribution After Pattern Photobleaching (FRAPP), fluorescence emission is used to detect the concentration of the molecules of interest as a function of position and time. The light bleaches, or irreversibly photochemically destroys, the fluorescence of dye molecules in the illuminated regions of the sample. The pattern decays by the diffusion of the remaining, unbleached molecules.

1.5. Characteristics of Transient Grating Techniques

Gratings can be created with dimensions comparable to the wavelength of light or longer. The distance over which the diffusion takes place, and thus the time scale of the measurement, can be adjusted for convenience. Time scales from nanoseconds upwards can be measured.

In the case of mass diffusion measurements, the concentration gradients are created in situ, without any mechanical manipulation of the samples. Thus artifacts which may arise due to surface phenomena are eliminated.

Diffusion is measured along a specified direction. For anisotropic samples, transport properties can be measured as a function of direction.

For mass diffusion experiments, it is required that the lifetime of the diffusing species must be much longer than the time required for diffusion. This requirement places a limitation on the use of thermally reversible photochromic dyes for slow diffusion. It is also necessary that the method of detection not significantly perturb the grating.

2. HISTORY

In 1971, Eichler, et al. of the I. Physikalisches Institut der Technischen Universität Berlin, reported the first detection of a transient grating.[1] They used a ruby laser to produce a thermal refractive index grating in dye-containing methanol. In 1972, they reported additional measurements, including the time decay of the grating due to thermal diffusion,[2] on a time scale of 1 ms.

In 1973, Pohl, et al. of the IBM Zurich Research Laboratory reported the measurement of the thermal diffusivity in NaF as a function of temperature.[3] The grating was created by a CO_2 laser and detected by its diffraction of a HeNe laser beam. The time scale was 100 μs. They called this technique "Forced Rayleigh Scattering".

In 1978, Salcedo, *et al.* at Stanford University reported the measurement of singlet electronic excitation transport in a molecular crystal.[4] The time scale of their measurement was 1 ns. Also in 1978, the first reports were published of the measurement of mass transport by transient grating techniques. Hervet, *et al.* of the Collège de France reported the translational diffusion coefficient of the photochromic dye methyl red in an aligned liquid crystal, as a function of direction.[5] The time scale was 0.1 s. Smith and McConnell at Stanford University used Fluorescence Redistribution After Pattern Photobleaching to measure the diffusion coefficient of dye-labeled phospholipid in oriented multibilayer films.[6] The time scale was 100 s.

Holographic techniques have been used to measure diffusion of polymer molecules in solution by Léger, *et al.* of the Collège de France,[7] by Coutandin and Sillescu of the Institut Für Physikalische Chemie der Universität Mainz,[8] and by Nemoto, *et al.* of the University of Wisconsin.[10] Smith, *et al.* at IBM Almaden Research Center have used FRAPP to measure the diffusion of polymer molecules in the melt.[9]

3. FLUORESCENCE REDISTRIBUTION AFTER PATTERN PHOTOBLEACHING

3.1. Principles

In FRAPP, an intense burst of laser light is used to irreversibly photochemically destroy (bleach) fluorescent dye molecules in the illuminated regions of the sample. FRAPP differs from Forced Rayleigh Scattering (FRS) in that fluorescence rather than diffracted beam intensity is used to detect the amplitude of the transient grating. Although other methods have been used, we will be concerned here with the creation and decay of the grating as described in sections 1.1 through 1.3. Since the species to be detected are destroyed by the initial burst of laser light, the concentration C as a function of position in the sample and time can be represented by:

$$C(x,t) = C_{ave} - \sum_{i=1}^{\infty} \exp(-D(ia)^2 t) H_i \sin iax \qquad 7$$

where C_{ave} is the average concentration of the fluorescent species, and H_i is the amplitude of each Fourier component present in the pattern.

After the pattern is created, the laser intensity is reduced by a factor between 10^{-3} and 10^{-6}. Thus the same wavelength of light, and the same pattern of illumination, which were used to create the grating are used to detect its decay. As fluorescent molecules diffuse from the dark regions of the sample into the illuminated regions, the total observed fluorescence intensity increases. The fluorescence intensity from each volume element of the sample is proportional to the concentration of the dye molecules and to the excitation light intensity. The total fluorescence intensity from the sample, I_F, is given by:

$$I_F(t) = \int_{x=0}^{2\pi/a} QC(x,t)I(x) = QAC_{ave} - \pi QBH_1 \exp(-Da^2 t) \qquad 8$$

where $C(x,t)$ is given by Equation 7, $I(x)$, A, and B by Equation 2, and H_1 is the amplitude of the first order Fourier component of $C(x)$. Q is the fluorescence yield of the dye molecule.

We see from equation 8 that the total fluorescence intensity from the sample is at a minimum at $t = 0$ (immediately after bleaching), and that the intensity increases with time towards a steady state value. This process is sometimes referred to as "fluorescence recovery". The sinusoidal pattern of illumination selectively detects the first order Fourier component of $C(x)$, and thus the recovery is described by a single exponential function. The recovery has a time constant of $1/Da^2$.

3.2. Comparison with Forced Rayleigh Scattering

The apparatus required for FRAPP and FRS are almost identical. The two techniques are applicable to the same types of samples. Even the same fluorescent dye molecules can be used. In principle, both types of detection could be used simultaneously in a single experiment. In general, one or the other technique will be easier to perform on a particular type of sample or with particular dye labeled molecules which are available.

3.2.1. Advantages of Forced Rayleigh Scattering. In FRS, the interference pattern is used only to create the grating, and a simple laser beam is used to detect it. Thus the requirements for mechanical stability of the optical system are less stringent for FRS than for FRAPP. In FRAPP, the interference pattern must remain in correct alignment on the sample during the entire course of the measurement. Motion of the pattern due to vibration or air convection currents causes noise in the FRAPP measurement. Since a diffracted laser beam is used for FRS detection, the light collection optics are very simple. It is also easy to direct the beams through small windows in a sample chamber. In FRAPP the fluorescence is emitted in all directions, and it is desirable to collect the light with high numerical aperture optics for maximum sensitivity.

If thermally reversible photochromic dyes are used for FRS, the same molecules can participate in repeated measurements, after a suitable relaxation time. Thus a small quantity of sample can be used many times to make measurements as a function of temperature or concentration.

3.2.2. Advantages of FRAPP. Fluorescence spectroscopy is a very sensitive technique, and very small numbers of molecules can be detected. FRAPP has been used to measure the diffusion coefficient of lipids in a single molecular monolayer in which only 1% of the lipid molecules were labeled.[11] Thus the signal-to-noise ratio of FRAPP measurements can be very high, and a single measurement (no averaging) can yield a precise value for the diffusion coefficient. Very small amounts of labeled material are required for FRAPP measurements. Laser light scattered by the sample or its container does not interfere with the detection of fluorescence, since the wavelengths are different. Both coherent and incoherent scattering can cause difficulties in FRS measurements.

3.3.3. Experimental problems in FRAPP. The most difficult part of FRAPP experiments on synthetic polymers lies in the preparation of well characterized, monodisperse, labeled molecules. In addition, the necessity for using labeled molecules can lead to difficulties in the measurements themselves. For instance, the fluorescent labels may aggregate in the sample, or the bleaching reaction may alter the sample in some way. Fortunately, it is often possible to avoid this type of problem by working at sufficiently low concentrations of the labeled molecules. One can verify the absence of label-induced effects by measuring the

diffusion coefficient as a function of the concentration of labeled molecules, and as a function of the amount of bleaching. Some dyes can bleach reversibly, and regain their fluorescence by chemical reaction. To rule out this possibility, one can make measurements over a range of pattern periods to see that the time constants change as the square of the pattern spacing (equation 8). It is also necessary to make sure that the light used for observation of the fluorescence does not produce a significant amount of bleaching.

4. POLYMER MELT DIFFUSION

A study of the diffusion of polymer molecules in the melt will be used to illustrate the application of FRAPP to polymer diffusion. In this example, the diffusion coefficient of poly(propylene oxide) has been measured as a function of concentration. The results are discussed in terms of the reptation model. A more complete description of the materials and methods may be found elsewhere.[12, 13]

4.1. Background

The reptation model plus scaling arguments have been used to predict the concentration dependence of the diffusion of polymer chains in solution for concentrations $C > C_E$ where C_E is the concentration for the onset of entanglement.[14, 15] These theories predict that the diffusion coefficient in good solvents is proportional to $C^{-7/4}$, in agreement with data from FRS measurements.[7] A number of measurements of the diffusion of polystyrene in good solvents, using a variety of experimental techniques, suggest that the power law $C^{-1.75}$ is observed only in a limited concentration range just above C_E.[16] Interpretation of the results for higher concentrations is complicated by changes in the magnitude of excluded volume effects, and by changes in the monomeric friction coefficient with concentration.

The interpretation of measurements of diffusion coefficients in the melt is not complicated by a concentration crossover between good and poor solvent behavior. An investigation of the dependence of the diffusion coefficient upon concentration in the melt was thus chosen as a test of the reptation model. For this study, poly(propylene oxide) (PPO) having a molecular weight ($M_W = 32\ 000$) which is approximately five times the critical molecular weight for entanglement ($M_C = 6000$) was diluted to various concentrations with PPO of much lower molecular weight ($M_W = 1000$). In order to keep the concentration of labeled molecules low (0.1% by weight), only a portion of the higher molecular weight molecules were labeled. Each sample contained this fixed amount of labeled molecules, and varying proportions of the higher and lower molecular weight PPO.

4.2. Measurements

The polymer melt samples were contained in fused silica cuvettes having dimensions 1 cm $\times$ 3 cm $\times$ 100 μm. The light source was an argon-ion laser producing 0.5 W at 488 nm. Total power incident on the sample for photobleaching was 0.2 W. Duration of the bleaching exposure ranged from 0.2 to 1.0 s, which produced initial reductions in the observed fluorescence intensity of from 25% to 75%. The total power of the two beams was reduced to 1 μW for observation of the fluorescence recovery. The illuminated region of the sample was 100 μm thick and approximately 0.5 mm in diameter. The pitch of the periodic pattern was varied within the range 4 μm to 28 μm by changing the angle of intersection of the beams. It is necessary that the diameter of the laser beams be much greater than the pattern pitch, so that equation 3, which assumes an infinite pattern, is applicable. Diffusion coefficients were deduced from a least-squares fit of a single exponential function to the fluorescence recovery data. All measurements were made at 25° C.

4.3. Materials

Fluorescent-dye-labeled PPO was synthesized with exactly one fluorophore in the polymer backbone. This was achieved by initiating the polymerization of propylene oxide with the dye 4-diethanolamino-7-nitrobenzofurazan with zinc hexacyanocobaltate as catalyst in tetrahydrofuran. A fraction of this dye-labeled polymer having molecular weight $M_N = 33\ 600$ and a narrow molecular weight distribution, $M_W/M_N = 1.1$, was selected by size exclusion chromatography. Two unlabeled PPO samples were used for these experiments. The lower molecular weight polymer, $M_W = 1000$, $M_W/M_N = 1.1$, was used as the solvent. The higher molecular weight polymer, $M_W = 32\ 000$, $M_W/M_N = 1.6$, had approximately the same molecular weight as the labeled material. Samples were prepared for diffusion measurements by mixing the three PPO components in dichloromethane solution and then removing the dichloromethane under vacuum at 50° C. The fluorescent-dye labeled PPO was always present in the sample at a concentration which produced an absorbance of 0.5 cm^{-1} at 488 nm.

4.4. Results

Diffusion coefficients of the fluorescent-dye labeled PPO are plotted in Figure 1 as a function of the total weight fraction of the high molecular weight PPO in the solution. The diffusion coefficients span almost two orders of magnitude. The extremities of this range are $D = 1.1 \times 10^{-10}$cm^{2}s^{-1} for the self-diffusion of labeled PPO in unlabeled PPO of the same molecular weight, and $D = 4.8 \times 10^{-9}$cm^{2}s^{-1} corresponding to the diffusion of labeled PPO in the low molecular weight solvent. The straight line in Figure 1 is the best fit by least squares to the five points corresponding to concentrations of the high molecular weight polymer exceeding 65% by weight. The slope of this line on the logarithmic scale is -2.2. Thus, the dependence of the diffusion coefficient on concentration at the limit of high concentration is given by $D \propto C^{-2.2}$.

4.5. Discussion

According to the reptation model, the center-of-mass diffusion coefficient will be inversely proportional to the terminal relaxation time, τ_t, which is the time required for complete renewal of the "tube" by reptation.[14]

$$D_{rep} \propto \frac{R_0^2}{\tau_t}$$

9

where R_0^2 is the mean squared end-to-end distance.

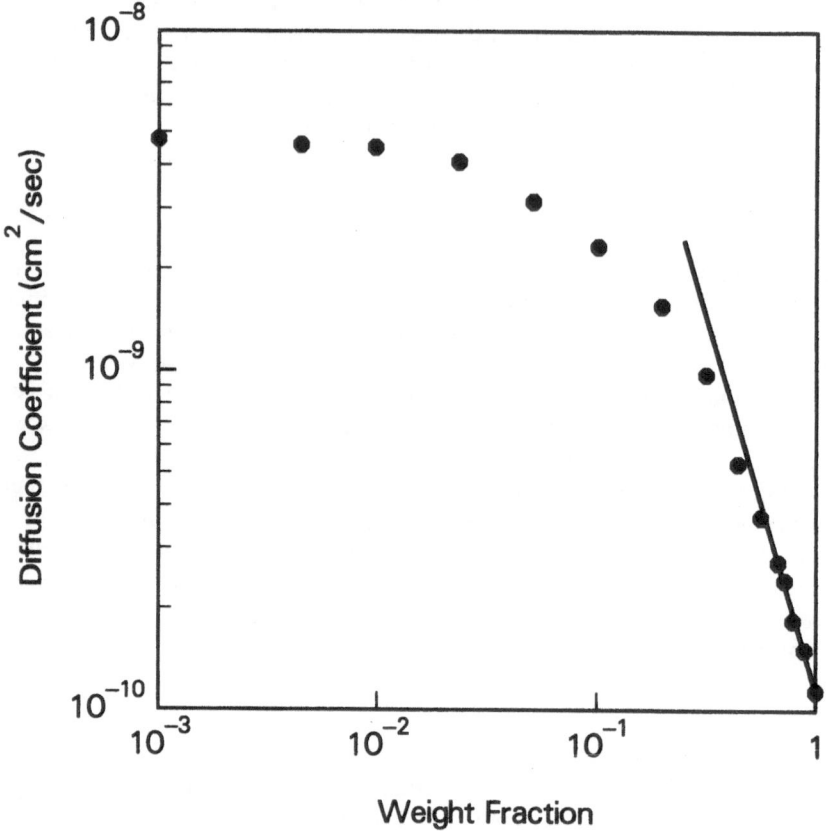

Figure 1. Diffusion coefficient (cm²sec⁻¹) of the labeled polymer as a function of the total weight fraction of the high molecular weight (M_W = 32 000) PPO in the lower molecular weight (M_W = 1000) polymer solvent. The solid line has a slope of -2.2. (Reprinted with permission from *Macromolecules*, in press. Copyright 1985, American Chemical Society.)

$$\tau_t = \frac{\tau N_b^3}{N_E}$$

10

τ is the shortest Rouse relaxation time, which is related to the local friction coefficient experienced by short chain segments. N_b is the number of bonds per chain, and N_E is the number of bonds per entanglement.

If we assume for the moment that this local friction coefficient in our polymer melt is independent of the concentration of the long chains, (as suggested by experimental data[12]) then the only parameter in Equation 10 which depends upon that concentration will be N_E. Theoretical arguments, based largely on dimensional analysis, have produced exponents ranging from -0.5 to -2 for the dependence of N_E upon concentration.[19, 20] Our data are consistent with the relationship: $D \propto C^{-2}$ (see Helfand, this volume).

Some other experimental data on the concentration dependence of the plateau modulus[21] and of diffusion [17] fall closer to the prediction $D \propto C^{-1}$. Thus the correct theoretical interpretation for this case remains an open question.

5. ACKNOWLEDGEMENTS

The polymer melt diffusion experiments were done in collaboration with Dr. Stephen J. Mumby, visiting scientist at IBM Research Laboratory, and with Dr. Edward T. Samulski and Li-Ping Yu of the Chemistry Department, University of Connecticut, Storrs, CT. I wish to thank Dr. Georgio Ronca and Dr. Robin Ball for helpful discussions.

REFERENCES

1.   H. Eichler, G. Enterlein, J. Munschau, and H. Stahl, 'Lichtinduzierte, thermische Phasengitter in absorbierenden Flüssigkeiten', *Zeitschrift für angewandte Physik* **31** , pp 1-4, 1971.
2.   H. Eichler, G. Enterlein, P. Glozbach, J. Munschau, and H. Stahl, 'Power requirements and resolution of real-time holograms in saturable absorbers and absorbing liquids', *Applied Optics* **11** , pp 372-375, 1972.
3.   D. W. Pohl, S. E. Schwarz, and V. Irniger, 'Forced Rayleigh scattering', *Physical Review Letters* **31** , pp 32-35, 1973.
4.   J. R. Salcedo, A. E. Siegman, D. D. Dlott, and M. D. Fayer, 'Dynamics of energy transport in molecular crystals: the picosecond transient-grating method', *Physical Review Letters* **41** , pp 131-134, 1978.
5.   H. Hervet, W. Urbach, and F. Rondelez, 'Mass diffusion measurements in liquid crystals by a novel optical method', *J. Chemical Physics*, **68** , pp 2725-2729, 1978.
6.   Barton A. Smith and Harden M. McConnell, 'Determination of molecular motion in membranes using periodic pattern photobleaching,' *Proc. National Academy of Sciences, USA* **75** , pp 2759-2763, 1978.
7.   L. Léger, H. Hervet, and F. Rondelez, 'Reptation in Entangled Polymer Solutions by Forced Rayleigh Light Scattering', *Macromolecules* **14**, 1981, pp 1732-1738.
8.   Jochen Coutandin and Hans Sillescu, 'Application of Holography to Measuring Slow Diffusion of Labeled Molecules', *Makromol. Chem., Rapid Commun.* **3**, 1982, pp 649-652.
9.   Barton A. Smith, Edward T. Samulski, L.-P. Yu, and Mitchell A. Winnik, 'Tube Renewal versus Reptation: Polymer Diffusion in Molten Poly(Propylene Oxide)', *Phys. Rev. Lett.* **52**, 1984, pp 45-48.
10.   Norio Nemoto, Michael R. Landry, Icksam Noh, Toshiaki Kitano, Jeffrey A. Wesson, and Hyuk Yu, 'Concentration Dependence of Self-Diffusion Coefficient by Forced Rayleigh Scattering: Polystyrene in Tetrahydrofuran', *Macromolecules*, **18**, 1985, pp 308-310.

11. Robert M. Weis, Krishna Balakrishnan, Barton A. Smith, and Harden M. McConnell, 'Stimulation of Fluorescence in a Small Contact Region between Rat Basophil Leukemia Cells and Planar Lipid Membrane targets by Coherent Evanescent Radiation', *J. Biological Chem.*, **257**, 1982, pp 6440-6445.

12. Barton A. Smith, Edward T. Samulski, Li-Ping Yu, and Mitchell A. Winnik, 'Polymer Diffusion in Molten Poly(Propylene Oxide)', *Macromolecules*, **18**, 1985, pp 1901-1905.

13. Barton A. Smith, Stephen J. Mumby, Edward T. Samulski, and Li-Ping Yu, 'Concentration Dependence of the Diffusion of Poly(propylene oxide) in the Melt', *Macromolecules*, in press.

14. P. G. de Gennes, *Scaling Concepts in Polymer Physics*, Cornell University Press, Ithaca, New York, 1979.

15. P. G. de Gennes and L. Léger, 'Dynamics of Entangled Polymer Chains', *Ann. Rev. Phys, Chem.*, **33** , 1982, pp 49-61.

16. Matthew Tirrell, 'Polymer Self-diffusion in entangled systems', *Rubber Chem. and Technology*, **57**, 1984, pp 523-556.

17. S. F. Tead and E. J. Kramer, 'Diffusion of Long Polystyrene (PS) Chains in Diluted PS Matrices', *Bulletin Amer. Physical Soc.*, **30**, 1985, p 499.

18. William W. Graessley, 'Entangled Linear, Branched and Network Polymer Systems — Molecular Theories', *Advances in Polymer Science*, **47**, Springer-Verlag, 1982, pp 67-116.

19. K. E. Evans and S. F. Edwards, 'Computer Simulation of the Dynamics of Highly Entangled Polymers. Part 2 — Static Properties of the Primitive Chain', *J. Chem. Soc., Faraday Trans. 2*, **77**, 1981, pp 1913-1927.

20. Michael Rubinstein and Eugene Helfand, 'Statistics of the entanglement of polymers: Concentration effects', *J. Chem. Phys.*, **82**, 1985, pp 2477-2483.

21. W. W. Graessley and S. F. Edwards, 'Entanglement interactions in polymers and the chain contour concentration', *Polymer*, **22**, 1981, pp 1329-1334.

CURRENT PROBLEMS IN UNDERSTANDING THE BEHAVIOR OF POLYMER GLASSES

J. J. Aklonis
Polymer Materials Research Center
Loker Hydrocarbon Research Institute
University of Southern California
Los Angeles, CA 90089, USA

ABSTRACT. Two general types of treatments have been proposed to explain the glass transition, thermodynamic theories and kinetic or ordering parameter models. The successes of the thermodynamic approaches are reasonably well known and are only alluded to here. In this paper some of the complexities of the phenomenological behavior of glasses such as time dependence, non linearity, asymmetry and memory effects are reviewed. In addition, an effort is made to make obvious the physical underpinings by which multiordering parameter models treat such complicated behavior. In terms of a broad overview, it becomes clear that the thermodynamic and kinetic treatments are complementary, treating different aspects of this complicated behavior. In the hope that shortcomings of theories point the direction for additional development, several perplexing problems which have surfaced in attempting to treat accurate kinetic data in terms of multiordering parameter models are discussed.

1. INTRODUCTION

Various types of glasses have been known and used for centuries and it should probably be expected that new types of glasses will continue to be discovered in the future as they have been in the recent past. (1) In spite of our familiarity with the vitreous state which arises from having and using so many types of glasses in varied applications, I think it is fair to say that our understanding of the vitreous state is far from being complete.

In this brief discussion, I intend to point out some of the aspects of the behavior of glasses which I find fascinating; I will also describe how theoretical efforts have been made to treat these phenomena. In accord with

M. A. Winnik (ed.), Photophysical and Photochemical Tools in Polymer Science, 407–427.

the title, I will emphasize what I feel to be some of the major problems which have yet to be addressed by any truly viable theory of the glass transition.

Theoretical treatments of the glass transition fall into two broad categories, thermodynamic (2-3) and kinetic.(4-13) This might seem to be a strange situation since we all know that kinetics and thermodynamics are very different fields, one dealing with whether particular transitions are or are not possible and the other is concerned with just how rapidly the possible transitions take place. In fact, separate schools have viewed the glass transition from the perspectives of each of these areas and both have made considerable success in rationalizing or explaining some of the complex behavior observed experimentally. However, as will be made clear as this discussion develops, these diverse treatments must, by the very nature of their scientific disciplines, treat different aspects of glass transition behavior; a truly complete theory which remains to be developed must encompass the concerns of both types of treatments.

2. THERMODYNAMIC TREATMENT OF THE GLASS TRANSITION

In Figure 1, I have plotted a schematic of the volume or enthalty behavior of a glass forming material in the glass transition region. For the time being, the particular sequence of experimental events by which this behavior is measured need not be emphasized but it will be of paramount importance later.   From the data, the glass transition temperature can be determined by extrapolating

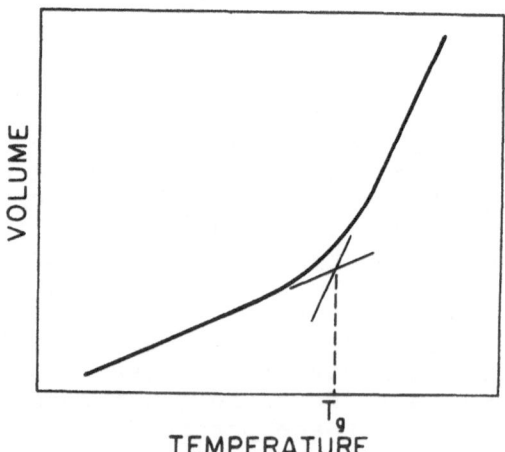

Figure 1.   Schematic volume versus temperature plot used
            to define Tg in the simplest possible fashion.

the sensibly straight line high and low temperature
behaviors to a point of intersection; the temperature at
which these two straight lines intersect is usually
defined as T_g. At a second order phase transition, a first
derivative of the Gibbs free energy shows a change in
slope. Since volume is a first derivative of G, the
behavior shown in figure 1, which is characteristic of the
glass transition, has often caused the glass transition
to be considered a second order phase change.

There is a considerable amount of work in the
literature based on this idea.(2,3) One of the most
successful theoretical treatments in this area is that of
Gibbs and DiMarzio.(3) Briefly, they write an expression
for the configurational partition function of a long
polymer molecule and are particularly interested in
the temperature dependence of the entropy calculated via
this partition function. They find that the number of
configurations accessible to the polymer molecule decreases
rapidly with temperature. In fact, only a single
configuration is probable at temperatures below a certain
temperature T_2, which is still well above absolute zero; it
has been argued that this transition temperature is related
to the glass transition temperature T_g.(14)

This theory has been successfully used to treat
the variation of T_g with molecular weight, (3a) with
pressure, (3a) with concentration of diluent, (3c) with
copolymer composition, (3b) etc. It is very impressive
that such a straightforward thermodynamic approach is so
successful in what will turn out to be such a complex area.

3.   ADDITIONAL PHENOMENOLOGICAL ASPECTS ASSOCIATED WITH THE GLASS TRANSITION

3.1.   Time Dependence

The behavior depicted in figure 1 which I have used to
start the discussion of the glass transition is observed
only under certain specific experimental conditions. If
one were to start with the glass former at temperatures
well above the glass transition temperature and then cool
the material at a fixed rate through the glass transition
domain one does, in fact, observe such behavior. A cooling
rate of $1^{\circ}C$ per minute would be appropriate here.

In figure 2, I have depicted what would happen if the
same experiment were done at a slower cooling rate,

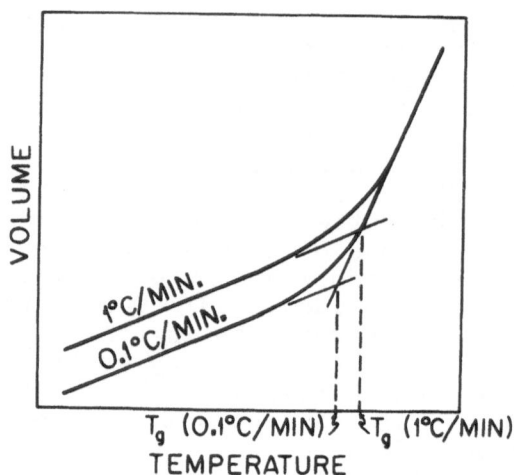

Figure 2.   Schematic volume versus temperature plots for
             experiments carried out at two different cooling
             rates.

perhaps, $0.1°C$ per minute. We see that similar behavior is
observed, however, extrapolating the high and low temperature straight line portions results in a lower glass transition temperature for the slower experiment than for the
more rapid experiment.
      Here it is tempting to assume that the faster experi-
ment was just "too rapid" and that the slower experiment
determines "the glass transition temperature". One might
base this choice on the argument that the slower transition
is more likely to have been carried under reversible
conditions and therefore much better approximates the type
of process which can be treated by thermodynamics. We have
considerable experience with systems which respond this
way. Glasses however are different. Experiments have
shown that the glass transiton phenomenon itself is
intrinsically a function of experimental rate (13) and, for
the type of experiments being discussed here, the glass
transition temperature changes by 3 to $4°C$ for each
factor of ten change in experimental rate.(15) All
experiments on glasses near the transiton region show
experimental rate dependencies.

3.2. Differences Observed upon Heating and Cooling;
Asymmetry

      Let us once again consider the experiment depicted in
figure 1. Here we monitored the volume of a glass former

while cooling at the fixed rate of $1^{\circ}C$ per minute starting at a temperature well above the glass transition region. This behavior is redrawn in figure 3 and looks like a second order thermodynamic transition. Some additional behavior is also sketched which is seen if the sample is heated, rather than cooled, at the same rate as soon as

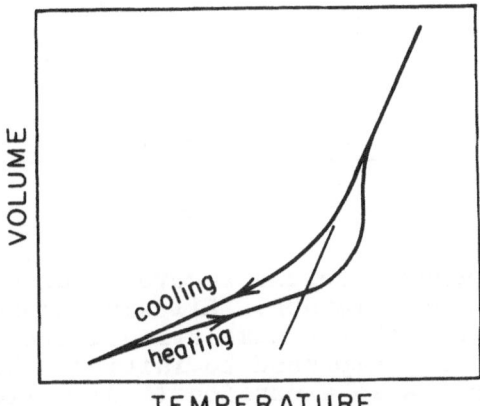

TEMPERATURE

Figure 3. Schematic volume versus temperature plots of behavior measured under heating and cooling conditions.

the lowest temperature is reached. The differences in behavior are striking. Upon heating, the volume temperature behavior actually cuts through the extrapolation of the high temperature liquid line and rejoins this line only at temperatures substantially above T_g as measured via cooling. To be sure, this is just a type of hysteresis. However it is important to analyze this behavior from the viewpoint of thermodynamics. Whereas the behavior upon cooling may look like a second order phase change, upon heating the observed behavior looks much more like a first order thermodynamic transition. Such dissimilar behavior upon heating and cooling has been called the asymmetry associated with the glass transition and this feature, too, is clear and obvious for experiments carried out in the glass transition range.

3.3. Non-Linear Behavior

Except in the limit of exceedingly small perturbations, the kinetic behavior of glasses is non linear. In

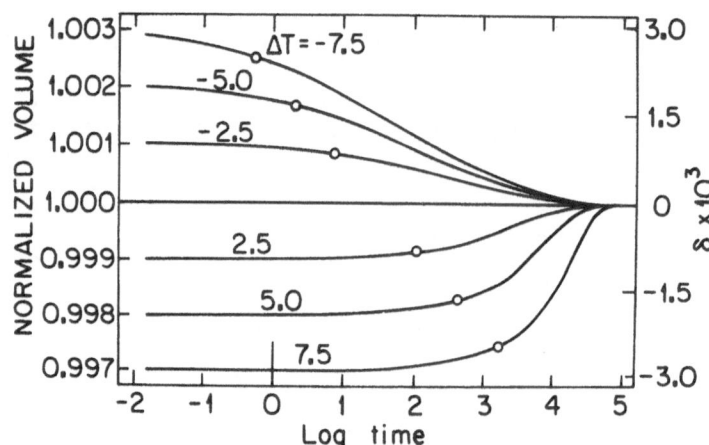

Figure 4. Time dependent volume recovery for simple
approach experiments. All experiments are
carried out at the same temperature. Δ T is the
temperature jump used to initiate the recovery.
Δ T>0 results in expansion (δ<0) while Δ T<0
results in contraction (δ>0).

figure 4 the results of a different kind of experiment are
depicted. To this point, we have been considering experiments in which constant temperature rate changes have been
applied to glass formers. Here, the system is imagined to
be at equilibrium and then subjected to a rapid change in
temperature. Experimentally one observes a rapid adjustment in properties such as the volume or enthalpy which is
then followed by a much slower change as equilibrium is
approached. In figure 4, the time dependence of this slow
approach to equilibrium is depicted for contraction
experiments caused by rapid temperature decreases and
expansion experiments caused by rapid temperature
increases.

In a linear system, the stress and strain are proportional
to one another so that doubling the stress results in
doubling the strain. This is clearly not the case here;
the points indicated on each curve represent 10% recovery
toward equilibrium. In a linear system, these points would
fall on a vertical line in expansion and also in contraction.

3.4. Memory Experiments

Glasses exhibit memory effects, which still further
complicate their kinetic character; typical behavior is
shown in figure 5. Here, experiment A depicts a normal

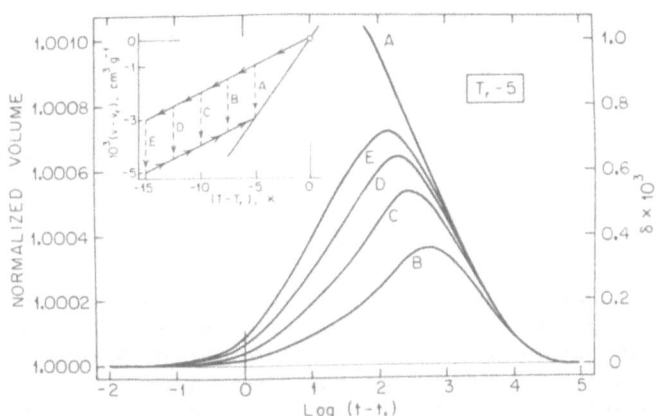

Figure 5. Memory behavior of glass formers. All experiments carried out at T_r-5°C. Preliminary thermal treatments shown in the insert.

contraction induced by a temperature jump of -5°C starting from the reference temperature T_r. The other four experiments depict multiple temperature jump situations. In experiment B, for example, a -7.5°C temperature decrease is used to perturb the system from equilibrium. Recovery is allowed to take place until the point is reached at which the application of a 2.5°C temperature increase results in the sample obtaining its equilibrium volume at the temperature T_r-5°C. Thus, immediately after the second temperature jump, the sample is apparently at equilibrium, having been brought to its equilibrium volume at the temperature T_r-5°C. If one monitors the volume after the attainment of this "equilibrium", one finds a spontaneous isothermal increase in volume followed by a subsequent decrease in volume back to the now real equilibrium condition. Different thermal pathways will result in different extents of memory as shown in the figure. If more complex temperature treatments are used, even more complex memory behavior will result.

3.5. Complex Tau Effective Behavior

    In the past there have been attempts to treat the kinetics of glass formers in terms of single ordering parameter models. (13) For experiments which measure volume, it is convenient to define a new parameter called delta as the normalized departure from equilibrium.

$$\delta = (V - V_\infty)/V_\infty \qquad (1)$$

For expansion experiments delta will be negative while for contraction experiments delta will be positive. At equilibrium delta equals 0. Single ordering parameter treatments suggest that the rate of approach to equilibrium (d δ/dt) is proportional to delta. Moreover, the proportionality constant can be given as a reciprocal relaxation time such that one writes

$$\frac{d\delta}{dt} = -\frac{\delta}{\tau} \qquad (2)$$

It is not difficult to appreciate that if delta is large and positive, the molecules are far apart and their motions can occur reasonably easily. On the other hand, at the same temperature, if delta is smaller, the molecules will be closer together and the time scale of all motions will be shifted to longer times. Thus the relaxation time which characterizes the rate of approach to equilibrium, itself depends upon how far the system is from equilibrium. In equation 3 this is emphasized by indicating that tau depends upon delta.

$$\frac{d\delta}{dt} = -\frac{\delta}{\tau(\delta)} \qquad (3)$$

The question now becomes just what particular functional dependence should tau have on δ(t)? Kovacs (13) decided that this question could be answered experimentally. He rewrote equation 3

$$\frac{1}{\tau_{eff}} = -\frac{1}{\delta}\frac{d}{dt} \qquad (4)$$

where τ_{eff} is the effective retardation time and carried out a series of simple temperature jump experiments in both expansion and contraction to measure it. He found the very complicated behavior shown in figure 6 for polyvinyl acetate and such complicated behavior seems to be characteristic of glasses in general.

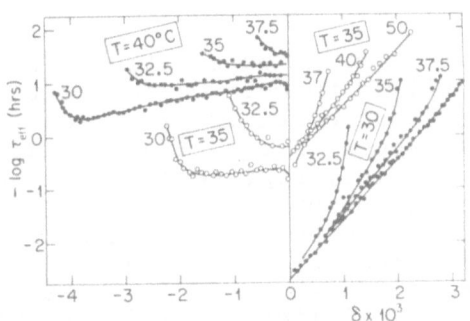

Figure 6. Log τ_{eff} versus δ for simple approach experiments in expansion ($\delta < 0$) and contraction
($\delta > 0$) at 45°C, 35°C and 30°C for
polyvinylacetate (after Kovacs) (13).

Since this tau effective data is so complex, it is
clear that there will be no simple functional dependence of
tau on delta. In fact, this analysis of the data by Kovacs
clearly shows that a more general theory must be developed.

4.0. MULTIORDERING PARAMETER MODELS

The attempt to describe the kinetic behavior of
glasses in terms of a single ordering parameter model is
analogous to trying to mimic the stress relaxation behavior
of a high molecular weight polymer in terms of a single
Maxwell element.(16) The dynamic behavior of a single
Maxwell element is just too simple to reproduce a stress
relaxation master curve. This arises from the fact that
various molecular motions with various relaxation times
contribute substantially to viscoelastic behavior.
Attempting to model this richness in terms of a single
relaxing element with a single relaxation time is doomed to
failure. In a similar manner, various molecular modes of
motion contribute to volume or enthalty adjustment in
glasses. Since it is reasonable to expect that these
various modes have different relaxation times, multiordering parameter models have been developed based on this
picture.(5-12) For simple approach experiments involving
single temperature jumps with systems initially at
equilibrium, one form of a multiordering parameter model
states

$$\frac{d\delta_i}{dt} = - \frac{\delta_i}{\tau_i(\delta)} \quad 1 \le i \le N \qquad (5)$$

$$\delta = \sum_{i=1}^{N} \delta_i \qquad (6)$$

$$\tau_i = \tau_{i,r} \, a_T a_\delta \tag{7}$$

Equation 5 is very similar to equation 3 except that a subscript i is affixed to both delta and tau. This subscript indicates that delta is partitioned into N individual pieces as is shown in equation 6. Each piece is called δ_i and ostensibly represents the fractional departure from equilibrium controlled by a particular mode of the molecule, that mode with a relaxation time τ_i. In equation 7, it is clear that ζ_i depends upon temperature through the multiplicative shift factor a_T and delta through the other multiplicative shift factor $a\delta$. $\tau_{i,r}$ is the retardation time of the ith mode of motion at the reference temperature at equilibrium, i.e., delta = 0. There are many expressions (5a) in the literature which can be used to state the particular functionalities for a_T and $a\delta$. We have used equation 8 (which has been shown to be consistent with expressions used earlier) to express these dependencies

$$\tau_i(\delta,T) = \tau_{i,r} \exp(-\Theta(T-T_r)) \exp\left(-(1-x)\right) \frac{\Theta\delta}{\Delta\alpha} \tag{8}$$

where Θ is a material constant equal to E_a/RT_r^2 which characterizes the temperature dependence of τ_i in equilibrium conditions, (E_a is an apparent activation energy and R the gas constant), x is a partition parameter ($0 \leq x \leq 1$) which determines the relative contributions of temperature and delta to each retardation time and $\Delta\alpha$ is the difference in coefficients of thermal expansion of the liquid and the glass. To use this simple model one must determine exactly how the total departure from equilibrium is partitioned among the various molecular modes of motion with their individual retardation times, i.e., one must determine the distribution function of δ_i versus τ_i. Here two separate treatments emerge.

One treatment makes use of the well known and very convenient Williams-Watts function. (6,7,8-12) This function is of the form

$$\phi(u) = \exp(-u^\beta) \tag{9}$$

where β is a constant ($0 < \beta < 1$). This function has been found to efficiently treat a large variety of data including mechanical, thermal and dielectric data.(17) In terms of this treatment, the distribution function is given as the rather complex expression (10)

$$G(s) = \frac{1}{\pi} \int_o^\infty e^{-z}\left\{e^{-(zs)^\beta \cos\pi\beta} \sin\left[(zs)^\beta \sin\pi\beta\right]\right\} dz \tag{10}$$

This distribution function is sketched in figure 7 for three values of beta as shown.(10) It must be kept in mind that one of the most important reasons for choosing a

distribution function of the Williams-Watts type is that
calculations within this formalism are reasonably easy to
carry out.

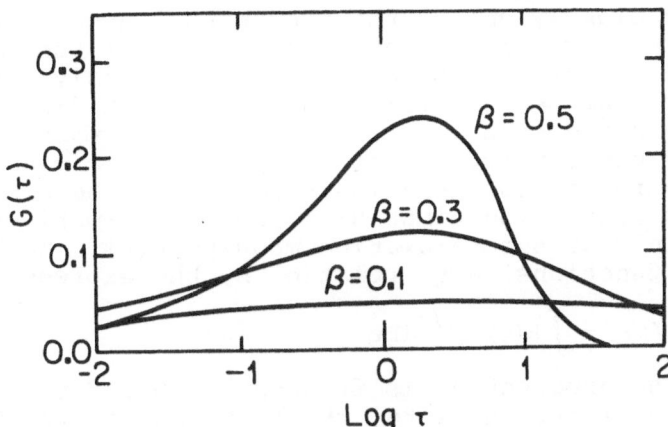

Figure 7.   Schematic of Williams-Watts distribution
             functions for three values of β according to
             equation (10). Reference (10).

     The other school has not assumed a form of the
distribution but rather has used trial and error methods to
determine the appropriate function which will mimic data
like that shown in figure 5.   In fact, evaluating the
credibility of a distribution function by comparing predicted and experimental tau effective behavior is a very
critical test of any particular distribution function since
tau effective is actually a derivative of the experimental
data.   In figure 8, one sees the two box distribution

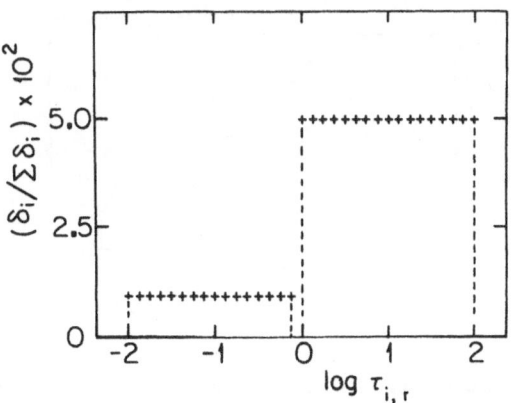

Figure 8.   Two box distribution function used in reference
             (5).

function which we have found to give very good agreement
between theory and experiment.

Before going further, I think it will be helpful if we
examine closely one of the fundamental underlying aspects
of the theory which gives rise to many of the important
kinetic features I have sketched previously. First,
however, the general relationship between distribution
functions and experimentally observed response functions
must be understood. As an example in this case, I will
briefly review the relationship between the distribution of
elastic relaxation times H( ) and the experimentally
determined stress relaxation modulus E(t). Mathematically
these functions are related by the expression

$$E(t) = \int_{-\infty}^{\infty} H(\tau)e^{-t/\tau} d\ln\tau \qquad (11)$$

where the exponential is frequently called the kernel. In
figure 9, I have depicted the distribution function as the
wedge and box which Tobolsky suggested and pioneered.(8)

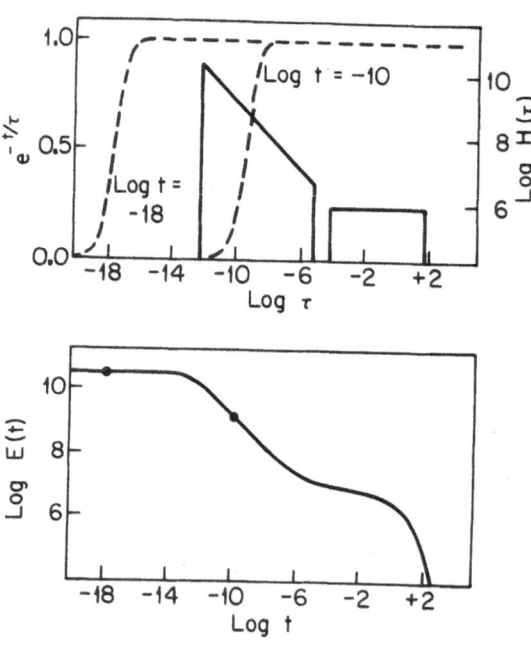

Figure 9.   Relationship between H(γ) and E(t) as expressed
             in equation (11).

To determine the stress relaxation modulus from this
distribution function, equation 11 suggests multiplication
of H(τ) by the kernel and subsequent integration. If t is
very small, the kernel will have a value of 1 for all
values of tau where H(τ) is significant and thus, at small
values of t, the modulus is merely the integral of the
entire distribution function. However, as time increases,
the kernel moves to the right on the log tau scale and   at
some later time, the kernel might look as depicted by the
second exponential function. Now it is clear that pointwise
multiplication followed by integration will yield a lower
value for the stress relaxation modulus. The actual
modulus related to the distribution function in figure
9 is depicted in the lower part of this figure.

     Let us consider what happens for volume relaxation
using a distribution function like that shown in figure 8
in simple expansion and contraction experiments. Once
again, it should be kept in mind that kinetic aspects of
the glass transition are highly asymmetric. Although we
have already discussed some of the vast differences in
behavior observed when heating and cooling a glass, here
it will be helpful to briefly consider one aspect of
expansion and contraction experiments just a bit more
carefully. In figure 4 concentrate on the -7.5°C
contraction curve and contrast this with the +7.5°C
expansion curve. It is important to remember that both
experiments are carried out at the same experimental
temperature. Only the initial conditions vary.
Remembering that the time scale is logarithmic, one
realizes the enormous asymmetry in this behavior. The
contraction curve begins to relax quickly. In fact, 10%
recovery has already occurred before one time unit (log t =
0) has elapsed. But the rate of this contraction slows and
slows as equilibrium is approached. In expansion, the
situation is very different. Virtually no recovery has
taken place at 1000 units (log t = 3) elapsed time; at a
time somewhat longer than this, the 10% recovery point is
reached and then a rapid acceleration to equilibrium is
observed. These are very different behaviors indeed.

     Still, it is easy and satisfying to understand how the
kind of kinetic models outlined above treat this asymmetry.

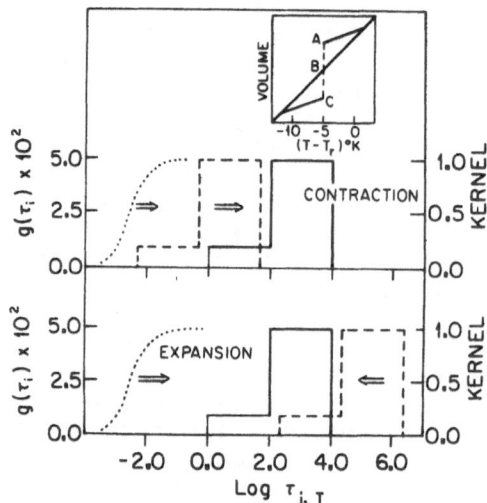

Figure 10.   Relative motion of the distribution function
              and the kernel which gives rise to the marked
              asymmetry in expansion and contraction experi-
              ments.

Figure 10 is similar to figure 9 in that we have a
distribution function and a kernel. The distribution
function is the one appropriate for volume relaxation as
sketched in figure 8 and the fact that the kernel is not
strictly exponential need not concern us here. Let us
first consider contraction which is represented on the top
half of the figure and also as the A to B transition in
the upper insert. The sample, in this case initially at
equilibrium at T_r, the reference temperature, is quenched
to $5^\circ C$ below T_r. This is shown in the insert. Since the
recovery experiment takes place at $T_r -5^\circ C$, the
distribution function in solid lines is at this
temperature.   It is the familiar two box distribution
function shown in figure 8. However, looking back at the
mathematical formalism given in equations 5 through 7, it
is clear that tau depends upon delta. The particular
functions used necessitate that for positive values of
delta, recovery times shorten which is in accord with our
intuitive ideas. Thus at point A in the insert,
immediately after quenching to the experimental temperature, the distribution function depicted by the solid lines
is not really appropriate for the contraction experiment.
It should be shifted to shorter times as indicated by the
dashed lines. The kernel once again moves to the right and
at reasonably short times will begin to overlap the
displaced distribution function; this accounts for

relaxation toward equilibrium, i.e., δ will diminish and
the volume will drop below A toward B in the insert.
However, as delta diminishes, the distribution function
must also move toward the right since, at equilibrium,
(indicated by the point B in the insert), the distribution
function must find itself in the position shown by the
solid lines. Thus, the kernel is attempting to catch a
receding distribution function. This accounts for the fact
shown in figure 4 that contraction experiments start early
and take a very long time to reach equilibrium.

The situation with expansion is very different. Now,
in the insert, the trajectory is from the point C through
B. In the lower half of the figure, once again the solid
line distribution function represents the position of this
function at equilibrium, i.e., point B. At point C, delta
is negative resulting in the distribution function being
displaced toward longer retardation times. Now the kernel
must move a very long distance indeed before any recovery
begins. This is the long non-relaxing initial plateau
observed in expansion as shown in figure 4. However, as
soon as recovery does start, delta diminishes which means
that the distribution function shifts toward the left,
i.e., toward the approaching kernel. In this situation, a
cooperative like transition occurs where we see a very
rapid approach toward equilibrium. Thus, the apparent
complicated asymmetry, and many other apparently
complicated features of the kinetics of the glass transition can be easily explained in terms of multiordering
parameter models with temperature and structure dependent retardation time spectra.

I would now like to emphasize some of the shortcomings
of these theories, starting by looking at the "big picture"
in terms of both thermodynamic and kinetic features and
then concentrating more fully on problems associated with
the kinetic theories.

5.0. PROBLEMS - GENERAL PERSPECTIVE

I have very briefly sketched a themodynamic view of glasses
and somewhat more fully, although certainly not completely,
outlined multiordering parameter or kinetic models of this
transition. Both of these techniques have strengths
and both have weaknesses. In fact, the strengths and
weaknesses are complementary. If we are faced with a new
polymer and want to predict its glass transiton
temperature, (or, more correctly, its glass transition
temperature range) multiordering parameter models offer
little help. However, by judicious application of the Gibbs
DiMarzio theory (3) or some of the subsequent developments of
this theory,(19) it is very likely indeed that a glass
transition temperature range can be predicted. On the

other hand, if one needs quantitative information about how the measured glass transition temperature of this new material might be affected by changing the heating rate used in the differential scanning calorimeter or why a heat capacity peak in this apparatus is seen on heating but never on cooling, don't expect thermodynamic treatments which assume equilibrium and reversibility to be of much help. It is clear that the glass transition is complicated. It has some features which can be treated using the formalism of thermodynamics and others which have been moderately successfully attacked by using the kinetic approach. Nevertheless, a completely viable theory of this most interesting area must encompass all aspects of the phenomenology within a single framework. The advent of such a treatment will be very exciting.

It is often the case that one learns more from situations where theories do not work than from where they are successful. With that thought in mind, I would like to concentrate on three areas where multiordering parameter models have not been completely successful. This emphasis should not be taken to indicate that the thermodynamic treatments are not equally vulnerable but rather merely reflects my own familiarity and interest in such treatments.

6.0. PROBLEMS ASSOCIATED WITH REPRODUCING EXPERIMENTAL DATA IN TERMS OF MULTIORDERING PARAMETER MODELS

6.1. General Agreement between Theory and Experiment

A considerable amount of effort has been invested in attempting to treat carefully gathered experimental data of high quality in terms of multiordering parameter models. (5,6,8-12) The vast majority of this work has been done utilizing the Williams-Watts function. As mentioned above, one of the most compelling reasons for using this particular formalism involves a considerable simplification of computational mathematics. I believe that the general conclusion of all of this work is that there seems to be no set of parameters (beta, x and theta) which completely satisfactorily reproduce the experimental data over a wide range of conditions. It has been stated that the parameter space is "very flat" which indicates that it is difficult to determine the best set of parameters, especially when one realizes that experimental error is present; work is continuing along these lines. Here, it should be realized, that the Williams-Watts formalism may not be flexible enough to be used with very high quality data. It may eventually be necessary to use other distribution functions such as the non analytic one shown in figure 8. However, it is clear that the use of such functions involves much more difficult calculations.

6.2. Approaches Toward Equilibrium

As mentioned earlier, Kovacs (13) has measured tau

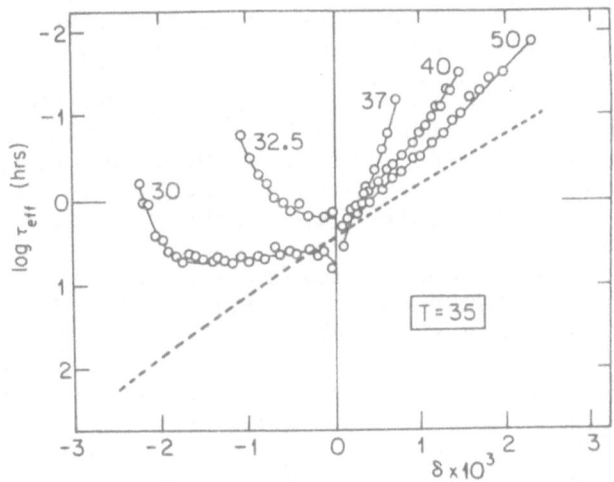

Figure 11. Selection of data from figure 6 emphasizing differences in approaches to equilibrium via expansion at miniscule departures from equilibrium.

effective for glassy samples and the immediately relevant section of his data is depicted in figure 11. This work was done on polyvinyl acetate at 35°C. What is particularly surprising is that the approaches of the two expansion curves to equilibrium are so different. At delta approaching -10^{-4}, Kovacs argues that he still has reasonable experimental accuracy. Nonetheless we find that the tau effective values differ by a factor of 5. From either a theoretical or an intuitive point of view, it is very difficult to understand how the sample so effectively stores knowledge of its initial condition when it is so close to equilibrium and then manifests behavior in terms of retardation times which are different by such a substantial factor. In terms of the model depicted in figure 10, there is no way at all to rationalize this behavior. Recently, Ngai and Rendell (7) have come up with a new formalism based on strength of coupling between the molecular modes of interest and those of the underlying thermal bath. Their theory is more flexible and ostensibly allows rationalization of such behavior. For the present,

however, it is not yet clear to me just what this
rationalization means.

6.3. Some Perturbing Experimental Facts

In dealing with the behavior of polymers, one gets
very accustomed to working with distribution functions.
Fundamentally this arises because there are many molecular
modes of motion which contribute to the relaxations
observed in polymers and these various modes can have
enormously different relaxation times. This potentially
complicated situation is often simplified by the fact that
the distribution functions which measure contributions of
these various motions to different types of behavior shift
more or less homogeneously when the sample is subjected to
change. For example, unless one looks very carefully, (20)
time temperature superposition or thermo-rheological
simplicity (21) are good approximations for most systems.
There is presently data emerging in the literature
(22,23) which clearly shows that the situation in terms of
the relaxation of glasses toward equilibrium is very much
more complicated. Let us consider a specific example
which will make this clear. Suppose one quenches
polystyrene from a temperature well above its glass
transition temperature to one well below this range. It is
now possible to monitor the approach of this sample toward
equilibrium using various techniques. Let us consider two
specific methods. The sample could be placed in a
differential scanning calorimeter and its enthalty measured
as a function of time after being quenched. One could
accurately analyze the DSC curves to monitor the approach
of the enthalpy toward equilibrium. In a similar manner,
one might imagine carrying out stress relaxation
experiments at various times after the thermal quench. In
these experiments one would be monitoring the approach
toward equilibrium by watching the rate at which the
distribution of relaxation times, one for mechanical
relaxation and the other for enthalty relaxation, move
with time. In each case we are interested in the
relaxation caused by the temperature jump. The progression
of sample adjustments to this perturbation would be
monitored by watching the enthalty or the distribution of
mechanical relaxation times move with time. What has been
seen experimentally (23) is that the two recovery processes
occur with enormously different relaxation times, a very
surprising effect. It seems clear from the literature that
while volume and enthalpy relax toward their equilibrium
values reasonably quickly after a thermal quench, density
fluctuations and mechanical relaxation recover very much
more slowly.(22,23)

From these experiments, we see that the individual distribution functions governing relaxation of various properties of glasses do not move in any simple straightforward fashion as was originally expected.(24) If such behavior turns out to be intrinsic in glasses and not just associated with some experimental difficulty or artifact which has been overlooked, it appears that the molecular theory which will be necessary to explain this kind of behavior must involve some feature of molecular interaction which has been totally overlooked in the theories proposed to date.

7.0. Conclusions

One of the first concepts we learned when introduced to polymer science is that of the glass transition. It all seemed so simple when we started. I have pointed out a few of the more dramatic features of the glass transition phenomenon and have shown a few examples of how such behavior is rationalized by kinetic theories. Although I have alluded to thermodynamic treatments, they have received only very brief coverage here since I was most concerned with time dependent behavior. Nevertheless, I have tried to emphasize the fact that thermodynamic and kinetic treatments in this area are complementary rather than overlapping and that a truly viable theory must eventually develop which will encompass the considerable successes of both of these areas within one framework. Finally, I have pointed out a few of the problems presently being encountered in attempting to apply multiordering parameter models to kinetic aspects of the glass transition with the hope that the limitations of the theory will offer clues as to how new and better formalisms should be developed.

8.0. ACKNOWLEDGMENT

This work was supported by Grant No. DAAG 29-82-K-0121 from the U. S. Army Office of Scientific Research.

9.0. REFERENCES

1. P. E. Duwez, Top. Appl. Phys., **46**, 19 (1981).

2. a) T. G. Fox and P. J. Flory, J. Appl. Phys., 21, 581 (1950); b) R. Simha ad R. F. Boyer, J. Chem. Phys., 37, 1003 (1962).

3. a) J. H. Gibbs and E. A. DiMarzio, J. Chem. Phys., 28, 373 (1958); b) E. A. DiMarzio and J. H. Gibbs, J. Polym.

Sci., 40, 121 (1959); c) E. A. DiMarzio and J. H. Gibbs, J. Polym. Sci., Al, 1417 (1963); d) E. A. DiMarzio, J. Res. Nat. Bur. Stand., A68, 611 (1964).

4. O. S. Narayanaswamy, J. Amer. Ceram. Soc., 54, 491 (1971).

5. (a) A. J. Kovacs, J. M. Hutchinson and J. J. Aklonis, "The Structure of Non-Crystalline Materials, P. H. Gaskel, Ed., Taylor and Francis, 1977, p. 153. b) A. J. Kovacs, J. J. Aklonis, J. M. Hutchinson and A. R. Ramos, J. Polymer Sci., 17, 1097 (1979). c) J. J. Aklonis and A. J. Kovacs, "Contemporary Topics in Polymer Physics", Vol. 3, M. C. Shen, Ed., Plenum, 1979, p. 209.

6. C. T. Moynihan, A. J. Eastel and M. A. Debolt, J. Am. Ceram. Soc., 59, 12 (1976).

7. a) K. L. Ngai, "Comments Solid State Physics", 9, 127 (1979). b) K. L. Ngai, "Comments Solid State Physics", 9, 141 (1980).

8. a) R. E. Robertson, J. Polymer Sci., Part C, 63, 175 (1978). b) R. E. Robertson, J. Polymer Sci. Polym. Phys. Ed. 17, 597 (1979).

9. Ryong-Joon Roe, J. Appl. Phys 48, 4084 (1977).

10. T. S. Chow and W. M. Prest, J. Appl. Phys. 53 (10), 6568 (1982).

11. I. M. Hodge and A. R. Berens, Macromolecules, 15, 762 (1982).

12. J. M. O'Reilly, J. Appl. Phys. 50, 6083 (1979).

13. A. J. Kovacs, Fortsche.   Hochpolym. Forsch. 3, 394 (1963).

14. G. Adam and J. H. Gibbs, J. Chem. Phys., 43, 139 (1965).

15. J. D. Ferry, Viscoelastic Properties of Polymers, 3rd Edition, John Wiley, New York (1980) p. 284.

16. J. J. Aklonis and W. J. MacKnight, Introduction to Polymer Viscoelasticity, 2nd Edition, John Wiley, New York (1983), Chapter 7.

17. N. G. McCrum, B. E. Read and G. Williams, Anelastic and Dielectric Effects in Polymeric Solids, John Wiley, New York (1967).

18. A. V. Tobolsky, Properties and Structure of Polymers, John Wiley, New York (1960), page 128.

19. For example: P. B. Couchman, Phys. Lett. 70A, 155 (1979).

20. D. J. Plazek, J. Phys. Chem. 69, 3480 (1965).

21. J. D. Ferry, Viscoelastic Properties of Polymers, 3rd Edition, Wiley, New York (1980), Chapter 11.

22. J. H. Wendorff, J. Polymer Sci., Polymer Letters, 17, 765 (1979).

FLUORESCENCE PROBES FOR POLYMER FREE-VOLUME

Rafik O. Loutfy
Xerox Research Centre of Canada
2660 Speakman Drive
Mississauga, Ontario L5K 2L1
CANADA

ABSTRACT. A series of fluorescence probes (p-(N,N-dialkylamino) benzylidene malononitriles) which belong to a class of organic compounds known as "molecular rotors" has been developed. The internal molecular rotation of these compounds can be slowed down by increasing the surrounding media rigidity, viscosity or decreasing the free-volume available for molecular relaxation. Inhibition of internal molecular rotation of the probe leads to a decrease in the non-radiative decay rate and consequently enhancement of fluorescence. This behavior can be used to study both the static and dynamic changes in free-volume of polymers as a function of polymerization reaction parameters, molecular weight, stereoregularity, crosslinking, polymer chain relaxation and flexibility. In addition, the dependence of the fluorescence emission maximum of these probes on media polarity allow continuous monitoring of the probes location in the polymer matrix. These fluorescence materials are capable of simultaneously probing the flexibility and polarity of the surrounding media.

1. INTRODUCTION

There has been a long history of theoretical and experimental papers which are concerned with the development of the free-volume concept to explain transport and diffusion in polymer systems. The original work is probably that of Batschinski (1913)[1]. who postulated that the viscosity of a liquid was inversely proportional to the amount of free space in the system. Much later, Doolittle (1951)[2] utilized an empirical exponential representation for the dependence of viscosity on free-volume to describe the temperature variation of viscosity for low molecular weight liquids. Bueche (1953)[3] derived an expression for polymer segmental mobility by considering free volume fluctuations. Fujita (1960)[4] formulated a free-volume description of diffusion in concentrated polymer solutions. Williams, Lendel and Ferry (WLF) (1955)[5] demonstrated that the temperature dependence of molecular mobility and relaxation in glass-forming liquids is controlled by the available free-volume not only in the melt region but also in the "glassy region". The WLF empirical formula relating the viscosity, η , at a given temperature T to those at the glass transition temperature Tg is entirely based on the

M. A. Winnik (ed.), Photophysical and Photochemical Tools in Polymer Science, 429–448.

relationship between viscosity and free-volume and is given as:

$$log \left(\frac{\eta}{\rho T} / \frac{\eta_g}{\rho_g T_g} \right) = - \frac{C_1 (T - T_g)}{C_2 + (T - T_g)} \tag{1}$$

The left-hand side is commonly expressed in terms of just η/η_g in as much as the product ρT is insensitive to temperature variations. Thus, we obtain the familiarly derived form:

$$log (\eta/\eta_g) = -17.44 (T - T_g)/[51.6 + (T - T_g)] \tag{2}$$

where the values of the universal constants C_1 and C_2 have been inserted. The WLF equation has been shown to describe the temperature dependence of viscosity and relaxation rate in many polymer systems, polymer solutions and for many glass-forming liquids. In these systems, the free-volume fraction shrinks with decreasing temperature to about 0.025 at the glass transition temperature, T_g, and both $\eta(T)$ and time-temperature shift factors are correlated in terms of free-volume expansion between T_g and T.

The objective of this work is to develop a method to study the distribution and changes in free-volume in polymer systems.

2. RESULTS AND DISCUSSION

In recent years the application of luminescence spectroscopy for the study of a diversity of phenomena occurring in synthetic and natural polymers has become widespread[6-11]. The versatility and sensitivity of luminescence as a technique in polymer chemistry stem from the multiple aspects of the interaction of electronically excited states with their immediate environment. We have recently described a novel phenomenon associated with the effect of media rigidity on the fluorescence intensities of a series of donor-acceptor dyes, [p-(N,N dialkylamino)-benzylidene]malononitriles 1-3[6]. The singlet-excited-state lifetime of these dyes is estimated to be 3-10 ps in solution, which corresponds to a nonradiative decay rate, k_{nr}, of the order of $10^{11} s^{-1}$. The extremely fast deactivation rate of the singlet state of these materials was attributed to rapid torsional relaxation. We have shown that environmental factors restricting the internal molecular rotation of these dyes lead to a decrease in k_{nr} and consequently an increase in fluorescence yield. These dyes are, therefore, excellent microscopic probes for measuring the torsional rigidity of the surrounding polymer media, and their fluorescence yield is very sensitive to dynamic structural changes occurring over a wide temperature range.

The fluorescence properties of these probes permits us to study the rotational relaxation in various polymers and even during polymerization reactions and thereby obtain information on the microscopic rigidity of the media. In the following discussion a description of the photophysical properties of the dyes 1-3 will be given, with particular emphasis on the excited-state conformational relaxation in various media. This will be followed by a discussion related to the application of these probes to study polymerization reactions, the effect of polymer molecular structure on free-volume, the dependence of polymer chain relaxation on molecular weight, and the effect of temperature on polymer conformation and free-volume.

2.1   Photophysical Properties Of The Fluorescent Probes

Figure 1 displays the absorption and emission spectra of dyes 1-3 in ethyl acetate at 25°C. The dyes exhibit an intense (ε_{max} = 5 x 10^4 M^{-1} cm^{-1}) absorption band in the blue and a weak fluorescence emission in the green region of the spectrum. The S_1 state of these dyes is a π,π^* state with a considerable amount of charge transfer (CT) character. The ground state, dipole moment, μ_g, is about 9 Debyes which increases to about 24 Debyes upon excitation[10]. The absorption maximum (λ_{max}) and fluorescence maximum (λ_F) of the dyes 1-3 shift to longer wavelength when the dielectric constant of the media increases, which is consistent with the high CT character of S_1 state. Actually, the fluorescence emission maximum (λ_F) correlate well with the solvent dielectric constant, as shown in Figure 2 for dye 1. This correlation is important because it can be used as tool to determine the polarity of the surrounding media.

Despite the similarity of the calculated radiative decay rate of dyes 1-3 ($\sim$ 3 x 10^8 S^{-1}), their quantum yields of fluorescence, ϕ_f, are different in any given media. In non-viscous media (ethyl acetate, vinyl monomers) ϕ_f of 1,2 and 3 is 0.89 x 10^{-3}, 2.1 x 10^{-3} and 3 x 10^{-3} respectively, at room temperature. ϕ_f increases as the dimension of the molecular probe increases. In more rigid media at RT, such as poly (methyl methacrylate), ϕ_f of 1,2 and 3 are 0.012, 0.057 and 0.12, respectively. Further experiments with these dyes at 77K in 2-methyl tetrahydrofuran glass matrix give ϕ_f of nearly unity, an increase of over 300 times. A media-dependent excited state relaxation must be proposed to explain these dramatic changes in ϕ_f of these dyes. Since these dyes exhibit very little triplet yield, the main pathway for non-radiative deactivation of the excited state is internal conversion. The absence of change in fluorescence emission maximum between room temperature and 77K, and the approach of ϕ_f to unity at 77°K, indicates that the emitting states must be those excited states which maintain a ground-state conformation. Previous work at Xerox[10-12] has shown that for molecular

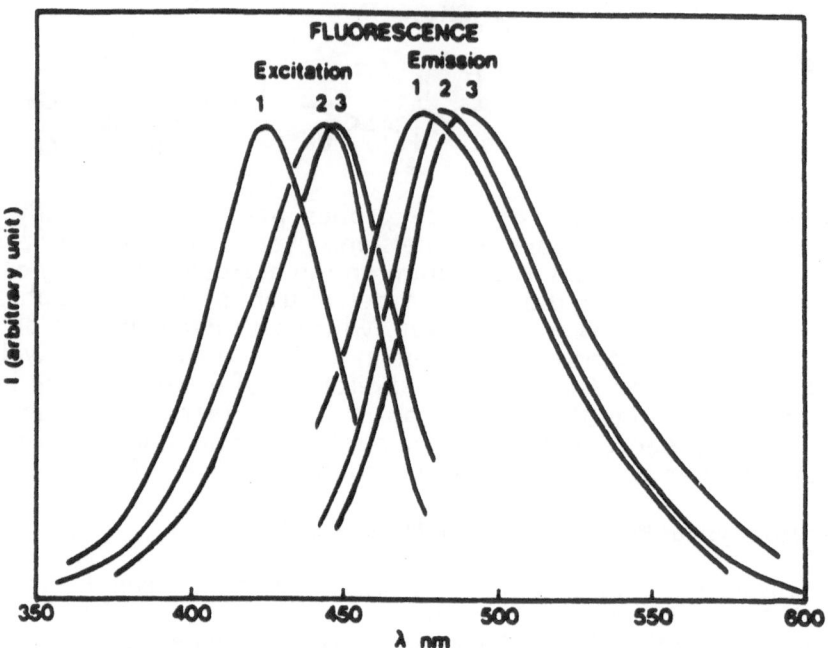

Figure 1. The fluorescence emission and excitation spectra of di-alkylaminobenzylidenemalononitriles. 1, 2, and 3 is ethyl acetate at room temperature.

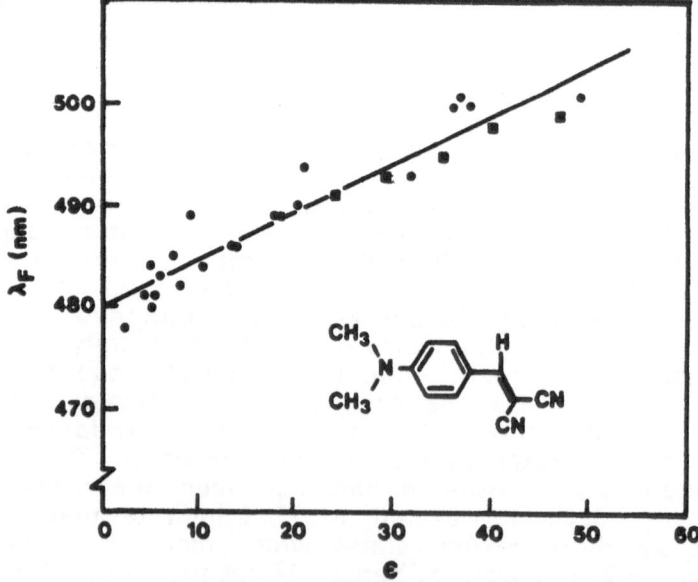

Figure 2. Correlation between the fluorescence emission maximum of dye 1 and the dielectric constant of the solvent.

TABLE I
Spectroscopic Data for Julolidine Malononitrile in Various Media

	Dye Absorption λ_{max} (nm)	Dye fluorescence				
		In monomer		In polymer ϕ_F	
		λ_F (nm)	ϕ_F ($\times 10^{-3}$)	70	23°C
methyl methacrylate	450	493	3.0	0.05	0.12
ethyl methacrylate	449	492	3.0	0.01	0.046
n-butyl methacrylate	448.5	486	2.8	0.005	0.02
styrene	454.5	485	3.0	0.014	0.04
vinyl acetate	449.5	490	2.9	⋯	⋯
ethyl acrylate	449	491	2.9	⋯	⋯
styrene/butyl methacrylate	450	490	2.9	⋯	⋯

rotors such as **1-3** torsional motion in the excited state is capable of inducing radiationless decay, $S_1 \rightarrow S_0$. It has also been suggested that the torsional motions responsible for inducing radiationless decay are hindered by the viscous drag of the solvent. Rotation of the aryl group in the excited state is considered to be the rate-determining step, leading to rapid internal conversion to the ground state. The effect of the medium is to hinder or slow down the torsional relaxation of molecular rotors, thus decreasing the radiationless decay rate, k_{nr}, and increasing ϕ_f.

2.2   *Fluorescence Probe 3 in Bulk Polymerization Reactions:*

We investigated the dependence of the fluorescence intensity of **3** on polymerization reactions. The polymerization reactions investigated were those of methyl methacrylate, ethyl methacrylate, n-butyl methacrylate, ethyl acrylate, styrene, and copolymerization of styrene/n-butyl methacrylate, with the fluorescent probe **3** simply dissolved in the monomer at 10^{-5} M concentration. Bulk polymerization was initiated using AIBN (0.5% by weight) at 70^{0}C. The fluorescence intensity of the probe (λ_{ex} 430 nm) was continuously monitored at 500 nm at the polymerization temperature. Figure 3 shows the change in fluorescence intensity (I_F) of **3** as a function of polymerization time for each of MMA, EMA and BMA. Figure 4 shows the dependence of I_F of **3** on the polymerization time of styrene. Similar results were obtained with ethyl acrylate and co-styrene (65%) /n-butyl methacrylate (35%). A curious behavior was observed. The fluorescence intensity remained almost constant in time until a critical moment is reached where a sharp rise in fluorescence intensity occurrs, followed by a leveling-off as the polymer limiting conversion is reached. The S shaped fluorescence intensity dependence on polymerization time is a common behavior to all polymerizations studied. However, the lag period, the slope of the fluorescence rise and the magnitude of fluorescence increase, all depend strongly on the rate of polymerization (temperature, initiator concentration and monomer reactivity) as well as on the particular polymer formed. In Table I the fluorescence yield, ϕ_f, of **3** at the limiting conversions are listed. The observed increase in ϕ_f going from the fluid monomer to the glassy polymer at 70^{0}C was a factor of 17, 3.4, 1.8 and 4.7 for MMA, EMA, n-BMA and styrene respectively. The polymerization region in which fluorescence intensity increases sharply appears to correspond to the increase of medium viscosity from fluid to rigid glass.

The importance of viscosity and free volume in the molecular relaxation processes of excited dyes has been well documented in the case of polymethines, di and triphenylmethanes and coumarine dyes[13-21]. For dyes in which rotation-dependent non-radiative decay (k_{nr}) links the excited state conformation to the media free-volume, V_f, one can express k_{nr} in terms of solvent free volume as:

$$k_{nr} = k_{nr}^0 \; \exp\left(-\beta \frac{V_0}{V_f}\right) \qquad (3)$$

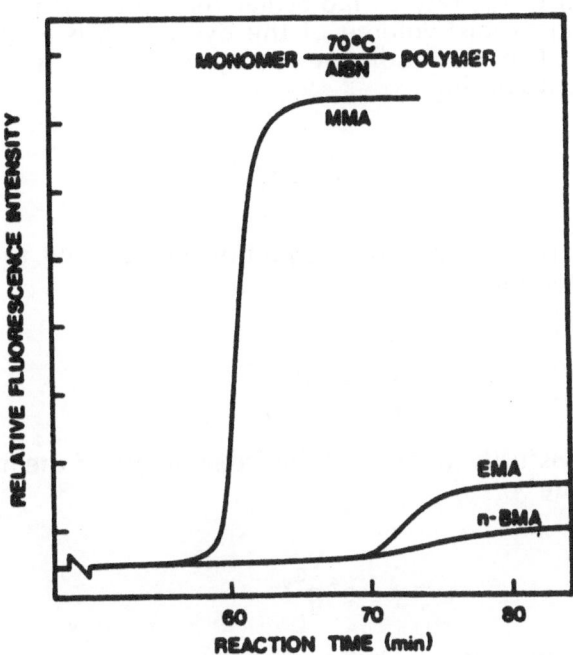

Figure 3. Dependence of 3 fluroescence intensity on the polymerization of methyl methacrylate (MMA), ethyl methacrylate (EMA), and n-butyl methacrylate (n-BMA).

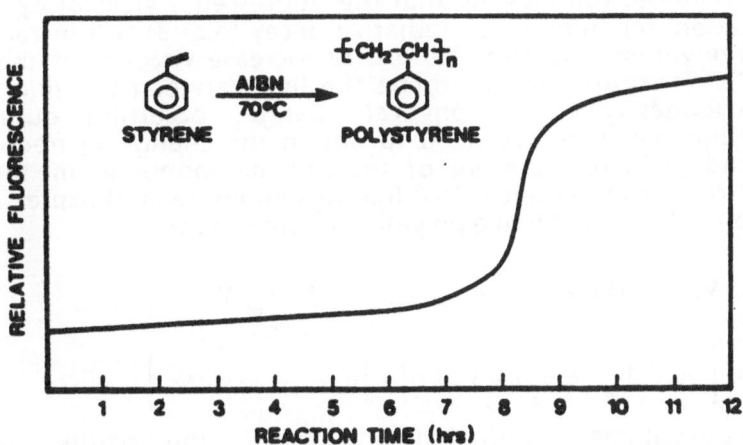

Figure 4. 3 fluorescence intensity change as a function of styrene polymerization time.

Here k_{nr}^0 is the intrinsic rate of molecular relaxation of the dye, V_0 is the occupied (Van der Waals) volume of the dye and β is a constant for the particular dye. The non-radiative decay rate, k_{nr} is related to the fluorescence yield according to:

$$k_{nr} = k_r \left(\frac{1}{\phi_f} - 1 \right)$$ (4)

Equation (4) can be substituted in Equation (3) to yield the fluorescence dependent free-volume.

$$\phi_f = \left(\frac{k_r}{k_{nr}^0} \right) exp \left(\beta \frac{V_0}{V_f} \right)$$ (5)

An expression of viscosity in terms of the free-volume of the media has been derived by Doolittle[14]:

$$\eta = A \exp \left(\frac{V_0}{V_f} \right)$$ (6)

Combining Equations (6) and (5), the relationship between the dye fluorescence quantum yield and viscosity can be derived as:

$$\phi_f = B \left(\frac{\eta}{T} \right) x$$ (7)

where $B = (k_r/k_{nr}^0)(T/A)x$. Here x is a constant between zero and one.

The above relationships indicate that the fluorescence yield of dyes which exhibit rotation-dependent non-radiative decay (e.g. 3) will increase with decrease free-volume Equation (5) and/or increase viscosity of the media Equation (7). Therefore, to determine the link between the fluorescence intensity changes and the physical changes occurring during the polymerization reactions, we need to obtain the change in free-volume, viscosity, and glass temperature of the polymer/monomer mixture as a function of conversion. Bueche[22,23] has developed general expressions for the variation of V_f, T_g and η of a polymer diluent system:

$$V_f = 0.025 + a_p(T - T_{gp}) V_p + a_d(T - T_{gd}) V_d$$ (8)

$$V_f = \left[a_p V_p T_{gp} + a_d(1 - V_p)T_{gd} \right] / \left[a_p V_p + a_d(1 - V_p) \right]$$ (9)

where V_p is the volume fraction of the polymer, η_g is the viscosity at the glass temperature and η is the polymer or glass forming liquid viscosity at temperature T. These relationships have been tested for several polymer-diluent systems and have been found to be reasonably accurate.[23]

Application of these relationships require the knowledge of the glass temperature of the polymer, T_{gp} and the diluent, T_{gd} the expansion coefficient of the polymer a_p and of the diluent, a_d. The value of a_p is very close to 4.8×10^{-4} per ^0C for most polymers and 10^{-3} per ^0C for most diluents. We studied, for example, the polymerization of MMA $\rightarrow$ PMMA. Here T_{gp} = 110^0C, T_{gd} = -102.8^0C.

From conversion-time measurements for MMA bulk polymerization initiated by AIBN at T = 70^0C we could determine the volume fraction of the polymer V_p. One can thus compute the change in V_f and η of MMA/PMMA mixtures as a function of conversion, using Equations (8) and (9) respectively.

Figure 5 shows the dependence of the fluorescence yield of **3** on the viscosity of PMMA/MMA systems. A gradual increase in fluorescence occurs as the viscosity increases from 0.1 to 2cp. This is followed by a sharp rise in fluorescence as the viscosity of the medium changes from 2 to 100 cp. The slope of this portion of the plot corresponds to $\phi_f a \eta^{\frac{2}{3}}$ which is in agreement with the Foster and Hoffmann[16] model, and also with Law's[11] recent results on a similar probe. Further increase in conversion leads to rapid increase in the macroscopic viscosity, as the glassy state is approached. Once T_g exceeds the reaction temperature no further change in fluorescence occurs; the fluorescence in this region levels off.

The viscosity dependent ϕ_f arises from the dependence of η on the free-volume for the media. At polymer conversion of 60% and less the free-volume is plentiful and only small changes in fluorescence is seen. However, as the bulk polymerization approaches the glassy state rapid reduction in the free-volume occurs, which leads to the observed rapid rise in fluorescence intensity. According to Eq. 5 a plot of $\ln\phi_f$ vs $1/V_f$ should give a straight line, the slope of which gives $V_0\beta$ and the intercept $\ln(k_r/k_{nr}^0)$. Figure 6 shows such a plot which gives a value of $V_0\beta = \frac{1}{2}$.

The phenomenon observed here, which is common to most polymerization reactions, demonstrates that polymers can interfere sterically with processes involving movement of parts of the guest molecules (fluorescent probes). As the polymer glassy state is approached, the relative free volume diminishes sharply, and the medium viscosity increases rapidly; mobility becomes restricted and the deactivation rate of the probe becomes controlled by the microscopic free volume provided by the polymer. This accounts for the abrupt increase in fluorescence until the limiting conversion is reached, at which point fluorescence levels off.

One can imagine the use of these dyes to monitor, on-line, the progress of bulk polymerization to prevent run-away reactions. Such an application could be of commercial importance.

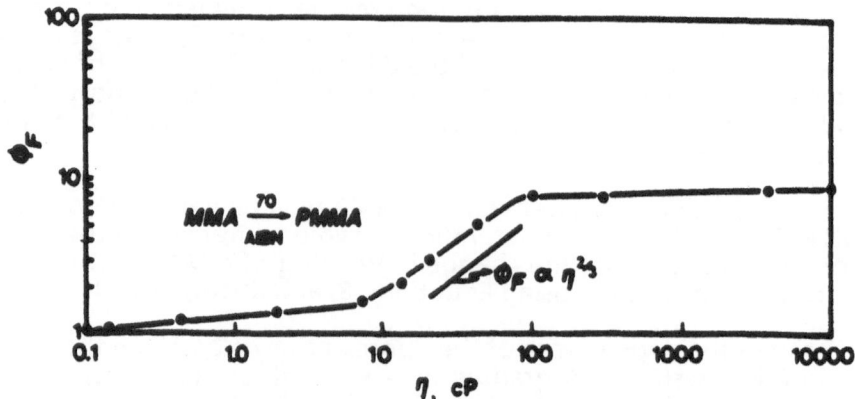

Figure 5. Dependence of fluorescence yield of 2 (at 70°C) on the viscosity of PMMA/MMA mixture.

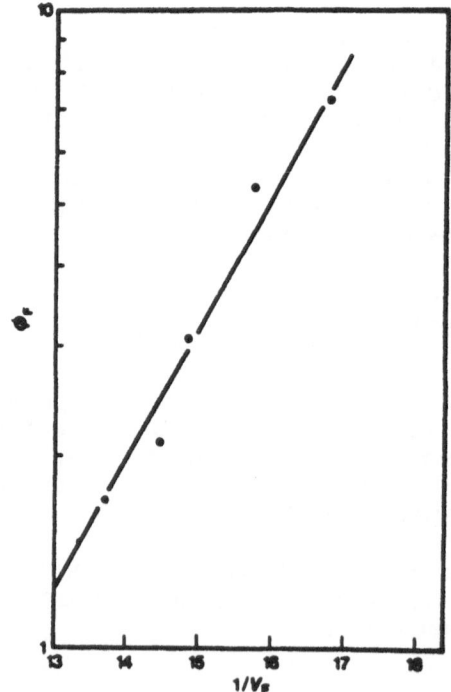

Figure 6. Dependence of fluorescence yield of 2 (at 70°C) on PMMA/MMA mixture free volume.

2.3 *Effect of Polymer Molecular Structure on the Fluorescence of the Molecular Rotors:*

We have investigated the effect of poly(alkylmethacrylate), PRMA, and poly(alkylacrylate), PRA, matrices on the fluorescence quantum yield (ϕ_f) of dye **2** to study the effect of polymer molecular structure on free-volume and to study polymer segmental relaxation processes.   Figure 7 shows the chemical structure of the polymers used. Table II lists the absorption and fluorescence spectral data of dye **2** in the various matrices. The glass transition temperatures of the polymers chosen varied from -54 to 105⁰C. No direct correlation between the glass transition relaxation temperature (T_g) of the polymer binder and the ϕ_f of **2** is observed. Instead, ϕ_f of **2** is related to the fluorescence emission frequency (v_f) of **2**, a parameter which has been shown to exert little influence on the ϕ_f values in organic solvents[11]. Since v_f is correlated to the polarity of a medium, our results suggest that **2** is located in different sites in various polymer matrices. The variation of ϕ_f with v_f is an indirect reflection of the difference in free volume (polymer chain flexibility) in these various sites of the polymers studied. This finding demonstrates the power of these probes and the utilization of the dual functionality, v_f to probe the location of the dye and ϕ_f to probe the   rigidity of that environment.

When the ϕ_f of **2** is plotted as a function of v_F (Figure 8), we observe two nearly parallel curves. One curve belongs to the PRMA family and the other curve belongs to the PRA family. The data for **2** in PVAc is incorporated into the PRA family because of their similarity in structure. The ϕ_f versus v_f curve of PRMA is always on the low frequency side of PRA. This implies that, for the same R group, **2** is located in a more polar microenvironment in PRMA polymers, a probable consequence of the conformational effect produced by the methyl groups on the polymer backbone.

The variation of v_f values shown in Table II indicates that **2** prefers different microenvironments within a family of polymers. In PMMA, PMA and PVAc matrices, **2** is basically located in the vicinity of the polymer backbone (the polar carboalkoxy group). This is confirmed by the low v_f values (20325-20408 cm⁻¹) observed in these three polymers, which have an effective dielectric constant of ~20. The ϕ_f of **2** in PMMA, PMA and PVAc are 0.0123, 0.013 and 0.011, respectively, despite their large differences in T_g values. The similar ϕ_f values in these three polymers suggests that the polymer chain flexibility (or free volume) around the polymer backbones of PRMA, PRA and probably poly(vinyl alkanoate) is probably very similar.

For PRMA and PRA polymers with longer alkyl side chains, the increase in v_f values observed within a family of polymers indicates that **2** tends to gradually penetrate deeper and deeper into the alkyl chain as its chain length increases. Using the v_f of **2** as an indicator for the location of **2** in these polymers, the change in ϕ_f values within a family of polymers will then give some information on the flexibility of segments of known locations along the alkyl side chain. For example, in PHDMA matrix, v_f of **2** is 20877

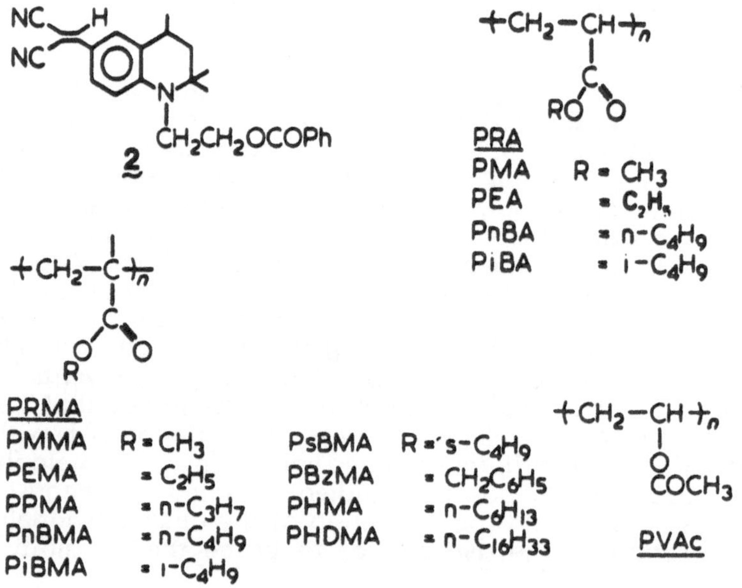

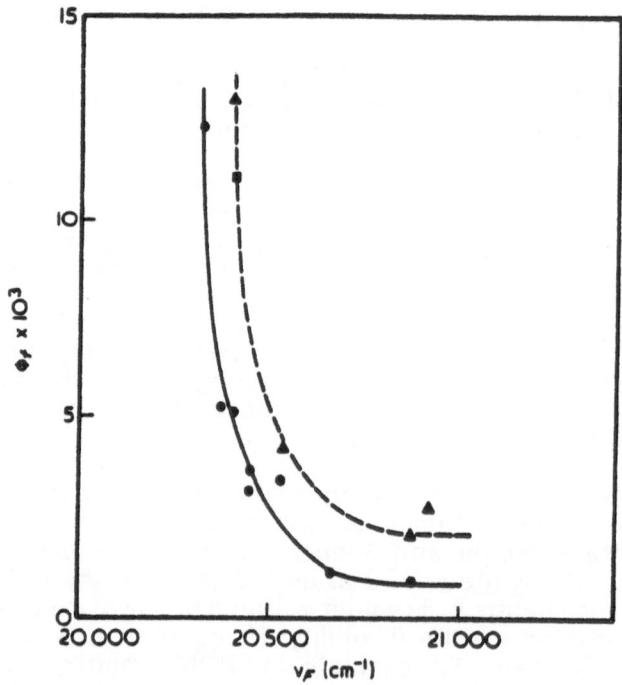

NC
H
NC
N
CH_2CH_2OCOPh
2
$+CH_2-CH+_n$
C
RO O
PRA
PMA R = CH_3
PEA = C_2H_5
PnBA = n-C_4H_9
PiBA = i-C_4H_9
$+CH_2-C+_n$
C
O O
R
PRMA
PMMA R = CH_3
PEMA = C_2H_5
PPMA = n-C_3H_7
PnBMA = n-C_4H_9
PiBMA = i-C_4H_9
PsBMA R = s-C_4H_9
PBzMA = $CH_2C_6H_5$
PHMA = n-C_6H_{13}
PHDMA = n-$C_{16}H_{33}$
$+CH_2-CH+_n$
O
$COCH_3$
PVAc

Figure 7. Structures of dye 2, poly(alkylmethacrylate), PRMA, poly(alkylacrylate), PRA and poly (vinyl acetate), PVAc.

15
10
5
0
$\phi_f \times 10^3$
20000
20500
21000
ν_F (cm^{-1})

Figure 8. Plot of ϕ_f of 1 as a function of ν_F of (○ PRMA; ▲ PRA and ■ PVAc).

Table II Absorption and fluorescence emission spectral data of 2 in PRMA, PRA and PVAc matrices

Polymer	$T_g{}^a$	$\nu_{abs}{}^b$	$\nu_F{}^c$	$\phi_f \times 10^3{}^d$
PRMA:				
PMMA	105	22341	20325	12.3
PEMA	65	22396	20450	3.6
PPMA	35	22426	20408	5.1
PnBMA	20			5.1
PiBMA	53	22533	20450	3.1
PsBMA	60	22426	20534	3.4
PBzMA	54	22163	20367	5.2
PHMA	−5	22426	20661	1.1
PHDMA	15	22532	20877	0.96
PRA:				
PMA	9	22222	20408	13.0
PEA	−24	22311	20533	4.2
PnBA	−54	22426	20876	2.1
PiBA	−24	22431	20920	2.8
PVAc	35	22222	20408	11.0

a Glass transition temperature, in °C, values either taken from J. Brandrup and E. H. Immergut, 'Polymer Handbook' 2nd Edition, John Wiley and Sons Inc., or specified by suppliers
b Absorption maximum frequency, in cm^{-1}
c Fluorescence emission maximum frequency, in cm^{-1}
d Better than ±10%

Table III
Dye Fluorescence in Polystyrene with Varying Molecular Weight

polystyrene		dye fluorescence
M_n	$T_g,{}^a$ °C	$\phi_f/10^{-3a}$
1.04×10^3	−130	3.0
8.0×10^2	27	4.9
2.0×10^3	49	6.8
4.0×10^3	59	7.5
5.0×10^3	64	7.6
9.0×10^3	83	8.2
1.75×10^4	94	11.1
5.0×10^4	100	15.0
1.0×10^5	104	18.5
2.23×10^5	105	19.6
1.8×10^6	107	19.3

aMeasured at 23°C; estimated error ± 10%.

cm^{-1}, indicating that **2** is basically located in a hydrocarbon environment (v_f of **2** in benzene is 20921 cm^{-1}). Information on the free volume (or chain flexibility) of the hydrocarbon chain in these polymers is then obtainable. The ϕ_f of **2** in PHDMA is 0.96 x 10^{-3}, indicating that the free volume in the hydrocarbon chain region in this polymer is very similar to those of low viscosity solvents[10].

These results indicate that the flexibilities of the polymer backbones of PRMA, PRA and PVAc are very similar. However, local polarity and local flexibility of polymer segments vary from site to site and from polymer to polymer. Great care should be exercised in using probes or labels in studying polymeric systems, especially when the location of the probe is difficult to determine.

2.4   Probing Polymer Chain Relaxation as a Function of Molecular Weight:

The effect of the molecular weight of monodispersed atactic polystyrene as non-fluorescent host polymer on the fluorescence yield of dye **2** was investigated. Films of dye **2** (0.5 wt.%) in atactic polystyrene (a-PS) of various molecular weight were prepared by solvent casting from 10% methylene chloride solution. The fluorescence data of dye **2** in polystyrene films with molecular weight from 10^2 to 10^6 are given in Table III and plotted in Figure 9. The fluorescence yield of dye **2** increases gradually with the increase of polystyrene molecular weight up to M_n = 10,000. Between M_n = 10^4 and 10^5, ϕ_f increases rapidly by about a factor of **2** to a plateau that extends to M_n = 1.8 x 10^6. This fluorescence behavior of the molecular probe is qualitatively very similar to that observed for poly(2-vinylnaphthalene) guest in PS solution.[24] However, the origin of the change in the fluorescence of the two probes with PS molecular weight must be different, since in our experiment a molecular rather than a polymeric probe was used. Probe miscibility with the host polymer is not an issue here.

The gradual increase in fluorescence efficiency of dye **2** with increase in the molecular weight of the host polystyrene to $M_n < 10^4$ can be attributed to inhibition of radiationless decay by rigidization of the probe by the local environment. This arises entirely from a decrease in the available free volume as the molecular weight of PS increases. Since the change in the glass transition temperature, T_g, with M_n is due to changes in free volume, a correlation between ϕ_f of **2** and T_g of the host polymer is expected and is observed, supporting the validity of the free-volume concepts as the controlling factor of torsional motion of the excited dye.

The situation is quite different for $M_n > 10^4$ and $M_n < 10^5$; the fluorescence of **2** abruptly increases by a factor of 2 and then levels off at $M_n > 10^5$. In that molecular weight range the glass transition temperatures of PS hardly changes. It should also be noted that all films prepared were optically clear; thus it might be argued that no phase separation has occurred.

The simplest possible explanation of the sudden rise in the fluorescence of the molecular probe **2** in PS above a critical molecular weight, $M_c \sim 10^4$, is due to an abrupt change in the morphology of the bulk polymer resulting from

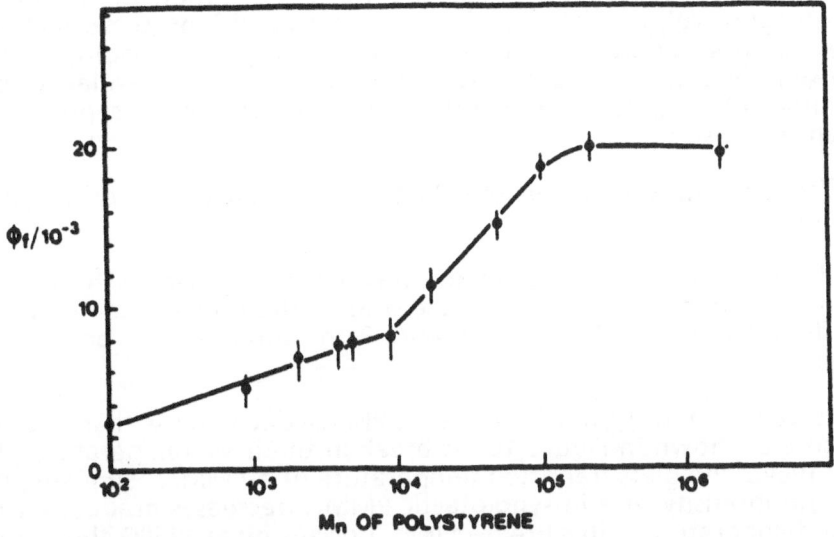

Figure 9. Variation of 2 fluorescence with polystyrene molecular weight, M_n.

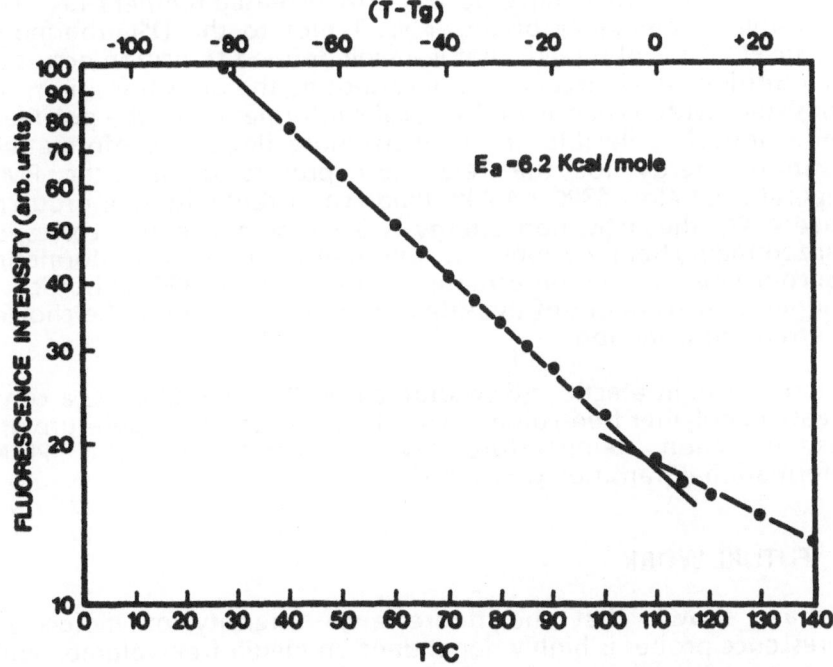

Figure 10. Dependence of fluorescence of 2 on temperature in a-PMMA.

chain contraction, or coiling. Since the torsional motion of the probe becomes progressively restricted by chain contraction, this process effectively decreases the non-radiative decay rate, k_{nr}, of the probe, leading to the sudden rise in fluorescence above that critical M_n. At PS molecular weight greater than 10^5 a polymer network will form and at this point the fluorescence levels off.

2.5   Effect of Temperature on the Probe Fluorescence in Solid Polymer Films.

To test the validity of the free-volume restriction imposed by a polymer matrix on the dye internal relaxation, we studied the effect of temperature on the fluorescence yield of dye 2 and 3 in stereoregular polymethyl methacrylates.

The fluorescence intensity, I_F, of 2 in atactic PMMA decreases with increasing temperature as shown in Figure 10. A break in the I_F vs. temperature plot occurs at 106°C, the glass transition temperature of a-PMMA. Similarly, the fluorescence intensity of 3 in syndiotactic PMMA decreases gradually with increasing temperature with a break in I_F vs. T occurring at 125°C, the T_g of S-PMMA as shown in Figure 11.

The fluorescence-temperature behavior of 3 in isotactic PMMA is radically different (See Figure 11). A sharp drop in the fluorescence occurs at ~43°C, followed by the normal decrease in I_F with increased temperature. There is a remarkable resemblance of the I_F vs. T plot to the DSC thermogram of isotactic PMMA is observed; that is a well defined transition occurs at 43°C. This transition which occurs at 43°C cannot be the glass transition; rather, it is consistent with a change in the local conformation of the polymer chains from a tight, less flexible, to an open, more flexible, conformation. The activation energy for the relaxation process of isotactic PMMA at temperatures below 43°C is 4.2 kcal/mol, consistent with side-group motion. Above 43°C, the activation energy is 8.5 kcal/mol, which corresponds to localized main-chain motions. On the basis of the above information, one must conclude that the conformational transition at 43°C for isotactic PMMA is triggered by rotations of the side groups, which change the choice of the preferred conformation.

To summarize, in atactic and syndiotactic PMMA, we observe a continuous increase in polymer free-volume with the increase of temperature up to the glass transition temperature, whereas with isotactic PMMA, a conformational transition was noted.

3.   FUTURE WORK

We have shown that the fluorescence intensity of molecular rotor fluorescence probes is highly dependent on media free-volume, while their fluorescence emission maxima are sensitive to media polarity. This dual functionality of these probes makes them useful in the study of a variety of polymer science problems such as curing of epoxy polymers,

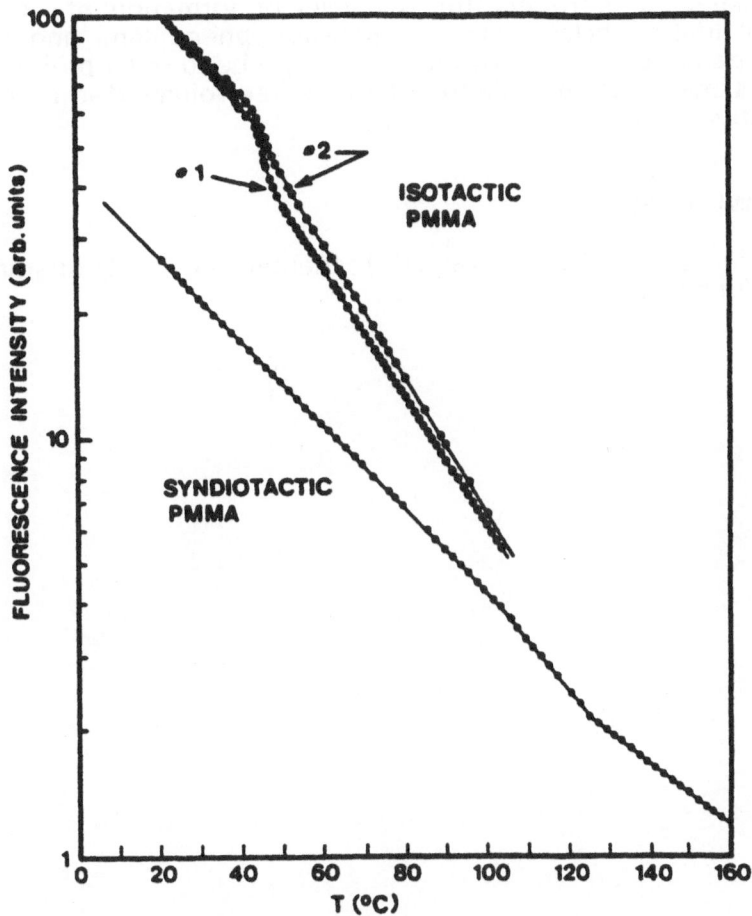

Figure 11. Effect of temperature on the fluorescence of 3 in syndiotactic and isotactic PMMA.

photocrosslinking of polymers, the dynamics of formation of sterically stabilized coloidal particles, polymer relaxation phenomena and many others. By expanding the series of fluorescence probe to cover probes with different sizes one should be able to determine free-volume distribution in polymeric systems.

4.   ACKNOWLEDGMENT

The author is indebted to Dr. K. Y. Law, D. Teegarden and P. R. Sundararajan for many useful discussions.

REFERENCES

1.   Batschinski A. J., 'Investigations of Internal Friction of Fluids' *J. Phys. Chem.*, **84**, 643(1913).

2.   Doolittle A. K., 'Studies in Newtonian Flow II. The Dependence of the Viscosity of Liquids or Free-Space' *J. Apl. Phys.*, **22**, 1471(1951).

3.   Bueche F., 'Segmental Mobility of Polymers Near Their Glass Temperature' *J. Chem. Phys.*, **21**, 1850(1953).

4.   Fujita H., *Trans. Faraday Soc.*, **56**, 424(1960).

5.   Williams M. L., Landel, R. F. and Ferry J. D., 'The Temperature Dependence of Relaxation Mechanisms in Amorphous Polymers and Other Glass Forming Liquids' *J. Am. Chem. Soc.*, **77**, 3701(1955).

6.   Beavan S. S., Hargreave J. S. and Phillips D., *Adv. Photochem.* **11**, 207(1979).

7.   Morawetz H., *Science (Washington, D.C.)* **203**, 405 (1979).

8.   Winnik M. A., Redpath T., Richard D. H., *Macromolecules*, **13**, 328 (1980).

9.   Semerak S. N., Frank C. W., *Macromolecules* 14, 443 (1981). G. E. Johnson, T. A. Good, *Macromolecules*, **15**, 409 (1982).

10.   Loutfy R. O., Law K. Y., *J. Phys. Chem.* **84**, 2804 (1980); *Macromolecules* **14**, 587 (1981). Loutfy R. O., Ibid. **14**, 270, (1981). *J. Polym. Sci., Polymer Phys. Ed.*, **20**, 825 (1982).

11.   Law K. Y., *Chem. Phys. Lett.*, **75**, 545 (1980); *Photochem. Photobiol.*, **33**, 799 (1981).

12.   Loutfy R. O. and Arnold B. A., *J. Phys. Chem.* **86**, 4205(1982).

13.   Oster G. and Nishijima Y. J., *Am. Chem. Soc.*, **78**, 1581 (1956).

14.   Doolittle A. K., *J. Appl. Phys.*, **23**, 236(1952).

15.   Johnson G. E., *J. Chem. Phys.*, **63**, 4047 (1975).

16.   Foster T. and Hoffmann G. Z., *Phys. Chemie NF* **75**, 63(1971).

17.   Tredwell C. J., and Keary C. M., *Chem. Phys. Letters*, **43**, 307, (1979).

18.   Taylor J. R., Adams M. C. and Sibbett W. , *Appl. Phys.* **102**, 847 (1980),

19.   Jaraudias J., *J. Photochem.*, **13**, 35 (1980).

20.   Baker, R. H., and Gratzel, M., *J. Am. Chem. Soc.*, **102**, 847(1980).

21.   Jones, G., Jackson, W. R., and Halpern, A. M. , *Chem. Phys. Letters,* **72**, 391(1980); Karsten, T. and Koh, K., J. Phys. Chem., **84**, 1871(1980).

22.   Bueche, F. and Kelley, F. N., *J. Polym. Sci.*, L, 549(1961).

23.   Bueche, F., 'Physical Properties of Polymers' *Interscience*, New York Chp. 5 (1962).

24.   Frank. C. W., Gashgari M. A., *Macromolecules* **12**, 163(1979); **14**, 1558(1981).

EXCIMER FLUORESCENCE AS A PROBE OF MOBILITY IN POLYMER MELTS

Liliane BOKOBZA, Lucien MONNERIE
Laboratoire de Physico-Chimie Structurale et Macromoléculaire
Ecole Supérieure de Physique et de Chimie Industrielles de
Paris
10, rue Vauquelin
75231 Paris - France

ABSTRACT. Intramolecular excimer formation of various probes dispersed in different elastomers has been investigated. The dynamic behaviour of the fluorescence probes is related to the mobility of the host matrix. The rate of conformational change appears to reflect the segmental chain motions involved in the glass-rubber transition. The temperature dependence of the correlation times of the rotational motions related to the excimer sampling mechanism follows the WLF equation.

I - INTRODUCTION

Molecular motions in bulk polymers have been investigated by various spectroscopic techniques such as dielectric relaxation, NMR, ESR and fluorescence anisotropy decay. These techniques yield the temperature dependence of the correlation times but do not give information about the amplitude of the motion performed by the labels or the probes. The excimer fluorescence technique can provide original information about this phenomenon.

Bichromophoric molecules of general formula :

$$CH_3 - \underset{\underset{Ar}{|}}{CH} - X - \underset{\underset{Ar}{|}}{CH} - CH_3 \qquad or \qquad Ar - CH_2 - X - CH_2 - Ar$$

$$X = CH_2 \text{ or } O$$

where the two aromatic groups (Ar) are separated by a three atom linkage give rise to intramolecular excimer formation. This intramolecular excimer formation involves rotation about the C - X bonds of the linkage to achieve a conformation in which the two chromophores overlap in a sandwich-like arrangement. Conformational energy calculations performed on the probes in order to determine the most stable ground state and excimer conformations should allow an estimation of the volume swept out during the conformational change required for excimer formation. Fluorescence lifetime measurements yield the time necessary

449

M. A. Winnik (ed.), Photophysical and Photochemical Tools in Polymer Science, 449–466.

for this conformational transformation. Analysis of the emission properties of an intramolecular excimer-forming probe, dispersed in a
polymer host matrix at a sufficiently low concentration to obtain solely
intramolecular interactions, can give evidence of a range of temperatures where the conformational change becomes efficient. In addition,it
might be possible to determine whether the rate of conformational change
required for excimer formation depends on the free volume induced by
the segmental motions occuring at the glass transition of the polymer.

     In this paper will be presented the results obtained in three fami-
lies of elastomers, using probes of various sizes which exhibit different photophysical behaviour and give rise to different excimer species.
The correlation times,ranging from 10^{-6} - 10^{-9} s,have been measured as
a function of temperature.

II - DETERMINATION OF THE RATE CONSTANT FOR INTRAMOLECULAR EXCIMER
     FORMATION

Analysis of the experimental data is performed according to the conventional kinetic scheme (1) :

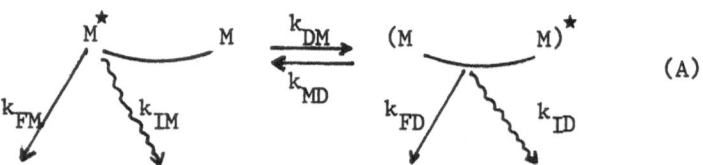

(A)

where k_{DM} and k_{MD} are the rate constant for intramolecular excimer formation and dissociation respectively, k_{FM} and k_{FD} are the rate constant
of fluorescence from the local excited state (monomer) and excimer, k_{IM}
and k_{ID} are the rate constant of non-radiative decay from the monomer
and the excimer. Solution of the rate equations appropriate to the kinetic scheme of equation (A) assuming a δ function excitation pulse
yields the following expressions for the time dependence of monomer
$I_M(t)$ and excimer $I_D(t)$ fluorescence decays :

$$I_M(t) = \frac{k_{FM}}{\beta_2 - \beta_1}[(X-\beta_1)e^{-\beta_2 t} + (\beta_2 - X)e^{-\beta_1 t}] \tag{B}$$

$$I_D(t) = \frac{k_{FD}k_{DM}}{\beta_2 - \beta_1}[e^{-\beta_1 t} - e^{-\beta_2 t}] \tag{C}$$

In these expressions :

$$\beta_{1,2} = \frac{1}{2}\{(X+Y)\mp[(X-Y)^2 + 4 k_{DM}k_{MD}]^{1/2}\} \tag{D}$$

and

$$X = \frac{1}{\tau_M} = k_{FM} + k_{IM} + k_{DM}$$

$$Y = \frac{1}{\tau_D} = k_{FD} + k_{ID} + k_{MD}$$

τ_M and τ_D represent the excited monomer and excimer lifetimes, respectively.

A direct access to the dynamic properties of the probe can be derived from the fluorescence decay measurements by the determination of the rate constant of excimer formation k_{DM}. Measurement of the monomer fluorescence decay alone enables the determination of k_{DM}. Besides the decay parameters β_1 and β_2 and their amplitude ratio $A = \frac{\beta_2 - X}{X - \beta_1}$, the monomer decay τ_o in the absence of excimer formation is needed. This monomer decay time τ_o can be defined as the fluorescence lifetime of an aromatic group moiety under the assumption that the probe does not undergo intramolecular excimer formation. It is commonly determined by measuring the fluorescence decay time of a model compound containing only one chromophore. In the absence of excimer dissociation, the monomer decay would be monoexponential with a decay function $\beta_2 = X = 1/\tau_M$ (equation D). This is the behaviour usually observed at low temperatures.

The most convenient experimental measure of excimer fluorescence under photostationary-state conditions is the ratio I_D/I_M, where I_D and I_M are the emission intensities of the excimer and monomer bands, respectively. The analysis of the kinetic scheme (equation A) leads to the following expression for the I_D/I_M ratio :

$$\frac{I_D}{I_M} = \frac{k_{FD}}{k_{FM}} k_{DM} \tau_D \qquad\qquad (E)$$

This ratio provides an approximate measure of the efficiency of the rotational process and hence of the mobility of the probe in the host matrix. Nevertheless I_D/I_M is directly proportional to k_{DM} only if τ_D is a constant over the range of temperature investigated. This assumption has often been used when the temperature-dependent fluorescence spectra are characterized by an isoemissive point. In somes cases, τ_D is temperature-dependent despite the existence of the isoemissive point (2). Thus, taking the I_D/I_M ratio as proportional to k_{DM} may lead to incorrect conclusions about the matrix mobility. Only transient measurements allow a quantitative determination of the dynamic properties of the probe.

III- CHARACTERISTICS OF THE HOST POLYMERIC MATRICES

One important feature of a polymeric material is its glass transition. The mobility of the polymer segments involved in the glass relaxation is described by the WLF equation using the free-volume concept:

$$\log a_T = \log \frac{\tau_C(T)}{\tau_C(Tg)} = -\frac{C_1(T-Tg)}{C_2+(T-Tg)}$$

where a_T is the shift factor, $\tau_C(T)$ the correlation time at a temperature T and $\tau_C(Tg)$ the correlation time at the reference temperature chosen as the glass transition temperature determined by DSC. $\tau_C(Tg)$ is the fitting parameter leading to a vertical shift, C_1 and C_2 are two constants depending on the chemical structure of the medium. The selected values are those given by Ferry (3) for elastomers of microstructures similar to those of the samples investigated in the present study. This WLF equation is commonly used for mechanical relaxation data in a lower frequency range ($f < 10^6$ Hz).

The characteristics of the matrices investigated are summarized in Table 1 and Table 2.

These matrices are :

- polybutadiene (PB) : $\text{-CH}_2\text{-CH=CH-CH}_2\text{-}$

random styrene-butadiene copolymers (SBR): $\text{-CH}_2\text{-CH-}_x\text{-CH}_2\text{-CH=CH-CH}_2\text{-}_{1-x}$

for which the set of parameters C_1 and C_2 of the WLF equation are C_1= 11.2 and C_2=60.5;

- polyisoprene (PI) : $\text{-CH}_2\text{-C(CH}_3\text{)=CH-CH}_2\text{-}$
and a random styrene-isoprene
copolymer (SIR) : $\text{-CH}_2\text{-CH-}_x\text{-CH}_2\text{-C(CH}_3\text{)=CH-CH}_2\text{-}_{1-x}$

for which C_1= 16.8 and C_2= 53.6;

- poly(dimethylsiloxane)(PDMS) :
$$\begin{bmatrix} & CH_3 & \\ & | & \\ & -Si-O- & \\ & | & \\ & CH_3 & \end{bmatrix}$$

and a poly(dimethyldiphenylsiloxane)(PDMPS):
$$\begin{bmatrix} & \bigcirc & \\ & | & \\ - & Si-O & - \\ & | & \\ & \bigcirc & \end{bmatrix}_x \begin{bmatrix} & CH_3 & \\ & | & \\ - & Si-O & - \\ & | & \\ & CH_3 & \end{bmatrix}_{1-x}$$

for which C_1 = 6.1 and C_2 = 69.0.

Polymer	% Styrene (in mass)	Microstructure of PI or PB phase			Tg (°C)
		% cis	% trans	% vinyl	
PB Diene 45 NF	0	37	51	12	- 91
SBR Solprene 1204	27	24	43	33	- 48
SBR Stereon 704	19	35	55	10	- 72
SBR Stereon 705	26	33	57	10	- 64
PI IR 307	0	92	5	3	- 58
SIR	20	50	31	19	- 27

Table 1 - Characteristics of the polybutadiene, the polyisoprene, and their respective copolymers with styrene

Polymer	$\overline{M}_W$	% diphenylsiloxyl groups	Tg (°C)
PDMS	5000	0	- 124.5
PDMPS	600 000	11.1	- 96.0

Table 2 - Characteristics of the poly(dimethylsiloxane) and of the poly(dimethyldiphenylsiloxane).

Among these samples, only the PDMS crystallizes (at - 40°C) which considerably reduces the field of the experiments. The interesting feature of the family of styrene-butadiene copolymers is the variation of the glass transition temperature Tg induced by changing either the microstructure of the butadiene phase or the styrene content.

The polydienes consist of films cured with dicumyl peroxide. These cross-linked samples were swollen with a solution of the probe in cyclohexane and dried by extensive pumping in vacuo. Films with raw polybutadiene were also prepared by casting onto a quartz plate a cyclohexane solution containing the probe.

The bulk samples of PDMS and PDMPS are fluids of widely differing macroscopic viscosity. The final probe concentration does not exceed 2×10^{-7} mol/g of polymer in the films and the absorbance in PDMS or PDMPS is about 0.1 at the wavelength of the excitation light. At these concentrations, only intramolecular interaction between the chromophores can occur.

IV - INVESTIGATION OF THE CHARACTERISTIC OF THE MATRIX MOLECULAR
MOBILITY THROUGH THE EXCIMER EMISSION OF 10,10'-DIPHENYL-BIS-9-
ANTHRYLMETHYLOXIDE (DIPHANT) (2,4).

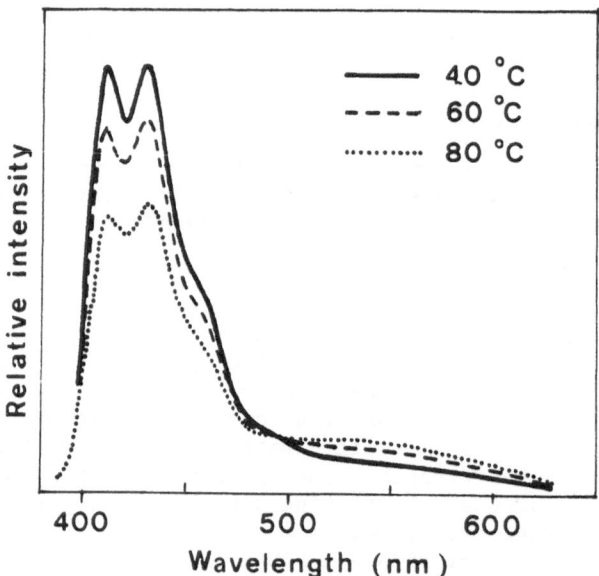

Diphant

As a first step we shall consider some typical features of the
emission behaviour of diphant dispersed in the various polymer host
matrices.

The temperature dependence of the excimer-monomer spectra of
diphant in PB is given in Fig. 1.

Fig. 1 - Temperature dependence of monomer and excimer fluorescence
intensities of diphant in polybutadiene

The excimer and monomer emission bands are well separated, which
reflects a large stabilization energy of the excimer complex. An isoemissive point is detected near 500 nm even at relatively high tempera-

tures. The I_D/I_M ratios significantly lower than those obtained in solution (5) are an indication of the restriction in the mobility of the probe itself in the relatively rigid polymer matrix. The evolution of the monomer and excimer fluorescence intensities as a function of temperature is shown in Fig. 2. The approximately constant value of I_M between - 100 and - 20°C indicates that the probe motions are still hindered. The non-zero value of I_D in this range of temperature means that the excimer is not formed by the usual process of conformational change but more likely arises from excitation of a chromophore pair preformed in the ground state. At higher temperatures, I_M is observed to decrease at the expense of I_D.

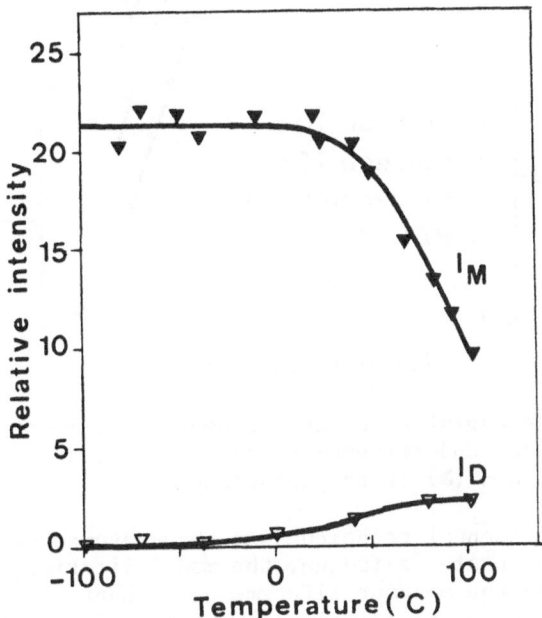

Fig. 2 - Temperature dependence of monomer and excimer fluorescence intensities of diphant in polybutadiene

The transient measurements are in agreement with the photostationary results. Presented in Fig. 3 are the lifetimes of monomer decay and of the model compound 9-(methoxymethyl)-10- phenylanthracene, as well as their temperature dependences, in a variety of polymer matrices.

The onset of mobility, detected at the beginning of the decrease of the monomer lifetime, occurs about - 20°C in PB, that is 70°C above the static reference temperature Tg measured at 1 Hz. This shift illustrates the time-temperature superposition principle : the motion is observed at the dynamic glass transition which corresponds to the fre-

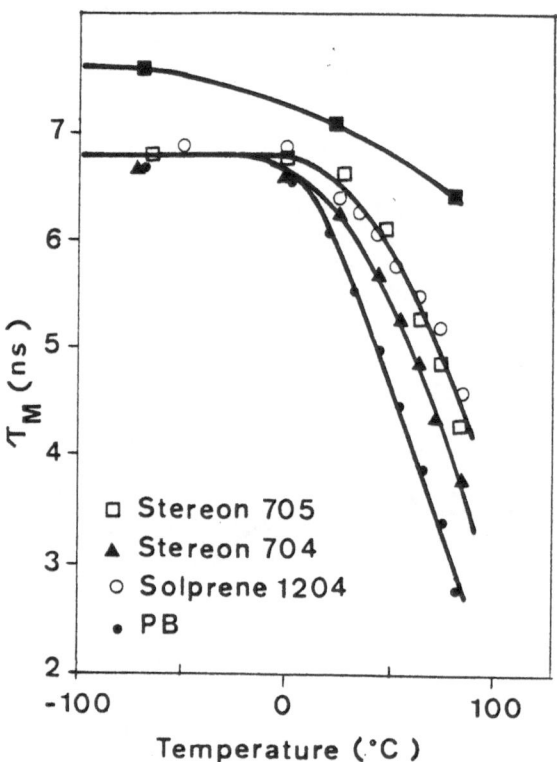

Fig. 3 - Temperature evolution of the monomer lifetime of diphant in
        polybutadiene and styrene-butadiene copolymers and of the
        model compound (■) in polybutadiene

-quency of the experimental technique. At temperatures where the rotational motion of the probe is frozen, the model lifetime is always
slightly larger than the monomer lifetime of diphant. A similar result
was noted by De Schryver et al. (5) in solution and explained by an intramolecular interaction between the chromophores of diphant in the
ground state. Assuming that the excited monomer lifetime of diphant
in the absence of excimer formation exhibits the same temperature dependence as that of the model compound, we can easily deduce the rate constant of intramolecular excimer formation K_{DM}.

     The data obtained for diphant in PB and its copolymers show clearly
that at a given temperature, the monomer lifetime, and consequently the
rate of conformational change of the probe, is affected by the host
matrix.

     Fig. 4 gives evidence of the temperature dependence of the excimer
lifetime τ_D of diphant even in the range of existence of the isoemissive
point. But contrary to τ_M, τ_D Plotted versus temperature exhibits little
scatter within the range of polymers studied. The similar values obtained in solution (5) show that τ_D is unaffected by the rigidity of the
matrix and is, indeed, and intrinsic characteristic of the fluorescence
probe.

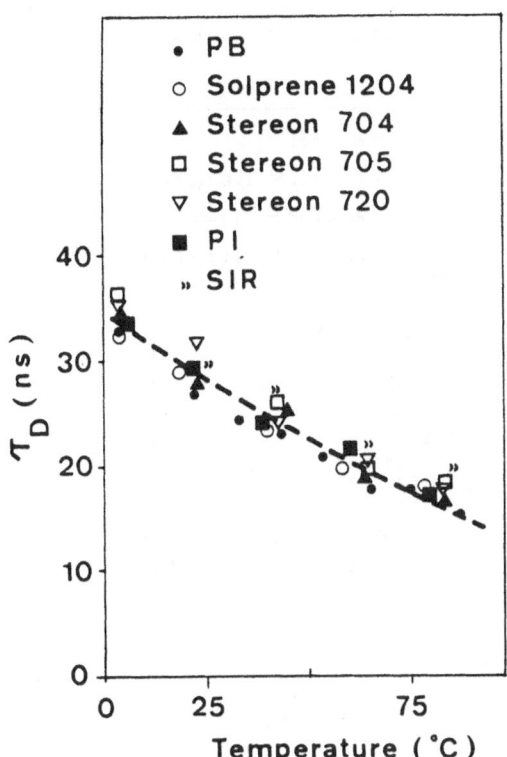

Fig. 4 - Temperature evolution of the excimer lifetime of diphant in
         polybutadiene, polyisoprene, and their respective copolymers
         with styrene.

As noted in the introductory section, our purpose is to use the
excimer fluorescence technique to study the flexibility of various
polymer chains in polymer matrices. One question which arises is to
know if the probe undergoes a conformational change via a free-volume
dependent molecular relaxation process related to the glass transition
of the polymer matrix. The free volume of a polymer at a given temperature should qualitatively relate to its Tg value.

The temperature dependence of the correlation time τ_C of the motion
involved in intramolecular excimer formation (τ_C is defined as the
reciprocal of the rate constant k_{DM} of this process) is compared to
the WLF equation :

$$\log a_T = \log \tau_C + constant$$

It is apparent from the plots of log τ_C versus (T-Tg) shown in
Fig. 5 that the rotational mobility of diphant reflects the glass-
rubber relaxation of the host matrix. The experimental data show little
scatter in the fit provided by the WLF equation for temperatures

higher than Tg + 100°C. At lower temperatures, the divergence
from this curve may be attributed to a loss of accuracy of the τ_C
calculation.

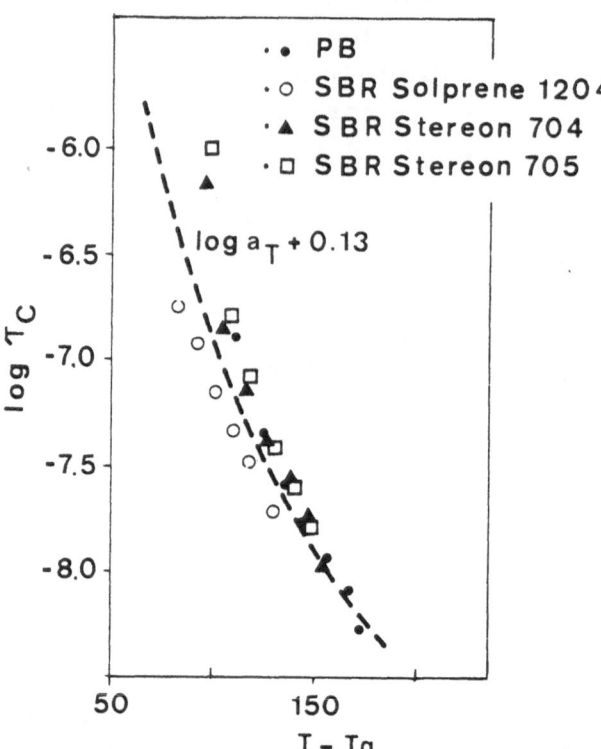

Fig. 5 - Logarithmic plot of the correlation time vs T-Tg for diphant
         dispersed in polybutadiene and styrene-butadiene copolymers.
         The dotted line represent the WLF equation.

     A more refined analysis reveals that, if the correlation times
precisely superimpose on the WLF curve for PB and Stereon 704 and 705,
their values are smaller in Solprene 1204. This change in mobility
can be interpreted by an additional effect of microstructure upon the
τ_C (Tg) vertical shift factor. Solprene 1204 differs from PB and its
other copolymers by its microstructure (Table 1). The change of the
elastomer phase in Solprene 1204, consisting of an increase of the
vinyl conformations to the detriment of the 1,4 cis or trans conformations would induce a diminution of τ_C (Tg). In other words, it appears
that the vertical shift factor τ_C(Tg) relative to the WLF equation is
slightly different in Solprene 1204 from its value in PB and Stereon
704 and 705. A further shift of the experimental correlation times
for the Solprene 1204 matrix indicates in Fig. 6 that the temperature
dependence in PB and its copolymers is well-fit by the WLF equation.

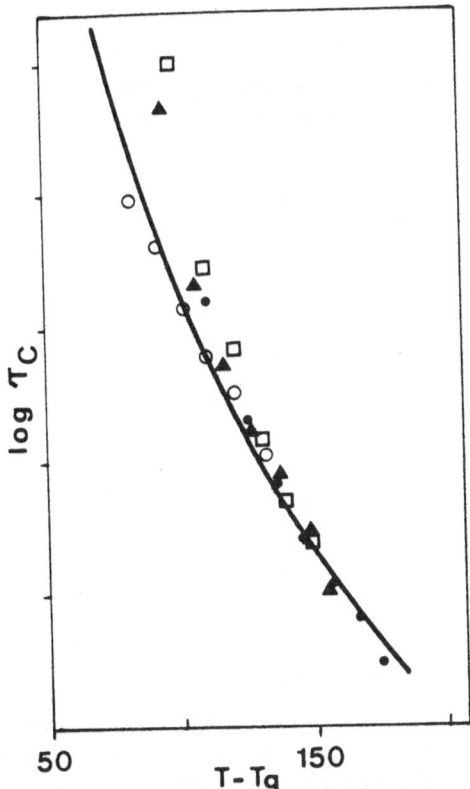

Fig. 6 - Master curve of the correlation time vs T-Tg for diphant dispersed in polybutadiene and styrene-butadiene copolymers

A master curve is obtained for each family of polymers (PB and SBR,
PI and SIR, PDMS and PDMPS) and the evolution of the absolute values
of the experimental correlation times follows the WLF equation.
In addition, we wish to emphasize that, at a constant (T-Tg) the
correlation times do not have the same value in each family of polymers
(Fig. 7).
This reveals the important effect of the chemical structure of the
matrix. The variation of the correlation time of the rotational motion
of the probe is an indirect reflection of the difference in free volume
offered by the polymer and accessible to the fluorescence probe to
achieve its conformational transition. The results would suggest that
the polymer chain flexibility (or free volume) around the polymer
backbone is different   in each family of matrices. But in each family
of polymers, the probe mobility is modulated by the same free volume
fluctuations as the polymer segments.

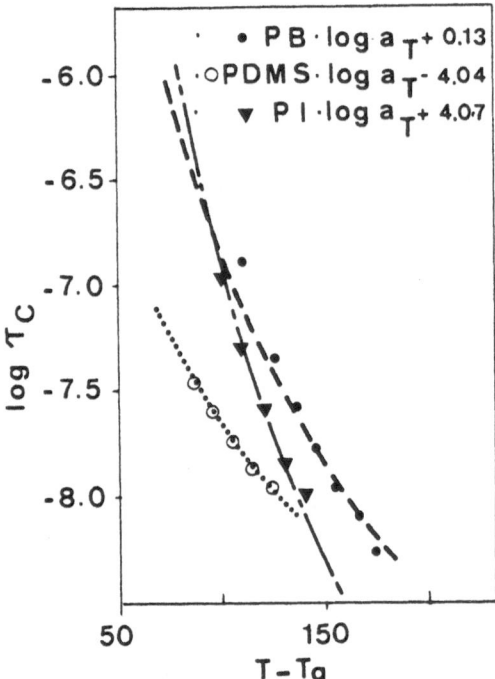

Fig. 7 - Logarithmic plot of the correlation time vs (T-Tg) for
       diphant dispersed in polybutadiene, polyisoprene and poly
       (dimethylsiloxane). The dotted lines represent the related
       WLF equations.

V - INFLUENCE OF THE SIZE OF THE CHROMOPHORE GROUP

In order to determine the influence of the physical properties and the
size of the fluorescent molecule on the observed mobility, we look at
the emission behaviour of another probe dispersed in some of the polymers used in the diphant probe studies.It would be of interest to investigate the intramolecular excimer formation in bis(9-anthryl) oxide,
less bulky than diphant. Unfortunately this compound is not photostable
and gives rise to photocyclomers. The present work has, therefore, been
performed with the 2,4-di(N-carbazolyl) pentane (meso-DNCzPe)(see formula) in which the aromatic group are of similar size to the bis-anthryl
derivative.
       The lifetime of the excited monomer of meso-DNCzPe in PB changes
from 10.5 ns at 0°C to 5.2 ns at 40°C while that of diphant passes from
6.6 ns to 5.1 at the same temperatures. The results obtained with the
meso-DNCzPe are summarized in Fig. 8.
       Once more, at the same (T-Tg), the conformational transition rate is
quite similar for polymer matrices belonging to the same family but large
differences occur from one family to another. The excellent agreement
obtained between the temperature dependence of the correlation

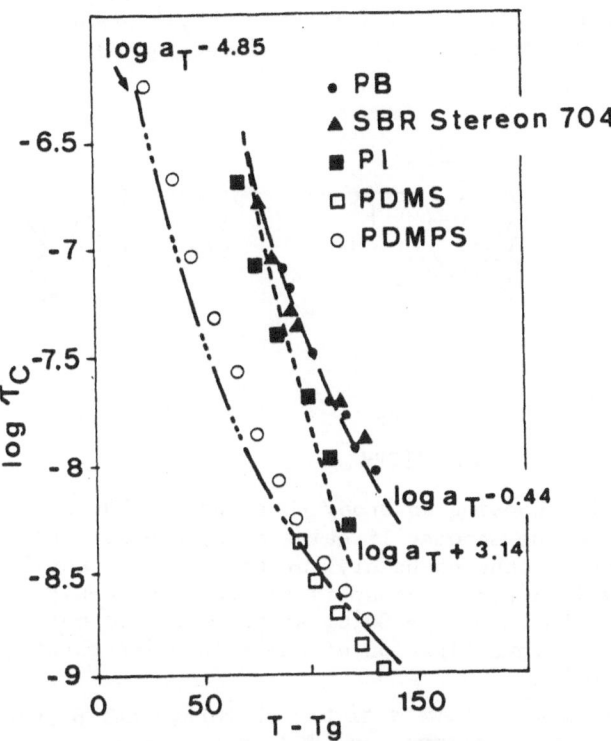

Fig. 8 - Logarithmic plot of the correlation time vs (T-Tg) for meso-
       DNCzPe. The dotted lines represent the WLF equations.

times and the one expected from the WLF equation, clearly indicates
that the intramolecular conformational change of the meso-DNCzPe probe
is controlled by the segmental motions of the polymer matrix which are
involved in the glass-transition phenomenon. It is worthwhile to note
that in all the matrices investigated, the onset of mobility in meso-
DNCzPe occurs at a temperature lower by about 20 degrees than that

observed in diphant while the monomer lifetimes of the two compounds
are of similar magnitude. This behaviour originates from a smaller
volume which is swept out during the intramolecular conformational
change associated to the excimer formation in DNCzPe. Conformational
energy calculations are actually in progress in order to estimate the
volumes swept out by the interacting chromophores during the intramolecular rotational motion.

VI - INFLUENCE OF THE EXCITED LIFETIME OF THE CHROMOPHORE GROUP

Intramolecular excimer formation of meso 2,4-di(1-pyrenyl)pentane
(meso-DIPP) and of meso bis[1-(pyrenyl) ethyl] ether(meso-B1PEE) dispersed in polybutadiene has also been investigated.

$$X = CH_2 \quad \text{for D1PP}$$
$$X = 0 \quad \text{for B1PEE}$$

Our objective in choosing to study these particular probes in polymer
matrices is that the monomer lifetime of pyrene is over twenty times
longer than that of the carbazolyl or the phenyl-anthryl group. The lifetime of the excited pyrene group is 184 ns in meso-DIPP and 235 ns in
meso-BIPEE in a PB film at - 90°C, so that the two molecules are appropriate probes for investigation of the relaxation phenomena in the time
range 10^{-9} - 10^{-6} s.

While meso-DNCzPe shows a simple photophysical pattern on account
of its single chain conformation (TG) in the ground state and its unique low energy excimer with total overlap of the carbazole groups (6, 7),
meso-DIPP and meso-BIPEE exhibit in solution a rather complex kinetic
behaviour. A recent conformational study on intramolecular excimer
formation in these models of poly(1-vinylpyrene) has been recently performed by De Schryver et al. (7, 8). The analysis of the fluorescence
decay in the monomer region for these two probes shows two excimer-forming components even at low temperatures interpreted as different rotamers of the TG conformation. The two components of the fluorescence
decay of the excimer emission are explained by the existence of two
excimer configurations (staggered and eclipsed).

The onset of mobility, accurately detected on account of the long
monomer lifetime, occurs at about - 50°C for both compounds dispersed

in PB. Temperature-dependent fluorescence spectra of meso-BIPEE in PB represented in Fig. 9 are characterized by an isoemissive point in the temperature range from - 50°C to 20°C.

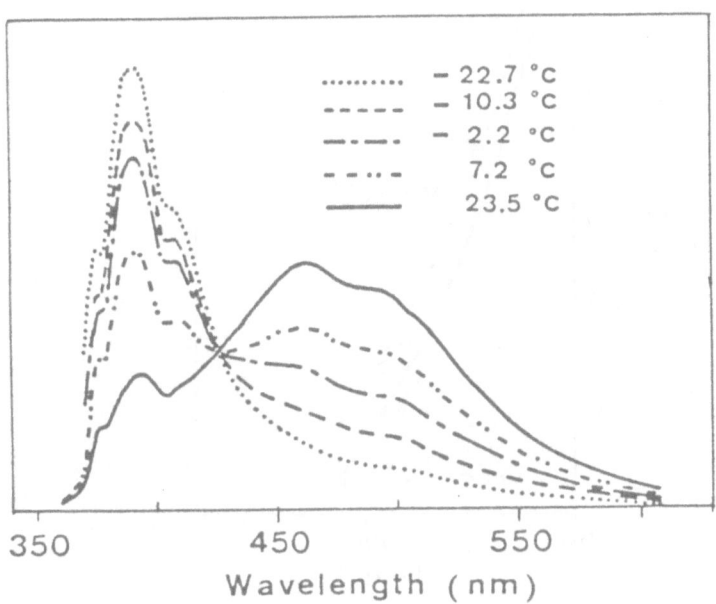

Fig. 9- Temperature-dependent fluorescence spectra of meso-BIPEE in polybutadiene

The "apparent" high mobility reflected by the large I_D/I_M ratios compared to diphant (Fig. 1) or to the carbazole derivative is in fact due to the small value of the radiative rate constant k_{FM} of the local excited pyrene group (equation E).

As is seen in Fig. 10, for both compounds, the temperature dependence of the experimental correlation times follow the WLF law.

VII - PROBE EFFECTS IN A POLYBUTADIENE MATRIX

Through this study, we have shown that the excimer fluorescence technique of a probe simply dispersed in a macromolecular medium provides detailed information about the molecular motions of the polymer.

Although the intramolecular rotational process reflects the glass transition of the host medium, within the same matrix, the absolute values of the experimental correlation times, differ for each probe. In Table 3

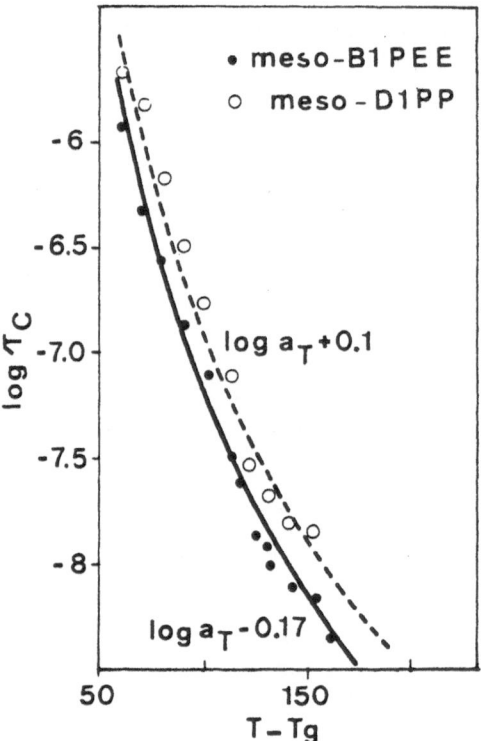

Fig. 10 - Logarithmic plot of the correlation time vs (T-Tg) for
meso-BIPEE and meso-DIPP in polybutadiene. The full and
the dotted lines represent the WLF equations

are reported the correlation times at 298 K in solution and in the
polybutadiene as well as the τ_c(Tg) values from the WLF equation
fitting parameter corresponding to the vertical shift factor.

It appears that in one medium, the correlation times are different
from one probe to another. As it is observed that the correlation
times of the motions involved in the excimer sampling mechanism follow
the temperature dependence of the WLF equation, such a behaviour can
only be ascribed to the prefactor term reflecting the efficiency of
collisions with the surrounding to yield the excimer state. In other
words, this reflects a different degree of coupling or cooperative
interaction between the probe and the host matrix.

Nevertheless, the correlation time of meso-DNCzPe is shorter than
that of the other probes. As already mentioned, this high efficiency
may be explained by the smaller volume swept out during the conformational change associated to the excimer formation.

It would be interesting to find small probe molecules for which the
excimer sampling is not controlled by segmental motions involved in the

	Diphant		Meso-DNCzPe		Meso-B1PEE		Meso-DlPP	
	τ_c (s)	$\tau_c(Tg)$	τ_c	$\tau_c(Tg)$	τ_c	$\tau_c(Tg)$	τ_c	$\tau_c(Tg)$
Solution	$2.5.10^{-9}(1)$		$5.10^{-10}(2)$		$6.67.10^{-10}(3)$ $5.56.10^{-10}(3)$		$3.45.10^{-9}(3)$ $2.08.10^{-9}(3)$	
PB	$7.05.10^{-8}(4)$	1.35	$1.66.10^{-8}(4)$	0.36	$3.18.10^{-8}(4)$	0.68	$6.21.10^{-8}(4)$	1.26

Table 3 ··

(1) In methylcyclohexane, at room temperature (from reference 5)
(2) In isooctane, at room temperature (from reference 7)
(3) In isooctane, at 298 K (from reference 9)
(4) At 298 K

glass-transition processes of the polymer host matrices.

Unfortunately, at the present time, the only available probes, 1,3-diphenylpropane and 2,4-diphenylpentane cannot be used because fluorescent impurities in the matrix are excited at the short wavelengths necessary to excite the probe.

VIII - CONCLUSIONS

The rotational mobility of intramolecular excimer-forming probes has been shown to reflect the glass-transition relaxation phenomena of the polymer host matrix, in agreement with the appropriate WLF equation.

In the case of labeled elastomeric chains (10, 11), probes with flexible tails (12) and large rigid probes (10, 11), fluorescence polarization studies have shown that the temperature dependence of the correlation time of the rotational motion agree with the temperature dependence of the WLF equation. On the other hand, smaller probes such as 9,10-dimethylanthracene dispersed in the same media, exhibit a different behaviour. In that latter case, the observed activation energies seem to be related to a secondary transition of the matrix.

The analysis of the emission properties of intramolecular excimer-forming probes of various sizes may afford a way to estimate the size of the region over which cooperative motion must occur so that the principles of free volume equilibration apply. With the use of small probe molecules, it should be possible to determine the lower limit of the size of the cooperatively rearranging region.

References

1. Birks, J.B.: 1970, "Photophysics of Aromatic Molecules", Wiley.

2. Pajot-Augy, E., Bokobza, L., Monnerie, L., Castellan, A., Bouas-Laurent, H. : 1984, Macromolecules, 17, pp. 1490-1496.

3. Ferry, J.D. : 1970, "Viscoelastic Properties of Polymers", Wiley.

4. Bokobza, L., Pajot-Augy, E., Monnerie, L., Castellan, A., Bouas-Laurent, H.: 1984, Polymer Photochem., 5, pp. 191-207

5. De Schryver, F., Demeyer, K., Huybrechts, J., Bouas-Laurent, H., Castellan, A.: 1982, J. Photochem., 20, pp. 341-354.

6. De Schryver, F.C., Vandendriessche, J., Toppet, S., Demeyer, K., Boens, N.: 1982, Macromolecules, 15, pp. 406-408.

7. Vandendriessche, J., Palmans, P., Toppet, S., Boens, N., De Schryver, F.C., and Masuhara, H.: 1984, J. Am. Chem. Soc., 106, pp. 8057-8064.

8. Collart, P., Demeyer, K., Toppet, S., De Schryver, F.C.: 1983, Macromolecules, 16, pp. 1390-1391.

9. Collart, P. Toppet, S., Zhou, Q.F., Boëns, N., De Schryver, F.C.: 1985, Macromolecules, 18, pp. 1026-1030.

10. Jarry, J.P., Monnerie, L.: Macromolecules, 12, pp. 927-932.

11. Queslel, J.P., 1982 : Thèse Docteur-Ingénieur, Université Pierre et Marie Curie, Paris.

12. Viovy, J.L., Frank, C.W., Monnerie, L.: Macromolecules, in press.

MASS DIFFUSION IN SOLID POLYMERS

J. E. Guillet
Department of Chemistry
University of Toronto
Toronto, Canada M5S 1A1

ABSTRACT. Mass diffusion in solid polymers occurs as the result of free volume created in the solid matrix. Experimental studies of molecular diffusion in solids have shown that there is an intimate relationship between diffusive transport and the types of molecular motion which can occur in the polymer, including the librational or rotational motion of side groups as well as larger scale motion of segments of the polymer backbone. As a consequence of this relationship, the onset of particular modes of motion in a solid polymer can be detected by observing the diffusive behavior of various types of molecular probes in the polymer. Because of the high sensitivity of luminescence procedures, it is often possible to observe diffusive processes using probes at very low concentrations, so that the bulk of the solid is not excessively perturbed by the presence of the chromophore. Luminescence measurements can thus be used both to detect "transitions" at which particular modes of molecular motion become observable, as well as providing sensitive methods for measuring diffusion and permeability in solid polymers. The principles of these procedures and a few practical examples of their application will be discussed in this chapter.

1. DIFFUSION AND MOLECULAR MOTION IN AMORPHOUS POLYMERS

Diffusion in solid polymers occurs as the result of molecular motion under an applied chemical or physical potential. Whereas in simple liquids diffusion occurs primarily as a result of translational displacements of small molecules, the situation is much more complex in a solid polymer matrix. Motions of the polymer molecule can involve a large variety of specific motions of atoms or groups of atoms within the molecule. The smallest scale motions are those of the atomic vibrations familiar to the infrared spectroscopist, while on the macroscopic scale polymer motion is observable in the phenomenon of fluid flow.

We can categorize three general types of polymer motion, as shown in Figure 1. These are (a) translational motion of the centre of mass of the polymer molecule, (b) segmental rotation derived from internal

M. A. Winnik (ed.), Photophysical and Photochemical Tools in Polymer Science, 467–494.
© *1986 by D. Reidel Publishing Company.*

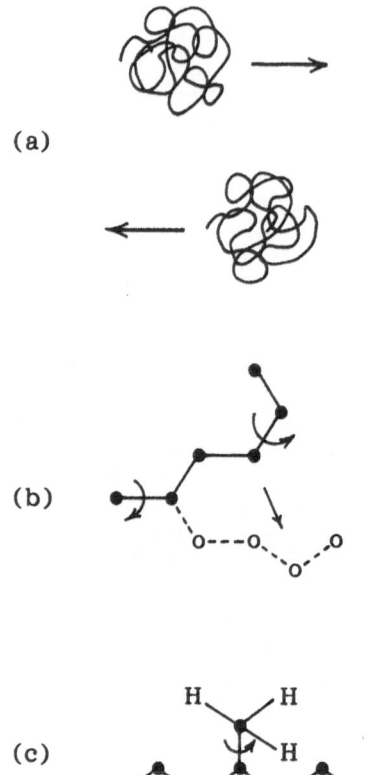

Figure 1.   Types of motion in polymers molecules: (a) translational;   (b) segmental rotation;   (c) group rotation.

rotation about covalent bonds contained in the backbone of a polymer chain, and (c) group rotation of functional groups attached to the side chain of the polymer backbone as a result of rotation about single bonds.   It is important to stress that although segmental rotation occurs around single bonds, the largest scale single bond segmental motion shown in Figure 1(b) cannot be identified with particular chemical groups or monomer units.   In particular, those associated with the glass transition temperature and with viscous flow of polymer melts often consist of units which may contain as many as 10 or 20 monomer units and except for very low molecular weight polymers will be smaller than the total chain length.

In the case of both segmental and group rotations it is not necessary that a full 360° rotation occurs around individual bonds.   It is often sufficient that a librational movement take place, giving rise to the observed mobility.   The first two of these motions, translational and segmental rotation, are usually considered to control the bulk physical properties of the macromolecular system, although there are

some exceptions to this, particularly in the case of very flexible molecules like polyethylene and poly(oxymethylene) where smaller-scale
motions can contribute as well. The hardness or stiffness modulus of
the typical polymer will change with temperature as thermal energy
is imparted to these different modes of motion of a polymer chain. However, as will be seen later, mass diffusion in polymers and particularly
the diffusive transport of small molecules in a polymer matrix can be
assisted by all the types of polymer motions shown in Figure 1.

    If a polymer containing flexible chains were cooled slowly to abso-
lute zero, one would expect that it would achieve its most stable geometry and closest packing. However, because of the very large macroscopic viscosity of the solid it seldom achieves this packing density in
finite times unless it undergoes crystallization at a relatively high temperature. Therefore, even at absolute zero the polymer matrix will
contain some free volume which, in this context, can be defined as
space unoccupied by atoms because the maximum packing density cannot
be achieved. At absolute zero, of course, this free volume is relatively
small and because there is no energy to activate the polymer molecules
they will not move to occupy the space. However, as the temperature is increased towards room temperature, thermal motion will begin
and the solid will expand. Additional free volume will be created by
this expansion and small-scale rotations or oscillations of the atoms will
begin (Figure 2). As the temperature increases, the free volume will

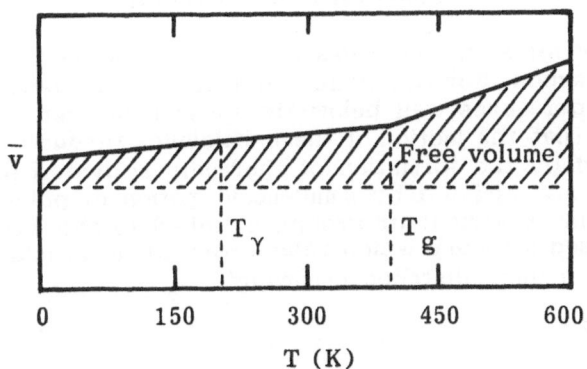

Figure 2. Specific volume and "free volume" of polymeric
material.

also increase and at particular temperatures, which are denoted transition temperatures, it will be possible to detect the motion of specific
groups within the polymer chain. However, the location of these so-
called transitions will depend on the method used to detect them and,
as shown later (Section 5), many of them show a strong dependence
on the frequency of the experimental procedure. Thus measurements
made at high frequencies detect motion in groups at higher temperatures than those involving lower frequencies. Motion of the smaller

groups within the polymer can be detected by a number of physical methods such as dielectric relaxation and thermal or mechanical deformation.[1,2]

As the temperature is raised, typical motions such as the rotational motion of the phenyl ring of polystyrene become observable. The point designated by T_γ in Figure 2 represents the temperature at which this motion is observable in polystyrene at a frequency of about 10 Hz. As the temperature is raised still further and additional free volume becomes available, a change in the slope of this specific volume curve is observed which is identified by T_g, the glass transition temperature. In most flexible polymer chains this transition is due to the occurrence of motion of very long segments of the polymer chain, often consisting of units of from 20 to 50 carbon atoms. The transitions which occur below T_g in solid polymers will vary with the chemical nature of the polymer. Specific motions which have been identified in polymer systems include the rotations of phenyl and methyl groups, rotations of carboxyl, carboxylic ester, nitrile and keto groups. Conformational transformations such as the transition from the boat to the chair form in cyclohexane rings can also be observed in a particular temperature region. Above T_g the motion of flexible chains becomes very rapid and, in fact, under these conditions the conformational mobility of polymer chains appears to be nearly as great as if they were dissolved at high dilution in a good solvent.

2.   DIFFUSION OF GASES IN AMORPHOUS POLYMERS

Diffusional motion of small molecules such as the permanent gases (oxygen, nitrogen, carbon dioxide, helium and argon) can occur in solid polymers at temperatures well below the glass transition temperature. It is believed that the motion of these molecules through the polymer matrix is assisted by the formation of thermally activated packets of free volume which are assisted by small-scale motion of polymer segments or the librational movement of groups attached to the side chains. Bueche [3] has proposed a theory which relates the diffusion constant D to a frequency ϕ and a jump distance δ through

$$D = 1/6\, \phi \delta^2 \tag{1}$$

It is expected that the jump distance δ would be relatively constant from one polymer to the next and that variations in the diffusion constant will be caused primarily by differences in the jump frequency ϕ. This frequency depends on the probability of collecting a suitable number of packets of free volume to give a hole volume V^* of sufficient size that the diffusing molecule can move into it. The creation of a hole in the liquid requires the expenditure of a certain quantity of energy E_h and the probability of forming such a hole can be calculated from Boltzmann statistics to be proportional to $\exp(-E_h/RT)$.

Application of this theory to both polymeric and small molecule liquids has given jump distances δ which correspond roughly to the dimensions of small molecules (i.e., from 2 to 20 Å). Typical jump frequencies range from 10^4 to 10^6 per second. These and other free volume theories of diffusion are described in more detail in Crank and Park [4]. The important concept, common to all free volume theories, is that diffusion occurs in polymers through free volume obtained by minor displacements of side groups or segments of the chain but without net translational movements of the centre of mass of the polymer.

3.   DIFFUSION ABOVE T_g

The large-scale segmental motions which occur in flexible polymers above their glass transitions impart relatively large amounts of free volume to the system. As a result, rates of diffusion of small molecules in polymeric materials above their glass transition are often only one or two orders of magnitude below those in ordinary liquids. Polymers are thus not a viscous medium in the conventional sense because the diffusing molecules are not affected by the motion of the centre of mass but only by segment diffusion. For example, measurements of the diffusion coefficients of naphthalene in polyethylene at 80°C by Heskins and Guillet [5] show an effective viscosity of about 1 poise, which is slightly higher than ethylene glycol but lower than the viscosity of glycerine. It is clear that the microviscosity of polymeric systems is many orders of magnitude lower than that which would be predicted from their macroscopic properties such as creep or viscous flow, which in the case of solid polyethylene at 80° could be of the order of 10^{10} poise. This fact has gone unrecognized in many studies of polymer photophysics and photochemistry.

As a polymer chain is crosslinked, the coordinated segment motion available for the creation of free volume is reduced with a consequent reduction in the diffusion constant. However, again the macroscopic properties of the polymer are affected much more by crosslinking than is diffusion. In fact, crosslinking has very little effect on diffusion until the crosslink density exceeds about 3%. [6,7] Further increases in crosslink density presumably then restrict the segmental motion responsible for the diffusion process.

Semicrystalline polymers such as polyethylene and polypropylene exhibit rather rapid diffusion processes for both small and intermediate size organic molecules at temperatures well below their melting points. It is now well established that diffusion in such polymers occurs exclusively through the amorphous regions of the semi-crystalline polymers. Until very high degrees of crystallinity are obtained in the polymer, it appears that the amorphous regions represent the contiguous phase which explains the rapid transport of small molecules through such polymers. This has important bearing on the chemical reactivity and aging characteristics of polyethylene and polypropylene and other semi-crystalline polymers.

4.   DETECTION OF SOLID PHASE TRANSITIONS BY
     LUMINESCENCE METHODS

4.1   Introduction

     The relationship between segmental and group motion in solid pol-
ymers and the rates of diffusion of small molecules suggests that diffusion measurements might be used to detect the transitions associated
with the occurrence of this type of motion, and indeed this turns out
to be the case.   One of the most sensitive measurements of the occurrence of solid-phase transitions is obtained by observing the quenching
of phosphorescent probes in solid polymers by the diffusion of oxygen.

4.2   Phosphorescent Probes for Low Temperature Solid Transitions

     The principle of the phosphorescence method has been developed
extensively by Somersall et al.[8]   In this work, groups capable of
phosphorescence emission such as keto- or naphthalene units, were copolymerized in small concentrations with a variety of polymers.   The
resulting polymers were coated as thin films and inserted between two
plates in a brass cell holder which could be placed in a phosphorimeter.
The cell was cooled with flowing nitrogen to temperatures of 70K and the
phosphorescence emission intensity was monitored as a function of temperature as the cell warmed to 25°C.   Whereas fluorescence emission
from these polymers showed very little change with temperature, the
phosphorescence intensity changed drastically.   Typical results are
shown in Figure 3.   Not only were large changes observed, but these
changes depended on the nature of the polymer.   When plotted in
Arrhenius form, experimental data on a large number of polymers show straight lines which intersect at points corresponding to the
known low-temperature transitions in the host polymer.   Results for a
copolymer of methyl methacrylate containing small amounts of naphthalene and phenyl ketone groups are shown in Figure 4.   The stronger
intensity of the naphthalene emission allows the detection of both the
γ-transition associated with the rotation of the α-methyl group and the
β-transition associated with the rotational motion of the methacrylate
ester group.   The lower intensity phosphorescence from the phenyl
ketone group was able to detect only the higher temperature β-transi-
tion.   Polymers whose repeating units exhibit even weak phosphorescence will also show such effects.   For example, the solid-phase transitions in poly(phenylacrylate) are shown in Figure 5.   The theory of
this effect has been described by Guillet [9] and is summarized below.

4.3   Theoretical Treatment for Phosphorescence in Solid Polymers.

     The experimentally determined rate constant for the decay of a
triplet chromophore attached to a polymer chain can be divided into
two terms $K_T = k_{PT} + k_{GT}$ in which k_{PT} is the rate constant for phosphorescence emission, which is generally regarded as being independent
of temperature over the temperature range of these experiments, while
the rate constant k_{GT} for non-radiative activation can be separated

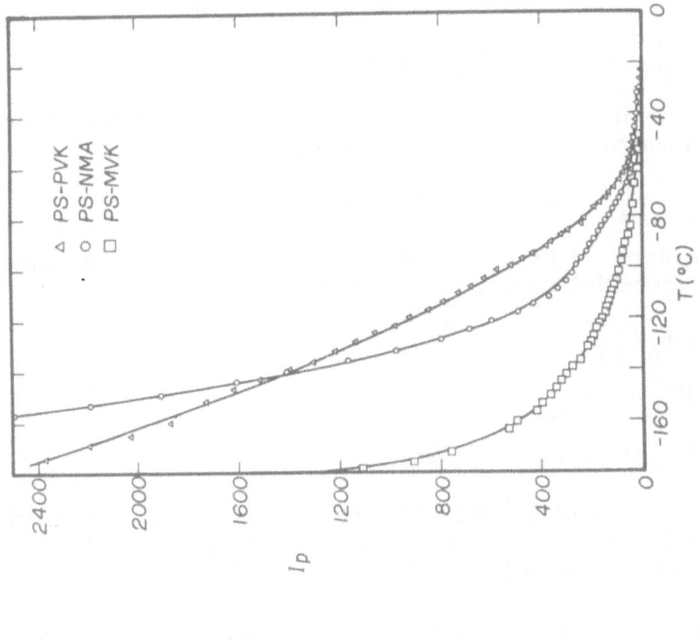

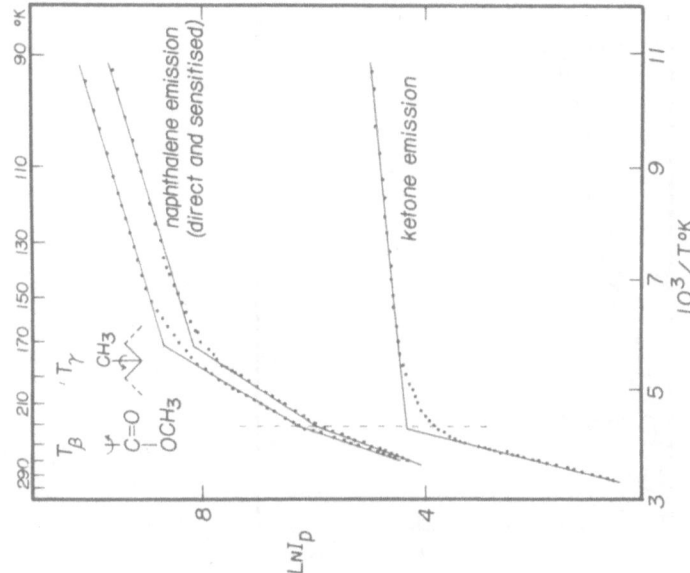

Figure 4. Arrhenius curves for phosphorescence of the copolymer MMA–NMA–PVK (80:10:10) [8].

Figure 3. Phosphorescence intensity as a function of temperature for styrene copolymers [8].

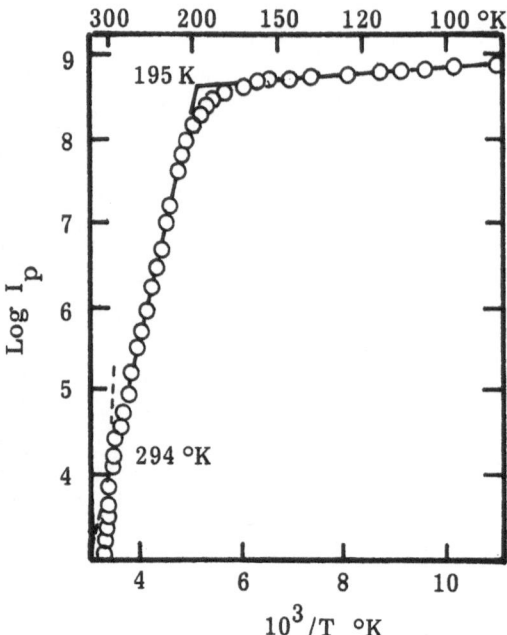

Figure 5.   Arrhenius plot for poly(phenylacrylate) phosphorescence.

itself into two terms, one of which is temperature-dependent and the other temperature-independent.   Thus we can write

$$k_T = k_{PT}^0 + k_{GT}^0 + A_{GT}' \exp[-\Delta E_{GT}/RT] \qquad (2)$$

Here the superscripted rate constants are now independent of temperature.   Rearrangement of this equation gives

$$\ln(k_T - k_T^0) = \ln A_{GT}' - \Delta E_{GT}/RT \qquad (3)$$

where k_T^0 is the value of k_T as $T \to 0$ K.   For steady-state illumination at low phosphorescence intensity, I_p, we have

$$I_p = BI_A \phi_{PT} \qquad (4)$$

$$= BI_A q_{PT} \phi_{TM} \qquad (5)$$

$$= BI_A (k_{PT}/k_T)(k_{TM}/k_M) \qquad (6)$$

where B is an instrumental factor assumed to be constant over the conditions of the experiment.   If we assume that k_T is the only temperature-dependent rate constant, it follows that

$$\ln[(1/I_p) - (1/I_p^0)] = \ln(A_{GT}' k_M / B k_{PT}^0 k_{TM} I_A) - \Delta E_{GT}/RT \qquad (7)$$

here ΔE_{GT} is the activation energy for the temperature-dependent part of the internal conversion rate constant k_{GM}. In the situation where I_P^0 is very much greater than I_P it is possible to approximate eq. (7) by the expression

$$\ln I_P = \text{const} + (\Delta E_{GT}/RT) \tag{8}$$

This explains the linear portions of the Arrhenius plots shown in Figure 4. This equation is, however, inadequate in cases where I_P and I_P^0 are of comparable magnitude. In these instances, use of eq. (8) will give activation energies which are too low. In some instances it is possible to approximate I_P^0 by the intensity of phosphorescence at 77 K, $I_{P(77)}$, which then can be substituted in eq. (7).

In the presence of a diffusible quencher such as oxygen, the total rate constant for decay of the triplet must include an additional term. Thus k_T will be given by

$$k_T = k_{PT} + k_{GT} + k_Q[Q] \tag{9}$$

where k_Q, the rate constant for quenching, is given by an expression of the form

$$k_Q = A_Q \exp[-\Delta E_Q/RT] \tag{10}$$

The expression for the intensity of phosphorescence now becomes

$$(1/I_P) - (1/I_P^0) = (k_M/BI_A k_{PT} k_{TM})(A_{GT}' \exp[-\Delta E_{GT}/RT] +$$
$$[Q]A_Q \exp[-\Delta E_Q/RT]) \tag{11}$$

When quenching becomes the dominant mechanism for triplet decay this equation can be approximated by

$$1/I_P = (k_M[Q]A_Q/BI_A k_{PT} k_{TM}) \exp[-\Delta E_Q/RT] \tag{12}$$

and hence a plot of the logarithm of the phosphorescence intensity versus $1/T$ according to

$$\ln I_P = \text{const} + (\Delta E_Q/RT) \tag{13}$$

will give a straight line with a positive slope equal to $\Delta E_Q/RT$. If the concentration of the diffusible quenching molecule is constant throughout the experiment, then the numerical value of ΔE_Q will be equal to the activation energy for diffusion ΔE_D. With gaseous quenchers such as oxygen, maintaining a constant partial pressure over a very thin film of the polymer does not assure that the concentration of the quencher is constant throughout the experiment and ΔE_Q will also include the temperature coefficient of the solubility of O_2 in the polymer.

4.4 Experimental Determination of Solid-Phase Transitions by Phosphorescence

   The detection of transition regions in polystyrene by this technique have been shown to be relatively independent of the nature of the phosphorescence probe.   For example, Figure 6 shows Arrhenius curves for a variety of polystyrene copolymers, which all show the γ-transitions in the region from around 180 K. In general, the higher the efficiency of emission from the phosphorescent probe, the more efficient it is at detecting the lower transitions in the polymer.   Table I summarizes transition temperatures detected by this technique on a number of polymers along with the assigned group motion associated with the transition.   It is particularly noteworthy that the activation energies associated with diffusion activated by a particular group motion increases with the size of the group. For example, Figure 7 shows that the activation energy associated with diffusion above particular transitions increases with the temperature of the transition.   This is in   good   agreement with the free-volume theory developed earlier for diffusive motion in solid polymers.

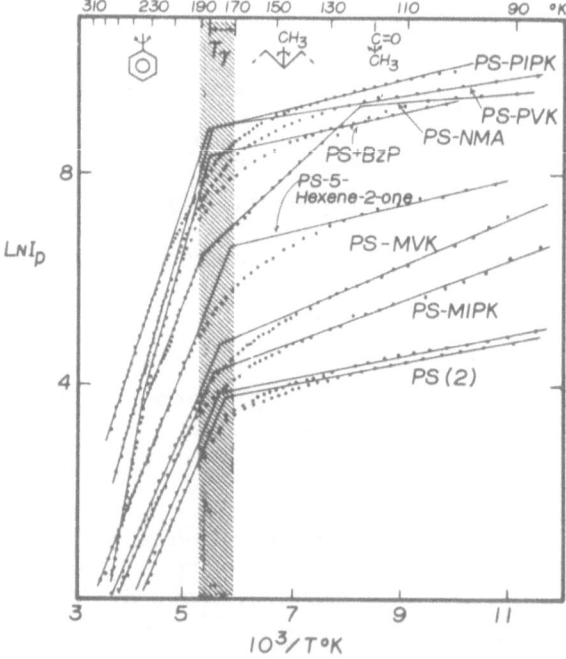

Figure 6. Arrhenius curves for polystyrene phosphorescence intensity in polystyrene and its copolymers containing keto or naphthalene groups [8].

TABLE I. Transition temperatures (T_{trans}(K)) and activation energies (E_a(kcal/mol)), from Arrhenius plots of polymer phosphorescence intensity

Polymers	C=O / CH$_3$		CH$_3$ (branched)		benzene ring		C=O / OCH$_3$		Chain segment rotation	
PS					178	5.2				
PS-PVK					182	7.0				
PS-PIPK					183	6.2				
PS-MVK					175	4.7				
PS-MIPK					183	4.9				
PS-5-hexen-2-one					175	5.4				
PS-NMA			120	2.0						
PE									163	4.3
PE-CO									163	4.5
PE-MVK									163	4.5
PE-MIPK									163	4.5
PMMA			~155	2.2						
PMMA-MVK			148	4.2			248	7.6		
PMMA-NMA			189	5.6			245	9.2		
PMMA-NMA-MVK (80:10:10)			173	3.0			236	6.0		
PVC (comm)									180	4.0
PVC (lab)									163	4.7
PVC-MVK									188	5.0
PVC-MIPK									163	4.2
PAN									~216	7.5
PAN-PVK									235	9.8
PAN-MVK	125	2.1							216	9.8
PAN-MA-MVK (75:20:5)	136	3.7							238	5.8
PMAN									234	7.6
PMAN-MVK									238	7.0
PMAN-MMA-MVK (75:20:5)									230	5.6
PMVK	145	5.8							230	12.2
PMIPK									233	7.2
PPVK					196	5.7				

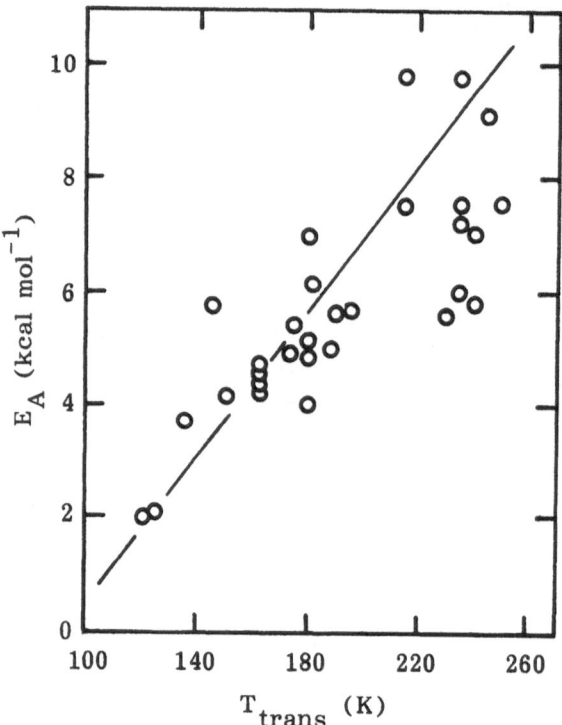

Figure 7. Activation energy (E_a) for various solid-phase transitions as a function of T_{trans}. (Data from Somersall et al. [8].)

Andrews [10] has shown that the simple Arrhenius form observed in the early experiments is not obeyed if the oxygen pressure is reduced to very low values. This can be explained by the theoretical treatment given in the previous section since if the concentration or partial pressure of oxygen is very low, the second term in eq. (11) will be small and will not become important until the diffusion rate becomes very high. The result of this will be that the data, when plotted in simple Arrhenius form, will give transition temperatures which are also too high. In the case of low oxygen pressures, the transition temperature will be identified with the first deviation from the lower temperature straight-line relationship. This is shown in Figure 8. Thus, when thick samples or low oxygen pressures are used, the transition can still be detected by this technique but the simple intersection of two straight-line Arrhenius plots will overestimate the correct value of the transition temperature.

5.   DETECTION OF THE GLASS TRANSITION BY LUMINESCENCE METHODS

The success of the phosphorescence procedure for determining low temperature transitions in solid polymers has led to a variety of

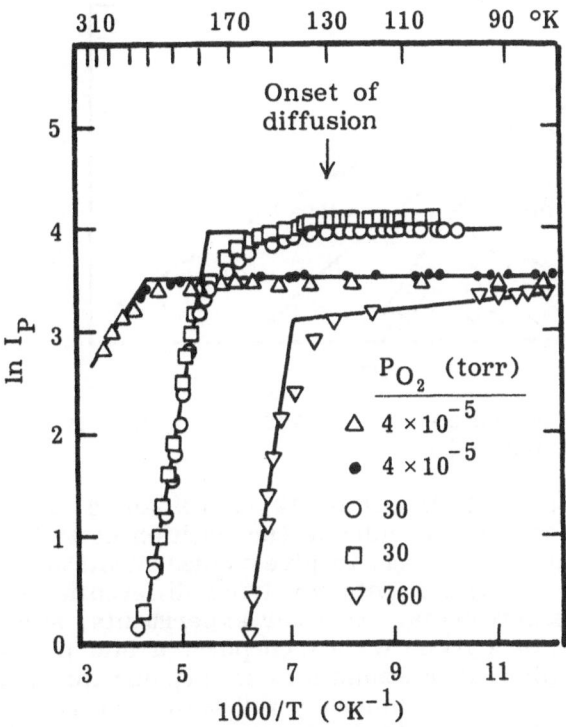

Figure 8.   Arrhenius dependence of PMMA-NMA phosphorescence at various oxygen partial pressures [10].

experiments to use luminescence probes to detect the glass transition.
For example, Pajot-Augy et al. [11] have used excimer emission from
diphant and other dimeric chromophores as a probe for determining
glass transition temperatures. However, in this work detection of excimer emission occurred at temperatures well above the normal glass transition temperature for the polymer matrix. The reason for this appears
to be that the formation of excimer takes place within periods of tens of
nanoseconds for an effective frequency of 10^7 or 10^8 per second. Similar results are also observed when molecular motion associated with the
glass transition is probed by NMR methods. It is well known that the
temperature at which particular forms of molecular motion are observed
is a strong function of the frequency of the measurements. Schematically, this is shown in Figure 9 which shows the Arrhenius diagram for
α, β and γ transitions in a typical polymer. Here the α transition is
the one associated with the glass transition, which involves very large-
scale motions of segments of the polymer chain. As pointed out in the
previous discussion and in Figure 9, the activation energy for the
occurrence of a particular type of motion increases with the amount of
free volume required.   The glass transition temperature is associated with a much larger activation energy than the lower temperature
transitions, and also a greater dependence on the frequency of the
measurement.

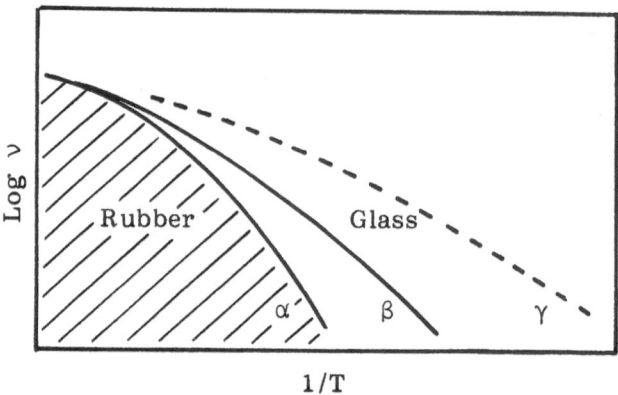

Figure 9. Arrhenius diagram for α, β and γ transitions in
a typical polymer [2].

Recently, Guillet and Sutherland [12] have found a fluorescence
measurement which does indeed indicate the position of the glass transition temperature, and which seems to give values identical to the so-
called "thermodynamic" measurements involving differential calorimetry
or specific volume measurements. In their experiments, small amounts
of a fluorophore such as pyrene were incorporated into the solid film
of a polymer along with similar amounts of an organic quencher such as
acetophenone. The film was annealed well above the glass transition
temperature, and cooled to room temperature or below. Then the
fluorescence intensity was monitored as a function of temperature as the
film was gradually heated to above its glass transition. Typical results
for polystyrene and poly(ethyl acrylate) are shown in Figure 10. It is
seen that the emission intensity decreases slowly below the glass transition temperature but decays drastically at T_g. Below T_g in the solid
matrix quenching of the pyrene fluorescence by the acetophenone (a
diffusion-controlled quencher) occurs slowly because of the relatively
high internal viscosity of the medium. The onset of the large-scale
segmental movements of the polymer chain at T_g permits the acetophenone to diffuse rapidly to the fluorophore and quench the emission.
As in the case of the lower temperature transitions discussed in the
previous section, the glass transition temperature can be determined in
these measurements by the onset of the more rapid diffusional motion
and the consequent drastic reduction in the fluorescence intensity.

Arrhenius plots of the data shown in Figure 11 also show straight-
line portions which intersect at the glass transition temperature. In
both Figures 10 and 11 the experimentally determined glass transition
temperatures for the polymers are indicated by the dotted lines. Experiments of this type show promise for the development of a rapid and
convenient method of measuring glass transitions by a luminescence
technique. It should also be possible, using the theoretical treatments
described in the next section, to determine the internal viscosity of the
polymer at various temperatures from such experiments.

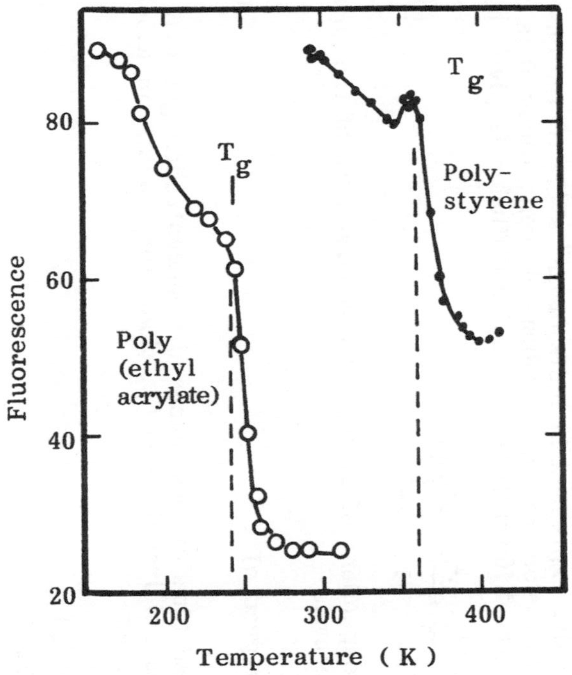

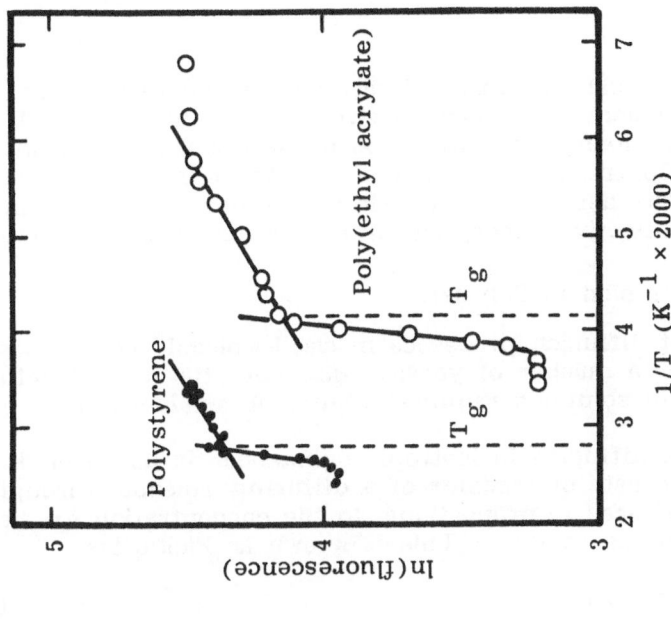

Figure 11. Arrhenius plots of fluorescence emission from polystyrene and poly(ethyl acrylate) films containing pyrene and aceto-phenone.

Figure 10. Fluorescence emission from poly-styrene and poly(ethyl acrylate) films containing traces of pyrene and acetophenone.

6. DIFFUSION AND PERMEABILITY IN POLYMERS

6.1 Introduction

Many photophysical and chemical reactions in polymers occur as a result of mass diffusion. This can be either diffusional motion of segments of a polymer chain or the diffusion of small molecules within the polymer matrix. Control of the diffusion mechanism is also of particular importance in the design of polymeric membranes for separation processes or for the selective absorption or desorption of chemical reagents.

6.2 Theory of Diffusion in Polymers

The theory of diffusion processes in small-molecule liquids has been well established for a number of years. However, the special nature of polymeric solids and solutions requires additional consideration.

The theory of diffusion in isotropic substances is based on the hypothesis that the rate of transfer of a diffusing molecule through unit cross-sectional area is proportional to the concentration gradient measured normal to the section. This is known as Fick's law:

$$F = -D(\partial c / \partial x) \tag{14}$$

where F is the flux (i.e., the rate of transfer per unit area of section), c is the concentration of diffusing substance, and x is the space coordinate measured normal to the section.   D is known as the diffusion coefficient, and in the simplest case (Fickian diffusion) is independent of c. If F and c are expressed in the same units, D has dimensions of $l^2 t^{-1}$ and is usually expressed in units of $cm^2\ s^{-1}$. In the case of three-dimensional diffusion through a volume element   where D is constant, the fundamental differential equation for diffusion is given by

$$\frac{\partial c}{\partial t} = D\left(\frac{\partial^2 c}{\partial x^2} + \frac{\partial^2 c}{\partial y^2} + \frac{\partial^2 c}{\partial x^2}\right) \tag{15}$$

Where D is not independent of c, the most general form of the fundamental equation becomes

$$\frac{\partial c}{\partial t} = \operatorname{div}(\operatorname{grad} c) \tag{16}$$

For diffusion across films or membranes the concentration gradient can often be treated as unidirectional (i.e., along the x axis), in which case eq. (15) becomes

$$\frac{\partial c}{\partial t} = D\left(\frac{\partial^2 c}{\partial x^2}\right) \tag{17}$$

This is known as Fick's second law.

For a plane sheet or membrane of thickness l whose surfaces are exposed to constant concentrations c_1 and c_2 of penetrant, a steady state will be reached where the rate of transfer F across all sections is constant and given by:

$$F = -D\left(\frac{dc}{dx}\right) = \frac{D(c_1 - c_2)}{l} \qquad (18)$$

In some cases in which a gas or vapor diffuses through a membrane, the surface concentrations may not be known and provided that Henry's law is obeyed, c will be proportional to the partial pressure of the gas, thus

$$c = Sp \qquad (19)$$

where p is the partial pressure and S is called the "solubility" or "solubility coefficient". The flux under these conditions is given by

$$F = \frac{P(p_1 - p_2)}{l} \qquad (20)$$

Where P is called the permeability, or permeability coefficient, and is given by

$$P = DS \qquad (21)$$

6.3 Solubility and Diffusion

It is important to note that the rates of reactions, whether photochemical or not, will not be controlled by the rate of diffusion alone, but also by the solubility. This also means that the rate of permeation will be proportional to both the diffusion constant and the solubility. For example, the diffusion constant for oxygen is quite large in many polymers, but usually the solubility is very low, and as a result, rates of oxidation can be quite small. Experimental values of the permeability (P) and diffusion constant (D) for various organic permeants and oxygen in low density polyethylene illustrate this point (Table II).

The concentration of diffusant c is given by

$$c = (P/D)p \qquad (22)$$

Thus at a given partial pressure p of the diffusing species, one can calculate that the concentration of benzene in polyethylene would be approximately 3×10^4 greater than that of oxygen, for example.

In the classical method of determining permeability and diffusion, a film is suddenly exposed to a constant partial pressure p_1 of the penetrant on one side and the partial pressure p_2 measured as a function of time on the other. A plot of p_2 versus time is of the form shown in Figure 12. Typically there will be a time lag before the diffusing species reaches the other side of the film, after which p_2 will increase with

TABLE II. Diffusion constants (D) and permeabilities (P) for small molecules in low density polyethylene

Permeant	$D \times 10^9$ (cm s^{-2})	$P \times 10^{10}$ (cm^3 (STP) cm cm^{-2} s^{-1} cm Hg^{-1})
Oxygen	460	2.9
Decane	3.5	~600
Benzene	8.2	5300
Hexane	--	2910
Carbon tetrachloride	--	3810
Ethyl alcohol	--	56
Ethyl acetate	--	513

Data taken from [13].

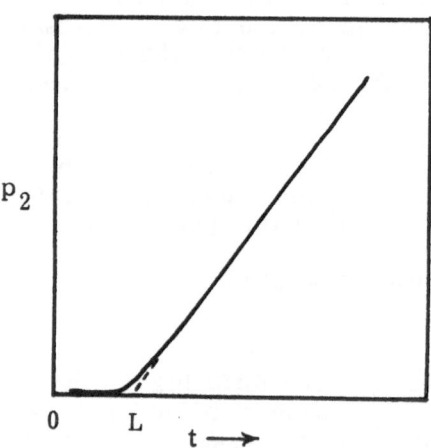

Figure 12. Rate of permeation of a penetrant across a thin film.

time at a rate which gives a straight-line relationship. The slope of this line gives the steady-state flux F which can be used to calculate the permeability P by

$$F = \frac{P(p_1 - p_2)}{l} \tag{23}$$

The intercept L of the straight-line portion with the time axis can be shown to give the diffusion constant D from the relation [4]

$$L = l^2/6D \tag{24}$$

This method works well when the polymer can be studied as a self-supporting film. However, in many cases, films cannot be easily obtained from polymers, or measurements are required at temperatures above the glass transition or in the melt, for example. In this case, a gas chromatographic procedure has been developed by Guillet and co-workers which is useful in measuring both solubilities and diffusion constants for small molecules [14–16]. Recently, luminescence procedures have been developed which supplement data obtained by more classical methods.

Since diffusion is a thermally activated process involving small scale translational motions of the penetrant and group or segment motion of the polymeric medium, a diffusion constant can be expressed in Arrhenius form as

$$D = D_A \exp[-E_d/RT] \tag{25}$$

where D_A and E_d are the pre-exponential factor and the activation energy of diffusion, respectively. Typical values for the activation energy E_d for the self-diffusion of small solvent molecules in conventional liquids are 8 to 12 kJ/mol, while diffusion in polymeric solids usually shows higher values, ranging from about 20 to 80 kJ/mol. This is clearly related to the greater energy required to displace large polymer segments to provide holes for the diffusing penetrant.

Typical values of diffusion constants and activation energies for gases in polyethylene are shown in Table III. It can be seen that as the size of the penetrant increases, the diffusion constant decreases but the activation energy increases substantially. As a result of the large value of the activation energy for diffusion in solid polymers, at high temperatures such as used in the processing of plastics in the molten state, the internal mobility of the polymer approaches that of small molecule liquids with respect to the permeation of small molecules.

Values for rates of gaseous diffusion in ordinary liquids are typically about two orders of magnitude greater than in polymers. For example, for oxygen in cyclohexane, $D = 5.3 \times 10^{-5}$ cm^2/s at 30°C, as compared with 4.6×10^{-7} in low density polyethylene.

TABLE III. Diffusion constants (D) and activation energies (E_d) for various diffusants in polyethylene at 25°C

Diffusant	$D \times 10^8$ $cm^2\ s^{-1}$	E_d $kJ\ mol^{-1}$
He	680	25
O_2	46	40
CO_2	37	38
CH_4	19	46
Benzene	0.82	--
n-Decane	0.35	--
n-Tetradecane (130°C)	1.0	67

7. MEASUREMENT OF DIFFUSION CONSTANTS BY LUMINESCENCE PROCEDURES

Most of the methods which have been developed to date to determine mass diffusion in polymer matrices by luminescence procedures have utilized an approach involving the quenching of a luminescence species by a mobile quencher. Among the earliest reports of such measurements was that of Oster et al. [17] who observed the quenching of a phosphorescent probe in glassy poly(methyl methacrylate) by the diffusion of oxygen. Many of these processes can be designed so that quenching occurs at virtually every collision between the reacting species. In this case, either transient or steady-state measurements of the intensity of luminescence can be used to determine the rates of mutual diffusion between the two particles. The theory is based on various modifications of Stokes' Law which gives the diffusion constant for a spherical particle as

$$D = kT / 6\pi\eta r \tag{26}$$

where k is the Boltzmann constant, T is the absolute temperature, η is the viscosity of the medium and r is the radius of the particle. Thus the rate of diffusion will be inversely proportional to the viscosity of the medium.

The rate constant for a diffusion-controlled reaction between two spherical particles is often described in terms of the modified Debye equation [18]

$$k_{diff} = \frac{1}{4}\left(2 + \frac{d_1}{d_2} + \frac{d_2}{d_1}\right)\frac{8\,RT}{3000\,\eta} \tag{27}$$

where k_{diff} is in units of liter/mol/s when η is in poise, $R = 8.31 \times 10^7$ erg/mol/deg and d_1 and d_2 are the diameters of the reacting species.

If the two particles are similar size, eq. (27) becomes

$$k_{diff} = \frac{8\,RT}{3000\,\eta} = 2.2 \times 10^5\,(T/\eta) \qquad (28)$$

This assumption is usually not valid where one of the species is part of a polymer chain. In this case a modified form of the Debye relation has been used by Heskins and Guillet [5]

$$k_{diff} = \frac{4\pi\rho'N}{1000}\,(D_1 + D_2) \qquad (29)$$

where ρ' is the sum of the radii of the colliding groups and D_1 and D_2 are the corresponding diffusion coefficients. In studies of the quenching of naphthalene fluorescence by both polymeric and small molecule ketones, this equation gave consistent results using a value of 4 Å for ρ', which is equivalent to assuming that only the C=O group is involved in the collision with naphthalene. If the second reactant is part of a copolymer chain, it seems likely that the segment diffusion rates will be more than an order of magnitude smaller than the translational diffusion of a small molecule and in this case eq. (29) reduces to

$$k_{diff} = \frac{4\pi\rho'ND}{1000} \qquad (30)$$

where D is now the diffusion constant for the small molecule reagent in the polymeric solvent. The diffusion constant at other than the experimentally determined temperature can be calculated if the viscosity of the solvent is known at that temperature from the simple relation

$$D_2 = D_1(T_2/T_1)(\eta_1/\eta_2) \qquad (31)$$

where η_1 and η_2 are the solvent viscosities at T_1 K and T_2 K, respectively.

As indicated previously, the viscosity of a polymeric medium can be related by free volume theories to the segmental motion of the polymer through eq. (1).

Experimental data on diffusion constants are usually derived from some variation of the quenching of luminescence in a polymer by a mobile quencher or mobile chromophore or both. In the static measurement, the intensity of luminescence is measured as a function of quencher concentration and expressed in the form of a Stern-Volmer relationship

$$\phi_0/\phi = 1 + k_q\tau[A] \qquad (32)$$

where [A] is the concentration of the quenching species. If the fluorescence lifetime in the absence of the quencher, τ, is known, and the quencher reacts at every collision, then $k_q = k_{diff}$, and combining eq. (1) with eq. (28) one obtains

$$\phi_0/\phi = 1 + 2.2 \times 10^5 \, (T/\eta) \, \tau[A] \tag{33}$$

where η is the "effective viscosity" of the medium. If one of the components, either the quencher or the luminescent chromophore, is attached to the polymer chain it is possible to assume that the segment diffusion is small compared to that of the small-molecule reagent and then the diffusion constant D of the moving small molecule can be obtained by the relation

$$\phi_0/\phi = 1 + (4\pi\rho'N/1000)D\tau[A] \tag{34}$$

The use of this relation assumes that there are no steric restrictions on the approach of the donor and acceptor species in polymer molecules, which are different in comparison to small molecule models. This assumption can usually be tested by selecting a range of small-molecule analogs for comparison. Even when quenching does not occur at every collision, it is still possible to apply eqs. (33) and (34), provided that an efficiency factor α is included. If this factor does not depend on the attachment of a chromophore to a polymer chain, then the procedure may still be valid.

Among the earliest studies of diffusion in polymers by fluorescence quenching is the work of Heskins and Guillet [5] who examined the quenching of naphthalene fluorescence in both small molecule ketones and in ethylene polymers containing carbonyl groups attached to the backbone of the chain (ethylene-CO). Both diffusion constants of the naphthalene in solid polyethylene and the efficiency of the energy transfer were investigated. It was shown that the efficiency factor α was about one-third as great when the carbonyl chromophore was contained in the polymer chain than for small-molecule model compounds. This was independent of whether or not the polymer was in solution or in the solid phase. The rate constant k_{diff} for naphthalene quenching in solid polyethylene at 80°C was found to be 1.0×10^8 L mol/s, which was in excellent agreement with the value determined for octadecane in polyethylene by a radio tracer procedure. Other workers [19-21] using a variety of polymeric chromophores, both in solution and in the solid phase, showed that this ratio ranges from between about 0.25 to 0.33. Presumably the presence of the polymer chain reduces the accessibility of the chromophore such that quenching occurs only with molecules moving with velocity vectors within a small solid angle of the "open" side of the chromophore.

8.   TRANSIENT METHODS FOR DIFFUSION MEASUREMENTS

Procedures described in the previous section utilize simple steady-state emission methods to estimate diffusion constants in polymers.

Transient procedures, though often more difficult to carry out, provide more information about diffusion mechanisms and generally more reliable data.

For example, studies by Andrews and Guillet [22] of the rate of phosphorescence quenching in solid polymer films have shown that it is possible to study very low rates of diffusion at cryogenic temperatures by the use of phosphorescence probes. Because of the longer lifetime of phosphorescence, much lower diffusion constants can be determined for diffusing quencher than by fluorescence. This procedure was first suggested by Oster et al. [17] and was used by Unterleiter and Hormats [23] and by Shaw [24] for measuring the diffusibility of oxygen in cylindrical samples of poly(methyl methacrylate). Andrews and Guillet studied the diffusion of oxygen in polystyrene containing small amounts of naphthylmethyl methacrylate copolymerized as the chromophore. The interpretation of the experiment requires a solution of the diffusion equation in one dimension (eq. 17).

In deriving solutions for this equation it is usually assumed that Henry's Law is valid so that the concentration c is proportional to the partial pressure of the gas surrounding the solid sample. General solutions to this type of equation are formally analogous to those obtained from the flow of heat in a solid bounded by parallel planes and developed by Carslaw and Jeager [25]. In this experiment, films of polymer were coated on polished aluminum plates so that diffusion occurred in only one direction from the front face. Phosphorescence was observed by reflection through the film. The concentration of gas absorbed by the film at any time t is given by the expression

$$c = lc_0 + l\left(\frac{c_2}{2} - c_0\right)\left(1 - \sum_{m=0}^{\infty} \frac{1}{(2m+1)^2} \exp\left[\frac{Dt(2m+1)^2\pi^2}{l^2}\right]\right)$$

(35)

and the intensity of phosphorescence will be given by

$$I_P(t) = I_A B \phi_P \sum_{m=0}^{\infty} \frac{1}{(2m+1)^2} \exp\left[\frac{-Dt\pi^2(2m+1)^2}{l^2}\right]$$

(36)

This series converges rapidly for times only slightly greater than zero and hence only a small error is incurred by retaining just the first term of the series which gives rise to the simple expression

$$I_P(t) = B' \exp\left[\frac{-Dt\pi^2}{l^2}\right]$$

(37)

Here B and B' are instrumental constants relating to the geometry of the sample and the optics of the phosphorimeter. The diffusion constant can be obtained from this equation by plotting the log of the phosphorescence intensity as a function of time, as shown in Figure 13

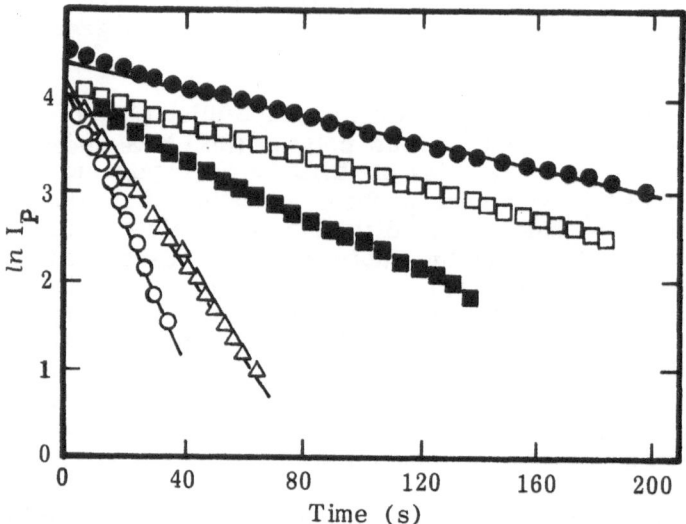

Figure 13. Semi-logarithmic plot of phosphorescence intensity versus time after exposure of polystyrene-naphthylmethyl methacrylate to oxygen [10]. Film thickness (cm ×10⁴) and temperature, respectively: O 8.0, -50.0°C; △ 8.0, -62.0°C; ● 17, -67.3°C; ■ 4.8, -82.2°C; □ 2.0, -104°C.

which includes data on films of varying thickness at different temperatures. The exponential nature of the decay is obvious from the linearity of these plots. The diffusion constants obtained from such data are shown in Table IV and a plot of these diffusion constants in Arrhenius form shows a linear relationship from -110°C to 25°C, with an activation energy of 28 kJ/mol for oxygen and polystyrene. The value agrees very well with those estimated from data of Somersall et al. [8] on phosphorescence quenching in polystyrene and similar to the activation energy obtained by Stannet [26] of 35 kJ/mol for oxygen in glassy polystyrene above 25°C. The advantage of this procedure is that it permits determination of very low diffusivities which would be nearly undetectable by conventional procedures. Utilization of this technique to measure diffusivities in polymers at cryogenic temperatures could provide much needed data on the correlation of gaseous diffusivities with molecular motion in solid polymers and also is of practical importance in determining the resistance in polymers to permeation by oxygen and other reactive gases.

For measurements at higher temperatures, MacCallum and Rudkin [27] developed a method for the determination of diffusion constants for oxygen in polystyrene and poly(methyl methacrylate). They studied the quenching of either the excimer emission of polystyrene or additive fluorescence in poly(methyl methacrylate) by oxygen in thin films after displacement of the nitrogen atmosphere over the sample by oxygen. The relationship between oxygen concentration at time t, $[O_2](t)$, and the diffusion coefficient D for a flat film of thickness l is given by

TABLE IV. Diffusivity of oxygen in films of PS–NMA of various thickness at different temperatures

Temperature, °C	Film thickness (cm)	Diffusion constant $(cm^2\ s^{-1})$
23.2 ± 0.1	2.8×10^{-2}	3.1×10^{-7}
3.0 ± 1.0	1.7×10^{-3}	1.2×10^{-7}
-50.0 ± 2.0	8.0×10^{-4}	4.3×10^{-9}
-62.0 ± 1.0	8.0×10^{-4}	3.2×10^{-9}
-67.3 ± 2.0	1.7×10^{-3}	2.1×10^{-9}
-82.2 ± 0.5	4.8×10^{-4}	3.3×10^{-10}
$-104\ \pm 1.0$	2.0×10^{-4}	3.6×10^{-11}

$$\frac{[O_2](t)}{[O_2](\infty)} = \frac{[I_N/I(t) - 1]}{[I_N/I_0)- 1]} = (4/l)(Dt/\pi)^{\frac{1}{2}} \tag{38}$$

where $I(t)$ is the emission intensity at time t after replacement of the nitrogen by oxygen, and I_N and I_0 denote steady-state values in the absence and presence of oxygen, respectively. Values for the diffusion coefficients of 1.0 and 3.8×10^{-8} cm^2/s were obtained at 20°C for polystyrene and poly(methyl methacrylate) [27].

9. DIFFUSION CONSTANTS BY FLUORESCENCE DEPOLARIZATION

It has been shown experimentally with small-molecule solvents that the depolarization of fluorescence of a fluorophore in a solvent can be related to the internal viscosity of the medium through

$$1/\rho = kT/3V\eta \tag{39}$$

where ρ is the rotational relaxation time for the particular chromophore. For ordinary solvents the viscosity of the medium can usually be taken as identical with the measured viscosity of the solvent. In this case, the effective molecular volume, V, of a molecule can be calculated from the slope of the plot of $1/\rho$ versus t/η for the solvent. In classic early work, Nishijima and his coworkers used this technique to determine the effective viscosity of both solutions and solid polymers [28-30] More recently, Jarry and Monnerie [31] have studied the fluorescence depolarization of large chromophores such as the diphenyl polyenes

dissolved in polyisoprene and other polymers. In principle, the diffusion constant for small molecules could be estimated using eq. (39) by
substituting the internal viscosity calculated from such measurements.
However, as yet no experimental data are available to test whether the
"internal viscosity" as determined from the rotational motion of large
fluorophores in solid solution will give identical values to the internal
viscosity as determined by translational diffusion of the same fluorophores. Should this turn out to be the case, then fluorescence depolarization measurements would be another weapon in the arsenal of
polymer photophysicists to determine the rates of mass transport of
small molecules in polymeric media.

10. REFERENCES

1. R. N. Haward, in *Molecular Behaviour and the Development of Polymer Materials*, ed. A. Ledwith and A. M. North. Chapman and Hall, London, 1975.
2. A. M. North, in *Molecular Behaviour and the Development of Polymer Materials*, ed. A. Ledwith and A. M. North. Chapman and Hall, London, 1975.
3. F. Bueche, *Physical Properties of Polymers*. Interscience, New York, 1962.
4. J. Crank and G. S. Park, ed., *Diffusion in Polymers*. Academic Press, London, 1968.
5. M. Heskins and J. E. Guillet, *Macromolecules*, $\underline{3}$, 224 (1970).
6. J. E. Guillet, *Polymer Photophysics and Photochemistry*. Cambridge University Press, Cambridge, 1985, p.60.
7. R. M. Barrer and G. Skirrow, *J. Polym. Sci.*, $\underline{3}$, 549 (1948).
8. A. C. Somersall, E. Dan and J. E. Guillet, *Macromolecules*, $\underline{7}$, 233 (1974).
9. J. E. Guillet, *op. cit.*, § 8.6.
10. M. Andrews, M.Sc. Thesis. University of Toronto, 1978.
11. E. Pajot-Augy, L. Bokobza, L. Monnerie, A. Castellan and H. Bouas-Laurent, *Macromolecules*, $\underline{17}$, 1490 (1984).
12. J. E. Guillet and D. G. J. Sutherland, unpublished work.
13. J. Brandrup and E. H. Immergut, ed., *Polymer Handbook*, Interscience, New York, 1965.
14. D. G. Gray and J. E. Guillet, *Macromolecules*, $\underline{6}$, 223 (1973).
15. J.-M. Braun, S. Poos and J. E. Guillet, *J. Polym. Sci., Polym. Lett. Ed.*, $\underline{14}$, 257 (1976).
16. J. E. G. Lipson and J. E. Guillet, *J. Polym. Sci., Polym. Phys. Ed.*, $\underline{19}$, 1199 (1981).
17. G. Oster, N. Geacintov and A. V. Khan, *Nature (London)*, $\underline{196}$, 1089 (1962).
18. P. Debye, *Trans. Electrochem. Soc.*, $\underline{82}$, 265 (1942).
19. A. C. Somersall and J. E. Guillet, *Macromolecules*, 5, 410 (1972).
20. P. Hrdlovic, J. C. Scaiano, I. Lukac and J. E. Guillet, submitted for publication.
21. B. Valeur and L. Monnerie, *J. Polym. Sci., Polym. Phys. Ed.*, $\underline{14}$, 29 (1976).
22. M. Andrews and J. E. Guillet, unpublished results.
23. F. C. Unterleitner and E. I. Hormats, *J. Phys. Chem.*, $\underline{69}$, 2516 (1969).
24. G. Shaw, *Trans. Faraday Soc.*, $\underline{63}$, 2181 (1967).
25. H. S. Carslaw and J. C. Jaeger, *Conduction of Heat in Solids*, Oxford University Press, Oxford, 1959.
26. V. Stannet, in *Diffusion in Polymers*, J. Crank and G. S. Park, eds., Academic Press, London, 1968.
27. J. R. MacCallum and A. L. Rudkin, *Eur. Polym. J.*, $\underline{14}$, 655 (1978).
28. Y. Nishijima and Y. Mito, *Rep. Progr. Polym. Phys. Jpn.*, $\underline{10}$, 139 (1967).
29. Y. Nishijima and M. Saito, *Rep. Progr. Polym. Phys. Jpn.*, $\underline{10}$, 135 (1967).

30. K. Hirota and Y. Nishijima, *Rep. Progr. Polym. Phys. Jpn.*, <u>23</u>, 555 (1980).

31. J. P. Jarry and L. Monnerie, *Macromolecules*, <u>12</u>, 927 (1979).

ELECTRONIC EXCITATION TRANSPORT AS A TOOL FOR
THE STUDY OF POLYMER CHAIN STATISTICS

Curtis W. Frank[*], Glenn H. Fredrickson[†], and Hans C. Andersen[‡]

Departments of [*]Chemical Engineering and [‡]Chemistry
Stanford University, Stanford, California 94305
[†]AT&T Bell Laboratories, Murray Hill, NJ 07974

ABSTRACT.   Recent theoretical studies on electronic excitation transport (EET) among chromophores attached to a polymer chain are reviewed
with the objective of making EET a quantitative probe of
macromolecular structure. One dimensional energy migration in an aryl
vinyl polymer containing single monomer and excimer states is shown to
lead to a transient fluorescence decay that is inherently nonexponential (or multiexponential). Two many body theories for three dimensional EET among a small concentration of chromophores attached to a
polymer chain are outlined. The first theory, developed for a random
distribution of chromophores along the chain contour, includes both
inter- and intramolecular excitation transport. It may be applied to
any type of chain statistics and to any transfer rate. Calculations
based upon Gaussian chains and Förster dipole-dipole interactions lead
to determination of the characteristic ratio C_∞, a measure of statistical chain flexibility, from fluorescence depolarization
measurements. The second theory, applicable to polymers containing
only two chromophores at the chain ends, allows determination of the
mean squared end-to-end distance, $\langle R^2 \rangle$, also from fluorescence
depolarization experiments.

1.0 INTRODUCTION

Electronic excitation transport (EET) describes the transfer of
electronic excitations among chromophores or dye molecules [1]. The
transfer process can proceed through several mechanisms. First, the
excitation energy might be released as radiative emission that the
adjacent chromophore could absorb. This energy can also be transferred nonradiatively through short range exchange interactions
induced by wave function overlap of the two chromophores. Finally,
long range electrostatic resonance interactions can cause the non-
radiative migration of an electronic excitation. In the subsequent
discussion we restrict our attention to singlet excitations for which
the nonradiative resonant EET mechanism is dominant. Radiative transport through emission and reabsorption can be minimized by performing
experiments at a low chromophore concentration.

M. A. Winnik (ed.), Photophysical and Photochemical Tools in Polymer Science, 495–522.

An expression for the rate of singlet excitation transport between two chromophores with long range resonant interactions was first derived by Förster [2]. He found that the rate of EET from a chromophore at position $\underline{r}_1$ to a chromophore at $\underline{r}_2$, W, is approximately given by

$$W = \frac{1}{\tau} (R_o/|\underline{r}_1 - \underline{r}_2|)^6. \tag{1}$$

The measured lifetime of the electronic excitation is denoted τ, and R_o is defined as the interchromophore separation at which the rate of EET is equal to the lifetime decay rate. Equation (1) has been found to describe the rate of EET accurately for a large number of dye molecules and chromophores [1,3].

When chromophores are placed in solution or in a solid material at finite concentration, an excitation localized on a particular chromophore can be transferred to any one of a large number of surrounding chromophores. This is a result of the long ranged, r^{-6}, transfer rate. Thus, a mathematical description of the EET process in such systems will involve the formulation and solution of a many body transport problem. Although such many body problems are difficult to solve, recent theoretical work on the transport of electronic excitations among randomly distributed sites in homogeneous systems by Gochanour, Andersen and Fayer (GAF) [4] and Loring, Andersen and Fayer (LAF) [5] has laid the foundation for the present study on inhomogeneous polymeric systems. In these papers self-consistent approximations are presented for the transport Green function, and connections are drawn with various fluorescence observables. We will make numerous references to this work in the following discussion.

The work of GAF and LAF shows that the strong distance dependence of the microscopic transfer rate will yield a collective or macroscopic rate of EET that is very sensitive to chromophore density, and to fluctuations in chromophore density. It is this feature of collective EET that has initiated the present investigation of EET as a probe of macromolecular structure. When chromophores are attached to polymer chains in a well defined manner, the EET dynamics will reflect ensemble averaged properties such as local chain segment density and flexibility, global configurational statistics, and intramolecular dimensions.

There are two types of fluorescence experiments that provide a direct measurement of the collective rate of EET [6]. The first of these is a trapping experiment in which two types of chromophores are incorporated in the polymer--donors and traps. The donor chromophores have a higher energy lowest excited state than the trap chromophores, which allows excitations to be transferred from a donor to a trap, but not from a trap to a donor chromophore. As a result, excitations initially on donor chromophores are eventually localized on traps, leading to trap fluorescence.

The total intensities of both donor and trap fluorescence are directly related to the time dependent, ensemble averaged probability that an excitation resides on a donor chromophore at time t [5,6].

We refer to this probability as $G^D(t)$ when it is calculated in the absence of fluorescence decay. $G^D(t)$ can be obtained theoretically by solving a many body EET problem, and depends strongly on the density and distribution of the donor and trap chromophores in the polymer. Fits of total fluorescence intensity data to theoretical expressions for $G^D(t)$ can provide information on macromolecular structure.

The second experiment that will allow the investigation of macro-molecular structure is a fluorescence depolarization study. When a collection of chromophores with randomly oriented absorption dipoles is illuminated with a plane polarized source, chromophores having dipoles aligned with the direction of polarization are preferentially excited. The subsequent fluorescence from these chromophores retains a high degree of polarization. However, if EET moves the excitations to surrounding chromophores with different orientations, the resulting fluorescence will be depolarized.

The extent of fluorescence polarization at a time t following a brief excitation pulse is directly related to the time dependent, ensemble averaged probability that an excitation resides on the chrom-ophore where it was created at t = 0 [4,6,7]. This probability, which will be designated $G^S(t)$, depends strongly on the density and dis-tribution of chromophores, and can be obtained from a theoretical analysis of the many body EET problem in the polymeric system. A fit of transient fluorescence depolarization data to a theoretical expres-sion for $G^S(t)$ provides parameter values that are directly related to chromophore distribution. $G^S(t)$ and $G^D(t)$ are portions of the total Green function for excitation transport.

The major objective of this paper is to introduce the results of recent many body analyses of EET in macromolecules. Although the general framework of the theoretical approach is discussed, we provide little detail on the specific calculations; these may be obtained from the thesis of G. Fredrickson [8] or the original references [9-13]. Rather, we concentrate on providing the appropriate background necessary to establish the significance of the results.

In Section 2.0 we examine trapping experiments in the aryl vinyl polymers, for which the trap is an excimer forming site (EFS). A one dimensional random walk model is used for analysis of the transient and photostationary fluorescence in dilute solution. In a second study, the three-dimensional results of LAF [5] are applied to trapping experiments in pure solid films of aryl vinyl polymers.

In Section 3.0 we consider fluorescence depolarization measure-ments in flexible polymers containing a small concentration of chromo-phores. Recent many body analysis of three dimensional excitation transport by Fredrickson, Andersen and Frank [9-13] are reviewed. For the case where the chromophores are distributed randomly along the contours of the chains, it is possible to determine the chain flexi-bility, as measured by Flory's characteristic ratio, C_∞. For the case where the chromophores are attached only at the chain ends, it is possible to determine the mean-squared end-to-end distance $\langle R^2 \rangle$.

2.0 TRAPPING EXPERIMENTS

Numerous studies of the photophysics of aryl vinyl polymers have shown that the monomer emission intensity can be empirically fit to a triple exponential [15-17]. The immediate conclusion that may be drawn from this observation is that the Birks kinetic scheme [1], which was developed for a collision-induced intermolecular process between small molecules in solution, is inapplicable to intramolecular excimer formation in macromolecules.

A quantitative explanation of this behavior has been somewhat difficult to obtain, however. There have been several proposals to identify the experimentally resolved exponentials with physical entities, such as multiple monomer states [17-20] or multiple excimer states [15,21]. It is quite probable that the kinetics of the aryl vinyl polymers are indeed more complex than the simple Birks scheme would allow. However, recent theoretical studies on electronic excitation transport in random systems of donors and traps have shown that the fluorescence decays are in general, nonexponential [4,5, 22-26]. A key feature of these analyses is that the trapping dynamics in one dimensional and quasi one dimensional polymeric systems require that a trapping rate <u>function</u> k(t) rather than a trapping rate <u>constant</u> be used. In Section 2.1 we give the relationship of k(t) to the observables in a trapping experiment and provide the connection with $G^D(t)$, which is obtained from theory.

In Section 2.2 we will demonstrate that the polymer configurations will determine a distribution of topologically distinct energy transport pathways from the monomer to the excimer forming site. Each pathway will have its own characteristic transport time and, thus, the observed decay will be an ensemble average over all possible migrative paths. We will present the simplest possible kinetic scheme with single monomer and excimer states in which EET plays a significant role. This work is to be distinguished from other studies on one-dimensional [28] and three dimensional [29] energy migration in miscible blends by the inclusion of segmental rotational motion.

In Section 2.3 a many body treatment developed by Loring, Andersen and Fayer [5], for three dimensional EET in homogeneous, randomly distributed systems of donors and traps is applied to pure aryl vinyl polymer films. Using this model, we will determine the EFS concentration in poly(2-vinyl naphthalene) and polystyrene films from photostationary fluorescence.

2.1 Connection Between Trapping Experimental Observables and EET Theory

A trapping experiment involves the excitation of the donor (or sensitizer) chromophore, and the subsequent observation of the fluorescence from a second chromophore type, the trap (activator, acceptor). Both photostationary and transient trapping experiments may be directly connected through the donor excitation function $G^D(t)$, which represents the probability that an excitation (created at t=0) resides on any donor chromophore at time t in the absence of other decay processes. When EET plays a significant role, it will be seen that $G^D(t)$ will be a nonexponential function. In such a case, a trapping rate <u>constant</u> formalism will be invalid. Rather, it is necessary to use an energy migration trapping rate <u>function</u> [30]

$$k_{EM}(t) = -\frac{d}{dt}\left[\ell n\ G^D(t)\right] \tag{2}$$

Transient donor and trap fluorescence intensities are given by [6]

$$i_M(t) = Q_{FM}\ k_M\ \exp(-k_M t)\ G^D(t) \tag{3}$$

$$i_E(t) = Q_{FE}\ k_E \exp(-k_E t)\ \int_0^t d\tau \exp(-k_M \tau)\left[-\frac{d}{d\tau}G^D(\tau)\right] \tag{4}$$

where k_M and k_E are the total rates of radiative and nonradiative decay of excited donors and traps (taken to be monomer and excimer states), respectively, while Q_{FM} and Q_{FE} are the fluorescence quantum efficiencies. Photostationary data may be expressed in terms of integrated trap to donor fluorescence intensities [6]

$$\frac{I_E}{I_M} = \frac{Q_{FE}}{Q_{FM}}\left[\frac{1-M}{M}\right] \tag{5}$$

in which M is the probability that an absorbed photon will decay through radiative or nonradiative processes from the donor (or monomer). This expression is general and does not assume Birks kinetics. M contains all of the morphological and dynamical information about the polymer chain that is necessary to establish the excitation path. M is directly proportional to the Laplace transform of $G^D(t)$

$$M = k_M\ \tilde{G}^D(\varepsilon = k_M) \tag{6}$$

where

$$\tilde{G}^D(\epsilon) = \int_0^\infty \exp(-\epsilon t) \ G^D(t) \ dt \tag{7}$$

in which ϵ is the Laplace variable.

Thus, both transient and photostationary trapping experiments may be interpreted quantitatively if an expression for the donor excitation function $G^D(t)$ is available. In Section 2.2 we outline the development of an expression for $G^D(t)$ for an aryl vinyl polymer in dilute solution. Here energy migration from a single monomer donor state to a single excimer trap state is analyzed by a one-dimensional model. We will present the analysis in more detail than the subsequent discussion of various many body theories because the treatment is straightforward and concise. This will allow the fundamental approach to be understood more clearly without the extensive mathematics required for the many-body treatment.

2.2 One Dimensional EET in the Aryl Vinyl Polymers

The kinetic scheme used here consists of a single excimer and monomer state. The excited monomer, M*, can proceed to the excimer either by segmental rotation with rate constant k_{rot} or by one-dimensional energy migration with a characteristic trapping rate function $k_{EM}(t)$. Alternatively, the excitation may decay with rate constant k_M. The analytical model is based on the following assumptions: 1) The equilibrium population of excimer forming site rotational dyads divides the chain into a series of finite-length segments of monomeric chromophores, 2) The excimer forming site traps terminating each chain segment are totally disruptive and will prevent any exciton migration out of the original chain segment, 3) Energy migration is fast compared with segmental rotation, 4) The rate of EET between two monomers is the same as that between a monomer and an excimer forming site.

For a system satisfying these assumptions, the excitation dynamics may be described by a Pauli master equation [14].

$$\frac{d}{dt} p_j(t) = W[p_{j+1}(t) - 2p_j(t) + p_{j-1}(t)], \qquad 2 \leqslant j \leqslant n-1$$

$$\frac{d}{dt} p_1(t) = W[p_2(t) - 2p_1(t)] \tag{8}$$

$$\frac{d}{dt} p_n(t) = W[p_{n-1}(t) - 2p_n(t)]$$

where

$$p_j(t) = p_j'(t) \exp(t/\tau) \tag{9}$$

is the probability that an excitation of infinite lifetime exists on chromophore j at time t, $p_j'(t)$ is the corresponding probability for an excitation with finite lifetime $\tau = (k_M + k_{rot})^{-1}$, and W is the rate constant for nearest-neighbor EET between two donor chromophores, averaged over the transition moment orientation and interchromophore separation associated with each rotational dyad.

The donor response function for a segment of length n is defined by

$$G_n^D(t) = \sum_{i=1}^{n} P_i(t) \tag{10}$$

and has been evaluated by Pearlstein [14]. The corresponding function for the complete polymer chain is obtained by averaging the finite segment result over all segment lengths, allowing for the initial excitation of excimeric sites [31]. The final averaged donor response function $G^D(t)$ represents the probability that a single excitation produced at $t = 0$ is still residing on a monomeric chromophore at time t, if the excitation has an infinite lifetime.

An expression for $G^D(t)$ valid for small EFS trap concentrations of $q \ll 1$ and long times, $Wt > 1$, has been obtained [30] using the t-matrix approach of Huber [25].

$$G^D(t) = (1-q)^2 \exp(4q^2 Wt) \, \text{erfc}[2q(Wt)^{1/2}] \tag{11}$$

where erfc(x) is the complementary error function. It is of interest to note that

$$G^D(t=0) = (1-q)^2 \tag{12}$$

$$= 1 - f_R \tag{13}$$

where f_R is the equilibrium fraction of chromophores in adjacent intramolecular EFS [32].

In order to relate $G^D(t)$ to the observables in a transient or photostationary fluorescence experiment, the results of Section 2.1 are used (c.f. eq. 3,4). The monomer response is given by

$$i_M(t) = Q_{FM} k_M (1-q)^2 \, \exp[(4q^2 W - k_M - k_{rot})t] \, \text{erfc}[2q(Wt)^{1/2}] \tag{14}$$

which is clearly a nonexponential result, in contrast to the commonly assumed triple-exponential summation. The transient behavior of the monomer is shown in Figure 1. The curves are plotted in terms of the dimensionless parameter $\alpha = Wt$. For small values of α, the monomer response is dominated by rotational and decay processes, and is nearly a single exponential. However, as α increases due to EET becoming rapid, the monomer behavior becomes highly nonexponential. We note that polystyrene (PS) in solution exhibits single exponential monomer behavior [33] while the poly(2-vinyl naphthalene) (P2VN) monomer

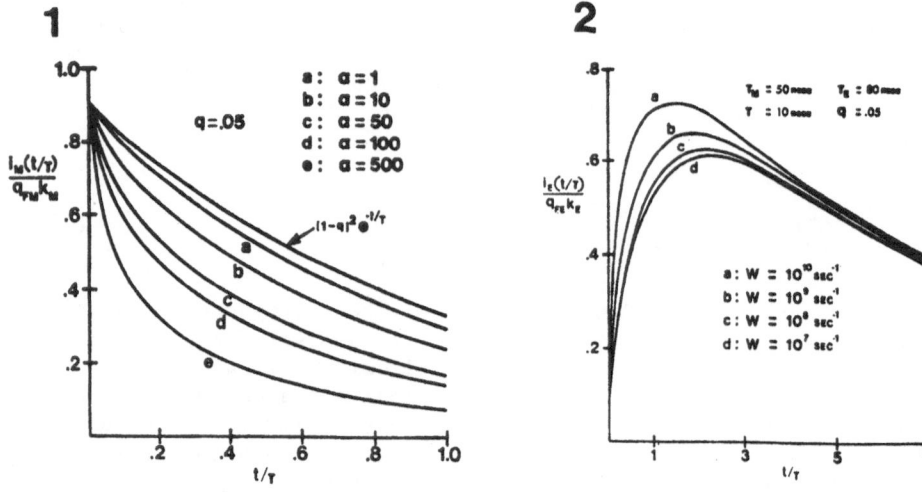

Figure 1. Transient fluorescence response of the monomer for an aryl vinyl polymer in solution. Taken from Figure 3 of reference [30].

Figure 2. Transient fluorescence response of the excimer for an aryl vinyl polymer in solution. Taken from Figure 4 of reference [30].

decay is multiexponential [16]. This is consistent with EET being slower in PS than in P2VN, a conclusion also drawn from analysis of polymer solid films [34,35].

Figure 2 presents the transient excimer response as a function of the conformationally averaged energy transport rate. The effect of increasing the EET rate is to move excitations to traps more readily. This leads to an increase in the intensity of excimer emission and a shift of the maximum intensity to shorter times.

Substitution of the Laplace transform of $G^D(t)$ into Equation (6) yields

$$M = \frac{k_M \tau (1-q)^2}{1 + 2q(W\tau)^{1/2}}$$

(15)

It is then of interest to examine the limiting behavior of Equation (5) using the expression for M given in Equation (15). If rotation is slow compared to migration, $W/k_M \gg k_{rot}/k_M > 1$ and

$W > \tau k_{rot}^2/(4q^2)$. Equation (5) then becomes

$$R \simeq \frac{Q_{FE}}{Q_{FM}} \left[\frac{2q}{(1-q)^2}\right]\left[\frac{Wk_{rot}}{k_M^2}\right]^{1/2} \tag{16}$$

If a Stokes law expression is assumed for k_{rot}, the viscosity dependence of the excimer to monomer ratio should be $R \sim \eta^{-1/2}$. This is consistent with photostationary fluorescence measurements on P2VN in toluene under hydrostatic pressure yielding an empirical expression of the form

$$R = A\eta^{-\beta} \tag{17}$$

with a best fit value of the exponent $\beta = 0.511$ [36].

When segmental rotation is extremely fast and EET is negligible, the $(W,q \to 0)$ limit of Equation (5) is

$$R \simeq \frac{Q_{FE}}{Q_{FM}} \left(\frac{k_{rot}}{k_M}\right) \tag{18}$$

which is identical with the low temperature diffusion controlled regime result obtained from the Birks scheme when the excimer dissociation rate is assumed to be much smaller than k_E.

Finally, in a miscible solid blend below the glass transition temperature k_{rot} is taken to be zero and

$$R = \frac{Q_{FE}}{Q_{FM}} \left\{\frac{2q\left[1 + \left(\frac{W}{k_M}\right)^{1/2}\right] - q^2}{(1-q)^2}\right\} \tag{19}$$

which shows that R is proportional to f_R for small trap concentrations, as was proposed some time ago [32].

In summary, this simple one dimensional scheme demonstrates that nonexponential (or multiexponential) fluorescence decays are to be expected even if there is only one excimer and one monomer species, as long as energy migration plays a significant role. The trapping rate constant formalism will be invalid unless the system is at very high temperatures in a nonviscous solvent, where rotational sampling and excimer dissociation may become the rate-determining steps to excimer formation.

2.3 Three Dimensional EET

Recent theoretical advances in the description of excitation transport in disordered, homogeneous materials have provided some useful theoretical tools for the analysis of many-body EET problems. Haan and Zwanzig [37,38] developed an elegant statistical mechanical

formulation of incoherent excitation transfer among randomly distributed points in a continuum. The problem was then solved approximately, utilizing a variety of theoretical methods [4,5,39-43]. One
of these methods is the self-consistent, graphical analysis of
Gochanour, Andersen, and Fayer [4]. The technique is based on a diagrammatic expansion of the Green function for excitation transport.
GAF obtain a class of self-consistent approximations for the Green
function that converge rapidly and that can be used to calculate all
transport properties of interest. The results are in excellent agreement with time dependent fluorescence depolarization experiments on
dye molecules in solution [7]. Loring, Andersen, and Fayer [5] subsequently extended the GAF method to the case of a two component homogeneous mixture of randomly distributed donor and trap chromophores.
The LAF theory is also in excellent agreement with trapping experiments performed in solution [44].

In Section 3.0 we will outline the method used to develop the
Green function for excitation transport among chromophores attached to
a polymer chain. The analysis is a generalization of the GAF and LAF
treatments to account for the correlations among chromophore positions
in polymeric systems. In the limit of very high polymer concentration, however, it may be possible to neglect such correlations. As
a result, it is worthwhile to apply the LAF analysis directly to films
of the aryl vinyl polymers.

An accurate approximate solution to the excitation transport
problem for a randomly distributed (i.e. uncorrelated) system of
donors and traps is the self-consistent "two body approximation"
derived by LAF. In this analysis the Laplace transform of $G^D(t)$ in
three dimensions is

$$\tilde{G}^D(\varepsilon) = \frac{[\tilde{G}^S(\varepsilon)]^2}{\tilde{G}^S(\varepsilon) - \tilde{\Delta}[\tilde{G}^S(\varepsilon)]} \qquad (20)$$

where

$$\tilde{\Delta}[\tilde{G}^S(\varepsilon)] = \frac{\pi}{2^{3/2}} C_D k_M^{1/2} [\tilde{G}^S(\varepsilon)]^{3/2} \qquad (21)$$

and

$$[\tilde{G}^S(\varepsilon)]^{1/2} = \frac{1}{2\varepsilon} \left\{ -\left[\frac{\pi}{2} k_M^{1/2} C_T + \frac{\pi}{2^{3/2}} k_M^{1/2} C_D\right] + \right.$$

$$\left. \left[\left(\frac{\pi}{2} k_M^{1/2} C_T + \frac{\pi}{2^{3/2}} k_M^{1/2} C_D\right)^2 + 4\varepsilon \right]^{1/2} \right\} \qquad (22)$$

The dimensionless monomer donor and excimer trap concentrations
are defined by

$$C_D = \frac{4}{3} \pi \left(R_o^{DD}\right)^3 \rho_D \tag{23}$$

$$C_T = \frac{4}{3} \pi \left(R_o^{DT}\right)^3 \rho_T \tag{24}$$

in which R_o^{DD} and R_o^{DT} are the Förster radii for donor-donor and donor-trap transfer and $\rho_D(\rho_T)$ is the bulk chromophore density for the donor (trap).

According to the LAF model, the probability M may be written as [34].

$$M = \frac{(1-q)X^2}{(1-YX)} \tag{25}$$

where $\qquad$ (26)

$$X = (Z^2 + 1)^{1/2} - Z \tag{26a}$$

$$Y = \frac{\sqrt{2} \, \pi^2}{3} n\left(R_o^{DD}\right)^3 (1-q) \tag{26b}$$

$$Z = \frac{\sqrt{2} \, \pi^2}{6} n\left(R_o^{DD}\right)^3 \left[(1-q) + q(1-0.5q)^{-1/2}\right] \tag{26c}$$

$$q_D = 1 - (1-q)^{1/2} \tag{26d}$$

Here n is the chromophore bulk density and q is the concentration of excimer forming sites.

The LAF analysis was applied to pure films of poly(2-vinyl naphthalene) and polystyrene [34]. When R_o^{DD} is assigned the value for transfer between 2-methyl naphthalene and itself (11.75 Å), the maximum trap concentration was found to be 0.072 for P2VN and 0.33 for PS. The interesting feature of these results is the relatively low value for P2VN, especially considering that the intramolecular EFS concentration is about 0.026 [34]. Apparently, it is much more unlikely to get extensive aromatic ring interaction in P2VN than in PS. In fact, Windle has suggested that preferential stacking of aromatic rings occurs in undiluted polystyrene films [45]. Our fluorescence results are consistent with this proposal.

3.0 FLUORESCENCE DEPOLARIZATION EXPERIMENTS

The excitation transport problem among chromophores attached to macromolecules is considerably more difficult than the corresponding EET problem in systems of randomly distributed chromophores. The lack of translational invariance and correlated polymer statistics increase the complexity of the theoretical analysis. Advances in the treatment of many-body EET in polymers have only recently been possible because of the emergence of powerful theoretical tools in the solution of the transport problem for randomly distributed chromophores.

Ediger and Fayer [46] have developed an approximate theory to describe EET among chromophores distributed randomly in a finite spherical volume. They utilized a density expansion technique similar to that of Haan and Zwanzig [38]. In a subsequent paper [47], Ediger and Fayer applied the finite volume theory to EET within a single polymer coil or aggregate. The polymeric entity was modeled as a sphere containing randomly distributed chromophores, and the Green function was obtained by ensemble averaging over a Gaussian distribution of sphere sizes. No excitation transport between chromophores in different spheres was allowed.

In Section 3.1 we provide background on the relationship between the theory of EET and fluorescence depolarization experiments. In Section 3.2 we outline the development of the master equation and the transport Green function. In Section 3.3 we briefly describe the results for polymers containing a small concentration of chromophores attached to the chain at random points along the contour. Finally, in Section 3.4, we consider a related calculation on polymer chains containing chromophores only at the chain ends.

3.1 Overview of Fluorescence Depolarization Theory

In this work we are interested only in the depolarization of fluorescence due to excitation transport and not to rotational relaxation. Hence, we consider viscous solutions or glasses. Furthermore, for simplicity we restrict consideration to polymers labeled exclusively with donor chromophores. Early workers have shown that the fluorescence polarization that is <u>retained</u> comes predominantly from the molecules that were initially excited [48,49]. The parallel and perpendicular fluorescence intensities have been shown by Gochanour [7] to be related to $G^S(t)$, which is the configurationally averaged probability that an excitation resides on its site of origin at time t.

$$I_{\parallel}(t) = \exp(-k_D t) \, (1 + 0.8 \; G^S(t)) \qquad\qquad (27)$$

$$I_{\perp}(t) = \exp(-k_D t) \, (1 - 0.4 \; G^S(t)) \qquad\qquad (28)$$

In general, it is more useful to work with the anisotropy, defined by

$$r(t) = \frac{I_{\parallel}(t) - I_{\perp}(t)}{I_{\parallel}(t) + 2I_{\perp}(t)} \tag{29}$$

In terms of the Green function

$$\frac{r(t;C_D)}{r(t;0)} = G^S(t) \tag{30}$$

Equation (30) has been normalized by the anisotropy at zero concentration. Thus, transient depolarization experiments provide a direct test of a transport theory, and useful information can be extracted from experiments performed at a single sample concentration. In terms of the steady state anisotropy $\bar{r}$ we may write

$$\frac{\bar{r}(C_D)}{\bar{r}(0)} = k_D \, \tilde{G}^S \, (\epsilon = k_D) \tag{31}$$

where $\tilde{G}^S$ is the Laplace transform of $G^S(t)$.

It is common for experimental data to be reported in terms of steady state measurements of the inverse polarization

$$P^{-1} = \frac{I_{\parallel} + I_{\perp}}{I_{\parallel} - I_{\perp}} \tag{32}$$

It can be shown [48] that for excitation by plane polarized light, P^{-1} is related to $\bar{r}$ by

$$\bar{r} = \frac{2}{3} \, (P^{-1} - \frac{1}{3})^{-1} \tag{33}$$

Thus, the inverse polarization is related to $\tilde{G}^S$ by

$$\frac{P^{-1}(C_D)}{P^{-1}(0)} = \frac{1}{6} \, (1 + \frac{5}{k_D \, \tilde{G}^S(\epsilon = k_D)} \,) \tag{34}$$

We thus find that in both trapping and depolarization experiments, the transient observables are related to a portion of the full Green function ($G^D(t)$ and $G^S(t)$, respectively), while the steady state observables are given in terms of the Laplace transform of these quantities.

3.2 Transport Dynamics and the Green Function

3.2.1 <u>Introduction</u>.   To investigate the use of EET as a quantitative tool for measuring chain flexibility, we consider linear
polymer chains. The chains have a small concentration of chromophores
(donors and traps) distributed randomly among their monomer units. In
such a system, interchromophore excitation transport will be dominated
by transfers between sites that are separated by large distances along
the chains, but not necessarily large distances in space. This type
of excitation transfer will be referred to as three-dimensional (3-D)
transport to distinguish it from the one-dimensional (1-D) transport
expected in polymers containing a large number of chromophores considered in Section 2.0.

The advantage of studying chains that contain a small number of
chromophores is twofold. First, the chromophore impurity is in low
enough concentration that the thermodynamic and statistical properties
of the polymer should not be affected. Secondly, it is clear that the
dynamics of 3-D excitation transport will be insensitive to the
detailed nature of the monomer units. This feature will allow a
single 3-D model to describe EET in a broad class of polymers containing a small number of attached chromophores.

The polymer chains considered in this section are assumed to be
monodisperse, and consist of n effective (Kuhn) statistical segments
[50,51] of length a. The Kuhn segments are introduced to remove short
range intramolecular interferences, and in the absence of excluded
volume effects the chains would be ideal [51]. There are correlations
among the positions of chromophores on the same chain, but for
simplicity we assume there are no correlations between the positions
of chromophores on different chains (i.e. the "correlation hole" is
not considered [52,60]). The chains have a small fraction of the
statistical segments containing donor or trap chromophores. The
chromophores are randomly distributed among the segments. We define
q_D as the average number of donors per segment, q_T as the average
number of traps per segment, and require
$q_D, q_T \ll 1$.   This condition leads to 3-D excitation transport as discussed above.

The theory presented here is limited to the transport of
incoherent excitations, describable with a Pauli master equation. In
addition, chromophore diffusion resulting from Brownian forces on the
chains is assumed to be negligible on the time scale of excitation
transport. However, no assumptions are made about the density of the
material, so the theory should be applicable to homogeneous melts and
to polymers dispersed in amorphous solids, as well as to chains in
solution.

3.2.2 <u>Many-Body Analysis</u> The polymeric system of interest for
the present discussion is assumed to consist of a polymer chain
containing N donor chromophores with locations dictated by

$$\{ \underline{R} \} = (\underline{r}_1, \underline{r}_2 \cdot \cdot \cdot \underline{r}_N) \tag{35}$$

For simplicity we do not include traps, although the original references do [9-13]. The probability that an excitation resides on the jth chromophore at time t, $p_j'(\{ \underline{R} \},t)$ satisfies the master equation

$$\frac{d}{dt} \, p_j'(\{ \underline{R} \},t) = \frac{-p_j'(\{R\},t)}{\tau} \, +$$

$$+ \sum_k \, W_{jk}\left[p_k'(\{ \underline{R} \},t) - p_j'(\{ \underline{R} \},t)\right] \qquad (36)$$

where W_{jk} is the transfer rate between sites j and k and τ is the measured lifetime of the excited species. The transport rate is assumed to be isotropic and of the form of Equation (1). It proves convenient to deal only with the excitation transport in the absence of the lifetime decay. This is accomplished by the substitution

$$p_j(\{ \underline{R} \}),t) = p_j'(\{ \underline{R} \},t) \, \exp(t/\tau) \qquad (37)$$

and the definition of a transfer matrix $\underline{\underline{Q}}$ given by

$$\cdot Q_{jk} = W_{jk} - \delta_{jk} \sum_\ell W_{k\ell} \qquad (38)$$

This allows the master equation to be rewritten as

$$\frac{d}{dt} \, \underline{p}(\{ \underline{R} \}),t) = \underline{\underline{Q}} \cdot \underline{p}(\{ \underline{R} \}),t) \qquad (39)$$

If we introduce the Laplace transform, the transformed state vector satisfies

$$\tilde{\underline{p}}(\{ \underline{R} \},\epsilon) = (\epsilon\underline{\underline{I}} - \underline{\underline{Q}})^{-1} \cdot \underline{p}(\{ \underline{R} \},0) \qquad (40)$$

where $\underline{\underline{I}}$ is the unit matrix and ϵ is the Laplace variable.

A Green function can be defined as the ensemble-averaged quantity

$$\tilde{\underline{\underline{G}}}(\epsilon) = \langle(\epsilon\underline{\underline{I}} - \underline{\underline{Q}})^{-1}\rangle \qquad (41)$$

where the ensemble average of a function $h(\{ \underline{R} \})$ is defined by

$$\langle h(\{ \underline{R} \})\rangle = \int d\underline{r}_1 \int d\underline{r}_2 \cdots \int d\underline{r}_N \, P(\{ \underline{R} \})h(\{ R \}) \qquad (42)$$

$P(\{ \underline{R} \})$ is the normalized multivariate distribution function for the

positions of the N chromophores. A key distinction between the
randomly distributed system of GAF and the polymer system is that in
the former case the pair correlation function simply involves the
number density of chromophores whereas the latter could involve the
Debye form [52] or be more complicated.

In a fluorescence depolarization experiment, $\tilde{G}^S(\epsilon)$, the Laplace
transform of the probability that an excitation resides on the site
where it was created at time t, is related to the diagonal elements of
the Green function

$$\tilde{G}^S(\epsilon) = \frac{1}{N} \sum \tilde{G}_{ii}(\epsilon) \tag{43}$$

The Green function may be expanded in powers of ϵ^{-1} and $\underline{Q}$ by use of
the identity

$$(\epsilon\underline{I} - \underline{Q})^{-1} = \epsilon^{-1}\underline{I} + \epsilon^{-1}\underline{Q} \cdot (\epsilon\underline{I} - \underline{Q})^{-1}$$

Thus it is possible to write

$$\hat{G}_{ii}(\epsilon) = \sum_{n=o}^{\infty} \epsilon^{-(n+1)} \langle (\underline{Q}^n)_{ii} \rangle \tag{44}$$

It is clear from Equation (44) that the diagonal portion of the
Green function consists of an infinite sum of products of ϵ^{-1} and W_{ij}
factors. One way to represent such a series is by a diagrammatic
representation in which each term is assigned a value according to a
particular set of rules [4,5,9,10]. Although space does not permit
extensive discussion, the general structure of the graphical analysis
is of interest. In this treatment the jth chromophore on the chain is
represented by a circle labeled j. Each factor of W_{ij} that appears in
a product term of Equation (44) is represented by a solid arrow from
circle i to circle j. This may be interpreted as an increase in probability on j due to transfer of excitation from i. Each factor of
$-W_{ij}$ is represented by a solid arrow from i to j followed by a
dashed arrow back to i. This corresponds to the decrease in probability on i due to transfer to j. Note that each diagram of this
graphical representation begins with a solid arrow leaving circle i,
proceeds through a continuous path of solid and dashed arrows representing a product of W_{ij} factors, and ends with a solid or dashed
arrow back to circle i.   Figure 3 illustrates several of the possible
diagrams in the infinite series for $\hat{G}_{ii}(\epsilon)$.

Similarly, the off-diagonal terms are given by

$$\tilde{G}_{ij}(\epsilon) = \sum_{n=o}^{\infty} \epsilon^{-(n+1)} \langle\langle \underline{Q}^n)_{ij} \rangle \qquad\qquad i \neq j \tag{45}$$

with typical diagrams shown in Figure 4. Note that in this case, each
diagram consists of a continuous path of solid and dashed arrows
beginning on circle i and ending on circle j. The value of a

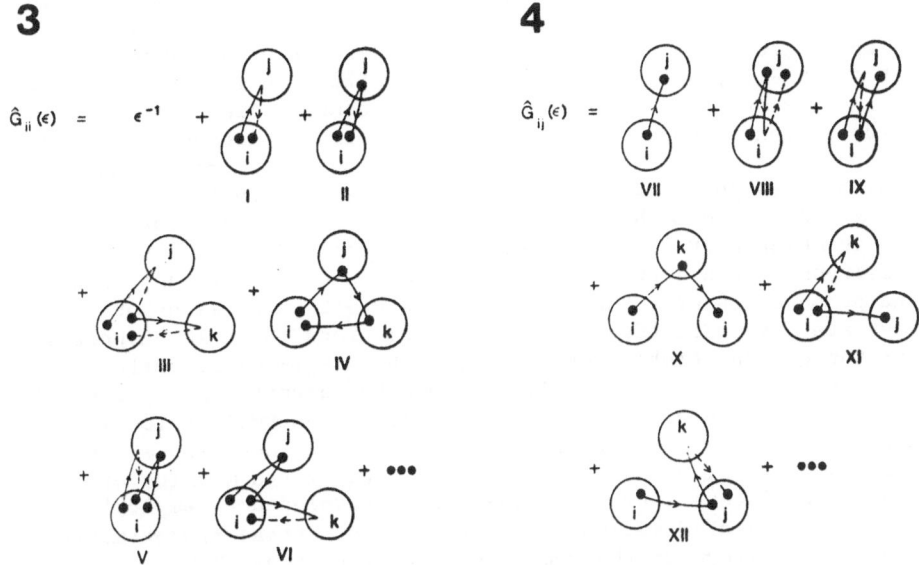

Figure 3.   Diagrammatic expansion of diagonal terms of transport Green function $\tilde{G}_{ii}(\epsilon)$ for an isolated polymer chain with randomly tagged chromophres. Taken from Figure 1 of reference [10].

Figure 4.   Diagrammatic expansion of off-diagonal terms of transport Green function $\tilde{G}_{ij}(\epsilon)$ for an isolated polymer chain with randomly tagged chromophores. Taken from Figure 2 of reference [10].

particular diagram in the off-diagonal $\tilde{G}_{ij}(\epsilon)$ expansion is calculated is in a manner analogous to the diagonal terms.

Diagrammatic or graphical representation of perturbation expansions are commonly used in many body physics. They are often more useful than the original perturbation expansions because various topological features of the individual diagrams can suggest useful ways to renormalize the series. The renormalization procedure is sometimes referred to as topological reduction [53] because the renormalized series typically contains a smaller number of diagrams than the original expansion. However, the rules for evaluating diagrams and the actual value of a graph are usually more complicated in the renormalized series.

There are several motivations for renormalization of a perturbation expansion. First, by suitable renormalization certain types of long-ranged or short-ranged (in time or position) divergences that are present in the original series can often be removed. Such divergences, which are not present in the exact solution of a given problem, often arise because of the particular mathematical formulation of the original expansion. A second use for renormalization is to extend the radius of convergence of a perturbation series. Another important use of topological reduction techniques is to derive formally exact relations between physical quantities of interest. Finally, topological reduction methods can often be used in conjunction with partial summation techniques to obtain self-consistent approximations. Such approximations are valuable because they ensure that one or more conservation laws that the exact physical quantity satisfies are also satisfied by the approximations for that quantity.

For the current problem of interest, the diagrammatic series for the Green function is topologically reduced by identifying portions of the diagrams that can be removed through vertex renormalization and the definition of a mass operator. (A vertex is a "dot" appearing in the diagrams of Figs. 3 and 4 and originally was given a value of ε^{-1}). Self-consistent approximations can be developed by employing partial summations of diagrams and requiring that these approximations can serve probability for the complete Green function. Two-body and higher order correlation functions for chromophore positions are required to utilize the various approximations.

3.3 Calculations for Randomly Tagged Chains

All of the time dependent and photostationary fluorescence observables can be obtained from the two functions $\widetilde{G}^S(\varepsilon)$ and $\widetilde{G}^D(\varepsilon)$. For simplicity we will consider here only $\widetilde{G}^S(\varepsilon)$, the Laplace transform of the probability that an excitation is on the initial site of absorption at time t. Several classes of approximations for $\widetilde{G}^S(\varepsilon)$ have been developed that incorporate all possible excitation transfers within clusters of two and three chromophores. The first class involves self consistent two-body and three-body equations for $\widetilde{G}^S(\varepsilon)$ that satisfy the requirement of conservation of probability, i.e.

$$\sum_j \widetilde{G}_{ij}(\varepsilon) = \varepsilon^{-1} \tag{46}$$

In a second class, a density expansion is used to obtain a series approximation for $\widetilde{G}^S(\varepsilon)$ that is accurate at short times and small concentrations. One approach is to construct a Padé approximant for $\widetilde{G}^S(\varepsilon)$ from this series that causes $G^S(t)$ to decay to zero at long times. A second approach is to construct cumulant approximants for the series by inverting the Laplace transform of the density expansion and re-expressing the series as a short-time expansion for $\ln[G^S(t)]$.

3.3.1 <u>Three Particle Padé Approximation</u>. Of the (highest order) three-body approximations, neither the self-consistent nor the three-body cumulant are well behaved at long times, showing divergences instead of the desired decay of $G^S(t)$ to zero. By contrast, the three-body Padé approximant is well behaved and is believed an accurate expression. It is possible, however, to remove the long time divergence of the three-body self-consistent approximation leading to the so-called three-body rational approximation, which is in good agreement with the three-body Padé [12]. Nevertheless, the three-body Padé remains a more compact expression, and we will employ it for parameterization studies.

Expressions for $\widetilde{G}^S(\varepsilon)$ have been determined for the special case of an infinitely long chain. As the number of chromophores goes to infinity, all of the sites on the chain become equivalent, and translational invariance along the chain contour is established. Under such conditions the values of the diagrams simplify, as does the structure of the renormalized series. For ideal, Gaussian statistics [51,52] the two and three-body correlation functions are known [10], and it is possible to evaluate the various two- and three-particle approximations for the Green function.

We first consider numerical calculations for an ideal Gaussian chain at infinite dilution with the isotropic Förster transfer rate [10]. The three-body Padé is given by

$$\widetilde{G}^S(\varepsilon) = \varepsilon^{-1}\left\{1 + \frac{2^{4/3}}{3^{1/2}}\pi\left[\frac{q^3}{\varepsilon\tau}\left(\frac{R_o}{a}\right)^6\right]^{1/3} + 4.123\left[\frac{q^3}{\varepsilon\tau}\left(\frac{R_o}{a}\right)^6\right]^{2/3}\right\}^{-1} \quad (47)$$

where a is the Kuhn segment length. The inverse Laplace transform of Equation (47) is shown in Figure 5 as a universal curve for $G^S(\theta)$, where

$$\theta = \left(\frac{tq^3}{\tau}\right)\left(\frac{R_o}{a}\right)^6 \quad (48)$$

As the chain concentration increases, energy migration will occur between different chains as well as on an intramolecular basis. To extend these results to non-zero concentration of labeled chains and to include traps, a three-particle Padé approximant has been evaluated [10] in terms of the dimensionless intramolecular donor and trap chromophore concentrations defined by

$$\bar{C}_D = \pi q_D\left(\frac{R_o^{DD}}{a}\right)^2 \quad (49)$$

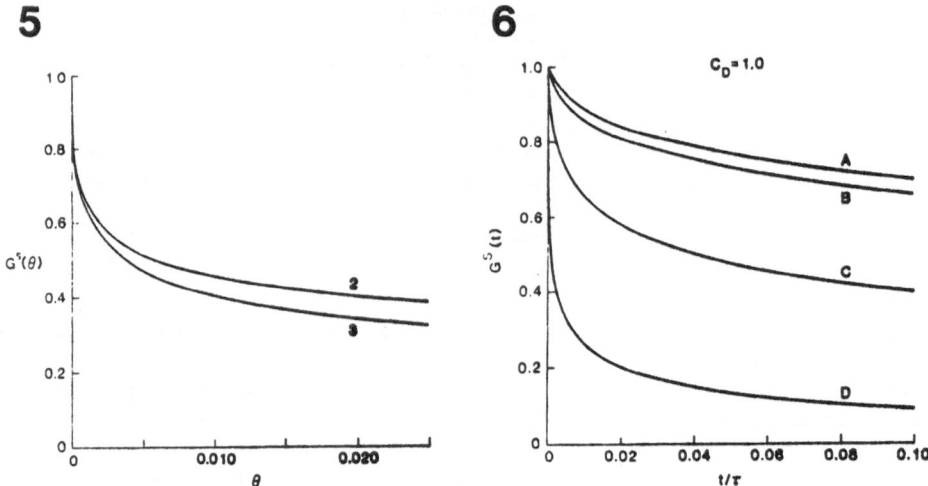

Figure 5. Inverse Laplace transform of the two and three-particle Padé approximants for $\tilde{G}^S(\varepsilon)$ for the randomly tagged polymer chain at high dilution. Taken from Figure 6 of reference [10].

Figure 6. Inverse Laplace transform of the three-particle Padé approximant for $\tilde{G}^S(\varepsilon)$ for the randomly tagged chain as a function of chain stiffness. In all curves $C_D = 1.0$. Curves A, B, C, and D correspond to $C_D = 0.01$, 0.10, $1,0$ and 5.0, respectively. Taken from Figure 1 of reference [11].

$$\bar{C}_T = \pi q_T \left(\frac{R_o^{DT}}{a}\right)^2 \tag{50}$$

and the dimensionless bulk chromophore concentrations defined by

$$C_D = \left(\frac{4\pi}{3}\right)\left(R_o^{DD}\right)^3 \rho_D \tag{51}$$

$$C_T = \left(\frac{4\pi}{3}\right)\left(R_o^{DT}\right)^3 \rho_T \tag{52}$$

where ρ_D and ρ_T are the bulk number densities of donors and traps in the system. These dimensionless concentrations are not independent, but are related by

$$\bar{C}_D C_T = \chi \, \bar{C}_T C_D \tag{53}$$

with

$$\chi = \frac{R_o^{DT}}{R_o^{DD}} \tag{54}$$

For reasons of space, we will not present the full expressions for $\tilde{G}^S(\varepsilon)$ and $\tilde{G}^D(\varepsilon)$, which may be found in the original reference [11]. The relationships are derived as expansions in fractional powers of $(\varepsilon\tau)$ with coefficients that have powers of C_D, $\bar{C}_C$, C_T, and $\bar{C}_T$ of total order 2. Numerical inversion of the Laplace transform may be obtained using the Stehfest algorithm [54].

Figures 6 and 7 show the results of numerical inversion for the case where there are no traps, i.e. $C_T = \bar{C}_T = 0$. Figure 6 illustrates the effect of statistical chain stiffness. As the Kuhn segment length a increases relative to R_o^{DD} the chain becomes stiffer and more expanded, causing $\bar{C}_D$ to decrease and the decay of the initial site excitation probability $G^S(t)$ to become slower. Figure 7 illustrates the effect of intermolecular excitation transport. An increase in C_D at fixed $\bar{C}_D$ corresponds to an increase in the chain density. This leads to much more rapid decay of $G^S(t)$.

The most significant feature of the random tag chain model is that fluorescence depolarization experiments will allow determination of the characteristic ratio [50,13] C_∞. This may be defined as

$$C_\infty = \lim_{n \to \infty} \left(\frac{n'a^2}{n\ell^2}\right) \tag{55}$$

where n' and n are the number of statistical segments and monomer units per chain, respectively; ℓ is the length of the monomer and a is the Kuhn segment length. We may write the number of chromophores per segment, q, as

$$q = \lim_{n \to \infty} \left(\frac{\sigma n}{n'}\right) \tag{56}$$

where σ is the average number of chromophores per monomer. This will be fixed by the synthesis of the polymer and may be measured using absorption spectroscopy. This allows the dimensionless intramolecular chromophore concentration to be expressed as

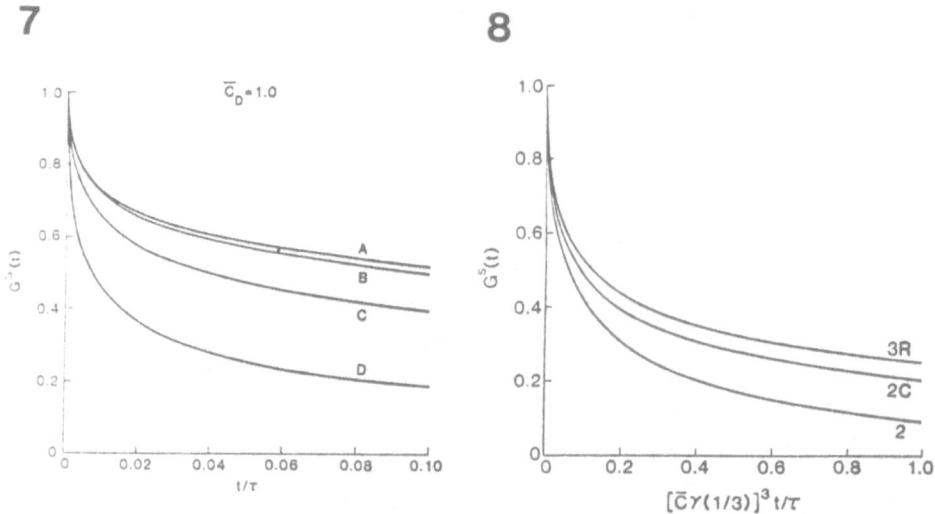

Figure 7. Inverse Laplace transform of the three-
particle Padé approximant for $\tilde{G}^S(\varepsilon)$ for the randomly tagged chain as
a function of chain density. In all curves $\bar{C}_D = 1.0$.   Curves A, B,
C, and D correspond to C_D = 0.01, 0.10, 1.0 and 5.0., respectively.
Taken from Figure 2 of reference [11].

Figure 8. Diagonal portion of the Green function for the randomly
tagged chain at high dilution. The universal curves labeled 3R, 2 and
2C are, respectively, the three-body rational self-consistent approximation, the two body self consistent approximation and the two-
particle cumulant. Taken from Figure 2 of reference [13].

$$\bar{C}_D = \frac{\pi\sigma\left(R_o^{DD}\right)^2}{C_\infty \ell^2}$$ (57)

Because R_o^{DD} and ℓ are known (or can be determined) independently,
a single parameter fit of time dependent depolarization data can yield
C_∞.

   3.3.2 Two-Particle Cumulant Approximation. The preceding discussion on the three particle Padé and the other approximations is not
limited to the case of ideal chain statistics, but it is only in this

case that the correlation functions are known exactly. There are numerous situations, however, in which non-ideal chain statistics are expected. Fortunately, it is still possible to detect deviations from ideal-chain statistics and to obtain approximations for nonideal-chain pair correlation functions through use of the two-particle cumulant approximant.

The general expression for the two-particle cumulant approximation for a translationally invariant polymeric material that contains a small chromophore concentration is [10,13]

$$G^S(t) = \exp\left\{\frac{-1}{32\pi^2} \int d\underline{r} \int d\underline{e}_1 \int d\underline{e}_2 \, g(\underline{r})\right.$$

$$\left. \times \left[1 - \exp(-2W(r,\underline{e}_1,\underline{e}_2)t)\right]\right\} \tag{58}$$

where $g(\underline{r})$ is the pair correlation function describing the average chromophore density at a displacement $\underline{r}$ from a particular chromophore averaged over all possible choices of that chromophore; $W(r,\underline{e}_1,\underline{e}_2)$ is a general excitation transfer rate depending on the distance of separation between two chromophores r and the unit orientation vectors of their transition dipoles $\underline{e}_1$ and $\underline{e}_2$. In addition, the chromophores are assumed to be motionless on the time scale of the excitation transport, and they are assumed to be randomly oriented in space.

We will consider several special cases for application of Equation (58) for Förster transfer and a general power law expression for the pair correlation function given by

$$g(\underline{r}) = \frac{A}{r^\alpha a^{3-\alpha}} \qquad (-3 < \alpha < 3) \tag{59}$$

For the case of EET among randomly distributed chromophores in an infinite volume with no correlations among chromophore positions $\alpha = 0$ and $g(\underline{r}) = A/a^3 = \rho$ where ρ is the number density of chromophores. This leads to

$$G^S(t) = \exp\left[-0.8452\left(\frac{\pi}{2}\right)^{1/2}C\left(\frac{t}{\tau}\right)^{1/2}\right] \tag{60}$$

where $C \equiv \frac{4}{3}\pi R_o^3\rho$. This problem was solved earlier by Gochanour, Andersen and Fayer [4]. Their three-body self consistent approximation agrees extremely well with Equation (60).

A second case of interest is when chromophores are attached randomly to an isolated linear ideal polymer chain. Here the pair correlation function is of the Debye form

$$g(\underline{r}) = \frac{3q}{\pi a^2 r} \tag{61}$$

where q is the average number of chromophores per statistical segment length. This yields

$$G^S(t) = \exp\left[-1.379\ \bar{C}\ \left(\frac{t}{\tau}\right)^{1/3}\right] \tag{62}$$

where

$$\bar{C} \equiv \pi q \left(\frac{R_o}{a}\right)^2$$

In Figure 8 we have plotted Equation (62) as well as the two body self-consistent approximation and the three-body rational self consistent approximation for this problem. The three-body rational approximation is believed to describe excitation transport on an ideal chain quite accurately [12]. Although the agreement between curves 3R and 2C is not as good as was observed for the case of randomly distributed chromophores in an infinite volume [13], the two body cumulant is bounded by the two-body and three-body curves. Thus, it should give a quantitative approximation to $G^S(t)$. It should be obvious that combination of Equation (57) for $\bar{C}$ with Equation (62) should lead to a straightforward determination of C_∞ from measurement of the transient fluorescence anisotropy.

One example of a non-ideal long range correlation that is of interest is that for a long, isolated chain containing randomly attached chromophores and placed in a good solvent. The appropriate correlation function is given by the Edwards scaling law [52]

$$g(\underline{r}) \sim r^{-4/3} a^{-5/3} \tag{63}$$

This yields an anisotropy decay of the form $\exp(-t^{5/18})$, corresponding to $\alpha = 4/3$.

3.4 Calculation for End Tagged Chains

A second class of many-body theories is for EET among chromophores attached to two specific sites on the polymer, such as at the chain ends. The treatment is similar to that treated earlier, and will not be discussed in any detail. In the original work [9] the analysis considers several cases: 1) the donors and traps are randomly distributed on the sites; 2) each macromolecule has exactly

one donor and one trap chromophore; and 3) each polymer chain has a
donor at each chain end. The formalism allows for intermolecular as
well as intramolecular EET, and has been shown to be exact in the
limit of isolated chains. At high chain densities or for large
intramolecular site separations, the self-consistent results of
Loring, Andersen and Fayer [5] are recovered.

In this review, we only present results for the case when the
chains contain only a single donor chromophore type, and no traps.
This is of interest because synthetic procedures have been developed
for placing identical chromophores on the chain ends. For example,
Winnik has prepared and examined several such types in his excimer
studies on cyclization dynamics, described earlier in these proceedings.

For simplicity, the polymer chain is assumed to have a Gaussian
distribution of intramolecular intersite distances with a distribution
function given by

$$f(\underline{r}) = (\frac{3}{2\pi <R^2>})^{3/2} \exp(\frac{-3r^2}{2<R^2>})$$ (64)

where $<R^2>$ is the second moment of the distribution, i.e. the mean-
squared end-to-end distance of the chain. Other distribution functions for short or stiff chains [55-59] have been utilized in a wide
variety of studies on biological systems, and could be employed when
appropriate. The general theory leads to expressions for $G^S(t)$ and
$G^D(t)$ in terms of the parameters: q_D, q_T, C_D, C_T, α_D, α_T and τ where

$$\alpha_D = \frac{(R_o^{DD})^2}{<R^2>}$$

$$\alpha_T = \frac{(R_o^{DT})^2}{<R^2>}$$ (65)

The parameters are related through the constraint

$$\frac{C_D}{C_T} = (\frac{\alpha_D}{\alpha_T})^{3/2} \frac{q_D}{q_T}$$ (66)

so that a decay curve for $G^S(t)$ or $G^D(t)$ will be specified by six
independent parameters.

The results for a chain containing only donor chromophores at the
chain ends are shown in Figures 9 and 10. In this case, $G^S(t/\tau)$
depends only on the parameters q_D, C_D and α_D. When $q_D = 1.0$ there
are exactly two donors on each chain. Figure 9 illustrates that as

the chains become more compact $\langle R^2 \rangle$ decreases and $G^S(t/\tau)$ decays to 1/2 more rapidly. This indicates that the excitation has randomized on the two donor chromophores where the excitation originated. Curve A in Figure 9 is the result for infinite chains having $\alpha_D = 0$, and is exactly the two-body self consistent solution of GAF. Thus, EET proceeds as if the chromophore positions were uncorrelated.

Figure 10 illustrates the effect of bulk chain density on $G^S(t/\tau)$ for fixed chain dimensions, i.e. constant $\langle R^2 \rangle$. As the chain density increases, $G^S(t/\tau)$ rapidly decays to zero. In fact, any non-zero value of C_D will lead to curves that decay to zero at long times. Intermolecular EET cannot be neglected unless α_D is at least an order of magnitude larger than C_D.

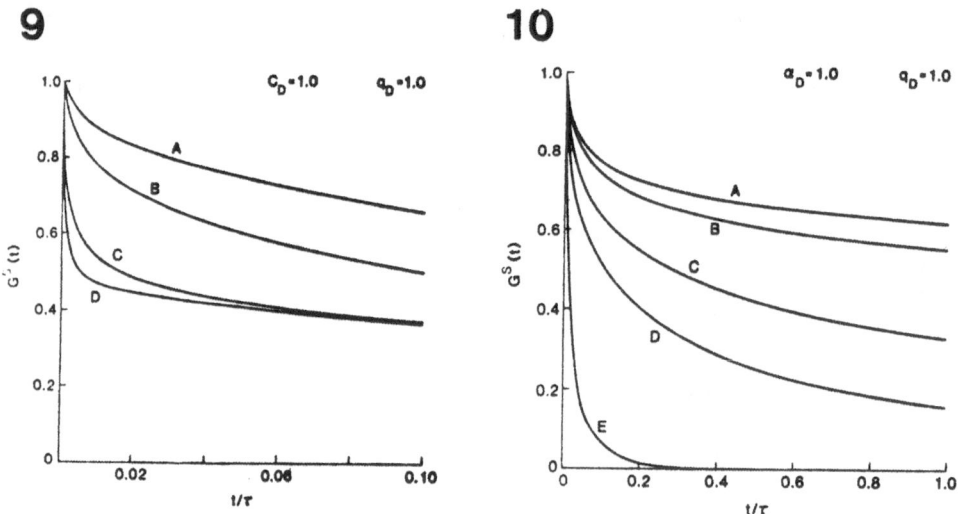

Figure 9. Inverse Laplace transform of $\tilde{G}^S(\epsilon)$ for an end tagged polymer with a Gaussian distribution in the absence of traps. Curves A, B, C, and D correspond to $\alpha_D = 0,1,5$ and 10, respectively. Taken from Figure 4 of reference [9].

Figure 10. Inverse Laplace transform of $\tilde{G}^S(\epsilon)$ for an end-tagged polymer with a Gaussian distribution in the absence of traps. Curves A, B, C, D, and E correspond to $C_D = 0, 0.1, 0.5, 1.0$ and 5.0, respectively. Curve A is the exact result for isolated chains. Taken from Figure 5 of reference [9].

ACKNOWLEDGEMENT

This work was supported by the NSF-MRL through the Center for Materials Research at Stanford University.

REFERENCES

(1) J.B. Birks, Photophysics of Aromatic Molecules. Wiley
     Interscience: N.Y., 1970.
(2) Th. Förster , Ann. Phys. (Leipzig), 1948, 2, 55.
(3) I.B. Berlman, Energy Transfer Parameters of Aromatic Compounds,
     Academic Press: New York, 1973.
(4) C.R. Gochanour, H.C. Andersen, and M.D. Fayer, J. Chem. Phys.
     1979. 70, 4254.
(5) R.F. Loring, H.C. Andersen, and M.D. Fayer, J. Chem. Phys. 1982,
     76, 2015.
(6) G. H. Fredrickson and C.W. Frank, Macromol. 1983, 16, 1198.
(7) C.R. Gochanour and M.D. Fayer, J. Phys. Chem. 1981, 85, 1989.
(8) G.H. Fredrickson, Ph.D. Thesis, Department of Chemical
     Engineering, Stanford University, Stanford, Calif., August,
     1984.
(9) G.H. Fredrickson, H.C. Andersen, and C.W. Frank, J. Chem. Phys.
     1983, 79, 3572.
(10) G.H. Fredrickson, H.C. Andersen, and C.W. Frank, Macromol. 1983,
     16, 1456.
(11) G.H. Fredrickson, H.C. Andersen, and C.W. Frank, Macromol. 1984,
     17, 54.
(12) G.H. Fredrickson, H.C. Andersen, and C.W. Frank, Macromol. 1984,
     17, 1496.
(13) G.H. Fredrickson, H.C. Andersen, and C.W. Frank, J. Polymer
     Sci.-Polymer Physics 1985, 23, 591.
(14) R.M. Pearlstein, J. Chem. Phys. 1972, 56, 2431.
(15) A.J. Roberts, C.G. Cureton, and D. Phillips, Chem. Phys. Lett.
     1980, 72, 554.
(16) D. Phillips, A.J. Roberts, and I. Soutar, Polymer 1981, 22, 427.
(17) D. Phillips, A.J. Roberts, and I. Soutar, J. Polym. Sci., Polym.
     Phys. Ed., 1982, 20, 411.
(18) D.A. Holden, P.Y.K. Wang, and J.E. Guillet, Macromol. 1980, 13,
     295.
(19) K. Demeyer, M. Van der Auweraer, L. Aerts, and F.C. De Schryver,
     J. Chim. Phys. 1980, 77, 493.
(20) D. Phillips, A.J. Roberts, and I. Soutar, Polymer 1981, 22, 293.
(21) S.E. Webber and P.E. Avots-Avotins, Macromol. 1981, 14, 105.
(22) R.D. Wieting, M.D. Fayer, and D.D. Dlott, J. Chem. Phys. 1978,
     69, 1991.
(23) D.D. Dlott, M.D. Fayer, and R.D. Wieting, J. Chem. Phys. 1978,
     69, 2752.
(24) R.F. Loring and M.D. Fayer, Chem. Phys. 1982, 70, 139.
(25) D. L. Huber, Phys. Rev. B 1979, 20, 2307.
(26) D. L. Huber, Phys. Rev. B 1979, 20, 5333.
(27) J. Klafter and R. Silbey, J. Chem. Phys. 1980, 72, 849.
(28) R. Gelles and C.W. Frank, Macromol. 1982, 15, 741.
(29) R. Gelles and C.W. Frank, Macromol. 1982, 15, 747.
(30) G.H. Fredrickson and C.W. Frank, Macromol. 1983, 16, 572.
(31) P.D. Fitzgibbon and C.W. Frank, Macromol. 1982, 15, 733.
(32) C.W. Frank and L.A. Harrah, J. Chem. Phys. 1974, 61, 1526.

(33) K.P. Ghiggino, R.D. Wright, and D. Phillips, J. Poly. Sci. Poly. Phys. Ed. **1978**, 16, 1499.

(34) S.N. Semerak and C.W. Frank, Canadian J. Chem. **1985**, 63, 1328.

(35) J.W. Thomas, Jr. and C.W. Frank, Macromolecules **1985**, 18, 1034.

(36) P.D. Fitzgibbon and C.W. Frank, Macromolecules **1981**, 14, 1650.

(37) S.W. Haan, Ph.D. Thesis 1977, University of Maryland, College Park, Maryland.

(38) S.W. Haan and R. Zwanzig, J. Chem. Phys. **1978**, 68, 1879.

(39) A. Blumen, J. Klafter, and R. Silbey, J. Chem. Phys. **1980**, 72, 5320.

(40) K. Godzik and J. Jortner, J. Chem. Phys. **1980**, 72, 4471.

(41) B. Movaghar, B. Pohlman, and W. Schirmacher, Solid St. Comm. **1980**, 34, 451.

(42) B. Movaghar, J. Phys. C.: Solid St. Phys. **1980**, 13, 4915.

(43) B. Movaghar, G.W. Sauer, J. Phys. C.: Solid St. Phys. **1980**, 13, 4933.

(44) R.J.D. Miller, M. Pierre, and M.D. Fayer, J. Chem. Phys. **1983**, 78, 5138.

(45) G.R. Mitchell and A.H. Windle, Polymer **1984**, 25, 906.

(46) M.D. Ediger and M.D. Fayer, J. Chem. Phys. **1983**, 78, 2518.

(47) M.D. Ediger and M.D. Fayer, Macromol. **1983**, 16, 1839.

(48) R.P. Hemenger and R.M. Pearlstein, J. Chem. Phys. **1973**, 59, 4064.

(49) A. Jablonski, Acta Phys. Pol. A, **1970**, 38, 453.

(50) P.J. Flory, Statistical Mechanics of Chain Molecules, Interscience Publishers: New York, 1969.

(51) H. Yamakawa, Modern Theory of Polymer Solutions, Harper and Row: New York, 1971.

(52) P.G. de Gennes, Scaling Concepts in Polymer Physics, Cornell University: Ithaca, NY, 1979.

(53) H.C. Andersen in Modern Theoretical Chemistry - Vol. 5: Statistical Mechanics A, B. Berne, Editor, 1977.

(54) H. Stehfest, Commun. Assoc. Comput. Mach. **1970**, 13, 47, 624.

(55) I.Z. Steinberg, Annu. Rev. Biochem. **1971**, 40, 83.

(56) A. Grinvald, E. Haas, and J.Z. Steinberg, Proc. Natl. Acad. Sci. **1972**, 69, 2273.

(57) E. Haas, M. Wilchek, E. Katchalski-katzir, and I.Z. Steinberg, Proc. Natl. Acad. Sci. **1975**, 72, 1807.

(58) E. Haas, E. Katchalski-katzir, and I.Z. Steinberg, Biopolymers **1978**, 17, 11.

(59) E. Katchalski-katzir, E. Haas, and I.Z. Steinberg, Ann. N.Y. Acad. Sci. **1981**, 366, 44.

(60) G. Fredrickson, Macromolecules, accepted for publication.

POLYMER BLEND THERMODYNAMICS: FLORY HUGGINS THEORY AND ITS
APPLICATION TO EXCIMER FLUORESCENCE STUDIES

Curtis W. Frank, Muhammad Amin Gashgari, and Steven N. Semerak

Department of Chemical Engineering, Stanford University,
Stanford, California 94305

Abstract - Polymer blends, which are multicomponent mixtures
of two or more homo- or copolymers, have attracted increasing attention as engineering materials because of the potential ease with which
the bulk properties may be controlled. A combination of theoretical
approaches and experiments has been employed to determine the degree
of mixing on the molecular scale. Although more recent thermodynamic
theories have been proposed, the classical Flory-Huggins treatment has
been shown to be particularly useful for polymer blends. In this
chapter the assumptions and basic features of this quasi-lattice
theory will be reviewed. In addition, excimer fluorescence will be
shown to be sensitive to changes in a number of parameters that are
expected to affect the Gibbs free energy of mixing: the solubility
parameters of the blend components, concentration, temperature and
molecular weight.

1.   INTRODUCTION

Although the impact of polymers on modern technology has unquestion-
ably been extensive, emphasis is passing from high volume-low cost
commodity materials to low volume-high cost specialty materials.
A major class of such high performance engineering plastics involves
polymer blends, or alloys, which are physical mixtures of two or more
homopolymers or copolymers. It has been estimated that 15% of all
engineering plastics are presently in the form of blends, and that the
United States production of blends for engineering materials will
triple by 1987 [1]. The major attraction of polymer blends is that
their properties may be controlled without requiring development of
new synthetic procedures [2-6]. However, a point of critical
importance to the optimization of the blend performance is the nature
of the mixing on the molecular level. Although the mechanical
properties may depend critically upon the blend morphology over the
distance scale of 100Å or less [7], a clear understanding of the
polymer-polymer interactions has proved elusive.

M. A. Winnik (ed.), Photophysical and Photochemical Tools in Polymer Science, 523–546.

Efforts to achieve a molecular structure-bulk property predictive capability have proceeded along both theoretical and experimental lines. The initial framework for analysis of the equilibrium thermodynamics was laid by Flory [8,9] and Huggins [10,11]. Although this has been appreciably extended in the equation of states treatment due to Flory and coworkers [12,13], and the lattice fluid theory of Sanchez [14], it remains a useful guide for the correlation of the experimental parameters that are significant to the control of blend miscibility. Experimental methods that have been employed to study the thermodynamics and kinetics of polymer blends encompasss virtually all of the major techniques of polymer science: electron microscopy; differential scanning calorimetry; dynamic mechanical and dielectric spectroscopy; pulsed nuclear magnetic resonance; and light, X-ray and small angle neutron scattering. Of special significance for this NATO Advanced Study Institute on Photochemical and Photophysical Probes in Polymer Science is that several fluorescence methods have been successfully used to study polymer blends. These include nonradiative energy transfer, described elsewhere in this volume by H. Morawetz, fluorescence depolarization described elsewhere in this volume by L. Monnerie, and excimer fluorescence, the subject of this chapter and a later one.

The objectives of this chapter will be twofold. Our primary purpose is to review the Flory-Huggins thermodynamic treatment, including the critical assumptions that must be understood in order to design and interpret experiments. A composition and temperature-dependent expression for the free energy of mixing of a polymer-polymer blend will be presented. Once this is in place, we will illustrate the application of excimer fluorescence as a probe of both enthalpic and entropic effects in blends containing poly(2-vinyl naphthalene) (P2VN). The treatment of the polymer photophysics will be phenomenological; it will only be necessary to assume that the experimentally observed ratio of excimer to monomer fluorescence is proportional to the local density of aromatic rings in the blend. More sophisticated treatments that incorporate a detailed consideration of energy migration will be examined in later chapters.

2.   THERMODYNAMICS OF POLYMER SYSTEMS

2.1   Entropy of Mixing

2.1.1 <u>Solutions of Small Molecules.</u> The ideal solution is the simplest solution model as it requires the solution components to be identical in chemical constitution and size. It follows that the energies of interaction between all molecular species then are the same and that the enthalpy of mixing, ΔH_M, is equal to zero. In addition, the entropy of mixing, ΔS_M, arises from purely combinatorial considerations. It is given by Eq. (1), which is referred to as the ideal entropy of mixing.

$$\Delta S_M = -k(n_1 \ln N_1 + n_2 \ln N_2) \ . \tag{1}$$

Here k is Boltzmann's constant, n_1 and n_2 are the number of molecules of solvent and solute, respectively, and N_1 and N_2 are their mole fractions.

Most real solutions do not behave ideally with deviations from ideality resulting from failure of either or both of the conditions describing an ideal solution: $\Delta H_M = 0$ and ΔS_M being given by Eq. (1). If ΔH_M is zero but ΔS_M is different from the ideal entropy of mixing, an athermal solution is obtained. A regular solution has the ideal entropy of mixing but the enthalpy of mixing is not zero. Finally, an irregular solution has both ΔS_M and ΔH_M different from their ideal values.

2.1.2 Flory-Huggins Theory for Polymer Solutions.

The first calculations of the entropy and enthalpy of mixing terms for a polymer/solvent system were carried out independently by Flory [8,9] and Huggins [10,11]. Both authors proceeded from the quasi-crystalline lattice model assuming that solvent molecules can interchange sites with the polymer chain segments, each of which is equal in size to a solvent molecule. It was also assumed that all polymer molecules are flexible and identical in size. The lattice was considered to consist of n_0 cells occupied by n_1 solvent molecules and n_2 polymer chains each consisting of x segments. Z was taken to be the lattice coordination number, defined as the number of cells that are nearest neighbors to a given cell.

The entropy of mixing was calculated by counting the number of possible configurations the chain can assume on the lattice. With the reference states taken as the pure solvent and the pure, perfectly ordered polymer, the configurational entropy of mixing, ΔS_c, is given by

$$\Delta S_c = -k\{n_1 \ln[n_1/(n_1 + xn_2)] + n_2 \ln[n_2/(n_1 + xn_2)]$$

$$-n_2(x-1)\ln[(Z-1)/e]\} \tag{2}$$

Equation (2) contains contributions from two processes: the disorientation of the perfectly ordered polymer molecules and the mixing of the disoriented polymer with solvent. In order to determine the configurational entropy of mixing an amorphous polymer with a solvent, ΔS_M^* , the disorientation entropy term, obtained from Eq. (2) by setting $n_1 = 0$, must be subtracted from ΔS_c. Thus

$$\Delta S_{disorientation} = kn_2 \{ \ln x + (x-1) \ln [(Z-1)/e] \}$$ (3)

For large values of x, $\ln x$ is negligible compared with the second term in Eq. (3) so that

$$\Delta S_M^* = -k(n_1 \ln \phi_1 + n_2 \ln \phi_2)$$ (4)

where ϕ_1 and ϕ_2 are the volume fractions of solvent and polymer, respectively.

$$\phi_1 = n_1/(n_1 + xn_2)$$

$$\phi_2 = xn_2/(n_1 + xn_2)$$ (5)

Comparison between Eqs. (1) and (4) reveals that the expression for the entropy of mixing of a pure amorphous polymer with solvent has the same form as that for an ideal solution except that volume fractions have replaced mole fractions of the components.

2.2   Enthalpy of Mixing

2.2.1 <u>Solutions of Small Molecules</u>. According to the lattice model, the formation of a regular solution involves the replacement of some of the contacts between like species in the pure solvent and solute with contacts between unlike species in the solution. There are three types of nearest neighbor contacts that are conveniently represented by {1,1}, {2,2} and {1,2}, with w_{11}, w_{22} and w_{12} representing the energies associated with these contacts, respectively. The overall energy change associated with the breaking of {1,1} and {2,2} type contacts and the formation of a {1,2} contact, which is called the interchange energy, Δw_{12}, is equal to the algebraic sum of the respective contact energies.

$$\Delta w_{12} = w_{12} - \frac{1}{2}(w_{11} + w_{22})$$ (6)

The enthalpy of mixing is directly related to this interchange energy according to

$$\Delta H_M = \Delta w_{12} P_{12}$$ (7)

where P_{12} is the average value of the total number of {1,2} contacts in the solution.

Scatchard [15] proposed a semi-empirical expression for the molar enthalpy of mixing given by

$$\Delta H_M = V_1 \, \phi_1 \phi_2 \left[(\frac{\Delta E_1}{V_1})^{1/2} - (\frac{\Delta E_2}{V_2})^{1/2} \right]^2 \qquad (8)$$

where ΔE_i is the molar energy of vaporization and V_i is the molar volume for component i. The ratio $(\frac{\Delta E}{V})$ is called the cohesive energy density (CED), and is a measure of the strength of the intermolecular forces holding the molecules together in the condensed state. Hildebrand [16] later introduced the solubility parameter, δ, defined as the square root of the CED, according to Eq. (9)

$$\delta \equiv (CED)^{1/2} \equiv (\frac{\Delta E}{V})^{1/2} \qquad (9)$$

which is evaluated at 298K. The molar change in internal energy, ΔE, is related to the molar heat of vaporization according to

$$\Delta E_i = \Delta H_{vaporization} - P\Delta V \simeq \Delta H_{vaporization} - RT \qquad (10)$$

and can be determined experimentally along with the molar volume V. Hence, the enthalpy of mixing given by Eq. (8) can be expressed as

$$\Delta H_M = V_1 (\delta_1 - \delta_2)^2 \, \phi_1 \phi_2 \qquad (11)$$

Hildebrand [16] has demonstrated that the solubility behavior of many non-polar solutes in non-polar solvents can be predicted satisfactorily from Eq. (11) for the enthalpy of mixing and Eq. (1) for the ideal entropy of mixing provided that volume fractions instead of mole fractions are used whenever the molar volumes of the components are not equal.

2.2.2 Flory-Huggins Theory for Polymer Solutions. The value of P_{12} in a polymer solution of a particular composition is given by the product of two terms: the probability that a particular site adjacent to a polymer segment is occupied by a solvent, and the total number of contacts between all polymer molecules and all of their neighbors. The first term is approximated by the volume fraction, ϕ_1, of solvent in the solution and the second term is given by $n_2[(Z-2)x+2]$. For large values of x the second term may be approximated satisfactorily by $n_2 Zx$, resulting in the following expression for P_{12}:

$$P_{12} = \phi_1 n_2 Zx \qquad (12)$$

According to Eq. (5), $\phi_1 n_2 x \equiv \phi_2 n_1$ so that the enthalpy of mixing the two components may be expressed as

$$\Delta H_M = Z \Delta w_{12} n_1 \phi_2 \tag{13}$$

The expression for ΔH_M may be recast in the form

$$\Delta H_M = kT \chi_{12} n_1 \phi_2 \tag{14}$$

where

$$\chi_{12} = BV_1/RT \tag{15}$$

and

$$B = Z \Delta w_{12}/V_s \tag{16}$$

χ_{12} is a dimensionless quantity that characterizes the interaction energy per solvent molecule divided by kT. V_1 is the molar volume of the solvent and V_s is the molecular volume of a polymer segment. Substitution of Eq. (15) into (14) yields the following expression for the molar enthalpy of mixing:

$$\Delta H_M = BV_1 \phi_1 \phi_2 \tag{17}$$

The successful application of the solubility parameter approach for regular non-polymeric solutions has led to its extension to polymeric solutions. Comparison of Eqs. (11) and (17) reveals that the term B, which represents the interaction energy density of the solvent-solute pair, corresponds to the square of the solubility parameter difference between the solvent and solute. Hence, the molar enthalpy of mixing of a low molecular weight solvent with a polymer may be expressed in terms of their solubility parameters according to Eq. (11). ΔH_M may also be written as

$$\Delta H_M = RT \chi_H \phi_1 \phi_2 \tag{18}$$

where

$$\chi_H = \frac{V_1}{RT} (\delta_1 - \delta_2)^2 \tag{19}$$

Since polymers generally cannot be volatilized without degradation, their solubility parameters must be determined by indirect methods. This may be done by measuring either the swelling of a slightly cross-linked sample or the intrinsic viscosity of the polymer in a series of solvents with known solubility parameters. The degree of swelling is considered to be proportional to the solvent quality with better solvents producing greater swelling. In addition, the intrinsic viscosity will increase with increasing polymer-solvent interaction as the polymer coils expand and interact. The degree of swelling or the intrinsic viscosity is usually plotted versus the solubility parameter of the solvent. The curve generally has a well defined maximum from which the solubility parameter of the polymer may be estimated.

The δ obtained from these methods is dependent on the nature of the solvents used. Both methods work best when only dispersive forces are active, which is the case when both polymer and solvents are non-polar and do not exhibit hydrogen bonding. Polar solvents and those with hydrogen bonding introduce additional interaction forces that usually lead to different values for δ [17]. As polar and hydrogen bonding forces are larger than those due to dispersion, their contributions to the cohesive energy should be taken into account. This has been accomplished by dividing the cohesive energy into three parts [18-20] according to Eq. (20).

$$\Delta E_i \equiv E_{cohesive} = E_d + E_p + E_h \tag{20}$$

where E_d, E_p and E_n are contributions from dispersion forces, polar forces, and hydrogen bonding respectively. The corresponding relation for the solubility parameter then is

$$\delta^2 = \delta_d^2 + \delta_p^2 + \delta_h^2 \tag{21}$$

where δ_d, δ_p and δ_h represent the solubility parameter components corresponding to the three types of interaction forces. Values of δ_d, δ_p and δ_h may be found in the recent extensive compilation by Barton [21].

Another method of estimating solubility parameters involves the concept of molar group additivity, first suggested by Dunkel [22]. In 1953, Small [23] introduced the molar attraction constant, F, defined as the square root of the product of the cohesive energy and the molar volume. Based on the available vapor pressure and heat of vaporization data in the literature for compounds with no hydrogen bonding, he assigned F values for the different chemical groups. In 1965, Van Krevelen [24] derived a set of atomic contributions to calculate F. In 1970, Hoy [25] re-examined a large amount of data on solvents and their solubility parameters and revised the F values of Small. He also proposed new group contributions and assigned F values for specific structural features that must be added to the group contributions. Solubility parameter values can be empirically calculated either from the F group molar attraction constants or from the cohesive energy group contribution constants that have been developed for polymers [26-29].

2.3   Free Energy of Mixing

2.3.1 Stability Criteria.

Given an expression for the free energy of mixing, the thermodynamic stability of binary blends requires that two conditions be satisfied [30]:

i) $\Delta G_M < 0$ and   ii) $\dfrac{\partial^2 \Delta G_M}{\partial \phi_1^2} > 0.$

A plot of $\Delta G_M/RT$ as a function of concentration provides a convenient way to describe when these conditions are met by a given polymer blend. Several free energy diagrams characteristic of polymer/polymer blends are shown schematically in Fig. 1. Curve A lies entirely in the negative ΔG_M region and is convex downward everywhere; hence, it corresponds to a thermodynamically stable blend. By contrast, curve D represents a totally immiscible blend with $\Delta G_M > 0$ over the entire composition range. More interesting behavior is observed for the two remaining cases. Curve C has $\Delta G_M < 0$ for a certain composition range and $\Delta G_M > 0$ elsewhere making the binary blend thermodynamically unstable. A blend whose free energy of mixing diagram is of this type will be miscible over a limited composition range outside of which the blend will undergo phase separation. Finally, Curve B, which also has $\Delta G_M < 0$, contains a section that is convex upward. This means that

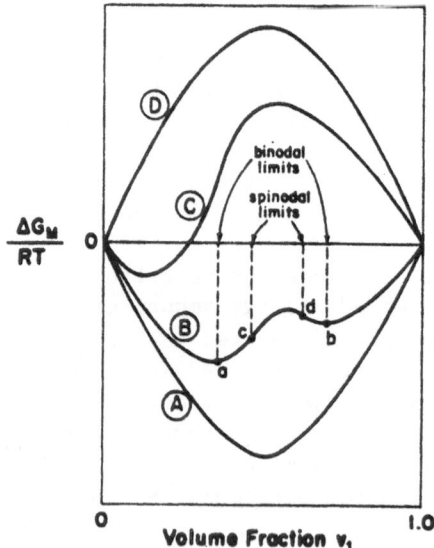

Figure 1.   Schematic concentration dependence of the free energy of mixing.

over that concentration region the solution will be thermodynamically
unstable and should separate at equilibrium into two phases of composition a and b, which are the two points of common tangency of a
straight line with curve B. Points a and b are temperature
dependent and represent the boundary between stable and metastable
compositions. The locus of such points as a function of temperature
is termed a binodal. The spinodal is defined as the locus of the
inflection points c and d, which represent the boundary between
metastable and unstable compositions. Hence, sections ac and bd
correspond to metastable phases and section cd corresponds to a
region where the system is completely unstable.

Figure 2 shows two typical phase diagrams obtained by plotting
the loci of points corresponding to the binodal and spinodal.
Figure 2a contains a lower critical solution temperature (LCST),
defined as the minimum temperature at which two phase behavior is
observed; an increase in the system temperature will lead to phase
separation. Figure 2b contains an upper critical solution temperature
(UCST), defined as the maximum temperature at which two phase behavior
is observed; an increase in the system temperature will lead to
miscibility. Polymer/solvent mixtures generally exhibit UCST
behavior and polymer blends generally show LCST behavior. However,
more complicated phase diagrams are possible: an LCST at high
temperature and a UCST at low temperature; an hourglass structure in
which the LCST and UCST have merged, leading to a miscibility gap; or
a closed two-phase region resulting from the UCST being greater than
the LCST.

2.3.2 Flory-Huggins Theory for Polymer Solutions. The assumptions
and approximations involved in the derivation of Eq. (4) for the
entropy of mixing impose certain limitations on its applicability to
real polymer solutions. One assumption made is that the pure solvent,
pure polymer and all intermediate compositions may be accommodated on
the same lattice. This requires that both solvent molecules and
polymer chain segments have identical size and spatial configuration,
which is seldom justified for real solutions. Another assumption made
implicitly in calculating the number of possible sites available for a
new polymer chain is that the segments of molecules already placed on
the lattice are distributed at random. Since a polymer chain occupies
a sequence of x consecutively adjacent cells, this assumption should
be valid for solutions of large polymer concentrations as polymer
chains will be intertwined with each other. However, this assumption
will be totally unrealistic for dilute solutions in which polymer
molecules are well separated from one another. Hence, Eq. (4) will
not be valid for dilute polymer solutions.

For concentrated solutions in which strong interactions due to
hydrogen bonding or dipole forces are absent, the free energy of

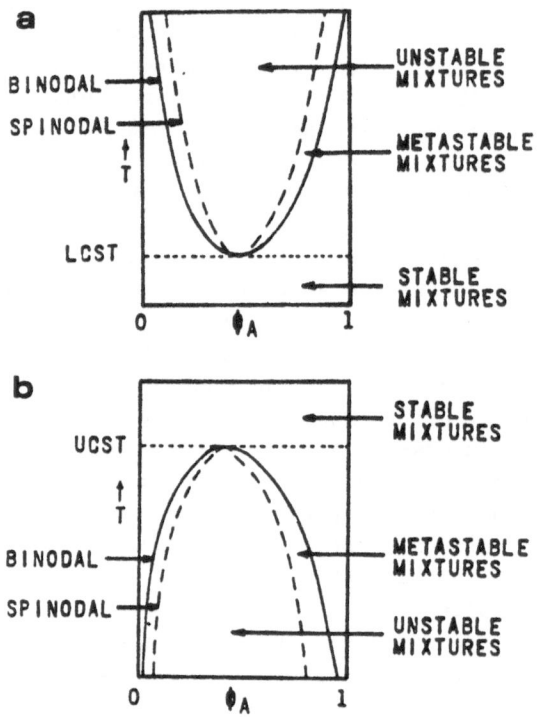

Figure 2. Typical phase diagrams for mixtures exhibiting a lower critical solution temperature (a) and an upper critical solution temperature (b).

mixing derived from Flory-Huggins theory has proven to be fairly successful in predicting the solubility behavior of polymers in low molecular weight solvents. This has suggested the extension to polymer-polymer mixtures. In fact, the more recent equation of state theories lead to the conclusion that Flory-Huggins theory should work much better for polymer/polymer than for polymer/solvent mixtures [5]. The first application of the Flory-Huggins theory of polymer solutions to mixtures of two polymers with or without added solvent was carried out by Scott [31] and Tompa [32] who derived expressions for the entropy, the enthalpy and the partial molar free energy of mixing. Scott [31] obtained the following expression for the total free energy of mixing:

$$\Delta G_M = \frac{RTV}{V_r} \left[\frac{\phi_1}{x_1} \ln \phi_1 + \frac{\phi_2}{x_2} \ln \phi_2 + x_{12} \phi_1 \phi_2 \right] \tag{22}$$

where V is the total volume of the mixture, V_r is a reference volume, taken to be the smaller of the repeat unit molar volumes for the two blend components, ϕ_i is the volume fraction of component i, x_i is the degree of polymerization of component i in terms of the reference volume V_r, and χ_{12} is the binary interaction parameter between the two polymers. χ_{12} is assumed to be independent of concentration, pressure and molecular weight and to have an inverse temperature dependence, as in Eq. (15).

The first two terms in Eq. (22) represent the combinatorial entropy contribution; it is negative and generally small for blends of high molecular weight polymers. Under such conditions the enthalpic contribution from the third term governs whether ΔG_M will be negative, a necessary but insufficient condition for a blend to be miscible.

The spinodal for a binary system is easily calculated by setting

$$\frac{\partial^2 \Delta G_M}{\partial \phi_1^2} = 0.$$ For the Flory-Huggins expression of the free energy of

mixing this yields the following equation:

$$(\chi_{12})_{spinodal} = \frac{1}{2} \left[\frac{1}{x_1(\phi_1)_{spinodal}} + \frac{1}{x_2(\phi_2)_{spinodal}} \right] \qquad (23)$$

The equations for the binodal are derived by setting the chemical potentials of polymer 1, defined as $\dfrac{\partial \Delta G_M}{\partial n_1}$, equal in the two coexisting phases. A similar procedure is used for polymer 2. The resulting binodal equations are

$$\ln \phi_1' + \left(1 - \frac{x_1}{x_2}\right) \phi_2' + x_1 \chi_{12}(\phi_2')^2 = \ln \phi_1'' +$$

$$\left(1 - \frac{x_1}{x_2}\right) \phi_2'' + x_1 \chi_{12}(\phi_2'')^2 \qquad (24)$$

$$\ln \phi_2' + \left(1 - \frac{x_2}{x_1}\right) \phi_1' + x_2 \chi_{12}(\phi_1')^2 = \ln \phi_2'' +$$

$$\left(1 - \frac{x_2}{x_1}\right) \phi_1'' + x_2 \chi_{12}(\phi_1'')^2 \qquad (25)$$

where primes designate one phase and double primes the other.

2.4   Alternative Thermodynamic Treatments

2.4.1 Empirical Description of the Binary Interaction Parameter. As
noted earlier, the most important contribution to ΔG_M for blends of
high polymers is due to enthalpic effects. Contrary to the
assumptions of the original Flory-Huggins treatment, there now is
experimental evidence that indicates that χ_{12} is dependent on
concentration and has a temperature dependence other than the inverse
relation of Eq. (15) [33].

A number of approaches have been considered to explain the
thermodynamic behavior of specific polymer mixtures. For example,
Tompa [34] suggested an empirical expression in which χ_{12} is a power
series in concentration:

$$\chi_{12} = \chi_1(T) + \chi_2 \phi_2 + \chi_3 \phi_2^2 \tag{26}$$

Here only the first term is assumed to be temperature dependent. Considerably more adjustable constants are available in the empirical
expression given by Koningsveld [35]

$$\chi_{12}(T, \phi_2) = \sum_{k=0}^{n} \chi_k(T) \, \phi_2^k \tag{27}$$

where   n   is the number of terms desired in the series and

$$\chi_k(T) = \chi_{k1} + \chi_{k2}/T + \chi_{k3}T + \chi_{k4} \ell n \, T \tag{28}$$

with the χ_{ki}'s being empirical constants. In Eq. (28) the χ_{ki}'s have
been shown to depend on measurable physical quantities such as heat of
mixing, specific heat and molecular weight [36-39].

In an analytical approach, Delmas et al. [39] applied the cell
theory of Prigogine [40] to a mixture of two polymers and arrived at
an expression for $\chi_{12}(T)$ that has only the second and third terms of
Eq. (28). The theory uses the concept of internal and external
degrees of freedom of the polymer molecule, the former related to the
strong intramolecular valence forces and the latter to the weak intermolecular Van der Waals forces. A smoothed potential of the Lennard-
Jones type is taken to represent the potential energy of interaction
between the two segments of polymer. χ_{k2} and χ_{k3} were found to depend
on the potential energy and the lattice coordination number.

One factor largely ignored in empirical approaches to fit the
experimental data is the possible existence of a non-combinatorial
component to the entropy of mixing. The classical combinatorial
entropic contribution to ΔG_M is calculated assuming complete randomness of orientation and rigid molecules [8-11]. The actual distribution of molecules will not be perfectly random, however, particularly

if there are any specific interactions. Moreover, for non-rigid molecules there is an influence of the surroundings on the average randomness of orientation of a segment in a polymer chain relative to the
orientation of the preceding segment. Experimental evidence has
clearly confirmed the existence of a non-combinatorial contribution to
the entropy of mixing [40]. In order to take this entropy into
account, Flory [41] suggested that Δw_{12} should be considered to
consist of two parts

$$\Delta w_{12} = \Delta w_H - T \Delta w_s \tag{29}$$

where Δw_H represents the change in heat content for the process and
Δw_s the non-combinatorial entropy change. Huggins, in his new theory,
accounted for these non-random factors by suggesting that the
interaction parameter should consist of an entropic part in addition
to the enthalpic part [42-44], as noted in Eq. (30).

$$\chi_{12}(T) = \frac{Z \Delta w_H}{kT} - \frac{S_{noncombinatorial}}{k}$$
$$= \chi_H + \chi_S \tag{30}$$

    A similar approach was used by Koningsveld [45] who obtained a
closed expression for the concentration dependence of χ_{12} for mixtures
of homopolymers given by

$$\chi_{12}(T,\phi_2) = \left[\alpha + \frac{\beta}{1 - \gamma \epsilon \, \phi_2} \right] \tag{31}$$

The significant aspect of Eq. (31) is that there are two contributions
to χ_{12}: an empirical entropy correction, α, and an enthalpic contribution, β, related to the heat of mixing. In the second term of
Eq. (31), $\beta = [(Z-2)\Delta w_{12}]/RT$; $\gamma = 2/(Z-2)$ relates the number of pairs
of unlike neighbors in the lattice to the volume fractions of the two
polymers and $\epsilon = \left[1/x_1 - 1/x_2\right]$. ϵ is related to the probability of
having unlike neighbors; it has an upper limit of unity for a polymer-
solvent mixture and a lower limit of zero for a mixture of two
infinitely high molecular weight polymers.
    It is of interest to note that if Z is between 6 and 12, γ has
a value of the order unity. In addition, for high molecular weight
polymers, ϵ is of the order of 10^{-3}. Thus, for low concentrations
with ϕ_2 below 0.10, the concentration dependent term $1-\gamma\epsilon\phi_2$ in
Eq. (31) is essentially unity. If, following Krause [29], β is taken
equal to the interaction parameter expected for regular solutions,
Eq. (31) then reduces to Eq. (30) where $\chi_S = \alpha$ and χ_H is given by
Eq. (19).

2.4.2 <u>Corresponding States Theories</u>. More rigorous treatments of
polymer solution thermodynamics are provided by the equation of state
theory (EST) developed by Flory and coworkers [12,13] and the lattice
fluid (LF) theory of Sanchez [14]. Both theories are corresponding
states approaches, in the spirit of the original development by
Prigogine [40]. Both approaches define the reduction parameters in
terms of physical properties of the pure components such as density,
coefficients of thermal expansion, isothermal compressibility and
thermal pressure, and prescribe methods to calculate the reduced
variables. These new theories are capable of predicting the temperature and pressure dependence of the interaction parameter, generally
considered to be beyond the powers of the original Flory-Huggins
theory. In addition, they incorporate a concentration dependence of
χ_{12} that is totally ignored in the original theory.

Although it might be more desirable to use either of these
approaches to analyze the phase equilibria, they require very accurate
physical property data that are unavailable for the blends of this
study. Rather, the approach taken is to use Flory-Huggins theory with
the interaction parameter suitably modified. Of prime importance is
the temperature dependence of χ_{12} contained explicitly in the RT
factor in χ_H as well as implicitly in the solubility parameter.
McMaster [46] has applied Flory's theory to mixtures of polystyrene
and poly(vinyl methyl ether) and has concluded that the thermal expansion coefficient plays an important role in determining polymer/
polymer miscibility; a small difference in pure component thermal
expansion coefficients is sufficient to cause blends of two high
molecular weight polymers to exhibit a lower critical solution
temperature. The temperature dependence of δ is estimated readily
using experimental measurements of the bulk density and of the thermal
expansion coefficient. The incorporation of the thermal expansion
coefficient to account for the temperature dependence of δ provides a
qualitative physical link betwen the original Flory-Huggins theory and
the EST method.

Even stronger justification for this approach comes from the work
of Biros, Zeman and Patterson [47]. They noted that the solubility
parameter method can incorporate dissimilarities in free volume,
important to the equation of state theories, as well as dissimilarities in contact energies, which form the basis for the classical
Flory-Huggins treatment, if the solubility parameters are assumed to
be temperature and pressure dependent. The proposed dependencies are
of the form

$$\left[\frac{\partial \ln \delta}{\partial T} \right]_p \simeq -\alpha_p \qquad\qquad (32)$$

$$\left[\frac{\partial \ln \delta}{\partial p} \right]_T = \beta_T \tag{33}$$

where α_p and β_T are the coefficients of thermal expansion and iso-thermal compressibility.

Biros, et al. [47] compared the predictions of both the equation of state and modified solubility parameter approaches for the dependence of the binary interaction parameter on temperature, pressure, solvent chain length, and polymer flexibility. Both treatments gave qualitatively similar predictions with values of χ_{12} based on solubility parameters always lower than the equation of state predictions. Quantitative agreement between the two approaches could be achieved if an entropic correction term, represented by α in Eq. (31), were employed. Thus, this modified Flory-Huggins approach should provide a good representation of the equilibrium thermodynamic state of the blends.

3. EXCIMER FLUORESCENCE STUDIES OF POLYMER BLENDS

3.1 Photophysics Summary

An excimer is an electronically excited molecular complex formed between two suitably oriented aromatic rings when one of them has been promoted to an excited state by absorption of energy. The normal characterization parameter is the ratio of the excimer fluorescence intensity to the monomer fluorescence intensity, R, where the monomer refers to the uncomplexed aromatic ring. Of importance for our work is that it is a common feature of the photophysical behavior of the aryl vinyl polymers, as described in a recent review [48]. The objective of this section is to demonstrate the sensitivity of excimer fluorescence to those variables expected to influence the free energy of mixing in polymer blends: solubility parameters of the two components, concentration, temperature and molecular weight.

In the aryl vinyl polymers, an excimer forming site (EFS) may result from interaction between rings on adjacent repeating units of the same chain, between vicinal rings located at widely separated points along the chain contour, and between rings on different chains. The first of these EFS is significant for studies of conformational chain statistics because only the trans, trans meso rotational dyad provides the appropriate geometry for ring overlap [49-51]. The second type of intramolecular EFS should, in principle, yield information about the overall chain conformation. For example, Winnik has conducted a series of studies in which excimer fluorescence from specially prepared polymers containing only two aromatic rings located at the ends of the chains has been used to study

chain cyclization dynamics [52]. Finally, the intermolecular EFS of
the aryl vinyl polymers is of considerable importance in studies on
blend phase morphology because it provides a measure of the local segment density [53-57].

Although the number of EFS traps will be an important factor
governing the observed fluorescence, a second factor of equal or sometimes greater importance is the phenomenon of electronic excitation
transport (EET) [51, 58-61]. This involves the radiationless transfer
of excitation energy, in the singlet state for compounds in this work,
from one aromatic chromophore to another. This process may be viewed
as a random walk with the rate of transfer between randomly oriented
absorption and emission dipoles being given at each step by

$$W = \tau_o^{-1}(R_o/a)^6 \qquad\qquad (34)$$

where τ is the lifetime of the chromophore in the absence of the
excitation transport; R_o is the Förster radius, which is the distance
at which the probabilities of transfer of the excitation and decay by
radiative or nonradiative means are equal; and a is the distance of
separation of the two chromophores.

From Eq. (34) it is clear that the EET process is strongly
dependent upon distance of chromophore separation. If an aryl vinyl
polymer chain is sufficiently isolated and expanded at high dilution
in a thermodynamically good polymer host matrix, the probability of
transfer of the excitation across a loop in the chain should be
minimized, and the EET may be modeled as a strictly one-dimensional
random walk [51,58,60]. On the other hand, at high concentration or
when aggregation of the aryl vinyl polymer has occurred as a consequence of unfavorable thermodynamics, the local density of aromatic
rings should be sufficiently high that the EET process becomes three
dimensional [59,61]. The key to photophysical structural analysis is
to relate the observed photostationary or transient fluorescence parameters to a specific morphological model that will then fix the
concentration of the EFS traps and the nature of the EET.

3.2   Thermodynamic Experiments

3.2.1 Effect of Solubility Parameter Differences. The polymer of
major interest for the initial blend research was poly(2-vinyl
naphthalene), an aryl vinyl polymer whose conformational energetics
are very similar to polystyrene but with a higher quantum yield. In
this first series of studies no attempt was made to develop a
quantitative description of the absolute value of R; rather, this
ratio was simply assumed to be proportional to the local segment
density of aromatic rings. An increase in the local segment
concentration of fluorescence guest polymer, due either to an increase

in the nominal bulk concentration in miscible blends or to phase
separation in immiscible blends, was expected to be reflected by an
increase in R.

The first experiment to demonstrate the utility of the excimer
probe was an investigation of P2VN dispersed at low concentration in a
homologous series of poly(alkyl methacrylates) having a range of solubility parameters that spanned that of the guest P2VN [53]. Since the
molecular weights of the guest and host polymers were high, the
combinatorial entropy of mixing was small and the enthalpic portion
was expected to dominate the free energy of mixing. The fluorescence
intensity ratio, $R = I_D/I_M$, determined from measurements of fluorescence envelope intensities, was observed to pass through a minimum as
a function of the host solubility parameter, as shown in Fig. 3.

Similar behavior was later observed for excimer fluorescence in
poly(4-vinyl biphenyl) and poly(acenaphthalene) [54]. Although the
poly(4-vinyl biphenyl) is expected to exhibit qualitatively similar
types of excimer forming sites as P2VN, the poly(acenaphthalene) (PAN)
will not allow intramolecular EFS to be formed between nearest
neighbor repeat units. As a result, the PAN data strongly suggest

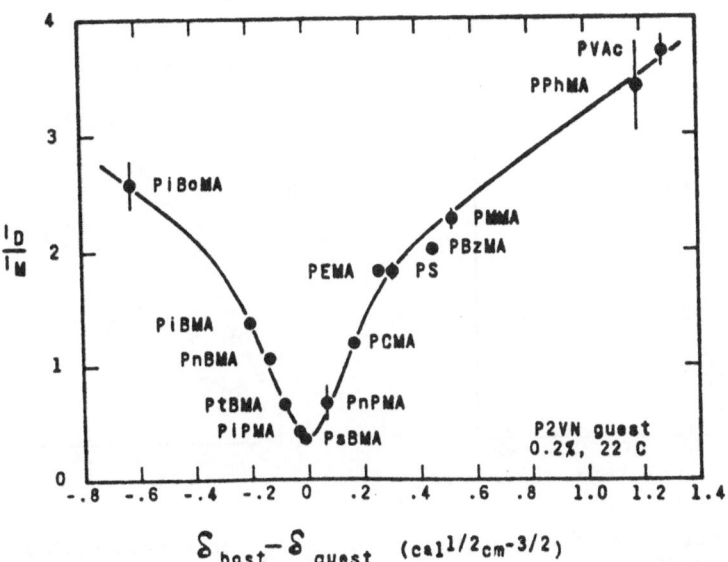

Figure 3. Dependence of observed excimer fluorescence in the P2VN
guest polymer on enthalpic interactions with the host polymer. Low
concentration (0.20%) P2VN films were prepared by solvent casting from
toluene at 22°C. The acronyms refer to the host polymer (primarily
poly(alkyl methacrylates), and are defined in reference 53. (Taken
from Figure 1 of reference 53.)

that intermolecular segmental interaction plays a significant role in
these blends. This could occur as a result of microphase separation
with the domain sizes too small to scatter light. Finally, analogous
results were observed for systems in which the fluorescent trap was
not an EFS but a covalently bound anthracene chromophore incorporated
at low concentration [62]. This is significant because there is no
ambiguity associated with the variability in types of EFS acting as
traps; the anthracene content is fixed by the synthesis. In all
cases, the results were interpreted in terms of a minimization of the
Flory interaction parameter when the solubility parameter of the P2VN
guest matched that of the host matrix.

3.2.2 <u>Effect of Concentration and Temperature.</u> In another series of
measurements the effect of concentration of the fluorescent polymer on
R was studied [54-57,63]. Typical results for blends of P2VN of
molecular weights 21,000 and 70,000 with PMMA (2500) and PS (2200) are

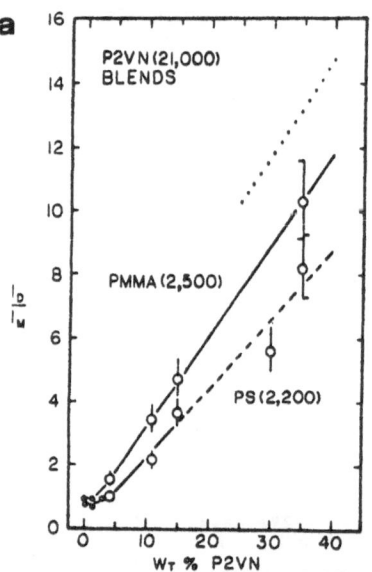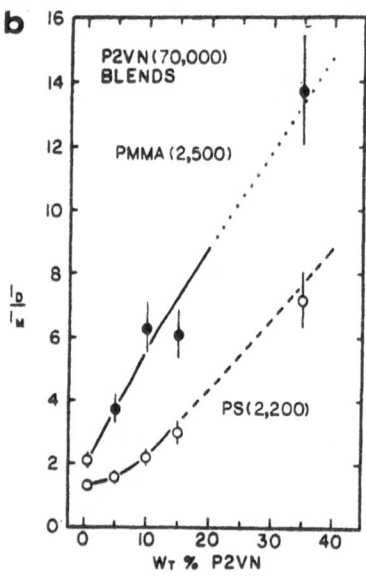

Figure 4. Effect of P2VN concentration on I_D/I_M for blends with PMMA
(2500) and PS (2200), where the values in parentheses refer to the
molecular weights. a: P2VN (21,000); b: P2VN (70,000). Open circles
represent films that were optically clear; solid circles represent
cloudy, phase separated films. The dashed line represents the data
from three molecular weights of P2VN (21,000; 70,000; 265,000) guests
in the PS(2200) host, while the dotted line represents the data from
P2VN (70,000); 265,000) guests in the PMMA(2500). (Taken from
Figures 4 and 5 of reference 63.)

shown in Fig. 4. An empirical clustering model was initially proposed [57] to correlate the influence of concentration, binary interaction parameter and molecular weight of the host polymer on R. This approach has been extended recently by the incorporation of scaling concepts leading to a parameter describing the average number of fluorescent polymer chains occupying the same volume in the blend [64]. Without going into these details, it is apparent from the higher I_D/I_M values that P2VN/PMMA blends are less thermodynamically compatible than P2VN/PS blends. In fact, the latter are optically clear over the concentration range examined while the former are phase separated and cloudy for all P2VN concentrations higher than .3%. Separate measurements of the glass transition temperatures by differential scanning calorimetry on toluene cast blends containing 35% P2VN confirmed that PMMA (2500) is immiscible with any molecular weight of P2VN greater than or equal to 21,000 [65,66]. The PS (2200) is miscible, however, with P2VN for molecular weight between 21,000 and 265,000.

One complication that was recognized at an early stage was that there is a real possibility that kinetic restrictions during solvent casting could affect the attainment of equilibrium since many of the blends containing P2VN are in the glassy state at 295 K [56]. For example, consider the situation for a binary blend at a composition such that equilibrium thermodynamics would predict that phase separation would occur. If the two phase boundary (binodal) is crossed at high solvent concentration upon solvent evaporation, there will be minimal diffusional barrier to phase separation. On the other hand, in cases for which the ternary system is miscible down to relatively low solvent concentration, the final morphology may be that of an apparently miscible alloy because there will be insufficient mobility in the blend to allow translational segmental diffusion. However, subsequent annealing of this nonequilibrium system at a temperature sufficiently above the glass transition could then lead to a phase separated structure.

These concepts were examined for the P2VN/poly(n-butyl methacrylate) (PnBMA) and P2VN/poly(methyl methacrylate) (PMMA) blends [56]. These systems were studied by varying the composition and the casting temperature, and by annealing the blends above the glass transition temperature. All of the qualitative observations on optical clarity for the P2VN/PnBMA blends prepared by solvent casting at temperatures greater than the T_g of the binary blend, approximately 295 K, could be satisfactorily explained using equilibrium Flory-Huggins lattice theory to obtain the entropy of mixing coupled with a modified regular solution approach to obtain the enthalpy of mixing. By contrast, the P2VN/PMMA blend exhibited significant deviations from equilibrium calculations with blends appearing to be more miscible than predicted. This was manifested both in the optical clarity and in the R values being lower than expected.

3.2.3 Effect of Molecular Weight. The entropic contribution to the free energy of mixing was examined in studies on the influence of

molecular weight in blends of P2VN with narrow molecular weight
distribution PS [55] and PMMA [63]. The solvent casting studies
demonstrated that the local segment concentration was not at its
equilibrium value in the high T_g blends such as these. Nevertheless,
the change in R for P2VN going from a low concentration miscible
value, ca. 1, to the pure film, ca. 45, is sufficiently large that it
is reasonable to expect that some change in R will be observed as the
molecular weight of the host is varied. Since there was no intent to
provide an explanation for the absolute value of ϕ_E/ϕ_M, this was
considered to be appropriate for the phenomenological approach.

Typical results for P2VN (21,000) and P2VN (70,000) in a series
of PMMA and PS host polymer matrices are shown in Fig. 5. In blends
that were believed to be miscible from the DSC measurements, such as
P2VN (21,000) in all molecular weight samples of PS, little change in
R was observed as the host molecular weight was increased, as shown in
Fig. 5a. In some systems, however, the DSC results indicated phase
separation at high P2VN guest concentration. This was generally confirmed to be the case at low P2VN concentrations (where DSC is
insensitive) by a rapid increase in R over a narrow range of host
molecular weight. This is observed for P2VN (21,000) in PMMA and for
P2VN (70,000) in both PMMA and PS, as seen in Fig. 5. In order to
explain the molecular weight and concentration results, the Flory
Huggins free energy of mixing was used to calculate the binodals for
P2VN (21,000) and P2VN (70,000) blended with the PMMA and PS host
series, shown in Fig. 6 and 7, respectively.

These binodal curves are plotted in a somewhat unusual manner,
with the interaction parameter being the variable quantity among the
different lines. Note that the PMMA system exhibits much steeper
binodals than for PS, consistent with the conclusion from all of the
experiments that PMMA is less compatible than PS. From these phase
diagrams it is possible to determine both the critical concentration
and the critical molecular weight at which immiscibility is expected
to set in, assuming, of course, that the equilibrium calculations are
valid for these glassy systems [55,63].

All of the fluorescence observations were successfully
interpreted in terms of the Flory-Huggins thermodynamics. This
indicates that, although the possible existence of kinetic restrictions must be recognized, they may be less significant if the only
objective is to explain the occurrence of <u>changes</u> in the photophysical
parameters. In a later chapter, we review work on blends of
polystyrene with poly(vinyl methyl ether) that permits interpretation
of the <u>absolute</u> values of the photophysical parameters.

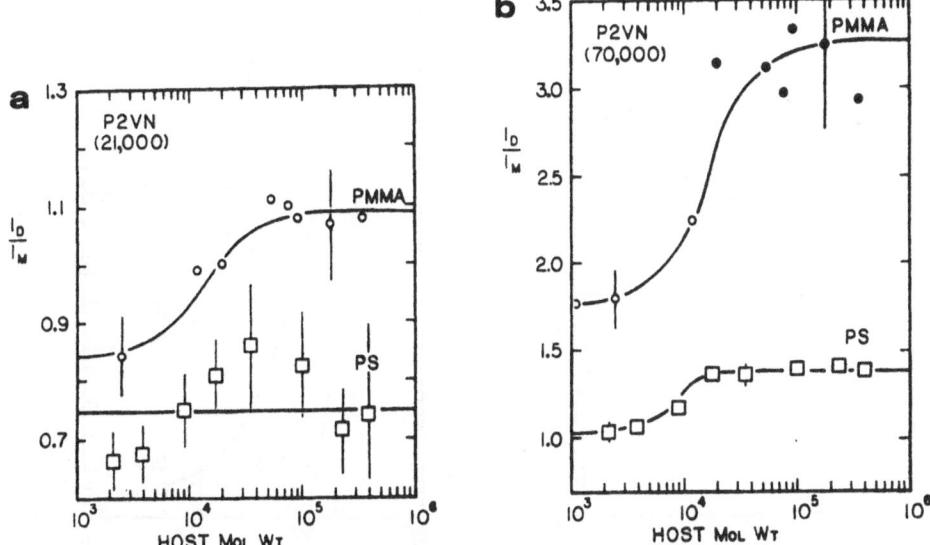

Figure 5. Effect of host molecular weight on I_D/I_M for 0.3% blends of P2VN with PMMA and PS. a: P2VN (21,000), b: P2VN (70,000). Filled plotting symbols indicate the film was visibly cloudy; open symbols denote clear films. (Taken from Figures 1 and 2 of reference 63.)

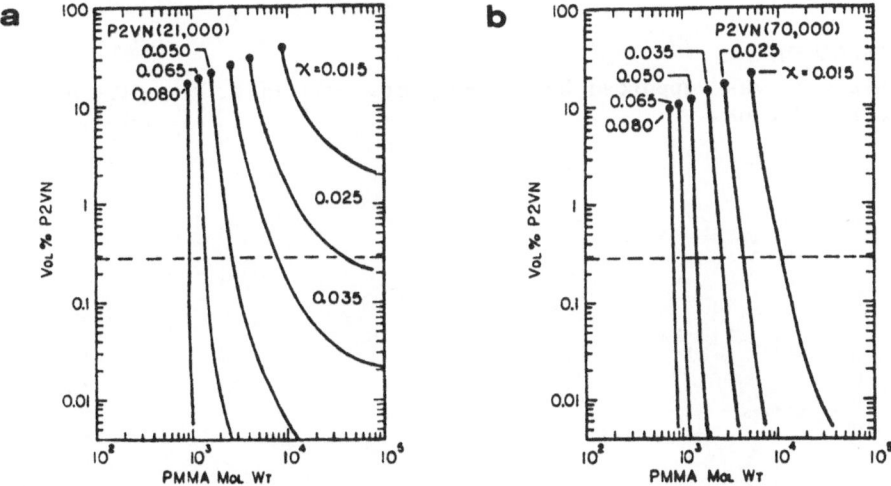

Figure 6. Binodal compositions calculated for P2VN/PMMA blends. a: P2VN (21,000); b: P2VN (70,000). Results for the P2VN-lean phase are shown as volume percent P2VN versus PMMA molecular weight. The interaction parameter for each curve is given and the critical point is indicated by the filled circle. The dashed line denotes 0.3 weight percent P2VN, the concentration for which the data of Figure 5 are presented. (Taken from Figures 9 and 10 of reference 63.)

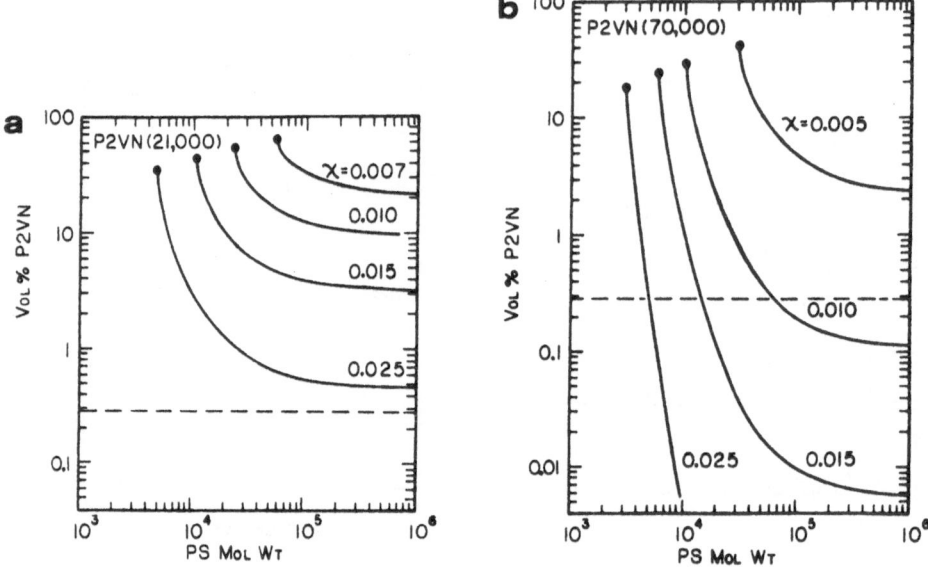

Figure 7.   Binodal compositions calculated for P2VN/PS blends.   See
Figure 6 for details.   (Taken from Figure 6 and 7 of reference 55.)

ACKNOWLEDGEMENT

        This work was supported by the Polymers Program of the National
Science Foundation.

REFERENCES

(1) Industrial Chemical News, **1983**, 4, No. 12, p. 35.
(2) N.A.J. Platzer, "Copolymers, Polyblends, and Composites": ACS Advances in Chemistry Series, No. 142, American Chemical Society: Washington, D.C., **1975**.
(3) D. Klempner, K.C. Frisch, Eds., Polym. Sci. Technol. **1977**, 10.
(4) L.H. Sperling, Ed., Recent Advances in Polymer Blends, Grafts and Blocks, Plenum Press: New York, **1974**.
(5) D.R. Paul and S. Newman, Eds., Polymer Blends, Academic Press: New York, **1978**, Vol. 1.
(6) O. Olabisi, L.M. Robeson, and M.T. Shaw, Polymer-Polymer Miscibility, Academic Press: New York, **1979**.
(7) A.J. Rudin, Macromol. Sci., Rev. Macromol. Chem. **1980**, C19(2), 267.
(8) P.J. Flory, J. Chem. Phys. **1941**, 9, 660.
(9) P.J. Flory, J. Chem. Phys. **1942**, 10, 51.
(10) M.L. Huggins, J. Chem. Phys. **1941**, 9, 440.
(11) M.L. Huggins, Ann. N.Y. Acad. Sci. **1942**, 43, 1.
(12) P.J. Flory, R.A. Orwoll and A. Vrij, J. Amer. Chem. Soc. **1964**, 86, 3515.
(13) P.J. Flory, J. Amer. Chem. Soc. **1965**, 87, 1833.
(14) I.C. Sanchez, "Statistical Thermodynamics of Polymer Blends," in Polymer Blends (D.R. Paul and S. Newman, Eds.), Academic Press: New York, **1978**, Vol. 1.
(15) G. Scatchard, Chem. Revs. **1931**, 8, 321.
(16) J.H. Hildebrand and R.L. Scott, The Solubility of Non-Electrolytes, Reinhold: New York, 3rd Ed., **1949**.
(17) H. Burrell, Polymer Handbook, (J. Brandrup and E.H. Immergut, Eds.) Part IV, 2nd Ed., **1975**, p. 337.
(18) C.M. Hansen, J. Paint Technol. **1967**, 39, 104, 511.
(19) C.M. Hansen, Ind. Eng. Chem., Prod. Res. Dev. **1969**, 8, 2.
(20) E.B. Bagley, T.P. Nelson and J.M Scigliano, J. Paint Technol. **1971**, 43, 35, 35.
(21) A.F. Barton, CRC Handbook of Solubility Parameters and Other Cohesion Parameters, **1983**.
(22) M. Dunkel, Z. Physik. Chem. **1928**, A138, 42.
(23) P.A. Small, J. Appl. Chem. **1953**, 3, 71.
(24) D.W. Van Krevelen, Fuel **1965**, 44, 235.
(25) K.L. Hoy, J. Paint Technol. **1970**, 42, 76.
(26) R.A. Hayes, J. Appl. Polymer Sci. **1961**, 5, 318.
(27) A.T. Di Benedetto, J. Polymer. Sci. **1963**, A1, 3459.
(28) R.F. Fedors, Polymer Engr. Sci. **1974**, 14, 147.
(29) S. Krause, J. Macromol. Sci., Rev. Macromol. Chem.. **1972**, C7, 251.
(30) M. Kurata, Thermodynamics of Polymer Solutions, Polymer Monograph Series, Vol. 1 (H.-G. Elias, Ed.), MMI Press: Harwood Academic Publishers, New York, **1982**.
(31) R.L. Scott, J. Chem. Phys. **1949**, 17, 279.
(32) H.Tompa, Trans. Faraday Soc. **1949**, 45, 1142.
(33) P.G. DeGennes, Scaling Concepts in Polymer Physics, Cornell University Press, Ithaca, NY, **1979**.
(34) H. Tompa, Polymer Solutions, Butterworth, London, **1956**.
(35) R. Koningsveld, Ph.D. Thesis, University of Leiden (1967).
(36) R. Koyama, J. Polym. Sci. **1949**, 35, 247.
(37) G. Rehage, Kunststoffe **1963**, 53, 605.

(38) A.J. Staverman in Handbuch der Physik, Springer-Verlag: Berlin and New
      York, 1962, Vol. 13, p. 456.
(39) G. Delmas, D. Patterson, and T. Somcynsky, J. Polym. Sci. 1962, 57, 79.
(40) I. Prigogine, The Molecular Theory of Solutions, North-Holland:
      Amsterdam, and Interscience: New York, 1957, Chapter 11 and Chapter 17.
(41) P.J. Flory, Principles of Polymer Chemistry, Cornell University Press:
      Ithaca, New York, 1953, Chapter 12.
(42) M.L. Huggins, J. Phys. Chem. 1970, 74, 371.
(43) M.L. Huggins, J. Phys. Chem. 1971, 75, 1255.
(44) M.L. Huggins, MTP Intern. Rev. Sci., Phys. Chem. Series 2, Vol. 8,
      Macromol. Sci., C.E.H. Bawn, Ed., Med. Techn. Publ. Co., London,
      1974, p. 123.
(45) R. Koningsveld and L.A. Kleintjens, J. Polymer Science: Polymer
      Symposium, 1977, 61, 221.
(46) L.P. McMaster, Macromolecules 1973, 6, 760.
(47) J. Biros, 1. Zeman, and D. Patterson, Macromolecules 1971, 4, 30.
(48) S.N. Semerak and C.W. Frank, Advances in Polymer Science 1983, 54, 31.
(49) C.W. Frank and L.A. Harrah, J. Chem. Phys. 1974, 61, 1526.
(50) P.D. Fitzgibbon and C.W. Frank, Macromolecules 1981, 14, 1650.
(51) P.D. Fitzgibbon and C.W. Frank, Macromolecules 1982, 15, 733.
(52) M.A. Winnik, Acc. Chem. Res. 1985, 18, 73.
(53) C.W. Frank and M.A. Gashgari, Macromolecules 1979, 12, 163.
(54) C.W. Frank, M.A. Gashgari, P. Chutikamontham and V. Haverly, Studies in
      Physical and Theoretical Chemistry 1980, 10, 187.
(55) S.N. Semerak and C.W. Frank, Macromolecules 1981, 14, 443.
(56) C.W. Frank and M.A. Gashgari, Macromolecules 1981, 14, 1558.
(57) C.W. Frank and M.A. Gashgari, Annals of the New York Academy of Sciences
      1981, 366, 387.
(58) R.G. Gelles and C.W. Frank, Macromolecules 1982, 15, 741.
(59) R.G. Gelles and C.W. Frank, Macromolecules 1982, 15, 747.
(60) J.W. Thomas, Jr. and C.W. Frank, Macromolecules 1985, 18, 1034.
(61) S.N. Semerak and C.W. Frank, Canadian J. Chem. 1985, 63, 1328.
(62) J.W. Thomas, Jr., C.W. Frank, D.A. Holden and J.E. Guillet, J. Polymer
      Sci.-Physics 1982, 20, 1749.
(63) S.N. Semerak and C.W. Frank, Macromolecules 1984, 17, 1148.
(64) J.W. Thomas, Jr.and C.W. Frank, manuscript in preparation.
(65) S.N. Semerak and C.W. Frank, Advances in Chemistry Series 1983,
      203, 757.
(66) S.N. Semerak and C.W. Frank, Advances in Chemistry Series 1984, 206, 77.

CHARACTERIZATION OF THE INTERPENETRATION OF CHAIN MOLECULES BY NONRADIATIVE ENERGY TRANSFER.

H. Morawetz
Department of Chemistry
Polytechnic Institutte of New York
333 Jay Street
Brooklyn, NY 11201

ABSTRACT. Emission spectra of blends of polymers tagged with fluorescent labels such that the emission spectrum of one overlaps the absorption spectrum of the other can be used to characterize polymer compatibility since interpenetration of donor and acceptor labeled chain molecules increases the efficiency of energy transfer. When solutions of a mixture of donor and acceptor labeled polymers are rapidly frozen and the solvent is removed by sublimation, the extent of chain entanglement which existed in the solution is retained and can be characterized by the reflectance fluorescence of pellets pressed from the material. Results were obtained for monodisperse polystyrenes in good and bad solvents as a function of solution concentration. Pellets obtained by rapid freeze drying of dilute solutions of a mixture of donor and acceptor labeled polymers were heated above the glass transition temperature and the interdiffusion of the polymer chains was followed by the emission spectrum reflecting an increasing energy transfer. The method allows a semiquantittatiye characterization of diffusion coefficients as small as 10^{-16} cm^2-s^{-1} .

1. Introduction.

Several important problems involving polymer systems depend on a characterization of the interpenetration of chain molecules. These concern (a) the compatibility of polymers in bulk, (b) the entanglement of chain molecules in semidilute solution and (c) the interdiffusion of polymers above the glass transition temperature. Studies carried out over the last few years have shown that emission spectra from systems containing polymers tagged with two fluorescent labels can yield information concerning each of these problems.

When a system contains two fluorescent species such that the emission spectrum of one (the "donor") overlaps the absorption spectrum of the other (the "acceptor"), excitation energy absorbed by the donor can be transferred by a dipole-dipole interaction mechanism to the acceptor, so that the acceptor emission is enhanced at the expense of emission from the donor molecules. The theory of the effect was derived

M. A. Winnik (ed.), Photophysical and Photochemical Tools in Polymer Science, 547–559.

by Förster (1) who showed that the sixth power of a characteristic distance R_o, at which the transfer efficiency is 50%, is proportional to the "overlap integral" J of the donor emission and the acceptor absorption spectra. With large values of J, R_o may be as large as 5 nm. This means that energy transfer can be studied in systems containing relatively low concentrations of fluorophores and that the phenomenon can also be observed in rigid systems.

It has long been known that most polymers are not miscible with one another. For instance, although we can mix styrene with methyl methacrylate, their polymers are quite incompatible and even such similar polymers as poly(methyl methacrylate) and poly(ethyl methacrylate) do not form stable blends. The reason for this is as follows: With compounds of low molecular weight an unfavorable mixing energy will rarely prevent mixing, since the large increase in entropy will render the free energy of the mixing process, $\Delta G = \Delta H - T\Delta S$ negative. However, when monomer residues are linked to long chain molecules, the entropy of mixing becomes negligible, although the energy of mixing is affected very little, so that polymers can be compatible only if their mixing is exothermic or athermal (2). Recent years have witnessed an increasing interest in the use of polymer blends and this has led to the development of a variety of methods for the characterization of polymer miscibility (3, 4). As we shall see, the use of emission spectra of blends of two polymers carrying donor and acceptor fluorophores, respectively, provides a powerful new approach to this problem.

When flexible chain molecules are dissolved in a good solvent medium, the molecular coils resist strongly mutual entanglement. A large contribution to this resistance is entropic and may be understood by the following reflection: If a single chain can exist in Z conformations, two chains separated from each other can assume Z^2 conformations. However, when the two molecular coils interpenetrate, their conformations become interdependent, since the space occupied by segments of one chain is excluded to segments of the other chain. Thus, the conformational entropy will decrease with increasing chain entanglement and such entanglement will become important only when the available volume can no longer accommodate the highly swollen polymer coils, all separate from each other. It has become customary to specify a concentration at which chain entanglement becomes important by c* (5) and much theoretical and experimental work has been concerned with changes in the properties of polymer solutions as this concentration is passed. Again, the measurement of energy transfer from donor- to acceptor-labeled polymers can enhance our understanding of this phenomenon.

Diffusion in polymer melts is extremely slow, particularly at temperatures close to the glass transition. If donor and acceptor labeled polymers diffuse into each other, energy transfer between the labels is enhanced, so that the emission spectrum can be used to monitor the process. As will be shown, this approach can lead to estimates of extremely small diffusion coefficients.

2. The compatibility of polymers in bulk.

A survey of studies of polymer compatibility (6) shows that most polymer pairs are incompatible. As we have seen, such a failure of polymers to produce stable blends follows necessarily from endothermic mixing. The theory of the heat of mixing was originally based on the cohesive energy density (CED) of the components of the blend; it was predicted that in the absence of specific interactions of the hydrogen bonding or acid-base type mixing would be endothermic or athermal in the unlikely case where the components of the blend have the same CED values (7). Later, a more sophisticated "corresponding states" theory was developed which led to a calculation of heats of mixing at any temperature T, pressure P and volume V if characteristic parameters T^*, P^* and V^* were known for the components (8). This approach allowed the prediction of the volume change characterizing the mixing process and this quantity could be important in the prediction of the temperature dependence of miscibility.

A number of techniques have been used in the past as criteria of polymer miscibility. Optical clarity has often been used for this purpose, but it can hardly be considered reliable, since a two-phase system will not scatter light if the refractive indices of the phases are sufficiently alike. Nevertheless, the simplicity of the method recommends it for use when it is applicable and determinations of cloud points have been used effectively for the study of phase separations of polymer blends on heating (9, 10). The most common method by which polymer blends are studied involves the determination of glass transitions by differential scanning calorimetry (DSC); here it is assumed that a two-phase system will exhibit the T_g of both components. However, when the composition of polymers to be blended was gradually changed, the appearance of two distinct T_g's was preceded by a broadening of the glass transition range, so that the point at which the system was considered to contain two phases was to some extent arbitrary (11).

In fact, any experimental method used for distinguishing one-phase and two-phase polymer systems will detect phase separation only above some characteristic dimension of phase domains. Electron microscopy can detect phase domains extending over as little as 2 nm (12). Neutron scattering from blends in which one component is deuterated furnishes an unambiguous test of polymer compatibility; if the apparent molecular weight derived from neutron scattering by a blend is identical with the light scattering molecular weight in solution, then the deuterated species must be molecularly dispersed in the blend (13). On the other hand, it has been reported (14) that even in a system in which electron microscopy revealed two phases with dimensions of 10-15 nm, no phase separation was indicated by the behavior of the glass transition.

As of now, theory is inadequate to make safe predictions concerning polymer miscibility. It has been suggested (15) that hydrogen bonding is responsible for the compatibility of poly(vinyl chloride) with polycaprolactone and the decrease in the carbonyl stretching frequency (16) is consistent with this interpretation. To account for the compatibility of polystyrene with poly(3,5-dimethyl p-phenylene oxide) is more difficult; it may be due to an increasing packing efficiency in the blend (17).

When two different polymers are labeled by a donor and an acceptor
fluorophore,respectively, the average donor-acceptor separation in a
blend of the two species will be much less in a one-phase system than
in a system in which the polymers are segregated into separate phases.
It has been demonstrated that this principle may be used to characterize polymer compatibility (18). A typical example of the data obtained
is shown in Figure 1 for blends of anthracene-labeled poly(methyl methacrylate) (PMMA) and naphthalene-labeled methyl methacrylate copolymers
with ethyl or butyl methacrylate. When both the donor and the acceptor

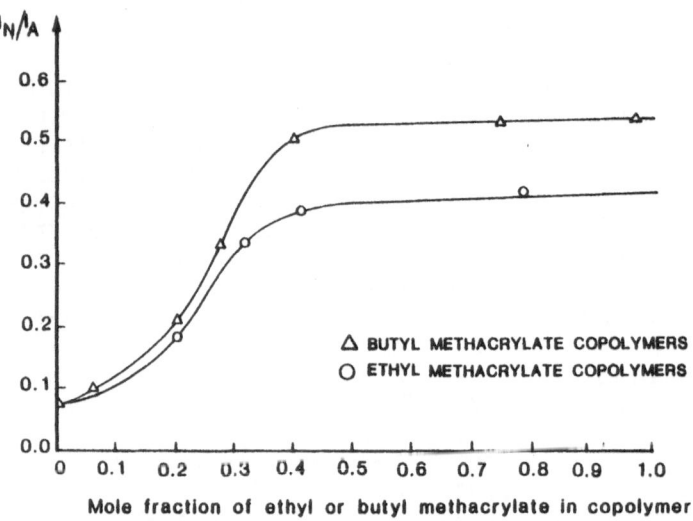

Figure 1. Ratio I_N/I_A of the emission intensity of the naphthyl donor
at 336 nm and the anthryl acceptor at 408 nm from blends containing
equal weights of methyl methacrylate copolymers containing 1.2 weight%
1-(2-naphthyl) ethyl methacrylate and PMMA containing 1.4 weight% of
1-(2-anthryl) ethyl methacrylate. Excitation was at 276 nm, where the
naphthalene label absorbs 80% of the incident light. The characteristic
distance R_0 at which 50% of the absorbed excitation energy is transferred from the naphthyl to the anthryl group is 2.3 nm.

are attached to PMMA, so that the two kinds of chain molecules are randomly intertwined, energy transfer is very efficient and I_N/I_A is small.
When PMMA is blended with copolymers with increasing ethyl or butyl
methacrylate content, I_N/I_A increases until it reaches a plateau characterizing well-separated two-phase systems. The most interesting feature
of the results is the gradual transition of the I_N/I_A ratio from that
corresponding to complete randomness of chain interpenetration to that
indicating complete phase separation. This may be interpreted as due to
a "fuzzy phase boundary" in systems not too far removed from compatibility, where chains can penetrate to some depth from one phase to the

other. Such "fuzzy" boundaries have been observed by electron microscopy (19) and phase contrast microscopy (20) and a theoretical analysis
has shown (21) how the mean depth of interpenetration shrinks as the
free energy of mixing of the two polymeric species assumes increasingly
positive values.

Emission spectra from blends of donor and acceptor labeled polymers seem to be extremely sensitive indicators of the uniformity of the
system in domains with dimensions comparable with R_o (i.e., about 2 nm).
It was found, for instance, that energy transfer from labeled PMMA to
labeled styrene-acrylonitrile copolymers (S/AN) indicates random mixing
if the copolymer contains 30-45 mole% acrylonitrile (18); when poly-
(ethyl methacrylate)(PEMA) was substituted fro PMMA, no composition of
the S/AN led to an equally efficient energy transfer, although energy
transfer was optimized with S/AN containing 35 mole% acrylonitrile (22).
Since both PMMA and PEMA yielded blends with S/AN copolymers exhibiting
a single glass transition (23) we conclude that the fluorescence technique is more sensitive in detecting a segregation of polymeric species
in a blend. This conclusion is also supported by a recent study of
blends of syndiotactic PMMA with poly(vinyl chloride) (24). It was found
that the energy transfer efficiency from the donor labeled PMMA to the
acceptor labeled PVC decreased smoothly with an increasing PMMA content
of the blend, although DSC data (26) exhibited a single glass transition
for all blends in which the vinyl chloride residues were in excess over
those of methyl methacrylate. The energy transfer results may be interpreted either as indicating an increasing fluctuation of composition
preceding phase separation, or as showing that systems with small phase
domains and fuzzy phase boundaries do not exhibit two glass transitions
although the departure from uniformity can be detected by a reduced
energy transfer.

Incompatible polymers may form a one-phase system on addition of
a block copolymer whose blocks are similar to the two homopolymers (25).
In that case, each of the homopolymers mixes with the similar block of
the block copolymer. This principle has been documented in an experiment in which a styrene-p-tert.butylstyrene block copolymer carrying
donor labels only in the styrene block was blended with acceptor-labeled
polystyrene or poly(p-tert.butylstyrene). As expected, energy transfer
was much more efficient in the polystyrene blends (27).

3. Use of energy transfer for the study of polymer association in solution.

The association of donor and acceptor labeled polymers in solution can
also be monitored by an increase in energy transfer. This has been demonstrated for aqueous solutions containing acidic and basic copolymers,
where addition of simple salts reduced, as expected, energy transfer
efficiency. The result was interpreted as due to a decrease in the polymer association with a decrease of their Coulombic interaction (28).
More recently, it was reported (29) that a graft copolymer carrying a
donor fluorophore in the chain backbone and acceptor groups at the end
of the side chains exhibited sharply increased energy tranfer when the
polymer associated to micelles.

4. Characterization of chain entanglements in solution.

Some attempts have been made in the past to develop a spectroscopic method for the study of polymer chain entanglements in solution. Thus, the
increase in the ratio of the excimer and "monomer" fluorescence intensity with an increase in the concentration of polystyrene has been interpreted as due to chain entanglements (30). In solutions containing a
mixture of a polymer carrying fluorescent residues and a polymer to
which quenching groups were attached, quenching was reported to commence
sharply at a critical solution concentration, corresponding presumably
to the onset of chain interpenetration (31). However, it may be pointed
out that both intermolecular excimer formation and fluorescence quenching depend on collision of the interacting groups; since chain mobility
is necessarily restricted by chain entanglements, neither of the above
experiments lends itself to a quantitative characterization of the de-
pendance of chain entanglements on solution concentration.

In a procedure which we have developed (32) it is possible to obtain quantitative data on the extent of chain entanglement as a function
of solution concentration. The method depends on the reasonable assumption that the extent of chain entanglement will not change when a polymer solution is rapidly frozen. Thus, if the solution contained a mixture of similar polymers labeled with donor and acceptor groups, respectively, the reflectance emission spectrum of a pellet pressed from
the material obtained after freeze-drying will depend on the original
concentration of the solution from which the pellet was derived. If the
solution concentration was well below the range in which chain entanglements would be expected, sublimation of the solvent from the freeze-
dried solution will lead to the collapse of the polymer chains to isolated globules (as had been observed by electron microscopy (33, 34))
and energy transfer between separate globules containing donor or acceptor residues will be minimal. At the other extreme, when a film is cast
from a solution containing a mixture of the donor and acceptor labeled
polymers, the energy transfer observed will correspond to perfect mixing
of the two species. Taking these two values as reference points, the
energy transfer in any freeze-dried sample can be interpreted in terms
of the fraction of the chains which were entangled with other chains.

Our original study (32) used a polydisperse polymer and this limited the conclusions which could be derived from the data. More recently, we have applied the method to monodisperse polystyrene to which the
appropriate labels were attached after light chloromethylation (35). It
was of particular interest to see how the interpenetration of flexible
chain molecules depends on the solvent power of the medium. With decreasing solvation, the polymer coils shrink, which should delay chain
entanglements until higher solution concentrations are attained and the
higher segment density within the molecular coils should also make chain
interpenetration more difficult. On the other hand, the mutual attraction between chain segments increases with a decreasing solvent power of
the medium and this should favor entanglements. To my knowledge, no ex-

perimental data existed before our work which would have illuminated the relative importance of these factors. Olaj and Pelinka **(36)** carried out a computer simulation of two chains of 50 links in a theta solvent. They found that the probability of finding the centers of gravity of the two chains at a separation larger than four chain links was enhanced, while the probability of finding them closer together decreased rapidly. Thus, the zero excluded volume was the result of a compensation of these two factors. While it is plausible that this conclusion is also valid for real chains of much greater length, it is obviously desirable to sub-stantiate this experimentally.

Results obtained with our freeze-drying technique using monodis-perse polystyrene (M=410,000) dissolved in two good solvents (benzene and dioxane at 25°C) and in cyclohexane at 50°C (a poor solvent; the theta temperature is (37) 34.4°C) are shown on Figure 2. The critical

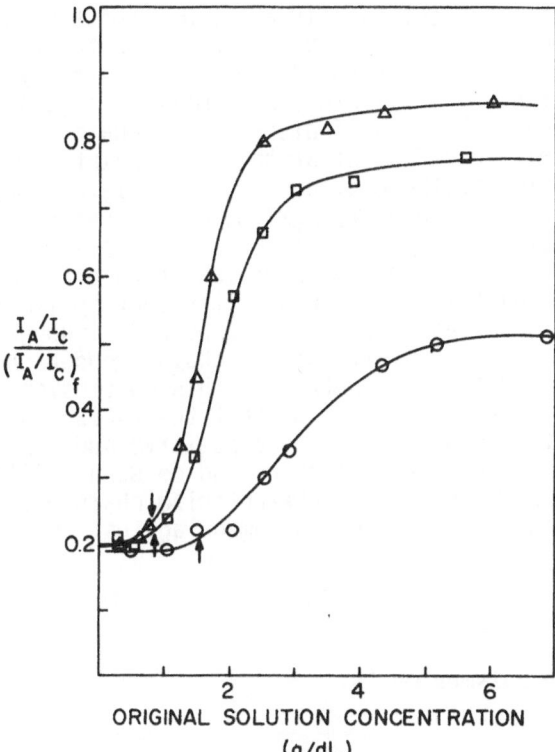

Figure 2. Ratio of anthracene and carbazole emission intensities (at 450 nm and 365 nm, respectively) from freeze-dried mixtures of polysty-renes (M = 410,000) carrying 0.0092 moles/kg of the labels, relative to this ratio of emission intensities of films cast from a mixture of these polymers. Excitation wavelength 294 nm. Original solution in benzene (**△**) dioxane (**□**) and cyclohexane (**○**). Arrows indicate $1/[\eta]$ in the various solvents.

concentration c* above which chain interpenetration changes the concentration dependence of various solution properties (e.g.,the osmotic pressure, the sedimentation coefficient) is of the order of m/s^3 where m is the mass of the chain molecule and s is the root-mean-square radius of gyration (5) or of the order of $1/[\eta]$. In Figure 2,$1/[\eta]$ is indicated for the three solvents investigated and it may be seen that it corresponds to the solution concentration at which our technique reveals incipient chain interpenetration. (Solutions in which the polymer concentration exceeds c* are referred to in the literature as "semidilute") The data lead, however, to another important insight: As the solution concentration is further increased, chain interpenetration apparently increases to level off at about $c = 3/[\eta]$ to a value far below that corresponding to random mixing obtained in films cast from mixtures of donor and acceptor labeled polymers.The location of the plateau decreases with a decreasing polymer solvation and for cyclohexane at 50°C (fairly close to the theta point) only about half of the polymer chains appear to be located in a volume occupied jointly with other chains.

In the model used by Flory and Krigbaum for a theory of the second virial coefficient,A_2,of flexible coil molecules (38) the polymer chain was represented by a cloud of disconnected chain segments and the vanishing second virial coefficient at the theta point corresponded to a free interpenetration of these clouds. Somewhat later, Orofino and Flory (39) wrote that "the quantity $A_2M/[\eta]$ may be regarded as a measure of non-interpenetration" implying that for $A_2 = 0$ the chains are completely interpenetrable. In Yamakawa's review of theories of the second virial coefficient (40)it is stated explicitly that "polymer molecules are completely interpenetrable at the theta point" although topological constraints due to the connectivity of the chains render this conclusion quite improbable. Our data also contradict this conclusion and support the suggestion of Olaj and Pelinka (36) that the vanishing second virial coefficient is due to a compensation of a positive and a negative part of the excluded volume integral. Bouchard and de Gennes (41) noted that considerable "self-knotting" should characterize chain molecules in theta solvents and this is a further argument against the concept of a "free interpenetration".

5. Self-diffusion in polymer melts.

According to generally accepted theories (42, 43) the self-diffusion coefficient D of polymer melts is inversely proportional to the square of the chain length. For long chains, this leads to extremely low values of D. Since the mean square displacement of a particle in a specified direction, $\langle x^2 \rangle$, in a time t is 2Dt, our ability to measure small D values is limited by the spatial resolution of the experimental technique. For instance, the first data on self-diffusion in a polymer melt were obtained following the diffusion of deuterated polyethylene into normal polyethylene by IR spectroscopy (44) and this technique could resolve only distances of 0.1 mm ; thus, experiments extending over a day could

characterize D values down to 10^{-9} cm^2-s^{-1}. This limited studies to
samples with molecular weights up to 23,000 at 176°C.

Obviously, it would be desirable to measure self-diffusion in much
longer polymer chains and to carry the studies to temperatures close to
the glass transition. We have shown (42) that significant data bearing
on this problem can be obtained by the following procedure: A dilute
solution containing a mixture of donor and acceptor labeled polymer is
rapidly frozen and the solvent is removed by sublimation so that the
chain molecules collapse into isolated globules (provided the polymer is
below its glass transition temperature). The resulting powder is pressed
into pellets and heated in an inert atmosphere for varying times to a
temperature above the glass transition. The reflectance fluorescence
spectrum at ambient temperature is followed as a function of heating
time and the increase in the ratio of acceptor and donor fluorescence
intensity is used to characterize the interdiffusion of the labeled polymer.

Figure 3 shows the results obtained with the use of this technique
and it can be seen that changes in the emission spectrum from labeled
poly(ethyl methacrylate) with a viscosity average molecular weight as
high as 500,000 could be observed in the temperature range of 120°C -
150°C over a time of less than 20 hours. These results demonstrate the
feasibility of the method,but in view of the steep dependence of the
self-diffusion coefficient on chain length, data obtained with a polydisperse polymer do not lend themselves to a reliable interpretation.
We are currently carrying out a similar study using monodisperse polystyrene to which fluorescent labels were attached after light chloromethylation.

Since donor and acceptor labeled polymer globules are randomly
distributed in the pellet,data obtained by this technique do not yield
a quantitative value for D. Nevertheless, its order of magnitude may be
estimated when it is assumed that transport over a distance comparable
to the radius r of the globule corresponds to half the change in the
ratio of acceptor and donor fluorescence intensities from its initial to
its equilibrium value, which takes place in a time $t_{1/2}$. Thus, with
$r = 6 \times 10^{-7}$ cm for a polymer with $M = 500,000$ and $t_{1/2} = 20,000$ s, we
may estimate $D \sim r^2/2t_{1/2} \sim 10^{-17}$ cm^2-s^{-1}. To appreciate the significance
of such a diffusion coefficient it should be pointed out that it corres-
ponds to a transport of 1 mm in 30 million years!

6. Concluding remarks.

In considering the significance of polymer compatibility it should be
pointed out that the concept of a "phase" in a system containing two or
more polymeric components is not as clearly defined as in the classical
chemistry of compounds of low molecular weight. If it is "a chemically
distinct, physically separable portion of the system", what are we to
make of a blend of two rather similar polymers where chains from each
"phase" penetrate to some depth into the other ? Is it reasonable to

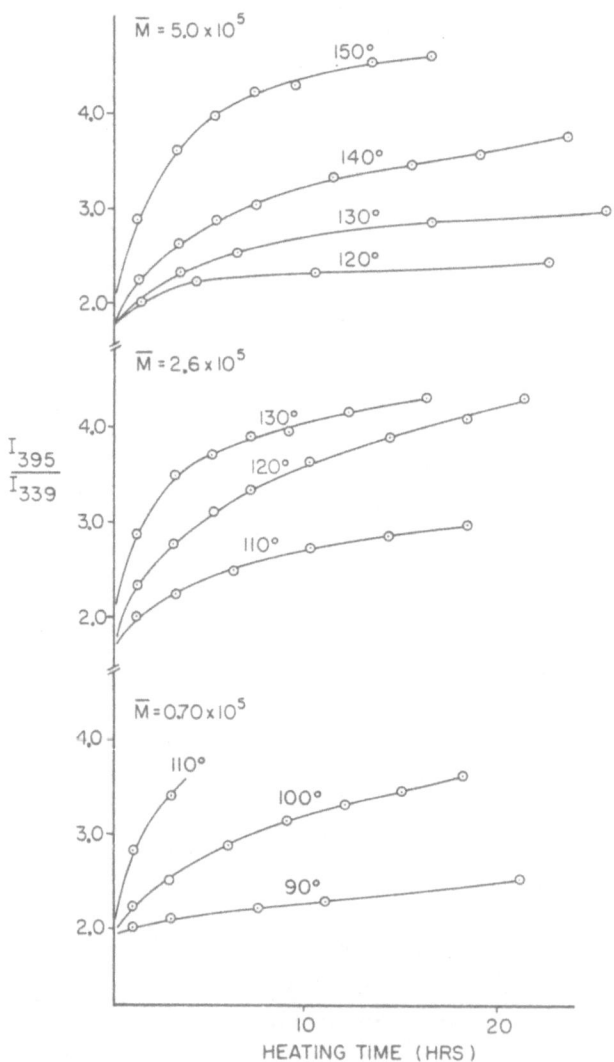

Figure 3. Dependence of the ratio of the emission intensities at 395 nm (the pyrene emission peak) and at 339 nm (the naphthalene emission peak) on heating time of pellets from freeze-dried mixtures of pyrene and naphthalene labeled poly(ethyl methacrylate).

make a clear-cut distinction between blends in which the composition is subject to increasing local fluctuations and systems with "fuzzy phase boundaries" ?

Another uncertainty in studies of polymer compatibility is due to the effect which the method of sample preparation may have on the result obtained. It is known that the properties of a film cast from a solution containing a mixture of two polymers may depend on the solvent used; in particular, if the solvent interacts much more strongly with one of the polymeric components, the solution may separate into two phases even though the two polymers are miscible with each other (43). This, no doubt, is the reason why polystyrene - poly(vinyl methyl ether) mixtures yield two-phase films when cast from chloroform, but single phase films when cast from toluene (44). For polymers blended in an extruder, it must be taken into account that compatible polymer pairs frequently exhibit phase separation at higher temperatures (9, 10); on the other hand, vigorous mechanical mixing may lead 'to quasi-homogeneous blends which may separate only very slowly into two phases.

As I have noted, the polymer concentration at which excimer emission begins to increase due to intermolecular polymer interaction (30) or where a polymer carrying quenching groups begins to reduce the emission from fluorescing groups attached to similar polymer chains (31) has been used previously to estimate the critical concentration c^* at which polymer coils begin to overlap. However, it should be pointed out that for $c > c^*$ these techniques are not suited for the estimation of the degree of chain interpenetration, since they depend on the collision of interacting groups and there is no way in which the reduced mobility of mutually entangled chains can be assessed. By contrast, the non-radiative energy transfer in freeze-dried samples does not depend on molecular motion. Also, because of the large decrease of volume, the freeze-dried sample requires a much smaller concentration of fluorescent labels on polymer chains to reveal their interpenetration than would be required if the emission spectrum were recorded from the solution. It is, of course, crucial to keep the label concentration low, so as to ensure that the labels do not modify significanly the properties of the polymers.

Although we have referred to the change in the emission spectrum of freeze-dried mixtures of donor and acceptor labeled polymers as characterizing self-diffusion, it should be pointed out that the "unfolding" of a globule containing a collapsed polymer chain to its equilibrium conformation is a very different process than the transport of the unfolded chain. Both processes are opposed by the viscosity of the medium and the conjecture that they are closely related seems reasonable, although data on their interrelationship are not available at this time.

A most powerful method for an unambiguous determination of self-diffusion coefficients in polymer melts , involving fluorescence recovery after photobleaching, has been developed in Sillescu's laboratory (45). Data were obtained for polystyrene up to M = 150,000 at 177°C, at which $D = 4 \times 10^{-13}$ $cm^2 \cdot s^{-1}$ was obtained. The high precision of the data permitted the authors to determine the effect of the labeling on the diffusion rate: Increasing the number of fluorescein labels from one to two on a polystyrene with M = 43,700 reduced D by about a third. It is obvious that our method is crude in comparison, but it recommends itself by its great experimental simplicity and the ability to study even slower processes.

References.

(1) Th. Förster, Disc. Faraday Soc.,27,7 (1959)
(2) G. Gee, Q. Rev. Chem. Soc., 1,265 (1947)
(3) O. Olabisi,L.M. Robeson and M.T. Shaw, "Polymer-Polymer Miscibility"
    Academic Press, New York, 1979
(4) D.R. Paul and S. Newman, Eds., "Polymer Blends", Academic Press,
    New York, 1978
(5) M. Daoud, J.P. Cotton, B. Farnoux, G. Janninck, G. Sarma, H. Benoit
    R. Duplessis, C. Picot and P.G. de Gennes, Macromolecules,8,804
    (1975)
(6) S. Krause, J. Macromol. Sci.,Rev. Macromol. Chem.,7,251 (1972)
(7) J.H. Hildebrand and R.L. Scott, "The Solubility of Nonelectrolytes",
    Am.Chem. Soc. Monograph No. 17, Reinhold,New York, 1950,Chapt. 7,8.
(8) P.J. Flory, J. Am. Chem. Soc.,86,1833 (1965)
(9) R.E. Bernstein, C.A. Cruz, D.R. Paul and W. Barlow, Macromolecules,
    10,681 (1977)
(10) T. Nishi and T.K. Kwei, Polymer ,16,285 (1975)
(11) J.R. Fried, F.E. Krause and W.J. McKnight, Macromolecules,11,150
    (1978)
(12) D.L. Handlin, W.J. McKnight and E.L. Thomas, Macromolecules,14,
    795 (1981)
(13) W.A. Kruse, R.G. Kirste, J. Haas, B.J. Schmidt and D.J. Stein,
    Makromol. Chem.,177,1145 (1976)
(14) M. Matsuo, S. Sagae and H. Asoi, Polymer,10,79 (1969)
(15) O. Olabisi, Macromolecules ,8,316 (1975)
(16) M.M. Coleman and J. Zarian, J. Polym. Sci.,Polym. Phys. Ed.,
    17,837 (1979)
(17) S.T. Wellinghoff, J.L. Koenig and E. Baer, J. Polym. Sci., Polym.
    Phys. Ed.,15,1913 (1977)
(18) F. Amrami, J.-M. Hung and H. Morawetz, Macromolecules,13,649 (1980)
(19) S.S. Voyutskii, A.D. Kaminskii and N.M. Fodiman, Kolloid-Z.,215,
    36, (1966)
(20) J. Letz, J. Polym. Sci.,Polym. Phys. Ed.,7,1987 (1969)
(21) E. Helfand and Y. Tagami, J. Polym. Sci.,B,9,741 (1971)
(22) H. Morawetz, Ann. N.Y. Acad. Sci.,366,404 (1981)
(23) D.R. Paul, private communication
(24) B. Albert, R. Jerome, Ph. Teyssie, G. Smyth, N.G. Boyle and V.J.
    McBrierty, Macromolecules,18,388 (1985).
(25) D.R. Paul, in D.R. Paul and S. Newman,Eds., "Polymer Blends",
    Academic Press, New York, 1978, Vol. 2, Chapter 12.
(26) J.W. Schurer, A. de Boer and G. Challa, Polymer,16,201 (1975)
(27) F. Mikeš, H. Morawetz and K.S. Dennis, Macromolecules,17,60 (1984)
(28) I. Nagata and H. Morawetz, Macromolecules,14,87 (1981)
(29) A. Watanabe and M. Matsuda, Macromolecules,18,273 (1985)
(30) (a) F. Nishihara and M. Kaneko, Makromol. Chem.,124,84 (1969);
    (b) J. Roots and B. Nystrom, Eur. Polym. J.,15,1127 (1979)
(31) Yu.E. Kirsh, N.R. Pavlova and V.A. Kabanov, Eur. Polym. J.,11,495
    (1975)
(32) J. Jachowicz and H. Morawetz, Macromolecules,15,828 (1982)

(33) M. Richardson, Proc. R. Soc. London,Sec.A,279,50 (1964)
(34) H. Reinlinger, Naturwiss. ,63,574 (1976)
(35) P. Chang, unpublished data
(36) O.F. Olaj and K.H. Pelinka, Makromol. Chem.,177,3413 (1976)
(37) Y. Miyake Y. Einaga and H, Fujita, Macromolecules,11,1180 (1978)
(38) P.J. Flory and W.R. Krigbaum, J. Chem. Phys.,18,1086 (1950)
(39) T.A. Orofino and P.J. Flory, J. Chem. Phys.,26, 1067 (1957)
(40) H. Yamakawa, "Modern Theory of Polymer Solutions",Harper and Row,
     San Francisco, 1971, p. 169.
(41) F. Brochard and P.-G. de Gennes, Macromolecules,10,1157 (1977)
(42) T. Y.-J. Shiah and H. Morawetz, Macromolecules,17,792 (1984)
(43) A. Robard and D.A. Patterson, Macromolecules,10,1021 (1977)
(44) M. Bank, J. Leffingwell and C. Thies, J. Polym. Sci.,A-2,10,1097
     (1972)
(45) M. Antonietti, J. Contandin, R. Grütter and H. Sillescu,
     Macromolecules,17,798 (1984)

ROTATIONAL DYAD STATISTICS AND ENERGY MIGRATION IN MISCIBLE AND IMMISCIBLE POLYSTYRENE/POLY(VINYL METHYL ETHER) BLENDS

Curtis W. Frank and Richard Gelles
Department of Chemical Engineering, Stanford University,
Stanford, California 94305

ABSTRACT. Excimer fluorescence is developed as a quantitative probe of isolated chain statistics and intermolecular segment density for miscible and immiscible blends of polystyrene (PS) with poly(vinyl methyl ether) (PVME). Rotational isomeric state calculations combined with a one-dimensional random walk model are used to explain the dependence of the excimer to monomer intensity ratio on PS molecular weight for 5% PS/PVME blends. A model for a three-dimensional random walk on a spatially periodic lattice is presented to explain the fluorescence of miscible PS/PVME blends at high concentrations. Finally, a simple two-phase morphological model is employed to analyze the early stages of phase separation kinetics.

1.0 INTRODUCTION

The growing role of polymer blends as engineering materials was outlined in an earlier chapter on Flory-Huggins thermodynamics and its application to excimer fluorescence studies. It was shown that this classical lattice theory gave quantitative explanation of changes in the excimer to monomer fluorescence intensity ratio that resulted from variation in several parameters expected to influence the Gibbs free energy of mixing. The objective of this chapter is to develop the photophysical analysis to the extent that the absolute value of the fluorescence observables may be interpreted in terms of the morphology of the blend.

The blend of interest is that formed between polystyrene (PS) and poly(vinyl methyl ether) (PVME). This has been examined by a variety of experimental methods including differential scanning calorimetry [1,2], dielectric relaxation [1,2], dilatometry [3], vapor diffusion [3], nuclear magnetic resonance [3,4], light transmission [4], and optical microscopy [4-9]. It has been shown that film casting PS/PVME from toluene or benzene leads to miscible blends [1-8]. Moreover, dielectric relaxation [1,2] and nuclear magnetic resonance studies [3,4] show that very small scale mixing is obtained. Using vapor diffusion, a negative interaction parameter has been measured [4], showing that enthalpic interactions between the two polymers favor miscibility.

M. A. Winnik (ed.), Photophysical and Photochemical Tools in Polymer Science, 561–588.

The advantages of the PS/PVME blend for photophysical studies are twofold. First, the low glass transition temperature of PVME (245 K) [1] ensures that a low concentration blend will be in the rubbery state over a wide temperature range. As a result, the conformational population of the PS chains may be assumed to be in equilibrium. Excimer formation due to rotational sampling should be unimportant provided the blends are not too high above T_g. Second, the attainment of random mixing by casting from toluene permits energy migration to be studied quantitatively in the limits of low and high PS concentration. The former allows modeling of the electronic excitation transport as a one dimensional random walk along an isolated PS chain while the latter permits three dimensional random walk models to be applied.

2.0 CONFIGURATIONAL STATISTICS OF POLYSTYRENE

The objective of this section is to demonstrate that intramolecular excimer fluorescence may be used in conjunction with rotational isomeric state theory to characterize aspects of the conformational structure of polystyrene. In Section 2.1 data on polystyrene and model compounds are analyzed to show the geometrical requirements for an excimer forming site (EFS) and to clarify the dissociation behavior. Rotational isomeric state theory is briefly described in Section 2.2 and then employed in Section 2.3 to calculate the EFS population for the isolated PS chain.

2.1 Photophysical Studies of Polystyrene and Related Model Compounds

Yanari and Bovey [10] were the first to report that the absorption spectra of atactic PS, isotactic PS, and copolymers of styrene and methyl methacrylate were almost identical to that of toluene. Vala [11] and Gargallo [12] later noted that ethylbenzene was a better single ring model compound for PS than toluene. Finally, Bokobza [13] found that isopropylbenzene gave an even better match than ethylbenzene. However, the racemic derivative of 2,4,6-triphenyl heptane had slightly higher molar extinction coefficients, which was attributed to the different local environments of the stereoregular dyads [14,15]. All these studies lead to the same conclusion: In PS and its model compounds, the only absorbing species is the single phenyl ring; the neighboring phenyl rings do not interact in the ground state.

The phenyl ring excimer was identified first by Ivanova in studies of benzene, toluene and p-xylene [16] and subsequently studied by others [12-19], but Yanari [10] was the first to identify excimer formation in polystyrene solutions and films. Shortly thereafter, Vala [11] observed that the ratio of excimer to monomer fluorescence was independent of PS concentration in dilute solution but increased after the PS content reached a certain value. Similar observations have been made in other studies [20-22]. The conclusion drawn from these experiments is that excimer formation may be intermolecular as well as intramolecular. This point has been utilized in the polymer blend research described in an earlier chapter.

A classic model compound study was performed by Hirayama [23] who examined compounds of the type $\phi-(CH_2)_n-\phi$, where n was varied from one to six. He found that excimer formation was only possible for n=3, showing that PS intramolecular excimer formation occurs between adjacent pendant phenyl rings. Other studies have shown that intramolecular excimer formation between non-adjacent rings is an unimportant process in PS. For example, in dilute solution no excimer emission is observed from head-to-head PS [24]. However, films of head-to-head PS do show excimer emission, proving that excimers can form intermolecularly in PS.

Förster showed in early work [25] that excimer formation in pyrene is only possible if the excited and ground state rings are close enough together to allow π orbital overlap. Since maximum overlap would occur in a sandwich type configuration, Hirayama concluded that the phenyl rings would also be in this geometry. Thus, the ring separation in the excited complex would have to be less than 3.2 Å, which is the thickness of the π orbital. He then noted that in the trans form of 1,3-diphenyl propane, the rings are 2.54 Å apart in the sandwich configuration, while in the case of the other 1,n-diphenyl alkanes, there are no stable conformations which would allow this geometry.

Additional information about the excimer geometry may be obtained from studies of the paracyclophanes in which two phenyl rings are bound through the para positions by methylene chains having n and m methylene units [11,26,27]. When n=m the most stable ground state conformation is one in which the two rings are in a parallel sandwich arrangement, while when n≠m, the rings are skewed. Cram has calculated ring separations for a number of these conformations [26]. For m=n=2 or 3, there is a distortion of the absorption spectrum, which is associated with existence of a ground state complex. In addition, only one fluorescence band at extremely long wavelengths is observed. This is expected since the rings are separated in the most stable conformation by distances of 1.54 Å and 2.52 Å, respectively. When n=m=4, there is no distortion of the absorption spectrum while the fluorescence spectrum is that of a typical phenyl ring excimer. Here, two ground state phenyl rings should be separated by 3.73 Å in the most stable conformation; if one ring is excited, only minimal molecular motion is required to bring about the excimer configuration. When n or m > 5, no excimer is observed; the fluorescence is characteristic of an isolated phenyl ring. Finally, when n=3, m=4 the separation between aromatic carbons on the two rings ranges from 2.84 Å to 3.41 Å. This compound shows intermediate characteristics with a distorted absorption spectrum, but a fluorescence spectrum more like the m=n=4 compounds than the other paracyclophanes which exhibit distorted absorption spectra. Morawetz and Wang [27] concluded from this that the perfect parallel sandwich arrangement might not be necessary in the excited complex.

It is also of interest to consider the effect of stereoregularity. Longworth [28] found ϕ_D/ϕ_M for isotactic PS to be 10.0
while atactic PS had a value of 2.3. Ishii [29] obtained a similar
result with the isotactic PS having a ratio of 15.4 and the atactic a
ratio of 4.6. Similar findings have been reported for PS model
compounds. Longworth and Bovey [30] were the first to note qualitatively that meso 2,4-diphenylpentane (DPP) had more excimer emission
than the racemic stereoisomer. Finally, Bokobza [13] found

$$\left(\frac{\phi_D}{\phi_M}\right)_{meso} \Big/ \left(\frac{\phi_D}{\phi_M}\right)_{racemic} \qquad \text{to vary from 7.2 for 2,4-DPP to 19.8 for}$$

2,4,6-triphenylheptane (TPH).

The photophysics of PS is greatly simplified because dissociation
of the excimer to excited monomer is unimportant over a wide temperature range. Hirayama [23] and Vala [11] noted that the presence of
dissolved oxygen quenched the excimer much more than the excited monomer. This can be explained if most of the excited monomer present
does not come from excimer dissociation. Moreover, for atactic PS in
solution, monomer fluorescence in the presence of a quencher obeys the
Stern-Volmer relationship over the temperature range 240 K –340 K
[31,32]. This is expected [32] provided the Birks kinetic scheme
holds and there is no thermal deactivation of the excimer to excited
monomer. However, this result is somewhat difficult to interpret
because Birks' scheme only works if all excited monomers are photophysically equivalent. This will not be true in general since
rotational sampling rates are different for rings in meso and racemic
dyads, leading to some complications in treating intramolecular
excimer formation.

For both atactic PS in solution [29] and the stereoregular
2,4-DPP's [33], the excimer fluorescence intensity remains constant,
then starts to decrease with temperature. This decrease starts at
about 300 K for the polymer and 330 K for the models. The monomer
fluorescence intensity, on the other hand, always decreases with
increasing temperature. Because no increase in monomer fluorescence
is observed when the excimer decreases, this has been interpreted as
showing that no dissociation to excited monomer occurs [29,33].

The issue has been finally settled by De Schryver and coworkers
[34] who have found that the monomer fluorescence decay of stereoregular 2,4-DPP's can be described by a single exponential over a wide
temperature range. Time resolved fluorescence studies of atactic PS
in dilute solution at room temperature have shown that at long times
only the excimer was present, while at short times there is both monomer and excimer [35]. This result is perhaps the strongest evidence
that no intramolecular excimer dissociation to excited monomer takes
place in atactic PS solutions at room temperature. Moreover, Gupta
and coworkers [36] have found recently that excimer dissociation to
excited monomer is also unimportant in pure PS films at room
temperature.

2.2 Rotational Isomeric State Theory

The rotational state of a given bond is defined in terms of interactions between chemical groups separated by three bonds. It is assumed that only the staggered conformations corresponding to local minima in energy-rotational space are allowed. The three possible conformations of the first internal bond in a dd meso dyad are shown in Figure 1. The Natta projections on the left have been drawn with the second pair of main chain bonds in the plane of the paper. The Newmann diagrams given on the right are obtained from the Natta projections by looking down the bond in question in the direction of the arrow.

The rotational states of the internal bond pair ϕ_i, ϕ_{i+1}, are defined relative to their states when all backbone bonds are in the same plane. For the tt meso dyad, $\phi_i = \phi_{i+1} = 0°$ when there is perfect staggering between groups separated by three bonds. Thus, $\phi_i = -120°$ and $+120°$ for the gt and $\bar{g}t$ dyads, respectively. However, it should be noted that, in general, higher order interactions cause the lowest energy state of a rotational dyad to deviate from perfect staggering of the groups separated by three bonds. Therefore, ϕ_i and ϕ_{i+1} are usually not exact multiples of $\pm120°$.

The experimental determination of dyad configuration in atactic PS by nuclear magnetic resonance (NMR) has proved difficult because of poor resolution [37]. However, much can be learned about PS dyad configuration from studies of stereochemical equilibria in the two- and three-ring model compounds DPP [37,39] and TPH [38]. These isomers have been separated and analyzed using gas chromatography by Flory and coworkers. The fraction of the meso form of DPP was found to be 0.48 [40]. Similarly, the probability that a dyad has the meso configuration, P_{meso}, in TPH was calculated to be 0.47 [41]. In the latter study, it was found that increasing the molecular weight decreases P_{meso}; P_{meso} for infinite molecular weight atactic PS was predicted to be between 0.40 and 0.43.

In all of these studies it was found that conformations involving $\bar{g}$ rotational states are present only in insignificant amounts. In addition, the tg (or gt, which is the same conformation by symmetry) state is strongly preferred for the meso dyad. At room temperature, less than 5% of the tt conformer of meso DPP was found [39,42], while gg states were present in even smaller amounts. In the all meso, or isotactic, PS, the tg conformation was also found to be strongly preferred [37]. Finally, two important conformations for the racemic dyad were found: the ground state tt and the gg conformation, which is only slightly higher in energy. These may be seen in Figure 2.

The conformational equilibria results obtained from model compounds can be used at least qualitatively to analyze the atactic polymer. However, in order to take into account the additional types of steric interactions resulting from both an increase in molecular weight and stereoirregularity, the rotational isomeric state (RIS) theory of conformational statistics is required. In this theory

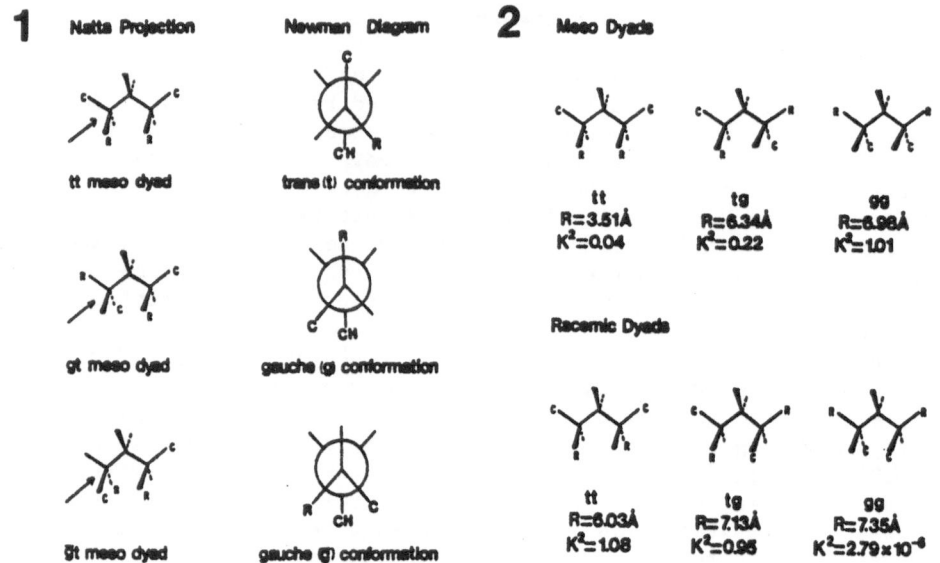

Figure 1. Natta projections and Newmann diagrams for a meso rotational dyad. In the Natta projection R refers to an aromatic ring and
C to one section of the remainder of the chain bonded to the dyad. In
the Newmann diagram CH corresponds to the right asymmetric carbon of
the Natta projection.

Figure 2. Natta projections of allowed meso and racemic dyads in the
two state model for the aryl vinyl polymers. The values listed for R
are the distances between the centers of mass of the substituent
phenyl rings and those for κ^2 are the orientation factors defined by
Equation 3.

[43,44], the rotational state of a backbone bond is considered to be a
discrete random variable distributed according to relative energy. It
is then assumed that the state of a given bond only depends on the
state of the previous bond, i.e., the collection of rotational states
along the chain backbone is a Markov chain [45] and matrix algebraic
techniques may be used for analysis. Once the rotational states of
the backbone bonds have been defined, statistical weights are assigned
to each conformation based on the type of interactions present. These
statistical weights are partition functions that are similar to the
transition probabilities used in the analysis of Markov chains, the
difference being that statistical weights are generally not normalized
to unity.
    In order to determine the averaged values of ϕ_i and ϕ_{i+1} for each
rotational dyad and the values of the statistical weights,

conformational energies as a function of the rotational states of the
two internal bonds are calculated taking all bonded and non-bonded
interactions into account. Ignoring solvent effects, Gorin and
Monnerie [46] determined the conformational energies for meso and
racemic DPP. However, their results do not agree with the
experimental findings discussed above in Section 2.1.

In a study of the conformational statistics of PS, Yoon,
Sundararajan and Flory (YSF) included solvent effects in the analysis
[47]. The influence of solvent on conformational energy is important
because the distance between chemical groups separated by four bonds
is large enough for solvent molecules to penetrate the space between
them and provide additional interactions. These results are in
complete agreement with the experimental studies of PS conformational
structure. For example, YSF found the $\bar{g}$ rotational states to be
insignificant due to severe steric interactions. In addition, the
inclusion of solvent effects led to the conclusion that the tt conformation is not the ground state of the meso dyad.

Finally, Stegan and Boyd [48] extended the PS and PS oligomer
analysis to include up to six bond interactions, and presented
numerical results from statistical weight calculations for all three
possible rotational states. Their results validated the neglect of $\bar{g}$
states in the determination of statistical weights by YSF. Since this
analysis does not significantly change the YSF results, the latter are
used here to reduce computation time. Thus, the only conformations
that are EFS in PS are the tt meso, $\bar{g}\bar{g}$ meso, and (t$\bar{g}$,$\bar{g}$t) racemic
dyads. However, since the $\bar{g}$ state is so high in energy that it is
insignificant in atactic PS of reasonable meso content, it may be
assumed that the only intramolecular EFS site is the tt meso dyad,
shown in Figure 2.

2.3 Calculation of Excimer Forming Site Concentration in Polystyrene

The intramolecular EFS concentration of atactic PS may be determined using RIS theory. Before proceeding with the calculation,
however, the computer code for the simulation was verified in a study
of end effects in stereoregular isotactic PS and 300 K. The results,
which give the dependence of the fraction of tt conformations on chain
length, are shown in Figure 3. The asymptotic value of the tt concentration determined by the computer study agrees with the analytical
result that P_{tt} = 0.134 for an infinite molecular weight isotactic PS
chain at 300 K.

For an atactic chain the fraction of tt meso dyads must be determined by a Monte Carlo simulation because the conformation of a given
dyad will depend on the configuration of neighboring dyads [49]. In
the calculation [50] the fraction of meso dyads and the molecular
weight were first fixed. The meso content of atactic PS was taken to
be 45%, a value consistent with the results from nuclear magnetic
resonance and stereochemical equilibria studies of PS and its oli-
gomers [40,41]. An ensemble of chains of the desired molecular weight
and tacticity was then generated randomly. The size of the ensemble

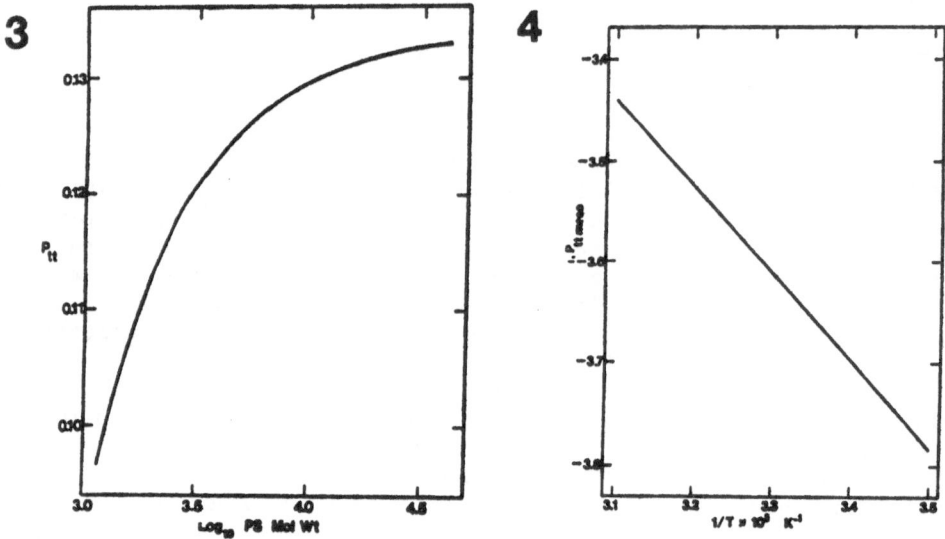

Figure 3. Dependence of the fraction of tt rotational dyads on molecular weight for polystyrene.

Figure 4. Dependence of the tt meso dyad concentration on temperature for infinite length atactic polystyrene chains containing 45% meso units.

needed to obtain reproducibility in $P_{tt,meso}$ to three significant figures decreased from 5,000 to less than 3,000 chains as the molecular weight increased from 2,200 to 17,500. For each chain in the ensemble, the probability that each dyad was in a tt meso state was calculated. Averaging over the ensemble gave the fraction of tt meso dyads, which is the concentration of dyads in which neighboring phenyl rings are in an EFS.

The results of the simulation are presented in Figure 4, which shows the temperature dependence of the tt meso dyad concentration for infinite length atactic chains. The Monte Carlo calculation shows that the presence of racemic dyads lowers the probability that a meso dyad is in the tt conformation. For example, at 300 K, a simulated atactic chain with 45% meso dyads has a tt meso concentration of about 2.6%. On the other hand, simple multiplication of the fraction of meso dyads by the tt dyad population for an isotactic chain yields a value of about 6%. For chains with 45% meso dyads, there was little dependence on molecular weight over the temperature and molecular weight range examined.

On the other hand, $P_{tt,meso}$ was found to be strongly temperature dependent, as shown by the Arrhenius-like temperature behavior of the curve in Figure 4. The effective activation energy of the tt conformation is 1.7 kcal/mole. It is of interest to note that while the presence of racemic dyads affects the conformational distribution of meso dyads, it has no significant influence on the effective activation energy of the tt meso state. For an isotactic, or all meso, chain, the effective activation energy of the tt state equals the equilibrium separation between the potential energy minima of the preferred tg and the tt conformation. This is 1.8 kcal/mole, very similar to that found for the atactic simulation.

3.0 SINGLET ENERGY MIGRATION IN POLYSTYRENE

The singlet monomer excitation in an aryl vinyl polymer such as PS is generally believed to have the ability to migrate between rings by a series of hops [36,51-53]. Recently, however, this has been called into question [54]. Unfortunately, the existence of singlet energy migration in PS cannot be proven directly, as is the case for triplet migration [52-53]. The objective of this section is to examine the experimental evidence related to energy migration. Transient, quenching and polarization measurements are considered, and it is concluded that explanation of the photophysical behavior of polystyrene must include energy migration.

3.1 Quenching Measurements

MacCallum and Rudkin [55] have recently studied the quenching of monomer fluorescence in styrene-methyl methacrylate copolymers of low styrene content, and the quenching of excimer fluorescence in PS. They analyzed their data with kinetics based on the Birks scheme and found that the quenching rate constant for the PS excimer, k_{QD}, was about the same as that for the copolymer monomer, k_{QM}. They concluded that this shows there is no energy migration in PS by the following reasoning: In the copolymers, which have low styrene content, there should be no migration and no excimer formation since all rings are isolated between the comonomer units. Thus, if there is energy migration in PS, k_{QD} for PS should be much larger than k_{QM} for the copolymer.

This premise is incorrect because k_{QM} simply describes the quenching of an isolated excited phenyl ring while k_{QD} always describes the direct quenching of the isolated excited complex. The possible contribution to excimer quenching caused by monomer quenching before the excitation can reach an EFS is already included in k_{QM}. Thus, k_{QM} for the copolymer monomer should be approximately equal to k_{QD} for the PS excimer. MacCallum and Rudkin's conclusion shows that it is often easy to misinterpret quenching experiments.

The enhanced quenching of excimer over monomer fluorescence in homopolymers has also been used to support the claim that singlet energy migration does take place [29,31,32,56], since excimer

formation and quenching are competing processes. If excimer formation
takes place mainly by energy migration to EFS, then the decrease in
excimer fluorescence would result from both direct quenching of the
excimer and quenching of the monomer.

In addition, there have been a number of studies of the steady
state fluorescence properties of styrene random copolymers in dilute
solution as a function of copolymer composition [57-60]. In these
studies, the conclusion that energy migration is important in PS has
generally been reached based on correlations between the ratio of
excimer-to-monomer fluorescence and various parameters that describe
copolymer composition.

One qualitative observation from the copolymer work is of
particular interest here. Hill [60] noted that the fluorescence ratio
is larger for acrylonitrile-styrene than when the comonomer is methyl
methacrylate. They offered two possible explanations. The first was
that, due to its large size, a methyl methacrylate residue presents a
partial barrier to energy migration. The other was that the conformational statistics of styrene dyads could be affected by the
comonomer. The former reason does seem reasonable and could possibly
explain why pure films of alternating copolymers of styrene and methyl
methacrylate do not show any excimer emission [61]. In these films,
the comonomer could also prevent intermolecular excimer formation,
which is the only type of excimer formation possible in the system.

Energy transfer studies, which are similar to quenching studies,
have also led workers to the conclusion that singlet energy migration
takes place in PS. Energy transfer refers to the process by which
singlet excitation is transferred non-radiatively by Förster transfer
to an impurity or dopant that can fluoresce. The dopant or impurity
may be either part of the polymer chain or dispersed in the system.
It is generally found in these experiments that in the presence of the
dopant or impurity, the dopant fluorescence is much stronger than
would be expected in the absence of migration [24,62-64].

3.2 Fluorescence Depolarization Measurements

The experimental method that has sparked the most controversy
with regard to energy migration in PS is fluorescence
depolarization. If a single ring with parallel emission and absorption dipoles is excited by light plane polarized an angle θ away from
the dipoles, then the observed fluorescence emission measured perpendicular and parallel to the direction of the plane polarized light will
depend on θ. If the ring does not move during the lifetime of the
excitation, the degree of polarization P will vary between -1 and +1
where

$$P = \frac{I_{\parallel} - I_{\perp}}{I_{\parallel} + I_{\perp}} \tag{1}$$

Here, $I_{\parallel}$ and $I_{\perp}$ are the fluorescence intensities measured parallel
and perpendicular to the direction of the exciting light.

Ghiggino and coworkers [65] have presented a very good review of both steady state and time-resolved depolarization measurements. They have pointed out that the emission anisotropy of a sample, r, is more useful than the degree of depolarization. Here

$$r = \frac{I_{\parallel} - I_{\perp}}{I_{\parallel} + 2I_{\perp}}$$ (2)

However, in studies discussed in the present chapter, only depolarization values have been reported. Provided that the excitation does not migrate between molecules and that no molecular motion takes place during the lifetime of the excitation, the depolarization of a perfectly random array of absorbing species equals 2 [65].

Studies of depolarization in solution are difficult to interpret because molecular motion is expected to cause depolarization. However, Reid and Soutar [59] have found that depolarization increases with increasing styrene content for styrene/methyl methacrylate copolymers in dilute solution. They found that the amount of depolarization correlated well with the mean sequence length of styrene residues and interpreted this as resulting from energy migration which occurs between neighboring phenyl rings in sequences separated by the comonomer. They measured a depolarization of over 100 for copolymers with a 94% styrene content. MacCallum [66], on the other hand, has reported that PS in solution shows no fluorescence depolarization. This is a rather surprising result, even if there were no energy migration in PS, because of the rapid motion that can take place in this system.

The results of studies of PS copolymers dispersed at low concentration in a rigid glass are similar to those obtained in fluid solution. For both methyl methacrylate and methacrylate comonomers, David [67] found that depolarization increased with increasing styrene content and reached values close to those found in fluid solution when the styrene content was large. David also examined these methacrylate and methyl methacrylate copolymers in pure film form [68]. Here they found that the fluorescence depolarization was greater than 50 when the styrene content was above 15% for methacrylate films and above 25% for the methyl methacrylate films. They concluded that energy migration between rings on different chains was very efficient in pure films.

MacCallum and Rudkin [69] once again obtained contrasting results in a study of the depolarization of PS films. They reported a measured value of 18 and concluded that energy migration does not occur in PS. Their results disagree with those of a recent study by Gupta [36], in which essentially complete depolarization of emission was found for atactic PS in both dilute solution and pure film form. It is of interest to note that in the latter work it was observed that the depolarization of pure PS decreases upon photodegradation. This could possibly explain the results of MacCallum.

From the above discussion, it appears that, except for the
results of MacCallum, fluorescence depolarization studies show that
energy migration takes place in PS. For the copolymers in glassy and
fluid dilute solution, the increase in depolarization with styrene
content could be explained in terms of the increase in the size of
blocks of consecutive styrene units and the fact that energy migration
occurs mainly between rings in the same block. The large depolarization obtained in copolymer films and pure PS may be interpreted in
terms of efficient migration between rings on different chains.

4.0 PHOTOPHYSICAL ANALYSIS OF POLYSTYRENE/ POLY (VINYL METHYL ETHER) BLENDS

In view of the strength of the preceding arguments it will be
assumed that singlet energy migration does take place in PS. Thus,
the process of excimer formation in PS can be viewed in terms of a
random walk of the excitation among the aromatic rings. At any step
in the walk, the excitation can be lost radiatively or non-radiatively
from the isolated excited ring, or monomer. Alternately, the excitation can be trapped at an EFS, leading to radiative or non-radiative
decay through the excimer.

The objective of this section is to present a quantitative
analysis of the photostationary state fluorescence of miscible and
immiscible PS/PVME blends. In Section 4.1 we develop the one-
dimensional random walk model that is used in conjunction with rotational isomeric state calculations to analyze low concentration miscible blends. In Section 4.2 we treat miscible blends having high PS
concentration using a spatially periodic three-dimensional random walk
model. Finally, in Section 4.3 we present a simple two phase morphological model and demonstrate how it may be used to monitor phase
separation kinetics.

4.1 One-dimensional Energy Migration in Miscible Blends

In this section we will consider the sensitivity of the energy
hopping rate to the distance of separation and orientation of the
phenyl rings. We then follow with the application of Levinson's one-
dimensional random walk model [70] suitably modified to account for
chain end effects to explain the dependence of I_D/I_M on PS molecular
weight for 5% PS/PVME blends. Finally, we present a critique of the
strictly one dimensional random walk model and note that it is only a
first approximation for the aryl vinyl polymers in which some cross-
loop energy migration is almost always possible.

4.1.1 Adjacent Phenyl Ring Separation and Relative Orientation.
Information about the separation and relative orientation
of adjacent phenyl rings is needed to understand the photophysics of
isolated PS chains. The absorption and emission dipoles are parallel
to the plane of the ring and perpendicular to the ring axis bisecting

the substituent [71]. A convenient way of representing relative ring orientation is with the orientation factor κ^2

$$\kappa^2 = (\cos \theta_T - 3 \cos \theta_D \cos \theta_A)^2 \tag{3}$$

where θ_T is the angle between the dipoles, while θ_D and θ_A are the angles between the vector connecting the centers of mass of the two rings and the emission and absorption dipoles, respectively. The calculated values of κ^2 and the distance between the centers of mass of the two rings, R, are given in Figure 2. These distances and orientations were calculated using the actual bond angles and distances determined for polystyrene [47]. Deviations from perfect staggering in the conformational angles ϕ_i and ϕ_{i+1} strongly affect κ^2. For example, when the tt meso dyad has $\phi_i = \phi_{i+1} = 0°$ then $\kappa^2 = 1$. When the gg racemic dyad has $\phi_i = -120° = -\phi_{i+1}$ then $\kappa^2 = 0.17$.

The rate constant for a single hop between two aromatic rings, W, can be related to the monomer lifetime, τ, through the Förster expression [72]

$$W\tau = \frac{3}{2} \kappa^2 \left(\frac{R_0}{R}\right)^6 \tag{4}$$

where R_0 is the Förster radius and $\tau = k_M^{-1}$. R_0 depends on the specific chemical groups under consideration and can be evaluated from spectroscopic data [73]. Its value for PS should be approximately equal to that for transfer between isopropylbenzene molecules, which has been calculated to be 6.47 Å [73].

In dilute systems where the vinyl aromatic polymer chains are isolated, energy migration should be primarily an intramolecular process. Fitzgibbon and Frank have studied intramolecular energy migration in low molecular weight poly(2-vinyl naphthalene) by Monte Carlo simulation on chains of 60 repeat units or less [74]. They found that approximately 97% of the transfers take place between adjacent pendant rings. In poly(2-vinyl naphthalene) the Förster radius is 11.75 Å [73], considerably larger than for PS. Thus, one would expect that in low molecular weight PS, even a larger percentage of the transfers would take place between nearest neighbors.

4.1.2 The Strictly One Dimensional Random Walk Model. The energy migration model for isolated chains was developed by Levinson [70] for infinite length chains and modified for the case of finite length chains by Fitzgibbon and Frank [74]. The following assumptions are made in the model: 1) The starting location of the excitation, or exciton, is completely random, i.e., all rings have the same absorption properties. 2) The exciton can only transfer between adjacent rings. 3) The probability that monomer decay takes place rather than transfer at a given ring, α, is constant along the chain. 4) EFS are distributed randomly with the probability that a ring pair is an EFS equal to q. 5) The excimer does not dissociate to excited monomer, i.e., EFS are traps of the excitation.

Because transfer only takes place between adjacent rings, chain ends are viewed as reflecting sites. An exciton on an end ring that is not in an EFS can either decay or be reflected back. On the other hand, rings in EFS act as absorbing sites since they are traps. Because the exciton cannot hop past chain ends and traps, the two types of sites serve as boundaries that subdivide the polymer into a series of sequences of consecutive non-trap rings. There are three different sequence types. One has chain ends at both ends, the second has a chain end and an EFS, while the third is bounded at both ends by EFS.

The approach used to determine $M(\alpha,q,N)$, the probability that an absorbed photon on a chain of N rings decays through the monomer, is to solve the problem for a specific starting location on a specific sequence type of given length, and then to average over all possibilities. Thus,

$$M(\alpha,q,N) = \sum_{i=1}^{3} \sum_{m} F_i(m) \, G_i(m,N) \tag{5}$$

where

$$F_i(m) = \frac{1}{m} \sum_{k=1}^{m} F_i(m,k) \tag{6}$$

Here, i stands for the sequence type and m the sequence length, which is the number of consecutive non-trap rings in the sequence. $F_i(m,k)$ is the conditional probability of eventual monomer decay given the exciton is absorbed at the kth ring from the start of a sequence of m consecutive non-trap rings with type i boundaries. Since there is equal probability that absorption occurs at any ring of the sequence, $F_i(m)$ is simply the averaged conditional probability of eventual monomer decay given the starting location is in a sequence of length m, type i. Finally, $G_i(m,N)$ is the probability that absorption takes place by a ring in a sequence of length m, type i on a chain of N rings.

The probability $F_i(m,k)$ can be determined with a finite difference equation since the random walk obeys the Markov property. In other words, if transfer takes place to ring k+1 from ring k, the probability of eventual monomer decay is the same as if ring k+1 were the point of initial absorption. Thus,

$$F_i(m,k) = \alpha + \frac{1}{2}(1-\alpha)[F_i(m,k-1) + F_i(m,k+1)] \tag{7}$$

The solution to Equation (7) is [70]

$$F_i(m,k) = 1 + C_1 \exp(2\sigma k) + C_2 \exp(-2\sigma k) \tag{8}$$

where

$$\sigma = \frac{1}{2} \ln \left[\frac{1 + \sqrt{2\alpha - \alpha^2}}{1 - \alpha} \right] \tag{9}$$

and C_1 and C_2 depend on the sequence type i and length m. The determination of C_1 and C_2 is left to the original reference [74]. Since each ring has two nearest neighbors, α can be related to the rate constants W and τ

$$\frac{1}{\alpha} = 1 + 2 \, W\tau \tag{10}$$

Equation (4) shows that $W\tau$ depends on the relative orientation and separation of the two rings. As shown in Section 2, different conformations result in different orientations and distances between adjacent aromatic rings. Thus, α is not constant at any single step in the walk but should be viewed as an averaged probability. We will return to this point later. It is possible to assume, however, that α is independent of molecular weight. This is because the concentrations of the various chain conformations depend little on molecular weight, as has been shown to be the case for the EFS dyad. Finally, Fitzgibbon and Frank [74] have found that Q_D/Q_M is also independent of molecular weight for poly(2-vinylnaphthalene) in fluid solution. It seems reasonable to assume that this is also true for PS dispersed in PVME. Thus, for a given temperature, the model has three parameters: α, q, and Q_D/Q_M.

4.1.3 Effect of PS Molecular Weight. The fluorescence results for 5% PS/PVME blends are shown in Figure 5. The ratio of excimer to monomer fluorescence intensities is given as a function of temperature for monodisperse PS molecular weights ranging from 2,200 to 390,000. The PVME was a polydisperse sample having $M_v \sim 47,000$.

Data at three temperatures are replotted in Figure 6 along with smooth curves showing the results of the fit to the 1-D model. Values of q, the trap concentration obtained from Monte Carlo simulations; Q_D/Q_M, the ratio of intrinsic quantum yields; and α, the emission probability parameter, for these three temperatures are as follows:

$$286 \text{ K} : 0.0227, \ 0.36, \ 6.01 \times 10^{-4};$$
$$308 \text{ K} : 0.0282, \ 0.53, \ 9.12 \times 10^{-4};$$
$$323 \text{ K} : 0.0321, \ 0.87, \ 1.36 \times 10^{-3};$$

where the data are listed as (T: q, Q_D/Q_M, α).

Qualitatively, the observed molecular weight dependence of the fluorescence ratio can only be explained by resorting to a photophysical model that includes energy migration. It has been previously postulated that the molecular weight effect in fluid solution derives from differences in the effectiveness of rotational sampling for bonds near the ends and in the center of a chain [75-77], but the explanation cannot be used for the PS/PVME blend because of its high viscosity. It has also been proposed [75] that the trap concentration

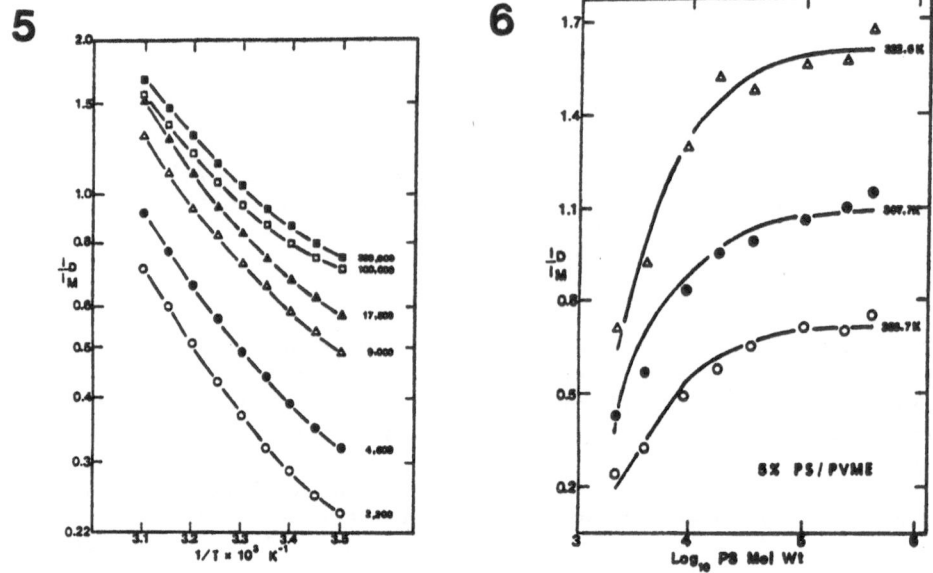

Figure 5. Temperature dependence of the photostationary excimer to monomer intensity ratio for 5% PS/PVME blends. The PS molecular weights are shown by each curve. Taken from Figure 1 of reference 50.

Figure 6. Comparison of experimental fluorescence results with the best fit of the one-dimensional random walk model for 5% PS/PVME blends. The temperatures of the blends are shown by each curve. Taken from Figure 4 of reference 50.

increases rapidly with PS molecular weight, but conformational calculations presented in Section 2 show that the trap concentration depends little on chain length.

This work demonstrates that the molecular weight dependence of the fluorescence ratio can be explained in terms of a one-dimensional random walk of the excitation. Due to the low trap concentration, very short chains may have no EFS at all. An increase in molecular weight allows the migrating excitation to "sample" more distinct rings so that the probability of finding an EFS trap will increase. Thus, I_D/I_M is expected to increase with molecular weight. This also explains why shorter chains exhibit a stronger temperature dependence than larger chains. Increasing the temperature will increase the trap concentration because the tt conformation is higher in energy than other meso dyads. Since shorter chains have less available rings for sampling, they should show a stronger temperature dependence, as has been observed here.

4.1.4   Critique of One-Dimensional Model. While the strictly
one-dimensional random walk model predicts that I_D/I_M should level off
at large chain lengths, the experimental data in Figure 6 show that
there is a persistence of the molecular weight effect at all
temperatures. Because the finite lifetime of the excitation limits
the number of Förster transfers that can be made, asymptotic behavior
at high molecular weights would be expected regardless of the dimensionality of the walk, as long as it is constant. However, the
observed behavior at high molecular weight may be explained by an
increase in dimensionality, which makes the EFS trapping process more
efficient since less rings are resampled. For example, the number of
distinct sites visited in n steps when there is no emission or
trapping is $O(n^{1/2})$ for a one-dimensional walk and $O(n)$ for a three-
dimensional walk [78].

A more quantitative argument that the walk is not strictly one-
dimensional over all or part of the molecular weight range can be made
by a comparison of the predicted value of the rate of one-dimensional
transfer with that obtained from the fluorescence data. In Figure 2
calculated values of ring separation R and orientation factor κ^2 have
been presented for the allowed conformational dyads of PS. From these
values and the Förster expression given in Equation 2, an order of
magnitude estimate of the rate of transfer relative to the rate of
monomer decay can be made for the case of strictly one-dimensional
hopping. The results of this calculation are shown in Table 1. As
noted in Section 2 the majority of non-EFS are either tg meso or tt
racemic dyads. Thus, an averaged value of $W\tau$ should be order one or
less.

TABLE 1

Predicted Dyad Single Hop Energy Migration Rates

Dyad	$W\tau$
tg meso	0.37
gg meso	0.96
tt racemic	2.47
tg racemic	0.80
gg racemic	1.95×10^{-6}

From Equation 3 and the values of α it is found that fits of the
one-dimensional random walk model to data yield $W\tau$ equal to 660, two
orders of magnitude larger than predicted. It is apparent that the
random walk is more efficient than if it were strictly one-
dimensional. It must be stressed, however, that any deviation from
the most probable conformational angles, as well as small torsional
motions of a phenyl ring about the bond connecting the ring to the
main chain, can change the orientation factor κ^2 significantly.
Nevertheless, since κ^2 can be at most four, $W\tau$ for a single hop is
still predicted to be order one or less. Thus, although the molecular

weight dependence of the fluorescence ratio can be fit reasonably well
with the strictly one-dimensional random walk model, there is strong
evidence that singlet energy migration in this system has at least
some three-dimensional character.

4.2 Three-Dimensional Energy Migration in Miscible Blends

At concentrations just above the dilute-semidilute transition for
an aryl vinyl polymer energy migration should be predominantly intramolecular with occasional hopping between chains when the excitation
reaches a ring which is close to the exterior of a random coil. As
the concentration is increased further, however, the situation becomes
simpler. When rings on different chains are as close as adjacent ring
pairs on a single chain, the rate of Förster transfer off chain will
be as fast as the down-chain rate. At this point, the presence of
interconnected chain segments is unimportant and the random walk
should be three-dimensional. In addition, formation of intermolecular
EFS should be facile.

In this section we examine the effect of PS concentration on the
fluorescence of miscible PS/PVME blends cast from toluene. We then
show how fluorescence data for miscible blends may be used to establish a "calibration" curve for M, the probability of eventual non-
radiative or radiative decay of the excitation by monomer fluorescence. Finally, we present a spatially periodic lattice model that
leads to reasonable predictions for Wτ for the three dimensional
energy migration process.

4.2.1   Spatially Periodic Lattice Model. The dependence of
I_D/I_M on PS volume fraction of 10μ thick films of PS/PVME blends cast
from toluene with PS molecular weights of 4000 and 100,000 is shown in
Figure 7 [79]. Figure 7 also shows the results of phase-separated
blends cast from tetrahydrofuran made with PS of molecular weight
100,000, to be considered in Section 4.3.

It is informative to note the general relationship between the
experimental observable I_D/I_M and M [74].

$$\frac{I_D}{I_M} = \frac{Q_D}{Q_M} \left[\frac{1}{M(\alpha,q)} -1 \right] \tag{11}$$

Both the one-dimensional random walk model described in Section 4.1
and the three dimensional model to be considered here are attempts to
provide analytical predictions of M and, hence, for the experimental
I_D/I_M results. We wish to emphasize, however, that it is possible to
obtain an experimental measurement of M through straightforward use of
Equation (11) and data for a miscible mixture [80]. This is shown in
Figure 8 for the 100 K PS/PVME blend cast from toluene.

In order to obtain an analytical expression for M in concentrated
PS/PVME blends, we utilize random walk theory. An interesting
parameter calculated by Montroll and reviewed by Barber [78], is the
number of distinct sites per step visited in n steps as n approaches

infinity. This calculation is for three dimensional hopping between neighboring sites on a periodic lattice without emission or trapping. The results given in Table 2 show that as the number of nearest neighbors increases, the probability that the walker returns to a site that was previously sampled decreases.

In a recent computer simulation Zumofen and Blumen treated the situation where hopping to all possible sites occurs with a distance dependence corresponding to the Förster mechanism and for the case of no emission or trapping [81]. Their results are shown in Table 3. It is apparent from a comparison of Tables 2 and 3 that when hopping is not limited to nearest neighbors, the probability that a site is resampled decreases. In addition, there is little site resampling in short walks.

In the spatially periodic model we incorporate the general features of these random walk analyses and assume that there is no ring resampling during the random walk of the excitation in PS/PVME

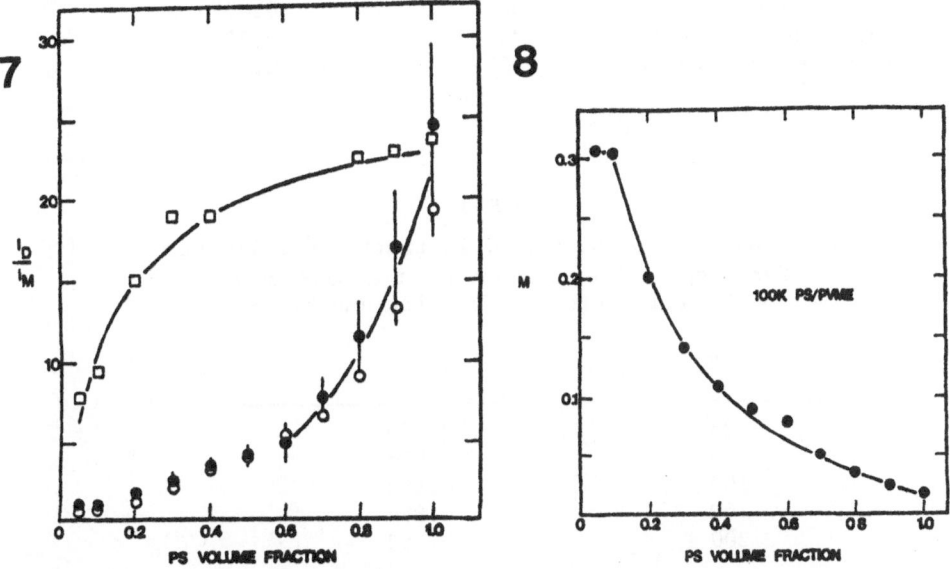

Figure 7. Concentration dependence of I_D/I_M for miscible PS/PVME blends cast from toluene (circles) and immiscible blends cast from tetrahydrofuran (squares). The solid line through the toluene cast film data is the best fit of the spatially periodic model, while the solid line through the THF results is the best fit of the two phase model. Taken from Figure 1 of reference 79.

Figure 8. Concentration dependence of the probability that an absorbed photon decays through radiative or nonradiative monomer processes for PS(100,000)/PVME blend. Taken from Figure 6 of reference 80.

blends. For blends with moderate PS concentrations, the number of
distinct sites visited per step will be close to unity, provided the
walk is short. Because the effect of EFS traps and monomer emission
will be to produce walks with a small number of steps, this simplifying assumption is reasonable at low concentration and should work
even better at higher concentrations. As a result, each step in the
random walk becomes independent and the probabilities of monomer decay
and trapping do not change at each step.

TABLE 2

The Number of Distinct Sites Visited Per Step at the Stationary
Distribution for Three-Dimensional Walks on Periodic
Lattices Without Emission and Traps

Lattice Type	Number of Nearest Neighbors	$\lim_{n \to \infty} S_n/n$
simple cubic	6	0.629
body-centered cubic	8	0.718
face-centered cubic	12	0.743

TABLE 3

The Dependence of Number of Distinct Sites Visited
Per Step on Step Number n for a Walk on a
FCC Lattice Without Emission and Traps

n	S_n/n
25	0.936
50	0.906
100	0.863
150	0.854
200	0.852

It is further assumed that if transfer to a ring in an EFS takes
place, the excitation will be trapped and will be lost from the system
through radiative or non-radiative decay from the excimer. Also, α
should be considered to be the expected value of the fraction of times
that monomer emission occurs before transfer. It may then be shown
[79] that

$$\frac{I_D}{I_M} = \frac{Q_D}{Q_M} \left[\frac{q}{1-q}\right] \frac{1}{\alpha} \tag{12}$$

To fit the concentration dependence of I_D/I_M, expressions for q and α are derived in terms of the PS volume fraction ϕ. A simple lattice approach is employed in which the size of the lattice site is taken equal to the size of the PS repeat unit; the PVME segments are then broken up to fit on the same lattice. Under the lattice assumption the separation between rings on adjacent lattice sites will be constant, although the number of rings next to a given ring will depend on concentration. It is assumed that transfer can only occur between rings that are nearest neighbors, that the rate of transfer between two neighbors is constant with composition, and that the sum of the rates of transfer to each of the nearest neighbors equals the net rate of transfer from a given ring. It is also assumed that the probability that a ring is in an EFS is the sum of the probabilities that a ring forms an excimer with each of the rings on nearest-neighbor sites.

The result of this analysis is that

$$
\frac{I_D}{I_M} = \frac{Q_D}{Q_M} \left\{ \frac{q_{intra} + (N-2)\zeta\phi}{1 - [q_{intra} + (N-2)\zeta\phi]} \right\} \tag{13}
$$

where ϕ is the volume fraction of polystyrene, q_{intra} is approximately twice the probability that a given dyad is a suitable excimer forming site, N is the coordination number and ζ is the probability that two rings occupying adjacent lattice sites will be in the appropriate geometry for excimer formation. Q_D/Q_M was found equal to 0.42 at room temperature from the low concentration PS/PVME blend study described in Section 4.1 [50].

This random walk model should only apply over the concentration range where rings on different chains are at least as close as adjacent ring pairs on the same chain. In Section 2 it was found that adjacent intramolecular ring pairs that are not EFS are between six and seven Å apart. If the intermolecular ring separation is assumed to be approximately equal to (1/ring concentration)$^{1/3}$, then the concentration at which this distance equals the average intramolecular separation of about 6.5 Å can be determined. This estimation gives a minimum concentration of $\phi_{PS} \approx 0.6$ for which the model is expected to work.

4.2.2   Effect of PS Concentration. The analysis was performed by fixing the value of $(N-2)\zeta$ and then determining the value of $NW\tau$ for each PS volume fraction from the average experimental ratio R_{exptl} for the 4000 and 100,000 molecular weight PS. These values were then averaged and used to calculate the fit result, R_{fit}. The best fit was chosen by minimizing $\Sigma(R_{exptl} - R_{fit})^2$. The best results were obtained for $(N-2)\zeta=0.61$ and $NW\tau = 25.9$. They are plotted as a solid line through the toluene-cast data in Figure 7.

In order to assess the significance of the results, it is necessary to determine whether the values of the parameters used to fit the data are reasonable. It is difficult to obtain a quantitative check on the parameter $(N-2)\zeta$. Nevertheless, since N should be between six and fifteen, the probability that two adjacent rings are in an intermolecular EFS is of order 10^{-2}. This is consistent with the strict geometric requirements necessary for excimer formation.

The value of $W\tau$ may be predicted more precisely using the theory of Förster [72]. For nonradiative energy transfer between randomly orientated chromophores, the factor κ^2 in Equation 3 has an average value of 2/3. In a manner consistent with the lattice approach used in the derivation of the energy migration model, it will be assumed that the ring separation equals that found in pure PS. This leads to $R = 5.5$ Å and $W\tau = 2.65$. From the best fit of the three-dimensional random walk model to the fluorescence results and the assumption that N should be between six and fifteen, the value of $W\tau$ is found to be between 1.7 and 4.3. This is reasonable agreement, considering the simple nature of the energy migration model.

4.3 Photophysical Analysis of Phase Separated PS/PVME Blends

Although the successful explanation of the fluorescence behavior for both low and high concentration miscible PS/PVME blends is of interest from a photophysical viewpoint, it is perhaps more technologically significant that fluorescence may also be used quantitatively to study immiscible blends. We begin this section with a consideration of a simple two phase morphological model that probably provides no more than a zeroth order approximation to the true morphology. Nevertheless, it yields quite reasonable results when applied to PS/PVME blends that are phase-separated by virtue of the solvent casting process. Next, we consider the study of the kinetics of phase separation and obtain a time dependent measurement of the concentration of PS in the rich phase. We suggest that measurements of this type may be useful in verifying recent theories of spinodal decomposition in polymer blends.

4.3.1 Morphological Model. We first assume that the blend can be described by binary equilibrium thermodynamics with two uniform phases in the blends, one rich in PS and one lean in PS. The volume fractions of PS in the rich and lean phases, ϕ_R and ϕ_L, respectively, are independent of the bulk PS composition ϕ_B. Thus, from a simple mass balance, the volume fraction of the rich phase in the blend, V_R, is given by

$$V_R = \frac{\phi_B - \phi_L}{\phi_R - \phi_L} \qquad (14)$$

Furthermore, the fraction of phenyl rings in the rich phase, X_R, equals the probability that a photon is absorbed by a ring in the rich phase

$$X_R = \frac{\phi_R V_R}{\phi_R V_R + \phi_L (1-V_R)} = \frac{\phi_R (\phi_B - \phi_L)}{\phi_R (\phi_B - \phi_L) + \phi_L (\phi_R - \phi_B)} \qquad (15)$$

To simplify the analysis, two assumptions about the photophysics of a two-phase blend are made. First, we assume that the photophysical behavior of one phase in a phase-separated system with concentration ϕ is identical to that for a miscible blend of the same composition, i.e., the phases are large enough so that there are no boundary effects. The second, more critical assumption, is that there is no energy migration between phases. This will be a good assumption when the lean phase concentration is low enough so that the PS coils are isolated and migration is intramolecular. At the other extreme, when the PS volume fraction is greater than 0.6, very few hops are made due to the large number of EFS. Thus, an excitation should not be able to make enough transfers without being trapped to escape a concentrated phase large enough to scatter light.

If M_R and M_L are the probabilities of eventual monomer decay given that the photon is absorbed by the rich or lean phase, respectively, then the ratio of excimer to monomer fluorescence is given by a simple weighted average of fluorescence contributions from the rich and lean phases:

$$\frac{I_D}{I_M} = \frac{Q_D}{Q_M} \left[\frac{X_R(1-M_R) + (1-X_R)(1-M_L)}{X_R M_R + (1-X_R)M_L} \right] \tag{16}$$

M_R and M_L are determined from the calibration curve, such as shown in Figure 8, for the rich and lean phase compositions.

An immediate test of this model is provided by the results for the phase separated PS/PVME blends cast from tetrahydrofuran, shown in Figure 7. The M_L and M_R values were determined for isolated PS chains and 98% PS, respectively. Using these values and an iterative procedure, the best fit of the experimental results was obtained for $\phi_L = 0.007$ and $\phi_R = 0.98$. The model is shown as the solid line through the square plotting symbols in Figure 7.

4.3.2   Phase Separation Kinetics.

There currently is considerable interest by the polymer physics community in the mechanism of phase separation in polymer blends. In an early study, Kwei, Wang and Nishi [4] employed pulsed NMR and found that the Cahn-Hilliard linearized non-statistical theory of spinodal decomposition [82] adequately described the process at short times: the average composition of the two phases changed exponentially with time while their volume fractions remained constant. In addition, it was observed that no change in phase volume fraction took place at longer times when the linearized model no longer was valid.

More recently, DeGennes [83] and Pincus [84] have extended the Cahn-Hilliard theory of spinodal decomposition to polymer blends in the melt. Their scaling relationships show that the kinetics of the early stages of the process should depend strongly on molecular weight. This analysis has been called into question by Ronca and Russell [85] who employ a very different form of the concentration gradient term.

Although a detailed test of these various theories is of
interest, for reasons of space we limit our treatment to presentation
of experimental results and an abbreviated analysis. Details are
provided in the original references [80,86]. The objective is to
demonstrate that the two phase morphological model may be used along
with a phase diagram to obtain a quantitative measure of the rich
phase PS concentration as a function of time during phase separation.

Typical fluorescence results are shown in Figure 9 for the 10% PS
(100,000)/PVME blend. The majority of the change in I_D/I_M occurs in
the first 10 minutes, with the ratios leveling off and becoming
constant for times greater than about 50 minutes. It is of interest
to compare these data with results of microscopy studies in which an
interconnected morphology was never observed for the 10% films but was
observed for the 50% PS films. This is consistent with Cahn's
demonstration that the minor phase must have a volume fraction greater
than 0.15 to allow connectivity. The more important observation, however, is that droplet coalescence begins to take place at the same
time that I_D/I_M levels off. This indicates that excimer fluorescence
is sensitive to changes in local concentration occurring during early
states of phase separation but not to the major morphological changes
that occur at long times.

From the NMR study of Kwei and coworkers [4] it appears reasonable to assume that the volume fraction of each phase remains constant
and equal to the equilibrium value at all times. From the M versus
ϕ_{PS} plots, such as shown in Figure 8, together with Equations (14-16),
it is possible to relate I_D/I_M for a phase separated blend to a unique
value of the rich phase composition ϕ_R.

Typical results are shown in Figure 10. If the linearized theory
of spinodal decomposition is used, the average rich phase composition
is predicted to change exponentially with time

$$\phi_R - \phi_0 \sim \exp\left(-t/\tau_{q(max)}^{-1}\right) \tag{17}$$

provided the Fourier Series solution to the diffusion equation can be
approximated by its maximum term. Here, $-\tau_{q(max)}^{-1}$ is the growth rate
of the dominant concentration fluctuation, and can be obtained from
the linear short time portion of the curves in Figure 10. It is clear
that the growth rate of the dominant concentration fluctuation is
slower for higher molecular weight PS. This is qualitatively as predicted; we consider the quantitative features elsewhere [86].

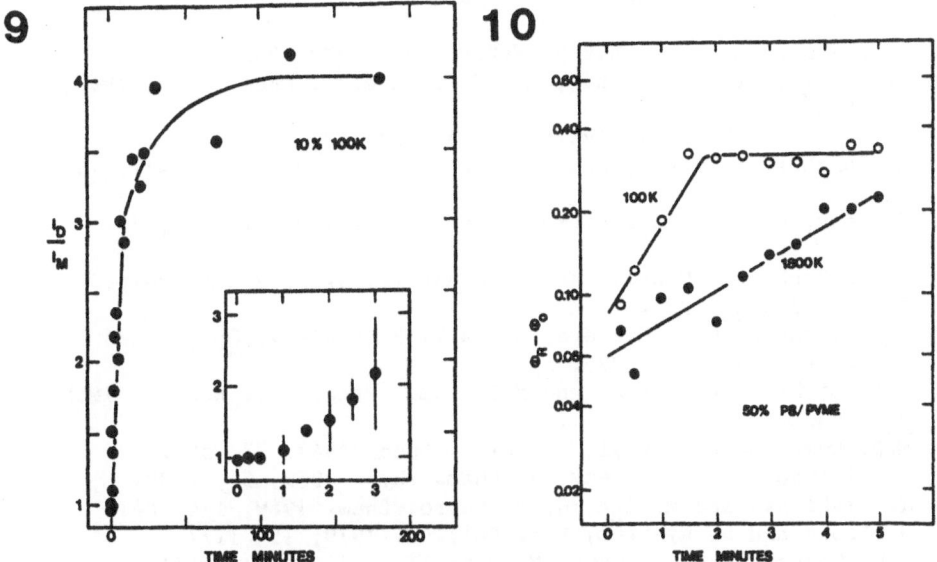

Figure 9.   Kinetics of phase separation in 10% PS/PVME blends annealed
for various times at 423 K, followed by a rapid quench to 295 K, where
I_D/I_M was measured.   Taken from Figure 8 of reference 86.

Figure 10.   Early stages of phase separation in 50% PS/PVME blends.
ϕ_R is the PS concentration in the rich phase and ϕ_0 is the nominal
bulk concentration.   Taken from Figure 13 of reference 86.

ACKNOWLEDGEMENT

This work was supported by the Army Research Office under
Contract DAAG 29-82-K-0019.

REFERENCES

(1)   M. Bank, J. Leffingwell, and C. Thies, Macromol. **1971**, <u>4</u>, 43.
(2)   M. Bank, J. Leffingwell, and C. Thies, J. Poly. Sci. **1972**. <u>10</u>,
      1097.
(3)   T.K. Kwei, T. Nishi, and R.F. Roberts, Macromol. **1974**, <u>7</u>, 667.
(4)   T. Nishi, T.T. Wang, and T.K. Kwei, Macromol. **1975**. <u>8</u>, 227.
(5)   L.P. McMaster, Macromol. **1973**, <u>5</u>, 760.
(6)   T. Nishi and T.K. Kwei, Polymer **1975**, <u>16</u>, 285.
(7)   N. Kuwahara, M. Ishizawa, S. Saeki, and M. Kaneko, Rep. Prog.
      Poly. Phys. Jap. **1976**, <u>19</u>, 9.
(8)   D.D. Davis and T.K. Kwei, J. Poly. Sci. Poly. Phys. Ed. **1980**,
      <u>18</u>, 2337.

(9)   S. Reich and Y. Cohen, J. Poly. Sci. Poly. Phys. Ed. **1981**, 19,
      1255.
(10) S.S. Yanari and F.A. Bovey, Nature **1963**, 200, 242.
(11) M.T. Vala, Jr., J. Haebig, and S.A. Rice, J. Chem. Phys. **1965**,
      43, 886.
(12) L. Gargallo, Scientia (Valparaiso) **1977**, 42, 11.
(13) L. Bokobza, B. Jasse, and L. Monnerie, Eur. Poly. J. **1977**, 13,
      921.
(14) F. Gény, C. Nöel, and L. Monnerie, J. Chim. Phys. **1974**, 71,
      1150.
(15) N.-L. Bach Van, C Nöel, and L. Monnerie, J. Poly Sci. Poly.
      Symp. **1975**, 52, 283.
(16) T.V. Ivanova, G.A. Mokeeva, and B.Y. Sveshnikov, Optics and
      Spectrosc. **1962**, 12, 325.
(17) J.B. Birks, C.L. Braga, and M.D. Lumb, Proc. Roy. Soc. A. **1965**,
      283, 83.
(18) M.D. Lumb and D.A. Weyl, J. Molec. Spec. **1967**, 23, 365.
(19) F. Hirayama and S. Lipsky, J. Chem. Phys. **1969**, 51, 1939.
(20) T. Nishihara and M. Kaneko, D. Makro. Chem. **1969**, 124, 84.
(21) J. Roots and B. Nyström, Eur. Poly. J. **1979**, 15, 1127.
(22) J.M. Torkelson, S. Lipsky, M. Tirrell, and D.A. Tirrell,
      Macromol., **1983**, 16, 326.
(23) F. Hirayama, J. Chem. Phys. **1965**, 42, 3163.
(24) W.E. Lindsell, F.C. Robertson, and I. Soutar, Eur. Poly. J.
      1981, 17, 203.
(25) T.H. Förster, Molecular Spectroscopy; Butterworths Scientific
      Publications, Ltd.: London, 1962.
(26) D.J. Cram, N.L. Allinger, and H. Steinberg, J. Am. Chem. Soc.
      1954, 76, 6132.
(27) Y.-C. Wang and H. Morawetz, D. Makro, Chem. Supp. **1975**, 1, 283.
(28) J.W. Longworth, Biopolymers **1966**, 4, 1131.
(29) T. Ishii, T. Handa, and S. Matsunaga, J. Poly Sci. Poly. Phys.
      Ed. **1979**, 17, 811.
(30) J.W. Longworth and F.A. Bovey, Biopolymers **1966**, 4, 1115.
(31) F. Heisel and G. Laustriat, J. Chim. Phys. **1969**, 66, 188.
(32) T. Ishii, T. Handa, and S. Matsunaga, Macromol. **1978**, 2, 40.
(33) L. Bokobza, B. Jasse, and L. Monnerie, Eur. Poly. J. **1980**, 16,
      715.
(34) F.C. DeSchryver, L. Moens, M. Van der Auweraer, N. Boens, L.
      Monnerie, and L. Bokobza, Macromol. **1982**, 15, 64.
(35) K.P. Ghiggino, R.D. Wright, and D. Phillips, J. Poly. Sci. Poly.
      Phys. Ed. **1978**, 16, 1499.
(36) M.C. Gupta, A. Gupta, J. Horwitz, and D. Kliger, Macromol. **1982**,
      15, 1372.
(37) F.A. Bovey, F.P. Hood, E.W. Anderson, and L.L. Snyder, J. Chem.
      Phys. **1965**, 42, 3900.
(38) H. Pivcova, M. Kolinsky, D. Lim, and B. Schneider, J. Poly.
      Sci.: Part C, **1969**, 22, 1093.

(39) T. Moritani and Y. Fujiwara, J. Chem. Phys. **1973**, 59, 1175.
(40) A.D. Williams, J.I. Brauman, N.J. Nelson, and P.J. Flory, J. Amer. Chem. Soc. **1967**, 89, 4807.
(41) A.D. Williams and P.J. Flory, J. Amer. Chem. Soc. **1969**, 91, 3111.
(42) B. Froelich, C. Noel, B. Jasse, and L. Monnerie, Chem. Phys. Lett. **1976**, 44, 159.
(43) P.J. Flory, Statistical Mechanics of Chain Molecules, Interscience Publishers: New York, 1969.
(44) P.J. Flory, Macromol. **1974**, 7, 381.
(45) P.G. Hoel, S.C. Port, and C.J. Stone, Introduction to Stochastic Processes, Houghton Mifflin Company: Boston, 1972.
(46) S. Gorin and L. Monnerie, J. Chim. Phys. Physicochim. Biol. **1970**, 67, 869.
(47) D.Y. Yoon, P.R. Sundararajan, and P.J. Flory, Macromol. **1975**, 8, 776.
(48) G.E. Stegan and R.H. Boyd, Polymer Preprints **1978**, 19, 595.
(49) P.J. Flory, Y. Fujiwara, Macromol. **1969**, 2, 315.
(50) R. Gelles and C.W. Frank, Macromol. **1982**, 15, 741.
(51) R.B. Fox, Pure and Appl. Chem. **1972**, 30, 87.
(52) W. Klöpffer, Spectr. Lett. **1978**, 11, 863.
(53) W. Klöpffer, N.Y. Acad. Sci. **1981**, 366, 373.
(54) J.R. MacCallum, Eur. Polym. J. **1981**, 17, 209.
(55) J.R. MacCallum and A.L. Rudkin, Eur. Poly. J. **1981**, 17, 953.
(56) T. Ishii, T. Handa, and S. Matsunaga, Makromol. Chem. **1977**, 178, 2351.
(57) C. David, M. Lempereur, and G. Geuskens, Eur. Poly. J. **1973**, 9, 1315.
(58) L. Alexandru and A.C. Somersall, J. Poly. Sci. Poly. Chem. Ed. **1977**, 15, 2013.
(59) R.F. Reid, I.S. Soutar, J. Poly Sci. Poly. Phys. Ed. **1978**, 16, 231.
(60) D.J.T. Hill, D.A. Lewis, J.H. O'Donnell, P.W. O'Sullivan, and P.J. Pomery, Eur. Poly. J. **1982**, 18, 75.
(61) R.B. Fox and T.R. Price, Macromol. **1974**, 7, 937.
(62) F. Hirayama, L.J. Basile, and C. Kikuchi, Molecular Crystals **1968**, 4, 83.
(63) C. David, M. Piens, and G. Geuskens, Eur. Poly. J. **1973**, 9, 533.
(64) W. Klöpffer, Eur. Poly. J. **1975**, 11, 203.
(65) K.P. Ghiggino, A.J. Roberts, and D. Philips, Adv. Poly. Sci. **1981**, 40, 69.
(66) J.R. MacCallum, Polymer **1982**, 23, 175.
(67) C. David, D. Baeyens-Volant, and G. Geuskens, Eur. Poly. J. **1976**, 12, 71.
(68) C. David, N. Putnam-de Lavareille, and G. Geuskens, Eur. Poly. J. **1977**, 13, 15.
(69) J.R. MacCallum and L. Rudkin, Nature **1977**, 266, 338.
(70) N. Levinson, J. Soc. Indust. Appl. Math. **1962**, 10, 442.

(71) H.H. Jaffé, M. Orchin, Theory and Application of Ultraviolet
      Spectroscopy, John Wiley and Sons, Inc.: New York, 1962.
(72) T.H. Förster, Disc. Far. Soc. 1959, 27, 7.
(73) I.B. Berlman, Energy Transfer Parameters of Aromatic Compounds,
      Academic Press: New York, 1973.
(74) P.D. Fitzgibbon and C.W. Frank, Macromol. 1982, 15, 733.
(75) T. Ishii, T. Handa, and S. Matsunaga, Macromol. 1978, 11, 40.
(76) Y. Nishijima, K. Mitani, S. Katayama, and M. Yamamoto, Prog.
      Polym. Phys. Jap. 1970, 13, 421.
(77) J.R. MacCallum, Eur. Poly. J. 1981, 17, 797.
(78) M.N. Barber, and B.W. Ninham, Random and Restricted Walks,
      Gordon and Breach: New York, 1970.
(79) R. Gelles and C.W. Frank, Macromol. 1982, 15, 747.
(80) R. Gelles and C.W. Frank, Macromol. 1982, 15, 1486.
(81) G. Zumofen and A. Blumen, Chem. Phys. Lett. 1981, 78, 131.
(82) J.W. Cahn, J. Chem. Phys. 1965, 42, 93.
(83) P.G. De Gennes, J. Chem. Phys. 1980, 72, 4756.
(84) P. Pincus, J Chem. Phys. 1981, 75, 1996.
(85) G. Ronca and T.P. Russell, Macromol. 1985, 18, 665.
(86) R. Gelles and C.W. Frank, Macromol. 1983, 16, 1448.

FLUORESCENCE QUENCHING OF ANTHRACENE-LABELLED POLYSTYRENE BY POLY(VINYL-METHYLETHER) : A NEW APPROACH FOR THE ANALYSIS OF PHASE SEPARATION PHENOMENA

Jean-Louis HALARY and Lucien MONNERIE
Ecole Supérieure de Physique et de Chimie Industrielles de la
Ville de Paris
10, rue Vauquelin, F-75231 PARIS Cedex 05 (France)

ABSTRACT - The phenomenon of static quenching by poly(vinylmethylether) of the fluorescence emission of anthracene-labelled polystyrene is evidenced and analyzed. It is applied to the investigation, on a molecular scale, of thermally-induced phase separation in polystyrene-poly (vinylmethylether) blends. Typical results thus obtained are presented, including binodal and spinodal determination, and dependence of phase diagram characteristics on polymer molecular weight, molecular weight distribution and isotopic substitution. Preliminary results on the phase separation kinetics are also given.

I - INTRODUCTION

Fifteen years ago, it was demonstrated that a small number of amorphous polymer mixtures are miscible at ordinary temperatures, but undergo thermally-induced phase separation at higher temperatures, because of the existence of a lower critical solution temperature (LCST). This phenomenon was investigated in the binary blend of polystyrene (PS) and poly(vinylmethylether) (PVME) using conventional techniques including differential scanning calorimetry, optical microscopy and light scattering (1 - 3).

As theoretical models were proposed to give an account for the spinodal decomposition for unstable one-phase compositions (4-6) and for the nucleation and growth mechanism for metastable one-phase compositions (7), the necessity arose for information on the earliest stages of phase separation. It is the reason why additional techniques including neutron scattering (8, 9), N.M.R. spectroscopy (10, 11) and fluorescence emission were used with the aim to detect the phase separation on a molecular scale.

Three fluorimetric approaches, based on different physical phenomena, have been yet reported and used in polymer blends. The first two have been developed by Morawetz et al (12) and by Frank et al (13) and involve the energy transfer between donnor and acceptor molecules, and the excimer emission, respectively. The third approach, which has been

M. A. Winnik (ed.), Photophysical and Photochemical Tools in Polymer Science, 589-610.
© 1986 by D. Reidel Publishing Company.

discovered in our Laboratory, applies to PS-PVME blends in which small amounts of anthracene-labelled polystyrene chains (PS*) have been added (14).

The aim of the present paper is to review both physical basis and recent applications of the latter method. We will successively discuss:

i / the determination of the equilibrium coexistence curve (14, 15)
ii / the determination of the spinodal (14, 15)
iii/ the effects of molecular weight and molecular weight distribution (14-16)
and iv/ the influence of polystyrene deuteration on the phase diagrams (16, 17)

II - RELATIONSHIP BETWEEN PHASE SEPARATION AND FLUORESCENCE EMISSION

In order to study the system PS-PVME, our procedure requires the addition in the blend of labelled polystyrene. The label PS*, which consists of PS chains containing anthracene fluorescent groups in their middle (Fig. 1), is suitable for this purpose. As previously reported (18) it is prepared by termination of living PS chains by 9,10 bis (bromomethyl)anthracene. PS* concentration is adjusted to be around 6 ppm of anthracenic units (i.e. less than 0.1 and 1 wt %

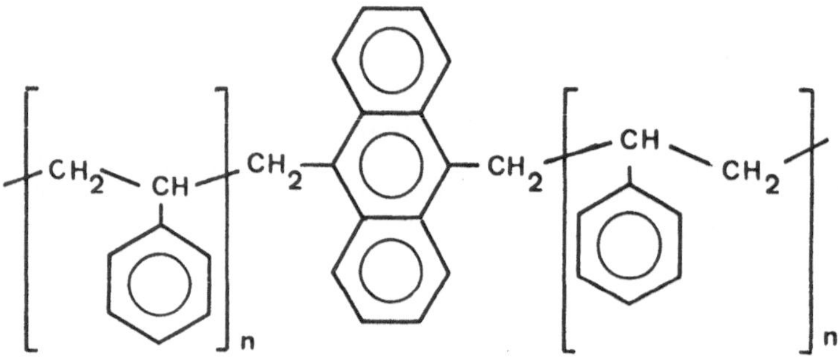

Fig. 1 - PS* formula.

of PS* of molecular weight 30,000 and 300,000, respectively). In these conditions, the PS/PS*/PVME ternary mixtures behave chemically like PS/PVME binary blends and the fluorescer concentration is low enough to prevent undesired intermolecular energy transfer. In consideration of the absorption and emission spectra of PS* (Fig. 2), the intensity of the fluorescence emission I_F may be conveniently determined by selecting excitation and fluorescence wavelengths around 365 and 440 nm, respectively. The experimental data reported below were obtained using a fluorescence microscope but such a device is not absolutely necessary. It allows one to study very small samples and to disregard the areas exhibiting some defects.

Heating of ternary blends of PS/PS*/PVME results in a sharp increase in the fluorescence intensity at a temperature T_F which depends

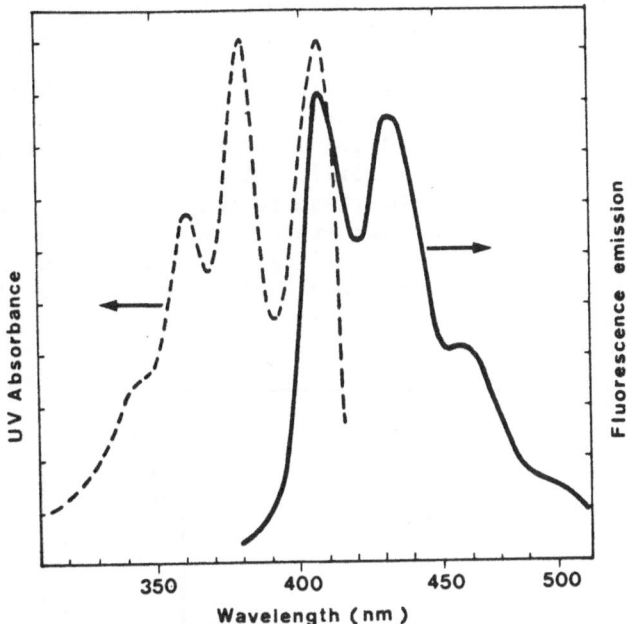

Fig. 2 - UV absorption (dashed curve) and fluorescence emission
        (solid curve) spectra of PS*.

on both the PS content and the polymer molecular weight. A typical
example of this phenomenon is given in Fig. 3 (T_F is marked by a vertical solid line). It stands to reason that T_F is related to the phase

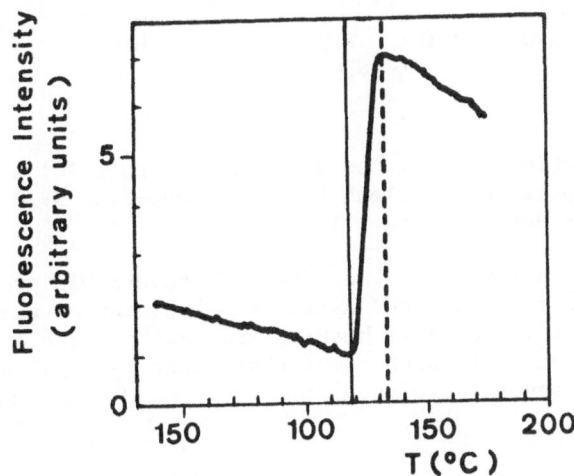

Fig. 3 - Typical recording of fluorescence intensity as a function of
        temperature. (Blend PS 233,000-PVME 99,000 33/67 wt %
        Heating rate : 10°C.min^{-1}).

separation since it is found a few °C below the cloud point temperature (marked by a vertical dashed line in Fig. 3), as determined from light scattering experiments (14) or by the ultimate decrease in fluorescence intensity owing to the appearance of cloudiness in the samples. As for the monotonic decrease in fluorescence intensity below T_F, it just results from the decrease of the label lifetime in the excited state as a consequence of its increased mobility above the glass transition temperature.Additionally, when the phase separation can be detected from small angle neutron scattering - for example in the case of PVME - per-deuterated polystyrene (PS-d_8) - a good agreement is observed between our fluorimetric determination (T_F) and the SANS cloud point (17), as shown in Fig. 4.

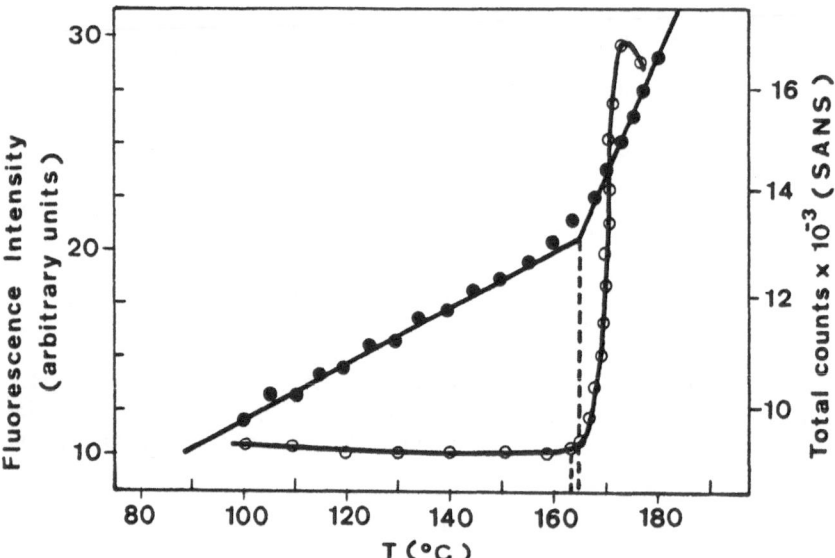

Fig. 4 - Phase separation as detected from fluorescence (O) and
          SANS (●) experiments. (Blend PS-d_8 119,000-PVME 99,000
          22/78 wt % - Heating rate : 1°C.min^{-1}).

Because phase separation is known to be a reversible process in the early stages, one may check the hysteresis behavior of the fluores-cence emission intensity. Repeated heating and cooling near T_c (while recording the variation of I_F with time) shows that I_F regains the initial value during several cycles, independent of the heating rate (Fig. 5).

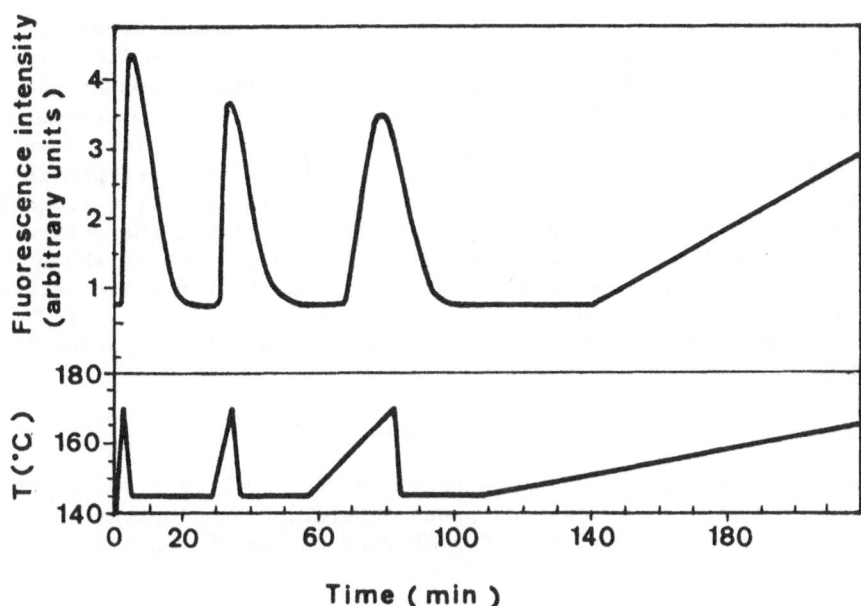

Fig. 5 - Typical recording of fluorescence intensity versus time on
repeated heating and cooling near the phase separation temperature. (Blend PS 67,000-PVME 99,000 19.7/80.3wt %).

III - EVIDENCE FOR FLUORESCENCE QUENCHING BY PVME

The origin of the change in fluorescence emission at the
temperature T_F lies in the photophysical process itself, i.e. in
quenching phenomena, the extent of which should depend on the nature of
the molecules next to the label. In fact, PS* is surrounded by a mixture of PS and PVME in the homogeneous blends, but mainly by PS molecules in the phase-separated systems. Thus, a simple explanation of
the change in fluorescence intensity would be that PS* emission is
greater in PS than in PS-PVME mixtures or, in other words, that PVME
is responsible for the quenching of the fluorescence emission.

III.1 - Origin of the quenching

Two types of fluorescence quenching may occur, dynamic (19)
and static (20). Dynamic quenching is characterized by an
energy transfer to non-radiative quencher molecules. In this process,
all the fluorescers are statistically affected and fluorescence intensity
and fluorescer lifetime in the excited state are quantities proportional with each other. On the other hand, static quenching results from
specific interactions between fluorescer and quencher molecules. Any
fluorescence emission is cancelled in the case of interactions, but

those fluorescers which do not interact keep their normal fluorescence characteristics. As a consequence, static quenching allows fluorescence intensity to decrease whereas fluorescer lifetime stands unchanged.

Fluorescence intensity and lifetime measurements on solutions of model compounds or directly on PVME/PS/PS* mixtures of various compositions (14) led to the conclusion that static quenching is responsible for the observed phenomena in this special case. It is only reasonable to place the corresponding specific interactions between the anthracenic units of PS* and the ether functions of PVME. Indeed, it was shown by Fourier-transform infra-red spectroscopy (21, 22) that similar interactions, although weaker, involve the ether lone-pair electrons of PVME with the benzene ring of PS and are responsible for the compatibility of both polymers.

III.2 - Sketch of phase separation phenomena

It remains now to explain why some static quenching occurs in compatible blends, but tends to cease as phase separation develops. As shown in Figure 6 a, the intimate mixing of PVME with PS and PS* chains which

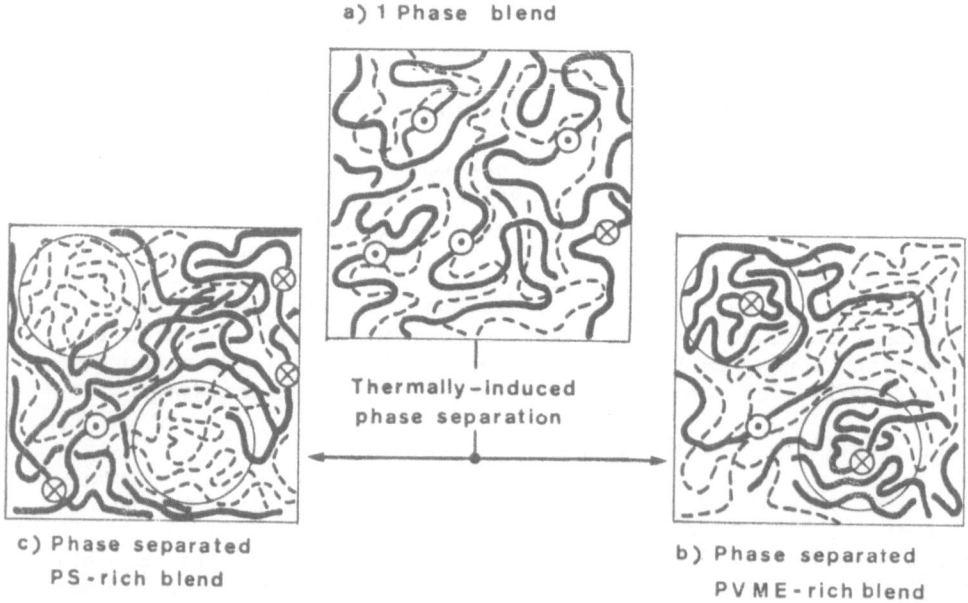

a) 1 Phase blend

Thermally-induced phase separation

c) Phase separated
PS-rich blend

b) Phase separated
PVME-rich blend

Fig. 6 - Sketch of phase separation phenomena. (dashed chains : PVME; solid chains: PS; (⊗) active label; (⊙) quenched label).

exists in the one-phase systems is favorable to specific interactions and thereby leads to a very small anthracene fluorescence emission. The situation is quite different in the two-phase systems because PS* tends to accompany PS. Suppose a thermally-induced phase separation to occur in a PVME-rich blend (Fig. 6 b) : PS-rich nodules segregate, inside which the probability for PS* to interact with PVME is very low. In the opposite situation of PVME-rich nodules formed from PS-rich blends (Fig. 6 c), PS* mainly remains in the bulk, almost surrounded by PS chains. In both cases, the overall probability for PS*-PVME interactions is decreased with respect to the one-phase state and the fluorescence intensity increases in proportion.

III.3 - Versatility of the label PS*

This method requires that a very small amount of PS* chains provide information on the phase separation in PVME-PS-PS* mixtures. We established that the observed changes in fluorescence intensity do not arise from the phase separation of PVME-PS* . Indeed, the observed phase separation temperature in some PVME-PS-PS* mixtures does not depend on the molecular weight of the label PS* in the range 30,000-300,000 (14). In addition, agreement was found between the SANS data on PVME/PS-d$_8$ and the fluorescence data on PVME/PS-d$_8$/PS* (Fig. 4). That is, PS* samples the PVME-PS mixture and is valid for the study of molecular weight as well as isotopic substitution influences on the phase diagrams.

IV - ANALYSIS OF PS-PVME PHASE DIAGRAMS

IV.1 - Determination of the equilibrium coexistence curves (binodals)

Determination of the binodals using our approach can be achieved with surprising accuracy over a very large range of polymer molecular weights ($\overline{M}_w$) : the most recent data cover PS-PVME blends with $\overline{M}_w$'s ranging from 20,400 to 1,660,000 for PS and from 45,000 to 1,330,000 for PVME (15). A few examples of such determinations are given in Fig. 7.

The experimental procedure for obtaining these equilibrium curves consists of heating up blends of various composition in a hot stage attached to the microscope and operated at a heating rate of 0.2°C.min^{-1}. Then, one must deduce T_F from the recordings of fluorescence intensity as a function of time, i.e. as a function of temperature (Fig. 3). In most cases, the temperature T_F does not exceed the equilibrium value by more than 1°C. Indeed, we have verified that 30 min isothermal experiments performed 1°C below T_F do not lead to any change in fluorescence intensity, confirming the absence of phase sepa-

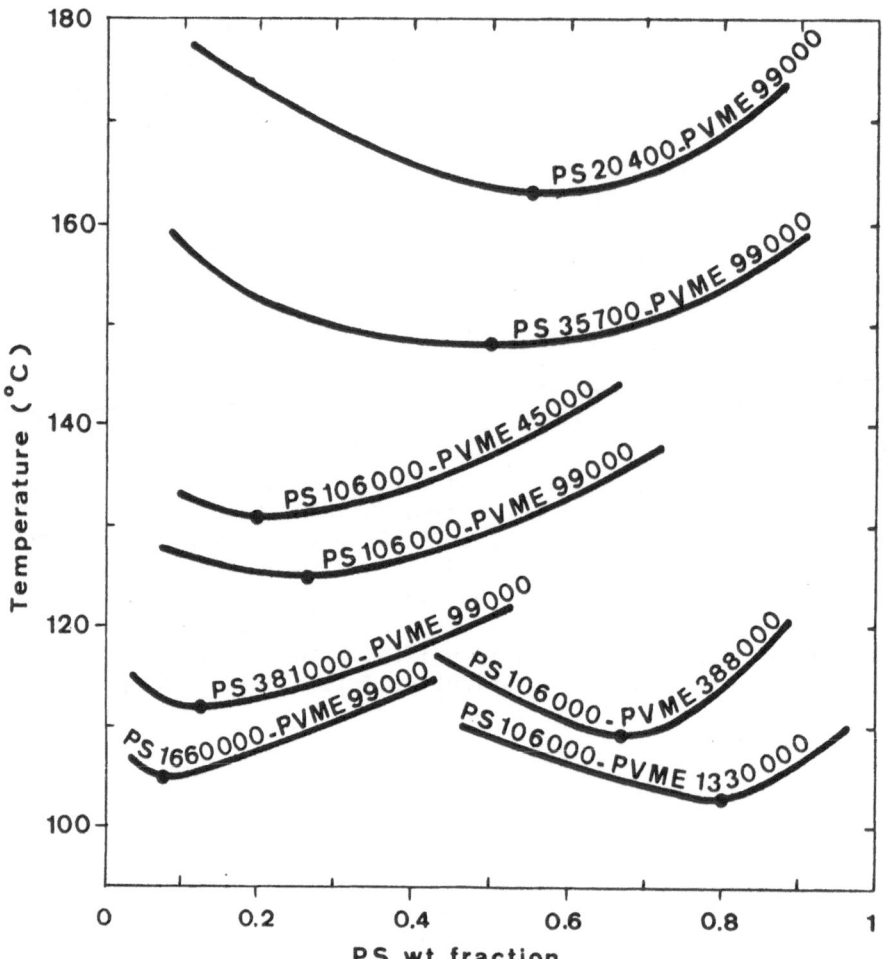

Fig. 7 - Equilibrium coexistence curves of various PS-PVME blends as
determined from fluorescence emission of PS*. (The numbers
indicate the polymer weight average molecular weights $\overline{M}_w$).

ration in such conditions. The only exceptions deal with blends containing one polymer of very large $\overline{M}_w$ (for instance, PS 1,660,000 or
PVME 1,330,000) or two polymers of large $\overline{M}_w$ (for instance, the blend
of PS 600,000 with PVME 388,000). In these cases, the heating rate of
$0.2°C.min^{-1}$ is too fast compared with the phase separation process. The
unique route towards binodal is to determine T_F at $0.2°C.min^{-1}$ and then
to perform isothermal experiments at temperatures increasingly lower
than T_F until no change in fluorescence intensity can be detected.

4.2 - Possible determination of the spinodals

4.2.1 - $\overline{M}_w \leqslant 100,000$

Let us concentrate first on the PS-PVME blends which contain polymers of moderate M_w (say, less than 100,000) and consider the influence of the heating rate (in the range $0.2\text{-}16°C.min^{-1}$) on the detected T_F values. Around the minimum of the coexistence curves, the heating rate is unimportant and the T_F values are identical within the confidence range (Fig. 8). However, for PVME or PS-rich compositions,

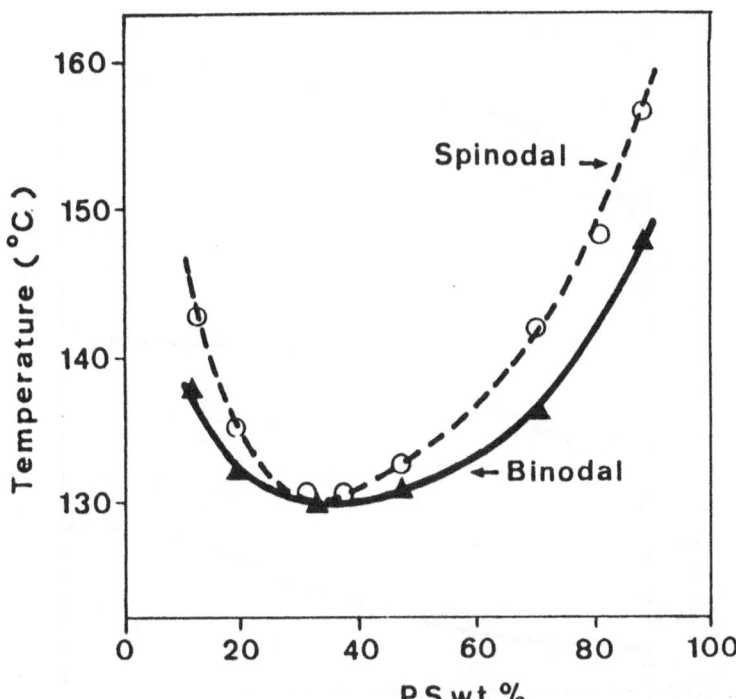

Fig. 8 - Experimental determination of binodal and spinodal curves. (Blend PS 67,000-PVME 99,000 - Heating rate ($°C.min^{-1}$) : (▲) 0.2; (O) 16).

modifying the heating rate from 0.2 to $16°C.min^{-1}$ allows the detected T_F to rise. The size of the gap increases with the departure from the minimum of the coexistence curve. This behavior has been attributed (14) to the different rates of the competing processes. Spinodal decomposition is the only process that takes place in the vicinity of the minimum in the phase diagram; depending upon the temperature, nucleation and growth mechanisms as well as spinodal decomposition may take place elsewhere. The spinodal decomposition can be immediately detected at a heating rate of 16°C.min^{-1}, as can be deduced from the absence of heating rate dependence around the minimum of the phase diagrams. This is because the spinodal decomposition is rapid relative to the heating rate in these mixtures. On the contrary, the nucleation and growth processes are probably too slow for detection under these conditions, while they are properly observed at a heating rate of 0.2°C.min^{-1}. Thus, it is possible to accurately measure the spinodal curve without any contribution from nucleation and growth mechanisms by a proper choice of experimental conditions.

    The data (Fig. 9) show that the temperature of the phase separation

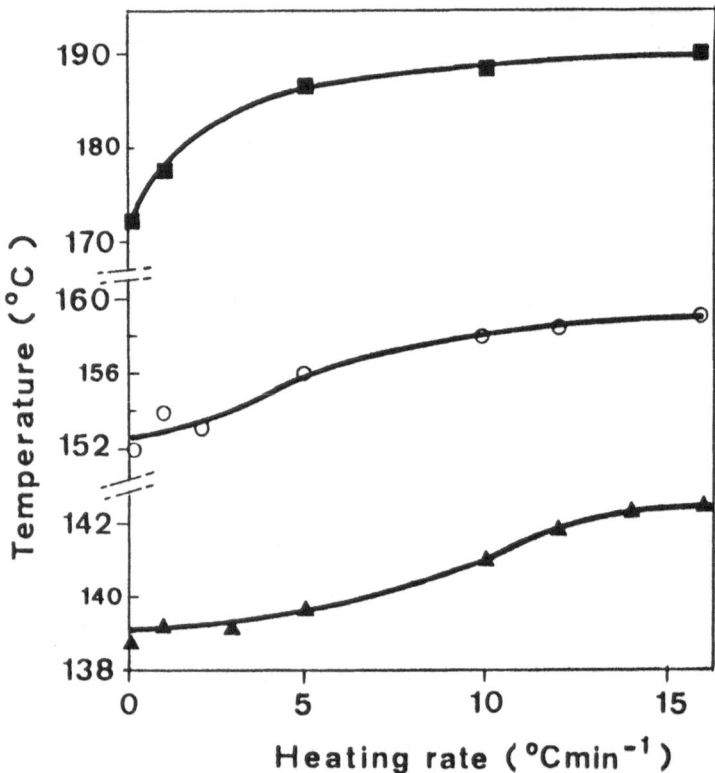

Fig. 9 - Examples of heating rate influence on observed phase separation temperature (composition far from w_{min}).

        (■) PS 20,400-PVME 99,000 21/79 wt %
        (O) PS 35,700-PVME 99,000 20/80 wt %
        (▲) PS 67,000-PVME 99,000 12/88 wt %

approaches a constant value as the heating rate increases for each mixture. The plateau value is reached at a heating rate value which depends on PS molecular weight and weight fraction but never exceeds $14°C.min^{-1}$.

4.2.2 - $\overline{M}_w$ > 100,000

The situation is less favorable for spinodal determination when the polymer molecular weights are larger than 100,000. Indeed, as above for the nucleation and growth process, conditions may be satisfied so that the applied heating rate matches the speed of phase separation. Consequently, a heating rate effect is observed over the entire composition range, including the composition w_{min} of the minimum of the phase diagram (Fig. 10). The extent of this effect may be characterized by

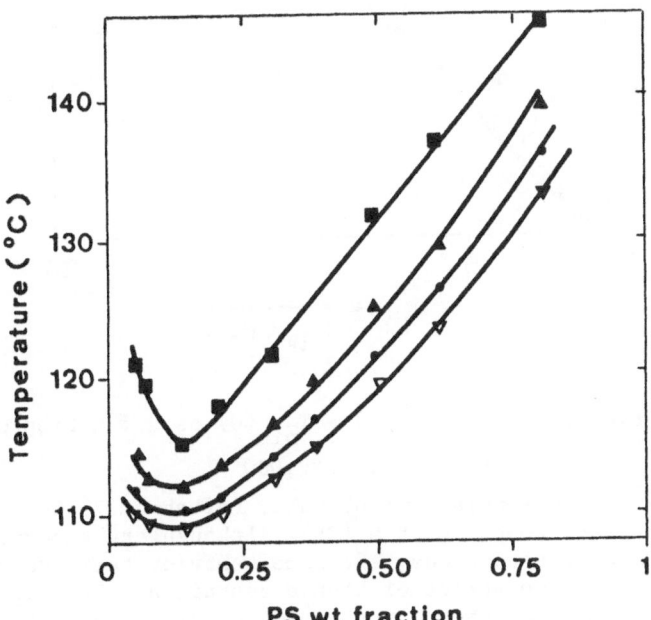

Fig. 10 - Influence of heating rate on observed phase separation temperatures. (Blend PS 600,000-PVME 99,000-Heating rate $(°C.min^{-1})$: (∇) 0.2; ($\bullet$) 1; ($\blacktriangle$) 5; ($\blacksquare$) 16).

the quantity $\Delta T = T_{min}(16) - T_{min}(0)$, which is the difference between the temperatures at which the phase separation is detected for the composition w_{min} in experiments at $16°C.min^{-1}$ and in isothermal experiments, respectively. As long as this process develops fast enough with respect to the heating rate, the temperature at which phase separation is detected coincides with T_{min} (0). On the other hand, when the speed of the process is dramatically slowed down, ΔT varies as the detection time lag, i.e. in inverse proportion to the speed of the process. The evolution of ΔT as a function of PS molecular weight for a series of blends containing the same PVME is illustrated in Fig. 11.

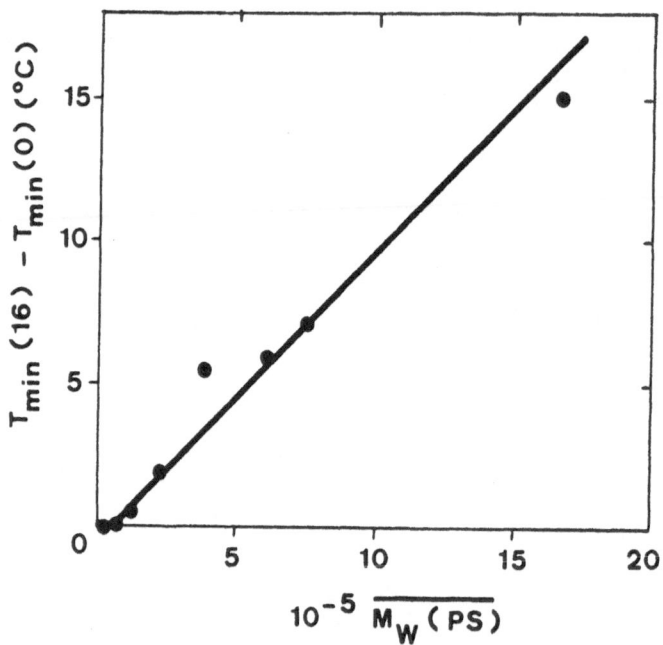

Fig. 11 - Evolution of $\Delta T = T_{min}$ (16) - T_{min} (0) as a function of $\overline{M_w(PS)}$ in mixtures PS-PVME 99,000

Above a critical molecular weight value (of about 70.000 in this case), ΔT increases linearly with $\overline{M_w}$ (PS). Although the experimental data available are less numerous, the same behavior seems to be encountered in the case of the series of blends containing the same PS and PVME's of variable $\overline{M_w}$, as shown by the plot of ΔT versus $\overline{M_w(PVME)}$ (Fig. 12).

This molecular weight dependence of ΔT is consistent with that of $-\tau_q^{-1}$, defined by Pincus (5) as the growth rate of the dominant concentration fluctuation. This quantity, indeed, should decrease with increasing molecular weight, as predicted from the theory (5) and qualitatively verified by Gelles and Frank (24).

From a practical viewpoint, the existence of large ΔT values prevents a precise determination of the spinodal curves. At best, the spinodal curve can be estimated by shifting the experimental curve at $16°C.min^{-1}$ to lower temperature by the amount ΔT, as shown

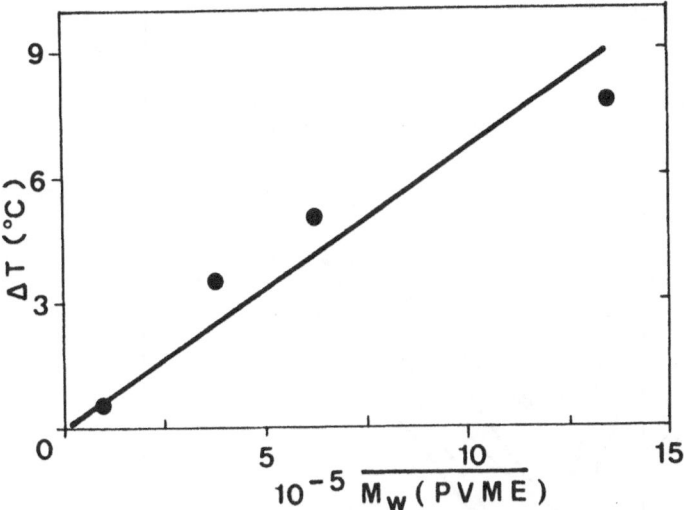

Fig. 12 - Evolution of $\Delta T = \Delta T_{min}$ (16) - T_{min}(0) as a function of

$\overline{M_w(PVME)}$ in mixtures PVME-PS 106,000

in Fig. 13. This procedure assumes that ΔT is independent of the
blend composition. However, this cannot be experimentally verified.

IV.3 - Analysis of the minima of the boundary curves

Let us go back on the phase diagrams in which both the binodal and spinodal are unambiguously determined with our fluorescence emission procedure (see for example figure 8). It is obvious that binodal and spinodal exhibit a common tangent around the minimum of the phase diagram
but nowhere else. This experimental observation would indicate that
the minimum of the temperature-composition diagram, as determined from
fluorescence experiments, is roughly identical with the thermodynamic
critical point, characteristic of the LCST. Bearing in mind that the
systems under consideration consist of PS's with a narrow molecular
weight distribution and of polydisperse PVME ($\overline{M_w}/\overline{M_n}$ = 2.12), this finding contrasts with M$_c$ Master's theoretical predictions (25). It would
mean, indeed, that the polydispersity of one of the blend components
does not force the critical point far away from the minimum of the
coexistence curve.
     The careful analysis of the influence of the polymer molecular
weights on the location of the phase diagram minimum (characterized
by a temperature T_{min} and a PS weight fraction w_{min}) is a way to corroborate this presumptive evidence. For this purpose, a suitable method
consists to pay attention to series of blends in which the molecular
weight of one component is kept constant whereas that,$\overline{M_w}(i)$,of the other

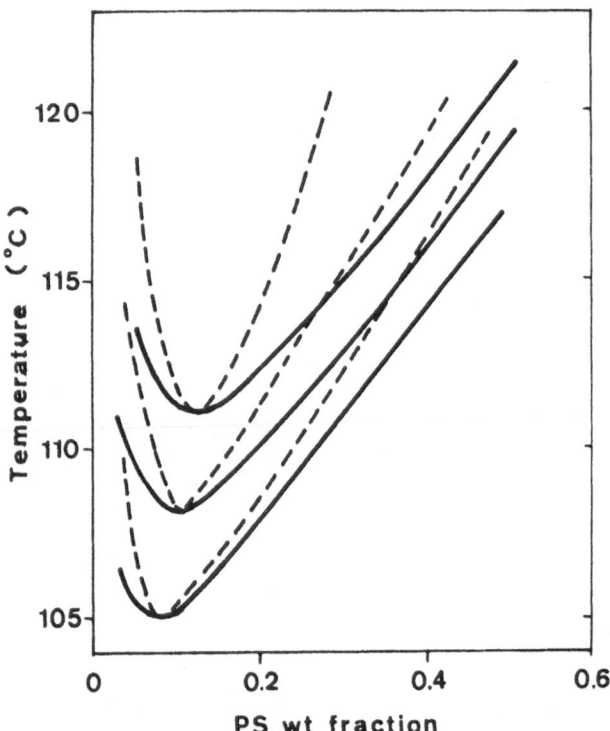

Fig. 13 - Sketch of experimental binodals (solid curves) and spinodals
estimated by a composition-independent $\underline{\Delta T}$ shift (dashed
curves) for mixtures PVME 99,000- high $\overline{M}_w$ PS. $\overline{M}_w$(PS) from
top to bottom : 381,000; 759,000; 1,660,000).

one is varied. Weight average molecular weights are condisered here, in
accordance with the current treatment of polydisperse systems (25);
however one may notice that the number average molecular weights
would be suitable as well, as long as the variable molecular weight
samples of a series exhibit roughly the same polydispersity. It is an
experimental evidence from our fluorescence data that a linear relationship exists between T_{min} and $[\overline{M_w(i)}^{-1/2}]$ (Fig. 14). Additionally,
T_{min} varies linearly with w_{min} (i), as shown in Fig. 15. As a consequence
of these two linear relationships, one may infer the existence of
a supplementary relation between $[w_{min}(i)]^{-1}$ and $[M_w(i)]^{1/2}$ (Fig. 16).
    It is worth noting that such a relation is consistent with one of
the critical conditions which are obeyed in Scott's treatment of polymer mixtures (26) by two isomolecular species A and B. Indeed, the
critical volume fraction of polymer A, ϕ_{cr}(A), satisfies the equation:

$$\phi_{cr}(A) = x_B^{1/2}/(x_A^{1/2} + x_B^{1/2}) \qquad [1]$$

where x_A and x_B are the degree of polymerization of polymer A and B in

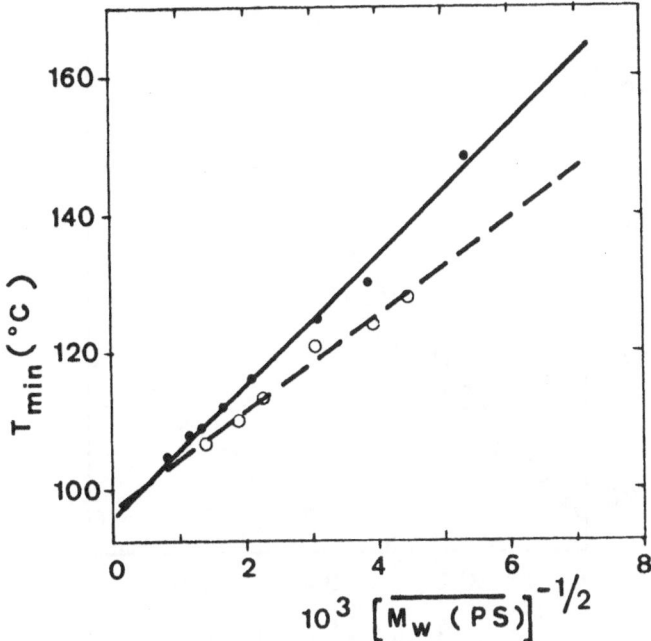

Fig. 14 - Plots of T_{min} versus $[\overline{M_w(PS)}]^{-1/2}$ for two series of blends PS-PVME 99,000. (Solid line : anionic PS's: dashed line: polydisperse PS's).

terms of a reference volume taken as the molar volume of the smallest polymer repeat unit. For each series of blends under consideration in our study, x_B is a constant, x_A is proportional to $\overline{M_w(i)}$ and the volume fraction $\phi_{cr}(A)$ is roughly equal to the weight fraction $w_{cr}(i)$ since PS and PVME have a comparable density; in such circumstances, equation [1] may be rewritten in the form :

$$[w_{cr}(i)]^{-1} = 1 + k(\overline{M_w(i)})^{1/2} \qquad [2]$$

where k is a constant.

In fact, the experimental data given in Fig. 16 fit correctly equation [2], in consideration of both the proportionality between $[w_{cr}(i)]^{-1}$ and $[M_w(i)]^{1/2}$ and the value of the intercept with the y-axis (0.94 as deduced from a least squares analysis). As a consequence, the experimental quantity $w_{min}(i)$ may be identified with the thermodynamic quantity $w_{cr}(i)$, despite the large polydispersity difference between the two components of the blend.

Furthermore, data such as those given in Fig. 14 and 15 provide us with values of the intercept with the y-axis . These quantities, so-called $(T_{min})_\infty$, are of particular interest because they represent the temperature at which phase separation should occur when the molecular

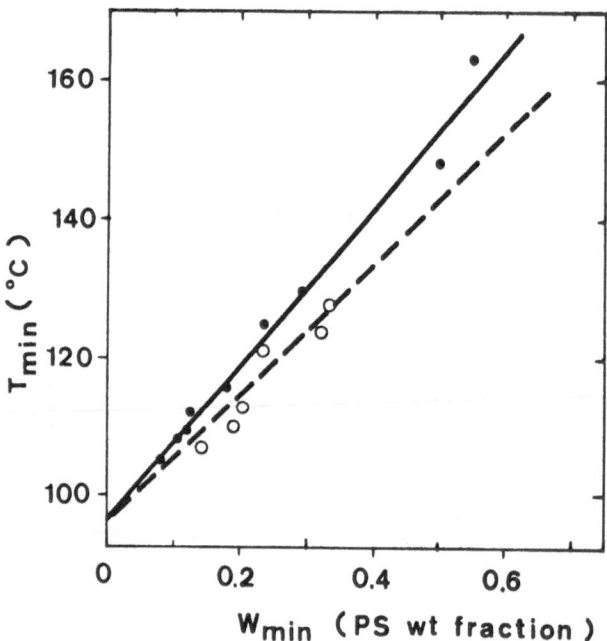

Fig. 15 - Plots of T_{min} versus w_{min} for two series of blends PS-PVME
99,000. (Solid line : anionic PS's; dashed line : polydis-
perse PS's).

weight of polymer i tends to be infinite. This situation is depicted
by $[\overline{M_w(i)}]^{-1/2}$ tending to zero and by $w_{min}(i)$ tending to zero as
well (equation [2].).The same values of $(T_{min})_\infty$ are found within the
experimental error (97 ± 2°C) (Table I):
i/ for a given series of blends, whatever the method of calculation and
ii/ whatever the nature and the molecular weight of the component which
is kept constant. In addition, the value of $(T_{min})_\infty$ is not modified

Series of blends	$(T_{min})_\infty$ deduced from plots of T_{min} :	
	versus $[\overline{M_w(i)}]^{-1/2}$	versus $w_{min}(i)$
PVME 99,000/i = PS	96.7°C	96.3°C
PS 106,000/i = PVME	97.0°C	94.3°C
PVME 388,000/i= PS	99.1°C	98.8°C

Table I - Values of $(T_{min})_\infty$ as determined from least squares curve
fitting.

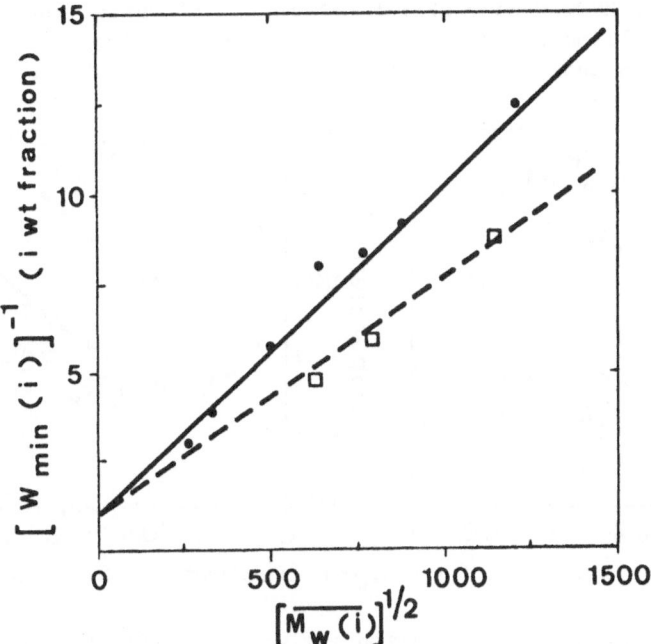

Fig. 16 - Plots of $[w_{min}(i)]^{-1}$ versus $[\overline{M_w(i)}]^{1/2}$ for two series of blends PS-PVME (Solid line : blends PVME 99,000 - PS(=i); dashed line : blends PS 106,000-PVME(=i).).

by an increase in PS polydispersity from 1.07 to about 1.8 (dashed lines in Fig. 14 and 15). All these observations suggest that $(T_{min})_\infty$ is an intrinsic characteristic of the pair PS-PVME, in agreement again with the theoretical description of polymer mixtures. Indeed, following the definition given by Nishi (27) :

$$(T_{min})_\infty = - B/AR \qquad [3]$$

where R is the ideal gas constant and A and B are also constants rela-ted to the temperature dependence of the interaction parameter χ bet-ween the two polymers by the relation :

$$\chi = A + B/RT \qquad [4]$$

IV.4 - Effects of PS deuteration on the phase diagram PS-PVME

As far back as figure 4, we have pointed out, on the basis of small angle neutron scattering data, that the perhydrogenous label PS* is suitable for PVME-perdeuterated polystyrene (PS-d_8) phase diagram deter-minations. Use was made of this versatility to cross-check previous photometric light scattering results (23) and confirm that the use of

PS-d₈ instead of hydrogenous PS(PSH) of identical chain length allows the LCST of the blend PS-PVME to rise by about 35°C (Fig. 17). Additionally, one can notice that no heating rate effect is detected around

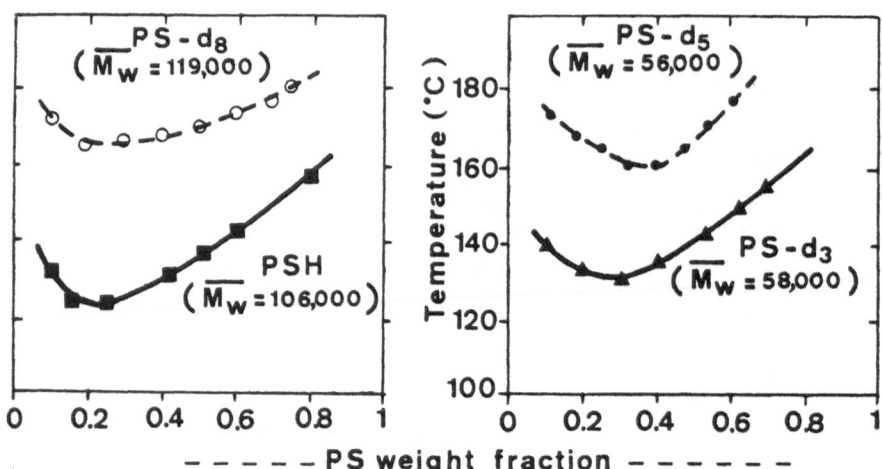

Fig. 17 - Examples of PS deuteration influence on PS-PVME phase
          diagrams

w_{min}, in good agreement with our previous findings about the blends containing an hydrogenous PS of moderate $\overline{M}_w$.

With the aim to go thoroughly into the origin of this isotope effect, we have determined the phase diagrams of PVME-selectively deuterated PS (16). For this purpose, we have taken an interest in poly(1-(pentadeuterophenyl)ethylene)(PS-d₅) and poly(1-phenyl trideuteroethylene)(PS-d₃) and shown that the blends PVME-PS-d₃ and PVME-PS-d₅ behave quite differently. As an example, the phase separation temperature T_{min} is about 30°C lower with PS-d₃ than with PS-d₅ of comparable chain length (Fig. 17).

A proper comparison of deuteration effects can be made thanks to the values of $(T_{min})_{\infty}$ which present the advantage to be independent of polymer molecular weight and molecular weight distribution. As shown in table II, PVME-PS-d₅ and PVME-PS-d₈ in one hand, and PVME-PSH and

Type of PS	$(T_{min})_{\infty}$ (°C)
hydrogenous (PSH)	97 ± 2
PS - d₃	96 ± 2
PS - d₅	128 ± 2
PS - d₈	129 ± 2

Table II - PS deuteration influence on the value of $(T_{min})_{\infty}$
           in the case of blends PS-PVME 99,000.

PVME-PS-d$_3$ in the other hand are series of blends which behave similarly. These results clearly show that the increase of the miscibility of PS and PVME upon deuteration is almost completely the result
of the deuteration of the phenyl group. Isotopic substitution of the
aliphatic chain atoms has a negligible effect. The phenyl rings of PS
must be involved in the specific interactions with the PVME chains
that contribute to the miscibility. These findings corroborate well
those of previous studies by Fourier-transformed infrared spectroscopy
(21, 22)

V - PHASE SEPARATION KINETICS

The fluorescence quenching procedure is also suitable for studying the
kinetics of phase separation. An example of experimental data obtained
in this way is given Fig. 18.
    As expected, the phase separation process develops farther.

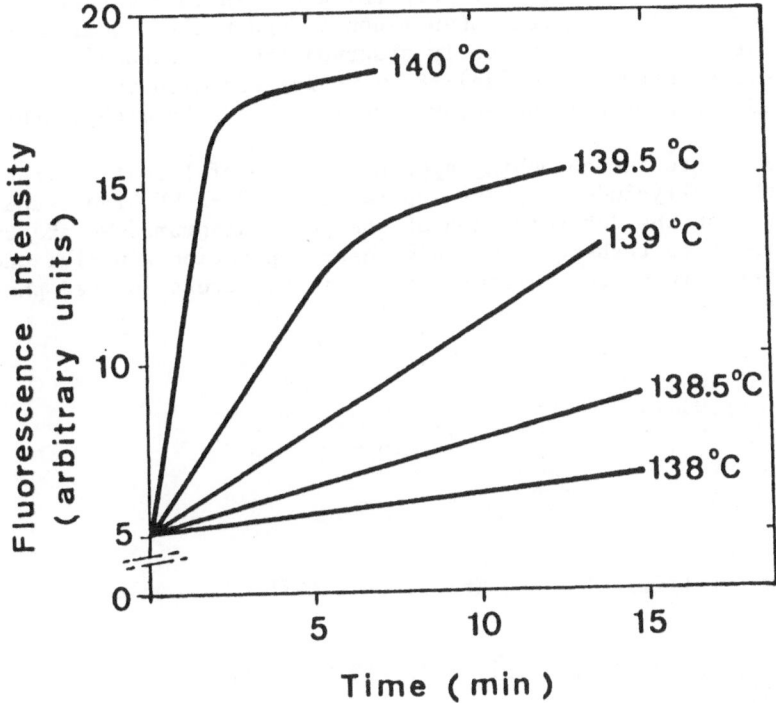

Fig. 18 - Example of phase separation kinetics in isothermal condi-
        tions. (Blend PS 67,000-PVME 99,000 12/88 wt %)

and farther as the temperature gap towards the equilibrium curve is increased. It is worth noting that the temperature gap may be very small in these experiments, thanks to the detection of the phenomena a molecular scale. This allows to detect the earliest stages of phase separation, as requested for a suitable check of the theoretical models (5-7). Unfortunately, polydispersity effects seem to stand in the way of this check. Indeed, Gelles and Frank (24) have advanced the PVME polydispersity as a possible reason for the discrepancy between Pincus's calculations and their excimer fluorescence experimental results. We have also performed some preliminary experiments which leave us to suspect a great influence of the molecular weight distribution on the speed of phase separation. Studies are currently in progress in our Laboratory which involve materials of narrow molecular weight distribution, obtained by fractionation of commercial PVME.

VI - CONCLUSION

Static quenching of PS* fluorescence emission has been discovered in PS-PVME blends. Although such phenomenon is generally uncommon in polymer mixtures, we have recently observed similar quenching effects in poly(dimethylphenylene oxide)-PS* mixtures and also in blends of chlorinated polyethylene and anthracene-labelled poly(methylmethacrylate).

   The fluorescence quenching approach is a useful tool for the study of thermally-induced phase separation in PS-PVME blends and is suitable for the determination of the phase diagrams. We are currently applying this technique to study phase separation kinetics and we plan to extend it to the molecular analysis of stress-induced phase separation.

REFERENCES

1. Bank, M., Leffingwell, J., Thies, C. : 1971, Macromolecules, 4, pp. 43-46.

2. Nishi, T., Wang, T.T., Kwei, T.K.: 1975, Macromolecules, 8, pp. 227-234.

3. Snyder, H.L., Meakin, P., Reich, S.: 1983 , Macromolecules, 16, pp. 757-762.

4. De Gennes, P.G.: 1980, J. Chem. Phys., 79, pp. 4756-4763.

5. Pincus, P.: 1981, J. Chem. Phys., 75, pp. 1996-2000.

6. Binder, K.: 1983, J. Chem. Phys., 79, pp. 6387-6409.

7. Binder, K., Heermann D.W. : 1985, in "Scaling phenomena in disordered systems", Plenum Press ed., New-York.(in press).

8. Jelenic, J., Kirste, R.G., Oberthür, R.C., Schmitt-Strecker, S., Schmitt, B.J.: 1984, Makromol. Chem., 185, pp. 129-156.

9. Yang, H., Shibayama, M., Stein, R.S., Han, C.C. : 1984, Polymer Bull., 12, pp. 7-13.

10. Kwei, T.K., Nishi, T., Roberts, R.F.: 1974, Macromolecules, 7, pp. 667-674.

11. Kaplan, S.: 1984, Polymer Preprints, 25, pp. 356-357.

12. Morawetz , H., Amrani, F.: 1978, Macromolecules, 11, pp. 281-282.

13. Semerak, S.N., Frank, C.W.: 1983, in "Polymer Characterization", ed. by C.D. Craver, A.C.S. series vol. 203.

14. Halary, J.L., Ubrich, J.M., Nunzi, J.M., Monnerie, L., Stein, R.S.: 1984, Polymer, 25, pp. 956-962.

15. Ubrich, J.M., Ben Cheikh Larbi, F., Halary, J.L., Monnerie, L., Han, C.C. : submitted to Macromolecules.

16. Ben Cheikh Larbi, F., Leloup, S., Halary, J.L., Monnerie, L.: submitted to Polymer,Communic.

17. Halary, J.L., Ubrich, J.M., Monnerie, L., Yang H., Stein, R.S. : 1985, Polymer,Communic., 26, pp. 73-76.

18. Valeur, B., Remmp, P., Monnerie, L.: 1974, C.R. Acad. Sci., 279 C, pp. 1009-1012.

19. Stern, O., Volmer, M.: 1919, Physik. Z., 20,pp. 183-188.

20. Perrin, F.: 1933, C.R. Acad. Sci., 196, pp. 485-487.

21. Lu, F.T., Benedetti, E., Hsu, S.L.: 1983, Macromolecules, 16, pp. 1525-1528.

22. Garcia, D.: 1984, J. Polym. Sci., Polym. Phys. Edn., 22, pp. 107-116 and pp. 1773-1779.

23. Yang H., Hadziioannou, G., Stein R.S. : 1983, J. Polym. Sci., Polym. Phys. Edn., 21, pp. 159-162.

24. Gelles, R., Frank, C.W. : 1983, Macromolecules, 16, pp. 1448-1456.

25. Mc Master, L.P.: 1973, Macromolecules, 6, pp. 760-773.

26. Scott, R.L.: 1949, J. Chem. Phys., 17, pp. 279-284.

27. Nishi, T.: 1980, J. Macromol. Sci., part B, 17, pp. 517-542.

LUMINESCENCE TECHNIQUES TO STUDY THE MORPHOLOGY OF PROTOTYPE INDUSTRIAL MATERIALS

Mitchell A. Winnik
Department of Chemistry and Erindale College
University of Toronto
Toronto, Ontario, Canada   M5S 1A1

ABSTRACT. Many commercial materials contain mixtures of incompatible polymers. During synthesis or processing, these polymers undergo phase separation to produce structures which presumably play an important role in the properties and applications of these materials. One would like to have analytical tools capable of elaborating these structures and exploring the interfaces and interphases within the system. Tools based upon fluorescence and phosphorescence spectroscopy show considerable promise here, particularly those based upon emission quenching processes. In this chapter some of these applications to a prototype material will be described.

1. INTRODUCTION

If techniques based upon fluorescence and phosphorescence spectro-scopy are going to gain acceptance as routine tools in the industrial laboratory, they must prove their merit by establishing that they can provide important information about typical polymeric industrial materials. Such materials are frequently prepared from recipes, where the recipes themselves have been optimized for product performance and not for structural simplicity. This means that one is dealing with complex materials composed of mixtures of homopolymers and various kinds of copolymers. These may generate microphase structures, inter-faces and interphase regions. One normally believes that the micro-scopic structure of the material and its dynamic response are somehow responsible for its desirable properties.

The molecular structures one ultimately obtains are rarely something engineered into the material during its design and development. Normally one knows little, if anything, about structure and dynamics at the 5 nm to 50 nm scale, distances on the order of polymer dimensions. The article of current faith is that if one had good analytical tools for determining morphology and dynamics at this level, then one could understand how they lead to material performance, and one could guide the

611

M. A. Winnik (ed.), Photophysical and Photochemical Tools in Polymer Science, 611–627.

synthetic chemist toward the rational design of high performance
materials.

Various approaches have been taken toward elucidating morphology
and dynamics at this level. Important information has become available
from experiments using solid state nmr,[1] electron microscopy,[2] neutron-
and x-ray scattering.[3] Luminescence spectroscopy should be able to
provide information complementary to these other techniques. An
important consideration is whether these luminescence techniques would be
useful in studying systems as complex as typical industrial polymer
materials.

My research group undertook as an objective to explore the utility
of these methods in such systems. To begin, we needed to examine some
complex material which might serve as a reasonable prototype for an
industrial polymer material. Such an initial choice is always
arbitrary. We were pursuaded by Dr. M.D. Croucher of the Xerox Research
Centre of Canada to become involved in a project involving non-aqueous
dispersions [NADs] of polymer particles.

Following a recipe taken originally from the patent literature,
we prepared dispersions of PMMA particles which themselves contained
a substantial amount of polyisobutylene [PIB, ca. 9% by weight]. We
modified the recipe by adding a naphthalene-containing comonomer during
one particular step, introducing naphthalene [N] groups into specific
components of the material. From this brief experience we learned an
important but general lesson: <u>Studies based upon fluorescent labels
require a synthetic chemist to work hand-in-hand with a spectroscopist
or an analytical chemist</u>.

Since that time we have carried out detailed studies of two NAD
systems.[4-7] Our concern was to explore what kind of information one
could learn about these materials from various different luminescence
quenching experiments. This chapter recounts some of these experiments
from the point of view of exploring various strategies and techniques for
studying complex multiphasic polymer materials with fluorescence and
phosphorescence labelling methods.

The chapter begins with a description of the materials studied,
followed by an analysis of five sets of experiments each designed to
explore a different property of the system.

2. MATERIALS

Non-aqueous dispersions [NADs][8] are said to be sterically stabilized. A
surface coverage of stabilizer polymer, normally very soluble in the
liquid medium but anchored into the particle surface, prevents close
approach of two particles. The coils of the stabilizer are swollen in
the solvent; osmotic and entropic forces resist interpenetration of these
coils.

To prepare polymer NADs, one polymerizes a soluble monomer whose
polymer is insoluble in the solvent. For example, the polymers of
methyl methacrylate [MMA] and vinyl acetate [VAc] are insoluble in
aliphatic hydrocarbons. The polymerization must be carried out in the

presence of a second polymer, either a preformed graft- or block-copolymer stabilizer, or a stabilizer-precursor [SP], which, through grafting, is transformed into the stabilizer during the particle synthesis step.

NADs and many aqueous dispersions are often referred to as core-shell materials. The belief is that phase separation occurs between the incompatible core polymer and stabilizer polymer. To minimize surface area, the latter is thought to form a spherical shell about the former. In fact rather little is known about particle morphology. As we shall see, other structures form, and few examples of true core-shell morphology are actually documented in the literature. Nevertheless, it is still convenient to refer to the insoluble, particle-forming polymer as the core polymer. I do so without prejudice as to the actual internal particle structure.

Fluorescence labelling experiments require that a dye be introduced into one component or the other of the system. This is most easily accomplished by introducing a dye-labelled comonomer during the synthesis of the SP or of the particle itself. In this way one can label specifically the stabilizer or core regions of the particle. Four useful comonomers are the methacrylate esters shown below. To help in understanding their spectroscopic behaviour, we found it useful to examine also the corresponding pivalate esters as models for the polymer-bound dye molecules (see above).

2.1.   Recipe 1.   PMMA particles

Degraded butyl rubber (PIB of M ~ 40,000 containing 1% unsaturation) (10 g), methyl methacrylate (MMA, 90 g), azobisisobutyronitrile (0.1 g) are heated at 90° for 8 hours, under nitrogen, in a hydrocarbon solvent such as isooctane. Spherical particles form. Labelled particles are obtained by adding a methyl methacrylate derivative of a fluorescent dye [e.g., dye-$CH_2O_2C(CH_3)C=CH_2$]. Particles are purified of all soluble reactants and by products by repeated centrifugation, decanting, and redispersion in fresh solvent. Our dye-labelled particles had a composition with an IB/methacrylate mole ratio close to 1/9. Particle diameters varied from 0.5 μm to 3 μm.

In this recipe, grafting of PMMA onto the sites of unsaturation in the PIB, either through addition across the double bond or allylic hydrogen abstraction by the growing PMMA chain, competes with homopolymerization of the methyl methacrylate. PIB and PMMA are immiscible and must phase separate. PIB is very soluble in hydrocarbons; PIB not anchored into the particle would be removed by the purification process. The particles would not form stable colloidal dispersions if there were not a surface layer of PIB to provide steric stabilization.

2.2.   Recipe 2.   PVAc particles

In the first step, an alkane-soluble methacrylate polymer or copolymer is prepared by classical free radical means using 2-ethylhexylmethacrylate EHMA or (EHMA + dye-CH_2-MA). This material serves as the stabilizer precursor. In the second step, the SP (10 g), vinyl acetate (90 g), AIBN (0.2 g), and isooctane (100 g) are mixed in an inert atmosphere and heated for 8 hours at 80°. Particles are purified as in recipe 1. Core-labelled particles are prepared by adding a trace (100 ppm) of dye-MA to the particle synthesis step. Stabilizer-labelled particles are prepared by introducing a labelled SP into the particle synthesis. Our particles had mean diameters of 200 nm to 300 nm and a monomer ratio of methacrylate/(vinyl acetate) of 1/9.

3.   PARTICLE MORHOLOGY VIA A MOBILE QUENCHER

One useful approach to the study of particle morphology employs a
labelled phase in the particles in conjunction with a quencher which is
free to diffuse in the system. For particle dispersions, the quencher
can be dissolved in the continuous phase. Here experiments are described
on samples of naphthalene [N] labelled PMMA particles. Samples referred
to as N10, N2 and N(0.01) contain, respectively, 10, 2, and 0.01 mol% N
groups in the methacrylate chains.[9] Dispersions of the particles in
cyclohexane showed typical N fluorescence (with very little excimer
emission in the N10 sample).

   1-Alkylnaphthalene derivatives undergo fluorescence energy transfer
to anthracene [A] by the Förster mechanism with an R_0 of 22 A.[10]   In

$$N^* \ + \ A \ \xrightarrow{\ k_{ET}\ } \ A^* \ + \ N \qquad\qquad (1)$$

solution in isooctane at 22°, N-Piv* transfers energy to A with a rate
constant $k_{ET} = 2 \times 10^{10}$ $M^{-1}s^{-1}$, a diffusion controlled process.[9a]   If N2
and N10 had a strict core-shell morphology, we would anticipate that only
N groups within 2 R_0 units of the PIB-PMMA interface could transfer
energy to anthracene molecules simply dissolved in the isooctane solvent
in which the particles were dispersed.

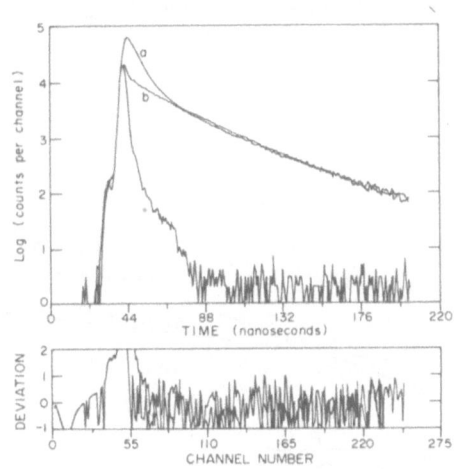

Figure 1. A plot of log I(t) vs t for N10 is isooctane containing 1 x 10⁻³ M anthracene. The lowermost curve is the lamp
excitation profile. Samples were excited at 280 nm. The
middle curve (a, N fluorescence) is measured at 337 nm. The
upper curve (b, A fluorescence) is measured at 450 nm.

Excitation of a dispersion containing 6 mg/ml N2 or N10 and 5 x 10⁻⁴ M
A at 280 nm, where the absorption of light by A is near a minimum and
that of N is a maximum, leads to fluorescence from both N and A. Energy
transfer can be proved by examination of the fluorescence decay [I(t)]
profile of N (337 nm) and A (450 nm). N^{*1} has a fluorescent lifetime
ca. 30–50 ns, which is concentration dependent because of the self-
quenching [$N^* + N \rightarrow 2N$]. In N10, Figure 1, it has a value of 39 nm.
Anthracene has a lifetime of 5.4 ns in isooctane. In Figure 1, the
decay of A* shows two components, one of 5.5 ns due to some direct
excitation of A groups and one of 39 ns. The latter decay must be due to
energy transfer. Note that at short times the I(t) for N* has faster
decaying components. The magnitude of these components increases with
the mole fraction of N and is consistent with N self-quenching. In this
section we ignore these components.

A key question about the energy transfer from N* to A is whether
it occurs by a static or dynamic process. In other words, can A diffuse
a distance comparable to R_0 during the 39 ns lifetime of N*? If this
were the case in homogeneous media, then varying the concentration of A
would affect the lifetime of N* according to the expression

$$\frac{1}{\tau_A} = \frac{1}{\tau_N} = \frac{1}{\tau_N{}^0} + k_{ET}\,[A] \qquad\qquad (2)$$

where $\tau_N{}^0$ is the lifetime of N* in the absence of A. In a core shell
model, most of the N* would be more than 50 Å from the PIB-PMMA interface
and ought to be protected from quenching by A.

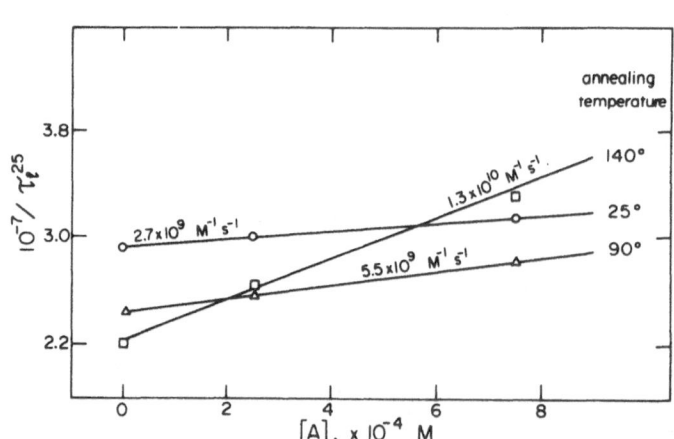

Figure 2. A plot of $(\tau_A)^{-1}$, measured at 25°, vs anthracene
concentration for N10 dispersions in isooctane. The slopes
labelled 90° and 140° refer to samples annealed at those
temperatures for three hours before cooling. The magnitudes of
the slopes [k_{ET}, eq. (2)] are indicated on the plot.

To our surprise, Figure 2, varying [$\underline{A}$] led to a change in $I_N(t)$. This suggests that virtually all the $\underline{N}$ groups in the dispersed particle are accessible to $\underline{A}$. From estimated diffusion coefficients, (10^{-16} $cm^2 sec^{-1}$ for $\underline{A}$ in PMMA at room temperature), we calculate that it should take $\underline{A}$ three years to diffuse 1 μm to the center of a pure PMMA parti- cle. Our observations are clearly inconsistent with this prediction. The slope of the 25° plot in Figure 2 is consistent with a diffusion constant for $\underline{A}$ some 10^{10} larger. Under these circumstances it would take only about a second for a species to diffuse 1 μm.

This result is also at odds with a core-shell structure for the polymer with a pure PMMA core. Some other structure must exist, which permits species such as $\underline{A}$ which can diffuse readily in isooctane and PIB to reach within 40 Å of most of the $\underline{N}$ groups. One such structure involves PMMA domains no larger than 100 Å thick, connected via inter- phase regions to channels of PIB which penetrate throughout the particle much as blood capillaries innervate muscle cells. This microphase model[11] is depicted in Figure 3. Such a structure could come about from

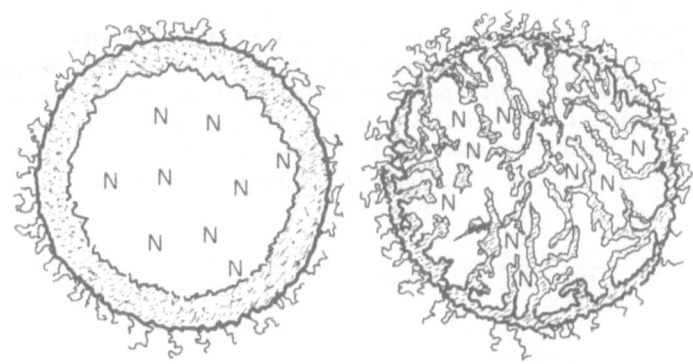

Figure 3. The "core-shell" and "microphase" models of polymer colloid structure. The hairy appendages represent the swollen coils of the stabilizer chains at the particle surface which provide steric stabilization against particle flocculation. The N groups are in the PMMA phase.

extensive grafting between PIB and PMMA or from coagulation of smaller particles in forming the 1 to 3 μm diameter particles we study. The PIB phase in this system would act as a conduit through which solvent and $\underline{A}$ molecules could diffuse into the deep reaches of the particle core.

Another view of particle structure is provided by the observa- tion of phosphorescence from $\underline{N}^{*3}$ in dispersions of N2 and N10 even at temperatures up to room temperature.[12] This result suggests that many N groups are buried in a glassy polymer environment, i.e., the PMMA phase. Curiously, preliminary studies by DSC show an endothermic process occurring in the 70°-90° region rather than at 110° expected for pure PMMA. The following results provide a spectroscopic view of this phenomenon.

When samples of N2 and N10 are heated for 3 hours at a given temperature and cooled back to 25°, changes in fluorescence decay properties of the <u>N</u> groups occur.[11] Glassy polymers are well known for their memory effects. Here, annealing in the presence of a hydrocarbon solvent has a number of interesting consequences. As the data in Figure 2 indicate, annealing increases the permeability of the particles to <u>A</u>, presumably by swelling and trapping solvent plus <u>A</u> in interphase regions upon cooling. A plot of k_{ET}^{25} (from measurements made at 25°) is sigmoidal in shape with an inflection at 75°. This type of behaviour is suggestive of some kind of a relaxation process which may be related to T_g for the system. The fact that this relaxation does not occur near 100° is perhaps a consequence of the extensive grafting between PIB and PMMA, and also the small size of the glassy PMMA domains.

4. CHARACTERIZING THE INTERPHASE WITH OXYGEN QUENCHING

In polymer blends the features one would like to characterize are the amount of interphase in the system and the nature of its properties. An interesting approach to this problem takes advantage of the different sorption properties of oxygen in the two polymers making up the blend. Sorption is the product of solubility times diffusion. Fluorescence and phosphorescence quenching provide a measure of sorption: For the process

$$ dye^* \ + \ O_2 \ \xrightarrow{k_q} \ dye \ + \ O_2 \tag{3} $$

the Stern-Volmer expression describes the sensitivity of the emission in intensity [I] and lifetime [ ] in the presence and absence [I^o, τ^o] of quencher:

$$ \frac{I^o}{I} \ = \ \frac{\tau^o}{\tau} \ = \ 1 \ + \ k_q \ \tau^o \ [O_2] \tag{4} $$

Quenching by O_2 is normally diffusion controlled; k_q is simply related to the sum of the diffusion coefficients of dye* and O_2. If [O_2] in the labelled phase is known, a plot of I^o/I or τ^o/τ vs [O_2] gives $k_q \tau^o$; τ^o is determined in a separate experiment. Normally, however, [O_2] is known only for the system as a whole and not for each individual phase. Under these circumstances one can determine $k_q \alpha$, where α is the partition coefficient for O_2 in the labelled phase of the system.

These principles are demonstrated in experiments on PVAc particles sterically stabilized with PEHMA.[6] Phenanthrene [Phe] is a particularly useful label for these studies. It does not form excimers; singlet self-quenching is inefficient. Its emission profile virtually always decays exponentially. Information about the interphase comes from comparing O_2 quenching data for stabilizer-labelled and core-shelled dispersions in cyclohexane.

The PEHMA stabilizer-precursor from which the particles were prepared contained ca. 1 Phe per chain on average. Figure 4 compares the sensitivity of Phe-MP, Phe-SP and the stabilizer-labelled particle Phe-S-coll. There are few surprises in the data. Note first that τ^o/τ and I^o/I give identical results.[6] Phe-Piv, a small molecule, is more mobile and quenched faster than the polymer-bound Phe-SP. Anchoring the stabilizer into the particle causes only another 20% reduction in quenching effectiveness. The only indication of subtle features of the system is that $I(t)$ for the particles is non-exponential when $[O_2]$ is present. This implies a distribution of environments for the Phe groups with somewhat different oxygen solubilities or diffusion rates. Even here, where we are obliged to calculate mean lifetimes, $\tau^o/\langle\tau\rangle$ data superimpose on the I^o/I data in Figure 4.

Quite different kinds of results are found when these experiments are repeated on the core-labelled particles,[6] Figure 5. Here the Stern-Volmer plot [equation (4)] is curved in a way that suggests that some Phe groups are harder to quench than others. These data fit a model, previously applied to proteins, which assumes that only a fraction f_a of the chromophores are accessible to quenching, the rest being buried. Plotting our data according to eq. 5, we find $f_a = 0.5$.

$$\frac{I}{I^o - I} = \frac{1}{f_a} + \frac{1}{f_a k_q [O_2] \tau^o} \tag{5}$$

Literature data are available on the solubility of oxygen in cyclohexane and in PVAc as well as on its diffusivity.[13] From this consideration we expect little quenching of O_2 of Phe groups in a pure PVAc phase. In the core-labelled particles, all the Phe groups are bound to PVAc chains. A sensible explanation of these results is that half the Phe groups are bound in an interphase region. Because of the presence of PEHMA, the interphase region should be swollen by solvent, enhancing both the local oxygen solubility and diffusion rate. If we presume that the Phe groups are passive markers for the presence of PVAc chains, we conclude that half the PVAc is in an interphase region.

From the slope in Figure 5 we calculate a value of $k_q = 9 \times 10^8$ $M^{-1}s^{-1}$ for the apparent oxygen quenching constant for Phe groups attached to PVAc in the interphase region. This value is half that obtained for Phe groups attached to the PEHMA. Some of this difference may be due to lower oxygen solubility in the interphase domain than in the total volume occupied by the stabilizer chains. A more important observation is that $k_q [O_2]$ is quite large in the interphase region. Oxygen appears to be highly mobile in this domain which may imply that in the dispersion, the interphase is swollen by the hydrocarbon solvent.

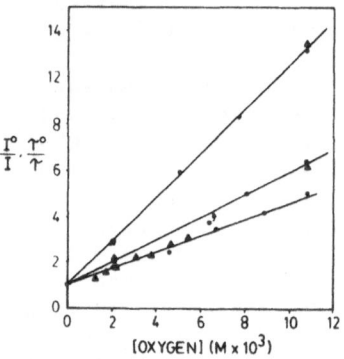

Figure 4. A plot of I^o/I and $^o/$ vs oxygen concentration in the solvent (cyclohexane) for Phe-Piv, the labelled polymer Phe-SP, and the PVAc particles labelled with Phe in the stabilizer, Phe-S-coll. For the particles, $I(t)$ is not exponential in the presence of O_2; the y-axis represents $\tau^o/\langle\tau\rangle$.

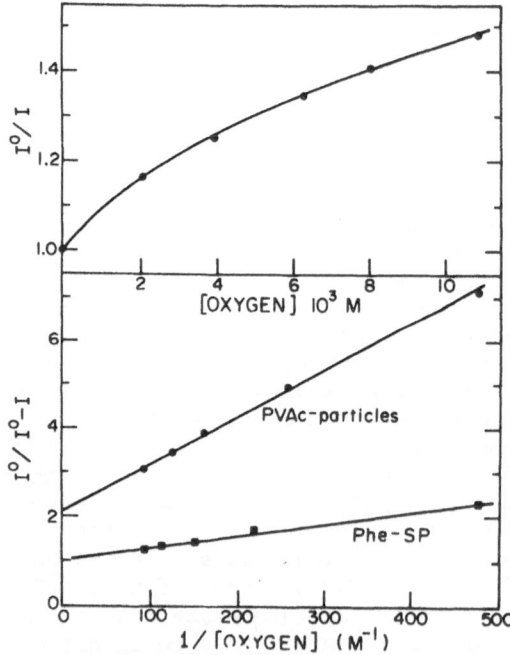

Figure 5. (a) A plot of I^o/I vs solvent concentration of $[O_2]$ for colloidal particles labelled with Phe in the PVAc phase; (b) A plot of $I/(I^o-I)$ vs $1/[O_2]$ (cf. eq. (5)) for the same data. The lower line is a replot of the data for Phe-SP from figure 4.

5. SWELLING AND CONTRACTION FROM SELF-QUENCHING

Fluorescence lifetime experiments on the PMMA particles N2 and N10
indicate that the fluorescence decay time is sensitive to the local
concentration of $\underline{N}$ groups. For example, the fluorescence decay time of
N10 is 39 ns, while that of N(0.01) is 55 ns. We attributed this
sensitivity to the process of self-quenching $[\underline{N}^* + \underline{N} -> 2\underline{N}]$. We do not
yet understand the mechanism of this process. It is probably related to
excimer formation but involving $\underline{N}/\underline{N}$ geometries that do not lead to
effective excimer emission. All these factors point to a rate constant
$k_{sq}(r)$ sensitive to the distance $\underline{r}$ between the $\underline{N}$ groups.

In the following paragraphs, this self-quenching is treated
qualitatively. Dexter's theory[14] of energy transfer by the exchange
mechanism can probably be fit to the data. The proper equations have
been developed by Inokuti and Hirayama.[15] In order to treat the data
properly, the R_0 value for the $\underline{N}^*/\underline{N}$ interactions should be determined
from model studies. This we have not yet done.

5.1. Fluorescence self-quenching

Fluorescence decays I(t) of the N groups in N2 and N10 are non-exponential. In the previous discussion we focussed only on the long-lived
component τ_1. These decay curves actually fit quite well to a sum of two
exponential terms (τ_{sh} is the short-lived component), although the
magnitude of the individual parameters is probably without significance.

$$I_N(t) = a_1 \exp(-t/\tau_1) + a_{sh} \exp(-t)/\tau_{sh}) \tag{6}$$

When samples of N2 or N10 are heated for three hours at a given
temperature and then cooled to room temperature, changes in $I_N(t)$ often
result.[16] For example, no changes occur, Figure 6, for samples heated up
to 60°. Annealing above that temperature causes an increase in τ_1 for
dispersions in a hydrocarbon solvent and a decrease in τ_1 for a
freeze-dried powder sample. These observations correspond to a swelling
of the PMMA phases in the presence of isooctane or hexadecane, leading to
an increase in the mean $\underline{N}/\underline{N}$ separation, and a contraction of the PMMA
phases in the dry powder.

If the $\underline{N}$ groups are statistically distributed along the PMMA
chains, as inferred from reactivity ratio of the two monomers, they
are likely to be statistically distributed spacially within the PMMA
phases of N2 and N10. The rigor of this assumption may be open to
question, but the point I wish to make is that we expect a distribution of $\underline{N}$ group pair separations about a mean value which depends on the
concentration of $\underline{N}$ groups in the phase. $\underline{N}$ pairs in close proximity
will lead to rapid quenching; those more distant, to slower quenching.
Isolated $\underline{N}$ molecules should maintain their 55 ns lifetime. From this
point of view, τ_{sh} samples the more closely spaced $\underline{N}$ pairs and a_{sh}/a_1
measures the fraction of $\underline{N}^*$ sampled by τ_{sh} as opposed to τ_1.

Swelling of the PMMA phases should lead to an increase in τ_{sh}
as well as an increase in a_1/a_{sh}. The distribution of N/N separations
should shift to larger r. Phase contraction should have the opposite
effect. Both of these changes occur for samples annealed above 60°. In
Figures 6, we note that τ_1 for the two dispersions pass through a maximum
for annealing at temperatures near 100°. Similar behavior is observed
for the ratio of a/a_{sh}. Both phenomena indicate that the extent of
swelling depends upon the temperature to which the sample is heated, and

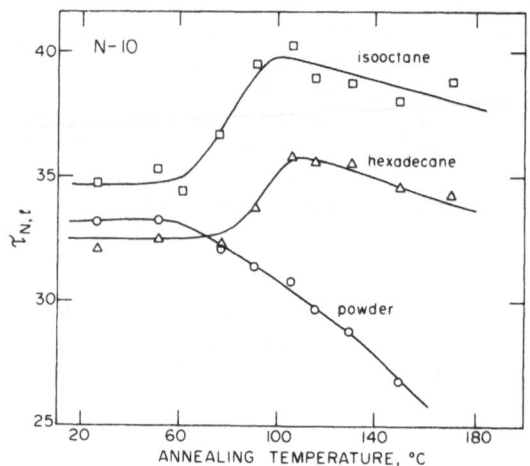

Figure 6. The $\tau_{N,1}$ values of N10 samples after annealing for
three hours at the temperatures indicated. The isooctane and
hexadecane dispersions contain 3 mg/ml. The powder samples
were heated under vacuum. The results are shown for samples
progressively treated by annealing at a given temperature,
cooling to room temperature where $I_N(t)$ was measured, and then
reheating to the next annealing temperature.

also, presumably, on the rate of cooling. These are complex phenomena,
particularly with the hydrocarbon solvents present. It is premature for
us to offer a detailed explanation for these observations. The important
point is that many types of industrial processes (e.g., melt extrusion)
involve treatment of polymer materials at elevated temperatures for
various periods of time. Fluorescence self-quenching provides a means of
studying changes occurring within individual microphases of blend and
composite materials.

5.2. Phosphorescence self-quenching

The phosphorescence lifetime of N groups at liquid nitrogen temperature
(77°K) is 2.3 sec.[9b,12] For chromophores at very low concentrations,
this decay time is independent of the solvent (frozen matrix). At

elevated [N̲], the phosphorescence decay times decrease, indicating
triplet self-quenching. This result is also consistent with the fact
that N2 and N10 have shorter triplet lifetimes than N(0.01) when powder
samples are cooled to 77°K.[9b]
      An important observation is that when powder samples of N2 and N10
are exposed to isooctane or methanol at room temperature and then cooled
to 77°K, the phosphorescence decay times are substantially longer than
the dry powder sample. This result points to solvent-induced swelling of
the PMMA phases during sample preparation leading to increased N̲/N̲ pair
separations. These differences are maintained when the sample is
cooled.[9b]
      What one needs to interpret self-quenching data properly is a cali-
bration curve relating mean decay time to local concentrations. More
specifically one should be able to fit I(t) curves to the form predicted
by Inokuti and Hirayama[15] to obtain a measure of the actual distribution
of dye/dye separations. Model studies are needed to demonstrate that
simple chromophores in homopolymer matrices show the appropriate
behavior. Once R_0 values are determined here, the way will be open to
more quantitative interpretation of swelling and contraction in complex
polyphasic polymer systems.

6. MOLECULAR RELAXATIONS IN THE GLASSY STATE

There is a substantial literature indicating that fluorescence and phosphorescence behavior of many dyes are sensitive to various relaxation
processes of the polymer matrices in which they are dissolved (17,18).
Large scale motions of a probe couple to chain motions responsible for
the glass transition and provide a measure of T_g. Smaller scale
rotations can sense matrix chain motions responsible for more localized
(β or γ) relaxation processes. It is also well known that diffusion of
gasses in glassy polymers is also very sensitive to these relaxations.[19]
If this diffusion leads to luminescence quenching, one expects the
emission intensity or decay time of a luminescent sensor to be sensitive
to these relaxation processes.
      The importance of these measurements is not only in their applica-
tion to simple homopolymers where other techniques abound, but in blend
and composite systems where one would like to understand the behavior of
each individual phase. In N2 and N10 the N̲ groups are sensors of the
PMMA phase.
      Two factors besides phosphorescence lead to deactivation of N̲*³:
The classical radiationless coupling of the excitation energy to vibrational modes of the matrix, which Jones and Siegel[20] investigated, and
a contribution due to impurity quenching. One can write the non-
radiative rate constant k_{nr} as a sum of a vibrational and a quenching
component

$$k_{nr} = k_{vib} + k_q \qquad (7)$$

Jones and Siegel show that when $k_q = 0$, the appropriate expression for treating the data is

$$\frac{\tau^0}{\tau} \; - \; 1 \; = \; a \; \exp(-E_a/RT) \tag{8}$$

where τ is the triplet lifetime, and τ^0 is the limiting low temperature lifetime, normally taken to be that at 77°K. The constants a and E_a are related to the vibrational levels of the matrix into which the excitation energy is dissipated.

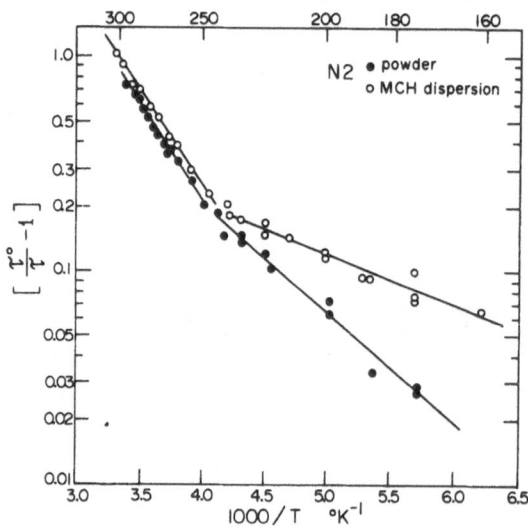

Figure 7. A plot of log $[(\tau^0/\tau)-1]$ vs 1/T for two samples of N2, a powder sample and a dispersion in methylcyclohexane [MCH]. Samples were cooled to 77°K and τ was measured at various times as the samples warmed to room temperature.

The rate constant k_q should also follow Arrhenius behavior.[18] Equation 8 will also apply in the presence of impurities. Here, however, a and E_a will give values characteristic of impurity diffusion if diffusion of the quencher is the rate-limiting process determining the triplet lifetime of N.

Presented in Figure 7 is an Arrhenius plot of log $[(\tau^0/\tau)-1]$ vs 1/T for samples of N2. Two samples are shown. The lower curve is a dry powder sample; the upper curve, a dispersion in methylcyclohexane. Both curves are conveniently fit to two straight lines intersecting at ca. -35°C. The break indicates a change in the rate-limiting step in the phosphorescence deactivation process.

The important question is what this data means. Particularly
helpful information is provided by two previous studies on pure PMMA.
Jones and Siegel[19] report that films of purified PMMA carefully outgassed
above T_g and annealed show no such break in the temperature dependence of
naphthalene phosphorescence. From a spectroscopic point of view they
identify the weak activation energy for decrease in the triplet lifetime
with increasing temperature to radiationless decay via coupling to
vibrational modes of the matrix. Their work helps identify the break in
the plot in Figure 8 to the presence of impurities, probably residual
oxygen, in the sample.

From these experiments we infer that the triplet lifetime of $\underline{N}$ in
our samples in Figure 7, between -35° and 0° is limited by the effectiveness of diffusion of oxygen and other traces of quenching impurities in
the sample. From the activation energy obtained from the Arrhenius
plot[12] [Figure 8] between -35° and 0°, we can identify the quenching
process with the α-methyl relaxation process in the PMMA phase. Oxygen
diffusion is coupled to the rotation of α-methyl groups in the polymer
backbone which create sufficiently rapid fluctuations of free volume for
this diffusion to compete with other processes in quenching $\underline{N}^{*3}$
phosphorescence.

It is tempting to identify the break point in Figure 8 as a characteristic temperature of the system. Guillet, in his chapter (18), makes
such assignments in several homopolymer systems. Indeed one finds in the
literature[21] a maximum at -35° corresponding to the α-methyl group rotation in an nmr relaxation experiment. In this instance the temperature
correspondence between the two sets of experiments must be accidental.
The relaxation temperature is frequency dependent. In the nmr experiment
the maximum corresponds to a methyl group rotational frequency of
10^6 Hz. In the phosphorescence quenching experiment, the triplet lifetimes are on the order of seconds. Impurity diffusion is coupled in an
yet unknown way to the methyl group rotation. From a kinetic point of
view, -35° is the temperature where the rate of impurity quenching
exceeds that of other processes in limiting the phosphorescence decay
time.

7. SUMMARY

Techniques based upon fluorescence and phosphorescence spectroscopy
provide a wealth of information about morphology and dynamics in complex
polymer systems. These are systems typically composed of several different polymer phases separated by sharp interfaces or diffuse interphases.
A particularly useful technique for studying such systems involves
labelling one individual component with a dye. Depending upon the dye
one can study the quenching of its fluorescence or phosphorescence by a
mobile probe added passively to the system. From the rates and efficiency of the quenching, one can learn about local morphology (with resolution of 10 Å to 100 Å); one can measure the extent of interphase formation; and one can study molecular relaxation processes in individual
phases of the material.

Under slightly different conditions, when the dye concentration reaches several percent, interaction between dyes allows one to study factors which promote swelling or contraction within individual phases of a polymer material.

8. REFERENCES

1. (a) L.A. Belfiore, F.C. Schilling, A.E. Tonelli, A.J. Lovinger and F.A. Bovey, Macromolecules, 1984, 17, 1561.

    (b) J.J. Dumais, L.W. Jelinski, L.M. Leung, I. Gancarz, A. Galambos and J.T. Koberstein, Macromolecules, 1985, 18, 119.

2. J.S. Trent, J.I. Scheinbeim and P.R. Couchman, Macromolecules, 1983, 16, 589.

3. L.H. Sperling, Polym. Eng. and Sci., 1984, 24, 1.

4. O. Pekcan, M.A. Winnik, L. Egan and M.D. Croucher, Macromolecules, 1983, 21, 1011.

5. M.A. Winnik, Pure Appl. Chem., 1984, 56, 1281.

6. L.S. Egan, M.A. Winnik and M.D. Croucher, Polym. Eng. and Sci., in press (1986).

7. M.A. Winnik, in 'Polymer Surfaces and Interfaces,' J. Feast and H. Munro, editors, Wiley-Interscience London, 1985.

8. (a) K.E.J. Barrett, editor, 'Dispersion Polymerization in Organic Media,' Wiley Interscience, London, 1975.

    (b) D.H. Napper, 'Polymeric Stabilization of Colloidal Dispersions,' Academic Press, New York, 1983.

9. (a) O. Pekcan, M.A. Winnik, L. Egan and M.D. Croucher, Macromolecules, 1983, 16, 699.

    (b) O. Pekcan, M.A. Winnik and M.D. Croucher, J. Coll. Interfac. Sci., 1983, 95, 420.

10. I.B. Berlman, 'Energy Transfer Parameters of Aromatic Compounds,' Academic Press (1973).

11. O. Pekcan, M.A. Winnik and M.D. Croucher, J. Polym. Sci. Polym. Lett., 1983, 21, 1011.

12. O. Pekcan, M.A. Winnik and M.D. Croucher, Can. J. Chem., 1985, 63, 129.

13. J. Brandrup and E.H. Immergut, editors, 'Polymer Handbook,' 2nd edition, Wiley Interscience, New York, 1975, p. 237.

14. D.L. Dexter, J. Chem. Phys., 1953, 21, 836.

15. M. Inokuti and F. Hirayama, J. Chem. Phys., 1965, 43, 1978.

16. M.A. Winnik, O. Pekcan and M.D. Croucher, Macromolecules, submitted (1985).

17. (a) K. Horie and I. Mita, Chem. Phys. Lett., 1982, 93, 61.

    (b) K. Horie, K. Morishita and I. Mita, Macromolecules, 1984, 17, 1746.

    (c) H. Rutherford and I. Soutar, J. Polym. Sci. Polym. Phys. Ed., 1980, 18, 1021.

    (d) F.E. El-Sayed, J.R. MacCallum, P.J. Pomery and T.M. Shepherd, J. Chem. Soc. Faraday Trans. II, 1979, 75, 79.

    (e) A.C. Sommersall, E. Dan and J.E. Guillet, Macromolecules, 1984, 7, 1746.

    (f) W.E. Graves, R.H. Hofeldt and S.P. McGlynn, J. Chem. Phys., 1972, 63, 129.

    (g) W.H. Melhuish, Trans. Faraday Soc., 1966, 62, 3384.

18. J.E. Guillet, chapter 21 in this volume.

19. D.R. Paul, Ber. Bungenges Phys. Chem., 1979, 83, 294.

20. P.F. Jones and S. Siegel, J. Chem. Phys. 1969, 50, 1134.

21. (a) T. Kawai, J. Phys. Soc. Jpn., 1961, 6, 1220.

    (b) J.G. Powles and P. Mansfield, Polymer, 1962, 3, 336.

CONTRIBUTORS

LIST OF PARTICIPANTS

Abel-Mottaleb M.S.
Department of Chemistry
Faculty of Science
Ain Shams University
Abbassia, Cairo
EGYPT

Aklonis J.J.
Department of Chemistry
University of Southern California
Los Angeles, California 90007
U.S.A.

Altomare A.
Dipartimento di Chimica e
Chimica Industriale
University of Pisa
Via Risorgimento 35
56100 Pisa
ITALY

Ausserre Dominique
College de France
11 Place M. Berthelot
75231 Paris Cedex 05
FRANCE

Bakir E.T.
School of Chemistry
University of Sussex
Falmer, Brighton
U.K.

Biddle Derek
Department of Physical Chemistry
University of Gothenburg
CTH
S-41296 Goteborg
SWEDEN

Bitai Irene
Institute for Physical Chemistry
University of Vienna
Wahringerstrasse 43
A-1090 Vienna
AUSTRIA

Boens N.
Laboratory for Molecular
Dynamics and Spectroscopy
Celestijnenlaan 200F
B-3030 Heverlee
BELGIUM

Bokobza Liliane
Laboratoire de Physico-Chimie
Structurale et Macromoleculaire
ESPCI
10, rue Vauquelin
75231 Paris Cedex 05
FRANCE

Carlini Carlo
Dipartimento di Chimica e
Chimica Industriale
University of Pisa
Via Risorgimento 35
56100 Pisa
ITALY

Ciardelli Francesco
Dipartimento di Chimica e
Chimica Industriale
University of Pisa
Via Risorgimento 35
56100 Pisa
ITALY

Cuniberti Carla
Istituto di Chimica Industriale
Corso Europa 30
Universita di Genova
16132 Genova
ITALY

De Schryver F.C.
Laboratory for Molecular
Dynamics and Spectroscopy
Celestijnenlaan 200F
B3030 Heverlee
BELGIUM

Ediger Mark D.
Department of Chemistry
University of Wisconsin
Madison, Wisconsin 53706
U.S.A.

Egan Luke S.
Department of Chemistry
University of Toronto
Toronto, Ontario M5S 1A1
CANADA

Ferreira Joao F. Alves
Physics Laboratory
Universidade do Minho
4700 Braga
PORTUGAL

Ferriera M. Isabel
Physics Laboratory
Universidade do Minho
4700 Braga
Portugal

Fibiger Richard
The Dow Chemical Company
Building 1702
Midland, Michigan 48674
U.S.A.

Fornes Raymond E.
Physics Department
402A Cox Hall
North Carolina State University
Raleigh, North Carolina 27695-8202
U.S.A.

Frank Curtis W.
Department of Chemical Engineering
Stanford University
Stanford, California 94305
U.S.A.

Gabor Gavriella
Israel Institute for Biological
Research
P.O.B. 19
Ness-Ziona 70450
ISRAEL

Gattiglia Ercole
Centro Studi Chimico-Fisici
di Macromolecole Sintetiche e
Naturali
Istituto di Chimica Industriale
Corso Europa 30
16132 Genova
ITALY

Geri Alessandro
C.N.R. - FRAE
Via Castognoli 1
40126 Bologna
ITALY

Goedeweeck Rudy
Laboratory for Molecular
Dynamics and Spectroscopy
Celestijnenlaan 200 F
B 3030 Heverlee
BELGIUM

Gonzales Machado Ana Maria
Max-Plank Institut für
Quantenoptik
8046 Garching-bei-Munchen
GERMANY

Granville Maryse
Laboratoire de Chimie
Macromoleculaire
et de Catalyse Organique
Universite de Liege
Sart-Tilman (B6), LE
4000 Liege
BELGIUM

Guillet James E.
Department of Chemistry
University of Toronto
Toronto, Ontario M5S 1A1
CANADA

Gupta Amitava
Mail Stop 67-201
Jet Propulsion Laboratory
Pasadena, California 91109
U.S.A.

Gusten Hans
Kernforschungszentrum Karlsruhe
Institut für Radiochemie
Postfach 3640
D-7500 Karlsrue
WEST GERMANY

Guven Olgun
Hacettepe University
Department of Chemistry
Beytepe, Ankara
TURKEY

Gysel Hermann
Institute for Applied Physics
University of Berne
Silderstrasse 5
CH-3012 Berne
SWITZERLAND

Haas Elisha
Department of Life Sciences
Bar-Ilan University
Ramat-Gan 52100
ISRAEL

Halary Jean-Louis
Laboratoire de Physico-Chemie
Structurale et Macromoleculaire
ESPCI
10, rue Vauquelin
75231 Paris Cedex 05
France

Helfand E.
AT & T Bell Laboratories
Murray Hill, New Jersey 07974
U.S.A.

Hofmann Manfred
CIBA-GEIGY AG
P & A Research Centre
WFM 190/161
CH-1701 Fribourg
SWITZERLAND

Horie Kazuyuki
Institute for Interdisciplinary
Research
Faculty of Engineering
The University of Tokyo
4-6-1 Komaba, Meguro-Ku
Tokio 153
JAPAN

Irie M.
The Institute of Scientific
and Industrial Research
Osaka University
Ibaraki, Osaka 567
JAPAN

Jovin T.
MPI für Biophys Chemie
Postfach 2841
D-3400 Gottingen
WEST GERMANY

King Mike
ICI, New Science Group
P.O. Box 11, The Heath
Runcorn, Cheshire WA7 4QE
U.K.

Levy R.L.
McDonnell Douglas Company
Research Division
P.O. Box 516
St. Louis, Missouri 63116
U.S.A.

Limm William
Department of Chemistry
University of Toronto
Toronto, Ontario M5S 1A1
CANADA

Locher Rene
Laboratorium für Physicalische
Chemie
ETH-Zentrum
CH-8092 Zurich
SWITZERLAND

Leger L.
College de France
11, Place Marcelin Berthelot
75231 Paris Cedex 05
FRANCE

MacCallum J.R.
Department of Chemistry
The University
St. Andrews, Scotland KY16 9ST
U.K.

Martinho Jose
Centro de Quimica Fisica Molecular
Complexo I - IST
Av. Rovisco Pais
1096 Lisboa Codex
PORTUGAL

Masuhara H.
Kyoto Institute of Technology
Department of Polymer Science
and Engineering
Matsugasaki, Kyoto 606
JAPAN

Mattice Wayne L.
Department of Chemistry
Louisiana State University
Baton Rouge, Louisiana 70803-1804
U.S.A.

Mendenhall G. David
Department of Chemistry and
Chemical Engineering
Michigan Technological University
Houghton, Michigan 49931
U.S.A.

Monnerie L.
Lab. de Physico—Chimie Structurale
et Macromoleculaire
10, rue Vauquelin
75231 Paris Cedex 05
FRANCE

Morawetz Herbert
Department of Chemistry
Polytechnic Institute of New York
333 Jay Street
Brooklyn, New York 11201
U.S.A.

Nealon Donald G.
Department of Chemistry
University of Texas
Austin, Texas 78712
U.S.A.

Neves Maria Edite
Laboratorio de Fisica
Av. Joao XXI
Universidade do Minho
4700 Braga
PORTUGAL

North Alastair M.
Asian Institute of Technology
P.O. Box 2754
Bangkok
THAILAND

Orrah Jacqueline
Chemistry Department
York University
Heslington
York, Yorkshire YO1 5DD
U.K.

Ors Jose
AT & T Technologies
ERC
Box 900
Princeton, New Jersey 08540
U.S.A.

Pekcan Onder
Department of Physics
Hacettepe University
Ankara
TURKEY

Perico A.
Istituto di Chimica Industriale
Corso Europa 30
Universita di Genova
16132 Genova
ITALY

Persoons Andre
Lab. Chemical and Biological
Dynamics
University of Leuven
Dept. of Chemistry
Celestijnenlaan 200D
B3030 Heverlee
BELGIUM

Phillips David
The Royal Institution
21 Albemarle Street
London W1X 4BS
U.K.

Pieroni Osvaldo
CNR - Istituto di Biofisica
Via S. Lorenzo 26
56100 Pisa
ITALY

Poole John A.
Department of Chemistry, CAS
Temple University
Philadelphia, Pennsylvania 19122
U.S.A.

Prieto Manuel Jose Estevez
Centro de Quimica Estrutural
Complexo I
Instituto Superior Tecnico
Av. Rovisco Pais
1096 Lisboa Codex
PORTUGAL

Scarlata Suzanne
AT & T Technologies
P.O. Box 900
Princeton, New Jersey 08540
U.S.A.

Schen Michael Alan
Lab. Chimie Macromoleculaire
U.S.T.L.
Place Eugene Bataillon
34060 Montpellier
FRANCE

Shea Kenneth J.
Department of Chemistry
University of Caalifornia
Irvine, California 92717
U.S.A.

Sinclair Andrew
Xerox Research Centre of Canada
2660 Speakman Drive
Mississauga, Ontario L5K 2L1
CANADA

Slomkowski Stan
Center of Macromolecular Studies
Polish Academy of Sciences
Boczna 5
90-362 LODZ
POLAND

Smith B.A.
IBM Almadeu Research Center K91/801
650 Harry Road
San Jose, California 95120-6099
U.S.A.

Solaro Roberto
Dipartimento di Chimica e Chimica
Industriale
University of Pisa
Via Risorgimento 35
56100 Pisa
ITALY

Soutar Ian
Department of Chemistry
Heriot-Watt University
Riccarton, Edinburgh EH14 4AS
U.K.

Squire David
European Research Office
USARDSG
Box 65
FRO New York 09510
U.S.A.

Stanton Deirdre
Department of Chemistry
University of Toronto
Toronto, Ontario M5S 1A1
CANADA

Sung C.P.S.
Institute of Material Science IMS
U-136
University of Connecticut
97 North Eagleville Road
Storrs, Connecticut 06268
U.S.A.

Tazuke S.
Tokyo Institute of Technology
4259 Nagatsuta, Midori-ku
Yokohama 227
JAPAN

Torkelson John M.
Department of Chemical Engineering
Technological Institute
Northwestern University
Evanston, Illinois 60201
U.S.A.

Unger Israel
Department of Chemistry
University of New Brunswick
Bag Service No. 45222
Fredericton, New Brunswick E3B 6E2
CANADA

Van Wonterghem Bruno
Department of Chemistry
University of Leuven
Celestijnenlaan 200D
B-3030 Heverlee
BELGIUM

Wang Francis
Polymer Division
National Bureau of Standards
Building 224, Room B320
Gaithersburg, Maryland 20899
U.S.A.

Webb John D.
Solar Energy Research Institute
1617 Cole Boulevard
Golden, Colorado 80401
U.S.A.

Webber Stephen E.
Department of Chemistry
University of Texas
Austin, Texas 78712
U.S.A.

Winnik Françoise
Xerox Research Centre of Canada
2660 Speakman Drive
Mississauga, Ontario L5K 2L1
U.S.A.

Winnik M.A.
Department of Chemistry
University of Toronto
Toronto, Ontario M5S 1A1
CANADA

SUBJECT INDEX